U0236274

斑马鱼行为与神经遗传学手册

Behavioral and Neural Genetics of Zebrafish

（加）R. T. 盖尔洛伊（Robert T. Gerlai）主编
李丽琴　石　童　等译

化学工业出版社
·北京·

Behavioral and Neural Genetics of Zebrafish
Robert T. Gerlai
ISBN: 978-0-12-817528-6

Authorized Chinese translation published by Chemical Industry Press Co., Ltd.

《斑马鱼行为与神经遗传学手册》（李丽琴 石童 等译）
ISBN: 987-7-122-43396-1

北京市版权局著作权合同登记号：01-2023-2726

图书在版编目（CIP）数据

斑马鱼行为与神经遗传学手册 /（加）R. T. 盖尔洛伊（Robert T. Gerlai）主编；李丽琴等译．—北京：化学工业出版社，2023.9
书名原文：Behavioral and Neural Genetics of Zebrafish
ISBN 978-7-122-43396-1

Ⅰ.①斑… Ⅱ.①R…②李… Ⅲ.①神经科学 - 医学遗传学 - 手册 Ⅳ.①Q189-62

中国国家版本馆 CIP 数据核字（2023）第 075034 号

责任编辑：傅四周　　文字编辑：陈小滔　刘洋洋
责任校对：宋　夏　　装帧设计：韩　飞

出版发行：化学工业出版社（北京市东城区青年湖南街 13 号　邮政编码 100011）
印　　装：北京建宏印刷有限公司
787mm×1092mm　1/16　印张 33¼　字数 771 千字　2024 年 1 月北京第 1 版第 1 次印刷

购书咨询：010-64518888　　售后服务：010-64518899
网　　址：http://www.cip.com.cn
凡购买本书，如有缺损质量问题，本社销售中心负责调换。

定　价：399.00 元　

纪 念

谨以此书纪念 Christine Hone-Buske 博士（1984 年 2 月 10 日—2019 年 5 月 22 日），她是我在多伦多大学密西沙加分校斑马鱼实验室的第一名本科生，后来深造成为研究生。

撰稿者名单

Rida Ansari Department of Cell and Systems Biology, University of Toronto, Toronto, ON, Canada; Neuroscience and Mental Health Program, Hospital for Sick Children, Peter Gilgan Centre for Research and Learning, Toronto, ON, Canada

Lior Appelbaum The Faculty of Life Sciences and the Multidisciplinary Brain Research Center, Bar-Ilan University, Ramat Gan, Israel

W. Philip Bartel Department of Neurology, University of Pittsburgh, Pittsburgh, PA, United States

Carla Denise Bonan Laboratório de Neuroquímica e Psicofarmacologia, Escola de Ciências da Saúde e da Vida, Pontifícia Universidade Católica do Rio Grande do Sul, Porto Alegre, Rio Grande do Sul, Brazil

Caroline H. Brennan School of Biological and Chemical Sciences, Queen Mary University of London, London, United Kingdom

Edward A. Burton Department of Neurology, University of Pittsburgh, Pittsburgh, PA, United States

Michael J. Carvan Ⅲ University of Wisconsin-Milwaukee, School of Freshwater Sciences, Milwaukee, WI, United States

Karl J. Clark Department of Biochemistry and Molecular Biology, Mayo Clinic, Rochester, MN, United States

Rosane Souza Da Silva Laboratório de Neuroquímica e Psicofarmacologia, Escola de Ciências da Saúde e da Vida, Pontifícia Universidade Católica do Rio Grande do Sul, Porto Alegre, Rio Grande do Sul, Brazil

Konstantin A. Demin Institute of Translational Biomedicine, St. Petersburg State University, St. Petersburg, Russia; Institute of Experimental Medicine, Almazov National Medical Research Centre, Ministry of Healthcare of Russian Federation, St. Petersburg, Russia

Flavia De Santis ZeClinics SL, IGTP (Germans Trias i Pujol Research Institute), Barcelona, Spain

Vincenzo Di Donato ZeClinics SL, IGTP (Germans Trias i Pujol Research Institute), Barcelona, Spain

Elena Dreosti Wolfson Institute for Biomedical Research, University College London, London, United Kingdom

Marc Ekker Department of Biology, University of Ottawa, Ottawa, ON, Canada

Amanda Facciol Department of Cell and Systems Biology, University of Toronto, Mississauga, ON, Canada

Gert Flik Department of Animal Ecology and Physiology, Institute of Water and Wetland Research, Faculty of Science, Radboud University, Nijmegen, the Netherlands

Robert T. Gerlai Department of Psychology, University of Toronto Mississauga, Mississauga, ON, Canada

Marnix Gorissen Department of Animal Ecology and Physiology, Institute of Water and Wetland Research, Faculty of Science, Radboud University, Nijmegen, the Netherlands

Fabrizio Grieco Noldus Information Technology BV, Wageningen, the Netherlands

Su Guo Department of Bioengineering and Therapeutic Sciences, Programs in Biological Sciences, Institute of Human Genetics, University of California, San Francisco, CA, United States

Amanda A. Heeren Department of Biochemistry and Molecular Biology, Mayo Clinic, Rochester, MN, United States

Dennis M. Higgs Department of Integrative Biology, University of Windsor, Windsor, ON, Canada

Ellen J. Hoffman Child Study Center, Program on Neurogenetics, Yale School of Medicine, Yale University, New Haven, CT, United States

Kerstin Howe Wellcome Sanger Institute, Cambridge, United Kingdom

Khang Hua Department of Biology, University of Ottawa, Ottawa, ON, Canada

Arnim Jenett TEFOR Paris Saclay, CNRS UMS2010, INRA UMS1451, Universite Paris Sud, Université Paris-Saclay, Gif-sur-Yvette, France

Allan V. Kalueff School of Pharmacy, Southwest University, Chongqing, China; Almazov National Medical Research Centre, St. Petersburg, Russia; Ural Federal University, Ekaterinburg, Russia; Granov Russian Research Center of Radiology and Surgical Technologies, St. Petersburg, Russia; Laboratory of Biological Psychiatry, Institute of Translational Biomedicine, St. Petersburg State University, St. Petersburg, Russia; Laboratory of Biopsychiatry, Scientific Research Institute of Physiology and Basic Medicine, Novosibirsk, Russia; ZENEREI Institute, Slidell, LA, United States

Konstantinos Kalyviotis Department of Bioengineering, Imperial College London, South Kensington Campus, London, United Kingdom

Justin W. Kenney Department of Biological Sciences, Wayne State University, Detroit, MI, United States

Anton M. Lakstygal Institute of Translational Biomedicine, St. Petersburg State University, St. Petersburg, Russia; Laboratory of Preclinical Bioscreening, Granov Russian Research Center of Radiology and Surgical Technologies, Ministry of Healthcare of Russian Federation, Pesochny, Russia

Han B. Lee Department of Biochemistry and Molecular Biology, Mayo Clinic, Rochester, MN, United States

Emmanuel Marquez-Legorreta School of Biomedical Sciences, The University of Queensland, Brisbane, QLD, Australia

Noam Miller Departments of Psychology & Biology, Wilfrid Laurier University, Waterloo, ON, Canada

Rodsy Modhurima Department of Biochemistry and Molecular Biology, Mayo Clinic, Rochester, MN, United States

Stephan C. F. Neuhauss Department of Molecular Life Sciences, University of Zurich, Zurich, Switzerland

Lucas P. J. J. Noldus Noldus Information Technology BV, Wageningen, the Netherlands; Department of Biophysics, Donders Institute for Brain, Cognition and Behavior, Radboud University, Nijmegen, the Netherlands

William H. J. Norton Department of Neuroscience, Psychology and Behaviour, University of Leicester, Leicester, United Kingdom

Periklis Pantazis Department of Bioengineering, Imperial College London, South Kensington Campus, London, United Kingdom

David M. Parichy Department of Biology and Department of Cell Biology, University of Virginia, Charlottesville, VA, United States

Marielle Piber School of Biomedical Sciences, The University of Queensland, Brisbane, QLD, Australia; School of Medicine, Medical Sciences & Nutrition, University of Aberdeen, Aberdeen, United Kingdom

John H. Postlethwait Institute of Neuroscience, University of Oregon, Eugene, OR, United States

David Pritchett School of Biological and Chemical Sciences, Queen Mary University of London, London, United Kingdom

Hanyu Qin Department of Biosystems Science and Engineering, ETH Zürich, Basel, Switzerland

Jason Rihel Department of Cell and Developmental Biology, University College London, London, United Kingdom

Domino K. Schlege Department of Molecular Life Sciences, University of Zurich, Zurich, Switzerland; Life Science Zurich Graduate School, Ph.D. Program in Molecular Life Sciences, Zurich, Switzerland

Ethan K. Scott School of Biomedical Sciences, The University of Queensland, Brisbane, QLD, Australia; The Queensland Brain Institute, The University of Queensland, Brisbane, QLD, Australia

Alexander Stankiewicz BioChron LLC, Worcester, MA, United States

S. J. Stednitz Institute of Neuroscience, University of Oregon, Eugene, OR, United States

Ruud A. J. Tegelenbosch Noldus Information Technology BV, Wageningen, the Netherlands

Javier Terriente ZeClinics SL, IGTP (Germans Trias i Pujol Research Institute), Barcelona, Spain

Steven Tran Department of Biology and Biological Engineering, California Institute of Technology, Pasadena, CA, United States

Benjamin Tsang Department of Psychology, University of Toronto Mississauga, Mississauga, ON, Canada

Sachiko Tsuda Graduate School of Science and Engineering, Saitama University, Saitama, Japan

Ruud van den Bos Department of Animal Ecology and Physiology, Institute of Water and Wetland Research, Faculty of Science, Radboud University, Nijmegen, the Netherlands

Victor S. Van Laar Department of Neurology, University of Pittsburgh, Pittsburgh, PA, United States

Monica Ryff Moreira Roca Vianna Laboratóriode Biologia e Desenvolvimento do Sistema Nervoso, Escola de Ciências da Saúde e da Vida, Pontifícia Universidade Católica do Rio Grande do Sul, Porto Alegre, Rio Grande do Sul, Brazil

P. Washbourne Institute of Neuroscience, University of Oregon, Eugene, OR, United States

Svante Winberg Department of Neuroscience, Uppsala University Uppsala, Sweden Jiale Xu Department of Bioengineering and Therapeutic Sciences, Programs in Biological Sciences, Institute of Human Genetics, University of California, San Francisco, CA, United States

David Zada The Faculty of Life Sciences and the Multidisciplinary Brain Research Center, BarIlan University, Ramat Gan, Israel

Irina V. Zhdanova BioChron LLC, Worcester, MA, United States

《斑马鱼行为与神经遗传学手册》
译者名单

主　译： 李丽琴

副主译： 石　童

其他翻译人员（按照姓氏笔画顺序排列）：

王　陈　石晶晶　朱思庆
刘东鑫　刘占彪　刘艳丽
李海芸　吴方晖　宋云扬
宋良才　张　毅　张瑞华
陈学军　宗星星　赵　丽
郭煊君　曹念念　靳　倩

致　谢

感谢我的家人支持包容我熬夜编写这本书，感谢我的学生和同事们，他们的才华与热情让我的学术生活充满了乐趣，感谢我的匿名推荐人，他们专业的意见帮助我提高了这本书的质量。我也感谢多伦多大学密西沙加分校为我提供了良好的基础设施与学术环境，同时也感谢爱思唯尔公司编辑部的工作人员，尤其是 Kristi Anderson 女士和 Natalie Farra 博士给予我的无私帮助与支持。最后要感谢辛勤与专注的作者们，是他们的辛苦与付出，才使这本书得以顺利出版。

译者的话

《斑马鱼行为与神经遗传学手册》是加拿大学者 Robert T. Gerlai 主编的有关“斑马鱼模式生物研究”主题的专著。该书一方面强调了斑马鱼具有的很多出色的特性使其成为神经科学研究的有用物种；另一方面也强调了斑马鱼作为除小鼠之外的另一种重要的模式生物，能进一步提高生物医学研究的人体转化相关性，尤其在帮助研究者更好地了解人类大脑的功能和功能障碍方面具有无可比拟的优势。全书共分为六个部分。第一部分简要介绍了斑马鱼，并讨论了有关该物种的自然生态和行为学以及斑马鱼的饲养和繁殖方法等问题。第二部分和第三部分分别介绍了斑马鱼仔鱼和成鱼的行为，包括知觉系统、运动功能、动机，以及一些复杂的行为，包括社会行为、睡眠行为、攻击性和反捕食行为以及学习和记忆的不同形式和方面等。第四部分介绍了一些为斑马鱼开发或采用的现代重组 DNA 方法，包括转基因报告基因系统和 CRISPR/Cas9 系统等。第五部分为对上述讨论的主题进行总结，并综述了如何利用斑马鱼模拟和研究人类的不同中枢神经系统疾病和问题，包括乙醇滥用相关疾病、帕金森病、癫痫、衰老和睡眠障碍、自闭症和冲动行为等。第六部分介绍了数据收集和生物信息学相关的主题，包括如何设计和进行高通量行为筛选、采用什么软件工具进行行为测试、如何利用生物信息学资源进行遗传分析以及斑马鱼大脑解剖结构探索等。

本书系统介绍了斑马鱼作为模式生物的行为和神经遗传学相关的最新研究方法和最新概念，描述了斑马鱼仔鱼和成鱼的神经行为，提供了如何将这些行为方法用于斑马鱼疾病模型的大量实例，讨论了与斑马鱼神经行为遗传学相关的生物信息学和“大数据”相关问题，展示了斑马鱼在转化医学方面的优势与应用前景，该书所提供的多种疾病模型也可为了解人类疾病的发生、发展及其分子机制等提供良好的参考。

国内尚未见有关斑马鱼的行为和神经遗传学系统研究方面的论著。本书的编著者均为各自领域的专家，他们来自世界著名的科研机构和大学，他们从事斑马鱼模式生物研究工作并搜集了大量关于斑马鱼的行为和神经遗传学研究的相关资料。

本书是读者全面系统了解斑马鱼在生物医学转化研究方面的进展情况的综合性参考资料。因此，本书对从事神经行为学、神经生物学、神经药理学、临床医学及相关专业的科学研究人员等具有较大的参考价值，并可作为研究生教学参考书。

本书由中国人民解放军军事科学院防化研究院相关研究人员负责翻译。李丽琴研究员担任主译，石童副研究员担任副主译。王陈博士、石晶晶博士、朱思庆博士、刘东鑫硕士、刘占彪硕士、刘艳丽博士、李海芸博士、吴方晖博士、宋云扬博士、宋良才硕士、张毅博士、张瑞华博士、陈学军博士、宗星星博士、赵丽博士（首都医科大学）、郭煊君硕士、曹念念硕士、靳倩博士分别对有关章节进行了翻译和审校。在此，对所有参与翻译和审校的工作人员表示最诚挚的谢意！

由于水平有限，疏漏难免，请读者批评指正！

译 者

2023年9月

原版前言：行为与神经遗传学中的斑马鱼

美国总统乔治·布什（George H. W. Bush）的总统公告（#6158）宣称：20世纪的最后10年为“大脑十年”，旨在“提升公众对于从大脑研究所获益处的认识”（https://www.loc.gov/loc/brain/proclaim.html）。进入21世纪后仍延续了这一主题。2013年，美国总统奥巴马（Barack H. Obama）宣布启动BRAIN（通过推进创新神经技术进行大脑研究）计划（https://obamawhitehouse.archives.gov/BRAIN），其目标是“彻底改变我们对人类大脑的理解”，政府机构［包括美国国家卫生研究院（NIH）https://braininitiative.nih.gov/］和私营企业都已经参与其中。大脑作为我们身体中最复杂的器官，作为科学研究的最前沿，长期以来使得政治家和工作者们为之着迷。在全球范围内已有数亿美元的资金被用于大脑研究，我们对大脑的认识也随着研究的拓展而迅速增长，包括宏观结构、微观结构、连接组、生理学、细胞生物学、突触功能，以及该极具吸引力的器官的分子生物学和生物化学等方面。其中一些研究并不是直接针对人类进行的，而是针对科学家们认为与人类相似的实验室生物进行的。在21世纪大多数实验室的神经科学研究都是采用单一的生物模型小白鼠（*Mus musculus domesticus*）进行，专注于单一模式生物研究也是符合情理的，对该物种生物组织的多个层面、在广泛实验室进行的无数独立研究中所获得的经验知识能更容易地结合在一起。另一种生物模型，如斑马鱼，能为已经积累的大量知识补充些什么呢？我们希望这本书能提供一个结论性的答案。

本书的大部分章节都首先强调了斑马鱼在行为和神经遗传学方面与其他模式生物相比的优势。事实上，斑马鱼的确有很多出色的特性使其成为神经科学研究的有用物种，我不会在这里重申这些优势。但是我将提请读者注意一个观点，该观点为使用该物种提供了强有力的理由，即比较法的强有力的作用。简而言之，当科学家们用来建模和分析人类功能或疾病机制时采用单一实验室生物模型可能并不够，在该项研究中增加采用其他物种可能会进一步提高研究的转化相关性，即帮助研究者更好地了解人类大脑的功能及其障碍。但是，为什么这种与人类和其他哺乳动物在生物进化上相隔约4亿年的鱼对转化或生物医学研究有用呢？

解释该答案的最简洁的方法是使用布尔代数（Boolean algebra）（图 0-1）。为此，请参考我们对生物进化的了解。我们知道物种之间是相互关联的，一般而言，它们之间的关系越密切，它们拥有共同祖先的时间就越短，也就越相似。问题是我们通常不知道如何量化、识别和 / 或解释“相似性”。例如，如果我们能够确定地知道两个物种之间看起来相似的特征或生物学功能具有共同的进化起源（我们在进化生物学中称为“同源性”的相似性），那么我们就可以挑选具有与我们人类特征足够相似的物种，只需研究这些物种的这些特征就可以外推到人体，并能完全避免用人体自身研究的复杂性。然而，问题是通常明显的相似性（即使我们可以精确地定义和量化其程度）并不能保证进化的同源性。看似相似的结构和功能可能有不同的进化起源，因此，通常正是由于这个原因，它们背后也有不同的机制，例如昆虫、鸟类或蝙蝠的翅膀。我们如何确定我们在人类和另一个物种之间发现的明显相似的特征在进化上确实是同源的，并且在机制上也是相似的呢？这就是比较方法学的用武之地。

根据布尔代数逻辑寻找相似性就是在图 0-1 中寻找各区域之间的重叠区域。例如，假设圆圈 A 代表人类记忆的所有分子机制，圆圈 B 代表小鼠相同大脑功能的机制，重叠区域表示两者的相似度。当然，问题是我们目前还不能确切地知道这个重叠区域是什么，也就是说，我们还不知道小鼠和人类记忆的所有机制，更为重要的是，我们还不知道在这些物种中无记忆的所有机制。那么，我们如何才能将注意力集中在真正构成记忆的相关机制上呢？如果增加另一个物种研究，那么我们现在就可以在其中识别潜在的记忆机制（圆圈 C），并寻找这三个物种之间存在的重叠区域，我们能确定这三个物种间共有的一组机制。鉴于物种间彼此相关，人们总是能够识别这种重叠，即共同机制。已确定的共性不仅代表了与特定目标表型（在我们的示例中为记忆）更相关的机制，而且它们也可能代表了该目标表型更为基础的方面，因为它们代表了所研究物种的进化保守方面。在进化上

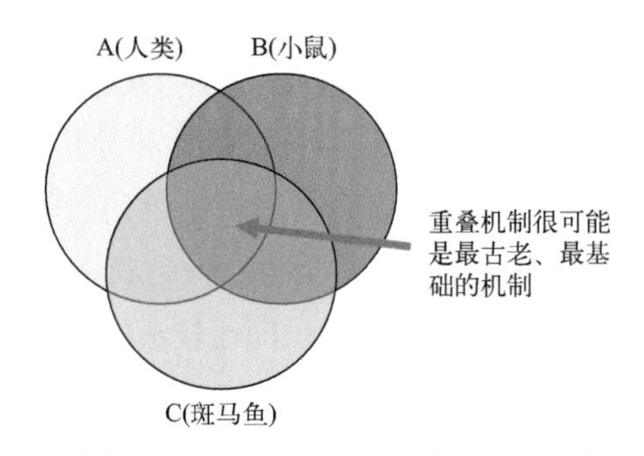

图 0-1　比较法的布尔代数逻辑解释了为什么使用多种物种可增加研究的转化相关性

圆圈 A（人类）、B（小鼠）、C（斑马鱼）分别代表针对这些物种的生物功能（例如记忆）相关机制。这三个圆圈之间的重叠区域代表了所有三个物种共有的机制。这些机制在进化上是最保守的，因此是研究生物学功能的最基础的核心部分

保守的特征、机制可能是我们生物学中最重要的核心方面。研究斑马鱼并将我们从该物种获得的结果与从小鼠和人类获得的结果进行比较，可能会增强我们对人类大脑功能和功能障碍的理解，即可能增加研究的转化相关性。为了能够进行这样的比较，我们需要积累关于第三种物种的研究数据，例如斑马鱼。本书首次尝试在神经行为遗传学领域内搜集此类数据方面做出一些努力。

本书并非旨在全面介绍斑马鱼神经和行为遗传学的方法和原理，这个快速发展的领域的相关知识已经相当广泛（图 0-2），本书并不能涵盖其所有的发展历程。我们的目标是通过对采用斑马鱼进行的一些最有趣和发展最快的神经和行为遗传学研究的部分研究成果来展示该物种。这些章节的作者都是各自领域的专家，其中一些是资深研究人员，另一些则是后起之秀，而且按照科学惯例，他们来自世界各地。

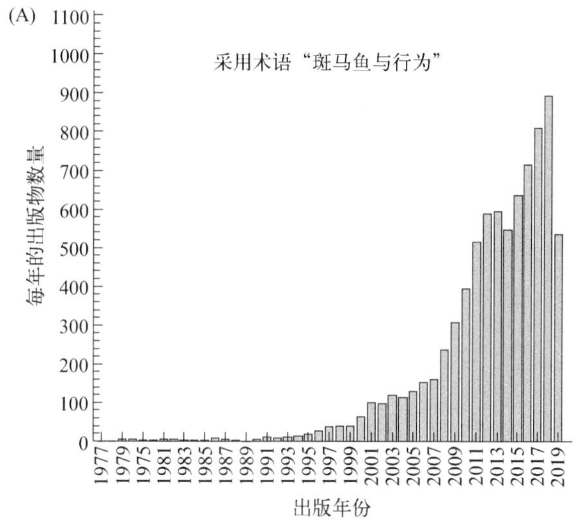

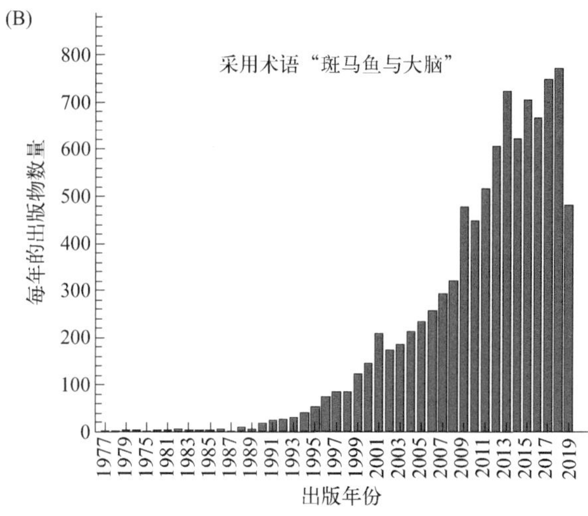

图 0-2

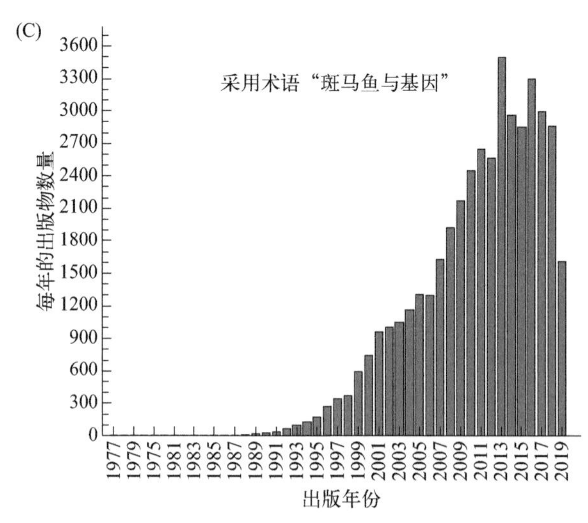

图 0-2　通过使用简单术语“斑马鱼与行为”（zebrafish AND behavior）[图（A）]、“斑马鱼与大脑”（zebrafish AND brain）[图（B）] 或“斑马鱼与基因”（zebrafish AND gene）[图（C）] 在 ISI Web of Knowledge 检索系统搜索（该图书馆搜索引擎中的所有数据库），获得的每年经同行评审的出版物数量随时间的变化情况

请注意这些出版物的数量呈现迅速增加趋势。另请注意，2019 年获得的值代表部分（不完整计数）结果，因为统计数据是在当年的年中（7 月）获得的

这本书是为那些有兴趣学习斑马鱼研究的神经科学家和愿意考虑在研究中使用遗传学、神经科学和行为学方法的斑马鱼专家编写的，这本书可能对那些在神经和行为遗传学研究方面感兴趣的本科生和研究生，以及对希望了解这一快速发展和激动人心的领域的新方法和方向的科研工作者有帮助。

R. T. 盖尔洛伊

（刘占彪翻译　李丽琴审校）

目　录

第三部分 109

斑马鱼成鱼的行为

第 一 部 分

斑马鱼概述：自然生态、行为学以及适宜的饲养繁殖条件

第一章

斑马鱼的生物和非生物环境

David M. Parichy[1], John H. Postlethwait[2]

1.1 引言

斑马鱼（*Danio rerio*）因其在发育遗传学上的潜在用途而被大家熟知（Chakrabarti et al., 1983; Kimmel, 1989; Grunwald and Eisen, 2002; Varga, 2018），该物种在我们理解行为和行为背后的神经生物学方面也变得越来越重要（Kalueff et al., 2013; Gerlai, 2014; Orger and de Polavieja, 2017; Zabegalov et al., 2019）。如果将斑马鱼简单地看作一种模型系统，对斑马鱼的深入了解可能有助于我们推广到对其他脊椎动物的认识；而如果将斑马鱼看作拥有自己的进化史，并对历史和现状具有选择机制的物种，那么认识它将可以帮助我们了解野生动物在行为、生活史特征、形态学、发育遗传以及生理方面是如何形成的。在斑马鱼的相关研究中，针对前一种观点的研究仅需具备斑马鱼和研究设施即可，而针对后一种观点的研究则需要将斑马鱼、研究设施与斑马鱼自然发展史的详细资料结合起来进行。

本章对斑马鱼的分布范围、进化起源以及它所处的非生物和生物环境进行简要概述。这些资料较为零散，而且至今为止实地考察研究范围也有限，因此本章的目的并不是详细描绘斑马鱼在自然界中的行为，而是收集一些重要的观察结果，为今后的工作提供一些建议。

1.2 地理分布与物种进化史

目前鲋属（*Danio*，也有称为短担尼鱼属）已确定有 26 个物种，分布于中南亚和东南

[1] Department of Biology and Department of Cell Biology, University of Virginia, Charlottesville,VA, United States.
[2] Institute of Neuroscience, University of Oregon, Eugene, OR, United States.

亚地区（Tang et al., 2010; McCluskey and Postlethwait, 2015; Froese and Pauly, 2019）。由于包含的地理区域较大且目前参与野外研究的研究人员相对较少，我们将来可能会发现更多的𫚥属新物种。实际上，近年来人们已经确认的该属的新物种数量明显增加［图 1-1（A）］。

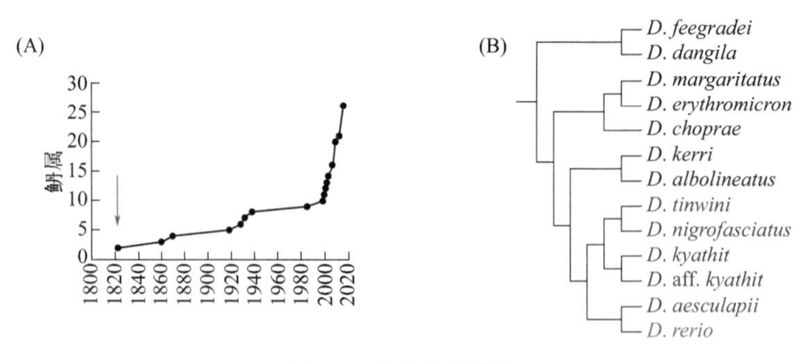

图 1-1 **斑马鱼的进化**

（A）𫚥属物种的累计数量，从 *Danio rerio* 开始累计（红色箭头）(Hamilton, 1822)。（B）几类𫚥属物种的系统进化关系，*Danio rerio* 以蓝色和红色字体表示。进化树只显示了几个已知进化系统中的一个。目前除了有对𫚥属物种方面的描述性文献（Fang, 1997b, 2000; 1997a; 1998; Fang and Kottelat, 1999, 2000; Roberts, 2007; Sen, 2007; Kullander and Fang, 2009b, 2009a; Kullander et al., 2009, 2015; Kullander, 2012, 2015; Kullander and Britz, 2015; Kullander and Noren, 2016），也有关于𫚥属物种的彩色图像和进化方面的文献（Quigley et al., 2004, 2005; Mills et al., 2007; Patterson et al., 2014; Spiewak et al., 2018; Lewis et al., 2019; Patterson and Parichy, 2019）

斑马鱼分布较广泛，近年来有分布于印度、尼泊尔、孟加拉国和缅甸等国的记载（Spence et al., 2006, 2007b, 2008; Engeszer et al., 2007; Whiteley et al., 2011; Arunachalam et al., 2013; Suriyampola et al., 2015; 以及其中的参考文献）［图 1-1（B）］。也有斑马鱼出现在巴基斯坦的报道（Daniels, 2002）。由于斑马鱼的界定条件不清，更不确定的是，形态相似的条纹𫚥属物种也有定义不清的情况，如黑带斑马鱼（*Danio quagga*）和花点斑马鱼（*Danio* aff. *kyathit*）（有可能 *Danio quagga* 和 *Danio* aff. *kyathit* 是同一物种），在印度东部各邦这些条纹𫚥属物种有可能被误认为是斑马鱼，而在缅甸，斑马鱼也有可能被误认为是条纹𫚥属物种（Fang, 1998; Quigley et al., 2005; Kullander et al., 2009; McCluskey and Postlethwait, 2015）。

与研究其他𫚥属物种的进化史类似，研究者已在不同条件下对斑马鱼的进化史进行了研究，使用不同方法其结论也不相同（Meyer et al., 1993, 1995; McClure, 1999; Parichy and Johnson, 2001; Sanger and McCune, 2002; Fang, 2003; Mayden et al., 2007; Fang et al., 2009; Tang et al., 2010）。最近有学者使用全基因组数据进行的分析指出，斑马鱼物种与其他𫚥属物种的亲缘关系很难辨别，因为在系统发育谱系中至少某种程度上有染色体片段渗入（McCluskey and Postlethwait, 2015）。由于𫚥属新物种的基因组测序已列入发展框架，进一步了解斑马鱼进化行为及其他特点的时机可能变得更加成熟。

1.3 非生物和生物环境的特征

表 1-1 总结了五项研究和未公开的印度和孟加拉国斑马鱼栖息地的现场观测记录（McClure et al., 2006; Spence et al., 2006; Engeszer et al., 2007; Arunachalam et al., 2013;

表 1-1 斑马鱼在野外的非生物条件

位置	海拔[①] /m	亲本水域[②]	深度[③] /cm	衬底[④]				水条件				研究		
				A	B	C	D	流速[⑤]	温度 /℃	pH	透明度[⑥]	来源[⑦]	来源地	月份
孟加拉国	~2	沟渠	80	×				静止	20	8	清澈	A	1	一
孟加拉国	~5	孤立河道小河	50	×				静止	22	8	中等	A	9	一
孟加拉国	~16	人工池塘	15	×				静止	21	8	清澈	A	15	一
孟加拉国	~16	池塘	40	×				静止	20	8	中等	A	16	一
孟加拉国	~16	池塘	103	×				静止	17	8	中等	A	17	一
孟加拉国	~18	池塘	96	×				静止	21	8	中等	A	22	一
孟加拉国	~18	水田沟	50	×				静止	23	8	清澈	A	23	一
孟加拉国	~18	池塘	65	×				静止	33	8	中等	A	23	一
孟加拉国	~18	灌溉渠	75	×				静止	33	8	中等	A	25	一
孟加拉国	~18	毗邻河流的小溪	120	×				流动	30	8	中等	A	26	一
印度西北、北、东北	~6,~129,~520	稻田，慢溪	16~57					慢速	27~34	7.9~9.2	清澈	B	2,4,5	九，十
印度东北	51	小溪二级支流边的沼泽	nd	×				慢速	25	6.8	清澈	C	1	七
印度东北	63	小溪二级支流	nd	×				中速	34	6.5	清澈	C	3	七
印度东北	63	小溪二级支流	nd	×				中速	33	6.4	中等	C	4	七
印度东北	63	小溪二级支流	nd	×				中速	32	6.3	清澈	C	5	七
印度东北	63	稻田	nd	×				非常慢速	39	7.3	清澈	C	6	七
印度东北	54	小溪二级支流	nd	×				慢速到中速	27	7.1	清澈	C	9	七
印度东北	74	沼泽	nd	×				静止	27	6.3	清澈	C	11	七
印度东北	50	小溪二级支流和水坑	nd	×				慢速到中速	28	6.1	清澈	C	12	七

续表

位置	海拔[1]/m	亲本水域[2]	深度[3]/cm	衬底[4]				水条件				研究		
				A	B	C	D	流速[5]	温度/℃	pH	透明度[6]	来源[7]	来源地	月份
印度东北	1323	小溪二级支流	nd	×				中速	27	6.5	清澈	C	14	七
印度东北	1323	小溪二级支流	nd	×				慢速	26	6.2	清澈	C	15	七
印度东北	1323	小溪二级支流，毗连的水坑	nd	×				静止到非常慢速	28	5.9	清澈	C	16	七
印度东北	1234	小溪二级支流	nd	×				快速	25	nd	清澈	C	18	七
印度东北	14	鱼塘	nd	×				静止	32	8.1	混浊	C	24	七
印度东北	14	鱼塘	nd	×				静止	31	7.8	混浊	C	26	七
印度东北	647	小溪一级支流	nd	×	×			慢速	20	6.3	清澈	D	11	十二
印度东北	675	小溪	nd	×	×			慢速	19	6.9	清澈	D	12	十二
印度东北	710	小溪	10~15	×	×			慢速	20	7.3	清澈	D	13	十二
印度东北	236	大河旁的池塘	nd	×	×			慢速	22	7.0	清澈	D	14	十一
印度东北	214	鱼塘	70		×	×		中速到快速	21	7.2	清澈	D	15	十一
印度东北	274	斜面	nd	×					22	7.3	混浊	D	16	十一
印度东北	342	小溪	10~60	×	×			中速	20	7.1	清澈	D	17	十一
印度东北	218	小河	20~60		×	×	×	中速到快速	19	6.3	清澈	D	18	十一
印度东北	396	小溪三级支流	nd		×			快速	21	7.3	清澈	D	19	十一
印度东北	341	小溪三级支流	1~15	×	×	×		慢速	20	6.5	清澈	D	20	十一
印度东北	1576	邻近小溪的水坑	nd						12	6.2	清澈	D	21	六
印度北部	268	小溪三级支流	30	×				中速	28	6.4	清澈	D	9	五
印度北部	656	养鱼场	nd	×				中速	24	7.5	清澈	D	10	五

续表

位置	海拔[①]/m	亲本水域[②]	深度[③]/cm	衬底[④]				水条件				研究		
				A	B	C	D	流速[⑤]	温度/℃	pH	透明度[⑥]	来源[⑦]	来源地	月份
印度东部	418	小溪四级支流	15~35		×	×		中速	27	6.6	清澈	D	5	五
印度东部	432	小溪三级支流	10~80		×	×		中速	27	6.4	清澈	D	6	五
印度东部	245	小溪二级支流	10~50	×				中速	27	7.2	混浊	D	7	八
印度东部	238	小溪二级支流	nd		×			中速	25	9.8	混浊	D	8	八
印度西南	328	小溪二级支流	10~35		×	×		中速	26	6.7	清澈	D	1	六
印度西南	680	河流	10~32		×	×		中速	26	7.1	清澈	D	2	三
印度西南	615	小溪	10~110		×	×	×	慢速	17	7.9	清澈	D	3	五
印度西南	599	小溪三级支流	30~45		×	×	×	慢速到中速	17	7.1	中等	D	4	五
印度东北	12	稻田	15	×				静止	27	7.7	nd	E	SV	十，十一
印度东北	12	灌溉渠	13	×				静止	27	7.4	nd	E	SN	十，十一
印度东北	41	小溪	10	×				快速	27	7.5	nd	E	FM	十，十一
印度北部中心	324	小溪	15	×				慢速	27	7.5	nd	E	FV	十，十一
印度东北	690	道路上的水坑	5		×			中速	37	nd	清澈	F		八
印度东北	1410	稻田	10~20	×				慢速	23	6.2	清澈	F		八

① ~表示从位置上通过谷歌卫星地图推断的数据。

② 在某些情况下，可能是在溪边的小水坑中收集到的鱼。

③ nd，表示未测量。

④ A，泥或淤泥；B，沙子；C，卵石、砾石或鹅卵石；D，大石块或基岩。

⑤ 静止，没有检测到流速；慢速，<6cm/s；中速，6～12cm/s；快速，>12cm/s。

⑥ 清澈，见底或30cm深；中等，<30cm深；混浊，肮脏、泥泞或<5cm深。nd，表示未测量。

⑦ A, Spence et al., 2006; B, McClure et al., 2006; C, Engeszer et al., 2007b; D, Arunachalam et al., 2013; E, Suriyanpola et al., 2015; F, Postlethwait et al., 实地观察。

Suriyampola et al., 2015）。尽管这些研究强调的是在不同环境下的不同特征，但也出现了一些共性。图 1-2 显示了一些斑马鱼栖息地。

图 1-2 **斑马鱼栖息地的例子**

（A）～（D）在印度梅加拉亚邦，附近有不流动的池塘或稻田的溪流（Engezer et al., 2007, 14 ～ 16 号地点）。（A）鸟瞰图，记录于 2009 年 11 月（谷歌卫星地图），显示了 2006 年 7 月拍摄的图片（B）、（C）和（D）的位置和方向。（B）河岸植被良好的小溪。（C）河上的稻田。（D）溪流附近的有水下植被的静止水，在季风季节被灌满水，但从空中看是干燥的。插图显示了一个小浅滩的两条斑马鱼，靠近水表面碎屑（箭头，在全图中）。（E）印度西孟加拉邦的一条小溪（Engeszer et al., 2007，3 号地点）。（F）印度奥里萨邦有斑马鱼和养殖鱼的混浊池塘（Engeszer et al., 2007，4 号地点）。（G），（H）印度梅加拉亚邦的现场，包括 37℃（北纬 25.740610 度，东经 91.806707 度）的淹没路面的水和邻近城镇的小溪（北纬 25.604767 度，东经 91.896118 度）。（G）图中的插图显示了浅水中的一小群斑马鱼和一条奇怪的具有鲜明特征的斑马鱼，其头部有反光虹膜斑点（箭头所指）（野外观察，Postlethwait, Raman, Chatterjee, Dey and Mylliemngap, 未出版）

斑马鱼通常出现在处在海平面至海拔1500m以上的浅溪流、死水区、稳定或间歇连接到溪流的盲渠或溪边的池塘。斑马鱼生存的水体底部通常是细泥、粉土或砂，也可以是卵石、较大的岩石或混合类的石块。斑马鱼不仅存在于天然水体中，还存在于稻田、灌溉沟渠和人工池塘中。无论是天然还是人工的水体，斑马鱼最常占据静止或缓慢移动的水域，有时也靠近流速相当快的主河道。河岸通常有丰富的悬垂和地表植被，且水下植物通常很丰富。斑马鱼利用植被提供的荫蔽和相当复杂的微环境进行产卵和仔鱼生长，以及避免被捕食仔鱼的（其他种类的）生物捕食。具有丰富植被的静水或近静水也确实发现有产卵的斑马鱼，也有仔鱼。然而，目前对野生斑马鱼的生命周期以及实验室观察结果与野外特定环境和行为的相关性仍知之甚少（Spence et al., 2007a, 2007b; Engeszer et al., 2008; Suriyampola et al., 2015）。

野生斑马鱼与其栖息地的结构和生物丰富性相适应，会以一些动物为食，特别是陆生和水生昆虫，但也会食用一些碎屑、藻类和高等植物等（McClure et al., 2006; Spence et al., 2007b; Arunachalam et al., 2013）。

斑马鱼所处的水域往往是清澈的，尽管偶尔也能在较混浊的环境中发现斑马鱼，这与降雨或其他干扰因素有关。遗憾的是，斑马鱼在水环境中所处的光谱和强度的细节还没有记载。这些数据对于解释竞争性互动、求爱、群聚以及躲避捕食等行为和潜在的相关电信号是至关重要的（Rosenthal and Ryan, 2005; Rosenthal, 2007; Engeszer et al., 2008; Kalueff et al., 2013; Lewis et al., 2019; Zabegalov et al., 2019）。

尽管实验室里斑马鱼的生存环境通常保持在28℃，但它们在自然界中生存的温度区间相当大，有记载它们生活在12～39℃的水域中。似乎可以肯定的是，一些斑马鱼种群也可承受低温的环境，在夏季的6月和7月，对温度最低的高海拔区域调查发现了斑马鱼。斑马鱼出没的水域的pH值为5.9～9.2。这些生存条件表明，不同斑马鱼种群之间可能存在有趣的热适应和其他生理适应性。明确这些参数在不同季节的均值和方差（包括在种群内部和种群之间）将为了解斑马鱼的生存环境特征提供重要的信息。

探寻斑马鱼栖息地的其他鱼类物种对了解斑马鱼形态和行为尤为重要。大量的物种现已被证明与斑马鱼出现在同一水体中（表1-2；图1-3）。其中包括潜在的竞争对手，如鳉鱼（*Aplocheilus*）、长须鲃（*Esomus*）、无须鲃（*Puntius*）等。此外，还存在卵和仔鱼的潜在捕食者，包括鳠鱼（*Mystus*）、泥鳅（*Lepidocephalus*）、刺鳅（*Mastacembelus*）和扁吻鱼（*Psilorhynchus*）。成年斑马鱼的潜在捕食者包括月鳢（*Channa*）、南鲈（*Nandus*）、针嘴鱼（*Xenentodon*）、大鲶鱼等。除了硬骨鱼外，有齿鱼类和其他水生无脊椎动物也有可能会捕食斑马鱼。目前尚不清楚包括涉水鸟或潜水鸟在内的陆生动物是否会捕食斑马鱼。探寻斑马鱼的捕食生物和与其有竞争的生物将为我们了解斑马鱼的选择机制以及整个生命周期阶段行为变化的生态情况提供有价值的启示（Engeszer et al., 2007a; Buske and Gerlai, 2011）。

表1-2 **与斑马鱼共生的硬骨鱼类**

科	属种
异鳉科	肉色青鳉，青鳉属物种
双边鱼科	肩斑玻璃鱼（玻璃鱼属）
攀鲈科	龟壳攀鲈（攀鲈属），无线棕鲈[①]，恒河毛足斗鱼[②]，条纹毛足鲈，拉利毛足鲈，印度副双边鱼[③]，兰副双边鱼[③]

续表

科	属种
鳗鲡科	孟加拉鳗鲡（鳗鲡属）
单唇鲻科	条纹鰕�鲻，蓝鰕鲻
棕鲈科	眼带棕鲈（棕鲈属），印度黛鲈（黛鲈属），棕鲈属物种
鲮科	布氏鳠，脂鳠，恒河鳠，条纹鳠（鳠属）
颌针鱼科	尖嘴扁颌针鱼
鳢科	宽额鳢，东方鳢，翠鳢，斯氏鳢，鳢属血斑1型，鳢属血斑2型
慈鲷科	尼罗罗非鱼（罗非鱼属）
鲱科	印度细齿鲱（细齿鲱属），印度小鳞鲥（小鳞鲥属）
鳅科	沙鳅，达林沙鳅，突吻沙鳅，安氏鳞头鳅，冈特似鳞头鳅，雀斑鳞头鳅，似鳞头鳅属1型，似鳞头鳅属2型
鲤科	磨齿钝齿鱼（钝齿鱼属），鳙鱼（鳙属），杰伊异鲴（异鲴属），巴纳低线鱲，本迪尔低线鱲，盖茨山低线鱲，漫游低线鱲（低线鱲属），沙昆鲤，条纹元宝鳊，印度元宝鳊，翼元宝鳊，卷须鲮，印度鲮，里巴鲮，彩花缌唇鲃，鲤鱼，大斑马鱼（鲋属），龙纹斑马鱼（鲋属），梅加拉亚斑马鱼（鲋属），波条鲋（大鲋属），阿萨姆鲋（大鲋属），大鲋，长须鲋（鲋属），安氏墨头鱼，裸吻墨头鱼，马耶墨头鱼，长须墨头鱼，鲢鱼，杰氏高须鲃，科卢斯高须鲃，巴塔野鲮，波加野鲮，博格野鲮，蓝野鲮，生野鲮，芒野鲮，露斯塔野鲮，小鳞光唇鱼，大头骨鳊，纹唇鱼，双斑无须鲃，多肉无须鲃，沼泽无须鲃，玫瑰无须鲃，条纹无须鲃，金色无须鲃，小无须鲃，萨拉无须鲃，沙利无须鲃，蝶无须鲃，斯氏无须鲃，单点无须鲃，真无须鲃，无须鲃属物种，波拉长嘴鱲，细波鱼，波鱼，鲑口波鱼，大眼鲑口波鱼，霍兰氏鲑口波鱼，孟加拉鲑口波鱼，库德里结鱼，摩苏尔马哈那狄卡斯结鱼，结鱼属物种
骨鲱科	里氏拉格鲱
虾虎鱼科	印度短虾虎鱼，舌虾虎鱼，鳍虾虎鱼
刺鳅科	长吻棘鳅，大吻棘鳅，大头吻棘鳅，吻棘鳅属物种（吻棘鳅属），白点刺鳅，大刺鳅（刺鳅属）
南鲈科	南鲈
条鳅科	沙棘鳅（沙棘鳅属），条鳅属物种，南鳅属物种
驼背鱼科	弓背鱼
烛焰鲶科	长须烛焰鲶
丝足鲈科（攀鲈亚目）	恒河毛足鲈（毛足鲈属），拟篦斗鱼（拟篦斗鱼属）
裸吻鱼科	孟加拉裸吻鱼，裸吻鱼，细尾裸吻鱼，裸吻鱼属物种
北非鲶科	巴基斯坦真热带鲶
鲶科	双斑绚鲶，叉尾鲶
鲱科	鲃，黑鲱，纹胸鲱属物种
合鳃科	山黄鳝
海龙科	印度腹囊海龙
四齿鲀科	睛斑鲀

① 无线棕鲈应属于棕鲈科棕鲈属。

② 恒河毛足斗鱼（原文拼写错误，应为*Colisa chuna*），以及条纹毛足鲈和拉利毛足鲈应属于攀鲈亚目丝足鲈科毛足鲈属。

③ 印度副双边鱼和兰副双边鱼应属于副双边鱼科。

注：汇总自 Spence et al., 2006; Engeszer et al., 2007b; Arunachalam et al., 2013; Suriyampola et al., 2015。

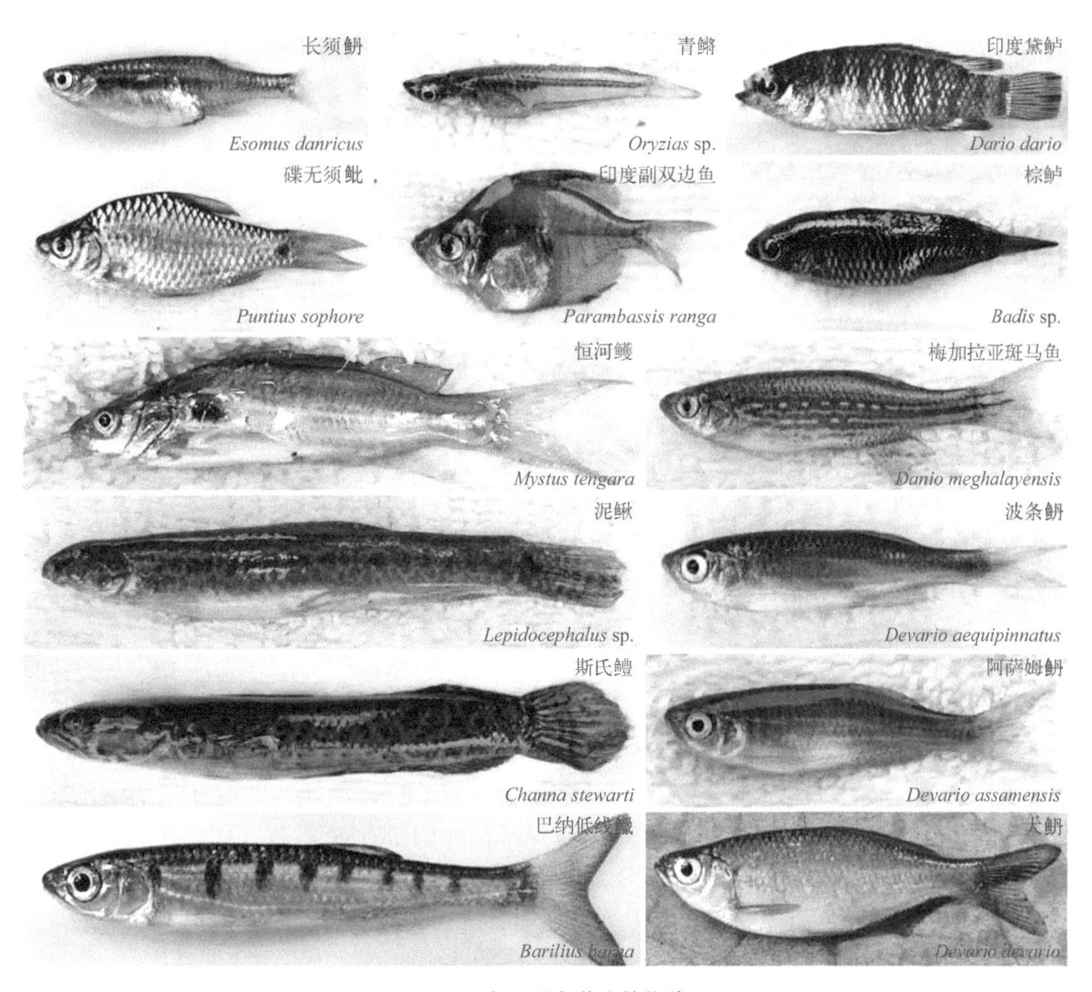

图1-3　**与斑马鱼共生的物种**

仅展示选定的示例。其他分类群的彩色图像可在文献中找到（Engeszer et al, 2007）。长须䲟属、青鳉属和其他属物种可能与斑马鱼有食物竞争关系，而鳞头鳅属、鳢属、鳠属、低线鱲属和其他属物种可能掠食斑马鱼卵、仔鱼或成鱼。体形较大的䲟属和大䲟属物种经常与体形较小的斑马鱼同时出现；其他体形较小的䲟属物种通常出现在斑马鱼分布范围外较远的东边（McCluskey and Postlethwait, 2015）

1.4　栖息地的环境变化

与其他物种相比，人类活动对野生斑马鱼的影响可能更大。已有文献标注出了斑马鱼栖息地的具体退化区域（Engeszer et al., 2007b; Arunachalam et al., 2013），之前有许多适合斑马鱼生存的水域不再适合它们。考虑到斑马鱼的地理分布的广泛性，以及基因组、发育和其他分析方法提供的大量信息，斑马鱼似乎是评估人类活动对其自然栖息地以及对其生态和进化响应的影响程度的一个有价值的模型。斑马鱼作为一种警示模型，可以提示以下信息：人类直接蚕食斑马鱼栖息地的情况，斑马鱼对生活、农业、工业污染物的接触情况，以及全球和区域气候变化对斑马鱼种群及其遗传学上产生的影响。获得这种了解的机会是需要那些在野外研究斑马鱼的研究人员持续发布高分辨率的全球定位系统数据，以便在未来几年探访斑马鱼的栖息地。

1.5 结论

斑马鱼已经成为发育生物学和发育神经生物学、进化生物学、基因组学和毒理学研究的一个良好系统。随着这一物种在行为神经科学或行为学研究中的应用越来越多，我们对其在自然世界中的了解不断深入，相信会提供更有价值的资料，并可能带来其他新的收获。尽管野生斑马鱼的生存环境与大多数实验室使用的斑马鱼相差甚远，但本文介绍的斑马鱼的研究和作者自己的见解表明，野外工作既易操作又具有启发性。如果我们对这些方面进行合作研究，那么这些努力还可以为全球的研究人员和学者搭建宝贵的科学和文化桥梁。

致谢

David M. Parichy 及其实验室人员感谢 NIH R35 GM122571 基金的支持。John H. Postlethwait 感谢 NIH R01 OD011116 基金支持，同时感谢 Rajiva Raman 教授（印度瓦拉纳西巴拿勒斯印度教大学动物学系细胞遗传学实验室）、Anupam Chatterjee 教授（印度东北希尔大学生物技术系主任）、Sankhadip Dey 和 Brandon Keith Mylliemngap（印度东北希尔大学）和 Manfred Schartl 教授（维尔茨堡大学）对实验室的支持，以及在他们熟练的帮助下为采集斑马鱼样本提供的便利。

（宗星星翻译　宋良才审校）

参考文献

Arunachalam, M., Raja, M., Vijayakumar, C., Malaiammal, P., Mayden, R.L., 2013. Natural history of Zebrafish (*Danio rerio*) in India. Zebrafish 10 (1), 1-14. https://doi.org/10.1089/zeb.2012.0803.

Buske, C., Gerlai, R., 2011. Shoaling develops with age in Zebrafish (*Danio rerio*). Progress in NeuroPsychopharmacology and Biological Psychiatry 35 (6), 1409-1415. https://doi.org/10.1016/j.pnpbp.2010.09.003.

Chakrabarti, S., Streisinger, G., Singer, F., Walker, C., 1983. Frequency of gamma-ray induced specific locus and recessive lethal mutations in mature germ cells of the zebrafish, *Brachydanio rerio*. Genetics 103 (1), 109-123.

Daniels, R.J.R., 2002. Freshwater Fishes of Peninsular India. Universities Press (India) Private Limited, Hyderguda, Hyderabad, India.

Engeszer, R., Dabarbiano, L., Ryan, M., Parichy, D., 2007a. Timing and plasticity of shoaling behaviour in the zebrafish, *Danio rerio*. Animal Behaviour 74 (5), 1269-1275. https://doi.org/10.1016/j.anbehav.2007.01.032.

Engeszer, R.E., Patterson, L.B., Rao, A.A., Parichy, D.M., 2007b. Zebrafish in the wild: a review of natural history and new notes from the field. Zebrafish 4 (1), 21-40. https://doi.org/10.1089/zeb.2006.9997.

Engeszer, R.E., Wang, G., Ryan, M.J., Parichy, D.M., 2008. Sex-specific perceptual spaces for a vertebrate basal social aggregative behavior. Proceedings of the National Academy of Sciences of the United States of America 105 (3), 929-933. https://doi.org/10.1073/pnas.0708778105.

Fang, F., 1997a. *Danio maetaengensis*, a new species of cyprinid fish from northern Thailand. Ichthyological Exploration of Freshwaters 8, 41-48.

Fang, F., 1997b. Redescription of *Danio kakhienensis*, a poorly known cyprinid fish from the Irrawaddy basin. Ichthyological Exploration of Freshwaters 7, 289-298.

Fang, F., 1998. *Danio kyathit*, a new species of cyprinid fish from Myitkyina, northern Myanmar. Ichthyological Exploration of Freshwaters 8, 273-280.

Fang, F., 2000. Barred *Danio* species from the Irrawaddy river drainage (Teleostei, Cyprinidae). Ichthyological Research 47 (1), 13-26. https://doi.org/10.1007/bf02674309.

Fang, F., 2003. Phylogenetic analysis of the Asian cyprinid genus *Danio* (Teleostei, Cyprinidae). Copeia (4), 714-728.

Fang, F., Kottelat, M., 1999. *Danio* species from northern Laos, with descriptions of three new species (Teleostei: Cyprinidae). Ichthyological Exploration of Freshwaters 10, 281-295.

Fang, F., Kottelat, M., 2000. *Danio roseus*, a new species from the Mekong basin in northeastern Thailand and northwestern Laos (Teleostei: Cyprinidae). Ichthyological Exploration of Freshwaters 11, 149-154.

Fang, F., Noren, M., Liao, T.Y., Kallersjo, M., Kullander, S.O., 2009. Molecular phylogenetic interrelationships of the south Asian cyprinid genera *Danio*, *Devario* and *Microrasbora* (Teleostei, Cyprinidae, Danioninae). Zoologica Scripta 38 (3), 237-256. https://doi.org/10.1111/j.1463-6409.2008.00373.x.

Froese, R., Pauly, D., 2019. FishBase. Retrieved from: www.fishbase.org.

Gerlai, R., 2014. Social behavior of zebrafish: from synthetic images to biological mechanisms of shoaling. Journal of Neuroscience Methods 234, 59-65. https://doi.org/10.1016/j.jneumeth.2014.04.028.

Grunwald, D.J., Eisen, J.S., 2002. Headwaters of the zebrafish—emergence of a new model vertebrate. Nature Reviews Genetics 3 (9), 717-724. https://doi.org/10.1038/nrg892.

Hamilton, F., 1822. An Account of the Fishes Found in the River Ganges and its Branches. Archibald Constable and Company, Edinburgh.

Kalueff, A.V., Gebhardt, M., Stewart, A.M., Cachat, J.M., Brimmer, M., Chawla, J.S., Craddock, C., Kyzar, E.J., Roth, A., Landsman, S., Gaikwad, S., Robinson, K., Baatrup, E., Tierney, K., Shamchuk, A., Norton, W., Miller, N., Nicolson, T., Braubach, O., Gilman, C.P., Pittman, J., Rosemberg, D.B., Gerlai, R., Echevarria, D., Lamb, E., Neuhauss, S.C., Weng, W., Bally-Cuif, L., Schneider, H., Zebrafish Neuroscience Research, C., 2013. Towards a comprehensive catalog of zebrafish behavior 1.0 and beyond. Zebrafish 10 (1), 70-86. https://doi.org/10.1089/zeb.2012.0861.

Kimmel, C.B., 1989. Genetics and early development of zebrafish. Trends in Genetics 5 (8), 283-288.

Kullander, S.O., 2012. Description of *Danio flagrans*, and redescription of D. choprae, two closely related species from the Ayeyarwaddy river drainage in northern Myanmar (Teleostei: Cyprinidae). Ichthyological Exploration of Freshwaters 23 (3), 245-262.

Kullander, S.O., 2015. Taxonomy of chain *Danio*, an Indo-Myanmar species assemblage, with descriptions of four new species (Teleostei: Cyprinidae). Ichthyological Exploration of Freshwaters 25 (4), 357-380.

Kullander, S.O., Britz, R., 2015. Description of *Danio absconditus*, new species, and redescription of *Danio feegradei* (Teleostei: Cyprinidae), from the Rakhine Yoma hotspot in south-western Myanmar. Zootaxa 3948 (2), 233-247. https://doi.org/10.11646/zootaxa.3948.2.5.

Kullander, S.O., Fang, F., 2009a. *Danio aesculapii*, a new species of danio from south-western Myanmar (Teleostei: Cyprinidae). Zootaxa 2164, 41-48.

Kullander, S.O., Fang, F., 2009b. *Danio tinwini*, a new species of spotted danio from northern Myanmar (Teleostei: Cyprinidae). Ichthyological Exploration of Freshwaters 20 (3), 223-228.

Kullander, S.O., Liao, T.Y., Fang, F., 2009. *Danio quagga*, a new species of striped danio from western Myanmar (Teleostei: Cyprinidae). Ichthyological Exploration of Freshwaters 20 (3), 193-199.

Kullander, S.O., Noren, M., 2016. *Danio htamanthinus* (Teleostei: Cyprinidae), a new species of miniature cyprinid fish from the Chindwin river in Myanmar. Zootaxa 4178 (4), 535-546. https://doi.org/10.11646/zootaxa.4178.4.5.

Kullander, S.O., Rahman, M.M., Noren, M., Mollah, A.R., 2015. *Danio annulosus*, a new species of chain *Danio* from the Shuvolong falls in Bangladesh (Teleostei: Cyprinidae: Danioninae). Zootaxa 3994 (1), 53-68. https://doi.org/10.11646/zootaxa.3994.1.2.

Lewis, V.M., Saunders, L.M., Larson, T.A., Bain, E.J., Sturiale, S.L., Gur, D., Chowdhury, S., Flynn, J.D., Allen, M.C., Deheyn, D.D., Lee, J.C., Simon, J.A., Lippincott-Schwartz, J., Raible, D.W., Parichy, D.M., 2019. Fate plasticity and reprogramming in genetically distinct populations of *Danio leucophores*. Proceedings of the National Academy of Sciences of the United States of America 116 (24), 11806-11811.

Mayden, R.L., Tang, K.L., Conway, K.W., Freyhof, J., Chamberlain, S., Haskins, M., Schneider, L., Sudkamp, M., Wood, R.M., Agnew, M., Bufalino, A., Sulaiman, Z., Miya, M., Saitoh, K., He, S., 2007. Phylogenetic relationships of *Danio* within the order Cypriniformes: a framework for comparative and evolutionary studies of a model species. Journal of Experimental Zoology Part B: Molecular and Developmental Evolution 308 (5), 642-654. https://doi.org/10.1002/jez.b.21175.

McClure, M., 1999. Development and evolution of melanophore patterns in fishes of the genus *Danio* (Teleostei: Cyprinidae). Journal of Morphology 241 (1), 83-105. https://doi.org/10.1002/(SICI)1097-4687(199907) 241:1<83::AID-JMOR5>3.0.CO;2-H.

McClure, M.M., McIntyre, P.B., McCune, A.R., 2006. Notes on the natural diet and habitat of eight danionin fishes, including the zebrafish *Danio rerio*. Journal of Fish Biology 69 (2), 553-570. https://doi.org/10.1111/j.1095-8649.2006.01125.x.

McCluskey, B.M., Postlethwait, J.H., 2015. Phylogeny of Zebrafish, a "model species," within *Danio*, a "model genus". Molecular Biology and Evolution 32 (3), 635-652. https://doi.org/10.1093/molbev/msu325.

Meyer, A., Biermann, C.H., Orti, G., 1993. The phylogenetic position of the zebrafish (*Danio rerio*), a model system in developmental biology: an invitation to the comparative method. Proceedings Biological sciences 252 (1335), 231-236. https://doi.org/10.1098/rspb.1993.0070.

Meyer, A., Ritchie, P.A., Witte, K.E., 1995. Predicting developmental processes from evolutionary patterns e a molecular phylogeny of the Zebrafish (*Danio Rerio*) and its relatives. Philosophical Transactions of the Royal Society B: Biological Sciences 349 (1327), 103-111. https://doi.org/10.1098/rstb.1995.0096.

Mills, M.G., Nuckels, R.J., Parichy, D.M., 2007. Deconstructing evolution of adult phenotypes: genetic analyses of kit reveal homology and evolutionary novelty during adult pigment pattern development of *Danio* fishes. Development 134 (6), 1081-1090. https://doi.org/10.1242/dev.02799.

Orger, M.B., de Polavieja, G.G., 2017. Zebrafish behavior: opportunities and challenges. Annual Review of Neuroscience 40, 125-147. https://doi.org/10.1146/annurev-neuro-071714-033857.

Parichy, D.M., Johnson, S.L., 2001. Zebrafish hybrids suggest genetic mechanisms for pigment pattern diversification in *Danio*. Development Genes and Evolution 211 (7), 319-328.

Patterson, L.B., Bain, E.J., Parichy, D.M., 2014. Pigment cell interactions and differential xanthophore recruitment underlying zebrafish stripe reiteration and *Danio* pattern evolution. Nature Communications 5, 5299. https://doi.org/10.1038/ncomms6299.

Patterson, L.B., Parichy, D.M., 2019. Zebrafish pigment pattern formation: insights into the development and evolution of adult form. Annual Review of Genetics 53. https://doi.org/10.1146/annurev-genet-112618-043741.

Quigley, I.K., Manuel, J.L., Roberts, R.A., Nuckels, R.J., Herrington, E.R., MacDonald, E.L., Parichy, D.M., 2005. Evolutionary diversification of pigment pattern in *Danio* fishes: differential fms dependence and stripe loss in *D. albolineatus*. Development 132 (1), 89-104. https://doi.org/10.1242/dev.01547.

Quigley, I.K., Turner, J.M., Nuckels, R.J., Manuel, J.L., Budi, E.H., MacDonald, E.L., Parichy, D.M., 2004. Pigment pattern evolution by differential deployment of neural crest and post-embryonic melanophore lineages in *Danio* fishes. Development 131 (24), 6053-6069. https://doi.org/10.1242/dev.01526.

Roberts, T.R., 2007. The "celestial pearl danio", a new genus and species of colourful minute cyprinid fish from Myanmar (Pisces: Cypriniformes). Raffles Bulletin of Zoology 55 (1), 131-140.

Rosenthal, G.G., 2007. Spatiotemporal dimensions of visual signals in animal communication. Annual Review of Ecology Evolution and Systematics 38, 155-178. https://doi.org/10.1146/annurev.ecolsys.38.091206.095745.

Rosenthal, G.G., Ryan, M.J., 2005. Assortative preferences for stripes in danios. Animal Behaviour 70, 1063-1066.

Sanger, T.J., McCune, A.R., 2002. Comparative osteology of the *Danio* (Cyprinidae: Ostariophysi) axial skeleton with comments on *Danio* relationships based on molecules and morphology. Zoological Journal of the Linnean Society 135 (4), 529-546.

Sen, N., 2007. Description of a new species of Brachydanio Weber and de Beaufort, 1916 (Pisces: Cypriniformes: Cyprinidae) from Meghalaya, north east India with a note on comparative studies of other known species. Records of the Zoological Survey of India 107, 27-31.

Spence, R., Ashton, R., Smith, C., 2007a. Oviposition decisions are mediated by spawning site quality in wild and domesticated zebrafish, *Danio rerio*. Behaviour 144 (8), 953-966. https://doi.org/10.1163/156853907781492726.

Spence, R., Fatema, M.K., Ellis, S., Ahmed, Z.F., Smith, C., 2007b. Diet, growth and recruitment of wild zebrafish in Bangladesh. Journal of Fish Biology 71 (1), 304-309. https://doi.org/10.1111/j.1095-8649.2007.01492.x.

Spence, R., Fatema, M.K., Reichard, M., Huq, K.A., Wahab, M.A., Ahmed, Z.F., Smith, C., 2006. The distribution and habitat preferences of the zebrafish in Bangladesh. Journal of Fish Biology 69 (5), 1435-1448. https://doi.org/10.1111/j.1095-8649.2006.01206.x.

Spence, R., Gerlach, G., Lawrence, C., Smith, C., 2008. The behaviour and ecology of the zebrafish, *Danio rerio*. Biological Reviews of the Cambridge Philosophical Society 83 (1), 13-34. https://doi.org/10.1111/j.1469-185X.2007.00030.x.

Spiewak, J.E., Bain, E.J., Liu, J., Kou, K., Sturiale, S.L., Patterson, L.B., Diba, P., Eisen, J.S., Braasch, I., Ganz, J.,Parichy, D.M., 2018. Evolution of Endothelin signaling and diversification of adult pigment pattern in *Danio* fishes. PLoS Genetics 14 (9), e1007538. https://doi.org/10.1371/journal.pgen.1007538.

Suriyampola, P.S., Shelton, D.S., Shukla, R., Roy, T., Bhat, A., Martins, E.P., 2015. Zebrafish social behavior in the wild. Zebrafish 13 (1), 1-8. https://doi.org/10.1089/zeb.2015.1159.

Tang, K.L., Agnew, M.K., Hirt, M.V., Sado, T., Schneider, L.M., Freyhof, J., Sulaiman, Z., Swartz, E., Vidthayanon, C., Miya, M., Saitoh, K., Simons, A.M., Wood, R.M., Mayden, R.L., 2010. Systematics of the subfamily Danioninae (Teleostei: Cypriniformes: Cyprinidae). Molecular Phylogenetics and Evolution 57 (1), 189-214. https://doi.org/10.1016/j.ympev.2010.05.021.

Varga, M., 2018. The doctor of delayed publications: the remarkable life of George Streisinger (1927e1984). Zebrafish 15 (3), 314-319. https://doi.org/10.1089/zeb.2017.1531.

Whiteley, A.R., Bhat, A., Martins, E.P., Mayden, R.L., Arunachalam, M., Uusi-Heikkila, S., Ahmed, A.T., Shrestha, J., Clark, M., Stemple, D., Bernatchez, L., 2011. Population genomics of wild and laboratory zebrafish (*Danio rerio*). Molecular Ecology 20 (20), 4259-4276. https://doi.org/10.1111/j.1365-294X.2011.05272.x.

Zabegalov, K.N., Kolesnikova, T.O., Khatsko, S.L., Volgin, A.D., Yakovlev, O.A., Amstislavskaya, T.G., Friend, A.J., Bao, W., Alekseeva, P.A., Lakstygal, A.M., Meshalkina, D.A., Demin, K.A., de Abreu, M.S., Rosemberg, D.B., Kalueff, A.V., 2019. Understanding zebrafish aggressive behavior. Behavioural Processes 158, 200-210. https://doi.org/10.1016/j.beproc.2018.11.010.

第二章

斑马鱼的饲养和繁殖：一些行为学和生态学方面的考虑

Benjamin Tsang[1], Rida Ansari[2,3], Robert T. Gerlai[1]

2.1 引言

斑马鱼应用于基础科学研究的态势发展迅速。大约在四五十年前由俄勒冈大学的George Streisinger博士发起后，斑马鱼研究迅速席卷科学界。斑马鱼作为一种动物模型，广泛应用于科学研究，并出现了指数级增长。斑马鱼适用于许多遗传和药理学操作，因此从发育生物学到分子遗传学，采用斑马鱼作为研究模型的研究人数呈指数级增长。不仅如此，斑马鱼在实验操作上还具有实用特性，包括体形小、繁殖能力强等特点。但关于斑马鱼饲养和繁殖基本原理的研究未呈现指数级增长，导致饲养和繁殖的行业标准（industry standards）欠佳，同时也会因为实验人员操作偏好、经验、能力不同而引起饲养和繁殖方法的不统一。

斑马鱼饲养方面缺乏系统研究的原因之一可能是在过去五十年里斑马鱼研究一直以发育生物学和分子遗传学研究为主（Carney and Mosimann, 2018; Fuentes et al., 2018; Patton and Tobin, 2019）。这些研究中有许多是关注基本发育过程的，专注于斑马鱼个体发育的最初几小时或几天，并研究有较大影响的基因。这些研究的主要目标是高效大量生产斑马鱼胚胎，因此相应地制定了斑马鱼饲养的“行业标准”。然而，正如我们所提到的，在过去的十年里，斑马鱼研究呈现出爆炸性增长，几乎表现在生物学的每个学科分支：从癌症

[1] Department of Psychology, University of Toronto Mississauga, Mississauga, ON, Canada.

[2] Department of Cell and Systems Biology, University of Toronto, Toronto, ON, Canada.

[3] Neuroscience and Mental Health Program, Hospital for Sick Children, Peter Gilgan Centre for Research and Learning, Toronto, ON, Canada.

到动物认知研究（Cayuela et al., 2019; Adams and Kafaligonul, 2018; Choo and Shaikh, 2018; Roper and Tanguay, 2018; Torraca and Mostowy, 2018; Dooley and Zon, 2000; Amatruda et al., 2002）。这些研究通常不仅采用胚胎或仔鱼，还采用成年斑马鱼。此外，这些研究通常调查的生物过程预计会受到许多影响力较小的基因和/或几种环境因素的影响，包括那些在实验室中斑马鱼的饲养和繁殖方法的影响。对于研究脊椎动物身体中最具可塑性的器官之一——大脑的功能神经科学家和行为科学家来说，对这些因素的认识尤为重要。因此，实验室饲养环境条件的优化已成为相关的问题。

我们调查相关文献和网上信息，结果表明不同实验室间，甚至同一实验室内所采用的饲养方法和环境条件都会有很大差异（Tsang et al., 2017）。在最近一次由美国国家卫生研究院组织的研讨会上，学者们也讨论了这一问题，题目是“斑马鱼和其他鱼类模型：严谨实验和可重复结果的外部环境因素”（https://orip.nih.gov/zebrafish-and-other-fish-models-extrinsic-environmental-factors-rigorous-experiments-and）。

本章中，我们将讨论当前采用的一些斑马鱼饲养方法，并在这里强调斑马鱼虽然在实验室中已经饲养了几十年，但通常缺乏对该物种的最佳饲养环境条件的系统和参数分析。我们还认为，为了开发这种优化的饲养和繁殖系统，需要考虑斑马鱼类物种繁衍生息的自然条件。但是并非所有人都同意这一观点。在建立斑马鱼饲养设施和设置斑马鱼饲养和繁殖的环境参数时，应考虑几个因素，其中包括效率、成本、占地面积以及设施维护的便利性。事实上，有些人认为，这些考虑因素否定了实验室环境应模拟斑马鱼在自然界的生活方式。然而，另一些人包括笔者在内的研究者认为，行为学（物种的自然行为）和生态学（物种进化和生活的自然环境条件）因素是至关重要的。在斑马鱼的饲养和繁殖过程中忽视物种进化和生活的自然环境条件因素可能会产生负面后果，例如，水压升高，斑马鱼可能更容易患病。在人工条件下，斑马鱼可能被迫在接近生理/体内平衡耐受阈值的情况下生存。因此，实验室环境中持续存在的条件波动，可能超过耐受阈值，进而引起斑马鱼种群中的方差（环境误差方差）显著增加。这种方差会降低统计检测能力，也就是说，无论研究者采用什么样的实验操作都很难得到有意义的结果。换句话说，根据这个论点，饲养斑马鱼时，不考虑其行为学和生态学因素，可能需要更大的样本量（成本）和/或导致假阴性结果。另一方面，斑马鱼在与其自然界条件相似的条件下饲养，可能会减少误差方差，结果的重复性和重现性更好（Gerlai, 2019）。值得注意的是，二十年前的小鼠神经行为遗传学研究也得出了类似的结论（Gerlai and Clayton, 1999a, 1999b）。在这一点上，还不清楚动物行为学和生态学良好的饲养条件是否具有优越性，或者更准确地说，我们不知道哪些方面的自然条件需要精确模拟，而哪些方面则可以在实验室忽略。由于缺乏对各种环境因素如何影响斑马鱼的系统分析，这些问题的答案仍然未知。

本章回顾了可能需要考虑的一些环境因素，讨论了“行业标准”和未来可能偏离“行业标准”的饲养方法。本章既不是对现有技术的综述，也不是对可能需要调查或修订的所有因素的系统评判。这只是一个简洁的，也许带有个人观点的总结，同时举例说明笔者认为的斑马鱼饲养标准目前或未来的发展方向。

2.2 标准化问题

在深入研究有关斑马鱼饲养的具体问题之前，简要讨论一下标准化本身的利弊。一般来说，这是一个重要的科学问题，在斑马鱼的饲养方面尤其重要。实验室内部和不同实验室之间饲养条件和程序的标准化可确保结果的可比性、重现性和重复性。然而，这种观点也并非所有人都同意。考虑到该物种的自然环境地域以及可能的自然遗传变异性，饲养野生斑马鱼或实验室培育的此类群体是否存在单一最佳饲养条件值得怀疑。此外，实验室品系可能经历过特定场所意外的人工选择和遗传漂变，即驯化，因此不论其来源如何，斑马鱼可能需要特殊的饲养和环境条件。关于标准化是否为最佳途径不属于理论问题，而属于经验问题，必须通过细致的分析加以探讨。通过实地研究已经揭示了斑马鱼在自然界生存的一系列条件。

斑马鱼是一种广温性的硬骨鱼类，原产于印度东部，分布于尼泊尔、巴基斯坦和喜马拉雅山脉，是一种天然耐寒的鱼类，栖息地广泛，能够适应和耐受各种环境条件（例如，Engeszer et al., 2007）。出于这个原因，养鱼爱好者认为斑马鱼是“初学者养的鱼”（beginner’s fish）。这种适应性强的特点或许也是斑马鱼在动物研究中备受欢迎的原因之一。但是这种强适应性以及对各种环境条件的耐受力是有代价的。斑马鱼养殖设施中采取了很多控制因素和方法，例如，水化学参数、饲养系统（housing systems）和繁殖方法，但没有考察所采用的设施是否对饲养特定斑马鱼种群为最优，或者这些条件是否对特定设施中斑马鱼的研究有影响。截然不同的设施和环境条件下饲养，可能会对结果的重现性和重复性产生影响，甚至意外的人工选择，可能导致场所特异性品系的发展。

然而，缺乏标准化对结果的重复性有负面影响还是正面影响，本身就是一个复杂的问题（Gerlai, 2019）。例如，人们普遍认为标准化和可控的实验室条件有助于实验室间结果的重复性，并减少环境误差方差。然而，也有一些人认为随着统计检测能力的提升，即使是很小的误差变化都会检测出来，因此实验操作因素对实验结果并不是非常重要。根据这一观点，从全方位的视角来看，标准化和减少表型误差方差实际上可能对我们有效的科学发现产生负面影响。与此相反的论点并不涉及标准化本身，而是关于如何以最佳方式饲养斑马鱼的问题。根据这一论点，如上所述，将斑马鱼饲养在不利条件下会在生理和 / 或心理上对其造成压力，因此会导致上述生物学误差方差增加，并对我们实验所需的任何操作产生显著偏差。综上，虽然这一观点并不提倡标准化，但却认同确定一套参数使生态学 / 生物学条件最佳来饲养斑马鱼是非常重要的。我们赞同后一种观点。我们认为，首先，必须研究什么才是饲养“标准”斑马鱼的最佳方法。接着，我们应该探究品系或种群在最佳饲养条件下是如何变化的。随后，我们应推动“全行业”采用一套准则，即斑马鱼饲养和繁殖的标准方案、程序、条件参数等。最终，我们可以基于特定斑马鱼实验室种群的具体特性对这些标准进行修改和调整。

由此，问题是我们是否能从自然界得到最佳饲养条件？尽管有关斑马鱼进化和生态学的知识有限，但最早的开创性研究已经开始收集有关斑马鱼自然栖息地的生物（活体）和非生物（非活体）方面 / 因素的数据（例如，Engeszer et al., 2007）。

2.3 自然界中的斑马鱼

首次发现斑马鱼是在印度东北部的比哈尔邦。在印度半岛的南部和西部，包括班加罗尔市，以及印度以外，北至巴基斯坦和尼泊尔，甚至孟加拉国和缅甸，都发现了斑马鱼。虽然斑马鱼自然种群的行为学和生态学数据很少，但一些田野调查记录了一些自然环境因素。例如，斑马鱼通常生活在软质、中性至碱性的热带淡水中，记录的盐度范围在 10 ～ 270μS 之间，pH 值在 7.0 ～ 7.7 之间，温度在 25 ～ 28℃之间。这些参数，尤其是盐度，应引起重视，因为它们与实验室通常采用的参数值有很大的不同。此外，斑马鱼的典型栖息地是浅而缓慢流动的溪流和静止的池塘，这些池塘是在雨季于溪流旁边形成的。这些溪流和池塘中的水通常是清澈的，但溪流或池塘底部可能有泥、沙或砾石。因此，雨水或暴风雨过后，水常常变得混浊。然而，在实验室里，我们把斑马鱼放在充满水的如水晶般清澈的鱼缸里，其底部是透明的，没有衬底（底部覆盖物）。通常在斑马鱼生活的自然水域中普遍存在水生植被，鱼有时可以通过悬垂植被或悬垂河岸获得掩蔽。这与实验室鱼缸环境非常不同，实验室鱼缸没有任何衬底、植物或其他藏身之处。

2.4 实验室饲养实践

目前斑马鱼饲养业和饲养系统开发的重点是实现斑马鱼高产繁殖，即快速繁殖大量后代，同时尽量减少所需的占用空间，在装置中最大限度地增加饲养动物的数量。斑马鱼繁殖、饲养方法和饲养装置有许多不同方案，几乎没有两个实验室有完全相同的做法。这种差异是实验室内和实验室间研究缺乏重复性的潜在原因。因此，重点是要考虑与当前育种和饲养装置有关的问题。本章我们将从后者开始探讨。

2.4.1 仔鱼和成鱼集中养殖与繁育系统设备

大多数实验室使用商用高密度系统机架（high-density system racks）。这些再循环系统机架是半自动的，能进行水过滤和消毒，包括化学（活性炭）、机械（海绵垫）和生物（生物球 / 转盘 / 流化床）过滤，紫外线消毒，以及每日自动换水。这些高密度系统的主要目标是提高效率，即最大限度地增加斑马鱼数量，同时尽量减少占用空间。因此，这些高密度系统兼具实用和经济的优势：少量的技术人员可完成大量的斑马鱼简单有效繁殖和饲养。但是，针对这些系统还需要注意以下方面。

2.4.2 从行为学角度看高密度系统的问题

2.4.2.1 物理空间

斑马鱼是位于中层至上层水体中快速游动的杂食性动物，并栖息在自然界相对开阔的水域。尽管目前包括笔者实验室在内的世界各地的斑马鱼实验室通常应用效率高且易管理的高密度机架，但这种形式的机架并没有考虑到动物行为学。一般是将包含 10 ～ 50 条鱼的斑马鱼群放到 2.5 ～ 10L 的水缸内。将斑马鱼束缚在如此小的鱼缸里是不符合自然规律的，这可能会导致许多负面后果。狭小的物理空间本身就是一个问题，因为快速移动的斑马鱼无法适应这种环境，可能会无法正常游动。此外，高密度饲养可能导致食物竞争加剧、攻击次数增加或异常，以及无法从更具支配性 / 攻击性的同类斑马鱼中逃离。高密度鱼房也可能导致异常环境条件，包括有机废物水平升高或异常波动、缺氧、盐度和 / 或 pH 值异常。尽管高密度机架系统的中央水槽中水的指标通常是经过测量和良好控制的，但进入特定水箱的流量、进水频率和投料量对系统内的各个水箱也会有很大的影响。

在自然界和实验室中，有机废物是通过氮循环自然去除的。斑马鱼以铵离子（NH_4^+）的形式排泄废物，再由硝化细菌（亚硝基单胞菌）将其转化为亚硝酸盐（NO_2），然后由硝化细菌和硝化螺菌科细菌进一步将其氧化为硝酸盐（NO_3^-），最后被植物作为营养物吸收。在实验室中，较小的水 / 鱼比意味着有环境条件波动的重大风险，这种环境条件可能破坏使铵离子和亚硝酸盐硝化的有益细菌群。由于铵离子和亚硝酸盐含量增加，斑马鱼会出现精神紧张和生病的症状，包括鳃发红和突出、换气过度（鳃盖运动频率增加）、焦虑样行为（如栖息在底部或表层）和其他异常行为（如瘙痒）。这些直接的反应也可能导致长期的后果，包括免疫功能降低，对病原体的易感性增加，甚至是影响许多生物功能的长期表观遗传变化。上述问题不仅仅是理论上的，而且已经有证据表明高密度饲养系统可能导致鱼的压力增大（Shams et al., 2017a, 2017b; Parker et al., 2012）。

例如，Shams 等（2017b）研究了长期隔离对斑马鱼的影响，预测与啮齿动物、灵长类动物和人类类似，隔离会增加焦虑并导致皮质醇水平升高，慢性隔离会导致成年斑马鱼的焦虑样相关行为、体内皮质醇和单胺类物质水平升高。然而预测的结果却与发现的结果不一致。他们发现与那些以标准方式饲养在拥挤的小水缸内的斑马鱼相比，被隔离的斑马鱼焦虑和皮质醇水平有所降低。另一项研究也发现了类似的惊人结果（Shams et al., 2017a）。这些发现是值得重视的，因为大量的已有研究已证实斑马鱼具有高度群居性（Miller and Gerlai, 2012, 2011, 2008, 2007），而且该物种是高度积极寻找同物种游动的（Al-Imari and Gerlai, 2008; Saif et al., 2013; Scerbina et al., 2012）。因此，当斑马鱼被群居安置时，焦虑和皮质醇水平的升高不可能是由群居安置所引起的，却很可能是源于上述讨论的小型高密度饲养系统所引起的负面特征。总而言之，目前高密度系统机架的行业标准，可能并非饲养斑马鱼的最佳方法。特别是对于斑马鱼的大脑功能和行为研究，或许需要增加水缸的大小或体积，以及降低鱼的密度。但理想的鱼缸大小是多少、应该在一个鱼缸里养多少条鱼，这些都是未来需要探索的经验性问题。

2.4.2.2 衬底

从饲养学的角度来看，除去衬底被认为是一个明显的优势，因为它可以保持水的洁净，防止碎屑、有机废物和厌氧菌的积聚。然而从行为学或生态学的角度看这是错误的。最近的研究表明，斑马鱼更喜欢有衬底的区域（Schroeder et al., 2014）。此外，当让斑马鱼在贫瘠或丰富的环境中进行选择时，与贫瘠的环境相比，斑马鱼在有植物、陶罐和其他衬底的环境中度过的时间更长（Lee et al., 2018）。即便衬底是沙子和砾石，与无衬底相比度过的时间也更长。有趣的是，当衬底以图像的形式呈现在水槽下时，发现存在对衬底种类的偏好性（Schroeder et al., 2014），这可以是一种模拟自然并使实验室环境丰富的简单方法。

2.4.2.3　植物群

在实验室里，饲养斑马鱼很少使用活的水生植物。然而，斑马鱼在自然界是生活在溪流和稻田中的，那里有丰富的植物种类。实验室里未使用水生植物有利也有弊。若水槽里有活植物，会使斑马鱼的生存环境变得丰富。与那些放置在裸槽里的斑马鱼相比，放置在有活植物环境中的斑马鱼体色更鲜艳，此外还表现出更自然的行为。在斑马鱼繁殖过程中提供人工或活的植物能使其产生更多的鱼卵，因为在斑马鱼求偶过程中，植物很可能为其提供休息的地方，并作为一种天然刺激物。这表明提供人工或活的植物的水槽接近自然繁殖的湖岸。研究发现，雄性和雌性斑马鱼都喜欢植物，不喜欢贫瘠的环境（Lee et al., 2018）。此外，活植物通过硝酸盐吸收在去除有机废物方面起着至关重要的作用。如果没有植物，则以定期换水手动清除废物的方式来取代氮循环中的这一步骤。然而在水箱中放置活植物，除上述好处外，也必须考虑到一些负面因素，即实际问题。斑马鱼自然栖息地中的水生植被，有营养丰富的以沙子、表土和砾石混合物组成的衬底，且植物可以接受强烈的自然光照。而在实验室环境添加砾石、沙子和土壤则是一项挑战，这需要掌握先进的共生体系知识。同时，为植物生长增加适当额定功率的照明也将大大增加运营成本，还可能诱发藻类的强劲生长，这是一种不希望出现的现象，藻类会覆盖水槽表面，使得观察斑马鱼的健康状况变得十分困难。尽管如此，实现实验室水槽中有活植物的潜在方案可能是使用需低光照和不需要土壤的根茎植物。养鱼爱好者和水生植物专业人士经常使用某些植物，例如，榕叶属（*Anubias*）和星蕨属（*Microsorum*）植物，但还没有在斑马鱼实验室饲养设施中进行探索。

2.4.2.4　水温

变温动物如鱼类，对温度变化表现出不同程度的耐受性，并且它们有最佳生长的狭窄温度范围。斑马鱼可以被归类为广温性鱼类，因为它们表现出对宽广温度范围的耐受性。斑马鱼种群其天然栖息地的水温范围为16.5～33℃，而在有温度控制的实验室研究表明，如果有足够长的适应时间，它们可以在6.7～41.7℃的温度下存活（Cortemeglia and Beitinger, 2005）。尽管斑马鱼具有很强的温度耐受性，但其胚胎发育的最佳温度范围却相当狭窄。尽管这一研究结果尚不明确，成年斑马鱼也可能存在一个狭窄的最佳生长温度范围。斑马鱼在28.5℃的恒温条件下生长最快。温度可能会影响多种行为方式，比如性别分化。靠近22～31℃范围的两端或以外的温度会导致斑马鱼群性别比例失衡。还有，较高温度可能导致斑马鱼生存能力下降，包括尾部、心血管系统、头部、胸鳍等结构和器官畸形，并可能导致某些器官水肿或血液异常积聚等。考虑到这些和其他潜在问题，有必要监测不同温度对生长、繁殖和压力等参数的影响，从而更清楚地阐明斑马鱼的热偏好。这些尚未形成系统的研究对于优化和标准化斑马鱼的饲养和繁殖条件是至关重要的。

2.4.2.5 盐度

确定斑马鱼饲养系统中水的盐度，目前的行业标准中有两个潜在的问题：（1）盐的浓度或盐度；（2）所用盐的组成成分。我们在这里先讨论后者。大多数实验室采用反渗透过滤法去除水中溶解的盐、微量元素以及潜在的毒素。然而，反渗透水（R/O 水）对鱼类是致命的。为了避免 R/O 水中缺乏有益的可溶物质，斑马鱼饲养装置中会重新添加盐，以达到所需的盐度水平。这种做法的理由是，除了清除斑马鱼饲养系统水中潜在有害毒素外，还实现了对盐度和含盐量的精确控制。大多数斑马鱼养殖装置都添加海盐将 R/O 水配制成盐度水平适当的水。在我们的装置中，同大多数斑马鱼装置一样，使用“Instant Ocean”（商品名），这是一种当地供应商（水族馆、宠物商店）提供的海盐混合物。然而，使用这种盐有两个重要问题。一是其盐的成分未经科学用途验证，即其成分可能会（或可能不会）随盐混合物的制造日期和供应商的来源而波动。但或许更令我们担忧的是第二点，即海盐的主要成分。斑马鱼饲养设施使用的所有海盐混合物，包括“Instant Ocean”，其主要成分均为氯化钠，可这种盐在斑马鱼的天然水域中几乎不存在。那我们为什么要用这种盐呢？一是氯化钠具有抑菌作用，因此可以较好地控制疾病；另一个原因是这就是“行业标准”，约定俗成，有助于实现结果的普遍性、可转化性和重复性。目前，据我们所知暂时还没有对这种盐与更自然的条件（例如，碳酸钙和 / 或镁盐）使用效果进行对比的系统研究。然而，我们认为这种比较会很有价值，甚至行业标准可能会相应改变。

现行行业标准的另一个问题是盐的浓度，即盐度。斑马鱼是一种淡水鱼，但能耐受各种盐浓度，这反映了斑马鱼对其自然环境中盐浓度变化的适应能力。对养殖的建议盐度通常在 1000μS 左右，甚至经常超过这个值。虽然允许盐度高于或低于这个水平，但这可能会影响生长、存活和繁殖。例如，在盐度为 2000μS 的水中孵化的 2 ～ 4 细胞期胚胎的孵化率降至对照组的 86%，甚至发现更高的盐度会致命；短时间（1 ～ 2h）暴露于高盐度（≥ 6000μS）也会有害，导致孵化率降低、卵裂球形成中断、孵出时间延长。尽管斑马鱼胚胎逐渐发展出渗透调节能力，但这些研究和其他研究表明，海水盐度的变化可能影响斑马鱼仔鱼的多种生物过程，包括与内分泌系统相关的基因表达的改变（Hoshijima and Hirose, 2007）。而不同盐度对成年斑马鱼的影响尚不清楚。尽管如此，第一批证据表明盐度对斑马鱼的行为和大脑功能有显著影响（Mahabir and Gerlai, 2017）。虽然还没有对斑马鱼生存的最佳盐度进行彻底和系统分析，但我们怀疑这类研究将会发现目前在 1000 ～ 2000μS 水中饲养斑马鱼的行业标准是次优的，因为在自然界中这些斑马鱼生活在盐度为上述值的 1/10 的水域中。

理想情况下，成年斑马鱼储水槽和繁殖槽的水盐度应与胚胎培养基的盐度相匹配。如果成年种鱼储水槽中水的盐度高于此首选水平（很可能是因为成年种鱼繁殖槽每天多次接收食物，且食物中含盐），我们建议实验人员将种鱼放置于水与储水槽的水相匹配的繁殖槽中，并缓慢改变水参数（即降低盐度），以接近胚胎培养基的首选参数。这要耗费几个小时，即在前一天晚上把种鱼放进繁殖槽和第二天早上开灯产卵之间花费的时间。因此，实际上这个实验室程序将模拟季风雨的到来，用软雨水稀释水，这标志着斑马鱼在自然界的活跃繁殖季节来临。上述建议的实验室步骤带来一个新的问题：我们调整水参数的速度是否重要？

2.5 绝对值与水参数变化速度

在斑马鱼养殖中，将偏离的水参数值恢复到期望值的速度往往不受重视，事实上经常被忽视。为了满足重复性要求，受欢迎的斑马鱼养殖设施对所需的水参数有严格的指导规范。尽管如上所述，不同饲养设施的这些参数可能有很大差异，但对相同设施，水参数应保持在一个明确的狭小范围内。特别是水化学性质，包括 pH 值和盐度，以及温度都要进行调整，以符合既定的指导规范。然而，这些规范通常不考虑在检测到某个水参数偏离期望值后该参数调整到期望值的合适速度。这个问题在我们的饲养设施中也存在。例如，在饲养中经常出现的现象，由于有机废物的积累、过滤器堵塞、斑马鱼饲养过多或进料过多，pH 值都可能会急剧下降。一旦 pH 值低于 6.5，报警系统会自动通知用户或由技术人员人工检测发现，这时必须采取纠正措施。这种情况下的纠正措施通常包括添加碳酸氢钠，这是一种能将 pH 值缓冲到高于中性（即高于 7.0）的盐。然而，pH 值从测量的 6.5 到达到期望值，例如 7.5（通常推荐用于斑马鱼），表明 OH^- 浓度增加了 10 倍，而使用碳酸氢钠这种纠正通常在几分钟内实现。进行此类调整的技术人员或学生会确信已采取了适当的纠正措施，并达到了最佳 pH 值，但在这个过程中被忽视的是，实现这种快速纠正会在鱼身上引起巨大的应激性。笔者的个人观察，加上几十年来养鱼爱好人士的经验都表明，对鱼类来说，超出期望值的范围所产生的应激性可能要比这种快速纠正小。水参数纠正速度值究竟达到多大代表太快？水参数偏离最佳范围时鱼类能存活多长时间？不幸的是，这些问题至今没有明确的答案，主要因为缺乏对这些问题的系统分析。

2.6 实验水槽和储水槽条件

另一个令人烦恼的问题是，如何保持实验水槽和储水槽中的水参数相同。如上所述，这些参数的任何突然变化都可能给鱼类带来应激性，在行为研究中这是一个特别重要的问题，用来设计检测大脑功能改变的微小变化。例如，在笔者的实验室中，没有专用加热装置的实验槽的水温可以迅速下降到 22 ～ 24℃，即使室温比其他实验室高出很多（通常为 26 ～ 27℃）。与栖息地水池的温度（28℃）相比，这是一个相当大的温差。采用水槽加热器，并不一定总是能适当加热实验水槽，特别是对于复杂的水槽，如用于学习研究的 plus 或 T 迷宫水槽。此外，监视实验斑马鱼的行为需要俯视观察，为了能更好地监测斑马鱼或录制斑马鱼视频，该类水槽没有全部被遮盖。当用于学习研究时，若不遮盖实验槽就意味着测试仪器中的水可能会蒸发，从而水槽中水被冷却，但同样重要的是，由于蒸发可能持续数天，也会导致盐度增加。重复使用实验水槽而不更换水，也会导致有机废物的积累以及存在先前斑马鱼留下的嗅觉信号。当然，这些技术问题都是可以解决的。比如可以采用使用水槽下加热垫、在每次做完斑马鱼实验后更换实验水槽中的水等方法。然而，大多数对斑马鱼进行的行为或心理药理学还有神经行为遗传学研究中都没有提到上述问题或类似问题是如何解决的，因此根本无法确定这些问题是否得到解决或有被考虑。

2.6.1 影响斑马鱼繁殖的因素

最后考虑的一个问题是斑马鱼繁殖的方法。繁殖是一个受多种因素影响的复杂过程。配偶的选择以及一般的交配行为，取决于视觉和可能的嗅觉信号、社交以及水参数，如盐度、pH 值、氧气水平、温度还有光线。产卵受斑马鱼的年龄、大小和品种（基因型）、产卵频率、饮食和健康状况的影响。

对斑马鱼在野外和实验室环境中繁殖的研究是非常重要的。斑马鱼品系在实验室的繁殖是基于特定的繁殖计划，其目标是不仅确保产生下一代，而且确保这些品系的活力得到保留。由于繁殖力和活力的下降通常与近交衰退有关，维持品系活力可能需要保持这些品系的遗传异质性，或产生遗传多样性的杂交品系。然而，需注意通过近亲繁殖品系培育出的斑马鱼也可以具有良好的生存能力和繁殖力，同时培育高度或完全近亲繁殖品系，这一方向没有得到适当的重视，有关这一点我们稍后讨论。

斑马鱼通常在小型产卵水槽中完成繁殖，底部与繁殖斑马鱼之间有一个带孔的隔板分开，这样可以收集卵并防止成年斑马鱼吃掉卵。在大型或商业养殖设施中，繁殖可在大容量圆柱体中进行，圆柱体的锥形底部将鱼卵收集到一个有细网覆盖的容器中，然后从该容器获取鱼卵。繁殖通常是把性成熟的繁殖个体放在产卵槽中然后关灯。在自然界产卵发生在日出时，因此在实验室斑马鱼产卵也是在第二天早上开灯时。在一些发育研究中，精确控制产卵时间可以通过在前一天晚上使用一个可拆卸的多孔透明屏障将雌雄鱼分割在产卵池的两侧，并在“日出”时移除屏障而实现。

2.6.1.1 性别比例

大多数实验室使用群体交配，即用多只雄性和多只雌性交配，这主要遵循了我们所知道的这个物种的相关自然行为。然而，不同的实验室甚至同一实验室内的不同研究，其性别比例以及用于繁殖的数量都不尽相同。最常用的方法是两雌三雄或一雌两雄的比例，因为有人认为雄性数量多可以提高受精效率。然而，产卵群中雄性比雌性多也会导致雄性的攻击性增加，从而干扰成功交配，因此我们建议用较少的雄性与更多的雌性交配。另外值得注意的是，当斑马鱼的密度较低时（单位体积的水中或单位表面积上的鱼较少），雄性往往会主动向雌性求爱，周期性地返回产卵地点；雌性则跟随雄性在那里产卵。视觉刺激和行为信号可能参与配偶选择中，从而限制了对繁殖后代有贡献的斑马鱼的数量。例如，特别占优势的雄性可能垄断产卵，导致后代的遗传多样性降低。因此，如果目标是保持品系的遗传多样性，建议建立更多的成对的而非更少的多个体群体产卵。

2.6.1.2 近亲繁殖或异交

在实验室野生型斑马鱼衍生品系的繁殖是基于特定的繁殖计划，其目标是不仅确保产生下一代的鱼，而且确保这些品系的活力得以保持。由于繁殖力和活力的下降通常与近交衰退有关，保持品系活力可能需要保持这些品系的遗传异质性或产生遗传多样性的杂交品系。

大多数实验室使用多个雄性和雌性的种群交配来模仿斑马鱼的自然行为。一些实验室使用的小型繁殖水槽只能繁殖出少量的鱼，而产卵量最高的繁殖鱼群通常是需要更大的繁殖水槽。

当从诱变筛选中发现的特定突变体需要繁殖时，或者当携带目标突变（例如转基因）的鱼需要繁殖时，单对交配是必要的。然而，同胞交配（在每一代交配同胞）或每一代只交配少量斑马鱼类将不可避免地导致基因组位点的纯合子率增加。这通常会导致所谓的近交衰退。近交衰退的特点是生育能力下降，甚至不孕；对疾病的易感性增加；通常伴有体形较小，以及产生许多其他遗传缺陷。亲缘关系最远的个体的系统交配将减少近交衰退。然而，除非有效群体规模（即繁殖个体的数量）非常大，否则近亲繁殖将会继续存在。避免近交衰退的一种方法是冷冻一个起始群体中几个雄性的精子，然后用冷冻的精子再现该群体的原始遗传组合；另一种减少近交衰退的方法是杂交斑马鱼品系，这一过程解决了由这个杂交产生的最初几代的近亲繁殖相关问题。

然而，避免近交衰退的另一个重要的可能方法是近亲繁殖本身。近交衰退是由于杂合性位点丢失，而在杂合性位点上“不太优”的等位基因恰好变得不易改变。因为遗传漂变导致的等位基因变异的丢失，在每个基因位点上都是随机的，所以“不太优”的等位基因有 50% 的概率是不易改变的。也就是说，对于整个基因组而言的有 50% 的先前的杂合子位点会因为劣势基因较少变成纯合子位点。一些等位基因会产生致命的影响，而另一些则不会产生有害的影响。不管怎样，这些负面影响的结果总和就是近交衰退，且通常是近亲繁殖系的损失。然而，在农业或实验室物种中采用近亲繁殖的研究已有多次报道，约 4% ～ 5% 的近亲繁殖系在近亲繁殖瓶颈下存活，即使经过 15 ～ 20 代同胞交配后仍然存活。经过 20 代的同胞交配，则认为该系是自交系，约 99.8% 的基因座将是纯合的形式。换言之，如果实验一开始用同胞交配法近交 100 个不同的斑马鱼品系，那么在 20 代之后，即大约经过 6 ～ 7 年的繁殖之后，可以预期有 4 ～ 5 个稳定的、完全近交的品系。某些自交系存活而其他自交系消亡的确切原因几乎不为人知，但有研究认为，跨多个基因座的某些遗传组合，即上位性效应，可以抵消劣势等位基因在纯合基因座上固定的有害影响。而这些组合，全基因组等位基因固定模式，能保证特定自交系的存活。重要的是，一旦一个自交系度过了近亲繁殖的瓶颈，它将永远保持稳定和生存，也就是说，直到有害的自发突变发生为止，但这是一个相当罕见的事件。不幸的是，我们还没注意到有大规模的斑马鱼近亲繁殖计划，已经尝试或计划尝试培育近交斑马鱼品种。此外，我们还听过一些斑马鱼专家的“轶事”，其声称斑马鱼不能近亲繁殖。然而，另一种鱼——天堂鱼（叉尾斗鱼，*Macropodus opercularis*）的近交已成功完成（Gerlai et al., 1990; Csányi, Gerlai, 1988），同时通过经典同胞交配成功产生的第一个近交斑马鱼品系也已问世（Shinya and Sakai, 2011）。

2.7 结语

在表 2-1 ～表 2-3 里，列出了一些饲养和繁殖的参数，包括笔者认为最佳的，以及其他人使用或斑马鱼可以耐受的参数。然而，所列信息只是未来系统分析的基础。美国国家卫生研究院最近组织了一个题为“斑马鱼和其他鱼类模型：严谨实验和可重复结果

表 2-1 **斑马鱼仔鱼和成鱼的水分参数**

	盐度	pH	温度	硝酸盐 / 亚硝酸盐 / 氨
推荐值	100 ～ 200μS	6.5 ～ 7.5	27 ～ 29℃	0/0/0
承受范围	10 ～ 9000μS （Tsang et al., 2017）； （Hernandez et al., 2018）	4.0 ～ 8.2 （Kwong et al., 2014）	6 ～ 38℃ （Spence et al., 2007）； （Engeszer et al., 2007）	未评估

表 2-2 **斑马鱼仔鱼和成鱼的水槽设置**

	仔鱼密度	仔鱼衬底	仔鱼富集
推荐值	10 ～ 30 条 /L	裸露底部	无
承受范围	10 ～ 100 条 /L （Ribas et al., 2017）；（Dabrowski and Miller, 2018）	裸露底部	无

	成鱼密度	成鱼衬底	成鱼富集
推荐值	1.5 ～ 3 条 /L	裸露底部 / 粗砂砾 / 平坦的河卵石	组织培养（TC）或根茎植物 / 同种
实验室方法	30 ～ 60 条 /L （Castranova et al., 2011）	砾石 / 裸露底部	塑料植物 / 同种

表 2-3 **繁殖和饲养选择**

	繁殖方法	仔鱼的第一次饲料（受精后第 1 ～ 2 周）	仔鱼饲料（受精后第 3 ～ 8 周）	成鱼饲料（9 周 +）
推荐值	单对（1 雌 1 雄）突变繁殖设置或转基因 3 雌 2 雄设置，以保持遗传多样性	轮虫（最小尺寸的活饲料，可选择肠道装载营养素）	卤虫、微丸（新鲜孵化的卤虫和 200 ～ 400μm 食品）	轮流使用：卤虫，缓慢下沉的 0.5 ～ 1mm 颗粒，螺旋藻片
实验室方法	单对交叉 / 多个体产卵群	微米级干食品、草履虫、滴虫和轮虫类	各种微丸、薄片和卤虫	各种颗粒、薄片和卤虫

的外部环境因素”的研讨会（https://orip.nih.gov/zebrafish-and-other-fish-models-extrinsic-environmental-factors-rigorous-experiments-and），对这些因素进行了更为深入的研讨。这次会议既体现了本章所讨论问题的严肃性，也体现了斑马鱼研究者解决这些问题的紧迫性。尽管本次研讨会的领导和与会者提出了许多重要问题，并讨论了斑马鱼养护的几个方面，但一份共识文件，甚至一份共识声明都没有达成。这可能是因为没有对斑马鱼饲养和繁殖的最佳环境因素进行系统分析。笔者的许多方法已使用几十年，但仍然是基于试验和错误的方法、轶事、个人印象和习惯之上的。可这也并不是说对如何饲养这种漂亮的小鱼是一无所知的。目前已有优秀的出版物总结了一些饲养和繁殖方法（例如，Westerfield, 2000）。然而，对缺乏探究最佳饲养条件的系统性科学研究的批评仍然存在。同样重要的是，是否应该使这些条件标准化还有待解答。不过不管人们是赞成还是反对标准化，笔者认为，对该物种自然栖息地、生态学和其行为学的了解应该作为设计旨在优化该物种维护条件的实验研究的指导原则（Gerlai and Clayton, 1999a, 1999b）。一旦确定了最佳的饲养条件和方法，斑马鱼饲养设施也采用了这些条件和方法，斑马鱼研究将具有更高的重复性和重现性（Kaflkafi et al., 2018; Gerlai, 2019）。

（宗星星翻译　宋良才审校）

参考文献

Adams, M.M., Kafaligonul, H., 2018. Zebrafishea model organism for studying the neurobiological mechanisms underlying cognitive brain aging and use of potential interventions. Frontiers in cell and developmental biology 6, 135.

Al-Imari, L., Gerlai, R., 2008. Sight of conspecifics as reward in associative learning in zebrafish (*Danio rerio*). Behavioural Brain Research 189 (1), 216-219.

Amatruda, J.F., Shepard, J.L., Stern, H.M., Zon, L.I., 2002. Zebrafish as a cancer model system. Cancer Cell 1 (3), 229-231.

Carney, T.J., Mosimann, C., 2018. Switch and trace: recombinase genetics in zebrafish. Trends in Genetics 34 (5), 362-378.

Castranova, D., Lawton, A., Lawrence, C., Baumann, D., Best, J., Coscolla, J., et al., 2011. The effect of stocking densities on reproductive performance in laboratory zebrafish (*Danio rerio*). Zebrafish 8 (3), 141-146.

Cayuela, M.L., Klaes, K., Godinho Ferreira, M., Henriques, C.M., van Eeden, F., Varga, M., Mione, M.C., 2019. The zebrafish as an emerging model to study DNA damage in ageing, cancer and other diseases. Frontiers in Cell and Developmental Biology 6, 178.

Choo, B.K.M., Shaikh, M.F., 2018. Recent Advances in Zebrafish Researches. IntechOpen. https://doi.org/10.5772/intechopen.74456.

Cortemeglia, C., Beitinger, T.L., 2005. Temperature tolerances of wild-type and red transgenic zebra danios. Transactions of the American Fisheries Society 134 (6), 1431-1437.

Csányi, V., Gerlai, R., 1988. Open-field behavior and the behavior-genetic analysis of the paradise fish (*Macropodus opercularis*). Journal of Comparative Psychology 102, 326-336.

Dabrowski, K., Miller, M., 2018. Contested paradigm in raising zebrafish (*Danio rerio*). Zebrafish 15 (3), 295-309.

Dooley, K., Zon, L.I., 2000. Zebrafish: a model system for the study of human disease. Current Opinion in Genetics & Development 10 (3), 252-256.

Engeszer, R.E., Patterson, L.B., Rao, A.A., Parichy, D.M., 2007. Zebrafish in the wild: a review of natural history and new notes from the field. Zebrafish 4 (1), 21-40.

Fuentes, R., Letelier, J., Tajer, B., Valdivia, L., Mullins, M., 2018. Fishing forward and reverse: advances in zebrafish phenomics. Mechanisms of Development 154, 296-308.

Gerlai, R., Crusio, W.E., Csányi, V., 1990. Inheritance of species-specific behaviors in the paradise fish (*Macropodus opercularis*): a diallel study. Behavior Genetics 20, 487-498.

Gerlai, R., Clayton, N.S., 1999a. Analysing hippocampal function in transgenic mice: an ethological perspective. Trends in Neurosciences 22, 47-51 (161).

Gerlai, R., Clayton, N.S., 1999b. Tapping artificially into natural talents. - Reply Trends in Neurosciences 22, 301-302.

Gerlai, R., 2019. Reproducibility and replicability in zebrafish behavioral neuroscience research. Pharmacology, Biochemistry and Behavior 178, 30-38.

Hernandez, R., Galitan, L., Cameron, J., Goodwin, N., Ramakrishnan, L., 2018. Delay of initial feeding of zebrafish larvae until 8 days postfertilization has no impact on survival or growth through the juvenile stage. Zebrafish 15 (5), 515-518.

Hoshijima, K., Hirose, S., 2007. Expression of endocrine genes in zebrafish larvae in response to environmental salinity. Journal of Endocrinology 193 (3), 481-491.

Kaflkafi, N., Agassi, J., Benjamini, Y., Chesler, E., Crabbe, J., Crusio, W., Eilam, D., Gerlai, R., et al., 2018. Reproducibility and replicability of mouse phenotyping in pre-clinical studies. Neuroscience & Biobehavioral Reviews 87, 218-232.

Kwong, R., Kumai, Y., Perry, S., 2014. The physiology of fish at low pH: the zebrafish as a model system. Journal of Experimental Biology 217 (5), 651-662.

Lee, C., Paull, G., Tyler, C., 2018. Effects of environmental enrichment on survivorship, growth, sex ratio and behaviour in laboratory maintained zebrafish Danio rerio. Journal of Fish Biology 94, 86-95.

Mahabir, S., Gerlai, R., 2017. The importance of holding water: salinity and chemosensory cues affect zebrafish behavior. Zebrafish 4, 444-458.

Miller, N., Gerlai, R., 2007. Quantification of shoaling behaviour in zebrafish (*Danio rerio*). Behavioural Brain Research 184 (2), 157-166.

Miller, N., Gerlai, R., 2008. Oscillations in shoal cohesion in zebrafish (*Danio rerio*). Behavioural Brain Research 193 (1), 148-151.

Miller, N., Gerlai, R., 2011. Shoaling in zebrafish: what we don't know. Reviews in the Neurosciences 22, 17-25.

Miller, N., Gerlai, R., 2012. From schooling to shoaling: patterns of collective motion in zebrafish (*Danio rerio*). PLoS One 7 (11), e48865.

Parker, M.O., Millington, M.E., Combe, F.J., Brennan, C.H., 2012. Housing conditions differentially affect physiological and behavioural stress responses of zebrafish, as well as the response to anxiolytics. PLoS One 7 (4), e34992.

Patton, E., Tobin, D., 2019. Spotlight on zebrafish: the next wave of translational research. Disease Models & Mechanisms 12 (3), dmm039370.

Ribas, L., Valdivieso, A., Díaz, N., Piferrer, F., 2017. Appropriate rearing density in domesticated zebrafish to avoid masculinization: links with the stress response. Journal of Experimental Biology 220 (6), 1056-1064.

Roper, C., Tanguay, R.L., 2018. Zebrafish as a model for developmental biology and toxicology. In: Handbook of Developmental Neurotoxicology. Academic press, pp. 143-151.

Saif, M., Chatterjee, D., Buske, C., Gerlai, R., 2013. Sight of conspecific images induces changes in neurochemistry in zebrafish. Behavioural Brain Research 243, 294-299.

Scerbina, T., Chatterjee, D., Gerlai, R., 2012. Dopamine receptor antagonism disrupts social preference in zebrafish: a strain comparison study. Amino Acids 43 (5), 2059-2072.

Schroeder, P., Jones, S., Young, I., Sneddon, L., 2014. What do zebrafish want? Impact of social grouping, dominance and gender on preference for enrichment. Laboratory Animals 48 (4), 328-337.

Shams, S., Amlani, S., Buske, C., Chatterjee, D., Gerlai, R., 2017a. Developmental social isolation affects adult behavior, social interaction, and dopamine metabolite levels in zebrafish. Developmental Psychobiology 60 (1), 43-56.

Shams, S., Seguin, D., Facciol, A., Chatterjee, D., Gerlai, R., 2017b. Effect of social isolation on anxiety-related behaviors, cortisol, and monoamines in adult zebrafish. Behavioral Neuroscience 131 (6), 492-504.

Shinya, M., Sakai, N., 2011. Generation of highly homogeneous strains of zebrafish through full sib-pair mating. G3: Genes, Genomes, Genetics 1, 377-386.

Spence, R., Gerlach, G., Lawrence, C., Smith, C., 2007. The behaviour and ecology of the zebrafish, *Danio rerio*. Biological Reviews 83 (1), 13-34.

Torraca, V., Mostowy, S., 2018. Zebrafish infection: from pathogenesis to cell biology. Trends in Cell Biology 28 (2), 143-156.

Tsang, B., Zahid, H., Ansari, R., Lee, R., Partap, A., Gerlai, R., 2017. Breeding zebrafish: a review of different methods and a discussion on standardization. Zebrafish 14, 561-573.

Westerfield, M., 2000. The Zebrafish Book. A Guide for the Laboratory Use of Zebrafish (*Danio rerio*), fourth ed. Univ. of Oregon Press, Eugene.

第二部分

斑马鱼仔鱼的行为

第三章

斑马鱼仔鱼的视觉系统和对视觉刺激的行为反应

Domino K. Schlege[1,2], Stephan C. F. Neuhauss[1]

3.1 斑马鱼仔鱼视觉的关联问题

斑马鱼个体的成长初期要面临许多严峻的考验。雄鱼对雌鱼排出的鱼卵进行体外受精，然后留下胚胎自生自灭。为了避免鱼卵被捕食，雌鱼更倾向于在浅水或靠近植物的地方产卵，这就是亲代斑马鱼保护数百个胚胎的唯一措施。这些特殊情况可能解释了斑马鱼胚胎为什么在强选择压力下快速发育。当面对选择压力特别高时，胚胎中枢神经系统和感官需要快速发育，以便能够对环境做出反应，避免被捕食，并在卵黄囊营养耗尽后开始捕食。

仔鱼（和成鱼一样）主要依靠视觉来逃避捕食者和捕获猎物，因此，发育中仔鱼视觉系统的迅速发育也就不足为奇了。

3.2 斑马鱼视觉系统的一般结构

脊椎动物视觉系统的结构非常保守，尤其是眼睛及其眼球后部的视网膜。这种感官结构是在大约 5 亿年前寒武纪脊椎动物谱系出现的时候建立起来的（Lamb et al., 2007）。

[1] Department of Molecular Life Sciences, University of Zurich, Zurich, Switzerland.

[2] Life Science Zurich Graduate School, Ph.D. Program in Molecular Life Sciences, Zurich, Switzerland.

斑马鱼仔鱼的视觉系统工作过程为：首先经眼球感知和处理光刺激，再经视网膜投射，将信息传递到主要由视顶盖组成的更高级脑中枢及大脑视觉中枢，然后进一步视觉处理并作出适当的行为反应。

3.2.1 眼睛和视网膜的结构

眼睛前部由角膜、晶状体、睫状体和虹膜角膜角组成。它的主要功能是将入射光聚焦到眼睛后部的感光视网膜上（Gestri et al., 2012）。与陆地动物相比，斑马鱼拥有一个高折射率的球状晶状体，眼睛的折射能力基本全部来自这种晶状体。没有证据表明斑马鱼的眼睛有自动调焦功能（Easter and Nicola, 1996a）。因此，斑马鱼的视觉系统很可能是在固定景深的情况下工作的，景深能被校正到无穷远。

与眼前部相反，斑马鱼的视网膜由间脑直接输出，因此属于中枢神经系统的一部分。在仔鱼期，其视网膜极大，约占中枢神经系统神经元总数的一半。

斑马鱼视网膜和所有典型脊椎动物视网膜一样，由三个核层组成，由两个丛状（突触）层分隔。外视网膜与视网膜色素上皮紧密相连，包含一种视杆细胞及四种视锥细胞。视杆细胞对光线最敏感，但只能传递亮度信息。在更明亮的光线条件下，四种视锥细胞能够产生四色视觉。不同的单个视锥细胞在形态和视觉色素含量上都有差异。红敏和绿敏的视锥细胞融合形成双锥细胞（double cones），而短单锥细胞（short-single cones）对紫外线（UV）敏感，长单锥细胞（long-single cones）含有蓝色敏感的视蛋白。在成鱼的视网膜中，光感受器排列成有序的镶嵌行，但这种排列方式并未出现在仔鱼的视网膜中（Allison et al., 2010）。

最近的一项研究表明，仔鱼视网膜具有高度的各向异性，其区域专门用于无色、四色和紫外线介导的视觉，以最大限度地利用环境提供的视觉信息（Neuhauss, 2018; Zimmermann et al., 2018）。成鱼视网膜是否也表现为各向异性还需要进一步研究。

谷氨酸是光感受器利用高度特异化的突触前结构分泌的唯一神经递质。这些带状突触结构优化后有利于紧张性神经递质的释放。

内核层由双极细胞（bipolar cells）和水平细胞（horizontal cells, HCs）的核，以及无长突中间神经元（amacrine interneurons）和 Müller 胶质细胞的胞体组成，其中双极细胞和水平细胞是与光感受器直接以突触接触的二级神经元。在外网状层（outer plexiform layer, OPL），光感受器和内视网膜之间形成突触连接。离晶状体最近的核层为神经节细胞层，该细胞层包含移动性无长突细胞（displaced amacrine cells）和神经节细胞的胞体，它们的轴突形成视神经和视束（进入大脑后）。内网状层（inner plexiform layer, IPL）是由内层视网膜和神经节层之间的突触连接形成的，其相对厚度提示了视网膜中的连接具有极大的复杂性。

仔鱼的视网膜在细胞类型上的复杂性已经超过哺乳动物视网膜。虽然对视网膜中不同细胞类型进行分类仍在进行，但已经确定的是，与哺乳动物相比，斑马鱼视网膜中细胞类型的亚类数量更多。斑马鱼的视网膜有 4 种不同的水平细胞，至少有 17 种双极细胞（Connaughton et al., 2004）以及多达 70 种无长突细胞（Connaughton et al., 2004）。

3.2.2 视网膜发育

斑马鱼的视网膜发育非常迅速。视网膜由间脑外壁向外突起形成，在 10hpf［受精后 10h］左右形成视叶。一般来说，视网膜的发育是由内到外的：神经节细胞先于内核细胞分化，其次是外层视网膜细胞和 Müller 细胞（Hu and Easter, 1999; Schmitt and Dowling, 1994, 1999）。

神经节细胞在 28 ～ 32hpf 左右分化，内层视网膜紧随其后，光感受器外节在 55 ～ 60hpf 左右出现。同时，光感受器突触终末形成（Hu and Easter, 1999; Schmitt and Dowling, 1994, 1999）。在 84hpf 左右，光感受器开始传输信号到二级神经元，并在 5dpf（受精后 5 天）后具备完全的信号传递功能（Biehlmaier et al., 2003）。视觉系统功能的快速成熟表现为视网膜电图（记录反映外视网膜功能的总场电位）（Branchek and Bremiller, 1984; brokerhoff et al., 1995a; Makhankov et al., 2004）和视觉引导行为的成熟。奇怪的是，尽管杆状光感受器在早期仔鱼的视网膜中已经存在（Bilotta et al., 2001），但它只在 15dpf 后的阶段才会对视网膜电图（可能也对行为）起作用。

3.3 视网膜投射

3.3.1 斑马鱼作为神经元连接模型

斑马鱼已被证明是一个有价值的模型系统，可通过视网膜顶盖投射研究神经元的发育和轴突导向。仔鱼的透明性及其神经系统的小体积和简单性，为研究人员研究视网膜顶盖投射发育的解剖学和时间尺度提供了可能性。在早期的研究中，通过注射辣根过氧化物酶（horseradish peroxidase, HRP）或荧光染料（如 DiO 和 DiI）来顺行和逆行标记视网膜神经节细胞（retinal ganglion cell, RGC）轴突。随后，通过分析发育过程特定时间点染色的轴突及其树状分支模式，可了解这些投射的形成（Stuermer, 1988; Stuermer et al., 1990）。然而，这些研究只提供了某一过程点，不能排除固定过程造成的假象。延时成像技术（time-lapse imaging techniques）的出现（Kaethner and Stuermer, 1992），不仅能在体内进行长时间的非侵入性观察，而且增加了观测的时间分辨率。斑马鱼的高繁殖力和易饲养性及其对基因修饰的适应性，是它另一显著的优势（Niklaus and Neuhauss, 2017）。这些特征以及高通量筛选方法的发展，为大规模筛选可识别导致视顶盖投射缺陷的突变株系提供了可能（Brockerhoff et al., 1995b；Trowe et al., 1996; Karlstrom et al., 1996; Baier et al., 1996）。使用 RGC 轴突中携带荧光标记的转基因株系，无须注射染料，即可鉴定额外的突变（Xiao et al., 2005）。最后，更先进的成像技术也不断促进斑马鱼模型系统的发展。表 3-1 总结了斑马鱼视觉系统的研究方法。

表 3-1 斑马鱼视觉系统的研究方法

方法	使用 / 例子	参考文献
辣根过氧化物酶、DiO 和 DiI 注射	视网膜神经节细胞顺行和逆行标记	Stuermer, 1988; Stuermer et al., 1990; Kaethner and Stuermer, 1992
行为分析与正向遗传学	识别与视觉系统发育和功能有关基因的遗传筛选	Brockerhoff et al., 1995b, 1998; Trowe et al.,1996; Karlstrom et al., 1996; Baier et al., 1996;Neuhauss et al., 1999; Muto et al., 2005; Neuhauss, 2003; Gross and Perkins, 2008
	视觉惊吓反应与视觉运动反应（VMR）	Easter and Nicola, 1996b; Emran et al., 2007,2008; Burgess and Granato, 2007; Burton et al., 2017; Gao et al., 2014
	眼动反应（OKR）	Brockerhoff et al., 1995b, 1998; Neuhauss et al., 1999; Muto et al., 2005; Easter and Nicola, 1996b; Emran et al., 2007; Roeser and Baier, 2003; Huang and Neuhauss, 2008
	视动反应（OMR）	Neuhauss et al., 1999; Roeser and Baier, 2003; Naumann et al., 2016; Orger and Baier, 2005; Maaswinkel and Li, 2003
	趋光性	Brockerhoff et al., 1995b; Orger and Baier, 2005; Wolf et al., 2017; Hartmann et al., 2018; Burgess et al., 2010
	视觉背景自适应（VBA）	Neuhauss et al., 1999; Muto et al., 2005; Fujii, 2000; Logan et al., 2006; Mueller and Neuhauss, 2014
转基因技术	对特定类型细胞荧光标记和活体成像，跟踪发育 / 细胞结局。引入活性标记物（如 GCaMP 和 iGluSnFR）来研究神经元活性 用于谱系或连接体分析的多色标记（Cre/Lox 系统） 细胞类型特异性消融 基因过表达的时空调控	Niklaus and Neuhauss, 2017
反向遗传学	TALENs、锌指核酸酶、CRISPR/Cas9：定点突变创建基因敲除模型，例如，视网膜疾病 导致患者特异性 DNA 变化的基因敲入（目前在斑马鱼中效率很低），或者用荧光报告基因或抗体标签标记内源性蛋白质的基因敲入	Niklaus and Neuhauss, 2017

3.3.2 视顶盖 - 视网膜拓扑映射图的建立

在仔鱼中，视网膜轴突在 34 ～ 36hpf 左右从眼睛开始向大脑延伸（Stuermer, 1988）。视网膜轴突在大脑内投射的主要结构是视顶盖，以及其他 9 个树枝状结构（Burrill and Easter, 1994）。在中线形成完整的视交叉后，第一批轴突在 46 ～ 48hpf 后进入视顶盖。轴突终止于 4 个不同的层：视神经层（stratum opticum, SO）、纤维层和浅表灰质层（stratum fibrosum et griseum superficiale, SFGS）、中央灰质层（stratum griseum centrale, SGC）以及中央白质层和脑室周围层（stratum album centrale, and stratum periventriculare, SAC/SPV）之间的投射区（Bartheld and Meyer, 1987）。一旦轴突到达目的地，它们便在 70 ～ 72hpf 时发出分支形成树状结构，建立视网膜拓扑映射图（Stuermer, 1988）。值得注意的是，神经支配的时间与视觉引导行为的开始是一致的（Easter and Nicola, 1996b, 1997）。

哺乳动物视网膜与再生的斑马鱼视网膜相比，发育中斑马鱼的顶盖轴突分支从一开始就非常精确（Stuermer, 1988; Stuermer et al., 1990）。关于这一点是如何实现的，目前存在几种假设。引导分子的梯度和生长锥上适当表达的受体，轴突竞争 / 相互作用，以及突触活动都与这一过程有关。虽然 RGC 活性并不是建立轴突末端树枝状结构所必需的（Stuermer et al., 1990; Kaethner and Stuermer, 1994），但在 72hpf 对视顶盖进行初始神经支配后，由于视网膜和顶盖的不同生长，在重排过程中 RGC 活性依赖的树状结构的稳定尤其重要（Gnuegge et al., 2001）。通过促进新的分支的生长并选择性使新的分支稳定，突触形成与视网膜轴突树样生长和顶盖细胞的树突状生长都有关系（Niell et al., 2004; Meyer and Smith, 2006）。此外，在顶盖神经支配 24h 后，观察到通过 NMDAR 活性依赖性的末端树状结构的微调（Schmidt et al., 2000）。另一项研究表明，活性依赖性轴突竞争在轴突生长过程中也是很重要的。利用单个 RGC 轴突活性被抑制的转基因斑马鱼，Hua 等人研究发现，当轴突的活性水平被抑制到低于相邻 / 竞争轴突的水平时，轴突树的生长就会受到抑制（Hua et al., 2005）。最后，轴突竞争也影响终末树状结构的大小和复杂性（Gosse et al., 2008）。因此，RGC 轴突最初的轴突寻路似乎不依赖于活性和竞争，很可能由分子信号（molecular cues）所支配。然而，活性依赖的机制以及轴突之间的相互作用使得在发育后期轴突分支形成且更加完善。

3.4 视觉过程

3.4.1 视网膜的视觉处理

视觉信息的一些特征在到达大脑之前就已经在视网膜回路中被提取出来。不同类型的 RGCs 选择性地对它们作出反应，并通过特异性途径将它们传递到更高级的大脑区域。OPL 是视觉信息处理的第一层，在这里光感受器细胞与水平细胞（horizontal cell, HCs）和双极细胞（BCs）形成多个突触。HCs 与光感受器和 BCs 的横向相互作用产生中心 - 周围感受野。这些回路可检测无色和彩色对比度。类似地，IPL 中 BC 轴突、无长突细胞（amacrine cell, AC）和 RGC 树突间形成的突触，负责神经节细胞中心 - 周围感受野和进一步的计算，例如检测物体形状、大小、运动和方向性。大多数关于 BC 和 HC 视觉处理的研究都是在成年斑马鱼身上进行的（Meier et al., 2018）。然而，笔者认为仔鱼的视网膜很可能也有这种回路。

Antinucci 等人最近描述了仔鱼中的定向选择性（orientation-selective, OS）视网膜回路，并发现它参与 OS 无长突细胞和神经节细胞的关闭或开 - 关反应特性（Antinucci et al., 2016）。除基底节细胞（水平和垂直细胞）外，OS 无长突细胞和神经节细胞也参与斜向调节。无长突细胞树突野的细长形态与细胞首选的刺激方向一致，产生了方向调节，这是斑马鱼视网膜方向选择性的基础。他们还确定了一种分子标记物即跨膜蛋白 teneurin-3，它可以调节无长突细胞的方向。他们认为，BCs 和 RGCs 方向选择性与来自过表达

teneurin-3 无长突细胞对垂直方向调节的抑制性输入有关。

很明显，视觉处理和与之相关的行为输出对于仔鱼在其自然环境中的生存至关重要。然而，它们的栖息地对仔鱼的视觉系统构成了哪些具体的视觉挑战？这些挑战是如何解决的？在 Zimmermann 及其同事最近的一项研究中对这些问题做出了解答（Zimmermann et al., 2018）。高光谱成像可以评估斑马鱼仔鱼自然栖息地的色素含量。结合对仔鱼的功能性双光子成像，他们发现视网膜不仅在形态上而且在功能上是各向异性的，从而精确地满足它们在自然环境中的需要。它们在上额部视野中建立了一个紫外线主导的攻击区，该区很可能与捕捉猎物有关。对四原色敏感的回路主要观察水平线及其以下的视野，这也是自然界中色彩信息最多的部分。另一方面，无色回路控制着上部视野。这些发现很有趣，因为成年斑马鱼的视网膜中有一个均匀分布的锥形体区域（cone mosaic）。因此出现的问题是，锥形体区域视网膜与具有高度特异化功能的各向异性视网膜相比具有什么优势，从而导致了发育过程中的这种转变？

3.4.2 视顶盖的视觉处理

视顶盖受大多数视网膜轴突和其他感觉系统和大脑区域的传入神经的支配，也和运动神经元连接。因此，视顶盖在感觉输入的处理和整合中起着重要作用。然而，令人惊讶的是，人们对视顶盖在这些过程中的具体功能知之甚少。

在早期的研究中，在成鱼中发现了四种不同的视顶盖细胞类型，它们具有独特的反应特性（Sajovic and Levinthal, 1982a, 1982b）。他们发现，顶盖细胞的感受野比视网膜感受野大，并提出侧向抑制机制至少参与了一些顶盖细胞类型的特性形成。他们认为，这对体形辨别至关重要，而体形辨别可能在觅食过程中发挥作用。有趣的是，他们没有发现任何方向选择性的顶盖细胞类型。

20 多年后，对斑马鱼仔鱼进行功能性双光子钙成像，可以识别出方向选择性顶盖细胞（Niell and Smith, 2005）。此外，该研究发现，顶盖的大多数反应特性是在视网膜轴突开始分支时形成的，与视觉经验无关，并且在仔鱼大脑中是生来就存在的。Abbas 等人提供了一个关于方向选择性如何在光学顶盖中编码的模型（Abbas et al., 2017）。结果表明，RGC 的三向运动调节继承了一些方向选择性，但通过从中间神经元到室周区顶盖细胞的抑制性输入，顶盖中产生了对嘴侧到尾侧运动的另外反应类型。

哺乳动物顶盖中上丘在眼球转动中发挥作用。值得注意的是，仔鱼能够处理二阶线索，例如相对而言的时空变化，这已通过视动反应（optomotor response, OMR）得到证明（Orger et al., 2000）。长期以来，人们一直认为只有具有皮质的动物才能处理这种非傅里叶运动。视顶盖是仔鱼大脑中整合视觉和其他感官输入的主要结构，因此似乎是这种处理的可能部位。然而，由于眼动反应（optokinetic response, OKR）和 OMR 的功能都独立于完整的视顶盖，仔鱼的视顶盖似乎对于跟踪眼球运动或非傅里叶运动检测并不是必不可少的（Roeser and Baier, 2003）。全脑成像技术为感知运动反应的神经回路提供了新的视角（Naumann et al., 2016;Wolf et al., 2017; Chen et al., 2018），并且与传统技术相结合后，将进一步加深对视觉引导行为的认识。

3.5 斑马鱼和哺乳动物视觉处理的比较

考虑到脊椎动物视网膜的保守结构，所有脊椎动物的视网膜信息处理可能有直接的可比性。然而，由于解剖结构的差异，大脑中的视觉处理区在物种之间可能有很大的差异：哺乳动物对视觉信息的处理主要限定在皮层结构，然而硬骨鱼类缺失这种皮层结构。上丘是一种与视顶盖同源的大脑结构，哺乳动物的上丘只负责某些方面视觉信息的处理。哺乳动物只有部分视觉信息是在上丘中处理的。另一方面，斑马鱼的视觉处理只有视顶盖和其他尚未被识别的大脑区域参与，而皮层未参与。然而，这些解剖上的差异并不一定意味着斑马鱼的视觉处理过程不那么复杂。如前文提到的斑马鱼处理复杂非傅里叶运动的能力就是一个很好的例子。最初人们认为这需要大脑皮层的参与，然而，从解剖学角度来看，尽管产生的输出是相同的，但涉及的各个回路可能有相当大的不同。新开发的技术可以测量行为动物的全脑神经元活动（Chen et al., 2018），这有助于进一步揭示在斑马鱼的大脑和哺乳动物的视觉皮层中是否使用了同等功能的神经回路。

3.6 视觉行为

斑马鱼仔鱼在发育的早期就表现出一些视觉引导行为，这些行为与视觉系统解剖学中的发育过程相匹配（如图 3-1 所示）。这些固定模式行为，加上强大的遗传学和最先进的成像技术，使斑马鱼仔鱼成为研究视觉系统发育和功能的优良模型。

3.6.1 视觉惊吓反应与视觉运动反应

仔鱼对视觉刺激的第一反应是视觉惊吓反应，这是一种光照减少时的突然运动。它最早出现在 68 ~ 79hpf，在机械惊吓发作之后，以及视动反射前不久（Easter and Nicola, 1996b）。视觉惊吓反应与非视网膜感光细胞无关，因为没有生长出眼睛的胚胎并没有表现出这种行为（Emran et al., 2008）。

最初人们认为视觉惊吓反应是一种防御行为（Easter and Nicola, 1996b），类似于捕食者逼近时成年动物产生的逃跑反应。然而，Burgess 等人发现，对暗闪光的惊吓反应会导致所谓的 O 形弯，不会使仔鱼改变位置并远离假定的捕食者（Burgess and Granato, 2007）。此外，Mauthner 细胞介导机械性和听觉性惊吓反应，但视觉惊吓反应与 Mauthner 细胞无关。与其他惊吓反应相比，视觉惊吓反应的潜伏期更长。总之，视觉惊吓这些特征使得它成为一种并不适合躲避捕食者的行为，而主要是一种导航行为，该行为有助于仔鱼处在光线充足的最佳进食环境中。与视觉惊吓反应相比，最近一项研究发现了一种新的行为，称为阴影视觉刺激逃跑反应，类似于成人逃避（Dunn et al., 2016）。因此，很明显仔鱼的视觉系统能够区分整个视线范围内的光线变化（视觉惊吓）和视觉刺激强度的突然增加（阴

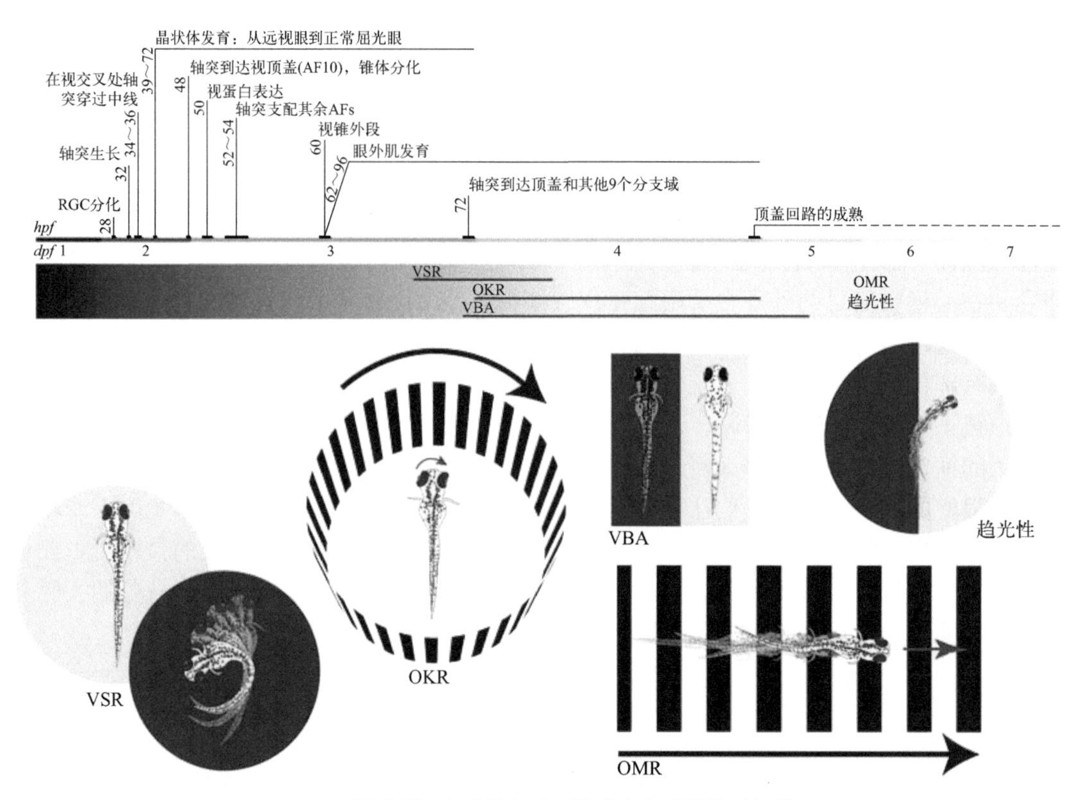

图3-1 **视觉引导行为的视觉系统发育和成熟的时间轴**

发育过程（时间线以上）和视觉引导行为的出现（时间线以下，图示）。视网膜神经节细胞（retinal ganglion cell, RGC）分化后，从28hpf开始，第一个RGC轴突在34～36hpf越过中线，到达视顶盖（48hpf），并在所有10个分支区域（arborization fields, AFs）中分支（72hpf）。视网膜视蛋白在受精50hpf左右开始表达，受精60hpf后在限定区域可见第一个视锥外段（cone outer segments）。随后，外段分布于整个视网膜（72hpf），并逐渐增大（96hpf）。从96hpf开始，视顶盖的树状结构和突触进一步细化。VSR：视觉惊吓反应（visual startle response），在68hpf后出现，反应外显率在79hpf时达到100%。OKR：眼动反应（optokinetic response），早在73hpf就可观察到，在80hpf强烈激发，在96hpf时完全成熟。VBA：视觉背景适应（visual background adaptation），从3dpf开始，5dpf后成熟（2dpf已经可以看到视网膜非依赖性黑素细胞扩散）。OMR：视动反应（optomotor response），在6dpf时首次完全成熟。趋光性（phototaxis）：在5dpf左右开始

影视觉刺激）。

在视觉惊吓反射的基础上，开发了一种不依赖运动的视觉分析方法来测试仔鱼感受光线简单变化的能力。这项实验被称为视觉运动反应（visual motor response, VMR）（Emran et al., 2007, 2008），包括一个自动跟踪系统，该系统用于监测放置于96孔板中的单个仔鱼在黑暗和光照期间的运动反应。除了上述光线递减时的行为，当灯光打开时，仔鱼的活动也会增加。此外，还可以追踪在变化刺激之间的行为。利用对光照和突然黑暗的反应来分别评估“开启”和“关闭”反应。由于该行为不是基于运动检测（可认为“色盲”）（Orger and Baier, 2005），因此它可用于评估光谱敏感性：Burton等人对比测试了窄波段和通常使用的广谱白光引起的刺激反应（Burton et al., 2017）。离散波段对VMR的各组成部分有不同的影响，例如典型的O形弯曲。因此，作者认为它们是由不同的光感受器亚群驱动的。他们的研究也对光遗传学与波段特异性VMR相结合的实验具有实践启示。最后，正如Gao等人使用VMR后所提出的，行为数据分析的机器学习算法可以实现更可靠的高通量筛选（Gao et al., 2014）。

3.6.2 眼动反应

眼动反应（optokinetic response, OKR）是描述鱼在感知到的运动方向上移动眼睛的偏好，这是仔鱼形成视觉需要的第一种视觉行为，可用于检测导致视觉缺陷的隐性突变（brokerhoff et al., 1995b）。与形态学筛查相比，这种行为方法可识别更细微表型的突变。此外，通过 OKR 检查还可发现导致视网膜外视觉系统缺陷的突变，而这种突变不太可能通过视网膜电图检查获得。使用 OKR 进行基因筛选，发现了大量导致视觉缺陷的突变（brokerhoff et al., 1995b, 1998; Neuhauss et al., 1999; Muto et al., 2005）。有关综述见文献（Neuhauss, 2003; Gross and Perkins, 2008）。

在这个实验中，仔鱼被固定在甲基纤维素或低熔点琼脂糖中，为其呈现一个移动的条纹滚筒或计算机生成模拟移动条纹的图像投影。追踪到的眼球运动通常表现为两个阶段：缓慢追踪移动物体，然后快速扫视，眼睛回归原位。Huang 和 Neuhauss 综述了跟踪和分析眼球运动的不同方法（Huang and Neuhauss, 2008）。

Easter 和 Nicola 是第一次将发育中的斑马鱼视觉系统的解剖特征与 OKR 关联起来的研究者，他们发现 73hpf 时 OKR 开始发生，直到 96hpf 时 OKR 达到成鱼的准确度（Easter and Nicola, 1996b）。有趣的是，该行为与视觉体验无关，因为饲养在黑暗环境中的斑马鱼也会引发 OKR。

有人认为，OFF 视网膜通路并不介导 OKR，因为在保持 OFF 视网膜通路完整的情况下，药理阻断 ON 双极细胞和 ON RGC 缺失的 *nrc* 突变株都会使 OKR 产生障碍（Emran et al., 2007）。因此，ON 通路被认为是驱动这种行为的原因。

未来研究的一个重要任务将是识别处理 OKR 刺激的脑回路，因为激光消融视顶盖只会减少扫视的频率，但对 OKR 的其他特性没有影响（Roeser and Baier, 2003）。

3.6.3 视动反应

视动反应（optomotor response, OMR）是另一种视觉引导行为，可用于前面提到的许多基因筛选。它与OKR类似，应用于运动检测。这项实验是基于仔鱼偏好游向感知到运动的方向，这可能是为了稳定自己在水中的位置。OMR 可以通过在测试箱下面向仔鱼呈现一个有移动边缘的屏幕来诱发（Neuhauss et al., 1999; Orger and Baier, 2005）。OMR 在 6 dpf 时就完全成熟，非常适合快速评估尚未形成集群行为的仔鱼的群体反应（Neuhauss et al., 1999）。向仔鱼呈现协调的或不太协调的等间距红绿格栅可分别导致较强的刺激和反应或较弱的刺激和反应（Orger and Baier, 2005）。因此，Orger 和 Baier 得出结论，在处理运动之前，来自红色和绿色光感受器的输入就汇集在一起，因此运动视觉被认为是“色盲”。这项研究还表明，与 OKR 相比，OMR 同时利用 ON 和 OFF 通路。用于筛查目的的可与趋光性特征区分开的 OMR 具有稳健性以及缓慢适应性等特征。尽管 OMR 可以用来评估运动的空间和时间频率（Maaswinkel and Li, 2003），但这种行为与顶盖的信息处理无关，因此不适合检查视顶盖中的缺陷（Roeser and Baier, 2003）。全脑成像与网络建模的结合，使人们首次了解了参与 OMR 的大脑区域和神经元回路（Naumann et al., 2016）。

3.6.4 趋光性

Brockerhoff等人首先描述了斑马鱼仔鱼的趋光性，并将其作为一种可能的视觉缺陷行为的筛查方法（Brokerhoff et al., 1995b）。仔鱼在适应光照或黑暗后，使其在更亮或更暗的隔间之间进行选择，这些仔鱼往往向最亮的方向游去。趋光性以紫外/蓝光敏感锥细胞为主，涉及部分红光敏感锥细胞，具有快速适应性特征（Orger and Baier, 2005）。此外，近红外光会诱发负趋光性（Hartmann et al., 2018）。Burgess等人认为，ON和OFF视网膜通路在这一行为中发挥了不同的作用，分别促进仔鱼向目标区域移动和远离非目标区域的重新定位（Burgess et al., 2010）。最后，Wolf等人提出了第一个斑马鱼仔鱼大脑回路驱动趋光行为的模型（Wolf et al., 2017）。

3.6.5 视觉背景适应

视觉背景适应（visual background adaptation, VBA）是一种行为反射。当仔鱼出现在明亮或黑暗的环境时，VBA分别牵涉到特殊黑素细胞中黑色素体收缩或弥散。鱼类体内的VBA受神经和激素系统的控制（Fujii, 2000; Logan et al., 2006），并依赖微管运输黑色素体（Logan et al., 2006）。黑色素体弥散从2dpf已经开始出现，可能是为了防止紫外线过度照射，因此其不依赖于功能性视网膜（Mueller and Neuhauss, 2014）。另一方面，从3dpf开始，出现光诱导黑色素体聚集，并一直持续到5dpf。这一部分VBA依赖于功能性视觉系统（Mueller and Neuhauss, 2014; Shiraki et al., 2010）。通过其他行为检测，几个缺乏VBA的突变体会表现出行为盲（Neuhauss et al., 1999）。因此，VBA缺陷往往与暗示视觉系统缺陷的行为盲（OKR, OMR）相关。随后筛选中发现，一些光感受器退化的突变体表现出正常的VBA, OKR/OMR缺陷与VBA之间的相关性较低（Muto et al., 2005）。这些发现可能说明是视网膜的非视觉部分而非光感受器部分参与了VBA，并发现运动视觉和背景适应可能对应不同的神经回路。

（宗星星翻译　赵丽审校）

参考文献

Abbas, F., Triplett, M.A., Goodhill, G.J., Meyer, M.P., 2017. A three-layer network model of direction selective circuits in the optic tectum. Frontiers in Neural Circuits 11, 88.

Allison, W.T., Barthel, L.K., Skebo, K.M., Takechi, M., Kawamura, S., Raymond, P.A., 2010. Ontogeny of cone photoreceptor mosaics in zebrafish. The Journal of Comparative Neurology 518 (20), 4182-4195.

Antinucci, P., Suleyman, O., Monfries, C., Hindges, R., 2016. Neural mechanisms generating orientation selectivity in the retina. Current Biology 26 (14), 1802-1815.

Baier, H., Klostermann, S., Trowe, T., Karlstrom, R.O., Nüsslein-Volhard, C., Bonhoeffer, F., 1996. Genetic dissection of the retinotectal projection. Development 123, 415-425.

von Bartheld, C.S., Meyer, D.L., 1987. Comparative neurology of the optic tectum in ray-finned fishes: patterns of lamination formed by retinotectal projections. Brain Research 420 (2), 277-288.

Biehlmaier, O., Neuhauss, S.C.F., Kohler, K., 2003. Synaptic plasticity and functionality at the cone terminal of the developing zebrafish retina. Journal of Neurobiology 56 (3), 222-236.

Bilotta, J., Saszik, S., Sutherland, S.E., 2001. Rod contributions to the electroretinogram of the dark-adapted developing zebrafish. Developmental Dynamics 222 (4), 564-570.

Branchek, T., Bremiller, R., 1984. The development of photoreceptors in the zebrafish, *Brachydanio rerio*. I. Structure. The Journal of Comparative Neurology 224 (1), 107-115.

Brockerhoff, S.E., Hurley, J.B., Janssen-Bienhold, U., Neuhauss, S.C., Driever, W., Dowling, J.E., 1995a. A behavioral screen for isolating zebrafish mutants with visual system defects. Proceedings of the National Academy of Sciences of the United States of America 92 (23), 10545-10549.

Brockerhoff, S.E., Hurley, J.B., Janssen-Bienhold, U., Neuhauss, S.C., Driever, W., Dowling, J.E., 1995b. A behavioral screen for isolating zebrafish mutants with visual system defects. Proceedings of the National Academy of Sciences of the United States of America 92 (23), 10545-10549.

Brockerhoff, S.E., Dowling, J.E., Hurley, J.B., 1998. Zebrafish retinal mutants. Vision Research 38 (10), 1335-1339.

Burgess, H.A., Granato, M., 2007. Modulation of locomotor activity in larval zebrafish during light adaptation. Journal of Experimental Biology 210 (Pt 14), 2526-2539.

Burgess, H.A., Schoch, H., Granato, M., 2010. Distinct retinal pathways drive spatial orientation behaviors in zebrafish navigation. Current Biology 20 (4), 381-386.

Burrill, J.D., Easter, S.S., 1994. Development of the retinofugal projections in the embryonic and larval zebrafish (*Brachydanio rerio*). Journal of Comparative Neurology 346 (4), 583-600.

Burton, C.E., Zhou, Y., Bai, Q., Burton, E.A., 2017. Spectral properties of the zebrafish visual motor response. Neuroscience Letters 646, 62-67.

Chen, X., Mu, Y., Hu, Y., et al., 2018. Brain-wide organization of neuronal activity and convergent sensorimotor transformations in larval zebrafish. Neuron 100 (4), 876-890.e5.

Connaughton, V.P., Graham, D., Nelson, R., 2004. Identification and morphological classification of horizontal, bipolar, and amacrine cells within the zebrafish retina. The Journal of Comparative Neurology 477 (4), 371-385.

Dunn, T.W., Gebhardt, C., Naumann, E.A., et al., 2016. Neural circuits underlying visually evoked escapes in larval zebrafish. Neuron 89 (3), 613-628.

Easter, S.S., Nicola, G.N., 1996. The development of vision in the zebrafish (*Danio rerio*). Developmental Biology 180 (2), 646-663.

Easter, J.S.S., Nicola, G.N., 1996. The development of vision in the zebrafish (*Danio rerio*). Developmental Biology 180 (2), 646-663.

Easter, S.S., Nicola, G.N., 1997. The development of eye movements in the zebrafish (*Danio rerio*). Developmental Psychobiology 31 (4), 267-276.

Emran, F., Rihel, J., Adolph, A.R., Wong, K.Y., Kraves, S., Dowling, J.E., 2007. OFF ganglion cells cannot drive the optokinetic reflex in zebrafish. Proceedings of the National Academy of Sciences of the United States of America 104 (48), 19126-19131.

Emran, F., Rihel, J., Dowling, J.E., 2008. A behavioral assay to measure responsiveness of zebrafish to changes in light intensities. Journal of Visualized Experiments 20.

Fujii, R., 2000. The regulation of motile activity in fish chromatophores. Pigment Cell Research 13 (5), 300-319.

Gao, Y., Chan, R.H.M., Chow, T.W.S., et al., 2014. A high-throughput zebrafish screening method for visual mutants by light-induced locomotor response. IEEE/ACM Transactions on Computational Biology and Bioinformatics 11 (4), 693-701.

Gestri, G., Link, B.A., Neuhauss, S.C.F., 2012. The visual system of zebrafish and its use to model human ocular diseases. Developmental Neurobiology 72 (3), 302-327.

Gnuegge, L., Schmid, S., Neuhauss, S.C.F., 2001. Analysis of the activity-deprived zebrafish mutant macho reveals an essential requirement of neuronal activity for the development of a fine-grained visuotopic map. Journal of Neuroscience 21 (10), 3542-3548.

Gosse, N.J., Nevin, L.M., Baier, H., 2008. Retinotopic order in the absence of axon competition. Nature 452 (7189), 892-895.

Gross, J.M., Perkins, B.D., 2008. Zebrafish mutants as models for congenital ocular disorders in humans. Molecular Reproduction and Development 75 (3), 547-555.

Hartmann, S., Vogt, R., Kunze, J., et al., 2018. Zebrafish larvae show negative phototaxis to near-infrared light. PLoS One 13 (11), e0207264.

Hu, M., Easter, S.S., 1999. Retinal neurogenesis: the formation of the initial central patch of postmitotic cells. Developmental Biology 207 (2), 309-321.

Hua, J.Y., Smear, M.C., Baier, H., Smith, S.J., 2005. Regulation of axon growth in vivo by activity-based competition. Nature

434 (7036), 1022-1026.
Huang, Y.-Y., Neuhauss, S.C.F., 2008. The optokinetic response in zebrafish and its applications. Frontiers in Bioscience 13, 1899-1916.
Kaethner, R.J., Stuermer, C.A., 1992. Dynamics of terminal arbor formation and target approach of retinotectal axons in living zebrafish embryos: a time-lapse study of single axons. Journal of Neuroscience 12 (8), 3257-3271.
Kaethner, R.J., Stuermer, C.A., 1994. Growth behavior of retinotectal axons in live zebrafish embryos under TTX induced neural impulse blockade. Journal of Neurobiology 25 (7), 781-796.
Karlstrom, R.O., Trowe, T., Klostermann, S., et al., 1996. Zebrafish mutations affecting retinotectal axon pathfinding. Development 123, 427-438.
Lamb, T.D., Collin, S.P., Pugh, E.N., 2007. Evolution of the vertebrate eye: opsins, photoreceptors, retina and eye cup. Nature Reviews Neuroscience 8 (12), 960e976.
Logan, D.W., Burn, S.F., Jackson, I.J., 2006. Regulation of pigmentation in zebrafish melanophores. Pigment Cell Research 19 (3), 206-213.
Maaswinkel, H., Li, L., 2003. Spatio-temporal frequency characteristics of the optomotor response in zebrafish. Vision Research 43 (1), 21-30.
Makhankov, Y.V., Rinner, O., Neuhauss, S.C.F., 2004. An inexpensive device for non-invasive electroretinography in small aquatic vertebrates. Journal of Neuroscience Methods 135 (1-2), 205-210.
Meier, A., Nelson, R., Connaughton, V.P., 2018. Color processing in zebrafish retina. Frontiers in Cellular Neuroscience 12, 327.
Meyer, M.P., Smith, S.J., 2006. Evidence from in vivo imaging that synaptogenesis guides the growth and branching of axonal arbors by two distinct mechanisms. Journal of Neuroscience 26 (13), 3604-3614.
Mueller, K.P., Neuhauss, S.C.F., 2014. Sunscreen for fish: co-option of UV light protection for camouflage. PLoS One 9 (1), e87372.
Muto, A., Orger, M.B., Wehman, A.M., et al., 2005. Forward genetic analysis of visual behavior in zebrafish. PLoS Genetics 1 (5), e66.
Naumann, E.A., Fitzgerald, J.E., Dunn, T.W., Rihel, J., Sompolinsky, H., Engert, F., 2016. From whole-brain data to functional circuit models: the zebrafish optomotor response. Cell 167 (4), 947-960.e20.
Neuhauss, S.C.F., 2003. Behavioral genetic approaches to visual system development and function in zebrafish. Journal of Neurobiology 54 (1), 148-160.
Neuhauss, S.C.F., 2018. Sensory biology: how to structure a tailor-made retina. Current Biology 28 (13), R737-R739.
Neuhauss, S.C., Biehlmaier, O., Seeliger, M.W., et al., 1999. Genetic disorders of vision revealed by a behavioral screen of 400 essential loci in zebrafish. Journal of Neuroscience 19 (19), 8603-8615.
Niell, C.M., Smith, S.J., 2005. Functional imaging reveals rapid development of visual response properties in the zebrafish tectum. Neuron 45 (6), 941-951.
Niell, C.M., Meyer, M.P., Smith, S.J., 2004. In vivo imaging of synapse formation on a growing dendritic arbor. Nature Neuroscience 7 (3), 254-260.
Niklaus, S., Neuhauss, S.C.F., 2017. Genetic approaches to retinal research in zebrafish. Journal of Neurogenetics 31 (3), 70-87.
Orger, M.B., Baier, H., 2005. Channeling of red and green cone inputs to the zebrafish optomotor response. Visual Neuroscience 22 (3), 275-281.
Orger, M.B., Smear, M.C., Anstis, S.M., Baier, H., 2000. Perception of Fourier and non-Fourier motion by larval zebrafish. Nature Neuroscience 3 (11), 1128-1133.
Regus-Leidig, H., Brandstätter, J.H., 2012. Structure and function of a complex sensory synapse. Acta Physiologica 204 (4), 479-486.
Roeser, T., Baier, H., 2003. Visuomotor behaviors in larval zebrafish after GFP-guided laser ablation of the optic tectum. Journal of Neuroscience 23 (9), 3726e3734.
Sajovic, P., Levinthal, C., 1982. Visual cells of zebrafish optic tectum: mapping with small spots. Neuroscience 7 (10), 2407-2426.
Sajovic, P., Levinthal, C., 1982. Visual response properties of zebrafish tectal cells. Neuroscience 7 (10), 2427-2440.
Schmidt, J.T., Buzzard, M., Borress, R., Dhillon, S., 2000. MK801 increases retinotectal arbor size in developing zebrafish without affecting kinetics of branch elimination and addition. Journal of Neurobiology 42 (3), 303-314.
Schmitt, E.A., Dowling, J.E., 1994. Early eye morphogenesis in the zebrafish, *Brachydanio rerio*. The Journal of Comparative Neurology 344 (4), 532-542.

Schmitt, E.A., Dowling, J.E., 1999. Early retinal development in the zebrafish, *Danio rerio*: light and electron microscopic analyses. The Journal of Comparative Neurology 404 (4), 515-536.

Shiraki, T., Kojima, D., Fukada, Y., 2010. Light-induced body color change in developing zebrafish. Photochemical and Photobiological Sciences 9 (11), 1498-1504.

Stuermer, C.A., 1988. Retinotopic organization of the developing retinotectal projection in the zebrafish embryo. Journal of Neuroscience 8 (12), 4513-4530.

Stuermer, C.A., Rohrer, B., Munz, H., 1990. Development of the retinotectal projection in zebrafish embryos under TTX-induced neural-impulse blockade. Journal of Neuroscience 10 (11), 3615-3626.

Trowe, T., Klostermann, S., Baier, H., et al., 1996. Mutations disrupting the ordering and topographic mapping of axons in the retinotectal projection of the zebrafish, *Danio rerio*. Development 123, 439-450.

Wagner, H.J., Wagner, E., 1988. Amacrine cells in the retina of a teleost fish, the roach (*Rutilus rutilus*): a Golgi study on differentiation and layering. Philosophical Transactions of the Royal Society of London B Biological Sciences 321 (1206), 263-324.

Wolf, S., Dubreuil, A.M., Bertoni, T., et al., 2017. Sensorimotor computation underlying phototaxis in zebrafish. Nature Communications 8 (1), 651.

Xiao, T., Roeser, T., Staub, W., Baier, H., 2005. A GFP-based genetic screen reveals mutations that disrupt the architecture of the zebrafish retinotectal projection. Development 132 (13), 2955-2967.

Zimmermann, M.J.Y., Nevala, N.E., Yoshimatsu, T., et al., 2018. Zebrafish differentially process color across visual space to match natural scenes. Current Biology 28 (13), 2018-2032.e5.

第四章

斑马鱼仔鱼的视觉逃逸：刺激、回路和行为

Emmanuel Marquez-Legorreta[1], Marielle Piber[1,2], Ethan K. Scott[1,3]

4.1 引言

动物能发现来袭的捕食者并避免其攻击的能力是进化选择行为的典型例子，这是因为失败的后果就是死亡。正因为如此，大多数动物通过进化形成了某种行为反应（behavioral responses），以增加逃离逼近危险的可能性。动物可应用多种感官模式探测和应对威胁，而视觉可以提供一整套关于捕食者的大小、形态、位置及运动的尤为丰富的信息，因此，视觉逃逸行为在整个动物界普遍存在就不足为奇了（Cooper and Blumstein, 2015; Sillar et al., 2016）。启发人们对包括在昆虫、甲壳类动物、鱼类、两栖动物、爬行动物、鸟类和哺乳动物（包括人类）等完全不同的生命系统中开展以下研究：视觉刺激和逃跑反应之间的关系研究，以及对刺激的感知、分析与触发反应相关的神经回路的研究（Schiff et al., 1962; Hayes and Saiff, 1967; Ewert, 1970; Sun and Frost, 1998; Carlile et al., 2006; Preuss et al., 2006; de Vries and Clandinin, 2012; Oliva and Tomsic, 2012; Landwehr et al., 2013; Yilmaz and Meister, 2013; De Franceschi et al., 2016; Fink et al., 2019）。

通常情况下，当动物在遇到视觉威胁时会朝着反方向做出快速运动，但是这种反应可依据不同情况和条件而变化。因此，虽然逃跑反应是一种强烈的先天性行为，但是，动物也可以根据其过去的经历或环境而做出调整。这种调整可能采取一定的形式，这些形式可以是有利于特定回避反应的形式，也可以是调节反应的速度和强度，或者是决定完全不反应。为了了解参与视觉逃避的核心回路以及调节反应的更为广泛的神经网络，必须研究各种威胁视觉刺激间的关系、一系列可能的输出行为，以及介于两者之间的大脑的活

[1] School of Biomedical Sciences, The University of Queensland, Brisbane, QLD, Australia.
[2] School of Medicine, Medical Sciences & Nutrition, University of Aberdeen, Aberdeen, United Kingdom.
[3] The Queensland Brain Institute, The University of Queensland, Brisbane, QLD, Australia.

动情况。

近年来，斑马鱼仔鱼已经发展成为研究此类问题的很有用的模型。早在受精1天的斑马鱼胚胎就能表现出应对抚摸刺激的逃逸行为（Eaton and Farley, 1973; Saint-Amant and Drapeau, 1998; Downes and Granato, 2006），而早在受精后4天的仔鱼就能执行视觉逃逸行为以逃避逼近的刺激（Yao et al., 2016）。斑马鱼是一种体外受精生物，意味着这种动物在早期发育阶段是很容易观测的。重要的是，斑马鱼仔鱼同胚胎一样均为光学透明体，这就意味着在不需要解剖、插管或者干扰仔鱼发育或神经活动的情况下就可以通过显微镜直接观察到其包括大脑在内的内部结构（Simmich et al., 2012）。

这些生物学属性为基因编码神经活性的荧光指示分子的应用（Broussard et al., 2014; Lin and Schnitzer, 2016），为双光子和选择性平面光学显微术的应用（selective plane illumination microscopy）（Ahrens et al., 2013; Wolf et al., 2015），为通过大型成像数据集中对全脑神经活动模式进行检测与建模的定量方法研究（Chen et al., 2018）等提供了近乎理想的研究平台。综合这些方法，可以在细胞分辨率下对斑马鱼仔鱼大脑的活动进行成像，并有助于理解组成大脑的成千上万个神经元的集体活动情况（Vanwalleghem et al., 2018）。最近已经扩展到可以对自由游动的斑马鱼仔鱼开展相关研究（Cong et al., 2017; Kim et al., 2017），同时也可以与用于相关环路功能测试（采用光遗传学或消融技术）和相关神经元形态描述（采用单细胞标记或免疫荧光技术）的强大工具进行整合研究（Leung et al., 2013; Feierstein et al., 2015; Dunn et al., 2016b; dal Maschio et al., 2017; Lovett-Barron et al., 2017; Helmbrecht et al., 2018; Tabor et al., 2018）。

总之，这些新方法与斑马鱼仔鱼的生物学特性相结合，使得其特别适合作为研究视觉逃逸的神经回路的模型。在本章中，我们将重点讨论斑马鱼仔鱼对视觉刺激的行为反应，以及驱动这些行为反应的神经回路，包括对触发视觉逃避的刺激类型以及刺激特征的描述，以及可能产生的各种逃避动作。我们将主要关注与斑马鱼仔鱼探测和应对视觉威胁相关的大脑区域、局部网络和细胞微环路方面的最新研究进展。最后，我们将提出一些未来的研究需求以更好地理解斑马鱼视觉逃避反应的神经回路和调控机制。

4.2 逼近刺激的基本特征（引起动物逃避的原因是什么？）

和大多数动物一样，斑马鱼有一种能对迎面而来的捕食者作出逃避反应的强大本能，这些反应已经在实验中得到了广泛的研究，在实验中使用的捕食者包括有生命的捕食者、屏幕显示的动画捕食者以及机器人模拟的捕食者（Bass and Gerlai, 2008; Saverino and Gerlai, 2008; Gerlai et al., 2009; Colwill and Creton, 2011; Ahmed et al., 2012; Luca and Gerlai, 2012; Ladu et al., 2015）。一般来说，斑马鱼对同域掠食者（sympatric predators）（斑马鱼可能在其自然栖息地遇到的掠食者）表现出比对异域掠食者（allopatric predators）或非掠食性鱼类更强的回避反应，这表明针对相关视觉威胁存在某种本能，使其在第一次遭遇危险时能够做出适应性行为（Bass and Gerlai, 2008; Ahmed et al., 2012）。从进化的角度看这种反应是非常有意义的，但是人们对于驱动逃跑行为的这些视觉刺激特性却知之甚少。

采用一种简化方法可解决这一问题，即将不同特征的刺激单独应用或组合应用，能衡量其在驱动逃避行为中的效力。例如，一些简单和非自然的刺激，如移动的矩形（moving rectangle）会引起强烈的反应（Ahmed et al., 2012）。另一些研究表明，将斑马鱼暴露于画有眼睛的各种简单形状的物体前都可以诱导逃避反应（Blaser and Gerlai, 2006）。食肉动物的身影（predatory silhouettes），包括鸟类（Dill, 1974）都能强烈地引起斑马鱼的反应。有趣的是，对这些掠食性身影的反应比对真正的捕食者的反应更加强烈，这表明斑马鱼不仅保存了掠食者的突出特性，而且可能会强化这种特性（Luca and Gerlai, 2012）。这些特性可能非常简单，因为进一步的实验表明，将复杂刺激减弱为简单移动的点也会引发逃避行为（Luca and Gerlai, 2012）。

最简单的刺激是在较亮的背景上出现一个逐渐扩大的黑暗圆圈［以下均称为“逼近”（loom）］，这种刺激已经成为研究模型（如斑马鱼仔鱼）的惊吓回路和行为的首选刺激物。然而，即使是这个简单的刺激，也包含了仔鱼在决定是否执行逃跑时需要处理的多种可能的线索（cue）。一种是亮度下降，当黑点接近斑马鱼时，它遮挡了视野中越来越大的部分。它还包含边缘信息，表示为不断变暗的圆圈与其较亮的背景之间的移动边界线［如图 4-1（A）～（C）所示］。一直以来，每个线索（变暗和移动边缘）的具体作用都成为人们研究的主题，但人们并未完全理解其在驱动逃逸行为中的确切作用。例如，对小鼠的研究表明，黑暗背景下的明亮刺激物不会引起惊吓，缺乏移动边缘的昏暗刺激物也不会引起惊吓（Yilmaz and Meister, 2013），表明这两种线索都必须存在才能使小鼠受到惊吓。在一项研究中，我们在斑马鱼仔鱼中发现相同的结果，该研究使用了明暗两种配置下的远去（后退 receding）和逼近（接近 approaching）的刺激以及黑暗刺激（dimming stimuli）（Temizer et al., 2015）。在这些实验中，明亮背景下的黑暗逼近刺激在 90% 以上的实验中引发了逃跑，而在黑暗背景上的明亮逼近刺激只在 20% ～ 30% 的实验中引发了惊吓。而渐渐远去的刺激效果甚至更弱，只有在不到 10% 的实验中引发了逃跑。为了测试亮度的变化是否可导致逃逸，实验时将全尺寸圆盘调暗，其调暗的时间动态与逼近的时间动态类似，实验结果表明这种调暗刺激并没有引起惊吓行为，表明亮度信息本身不足以驱动逃逸行为。“移动边缘”在逃避行为中的作用可以通过在等亮度的灰度背景上使用扩展的棋盘刺激来探索。这种移动信息类似于逼近刺激，但不包含昏暗刺激。利用这种方法，Dunn 等人（2016a）发现棋盘刺激驱动逃跑行为的概率与逼近刺激相似，而 Heap 等人（2018b）发现棋盘刺激引起的逃跑概率比逼近刺激低得多。正如 Heap 等人（2018b）所讨论的，棋盘刺激细节上的差异可能是导致这些不同结果的原因。结合这些研究结果，发现“移动边缘”似乎对诱导逃避行为是必要的，且亮度变化的方向（变亮或变暗）是很重要的，事实上，当移动边缘存在时，调暗可能有助于逃避行为的发生。

逼近刺激的时间和空间特性也对随后的逃逸反应的概率和时间有影响。不同模型系统的研究人员使用一种共同的语言来描述这些刺激的数学特性。在距离为（d）时，扩大的逼近刺激在斑马鱼的视网膜上投射出一个黑色圆盘，且其在视网膜上的角度大小（θ），随时间增加而增加［图 4-1（D）］。三角函数中，θ 是一半的物体大小（l）的反正切函数值的两倍。l 和物体接近速度（v）之间的关系，表示为比值（l/v），描述了物体在视网膜上大小的变化率。通过改变这一比值（l/v），研究人员可以操纵逼近刺激的时间动力学，以研究动物是如何运用这些信息来指导其行为的。

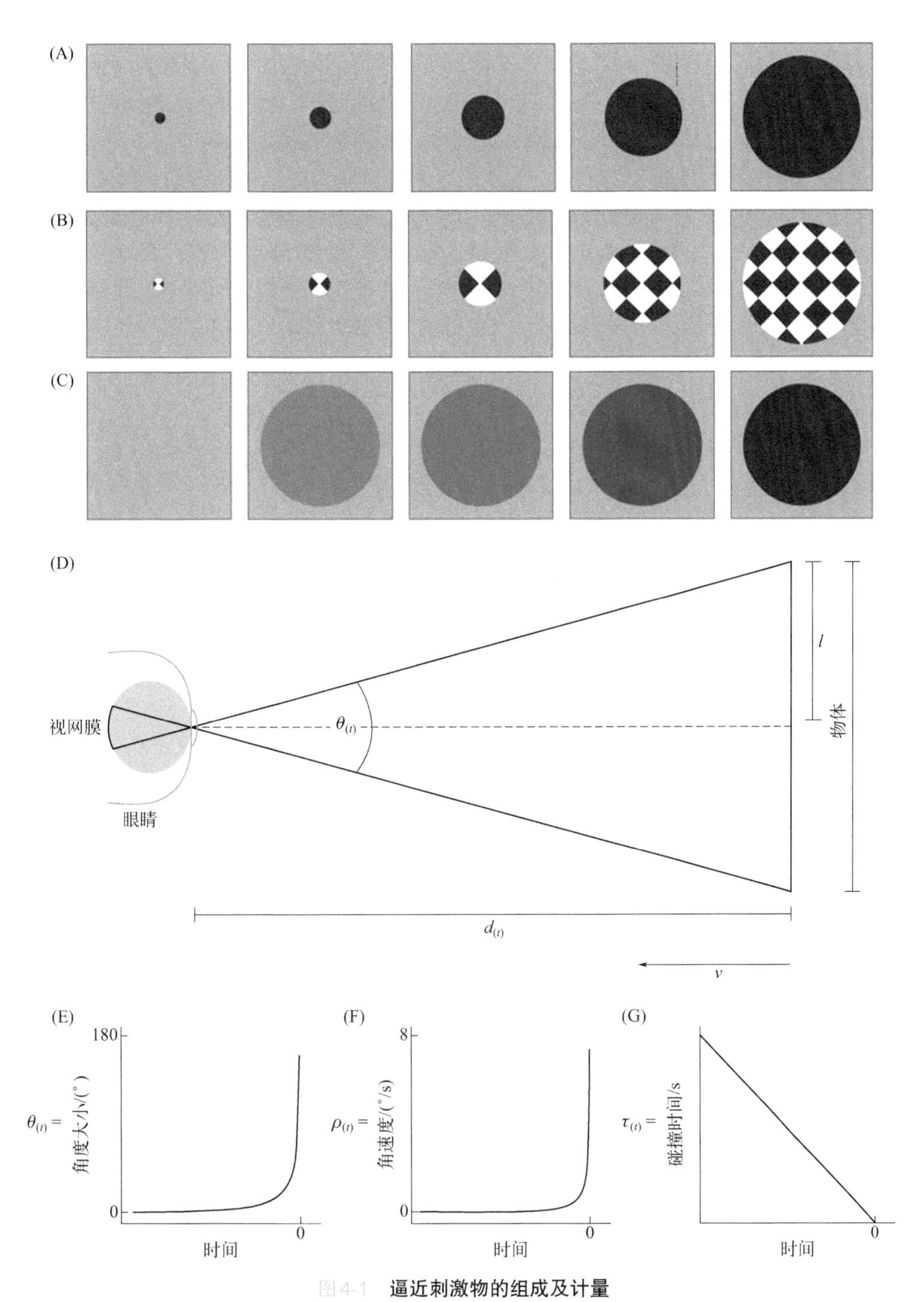

图4-1 **逼近刺激物的组成及计量**

(A)～(C)：分别表示逼近刺激物（A）、棋盘图案刺激物（B）和昏暗图案刺激物（C），刺激物从开始（左）到最后一帧（右）。(D)：示意图表示在观察者眼睛的视角 θ 处可观察到大小为l的物体的一半。随着物体以速度 v 接近，刺激物体逼近距离随着时间而缩短。如果刺激物接近速度是恒定的，角度 $\theta_{(t)}$ 中的刺激物和其增长率 [$\rho_{(t)}$=$\theta_{(t)}$] 都将随时间呈非线性增长。(E)～(G)：显示逼近过程中逼近刺激物的运动学参数。$\theta_{(t)}$ 为逼近刺激物的角度大小，$\rho_{(t)}$ 为逼近刺激物的角速度，$\tau_{(t)}$ 为碰撞时间（视觉刺激时间）。摘录自（Peek and Card, 2016）

对于以恒定速率接近的模拟捕食者，视网膜上的图像会以不断增加的速度增大［图 4-1（E）］。当单眼检测时，可感知 θ 的增长及其相应的增长率（ρ）的变化［图 4-1（F）］，包括碰撞时间（τ）［图 4-1（G）］（Hatsopoulos et al., 1995）。从理论上讲，这为动物提供了三个关键的光学参数，即物体的大小、物体逼近速度、物体增大速率这些参数，动物通过这三个关键参数可以探测到威胁。自然地，这引发了对大脑中编码这些参数的神经元的研究。

事实上，在蝗虫中的确发现了依赖于 θ 和 ρ 信息的刺激响应神经元和逃跑执行神经元（Fotowat et al., 2011）。这些神经元的信号转导随着逼近物体的增大而增加，在一个给定的角度阈值条件下，达到峰值，然后下降，其中这个阈值也与逃逸反应的启动相关。换句话说，当接近的物体达到某个角度大小时，神经元就会做出反应。一系列（l/v）比值范围的系统测试表明，是 θ 值决定了响应的时间。然后通过将这些信息与逼近刺激的其他属性信息相结合，使得蝗虫能够在适当时机逃生（Gabbiani et al., 1999）。

有趣的是，在鸽子中，图像扩展的所有三个光学变量分别与逼近敏感神经元的不同类别相对应（Sun and Frost, 1998）。首先，与物体的大小和速度无关，τ 细胞在碰撞前的特定时间开始活动，从而引发行为上的逃逸反应。其次，一旦超过了角速度的阈值，ρ 细胞就会做出反应。它们表示视角 θ' 变化率和物体绝对增大率。最后，η 细胞的放电率在一个临界刺激大小阈值（critical stimulus size threshold）时达到峰值，与接近物体的速度无关。最近一组斑马鱼仔鱼的研究暗示了视顶盖中存在 η 神经元（Temizer et al., 2015; Dunn et al., 2016a; Bhattacharyya et al., 2017）。然而，在这些研究中，可以依据对具有不同大小和增长速度的刺激反应时间，估算出阈值角度大小，结果发现在 20°（Temizer et al., 2015）、35°（Bhattacharyya et al., 2017）和 72°（Dunn et al., 2016a）分别作为响应发生角度时，阈值差异很大。这种差异性可以用实验方法的差异来解释，其中，Temizer 等人（2015）为头部嵌入式仔鱼（head-embedded larvae）呈现水平向的刺激，Bhattacharyya 等人（2017）为自由游动的仔鱼呈现一系列更自然的水平向刺激，Dunn 等人（2016a）从下方对自由游动的仔鱼进行逼近刺激。这些研究一致认为，存在触发斑马鱼仔鱼逃避行为的角度大小的阈值，但阈值的差异提醒我们，许多因素与刺激（大小、形状、生长速度和方向）和动物所处情况有关（在这种情况下，是头部嵌入或是自由游泳），这会影响动物的行为决定。这加强了视觉逃逸的基本属性：这是一种强大的、与生俱来的行为，但仍然受制于不可预知的外部世界的变化，以及动物的内部状态的调节。

4.3 逃避行为的执行

威胁性的视觉刺激可能导致以下多种行为中的任意一种。对于完全发育成熟的成年斑马鱼，这些可能包括惊吓、僵直和一系列回避行为（拍打、之字形、跳跃和潜水）（Luca and Gerlai, 2012; Bishop et al., 2016），这可能是根据捕食者的特征和动物所处的环境选择的。例如，通常在浅水中进行拍打，然后是静止不动，这一行为被认为是搅浑水域并伪装鱼类免受潜在捕食者的侵袭。（Bass and Gerlai, 2008; Luca and Gerlai, 2012）。同样，动物对特定行为的选择也可能受到其可利用的遮盖物（如阴影或植被）的影响（Hein et al., 2018），这是一种有趣而微妙的空间，但只是与斑马鱼仔鱼的惊吓行为有一定的关系，所以我们不在这里重点讨论。

包括斑马鱼仔鱼在内的鱼类最常见的逃逸行为是C形启动，它可以由触摸（Fetcho, 1991）、包括听觉线索在内的振动（Burgess and Granato, 2007）、水流（Stewart et al., 2013）和视觉（Eaton et al., 1977）而触发。逃逸潜伏期非常短，成年斑马鱼在刺激开始后5～10ms开始反应（Eaton et al., 1977）。逃逸行为涉及快速、有力地激活头部和身体肌肉，单侧收缩躯干和尾部的轴向肌肉，从而将身体弯曲成C形（Eaton et al., 1977）。这个动作可以分为三个阶段，经历15～20ms（Eaton et al., 1977）。上述身体的大范围收缩（第一阶段）通常发生在5～8ms后，随后身体的另一侧进一步收缩（第二阶段），在此期间尾部伸直。这两个阶段导致鱼快速远离捕食者，通常在大约20 ms内完成。在此之后，鱼可能会表现出各种行为，包括滑翔、减速或继续游泳（Eaton et al., 1977）。仔鱼的反应遵循与成鱼相同的运动过程，但在行为的速度、运动学和时间进程方面存在一些差异（Kimmel et al., 1974; Nair et al., 2015）。仔鱼的反应潜伏期略长，大于15 ms，达到完全收缩的速度比成鱼慢，但头部相对于体长的位移速度是成鱼的3倍（这是因为它们有较深的C形弯曲）（Eaton et al., 1977）。

涉及检测和处理视觉威胁的回路将在后面部分进行讨论，以下我们将简要讨论与惊吓相关的运动回路。短延迟C形启动是由Mauthner细胞（M细胞）介导的，这是一对双侧网状脊髓神经元，是支配整个脊柱的运动神经元。一个M细胞的单一动作电位会引发身体另一侧的C形弯曲（Zottoli, 1977; Eaton et al., 1982）。然而，如果不是激活M细胞，而是激活M细胞同源细胞，也会发生逃逸行为（Kimmel et al., 1980; Liu and Fetcho, 1999; Kohashi and Oda, 2008）。这种行为的C形起动具有长延迟特征，即需要更长的时间来启动C形弯曲，仔鱼的平均逃逸潜伏期为28.3ms（Liu and Fetcho, 1999; Budick and O'Malley, 2000; Burgess and Granato, 2007）。它们往往是由缓慢接近的逼近刺激引起的，不像短延迟的C形启动，涉及胸鳍的使用，这可能允许精准调节逃跑运动（McClenahan et al., 2012; Bhattacharyya et al., 2017）。相比之下，更快的逼近刺激接近速度通常会引起短延迟的C形启动（Bhattacharyya et al., 2017）。因此，短延迟和长延迟C形启动根据威胁的紧迫性提供不同的行为选择。M细胞驱动的短延迟C形启动提供了对即时威胁的快速和突然的响应，而未涉及速度和力量的调整与控制。长延迟的C形启动较慢，易变，运动学上更复杂，可能允许控制逃逸的速度或方向。

短延迟和长延迟的C形启动的这些不同属性，以及它们与动物面临威胁的性质之间的联系，只占鱼的广泛的行为表型和环境背景的一小部分。Feldman Barrett和Finlay（2018）认为防御行为是情境依赖的、有目的的、灵活的行动。只使用短延迟的C形弯曲会降低效率，而且会使捕食者可以预测其逃跑行为。因此，适当估计威胁水平的能力，并有一系列可能的反应，能够提高生存的可能性（Nair et al., 2017）。在较长的时间内，它们可以将环境的复杂细节例如躲藏的地方等集成到自身的逃跑决策中，为自身逃跑提供另一种复杂水平的决策依据。（Hein et al., 2018）。动物内部状态，如进食、交配或保卫领地的动机，也会影响其风险偏好，从而影响其对威胁的反应。核心逃脱回路提供一个可靠的、快速和强大的应激行为以应对紧急威胁，但由负责编码过去经验、环境背景和内部状态的其他回路所调控的核心回路可以提供正确的适应性行为以应对各种威胁。在接下来的两节中，我们将讨论斑马鱼仔鱼的核心逃逸回路，然后将提供一些证据以证明该回路是由大脑其他区域调控的。

4.4 视觉逃逸的核心回路

到目前为止，我们已经讨论了逼近刺激的关键特征以及由此产生的视觉诱发的逃避反应的特征。当然，这些刺激与反应的关系依赖于神经回路，这些神经回路负责感知和整合感官信息，计算威胁的属性或类型，并启动适当的运动回路来引发适应性反应。几十年以来这些回路的组成部分已经在一系列模型系统中被鉴定识别，但是直到现在，整个核心回路才开始受到关注，这在很大程度上有赖于在引言中所描述的近年来出现的新的光生理学和光遗传学方法。这些新技术方法能对大量神经元活动进行成像，从而使得斑马鱼仔鱼成为研究视觉惊吓回路的特别有价值的模型，并且在整个仔鱼大脑中可观察到逼近刺激的回路。

对视觉刺激做出反应的第一批神经元位于视网膜，视觉信息随后通过视网膜神经节细胞（retinal ganglion cell, RGC）轴突发送到大脑中的多个树枝状区域（arborization fields, AF）。大多数 RGC 延伸到顶盖的神经纤维 AF10（Burrill and Easter, 1994; Robles et al., 2014）（见图 4-2 和图 4-3 中顶盖的红色投影）。这些末端按精密的等高线图排列（横穿喙尾轴和内侧轴），并在顶盖神经纤维中不同的背侧（浅层）到腹侧（深层）层中复制，这种方式在大多数鱼类中高度保守（Vanegas and Ito, 1983; Vonbartheld and Meyer, 1987; Baier et al., 1996; Robles et al., 2013; Kita et al., 2015）。不同亚型的 RGC，从视网膜传递不同类型的视觉信息，靶向到特定的顶盖神经纤维的背腹侧层和其他 AF（Robles et al., 2013, 2014）。从视网膜产生的视觉信息的性质是重要的，因为它提供了顶盖和其他地方的回路解码和解释视觉刺激（如逼近）的资料。近年来，研究表明，RGC 亚型可以向顶盖传递特定目标、方向和大小的相关信息（Gabriel et al., 2012; Nikolaou et al., 2012; Lowe et al., 2013; Preuss et al., 2014; Semmelhack et al., 2014; Barker and Baier, 2015），包括顶盖接收视网膜中由特定 RGC 输入的逼近刺激的可能性（Temizer et al., 2015）。

图 4-2 总结了 RGC 传递给神经纤维的与逼近刺激相关的信息，包括神经纤维层和基底层（sublaminae）的细节，这些细节可能在后续处理中发挥特定的作用。携带运动信息和调暗信息的 RGC 主要投射到纤维层和浅表灰质层（stratum fibrosum et griseum superficiale, SFGS），也投射到中央灰质层（stratum griseum centrale, SGC）（Temizer et al., 2015）。调暗信息主要激活支配 SFGS6 深层的轴突，而 SFGS2-SFGS5 输入在接受逼近刺激时比接受调暗刺激时更活跃。支配 SFGS2 ～ 5 层的 RGC 对逼近物体的反应比对昏暗物体（亮度变化）的反应更强烈，这表明它们正在处理与移动边缘相关的信息。有趣的是，这最后一组 RGC 轴突可选择性地支配顶盖，而非其他的 AF（Robles et al., 2014）。逼近刺激信息通过顶盖的大量传递，表明该结构在检测视觉威胁方面发挥了重要作用，这一观点得到了实验的支持，即斑马鱼仔鱼顶盖神经纤维损伤显著降低逃逸反应（Temizer et al., 2015）。

其他对顶盖神经纤维层状结构的研究表明，顶盖可以区分逼近和其他类型的视觉刺激。除了亮度变化和移动边缘信息，顶盖还处理与移动物体的大小和方向相关的细节。似

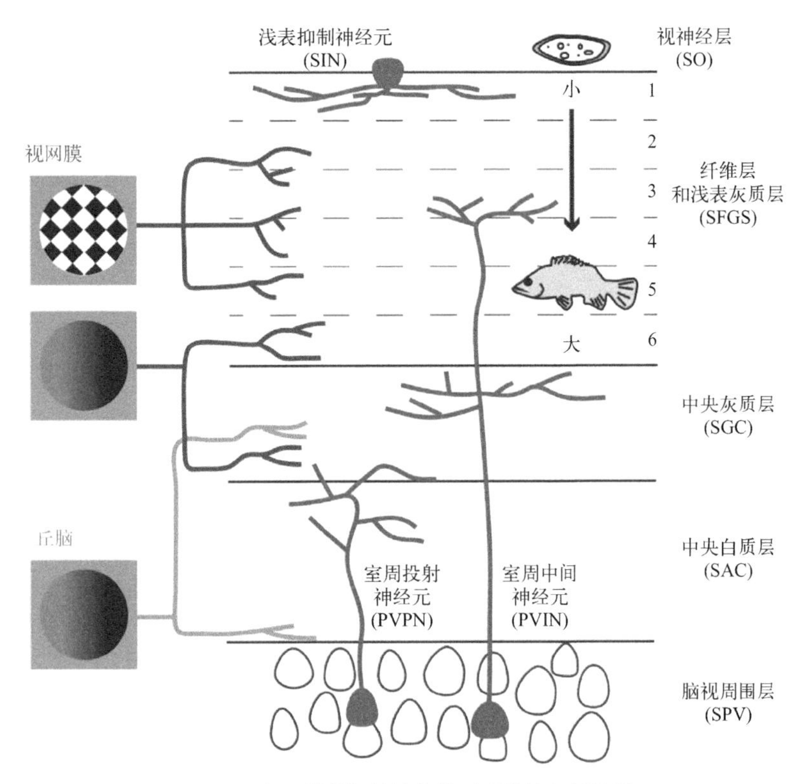

图 4-2 **逼近刺激相关信息传到顶盖神经纤维网**

左：RGC 轴突将运动信息传递到纤维层和浅表灰质层（SFGS）的 2 ～ 5 层，将调暗信息（dimming information）传递到 SFGS 的 6 层和中央灰质层（SGC）的深层（Temizer et al., 2015）。丘脑轴突将调暗信息传递到 SGC 的深层区域以及中央白质层（SAC）的深层部分（Heap et al., 2018a, b）。右：较小的物体优先在 SFGS 的视神经层（SO）和浅层进行处理，而较大的物体大多在 SFGS 的更深层进行处理（Preuss et al., 2014; Semmelhack, et al., 2014）

乎 SFGS 的深层更容易被较大的物体激活，而浅层和视神经层（stratum opticum, SO）对较小的物体的响应最为活跃（如图 4-2）（Preuss et al., 2014; Semmelhack et al., 2014）。这种筛选是由位于 SO 中的顶盖浅表抑制神经元（superficial inhibitory neurons, SIN）所介导的，它编码大小和方向信息（Del Bene et al., 2010; Preuss et al., 2014; Barker and Baier, 2015; Abbas et al., 2017; Yin et al., 2019）。调谐到特定刺激参数的 SINs 可调节其所在区域的室周中间神经元（periventricular interneurons, PVIN）的活动，PVIN 反过来激活室周投射神经元（periventricular projection neurons, PVPN），后者将轴突发送到运动前区（Del Bene et al., 2010; Helmbrecht et al., 2018）。通过这种方式，SIN 可能会控制从 RGC 靶向室周神经元（periventricular neurons, PVN）的信息，使具有特定特征（如逼近）的刺激更加显著。Dunn 等人（2016a）在模拟这种相互作用时发现，SINs 和 RGC 轴突的活动可以解释 PVN 对逼近刺激的反应。Barker 和 Baier（2015）的研究表明，将 SINs 消融可改变动物对大物体的回避反应，该研究为这一观点提供了支持，尽管在另一项类似的消融研究中没有观察到这种效果（Yin et al., 2019）。

这提供了一种貌似合理的机制，即顶盖可以根据视网膜输入信息来识别逼近刺激，但它没有考虑到视网膜信息也能传递给其他 9 个 AF，以及大脑其他视觉区域可能也参与逃避行为的反应。值得注意的是，最近的一项研究揭示了丘脑在逼近信息加工中的特定和重

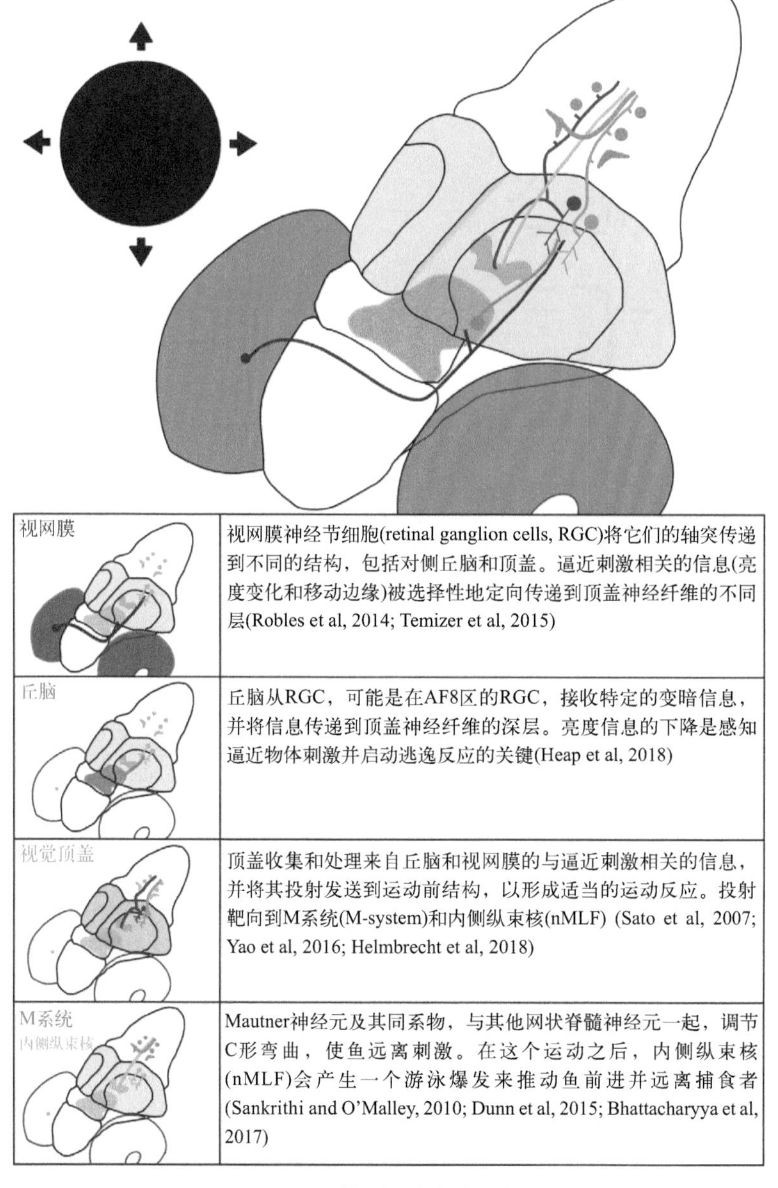

视网膜	视网膜神经节细胞(retinal ganglion cells, RGC)将它们的轴突传递到不同的结构，包括对侧丘脑和顶盖。逼近刺激相关的信息(亮度变化和移动边缘)被选择性地定向传递到顶盖神经纤维的不同层(Robles et al, 2014; Temizer et al, 2015)
丘脑	丘脑从RGC，可能是在AF8区的RGC，接收特定的变暗信息，并将信息传递到顶盖神经纤维的深层。亮度信息的下降是感知逼近物体刺激并启动逃逸反应的关键(Heap et al, 2018)
视觉顶盖	顶盖收集和处理来自丘脑和视网膜的与逼近刺激相关的信息，并将其投射发送到运动前结构，以形成适当的运动反应。投射靶向到M系统(M-system)和内侧纵束核(nMLF) (Sato et al, 2007; Yao et al, 2016; Helmbrecht et al, 2018)
M系统 内侧纵束核	Mautner神经元及其同系物，与其他网状脊髓神经元一起，调节C形弯曲，使鱼远离刺激。在这个运动之后，内侧纵束核(nMLF)会产生一个游泳爆发来推动鱼前进并远离捕食者(Sankrithi and O'Malley, 2010; Dunn et al, 2015; Bhattacharyya et al, 2017)

图 4-3 **核心视觉逃脱回路**

检测和响应视觉逼近刺激的大脑区域示意图

要作用（Heap et al., 2018b）。在这项研究中，Heap 和同事们描述了丘脑也参与视觉反应，他们发现丘脑对逼近刺激有反应。之后，他们鉴定了从丘脑到顶盖神经纤维的投射（主要位于其深层，在图 4-2 和图 4-3 中标记为橙色），并发现这些投射在逼近刺激期间特别活跃。为了评估这些投射的功能贡献，他们切除了丘脑 - 顶盖束，发现顶盖对逼近刺激的反应选择性降低，但对其他视觉刺激没有反应，这意味着丘脑为顶盖的视觉处理传递了逼近刺激相关的信息。丘脑 - 顶盖束的切除对逃逸行为也有较大的影响。这些切除不仅降低了逼近刺激逃跑的可能性，而且还改变了它们的方向，使仔鱼朝向而不是远离逼近刺激。为了测量丘脑逼近反应的刺激属性，他们试验了棋盘格刺激和调暗刺激，并发现亮度下降（调暗刺激）是丘脑逼近刺激响应的唯一原因。这种亮度信息很可能是投射到 AF8 的 RGC 轴突传递到丘脑的（Temizer et al., 2015）。在对这些刺激的进一步的行为实验中，Heap 和同事们发现，棋盘逼近类刺激比黑暗逼近类刺激的惊吓率低，但昏暗刺激无法引起反应，这与之前的研究一致（Temizer et al., 2015; Dunn et al., 2016a）。这表明丘脑亮度信息不能单独驱动逃逸反应，但增加了逃逸反应的概率并可控制其方向。为了证实亮度和丘脑在视觉逃逸中所扮演的角色，Heap 和同事们展示了一个双向刺激，其中一只眼睛可看到昏暗的物体，而另一只眼睛可看到扩大的棋盘格。这一刺激比单独昏暗刺激或单独棋盘逼近刺激更有可能导致逃离，并且出现面向棋盘逼近刺激而不是远离它的概率更高。这项研究的结果支持以下模型：在这个模型中，关于移动边缘的信息，直接从 RGC 传递到顶盖，这对于逃逸反应是必要的，但亮度信息通过丘脑传到顶盖是必要的，它可以指导动物逃离捕食者（亮度下降更早或更剧烈的那一边）。尽管这个模型与视网膜 - 顶盖视觉逃逸的基本模型有差异，但它仍然认为顶盖是逼近刺激信息加工的中心。作为移动边缘信息和亮度信息的交汇点，它可以很好地执行对视觉捕食者进行检测所需的关键计算。

顶盖微回路与这些信息流的整合细节在很大程度上仍然是未知的，但这些计算的结果被传送到神经纤维深层的室周投射神经元（periventricular projection neuron, PVPN）树突（Scott and Baier, 2009; Robles et al., 2011）（图 4-2），然后 PVPN 将这些计算结果传递给运动前神经元，形成运动反应（图 4-3）（Zottoli et al., 1987; Sato et al., 2007; Yao et al., 2016）。最近的研究结果表明，PVPN 的两个亚群是相关的。一个投射到对侧网状脊髓神经元和内侧纵束核（nucleus of the medial longitudinal fasciculus, nMLF），另一个则支配同侧网状脊髓神经元（Helmbrecht et al., 2018）（图 4-3 中分别标记为深蓝和浅蓝色的神经元）。这些 PVPN 亚型可能允许对侧 M 系统（驱动 C 形弯曲）的激活，以达到与其他姿势调整和抑制同侧脊髓运动回路相协调（Fetcho and Faber, 1988; Fetcho, 1991; Song et al., 2015）（综述见 Eaton et al., 1991; Korn and Faber, 1996; Eaton et al., 2001; Korn and Faber, 2005; Medan and Preuss, 2014; Hale et al., 2016）。有趣的是，对 M 细胞的顶盖投射在闪烁刺激和逼近刺激的时候都是一样的，但是 M 细胞对逼近刺激的响应更有特异性（Yao et al., 2016）。这些结果说明 M 细胞可能需要进一步的特殊信号才能达到阈值，或者需要一种抑制机制，这种抑制机制可以在其他刺激下抑制 M 细胞。如上所述，逃逸反应在 M 细胞消融后仍然会发生，这表明这些反应可以由多类网状脊髓神经元产生和调节（Eaton et al., 1982; Liu and Fetcho, 1999; Dunn et al., 2016a; Naumann et al., 2016），M 细胞介导短延迟 C 形弯曲，其他网状脊髓神经元驱动长延迟 C 形弯曲（Kohashi and Oda, 2008; Bhattacharyya et al., 2017）。这种运动回路的差异很重要，因为它是前面描述的快速、不变的逃逸反应（通常用于快速逼近和即将发生的碰撞）和较慢的、可调节的逃逸反应（用于较不紧急的威胁）之间的选择的基础（Bhattacharyya et al., 2017）。

在 C 形弯曲产生的初始重新定位后，逃逸反应包括游泳爆发，推动鱼向前并远离威胁。根据以往的研究结果，该运动反应很可能是由 nMLF（图 4-3 中浅绿色中心）产生的，因为它参与了逃逸反应和前游行为，并具有视觉反应神经元（Gahtan et al., 2002; Orger et al., 2008; Sankrithi and O'Malley, 2010; Severi et al., 2014; Thiele et al., 2014; Wang and McLean, 2014）。

4.5 核心逃逸回路的调控

C 形弯曲逃跑易造成混乱，而且耗费能量，因此，虽然在某些情况下 C 形弯曲逃跑对生存至关重要，但在不需要时，动物能够调整或防止这些行为也很重要。在本节中，我们将重点介绍内部状态、最近的实验和其他类型的环境对视觉逃逸行为的影响。我们还将回顾可能负责编码这些因素的大脑区域和回路，以及这些信息到达和影响核心逃逸回路活动的方式，以产生适合动物所面临环境的行为。这些途径的总结见图 4-4。

Lovett-Barron 等人（2017）最近提供了一个可以影响视觉逃逸的内部状态的研究案例，他们将钙成像和视觉逃逸分析结合起来研究，试图找到与快速反应相关的兴奋性活动神经元（他们以这些活动性神经元作为警惕性的代表）。通过免疫组化并将这些信息与钙成像结果进行对比，他们确定了在快速逃逸之前活动的神经元的特定功能特性。通过该方法鉴定了以下活动性神经元：蓝斑去甲肾上腺素能神经元，被盖中可卡因和安非他明调控的转录物（CART）和胆碱能神经元，以及下丘脑多巴胺能、5- 羟色胺能和神经肽 Y 神经元（图 4-4 中的紫色结构）。在使用咖啡因后，大脑结构中也发现了这些类似的活动性神经元，这也导致减少了反应时间。另一方面，下丘脑生长抑素神经元的活动与反应时间较长有关。这些结果启发 Lovett-Barron 和他的同事们得出以下结论：短反应时间之前的神经递质结构神经元活性参与了警觉，动物的这种内在状态缩短了其面对威胁刺激的反应时间，然而这种影响核心视觉逃脱回路的机制仍不清楚。

Yao 等人（2016）最近发现了视觉逃逸中感觉运动门控的一个重要例子。在他们的研究中，他们展示了从顶盖接收信息的下丘脑多巴胺能神经元是如何在非威胁性的视觉刺激（闪烁、后退的圆圈或明亮的逼近刺激）出现时更加活跃的。这种活动通过后脑的抑制性甘氨酸能中间神经元（图 4-4 中的黑色神经元）抑制顶盖向 Mauthner 神经元的传输。当一个逼近刺激出现时，这些多巴胺能神经元并不激活抑制性中间神经元，而是允许顶盖和网状脊髓神经元之间的自由传递，并允许逃逸反应。这种回路提供了一种安全性保障，以防止不必要的 C 形弯曲，即使这些视觉刺激包含一些与威胁刺激相同的特性（运动或边缘刺激）。

Filosa 和同事进行的第三项研究涉及视觉刺激、逼近和逃避行为与下丘脑 - 垂体间轴（hypothalamic-pituitary-interrenal, HPI）之间的关系（Filosa et al., 2016）。在寻找食物的压力下，饥饿的仔鱼变得不那么规避风险，对视觉刺激的惊吓响应降低，与进食相关的接近行为增加。这种状态与皮质醇减少和 HPI 轴活动减少有关。然后，他们发现饥饿的仔鱼投射到顶盖的中缝 5- 羟色胺能神经元（图 4-4 和图 4-5 中标记为粉红色）的活动也增加了，通过控制这些神经元的活动，他们可以调节饥饿与进食的仔鱼的行为。最后，他们发现 HPI 轴和 5- 羟色胺能神经元的活动以一种与行为结果相匹配的方式改变顶盖对不同大小的视圈的反应（大的刺激通常会引起惊吓，小的刺激会引起接近行为）。他们的研究表明，这些 5- 羟色胺能神经元的活动可调节顶盖回路的敏感性，从而使下游运动前区的平衡向回避或接近方向倾斜，这取决于动物的即时生存压力。

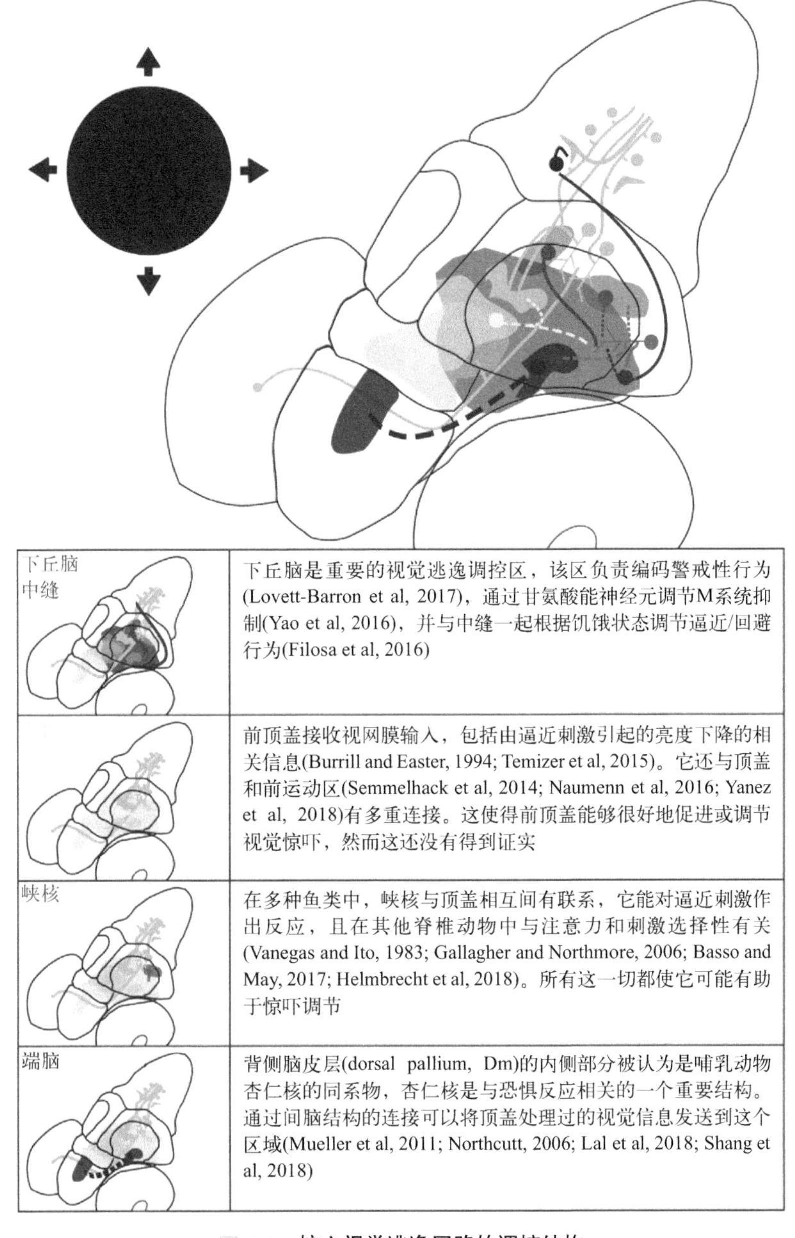

下丘脑 中缝	下丘脑是重要的视觉逃逸调控区，该区负责编码警戒性行为(Lovett-Barron et al, 2017)，通过甘氨酸能神经元调节M系统抑制(Yao et al, 2016)，并与中缝一起根据饥饿状态调节逼近/回避行为(Filosa et al, 2016)
	前顶盖接收视网膜输入，包括由逼近刺激引起的亮度下降的相关信息(Burrill and Easter, 1994; Temizer et al, 2015)。它还与顶盖和前运动区(Semmelhack et al, 2014; Naumenn et al, 2016; Yanez et al, 2018)有多重连接。这使得前顶盖能够很好地促进或调节视觉惊吓，然而这还没有得到证实
峡核	在多种鱼类中，峡核与顶盖相互间有联系，它能对逼近刺激作出反应，且在其他脊椎动物中与注意力和刺激选择性有关(Vanegas and Ito, 1983; Gallagher and Northmore, 2006; Basso and May, 2017; Helmbrecht et al, 2018)。所有这一切都使它可能有助于惊吓调节
端脑	背侧脑皮层(dorsal pallium, Dm)的内侧部分被认为是哺乳动物杏仁核的同系物，杏仁核是与恐惧反应相关的一个重要结构。通过间脑结构的连接可以将顶盖处理过的视觉信息发送到这个区域(Mueller et al, 2011; Northcutt, 2006; Lal et al, 2018; Shang et al, 2018)

图 4-4 **核心视觉逃逸回路的调控结构**

已知调控（实线）或可能调控（虚线）核心视觉逃逸回路的大脑区域结构的示意图

这些研究均表明下丘脑参与了核心逃逸回路，它似乎对逃逸行为有不同的影响，并在不同的条件下参与核心逃逸回路的不同部分。由于下丘脑也直接向顶盖神经纤维的深层发送投射（图 4-5 中的紫色轴突）（Heap et al., 2018a），它可能以某种方式影响逃逸行为，这种方式目前还未被研究。这使得下丘脑及其与大脑中逼近刺激敏感区域的回路间的相互作用成为进一步研究如何调节视觉惊吓的有趣课题。

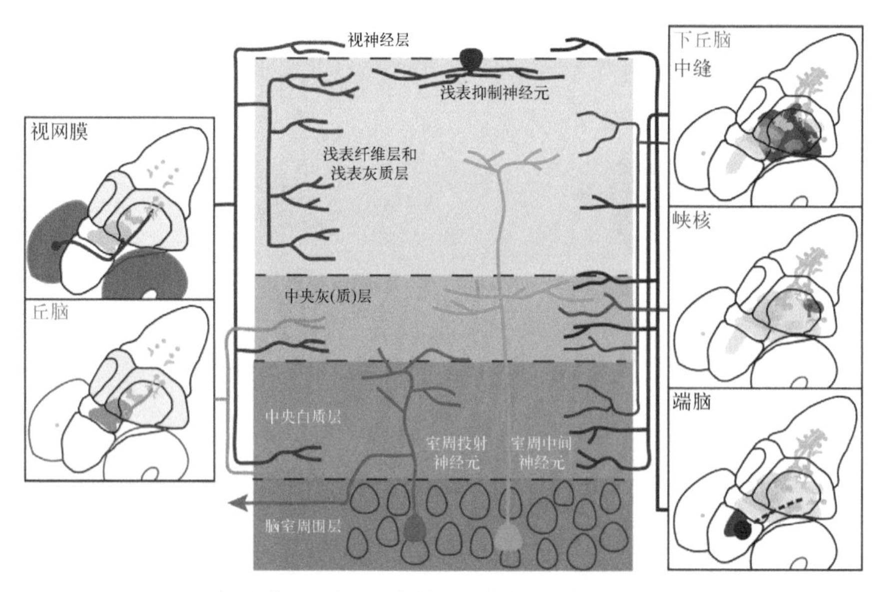

图 4-5　**视网膜、丘脑和顶盖神经纤维中调控信息的汇总示意图**

左：顶盖神经纤维的两个核心视觉传入神经分别是 RGC（红色）和来自丘脑的神经投射（橙色）（Temizer et al., 2015; Heap et al., 2018a,b）。右图：已知的顶盖神经纤维传入结构，以及可能调节逼近信息处理的结构，包括下丘脑（紫色）、中缝（粉红色）、峡核（绿色）和端脑（棕色）（Heap et al., 2018a,b; Filosa et al., 2016; King and Schmidt, 1993; Luiten, 1981; Folgueira et al., 2004）。中间：每一种结构的投射神经元均在神经纤维的刻板层（stereotyped laminae）和亚层（sublaminae）中终止，视觉（特别是视网膜）输入神经元投射到浅层区域，而调控性输入神经元投射到更深层的区域。视网膜信息由浅表抑制神经元和室周中间神经元接收和处理，室周中间神经元随后支配室周投射神经元，后者将轴突发送到运动前结构（Del Bene et al., 2010; Nevin et al., 2010）。PVN 和浅表抑制神经元的草图基于 Nevin 等人（2010）的研究

4.6　关于逼近回路的开放性问题

前面，我们概述了核心视觉逃逸回路和其他参与调节的大脑区域。然而，以上关于逼近刺激信息处理过程的描述示意图在两个方面仍然是不完整的。首先，目前人们只了解大脑各区域，对这些区域内负责处理信息的微回路知之甚少。虽然我们所描述的一些研究在钙成像过程中分析了这些区域内的单个神经元，但通常没有关于这些神经元形态或功能特征的信息。这一缺陷可以从我们目前关于顶盖对视觉逃逸贡献的认识中得到说明。顶盖直接从视网膜接收与逼近刺激相关的信息输入，以及来自丘脑的亮度信息，可能还有来自下丘脑和其他区域的调节信息。接收这些信息的顶盖神经元的结构和连通性，以及这些不同信息之间的联系和相互作用网络的功能结构等大多数仍然是未知的。在顶盖和其他与视觉惊吓有关的结构中，这些微回路的绘图（通过解剖和组织化学研究）和它们之间关系的功能测试（使用光遗传学）将是必要的，以便记录特定类型的神经元处理信息的类型和它们在更广泛网络中的作用。

我们的目标是将基于局部脑区的模型细化为一种描述微回路和调节视觉逃逸的神经网络模型，而一些有应用前景的研究已经开始在其他视觉通路中应用这种方法。这些研究

工作集中在神经元群体规模的功能成像（采用钙或电压指标）、解剖学标记（Forster et al., 2017a, 2018）和光学标记，以产生特定的大脑全息图像（Favre-Bulle et al., 2015; Accanto et al., 2019）。这些技术的结合使得对刺激有特殊反应的神经元（通过钙成像判断）可以通过在相关神经元上（dal Maschio et al., 2017; Kramer et al., 2019）使用靶向全息图来标记光激活 GFP（Patterson and Lippincott-Schwartz, 2002），从而将神经元形态与其活动联系起来。类似的方法可以用于选择性激活或沉默特定功能类别的神经元，在此过程中可使用光刺激和光遗传学执行器（actuators）（dal Maschio et al., 2017）。在神经元功能特征方面，体内钙成像和随后免疫组化的检测结果揭示了在感觉处理过程中具有特定活动模式的神经元的神经化学特性（Lovett-Barron et al., 2017）。在类似的方法中，转基因标记系统与钙成像相结合，可实时提供活动神经元的进一步信息（Dunn et al., 2016a; Yao et al., 2016; Heap et al., 2018b）。转基因标记系统的广泛文库（Davison et al., 2007; Scott et al., 2007; Scott and Baier, 2009; Satou et al., 2013; Kimura et al., 2014; Kawakami et al., 2016; Forster et al., 2017b; Tabor et al., 2019）将在未来成为一种灵活而有力的方法。

脑区局部模型的第二个不完善之处就是，该模型忽略了整个大脑中负责决策和执行视觉逃逸的区域。潜在的大脑区域包括额外的快速反应区域，如前顶盖（pretectum）、中脑被盖（midbrain tegmentum）和前脑（Dunn et al., 2016a; Chen et al., 2018），其功能尚未被彻底研究。在这些结构中，前顶盖（图 4-4 中标记为黄色）接受视网膜的直接输入（Burrill and Easter, 1994; Yanez et al., 2018），包括 AF6 中的亮度信息（Temizer et al., 2015）。此外，它与顶盖有交互联系（Yanez et al., 2018），并与斑马鱼的其他视觉诱发行为有关联（Kubo et al., 2014; Portugues et al., 2014; Semmelhack et al., 2014; Naumann et al., 2016）。中脑被盖是另一个潜在的反应区，它包括一个同样可能调节逃逸反应的核心。对金鱼和太阳鱼的研究表明，峡核（nucleus isthmi, NI）对逼近刺激比对新的视觉刺激更敏感，并与顶盖的深层（图 4-4 中的绿色区域）有交互联系（Vanegas and Ito, 1983; Striedter and Northcutt, 1989; King and Schmidt, 1993; Northmore and Gallagher, 2003; Gallagher and Northmore, 2006）。这增加了峡核 - 顶盖环参与处理逼近物体信息的可能性（Northmore and Graham, 2005; Graham and Northmore, 2007）。这种视觉反馈回路在大多数脊椎动物中是保守的（Gruberg et al., 2006），它的作用也已经在鸟类中被研究过。对猫头鹰、鸡和鸽子的一些研究表明，峡核可调控顶盖对突出视觉刺激作出反应，并与视觉注意有关（Marin et al., 2007; Asadollahi et al., 2010; Mysore and Knudsen, 2013, 2014; Goddard et al., 2014; Basso and May 2017）。

另一个可能参与调节对捕食者响应的结构是背侧脑皮层的内侧部分（dorsal pallium, Dm，图 4-4 中端脑的棕色结构）。这种前脑结构被认为是哺乳动物杏仁核的同源物（Wullimann and Mueller, 2004; Yamamoto et al., 2007; Mueller et al., 2011），因此可能参与斑马鱼对厌恶刺激的反应（von Trotha et al., 2014; do Carmo Silva et al., 2018; Lal et al., 2018）。Dm 的大部分视觉信息可能来自球前核（preglomerular nucleus）（图 4-4 中棕色传输结构）。根据弱电鱼、虹鳟、鲤鱼和金鱼的追踪实验，间脑结构接收来自顶盖的投射，并与背侧大脑皮层建立连接，后者通过向顶盖发送轴突来关闭环路（图 4-5 中的棕色部分）（von Trotha et al., 2014; do Carmo Silva et al., 2018; Lal et al., 2018）。这些连接让人联想到小鼠中逼近刺激触发的视觉逃逸反应所涉及的视觉通路（Mueller, 2012; Zhao et al., 2014; Carr, 2015; Wei et al., 2015; Yuan and Sus, 2015; Perathoner et al., 2016; Pereira and Moita, 2016; Shang et al., 2018; Zhou et al.,

2019），这可能在人类对威胁性视觉刺激的处理中很重要（McFadyen et al., 2017）。鉴于与哺乳动物的杏仁核连通性的相似之处，很容易让人想到 Dm 可能在鱼类中也扮演着类似的角色。

参与视觉逃逸反应的回路可以通过视网膜 - 丘脑 - 顶盖 - 后脑回路（图 4-3）或与多个调节结构的更复杂的相互作用（图 4-4）来达到其最小化形式。无论从哪种角度看，顶盖都是来自视网膜、丘脑、峡核、端脑和其他结构的大多数视觉信息汇聚的结构（图 4-5），也是逼近刺激的单独特征被整合的地方。因此，顶盖似乎充当了一个视觉运动中枢，在这里收集和理解视觉信息，并产生逃逸反应的预备运动信号。

这些微回路的框架如图 4-5 所示，其中我们总结了顶盖神经纤维的层状结构，它与各种视觉和调节信息输入有关。这凸显出视网膜输入调控 SFGS 的事实，即在这里，有关运动、边缘和图形的信息被传递，为视觉处理提供原始材料。在更深层次上，来自其他区域的输入（与亮度、背景和内部状态相关）更有可能提供了对方向、运动学或惊吓决策相关的进一步细节。顶盖的解剖结构（表层视网膜层 SINs，在浅表神经纤维中有树突的 PVINs，以及在深层神经纤维中有树突的 PVPN）展示了一个结构框架，当信息从浅表神经纤维到深层神经纤维传递，然后再传递到下游结构时，可以通过这个框架进行处理和调节。阐明执行这些处理的单个细胞类型和顶盖微回路将对未来的工作特别有意义，这是因为它们都将提供传递重要信息的精细功能图，并将揭示整合调控信息的回路机制。

（李海芸翻译　李丽琴审校）

参考文献

Abbas, F., Triplett, M.A., Goodhill, G.J., Meyer, M.P., 2017. A three-layer network model of direction selective circuits in the optic tectum. Frontiers in Neural Circuits 11.

Accanto, N., Chen, I.W., Ronzitti, E., Molinier, C., Tourain, C., Papagiakoumou, E., Emiliani, V., 2019. Multiplexed temporally focused light shaping through a gradient index lens for precise in-depth optogenetic photostimulation. Scientific Reports 9.

Ahmed, T.S., Fernandes, Y., Gerlai, R., 2012. Effects of animated images of sympatric predators and abstract shapes on fear responses in zebrafish. Behaviour 149, 1125-1153.

Ahrens, M.B., Orger, M.B., Robson, D.N., Li, J.M., Keller, P.J., 2013. Whole-brain functional imaging at cellular resolution using light-sheet microscopy. Nature Methods 10, 413.

Asadollahi, A., Mysore, S.P., Knudsen, E.I., 2010. Stimulus-driven competition in a cholinergic midbrain nucleus. Nature Neuroscience 13, 889-895. U138.

Baier, H., Klostermann, S., Trowe, T., Karlstrom, R.O., NussleinVolhard, C., Bonhoeffer, F., 1996. Genetic dissection of the retinotectal projection. Development 123, 415-425.

Barker, A.J., Baier, H., 2015. Sensorimotor decision making in the zebrafish tectum. Current Biology 25, 2804-2814.

Bass, S.L.S., Gerlai, R., 2008. Zebrafish (*Danio rerio*) responds differentially to stimulus fish: the effects of sympatric and allopatric predators and harmless fish. Behavioural Brain Research 186, 107-117.

Basso, M.A., May, P.J., 2017. Circuits for action and cognition: a view from the superior colliculus. Annual Review of Vision Science 3, 197-226.

Bhattacharyya, K., McLean, D.L., Maciver, M.A., 2017. Visual threat assessment and reticulospinal encoding of calibrated responses in larval zebrafish. Current Biology 27, 2751.

Bishop, B.H., Spence-Chorman, N., Gahtan, E., 2016. Three-dimensional motion tracking reveals a diving component to visual and auditory escape swims in zebrafish larvae. Journal of Experimental Biology 219, 3981-3987.

Blaser, R., Gerlai, R., 2006. Behavioral phenotyping in zebrafish: comparison of three behavioral quantification methods. Behavior Research Methods 38, 456-469.

Broussard, G.J., Liang, R.Q., Tian, L., 2014. Monitoring activity in neural circuits with genetically encoded indicators. Frontiers

in Molecular Neuroscience 7.
Budick, S.A., O'Malley, D.M., 2000. Locomotor repertoire of the larval zebrafish: swimming, turning and prey capture. Journal of Experimental Biology 203, 2565-2579.
Burgess, H.A., Granato, M., 2007. Sensorimotor gating in larval zebrafish. Journal of Neuroscience 27, 4984-4994.
Burrill, J.D., Easter, S.S., 1994. Development of the retinofugal projections in the embryonic and larval zebrafish (*Brachydanio rerio*). Journal of Comparative Neurology 346, 583-600.
Carlile, P.A., Peters, R.A., Evans, C.S., 2006. Detection of a looming stimulus by the Jacky dragon: selective sensitivity to characteristics of an aerial predator. Animal Behaviour 72, 553-562.
Carr, J.A., 2015. I'll take the low road: the evolutionary underpinnings of visually triggered fear. Frontiers in Neuroscience 9.
Chen, X.Y., Mu, Y., Hu, Y., Kuan, A.T., Nikitchenko, M., Randlett, O., Chen, A.B., Gavornik, J.P., Sompolinsky, H., Engert, F., Ahrens, M.B., 2018. Brain-wide organization of neuronal activity and convergent sensorimotor transformations in larval zebrafish. Neuron 100, 876.
Colwill, R.M., Creton, R., 2011. Imaging escape and avoidance behavior in zebrafish larvae. Reviews in the Neurosciences 22, 63-73.
Cong, L., Wang, Z.G., Chai, Y.M., Hang, W., Shang, C.F., Yang, W.B., Bai, L., Du, J.L., Wang, K., Wen, Q., 2017. Rapid whole brain imaging of neural activity in freely behaving larval zebrafish (*Danio rerio*). Elife 6.
Cooper, W.E., Blumstein, D.T., 2015. Escaping from Predators : An Integrative View of Escape Decisions. dal Maschio, M., Donovan, J.C., Helmbrecht, T.O., Baier, H., 2017. Linking neurons to network function and behavior by two-photon holographic optogenetics and volumetric imaging. Neuron 94, 774.
Davison, J.M., Akitake, C.M., Goll, M.G., Rhee, J.M., Gosse, N., Baier, H., Halpern, M.E., Leach, S.D., Parsons, M.J., 2007. Transactivation from Gal4-VP16 transgenic insertions for tissue-specific cell labeling and ablation in zebrafish. Developmental Biology 304, 811-824.
De Franceschi, G., Vivattanasam, T., Saleem, A.B., Solomon, S.G., 2016. Vision guides selection of freeze or flight defense strategies in mice. Current Biology 26, 2150-2154.
de Vries, S.E.J., Clandinin, T.R., 2012. Loom-sensitive neurons link computation to action in the *Drosophila* visual system. Current Biology 22, 353-362.
Del Bene, F., Wyart, C., Robles, E., Tran, A., Looger, L., Scott, E.K., Isacoff, E.Y., Baier, H., 2010. Filtering of visual information in the tectum by an identified neural circuit. Science 330, 669-673.
Demski, L., 2003. In a fish's mind's eye: the visual pallium of teleosts. In: Collin, S., Marshall, N.J. (Eds.), Sensory Processing in Aquatic Environments. Springer, New York, pp. 404-419.
Demski, L.S., 2013. The pallium and mind/behavior relationships in teleost fishes. Brain, Behavior and Evolution 82, 31-44.
Dill, L.M., 1974. The escape response of the zebra danio (*Brachydanio rerio*) I. The stimulus for escape. Animal Behaviour 22, 711-722.
do Carmo Silva, R.X., Lima-Maximino, M.G., Maximino, C., 2018. The aversive brain system of teleosts: implications for neuroscience and biological psychiatry. Neuroscience and Biobehavioral Reviews 95, 123-135.
Downes, G.B., Granato, M., 2006. Supraspinal input is dispensable to generate glycine-mediated locomotive behaviors in the zebrafish embryo. Journal of Neurobiology 66, 437-451.
Dunn, T.W., Gebhardt, C., Naumann, E.A., Riegler, C., Ahrens, M.B., Engert, F., Del Bene, F., 2016a. Neural circuits underlying visually evoked escapes in larval zebrafish. Neuron 89, 613-628.
Dunn, T.W., Mu, Y., Narayan, S., Randlett, O., Naumann, E.A., Yang, C.T., Schier, A.F., Freeman, J., Engert, F., Ahrens, M.B., 2016b. Brain-wide mapping of neural activity controlling zebrafish exploratory locomotion. Elife 5.
Eaton, R.C., Farley, R.D., 1973. Development of mauthner neurons in embryos and larvae of zebra fish, *Brachydanio-rerio*. Copeia 1973, 673-682.
Eaton, R.C., Bombardieri, R.A., Meyer, D.L., 1977. The Mauthner-initiated startle response in teleost fish. Journal of Experimental Biology 66.
Eaton, R.C., Lavender, W.A., Wieland, C.M., 1982. Alternative neural pathways initiate fast-start responses following lesions of the mauthner neuron in goldfish. Journal of Comparative Physiology 145, 485-496.
Eaton, R.C., Didomenico, R., Nissanov, J., 1991. Role of the mauthner cell in sensorimotor integration by the brainstem escape network. Brain, Behavior and Evolution 37, 272-285.
Eaton, R.C., Lee, R.K.K., Foreman, M.B., 2001. The Mauthner cell and other identified neurons of the brainstem escape

network of fish. Progress in Neurobiology 63, 467-485.

Echteler, S.M., Saidel, W.M., 1981. Forebrain connections in the goldfish support telencephalic homologies with land vertebrates. Science 212, 683-685.

Ewert, J.P., 1970. Neural mechanisms of prey-catching and avoidance behavior in toad (Bufo-bufo L). Brain, Behavior and Evolution 3, 36.

Favre-Bulle, I.A., Preece, D., Nieminen, T.A., Heap, L.A., Scott, E.K., Rubinsztein-Dunlop, H., 2015. Scattering of sculpted light in intact brain tissue, with implications for optogenetics. Scientific Reports 5.

Feierstein, C.E., Portugues, R., Orger, M.B., 2015. Seeing the whole picture: a comprehensive imaging approach to functional mapping of circuits in behaving zebrafish. Neuroscience 296, 26-38.

Feldman Barrett, L., Finlay, B.L., 2018. Concepts, goals and the control of survival-related behaviors. Current Opinion in Behavioral Sciences 24, 172-179.

Fetcho, J.R., 1991. Spinal network of the mauthner cell. Brain, Behavior and Evolution 37, 298-316.

Fetcho, J.R., Faber, D.S., 1988. Identification of motoneurons and interneurons in the spinal network for escapes initiated by the mauthner cell in goldfish. Journal of Neuroscience 8, 4192-4213.

Filosa, A., Barker, A.J., Dal Maschio, M., Baier, H., 2016. Feeding state modulates behavioral choice and processing of prey stimuli in the zebrafish tectum. Neuron 90, 596-608.

Fink, A.J.P., Axel, R., Schoonover, C.E., 2019. A virtual burrow assay for head-fixed mice measures habituation, discrimination, exploration and avoidance without training. Elife 8.

Folgueira, M., Anadon, R., Yanez, J., 2004. Experimental study of the connections of the telencephalon in the rainbow trout (*Oncorhynchus mykiss*). II: dorsal area and preoptic region. Journal of Comparative Neurology 480, 204-233.

Forster, D., Dal Maschio, M., Laurell, E., Baier, H., 2017a. An optogenetic toolbox for unbiased discovery of functionally connected cells in neural circuits. Nature Communications 8.

Forster, D., Kramer, A., Baier, H., Kubo, F., 2018. Optogenetic precision toolkit to reveal form, function and connectivity of single neurons. Methods 150, 42-48.

Forster, D., Arnold-Ammer, I., Laurell, E., Barker, A.J., Fernandes, A.M., Finger-Baier, K., Filosa, A., Helmbrecht, T.O., Kolsch, Y., Kuhn, E., Robles, E., Slanchev, K., Thiele, T.R., Baier, H., Kubo, F., 2017b. Genetic targeting and anatomical registration of neuronal populations in the zebrafish brain with a new set of BAC transgenic tools. Scientific Reports 7.

Fotowat, H., Harrison, R.R., Gabbiani, F., 2011. Multiplexing of motor information in the discharge of a collision detecting neuron during escape behaviors. Neuron 69, 147-158.

Gabbiani, F., Laurent, G., Hatsopoulos, N., Krapp, H.G., 1999. The many ways of building collision-sensitive neurons. Trends in Neurosciences 22, 437-438.

Gabriel, J.P., Trivedi, C.A., Maurer, C.M., Ryu, S., Bollmann, J.H., 2012. Layer-specific targeting of direction-selective neurons in the zebrafish optic tectum. Neuron 76, 1147-1160.

Gahtan, E., Sankrithi, N., Campos, J.B., O'Malley, D.M., 2002. Evidence for a widespread brain stem escape network in larval zebrafish. Journal of Neurophysiology 87, 608-614.

Gallagher, S.P., Northmore, D.P.M., 2006. Responses of the teleostean nucleus isthmi to looming objects and other moving stimuli. Visual Neuroscience 23, 209-219.

Gerlai, R., Fernandes, Y., Pereira, T., 2009. Zebrafish (*Danio rerio*) responds to the animated image of a predator: towards the development of an automated aversive task. Behavioural Brain Research 201, 318-324.

Giassi, A.C., Ellis, W., Maler, L., 2012a. Organization of the gymnotiform fish pallium in relation to learning and memory: III. Intrinsic connections. The Journal of Comparative Neurology 520, 3369-3394.

Giassi, A.C., Duarte, T.T., Ellis, W., Maler, L., 2012b. Organization of the gymnotiform fish pallium in relation to learning and memory: II. Extrinsic connections. The Journal of Comparative Neurology 520, 3338-3368.

Goddard, C.A., Mysore, S.P., Bryant, A.S., Huguenard, J.R., Knudsen, E.I., 2014. Spatially reciprocal inhibition of inhibition within a stimulus selection network in the avian midbrain. PLoS One 9.

Graham, B.J., Northmore, D.P.M., 2007. A spiking neural network model of midbrain visuomotor mechanisms that avoids objects by estimating size and distance monocularly. Neurocomputing 70, 1983-1987.

Gruberg, E., Dudkin, E., Wang, Y., Marin, G., Salas, C., Sentis, E., Letelier, J., Mpodozis, J., Malpeli, J., Cui, H., Ma, R., Northmore, D., Udin, S., 2006. Influencing and interpreting visual input: the role of a visual feedback system. Journal of Neuroscience 26, 10368-10371.

Hale, M.E., Katz, H.R., Peek, M.Y., Fremont, R.T., 2016. Neural circuits that drive startle behavior, with a focus on the mauthner cells and spiral fiber neurons of fishes. Journal of Neurogenetics 30, 89-100.

Hatsopoulos, N., Gabbiani, F., Laurent, G., 1995. Elementary computation of object approach by wide-field visual neuron. Science 270, 1000e1003 (New York, NY).

Hayes, W.N., Saiff, E.I., 1967. Visual alarm reactions in turtles. Animal Behaviour 15, 102.

Heap, L.A., Vanwalleghem, G.C., Thompson, A.W., Favre-Bulle, I., Rubinsztein-Dunlop, H., Scott, E.K., 2018a. Hypothalamic projections to the optic tectum in larval zebrafish. Frontiers in Neuroanatomy 11.

Heap, L.A.L., Vanwalleghem, G., Thompson, A.W., Favre-Bulle, I.A., Scott, E.K., 2018b. Luminance changes drive directional startle through a thalamic pathway. Neuron 99, 293.

Hein, A.M., Gil, M.A., Twomey, C.R., Couzin, I.D., Levin, S.A., 2018. Conserved behavioral circuits govern highspeed decision-making in wild fish shoals. Proceedings of the National Academy of Sciences of the United States of America 115, 12224-12228.

Helmbrecht, T.O., dal Maschio, M., Donovan, J.C., Koutsouli, S., Baier, H., 2018. Topography of a visuomotor transformation. Neuron 100, 1429.

Kawakami, K., Asakawa, K., Hibi, M., Itoh, M., Muto, A., Wada, H., 2016. Gal4 driver transgenic zebrafish: powerful tools to study developmental biology, organogenesis, and neuroscience. Advances in Genetics 95, 65-87.

Kim, D.H., Kim, J., Marques, J.C., Grama, A., Hildebrand, D.G.C., Gu, W.C., Li, J.M., Robson, D.N., 2017. Pan-neuronal calcium imaging with cellular resolution in freely swimming zebrafish. Nature Methods 14, 1107.

Kimmel, C.B., Patterson, J., Kimmel, R.O., 1974. The development and behavioral characteristics of the startle response in the zebra fish. Developmental Psychobiology 7, 47-60.

Kimmel, C.B., Eaton, R.C., Powell, S.L., 1980. Decreased fast-start performance of zebrafish larvae lacking mauthner neurons. Journal of Comparative Physiology 140, 343-350.

Kimura, Y., Hisano, Y., Kawahara, A., Higashijima, S., 2014. Efficient generation of knock-in transgenic zebrafish carrying reporter/driver genes by CRISPR/Cas9-mediated genome engineering. Scientific Reports 4.

King, W.M., Schmidt, J.T., 1993. Nucleus isthmi in goldfish e invitro recordings and fiber-connections revealed by HRP injections. Visual Neuroscience 10, 419-437.

Kita, E.M., Scott, E.K., Goodhill, G.J., 2015. Topographic wiring of the retinotectal connection in zebrafish. Developmental Neurobiology 75, 542-556.

Kohashi, T., Oda, Y., 2008. Initiation of mauthner- or non-mauthner-mediated fast escape evoked by different modes of sensory input. Journal of Neuroscience 28, 10641-10653.

Korn, H., Faber, D.S., 1996. Escape behavior e brainstem and spinal cord circuitry and function. Current Opinion in Neurobiology 6, 826-832.

Korn, H., Faber, D.S., 2005. The mauthner cell half a century later: a neurobiological model for decision-making? Neuron 47, 13-28.

Kramer, A., Wu, Y., Baier, H., Kubo, F., 2019. Neuronal architecture of a visual center that processes optic flow. Neuron 103 (1), 118-132.

Kubo, F., Hablitzel, B., Dal Maschio, M., Driever, W., Baier, H., Arrenberg, A.B., 2014. Functional architecture of an optic flow-responsive area that drives horizontal eye movements in zebrafish. Neuron 81, 1344-1359.

Ladu, F., Bartolini, T., Panitz, S.G., Chiarotti, F., Butail, S., Macrì, S., Porfiri, M., 2015. Live predators, robots, and computer-animated images elicit differential avoidance responses in zebrafish. Zebrafish 12, 205-214.

Lal, P., Tanabe, H., Suster, M.L., Ailani, D., Kotani, Y., Muto, A., Itoh, M., Iwasaki, M., Wada, H., Yaksi, E., Kawakami, K., 2018. Identification of a neuronal population in the telencephalon essential for fear conditioning in zebrafish. BMC Biology 16.

Landwehr, K., Brendel, E., Hecht, H., 2013. Luminance and contrast in visual perception of time to collision. Vision Research 89, 18-23.

Leung, L.C., Wang, G.X., Mourrain, P., 2013. Imaging zebrafish neural circuitry from whole brain to synapse. Frontiers in Neural Circuits 7.

Lin, M.Z., Schnitzer, M.J., 2016. Genetically encoded indicators of neuronal activity. Nature Neuroscience 19, 1142-1153.

Liu, K.S., Fetcho, J.R., 1999. Laser ablations reveal functional relationships of segmental hindbrain neurons in zebrafish. Neuron 23, 325-335.

Lovett-Barron, M., Andalman, A.S., Allen, W.E., Vesuna, S., Kauvar, I., Burns, V.M., Deisseroth, K., 2017. Ancestral circuits for the coordinated modulation of brain state. Cell 171, 1411.

Lowe, A.S., Nikolaou, N., Hunter, P.R., Thompson, I.D., Meyer, M.P., 2013. A systems-based dissection of retinal inputs to the zebrafish tectum reveals different rules for different functional classes during development. Journal of Neuroscience 33, 13946-13956.

Luca, R.M., Gerlai, R., 2012. In search of optimal fear inducing stimuli: differential behavioral responses to computer animated images in zebrafish. Behavioural Brain Research 226, 66-76.

Luiten, P.G., 1981. Afferent and efferent connections of the optic tectum in the carp (*Cyprinus carpio* L.). Brain Research 220, 51-65.

Marin, G., Salas, C., Sentis, E., Rojas, X., Letelier, J.C., Mpodozis, J., 2007. A cholinergic gating mechanism controlled by competitive interactions in the optic tectum of the pigeon. Journal of Neuroscience 27, 8112-8121.

McClenahan, P., Troup, M., Scott, E.K., 2012. Fin-tail coordination during escape and predatory behavior in larval zebrafish. PLoS One 7, e32295.

McFadyen, J., Mermillod, M., Mattingley, J.B., Halasz, V., Garrido, M.I., 2017. A rapid subcortical amygdala route for faces irrespective of spatial frequency and emotion. Journal of Neuroscience 37, 3864-3874.

Medan, V., Preuss, T., 2014. The mauthner-cell circuit of fish as a model system for startle plasticity. Journal of Physiology 108, 1290-140.

Mueller, T., 2012. What is the thalamus in zebrafish? Frontiers in Neuroscience 6.

Mueller, T., Dong, Z., Berberoglu, M.A., Guo, S., 2011. The dorsal pallium in zebrafish, *Danio rerio* (Cyprinidae, Teleostei). Brain Research 1381, 95-105.

Mysore, S.P., Knudsen, E.I., 2013. A shared inhibitory circuit for both exogenous and endogenous control of stimulus selection. Nature Neuroscience 16, 473-478. U143.

Mysore, S.P., Knudsen, E.I., 2014. Descending control of neural bias and selectivity in a spatial attention network: rules and mechanisms. Neuron 84, 214-226.

Nair, A., Azatian, G., McHenry, M.J., 2015. The kinematics of directional control in the fast start of zebrafish larvae. Journal of Experimental Biology 218, 3996e4004.

Nair, A., Nguyen, C., McHenry, M.J., 2017. A faster escape does not enhance survival in zebrafish larvae. Proceedings of the Royal SocietyBeBiological Sciences 284.

Naumann, E.A., Fitzgerald, J.E., Dunn, T.W., Rihel, J., Sompolinsky, H., Engert, F., 2016. From whole-brain data to functional circuit models: the zebrafish optomotor response. Cell 167, 947.

Nevin, L., Robles, E., Baier, H., Scott, E.K., 2010. Focusing on optic tectum circuitry through the lens of genetics. BMC Biology 8, 126.

Nikolaou, N., Lowe, A.S., Walker, A.S., Abbas, F., Hunter, P.R., Thompson, I.D., Meyer, M.P., 2012. Parametric functional maps of visual inputs to the tectum. Neuron 76, 317-324.

Northcutt, R.G., 2006. Connections of the lateral and medial divisions of the goldfish telencephalic pallium. The Journal of Comparative Neurology 494, 903-943.

Northmore, D.P.M., Gallagher, S.P., 2003. Functional relationship between nucleus isthmi and tectum in teleosts: synchrony but no topography. Visual Neuroscience 20, 335-348.

Northmore, D.P.M., Graham, B.J., 2005. Avoidance behavior controlled by a model of vertebrate midbrain mechanisms. Lecture Notes in Computer Science 3561, 338-345.

Oliva, D., Tomsic, D., 2012. Visuo-motor transformations involved in the escape response to looming stimuli in the crab Neohelice (¼Chasmagnathus) granulata. Journal of Experimental Biology 215, 3488-3500.

Orger, M.B., Kampff, A.R., Severi, K.E., Bollmann, J.H., Engert, F., 2008. Control of visually guided behavior by distinct populations of spinal projection neurons. Nature Neuroscience 11, 327-333.

Patterson, G.H., Lippincott-Schwartz, J., 2002. A photoactivatable GFP for selective photolabeling of proteins and cells. Science 297, 1873-1877.

Peek, M.Y., Card, G.M., 2016. Comparative approaches to escape. Current Opinion in Neurobiology 41, 167-173.

Perathoner, S., Cordero-Maldonado, M.L., Crawford, A.D., 2016. Potential of zebrafish as a model for exploring the role of the amygdala in emotional memory and motivational behavior. Journal of Neuroscience Research 94, 445-462.

Pereira, A.G., Moita, M.A., 2016. Is there anybody out there? Neural circuits of threat detection in vertebrates. Current Opinion in Neurobiology 41, 179-187.

Portugues, R., Feierstein, C.E., Engert, F., Orger, M.B., 2014. Whole-brain activity maps reveal stereotyped, distributed networks for visuomotor behavior. Neuron 81, 1328-1343.

Preuss, S.J., Trivedi, C.A., vom Berg-Maurer, C.M., Ryu, S., Bollmann, J.H., 2014. Classification of object size in retinotectal microcircuits. Current Biology 24, 2376-2385.

Preuss, T., Osei-Bonsu, P.E., Weiss, S.A., Wang, C., Faber, D.S., 2006. Neural representation of object approach in a decision-making motor circuit. Journal of Neuroscience 26, 3454-3464.

Robles, E., Smith, S.J., Baier, H., 2011. Characterization of genetically targeted neuron types in the zebrafish optic tectum. Frontiers in Neural Circuits 5.

Robles, E., Filosa, A., Baier, H., 2013. Precise lamination of retinal axons generates multiple parallel input pathways in the tectum. Journal of Neuroscience 33, 5027-5039.

Robles, E., Laurell, E., Baier, H., 2014. The retinal projectome reveals brain-area-specific visual representations generated by ganglion cell diversity. Current Biology 24, 2085-2096.

Saint-Amant, L., Drapeau, P., 1998. Time course of the development of motor behaviors in the zebrafish embryo. Journal of Neurobiology 37, 622-632.

Sankrithi, N.S., O'Malley, D.M., 2010. Activation of a multisensory, multifunctional nucleus in the zebrafish midbrain during diverse locomotor behaviors. Neuroscience 166, 970-993.

Sato, T., Hamaoka, T., Aizawa, H., Hosoya, T., Okamoto, H., 2007. Genetic single-cell mosaic analysis implicates ephrinB2 reverse signaling in projections from the posterior tectum to the hindbrain in zebrafish. Journal of Neuroscience 27, 5271-5279.

Satou, C., Kimura, Y., Hirata, H., Suster, M.L., Kawakami, K., Higashijima, S., 2013. Transgenic tools to characterize neuronal properties of discrete populations of zebrafish neurons. Development 140, 3927-3931.

Saverino, C., Gerlai, R., 2008. The social zebrafish: behavioral responses to conspecific, heterospecific, and computer animated fish. Behavioural Brain Research 191, 77-87.

Schiff, W., Caviness, J.A., Gibson, J.J., 1962. Persistent fear responses in rhesus monkeys to optical stimulus of looming. Science 136, 982.

Scott, E.K., Baier, H., 2009. The cellular architecture of the larval zebrafish tectum, as revealed by Gal4 enhancer trap lines. Frontiers in Neural Circuits 3.

Scott, E.K., Mason, L., Arrenberg, A.B., Ziv, L., Gosse, N.J., Xiao, T., Chi, N.C., Asakawa, K., Kawakami, K., Baier, H., 2007. Targeting neural circuitry in zebrafish using GAL4 enhancer trapping. Nature Methods 4, 323-326.

Semmelhack, J.L., Donovan, J.C., Thiele, T.R., Kuehn, E., Laurell, E., Baier, H., 2014. A dedicated visual pathway for prey detection in larval zebrafish. eLife 3.

Severi, K.E., Portugues, R., Marques, J.C., O'Malley, D.M., Orger, M.B., Engert, F., 2014. Neural control and modulation of swimming speed in the larval zebrafish. Neuron 83, 692-707.

Shang, C.P., Chen, Z.J., Liu, A.X., Li, Y., Zhang, J.J., Qu, B.L., Yan, F., Zhang, Y.N., Liu, W.X., Liu, Z.H., Guo, X.F., Li, D.P., Wang, Y., Cao, P., 2018. Divergent midbrain circuits orchestrate escape and freezing responses to looming stimuli in mice. Nature Communications 9.

Sillar, K.T., Picton, L., Heitler, W.J., 2016. The Neuroethology of Predation and Escape.

Simmich, J., Staykov, E., Scott, E., 2012. Zebrafish as an appealing model for optogenetic studies. Optogenetics: Tools for Controlling and Monitoring Neuronal Activity 196, 145-162.

Song, J.R., Ampatzis, K., Ausborn, J., El Manira, A., 2015. A hardwired circuit supplemented with endocannabinoids encodes behavioral choice in zebrafish. Current Biology 25, 2610-2620.

Stewart, W.J., Cardenas, G.S., McHenry, M.J., 2013. Zebrafish larvae evade predators by sensing water flow. Journal of Experimental Biology 216, 388-398.

Striedter, G.F., Northcutt, R.G., 1989. Two distinct visual pathways through the superficial pretectum in a percomorph teleost. The Journal of Comparative Neurology 283, 342-354.

Sun, H.J., Frost, B.J., 1998. Computation of different optical variables of looming objects in pigeon nucleus rotundus neurons. Nature Neuroscience 1, 296-303.

Tabor, K.M., Smith, T.S., Brown, M., Bergeron, S.A., Briggman, K.L., Burgess, H.A., 2018. Presynaptic inhibition selectively gates auditory transmission to the brainstem startle circuit. Current Biology 28, 2527.

Tabor, K.M., Marquart, G.D., Hurt, C., Smith, T.S., Geoca, A.K., Bhandiwad, A.A., Subedi, A., Sinclair, J.L., Rose, H.M., Polys, N.F., Burgess, H.A., 2019. Brain-wide cellular resolution imaging of Cre transgenic zebrafish lines for functional circuit-mapping. eLife 8.

Temizer, I., Donovan, J.C., Baier, H., Semmelhack, J.L., 2015. A visual pathway for looming-evoked escape in larval zebrafish. Current Biology 25, 1823-1834.

Thiele, T.R., Donovan, J.C., Baier, H., 2014. Descending control of swim posture by a midbrain nucleus in zebrafish. Neuron 83, 679-691.

Vanegas, H., Ito, H., 1983. Morphological aspects of the teleostean visual system: a review. Brain Research 287, 117-137.

Vanwalleghem, G.C., Ahrens, M.B., Scott, E.K., 2018. Integrative whole-brain neuroscience in larval zebrafish. Current Opinion in Neurobiology 50, 136-145.

von Trotha, J.W., Vernier, P., Bally-Cuif, L., 2014. Emotions and motivated behavior converge on an amygdala-like structure in the zebrafish. European Journal of Neuroscience 40, 3302-3315.

Vonbartheld, C.S., Meyer, D.L., 1987. Comparative neurology of the optic tectum in ray-finned fishes—patterns of lamination formed by retinotectal projections. Brain Research 420, 277-288.

Wang, W.C., McLean, D.L., 2014. Selective responses to tonic descending commands by temporal summation in a spinal motor pool. Neuron 83, 708-721.

Wei, P.F., Liu, N., Zhang, Z.J., Liu, X.M., Tang, Y.Q., He, X.B., Wu, B.F., Zhou, Z., Liu, Y.H., Li, J., Zhang, Y., Zhou, X.Y., Xu, L., Chen, L., Bi, G.Q., Hu, X.T., Xu, F.Q., Wang, L.P., 2015. Processing of visually evoked innate fear by a non-canonical thalamic pathway. Nature Communications 6, 6756.

Wolf, S., Supatto, W., Debregeas, G., Mahou, P., Kruglik, S.G., Sintes, J.M., Beaurepaire, E., Candelier, R., 2015. Whole-brain functional imaging with two-photon light-sheet microscopy. Nature Methods 12, 379-380.

Wullimann, M.F., Mueller, T., 2004. Teleostean and mammalian forebrains contrasted: evidence from genes to behavior. Journal of Comparative Neurology 475 (2), 143-162.

Yamamoto, N., Ito, H., 2005. Fiber connections of the anterior preglomerular nucleus in cyprinids with notes on telencephalic connections of the preglomerular complex. Journal of Comparative Neurology 491, 212-233.

Yamamoto, N., Ito, H., 2008. Visual, lateral line, and auditory ascending pathways to the dorsal telencephalic area through the rostrolateral region of the lateral preglomerular nucleus in cyprinids. The Journal of Comparative Neurology 508, 615-647.

Yamamoto, N., Ishikawa, Y., Yoshimoto, M., Xue, H.G., Bahaxar, N., Sawai, N., Yang, C.Y., Ozawa, H., Ito, H., 2007. A new interpretation on the homology of the teleostean telencephalon based on hodology and a new eversion model. Brain, Behavior and Evolution 69, 96-104.

Yanez, J., Suarez, T., Quelle, A., Folgueira, M., Anadon, R., 2018. Neural connections of the pretectum in zebrafish (*Danio rerio*). Journal of Comparative Neurology 526, 1017-1040.

Yao, Y.Y., Li, X.Q., Zhang, B.B., Yin, C., Liu, Y.F., Chen, W.Y., Zeng, S.Q., Du, J.L., 2016. Visual cue-discriminative dopaminergic control of visuomotor transformation and behavior selection. Neuron 89, 598-612.

Yilmaz, M., Meister, M., 2013. Rapid innate defensive responses of mice to looming visual stimuli. Current Biology 23, 2011-2015.

Yin, C., Li, X., Du, J., 2019. Optic tectal superficial interneurons detect motion in larval zebrafish. Protein and cell 10 (4), 238-248.

Yuan, T.F., Sus, H.X., 2015. Fear learning through the two visual systems, a commentary on: "A parvalbumin-positive excitatory visual pathway to trigger fear responses in mice". Frontiers in Neural Circuits 9.

Zhao, X.Y., Liu, M.N., Cang, J.H., 2014. Visual cortex modulates the magnitude but not the selectivity of loomingevoked responses in the superior colliculus of awake mice. Neuron 84, 202-213.

Zhou, Z., Liu, X., Chen, S., Zhang, Z., Liu, Y., Montardy, Q., Tang, Y., Wei, P., Liu, N., Li, L., Song, R., Lai, J., He, X., Chen, C., Bi, G., Feng, G., Xu, F., Wang, L., 2019. A VTA GABAergic neural circuit mediates visually evoked innate defensive responses. Neuron 103 (3), 473-488.

Zottoli, S.J., 1977. Correlation of the startle reflex and mauthner cell auditory responses in unrestrained goldfish. Journal of Experimental Biology 66, 243-254.

Zottoli, S.J., Hordes, A.R., Faber, D.S., 1987. Localization of optic tectal input to the ventral dendrite of the goldfish mauthner cell. Brain Research 401, 113-121.

第五章

斑马鱼听觉功能的研究进展

Dennis M. Higgs[1]

5.1 水下听觉环境及其与声音探测的相关性

在讨论鱼如何听到声音和能听到什么声音之前，首先对水下声音性能的特性进行基本描述。在水下，声音具有大量的粒子运动和压力分量，从声源发出的声音可以传播一段距离，而绝对传播距离取决于声音的频率、水深以及水本身的其他物理特性（Rogers et al., 1988）。在描述真实的水生环境中的声音传播时，没有完美的模型，但从理想的传播模型中可以确定一些基本原则，这些基本原则对目前的综述来说是足够的。在理想的环境中，声源的粒子运动分量将在靠近声源的地方占主导地位，同样来自声源的压力波将使声音传播得更远，而在含有鱼的野生环境中很少有这种理想环境。传统上，声音的粒子运动方面占主导地位的区域被称为近场，而压力占主导地位的区域被称为远场，近场 / 远场的边界理想化为波长的函数（Nedelec et al., 2016）。这些特性与当前讨论有关的原因是：鱼耳只对声音的粒子运动部分做出直接反应，因为这种运动是使毛细胞弯曲并将声音刺激转换为神经脉冲所必需的。从功能上来说，这意味着鱼耳只能探测到离声源距离较近的频率较低的声音（高至 800Hz），这是因为只有这种组合可以提供足够的粒子运动能量，使得毛细胞相对于耳石移动。为了探测到更高的频率和距离声源更远的声音，声波中的压力分量必须通过压力传感器和鱼的内耳之间的某种耦合转换为粒子运动。一条鱼要将声压转化为粒子运动，它必须有一些充满气体的结构，这些充气结构要么与耳朵物理耦合，要么离耳朵足够近，从而将充气结构中的振动传递给内耳毛细胞（Popper et al., 1999）。在斑马鱼和其他所有骨鳔总目鱼体内，这种充满气体的结构会随着压力变化而振动，鱼鳔是通过鳔骨物理连接到内耳，鳔骨是一种修饰过的骨骼组分，在物理上将鱼鳔连接到听软骨囊（vonFrisch, 1938;

[1] Department of Integrative Biology, University of Windsor, Windsor, ON, Canada.

Chardon and Vandewalle, 1997）。当声波的压力分量击中鱼鳔时，它会引起充气结构的振动，这种振动（粒子运动）沿着鳔骨传递到听囊。这种耦合不仅使斑马鱼的内耳能够检测到更高的频率，而且还提高了耳朵对不同频谱声音的敏感度（Higgs et al., 2002, 2003）。

5.2 斑马鱼耳的形态

一旦声音刺激以粒子运动的形式直接传达到内耳，或者先将压力波转化为粒子运动后再到达内耳，就必须使毛细胞弯曲以传导为神经脉冲。由于鱼体的密度与周围的水大致相同，在水和毛细胞之间不会产生相对运动而导致弯曲，但在每块听上皮上覆盖着致密的钙质耳石，它抵抗相对于水的运动［图 5-1（B）］。当鱼的身体随着水分子移动时，耳石滞后，导致毛细胞和上面的耳石之间的相对运动。毛细胞位于听上皮内，有多个微绒毛（又称静纤毛）和一个纤毛（动纤毛），这些纤毛从毛细胞的顶端伸出直到听内淋巴［图 5-1（C）］。多个静纤毛在靠近动纤毛时呈高度逐渐增加阶梯状六角形排列，这种关系决定了每个毛细胞的功能极性（Pickles and Corey, 1992）。静纤毛和动纤毛都通过顶连接（tip link）连接到一个非选择性阳离子通道（Pickles and Corey, 1992）。如果动纤毛弯曲远离静纤毛，这将导致顶连接拉伸并打开机械门控通道，引起感觉信号转导；如果动纤毛向静纤毛弯曲，这就阻断了通道的激活，使每个毛细胞的反应具有内在的方向性。毛细胞是一种非神经受体细胞，因为它没有轴突，而是释放谷氨酸与传入神经元结合，将刺激发送到脑干的神经中枢［图 5-1（C）］。整个上皮细胞层包含数千个独立的毛细胞（Bang et al., 2001），它们按相似极性排列成一组（Platt, 1993），因此整个上皮细胞层充当一个方向传感器，帮助编码声源的方向。

斑马鱼在成年期和整个发育过程中耳的形态都已被全面地研究（Bang et al., 2001; Platt, 1993; Haddon and Lewis, 1996）。与其他硬骨动物一样，成年斑马鱼的耳朵包括三个参与平衡调节的半规管和三个囊，即椭圆囊（utricle, U）、球囊（saccule, S）和听壶（lagena, L），含感觉上皮和覆盖其上的钙质耳石（Platt, 1993）（图 5-1 A）。每个耳石都有一个独特的形状：椭圆囊状的耳石（称为微耳石）是一个大的圆形耳石，球囊状耳石（称为矢耳石）是细长的耳石，而听壶耳石（称为星耳石）也是圆形的。由于碳酸盐颗粒增生形成耳石结构，下面的感觉上皮与它们各自的耳石形状相同（Dunkelberger et al., 1980; Campana, 1999），独特的形状可能影响耳石作用与刺激下层感觉毛细胞的方式（Lychakov and Rebane, 2000, 2005）。耳石与下面的感觉上皮通过一层胶状膜连接在一起。椭圆囊和半规管形成上面的更靠背侧的部分，而球囊和听壶形成下面的更靠腹侧的部分，通过椭圆囊上的导管与上部背侧部分连接。在三个末端器官中，感觉上皮纤毛束的形态和极性各不相同，但在这三个末端器官中，每块上皮的不同区域都有不同的极性组合。椭圆囊有两个不同的毛细胞组分，弧形微纹（striola）和毛状体（cotillus）区，其中弧形微纹区域的毛细胞密度大约是毛状体区域的两倍。球囊和听壶区的毛细胞密度无明显差异，但这两个末端器官的毛细胞密度也可能存在区域性差异（Platt, 1993）。半规管与椭圆囊（共同构成上部）主要负责鱼类的前庭反应，而球囊和听壶（共同构成下部）可能更多地参与将声音转

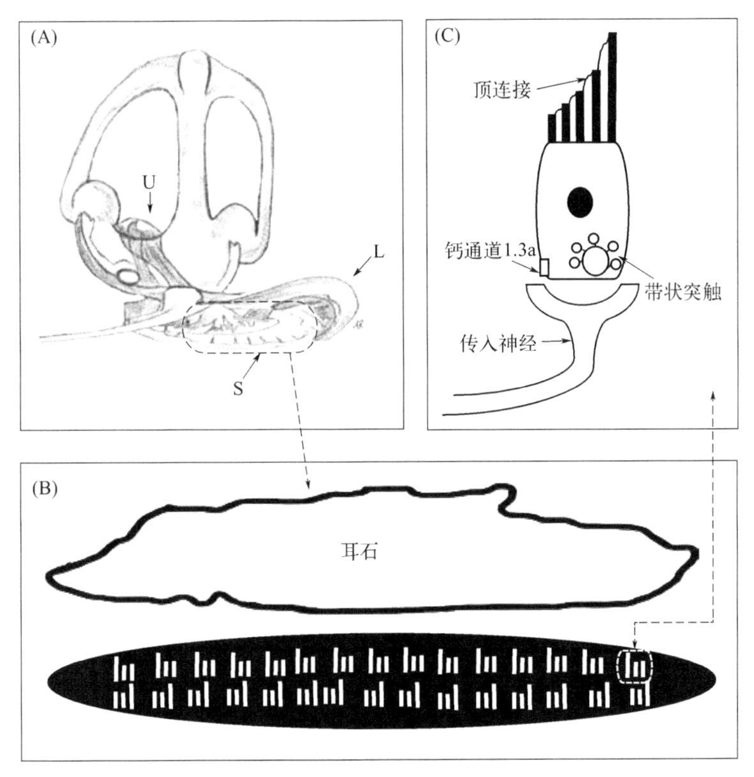

图5-1 **硬骨鱼耳朵的形态**

所有硬骨鱼的耳都由上下两部分组成［图（A）］，上部由半规管（semicircular canal）和椭圆囊（utricle, U）组成，主要用于平衡信息；下部由每只耳朵上的一个球囊（saccule, S）和听壶（lagena, L）组成，主要用于听觉。在每一个椭圆囊、球囊和听壶内都有一个耳石覆盖在感觉上皮上，感觉上皮包含数千个不同极性的毛细胞［图（B）］。极性是由纤毛的排列决定的，并决定了每个毛细胞的方向敏感性。单个毛细胞［图（C）］由静纤毛和一个动纤毛组成，动纤毛通过顶连接与细胞顶端表面的阳离子通道相连。在基底表面，每个毛细胞都有一个带状突触，其中含有充满谷氨酸的突触囊泡，这些突触囊泡在Cav1.3a电压门控通道的影响下释放神经递质

换为神经脉冲（Popper et al., 1999）。每个器官上均有成千上万的独特的毛细胞，它们横跨整个上皮中线并在相反的方向上具有方向敏感性（Platt, 1993），最接近于“垂直”方向模式（Popper and Coombs, 1982）。耳基板在受精后迅速发育，椭圆囊状耳石和球囊状耳石（分别为微耳石和矢耳石）在受精后约 20h 内形成，第一感觉毛细胞在受精后约 24h 内发育形成（Haddon and Lewis, 1996）。受精后 72h 形成的半规管作为听软骨囊的前庭出口（outpockets），而到受精后1周，内耳所有基本结构都很明显，包括传入神经分布（Haddon and Lewis, 1996）。半规管是由导管组织发生柱形态改变而形成的，至少部分受到调节细胞外基质的 G 蛋白偶联受体基因 *gp2 126* 的控制，部分受 *sox10* 表达水平的调控（Geng et al., 2013）。Sox10 突变体在耳基板（otic placode）形成中表现出明显的异常，这表明在耳分化中 sox 基因组具有很强的早期调控作用（Dutton et al., 2009）。虽然参与早期形态发生的基因调节机制仍不清楚（Alsina and Whitfield, 2017），但 *fgf* 表达（Hammond and Whitfield, 2011; Hartwell et al., 2019）和 Hedgehog 信号通路（Hammond et al., 2003）很显然在斑马鱼听软骨囊（otic capsules）的形成早期模式中发挥了重要作用。有研究表明，内耳发育过程中毛细胞和支持细胞的命运似乎受 *fgf* 和视黄酸（retinoic acid, RA）之间的反

馈回路控制，*fgf* 的高表达水平决定支持细胞的命运，而 RA 的高水平决定毛细胞的命运（Maier and Whitfield, 2014）。毛细胞一旦发育，就会表现出固有的极性（见上文），在发育过程中，毛细胞的极化方向是由基因 *vangl2* 和转录因子 *Emx2* 共同驱动的（Pickett and Raible, 2019）。内耳和神经丘感觉上皮中发育成熟的毛细胞的存活受到基因 *eye1a* 表达水平的部分调控（Kozlowski et al., 2005）。

5.3 关于耳与侧线神经丘作用的说明

在细胞水平上，耳朵和侧线系统（lateral line system）中将机械能转换为神经脉冲的感觉细胞是相同的；这两种系统都使用了毛细胞，这些毛细胞会对水的运动做出偏转反应并打开离子通道。这两种系统在毛细胞的感觉结构上有所不同，而正是这种不同从根本上决定了毛细胞会对什么频率的运动做出反应。耳毛细胞分布在覆盖着耳石的广阔的听觉上皮中，而侧线系统的毛细胞则更紧密地分布在单个的神经丘中，在静纤毛束和运动纤毛束的顶部形成一个凝胶状的杯状突起（Coombs and Montgomery, 1999）。每个神经丘包含多个毛细胞（Webb and Shirey, 2003），并且在神经丘内毛细胞以相反的极性排列，使整个结构具有最大的敏感性（Webb et al., 2008）。神经丘可分为管状和浅表两种，同样，神经丘的位置决定了它们最佳反应频率的基本差异。管状神经丘在鱼鳞下方，位于连接外部的毛孔之间，是骨化的管状组织，由此而得名。管状组织可以在鱼的头部和 / 或躯干上形成各种各样的图案（Webb et al., 2008），而每个管状组织则可以容纳数十到数百个神经丘（Coombs and Montgomery, 1999）。浅表神经丘不是在管状组织中形成的，而是直接位于鱼的外表面，其排列远不如管状神经丘有规律。管状和浅表神经丘的不同排列规则也决定了每种类型的反应敏感性；浅表神经丘对 50Hz 以下的水运动频率最敏感（Montgomery et al., 1994; M0ünz, 1985），而管状神经丘在频率达到 100 ～ 200Hz 时反应良好（Münz, 1985; Coombs and Janssen, 1990）。一般认为，浅表神经丘起着速度传感器的作用，可能是趋流性（rheotaxis，朝向水流的方位）所需要的（Montgomery et al., 1997），而管状神经丘是加速传感器，可能更多地用于检测与鱼群聚集、捕食或进食事件相关的水流变化（Coombs and Montgomery, 1999）。虽然只有在涉及耳朵的情况下才恰当地将一种感觉称为“听觉”，但听觉和神经丘毛细胞在探测声源附近的低频声音中可能有重叠的作用（Higgs and Radford, 2013）。

5.4 对斑马鱼在听觉研究中日益增加的重要性的分析

第一篇明确描述斑马鱼内耳各方面的论文似乎是 Platt（Platt, 1993）对成年斑马鱼的研究，随后是 Haddon 和 Lewis（1996）对耳朵发育的研究。随着斑马鱼作为一个重要模型系统的使用越来越广泛（Eisen et al., 1986; Westerfield, 1989），在整个 20 世纪 90 年代，人们对将斑马鱼作为听觉模型的研究兴趣逐渐增加（图 5-2），而进入 21 世纪后，随着现有分子手段和基因技术的日益成熟，人们对斑马鱼听觉模型的兴趣迅速增加（图 5-2）。自 2015 年以来，每年都有超过 100 篇论文研究斑马鱼听觉系统的某些方面（图 5-2），预计斑马鱼作为听觉模型的运用会不断增加，特别是在作为哺乳动物听觉功能障碍的模型这一方面（Yang et al., 2017）。

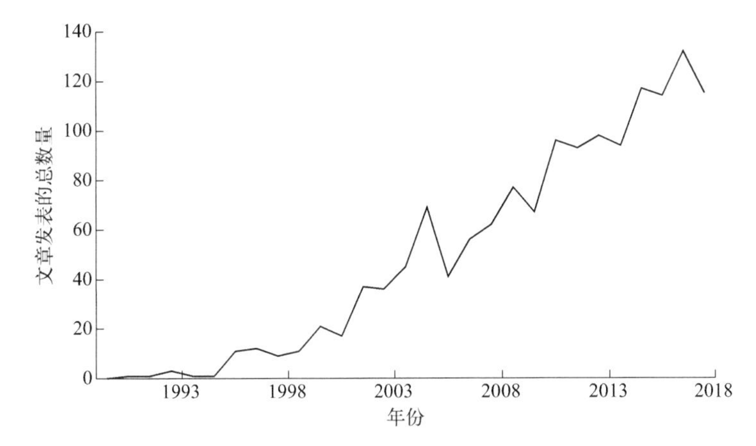

图 5-2　对截至 2018 年发表的关于斑马鱼耳或斑马鱼听力的所有论文进行 meta 分析（于 2019 年 8 月用谷歌学术搜索完成）

5.5　斑马鱼的听觉以及我们是如何知道的

对鱼的听力进行了行为和生理两方面的测试。一种未被充分利用的行为技术是使用经典的条件反射技术，首先训练斑马鱼对有食物奖励的声音做出反应，然后利用这种反应开发出所有频率的行为听觉图（Cervi et al., 2012）。虽然这种技术可以提供对声音的意志性反应的真实测定，但训练鱼需要时间就意味着这种方法对于研究发育中的鱼并不是一种可行的策略，因此只能研究成年动物。一种不需要训练时间的快速行为方法是利用鱼对快速出现的刺激信号表现出惊吓反应的倾向，从而量化听觉惊吓反应指标（Bang et al., 2002; Zeddies and Fay, 2005; Bhandiwad et al., 2013; Liu et al., 2018）。当受到惊吓时，许多鱼类会表现出一种被称为 C 形启动的典型逃逸反应，最终由称为 Mauthner 细胞的大型运动神经元驱动（Eaton et al., 1991; Medan and Preuss, 2014），这可以在斑马鱼受精后 6 天（6dpf）通过声音刺激诱导（Kimmel et al., 1974）。由于这种响应行为的刻板性动作，可以采用自动化视频分析，用于较宽频率和强度范围的高通量分析（Yang et al., 2017; Liu et al., 2018）以快速评估大量鱼类的听力能力和 / 或缺陷。斑马鱼的一个突变系减弱了声音惊吓反应，至少部分处理途径似乎与 *cyfip2* 基因的突变有关，该基因是脆性 X 类蛋白质编码基因的一部分（Marsden et al., 2018）。最后一种常见的行为分析方法是建立在逃逸反应的基础上，但使用了前脉冲抑制（prepulse inhibition, PPI）指标（Bhanddiwad et al., 2013）来评估听觉能力。通过用 PPI 测试声音刺激抑制听觉惊吓反应的能力，发现其产生的行为阈值比惊吓反应更敏感，这可能是因为引发真正惊吓行为所需的Mauthner神经元作出反应所需的阈值刺激更高。生理测试可以通过听觉诱发电位法得到整个脑干反应（Higgs et al., 2002, 2003; Monroe et al., 2016）或更精细的可使用听觉微音学方法（Lu and DeSmidt, 2013; Yao et al., 2016）。听觉诱发电位法是一种全脑反应法，通常将记录电极放置在脑干上，将参考电极放置在鱼的鼻子上，以得到对声音刺激的整个大脑的神经反应的总和。这种方法可以用于各种大小的鱼类，每条鱼大约1h就可以得到完整的听觉图，现在已经成为鱼类听觉生理学的标准方法（Ladich and Fay, 2013; Sisneros et al., 2016）。听觉诱发电位方法的好处是：它能测量中枢神经系统对刺激的反应，因此结果的综合性更强。但其缺点是：它不能区分由耳朵驱动的反应和由浅表或管状神经丘驱动的反应（Higgs and Radford, 2013; Higgs et al., 2015），因此很难识别反应的神经驱动因素。为了真正确定耳朵和耳朵内单个上皮细胞的作用，研究人员通过将电极直接引入单个听觉末端器官，然后通过各种振动刺激来刺激耳朵，使用了真正的微音记录（Lu and DeSmidt, 2013; Yao et al., 2016）。微音记录方法具有从耳朵获得直接反应的好处，但往往对振动刺激效果最好，故而被用于测试斑马鱼听力范围的低频部分。每一种方法都有其优点和缺点，因此最佳技术的使用取决于所提出的研究问题。

成年斑马鱼可以听到100～4000Hz的频率，最佳灵敏度在800Hz左右（Higgs et al., 2002）。一项研究（Wang et al., 2015）报告了频率上限为12000Hz，但这项研究是在一个非常小的水箱中进行的，众所周知，这个水箱的声学性能存在问题（Parvulescu, 1967; Rogers et al., 2016; Akamatsu et al., 2002），因此目前尚不清楚斑马鱼是对如此高的频率还是对鱼缸里的人工制品做出了反应。斑马鱼的听觉能力与金鱼相似（Popper, 1971; Higgs, 2002），并可能在鲤形目其他成员中相对保守，因为它们都通过鳔骨连接鱼鳔和内耳（Chardon and Vandewalle, 1997）。在成年鱼中，由行为条件作用形成的听觉图的形状依赖于训练刺激，但斑马鱼对至少200～1000Hz的纯音调表现出强烈的行为反应（Cervi et al., 2012）。成鱼对声音刺激的反应可能主要是由囊状上皮介导的（Zeddies and Fay, 2005），尽管它们也可能接受来自听壶和囊状毛细胞的刺激。绝对听觉灵敏度会因测量听觉能力的技术而不同（Rogers et al., 2016），但与其他物种相比，斑马鱼确实对其能听到的频率范围内的声音非常敏感。斑马鱼的听力在发育过程中会发生变化，检测到的频率范围受到鱼鳔和鱼耳之间连接的发育限制（Higgs et al., 2003）。根据行为范式，斑马鱼直到5dpf才对声音作出反应，但在那之后，斑马鱼对90～1070Hz的响亮声音表现出惊吓反应，在最低频率下反应最佳（Bhandiwad et al., 2013）。

虽然成年斑马鱼的听觉频率范围似乎已经确定，但必须注意的是它的听觉灵敏度。不同技术之间的听觉灵敏度测量差异很大（Rogers et al., 2016），甚至在使用相同技术的不同实验室之间（Ladich and Fay, 2013; Higgs, 2002）也存在差异，所以绝对听觉阈值没有确定值。在一个研究小组中，在相对意义上可以相信阈值测量，但现在看来，至少在斑马鱼中，品种差异也可以影响表观听觉表现。特异性表达GFP的转基因型与其背景品系相比，前者会导致听觉阈值升高（Monroe et al., 2016），甚至非转基因背景株也会彼此不同，与野生株也不同（Monroe et al., 2016）。在影响耳石形成的突变株中PPI是缺乏的，在其他选择突变中也可以消除（Burgess and Granato, 2007）。虽然目前还没有对鱼群中听力的遗传性进行研究，但很明显，不同品系的斑马鱼在听力上存在明确的差异性，转基因技术的发展，至少对于 GFP 的表达可能对听觉能力有明显的影响，所以在解释不同基因来源的斑马鱼种群之间的听力能力时应谨慎使用该技术。

由于对压力信息不敏感，斑马鱼仔鱼和成鱼的听觉能力在很多方面都有明显的差异，这种差异可以用来更好地理解声音输入的转导机制和整合。在受精后的第一周，早在鳔骨形成以传递压力信息之前（Higgs et al., 2003），听觉毛细胞只会对较低频率的刺激作出反应，而直接测量这些反应是毛细胞参与转导的有力指标。微音实验表明，斑马鱼在 20 ～ 400Hz 的直接刺激下表现出清晰的响应，响应阈值随着频率的降低而降低（Lu and DeSmidt, 2013; Yao et al., 2016）。有趣的是，野生型和 Et（krt4：GFP）品系斑马鱼之间的响应性没有差异（Yao et al., 2016），这表明该品系至少在发育早期是一个有效的模型。在大脑水平上，频率为 100 ～ 400Hz 的音频刺激可导致内侧八头肌核（medial octavolateralis nuclei, MON）和其他一些后脑区域的活动持续增加，丘脑和半规隆凸也会出现额外的但一致性较差的反应（Vanwalleghem et al., 2017），这些区域被认为对感觉整合很重要（Feng et al., 1999; Moller, 2002; Edds-Walton and Fay, 2005）。有证据表明，沿背神经轴的 MON 存在广泛的张力失调，刺激强度似乎是神经放电频率的函数（Vanwalleghem et al., 2017）。

5.6 毛细胞动力学

当声音传递到鱼耳上时，它可能通过毛细胞相对于上面的耳石的不同运动引起感觉传导。毛细胞的极性将决定最终的传导，如果这一运动导致静纤毛向动纤毛弯曲，顶连接将被拉伸，阳离子通道将打开，但如果声音来自相反的方向，机械传导通道将关闭（Kindt et al., 2012）。在激活的情况下，顶连接与非选择性离子通道耦合，但原位 K^+ 是主要的离子，所以打开通道导致大量的高浓度的 K^+ 离子内流，同时伴随着一些 Ca^{2+} 内流，从而引起毛细胞去极化（Pickles and Corey, 1992）。去极化导致膜电位的逐渐变化，从而导致毛细胞基底端电压门控钙通道的打开，最终导致突触神经递质释放增强。在脊椎动物中，由于大电导钙激活钾离子通道（large-conductance calcium-activated potassium channels, BK 通道，Fettiplace and Fuchs, 1999）介导的电共振，单个毛细胞具有一定程度的频率敏感性。在斑马鱼中，敲低编码这些 BK 通道的孔形成结构域的 2 个基因 *slo1a* 和 *slo1b* 可导致听觉阈值增加（Rohmann et al., 2014）。在“门控 - 弹簧”模型（“gating-spring” model）中顶连接发挥其作用（Corey, 2006），在该模型中，静纤毛的偏转导致顶连接基部的弹性元件拉伸并物理性地打开邻近的静纤毛上的离子通道，以增加其处于开放状态的可能性（Gillespie and Walker, 2001）。相反方向的移动将导致顶连接松弛，从而关闭通道（Gillespie and Walker, 2001）。在静止状态下，阳离子通道结构上是开放的，因此激发只会增加通道开放的时间，从而增加去极化率。

许多被归为 *circler* 基因突变的突变体最初被认为具有听觉和前庭功能缺陷，正是这些突变体使得我们第一次了解了毛细胞功能背后的基因作用（Nicolson et al., 1998）。组成传导通道本身的蛋白质仍存在争议（Pickett and Raible, 2019），但可能与 TRP 家族蛋白有关

（Corey, 2006; Sidi et al., 2003）。顶连接本身的结构是由钙黏素 23（cadherin23，Cdh23）和原钙黏附因子 15（protocadherin15，Pcdh15）编码的蛋白质组成的复合物，它们构成顶连接的结构元件（Ahmed et al., 2006; Kazmierczak et al., 2007; Maeda et al., 2017）。带有 *sputnik* 突变的斑马鱼（Pickett and Raible, 2019; Söllner et al., 2003）在 *Cdh23* 基因上有突变，不仅破坏传导束本身的结构完整性，而且在听觉和平衡功能的机械传导和行为缺陷方面也显示出明显的缺陷；证明了 *Cdh23* 在传导束功能中的重要性。在顶连接的下端，两个跨膜结构蛋白 Tmc1 和 Tmc2a 都在内耳毛细胞中特异性表达，并且都与 pdch15 的 C- 末端结构域结合（Maeda et al., 2017）。虽然还不完全清楚这些相互作用发挥了什么作用，但似乎很清楚的是，Pcdh15 和 TMC 家族蛋白之间的连接是斑马鱼毛细胞中传导机制的重要组成部分（Maeda et al., 2017），TMC 家族是机械传导所必需的阳离子选择性通道孔形成区域的潜在候选蛋白（Nicolson, 2017; Erickson et al., 2017）。穿膜内耳表达因子 tmie 蛋白也参与机械传导，由于缺乏功能性 *tmie* 基因的突变体会缺少顶连接，并且在斑马鱼仔鱼中也表现出了微音器电位（microphonic potential）的缺失（Gleason et al., 2009），而作用于 *tmie* 的吗啉类药物（morpholinos）可使毛细胞数量减少、半规管发育中断（Shen et al., 2008）。另一种跨膜蛋白 LHFPL5 可能在将 PCDH15 定位在静纤毛的位置中发挥重要作用，也可能在传导通道的形成中发挥作用（Xiong et al., 2012）。LHFPL5 的突变确实导致了仔鱼的听力和平衡障碍（Obholzer et al., 2008）。Barta 等人（2018）在比较斑马鱼听觉上皮毛细胞和非感觉支持细胞的转录组时发现，毛细胞中有 17 个基因特异性上调，但其中一些基因的功能仍有待阐明。

在顶连接的上端，也就是单纤维（filament）与下一个静纤毛连接的地方，也有一些蛋白质对这种连接的结构完整性是必需的，并且可能对保持顶连接的张力很重要（Gillespie and Walker, 2001）。*circler* 突变体 *sputnik* 和 *mariner* 在保持身体平衡和正常的游泳运动方面出现困难，并且都显示出分开的毛细胞束（Nicolson et al., 1998）。Myo7aa 蛋白是一种非常规的肌球蛋白，它与斑马鱼的顶连接中的 Cdh23 和支架蛋白 Ush1c 相互作用，*myo7aa* 基因突变导致毛细胞束分开（Ernest et al., 2000; Blanco-Sánchez et al., 2014）。Myo7aa 可能与内质网中的 Cdh23、Harmonin 以及 Ift88 蛋白形成复合物，该复合物的破坏不仅会导致传导束形态方面的问题，还可能导致斑马鱼毛细胞的凋亡（Blanco-Sánchez et al., 2014）。

最近，突触前水平的突触传递遗传学已经得到了充分的研究和全面的回顾（Nicolson, 2015）。激活后，位于带状突触的突触小泡释放谷氨酸。发生 *asteroid* 突变的斑马鱼对前庭听觉刺激的反应降低，这是由于编码囊泡谷氨酸转运体 3（vesicle glutamate transporter 3，vglut3）的基因发生突变，该基因现在被认为在突触囊泡形成和 / 或突触束（synaptic tethering）中很重要（Obholzer et al., 2008）。Ribeye 家族蛋白对脊椎动物突触的形成和功能至关重要（Nicolson, 2015），而 *ribeye a* 基因和 *ribeye b* 基因缺失会导致斑马鱼带状突触形成的中断（Sheets et al., 2011; Wan et al., 2005）。Ribeye 突变体可能通过与脊椎动物轴突活动区常见的 Rab3 束的相互作用（Nicolson, 2015），导致基底钙通道 Cav1.3a 的减少（Sheets et al., 2012）。快速释放谷氨酸对毛细胞响应声波刺激的锁相放电（phase-locked firing）是至关重要的（Einhorn et al., 2012），Rab3α 和囊泡上 ATP 合酶之间的相互作用对囊泡的快速装载是至关重要的，这是由于Rab3α的缺失会减少传入放电活性并阻断锁相放电（Einhorn et al., 2012）。突触囊泡一旦耗竭，就必须被重新循环利用，而*snyj*的缺失已被证明可以阻止突触膜对耗竭的囊泡的重吸收（Trapani et al., 2009）。

5.7　鱼类毛细胞死亡的原因

暴露于耳毒性化学物质和暴露于强烈或持续的声源会损害鱼的毛细胞（Brignull et al., 2009; McCauley et al., 2003; Smith et al., 2016）。对于化学污染，重金属如铜（Linbo et al., 2006）和镉（Faucher et al., 2006, 2008），以及氨基糖苷类抗生素（Song et al., 1995; Van Trum et al., 2010; Coffin et al., 2016）是研究毛细胞死亡机制和影响的最常用物质。许多噪声源已经被证明会导致硬骨鱼的毛细胞损失（McCauley et al., 2003; Popper and Hastings, 2009），慢性和急性暴露导致的细胞死亡依赖于音量（Smith et al., 2006; Hastings et al., 1996）。对于化学物质和噪声导致的毛细胞损失，已经证明涉及坏死和凋亡机制（Wiedenhoft et al., 2017），具体机制可能取决于声学损伤的程度。

5.8　毛细胞再生

损伤后，鱼类有能力完全再生听觉和神经丘毛细胞（Smith et al., 2006; Harris et al., 2003），然而这种再生机制仍然不清楚（Monroe et al., 2016; Lush et al., 2019）。由于容易获得神经丘，在合理假设听觉和神经丘毛细胞之间的机制具有共性的前提下，对神经丘毛细胞进行了更多的再生研究（Lush et al., 2019; Nicolson, 2005）。出于这个原因，本文将对这两种类型的毛细胞同等对待，以阐明目前我们对鱼类毛细胞再生的认识，但也应认识到这两类毛细胞之间如果有一些差异则可能会存在不同的毛细胞再生和修复的确切机制（Monroe et al., 2016）。鱼类毛细胞的再生是通过两组支持细胞的分化来完成的：一组形成新的支持细胞，另一组分化为两个新的毛细胞（Lush et al., 2019; Romero-Carvajal et al., 2015; ViaderLlargués et al., 2018）。采用$CuSO_4$对神经丘毛细胞的消融显示，一组基因在被称为肥大细胞的毛细胞前体细胞中快速分化表达（Steiner et al., 2014）。消融后肥大细胞中*arpc1a*、*atg2bl*、*btr04*、*fat2*、*klf3*、*lgals1l*和*prom2*基因均显著上调，而*fgfr1a*、*fndc7*和*tspan1*基因均显著下调（Steiner et al., 2014）。上调的基因都编码跨膜蛋白，参与细胞增殖、轴突投射、肌动蛋白组装等活动，但它们在肥大细胞分化中的确切作用尚不清楚（Steiner et al., 2014）。采用噪声消融斑马鱼囊状毛细胞，然后仔细分析其恢复的时间过程，发现消融后基因表达发生了大量变化，但是超过50个基因组成的核心组是*stat3/soc3*通路的一部分，显示了该通路中基因在毛细胞再生过程中的重要性（Liang et al., 2012）。在神经丘毛细胞中，化学消融后，Notch信号通路快速下调，基因*stat3*立即上调，然后是Wnt/βcatenin信号的上调（Jiang et al., 2014），而Notch和fgf3可能协同作用而阻断Wnt介导的再生

（Lush et al., 2019）。用新霉素对神经丘毛细胞进行化学消融，导致毛细胞初期完全丧失，在24h内开始再生，并在消融后72h完成再生（Ma et al., 2008）。再生来自支持细胞的分裂，并与*deltaA*、*notch3*和*atoh1*信号的上调相一致（Jiang et al., 2014; Ma et al., 2008），*notch3*在支持细胞中表达，*deltaA*在假定的毛细胞前体中表达（Ma et al., 2008）。*atoh1*的表达被认为是感觉上皮发育的必要和充分条件（Millimaki et al., 2007），而*atoh1*的错误表达导致毛细胞上皮异位发育（Sweet et al., 2011）。Pax-2蛋白的表达导致了球状毛细胞和囊状毛细胞的早期分化（Riley et al., 1999; Kwak et al., 2006）。至少在发育早期阶段，*Fgf*（Millimaki et al., 2007）和*sox2*（Sweet et al., 2011）的表达可能促进了*atoh1*的表达。在用耳毒性抗生素消融神经丘毛细胞后，Notch信号立即下调，但在消融约3～5h后再次上调，这可能是为了决定细胞命运（Jiang et al., 2014）。Notch和Fgf蛋白的下调共同启动再生，并可能抑制Wnt信号通路（Lush et al., 2019），特定的基因簇在初始再生和后期成熟过程中同时上调和下调（Lush et al., 2019）。Jak1/Stat3和BMP信号通路在神经丘毛细胞再生过程中也上调，Wnt/β-catenin信号通路在后期增殖阶段上调（Jiang et al., 2014），这可能是支持细胞分化的触发器（Head et al., 2013）。在对成熟内耳毛细胞和支持细胞进行比较时，*s100t*、*capb2b*、*otofb*、*tmc2a*、*cacnb3b*、*dnajc5b*、*tmc1*和*kcnq5a*基因在成熟内耳毛细胞中较支持细胞中的q-PCR值变化大，因此，这些基因对成熟内耳毛细胞的功能具有重要作用（Barta et al., 2018）。

5.9 斑马鱼耳石的形成与生长

在硬骨鱼中，每个听觉上皮（又称黄斑）与发育早期分泌的致密钙质耳石密切相关（Riley et al., 1997），并随着感觉上皮的生长而在整个生命周期中继续生长（Lombarte, 1992; Aguirre and Lombarte, 1999）。耳石对鱼类的听力是不可或缺的（见上文），并通过耳石膜与毛细胞上皮结合（Dunkelberger et al., 1980; Popper, 1977）。硬骨鱼的耳石主要是由感觉黄斑细胞分泌的蛋白质基质与碳酸钙融合而成（Campana, 1999）。现在已经发现了许多破坏耳石形成的突变体（Riley et al., 1997; Whitfield et al., 1996; Schibler and Malicki, 2007; Lundberg et al., 2015），也可能破坏听力，但这种破坏背后确切的基因机制仍在研究中。在这篇综述中，并未列出所有已知的突变体，而是将重点放在目前对耳石形成相关基因机制的理解上，并指导读者阅读上述综述中所参考已发表的文献。*kei*和*bks*基因的突变导致胚胎完全缺乏耳石，而*eis*基因的突变只形成一个耳石（Whitfield et al., 1996）。耳石形成的第一步是耳石种子颗粒进入听软骨囊（Riley et al., 1997; Tanimoto et al., 2011）。最初的蛋白质分泌可能受到*otopretin1*的调控（Söllner et al., 2004），最初的传导束（tethering）受*otog*表达的影响，发育后期传导束的维持也受*tecta*功能的影响（Stooke-Vaughan et al., 2015）。碳酸酐酶的体内循环水平也影响耳石生长（Anken et al., 2004），可能是通过调节内淋巴HCO_3^-水平（Lundberg et al., 2015）而调节耳石生长。*rst*突变首先导致正常的耳石分泌，但由此产生的耳石并没有固定在黄斑上，可能是由于耳石膜的形成缺陷（Whitfield et al., 1996）。耳石一旦开始形成，就会随着鱼的生长而继续生长，用年轮（通常是每日轮）来标记生长，仔细分析后可以表明特定鱼的年龄（Campana et al., 1992）。斑马鱼耳石的生长受 $CaCO_3$ 晶体中的有机基质相关的蛋白质控制，尽管在其他硬骨鱼类中可能没有发现相同的蛋白质（Weigele et al., 2016）。伴侣蛋白 GP96 的敲除可导致耳石前体颗粒无法附着在传导束细胞（tether cells）上，从而阻止耳石生长（Sumanas et al., 2003），而 *starmaker* 基因表达的减少导致正常耳石形状的破坏（Söllner et al., 2004）。Starmaker 蛋白的分泌可能受到 *otopretin 1* 的调控，因为该基因的突变可能是 *backstroke* 突变斑马鱼（Whitfield et al., 1996）完全缺乏耳石的原因（Söllner et al., 2004）。耳石基质蛋白（otolith matrix proteins, OMPs）是耳石生长所必需的（Murayama et al., 2000, 2002），而 *zOMP-1* 是正常耳石生长所必需的，相关蛋白耳石蛋白 1（Othlin 1）则是耳石在听软骨囊内的正确排列所必需的（Murayama et al., 2005）。耳石的持续生长受骨形态发生蛋白 Sparc 的控制，可能还与耳石蛋白 1 和另外两种耳石蛋白，即前小脑肽样蛋白（precerebellin-like）和神经源的丝氨酸蛋白酶抑制剂（neuroserpin）有关（Kang et al., 2008）。钙黏蛋白被分泌到听觉内淋巴，也可能促进耳石的持续生长，这是因为钙黏蛋白 11 基因（*cadherin11*）的敲减可导致耳石体积的减小（Clendenon et al., 2009）。

5.10 斑马鱼听觉刺激的传导

毛细胞活化传递到中枢神经系统是通过改变释放谷氨酸到毛细胞和相关传入神经元之间的突触间隙的速率。在没有感官刺激的情况下，在静息电位下会自发释放谷氨酸，导致感觉传入神经的自发活动（Sewell, 1984），这种释放的增加或减少取决于极化毛细胞立体纤毛阵列的激活方向（见上文）。当被声音刺激激活后，斑马鱼毛细胞突触前膜上钙通道的开放增加，导致带状突触上的突触囊泡融合。带状突触是感觉毛细胞的基本特征（Nicolson, 2015; Moser et al., 2006），由一层连接在毛细胞活动区突触体上的突触囊泡组成（Moser et al., 2006; Lenzi et al., 1999; Matthews and Fuchs, 2010）。突触带下方有无约束（untethered vesicles）的囊泡，允许快速释放神经递质（Mennerick and Matthews, 1996），而传导束的囊泡（tethered vesicles）在激活时负责缓慢并持续地释放神经递质（Moser et al., 2006; Lenzi et al., 1999）。谷氨酸释放后，与突触后传入的 AMPA 受体结合而将其激活（Glowatzki and Fuchs, 2002）。虽然斑马鱼突触连接的详细驱动因素仍有待阐明，但我们已开始了解参与这一过程的一些分子机制。在斑马鱼中，*ribeye* 是带状突触形成的重要组织者，因为在斑马鱼仔鱼中，敲除 *ribeye* 会导致带状突触更小或完全缺乏（Wan et al., 2005; Sheets et al., 2012）。*ribeye* 在基底外侧钙通道的聚集中也有作用，因为其表达水平的改变也会影响通道的数量和位置（Sheets et al., 2012）。斑马鱼带状突触的大小和形状也受到钙通道活动的影响（Sheets et al., 2012），特别是 Cav1.3a 通道，因为 *cav1.3a* 突变体具有增强的带状突触（Sheets et al., 2012），细胞内钙水平也有助于形成这些囊泡聚集（Nicolson, 2015）。*clarin1* 的敲除也影响了 ribeye 蛋白在带状突触形成中的作用（Ogun and Zallocchi, 2014），然而这种相互作用的机制尚不清楚（Nicolson, 2015）。有助于将囊泡锚定在突触前膜上的蛋白质包括小突触囊泡蛋白（synaptobrevin）和突触融合蛋白 3（syntaxin 3）上，它们的功能受到 WRB 蛋白的影响，如 *pinball wizard*（*pwi*）突变株所示（Lin et al., 2016）。与斑马鱼毛细胞的传入连接早在 27hpf 时便形成，而声音诱发电位直到 40hpf 才出现（Tanimoto et al., 2009），这表明突触在最初发育后需要额外的时间才能正常发挥功能。突触一旦形成，谷氨酸的突触释放依赖于与跨膜蛋白 Otoferlin 结合的 Cav1.3a 钙通道的钙内流（Sidi et al., 2003; Sheets et al., 2012）（Goodyear et al., 2010; Chatterjee et al., 2015）。*pinball wizard*（*pwi*）突变体的听力缺陷可归因于突触囊泡释放中断（Liu et al., 2018），并降低了 Rab3 和 Dnacj5 蛋白（两种已知的突触囊泡蛋白）的表达。囊泡循环至少部分依赖于脂质磷酸酶 Synaptojanin 1，因为以 *synj1* 为靶点的 *comet* 突变在重复刺激后会快速耗尽突触活性，并减少突触储备囊泡的数量（Trapani et al., 2009）。在被激活后，感觉信息沿着传入的第 8 传入通路（Ⅷ th afferent）先到达后脑内侧和背侧八掌外侧核（dorsal octavolateralis nuclei），然后在丘脑和半规隆凸（torus semicircularis）进一步加工（Higgs et al., 2006; Fay et al., 2008）。即使是 6dpf 的斑马鱼也有这种基本的神经支配模式，然而它们的频率响应和神经张力相对于成年斑马鱼来说有所减弱（Vanwalleghem et al., 2017）。

（李海芸翻译　李丽琴审校）

参考文献

Aguirre, H., Lombarte, A., 1999. Ecomorphological comparisons of sagittae in *Mullus barbatus* and M. *surmuletus*. Journal of Fish Biology 55 (1), 105-114.

Ahmed, Z.M., Goodyear, R., Riazuddin, S., Lagziel, A., Legan, K., Behra, M., Burgess, S.M., Lilley, K.S., Wilcox, E.R., Riazuddin, S., Griffith, A.J., 2006. The tip-link antigen, a protein associated with the transduction complex of sensory hair cells, is protocadherin-15. Journal of Neuroscience 26 (26), 7022-7034.

Akamatsu, T., Okumura, T., Novarini, N., Yan, H.Y., 2002. Empirical refinements applicable to the recording of fish sounds in small tanks. Journal of the Acoustical Society of America 112 (6), 3073-3082.

Alsina, B., Whitfield, T.T., 2017. Sculpting the labyrinth: morphogenesis of the developing inner ear. Seminars in Cell and Developmental Biology 65, 47-59. Academic Press.

Anken, R.H., Beier, M., Rahmann, H., 2004. Hypergravity decreases carbonic anhydrase-reactivity in inner ear maculae of fish. Journal of Experimental Zoology 301, 815-819.

Bang, P.I., Sewell, W.F., Malicki, J.J., 2001. Morphology and cell type heterogeneities of the inner ear epithelia in adult and juvenile zebrafish (*Danio rerio*). The Journal of Comparative Neurology 438 (2), 173-190.

Bang, P.I., Yelick, P.C., Malicki, J.J., Sewell, W.F., 2002. High-throughput behavioral screening method for detecting auditory response defects in zebrafish. Journal of Neuroscience Methods 118 (2), 177-187.

Barta, C., Liu, H., Chen, L., Giffen, K.P., Li, Y., Kramer, K.L., Beisel, K.W., He, D.Z., 2018. RNA-seq transcriptomic analysis of adult zebrafish inner ear hair cells. Scientific Data 5, 180005.

Bhandiwad, A.A., Zeddies, D.G., Raible, D.W., Rubel, E.W., Sisneros, J.A., 2013. Auditory sensitivity of larval zebrafish (*Danio rerio*) measured using a behavioral prepulse inhibition assay. Journal of Experimental Biology 216 (18), 3504-3513.

Blanco-Sánchez, B., Clément, A., Fierro, J., Washbourne, P., Westerfield, M., 2014. Complexes of Usher proteins preassemble at the endoplasmic reticulum and are required for trafficking and ER homeostasis. Disease Models and Mechanisms 7 (5), 547-559.

Brignull, H.R., Raible, D.W., Stone, J.S., 2009. Feathers and fins: non-mammalian models for hair cell regeneration. Brain Research 1277, 12-23.

Burgess, H.A., Granato, M., 2007. Sensorimotor gating in larval zebrafish. Journal of Neuroscience 27 (18), 984-4994.

Campana, S.E., 1999. Chemistry and composition of fish otoliths: pathways, mechanisms and applications. Marine Ecology Progress Series 188, 263-297.

Campana, S.E., Jones, C.M., 1992. Analysis of otolith microstructure data. In: Stevenson, D.K., Campana, S.E. (Eds.), Otolith Microstructure Examination and Analysis Can Spec Publ Fish Aquat Sci, vol. 117, pp. 73-100.

Cervi, A.L., Poling, K.R., Higgs, D.M., 2012. Behavioural measure of frequency detection and discrimination in the zebrafish, *Danio rerio*. Zebrafish 9, 1-7.

Chardon, M., Vandewalle, P., 1997. Evolutionary trends and possible origins of the Weberian apparatus. Netherlands Journal of Zoology 47, 383-403.

Chatterjee, P., Padmanarayana, M., Abdullah, N., Holman, C.L., LaDu, J., Tanguay, R.L., Johnson, C.P., 2015. Otoferlin deficiency in zebrafish results in defects in balance and hearing: rescue of the balance and hearing phenotype with full-length and truncated forms of mouse otoferlin. Molecular and Cellular Biology 35, 1043-1054.

Clendenon, S.G., Shah, B., Miller, C.A., Schmeisser, G., Walter, A., Gattone, V.H., Barald, K.F., Liu, Q., Marrs, J.A., 2009. Cadherin-11 controls otolith assembly: evidence for extracellular cadherin activity. Developmental Dynamics 238 (8), 1909-1922.

Coffin, A.B., Ramcharitar, J., 2016. Chemical ototoxicity of the fish inner ear and lateral line. In: Sisneros, J.A. (Ed.), Fish Hearing and Bioacoustics. Springer, New York, pp. 419-437.

Coombs, S., Janssen, J., 1990. Behavioral and neurophysiological assessment of lateral line sensitivity in the mottled sculpin, Cottus bairdi. Journal of Comparative Physiology A 167 (4), 557-567.

Coombs, S., Montgomery, J.C., 1999. The enigmatic lateral line system. In: Comparative Hearing: Fish and Amphibians. Springer, New York, NY, pp. 319-362.

Corey, D.P., 2006. What is the hair cell transduction channel? The Journal of Physiology 576 (1), 23-28.

Dunkelberger, D.G., Dean, J.M., Watabe, N., 1980. The ultrastructure of the otolithic membrane and otolith of juvenile mummichog. Journal of Morphology 163, 367-377.

Dutton, K., Abbas, L., Spencer, J., Brannon, C., Mowbray, C., Nikaido, M., Kelsh, R.N., Whitfield, T.T., 2009. A zebrafish model for Waardenburg syndrome type IV reveals diverse roles for Sox10 in the otic vesicle. Disease Models and Mechanisms 2 (1e2), 68-83.

Eaton, R.C., DiDomenico, R., Nissanov, J., 1991. Role of the Mauthner cell in sensorimotor integration by the brain stem escape network. Brain, Behavior and Evolution 37, 272-285.

Edds-Walton, P.L., Fay, R.R., 2005. Projections to bimodal sites in the torus semicircularis of the toadfish, Opsanus tau. Brain, Behavior and Evolution 66 (2), 73-87.

Einhorn, Z., Trapani, J.G., Liu, Q., Nicolson, T., 2012. Rabconnectin3a promotes stable activity of the Hþ pump on synaptic vesicles in hair cells. Journal of Neuroscience 32 (32), 11144-11156.

Eisen, J., Myers, P.Z., Westerfield, M., 1986. Pathway selection by growth cones of identified motoneurons in live zebra fish embryos. Nature 320, 269-271.

Erickson, T., Morgan, C.P., Olt, J., Hardy, K., Busch-Nentwich, E., Maeda, R., Clemens, R., Krey, J.F., Nechiporuk, A., Barr-Gillespie, P.G., Marcotti, W., Nicholson, T., 2017. Integration of Tmc1/2 into the mechanotransduction complex in zebrafish hair cells is regulated by Transmembrane O-methyltransferase (Tomt). eLife 6, e28474.

Ernest, S., Rauch, G.J., Haffter, P., Geisler, R., Petit, C., Nicolson, T., 2000. Mariner is defective in myosin VIIA: a Zebrafish model for human hereditary deafness. Human Molecular Genetics 9 (14), 2189-2196.

Faucher, K., Fichet, D., Miramand, P., Lagardère, J.P., 2006. Impact of acute cadmium exposure on the trunk lateral line neuromasts and consequences on the "C-start" response behaviour of the sea bass (*Dicentrarchus labrax* L.; Teleostei, Moronidae). Aquatic Toxicology 76 (3e4), 278-294.

Faucher, K., Fichet, D., Miramand, P., Lagardere, J.P., 2008. Impact of chronic cadmium exposure at environmental dose on escape behaviour in sea bass (*Dicentrarchus labrax* L.; Teleostei, Moronidae). Environmental Pollution 151 (1), 148-157.

Fay, R.R., Edds-Walton, P.L., 2008. Structures and functions of the auditory nervous system of fishes. In: Webb, J.F., Fay, R.R., Popper, A.N. (Eds.), Fish Bioacoustics. Springer, New York, NY, pp. 49-97.

Feng, A.S., Schellart, N.A., 1999. Central auditory processing in fish and amphibians. In: Fay, R.R., Popper, A.N. (Eds.), Comparative Hearing: Fish and Amphibians. Springer, New York, NY, pp. 218-268.

Fettiplace, R., Fuchs, P.A., 1999. Mechanisms of hair cell tuning. Annual Review of Physiology 61, 809-834.

Geng, F.-S., Abbas, L., Baxendale, S., Holdsworth, C.J., Swanson, A.G., Slanchev, K., Hammerschmidt, M., Topczewski, J., Whitfield, T.T., 2013. Semicircular canal morphogenesis in the zebrafish inner ear requires the function of gpr126 (lauscher), an adhesion class G protein-coupled receptor gene. Development 140 (21), 4362-4374.

Gillespie, P.G., Walker, R.G., 2001. Molecular basis of mechanosensory transduction. Nature 413, 194-202.

Gleason, M.R., Nagiel, A., Jamet, S., Vologodskaia, M., López-Schier, H., Hudspeth, A.J., 2009. The transmembrane inner ear (Tmie) protein is essential for normal hearing and balance in the zebrafish. Proceedings of the National Academy of Sciences of the United States of America 106 (50), 21347-21352.

Glowatzki, E., Fuchs, P.A., 2002. Transmitter release at the hair cell ribbon synapse. Nature Neuroscience 5 (2), 147.

Goodyear, R.J., Legan, P.K., Christiansen, J.R., Xia, B., Korchagina, J., Gale, J.E., Warchol, M.E., Corwin, J.T., Richardson, G.P., 2010. Identification of the hair cell soma-1 antigen, HCS-1, as otoferlin. The Association for Research in Otolaryngology 11, 573-586.

Haddon, C., Lewis, J., 1996. Early ear development in the embryo of the zebrafish, *Danio rerio*. The Journal of Comparative Neurology 365 (1), 113-128.

Hammond, K.L., Whitfield, T.T., 2011. Fgf and Hh signalling act on a symmetrical pre-pattern to specify anterior and posterior identity in the zebrafish otic placode and vesicle. Development 138 (18), 3977-3987.

Hammond, K.L., Loynes, H.E., Folarin, A.A., Smith, J., Whitfield, T.T., 2003. Hedgehog signalling is required for correct anteroposterior patterning of the zebrafish otic vesicle. Development 130 (7), 1403-1417.

Harris, J.A., Cheng, A.G., Cunningham, L.L., MacDonald, G., Raible, D.W., Rubel, E.W., 2003. Neomycin-induced hair cell death and rapid regeneration in the lateral line of zebrafish (*Danio rerio*). The Association for Research in Otolaryngology 4 (2), 219-234.

Hartwell, R.D., England, S.J., Monk, N.A., Van Hateren, N.J., Baxendale, S., Marzo, M., Lewis, K.E., Whitfield, T.T., 2019. Anteroposterior patterning of the zebrafish ear through Fgf-and Hh-dependent regulation of hmx3a expression. PLoS Genetics 15 (4), e1008051.

Hastings, M.C., Popper, A.N., Finneran, J.J., Lanford, P.J., 1996. Effects of low-frequency underwater sound on hair cells of the inner ear and lateral line of the teleost fish *Astronotus ocellatus*. Journal of the Acoustical Society of America 99 (3), 1759-1766.

Head, J.R., Gacioch, L., Pennisi, M., Meyers, J.R., 2013. Activation of canonical Wnt/b-catenin signaling stimulates proliferation in neuromasts in the zebrafish posterior lateral line. Developmental Dynamics 242 (7), 832-846.

Higgs, D.M., 2002. Development of the fish auditory system: how do changes in auditory structure affect function? Bioacoustics 12, 180-182.

Higgs, D.M., Radford, C.A., 2013. The contribution of the lateral line to "hearing" in fish. Journal of Experimental Biology 216, 1484-1490.

Higgs, D.M., Souza, M.J., Wilkins, H.R., Presson, J.C., Popper, A.N., 2002. Age-related changes in the inner ear and hearing ability of the adult zebrafish (*Danio rerio*). The Association for Research in Otolaryngology 3, 174-184.

Higgs, D.M., Rollo, A.K., Souza, M.J., Popper, A.N., 2003. Development of form and function in peripheral auditory structures of the zebrafish (*Danio rerio*). Journal of the Acoustical Society of America 113, 1145-1154.

Higgs, D.M., Lui, Z., Mann, D.A., 2006. Hearing and mechanoreception. In: Evans, D.H. (Ed.), The Physiology of Fishes, third ed. CRC Press, pp. 391-429.

Higgs, D.M., Radford, C.A., 2015. The potential overlapping roles of the ear and lateral line in driving "acoustic" responses. In: Sisneros, J.A. (Ed.), Fish Hearing and Bioacoutistics. Springer-Verlag, New York, pp. 255-270.

Jiang, L., Romero-Carvajal, A., Haug, J.S., Seidel, C.W., Piotrowski, T., 2014. Gene-expression analysis of hair cell regeneration in the zebrafish lateral line. Proceedings of the National Academy of Sciences of the United States of America 111 (14), E1383-E1392.

Kang, Y.J., Stevenson, A.K., Yau, P.M., Kollmar, R., 2008. Sparc protein is required for normal growth of zebrafish otoliths. The Association for Research in Otolaryngology 9 (4), 436-451.

Kazmierczak, P., Sakaguchi, H., Tokita, J., Wilson-Kubalek, E.M., Milligan, R.A., Müller, U., Kachar, B., 2007. Cadherin 23 and protocadherin 15 interact to form tip-link filaments in sensory hair cells. Nature 449 (7158), 87.

Kimmel, C.B., Patterson, J., Kimmel, R.O., 1974. The development and behavioral characteristics of the startle response in the zebra fish. Developmental Psychobiology 7 (1), 47-60.

Kindt, K.S., Finch, G., Nicolson, T., 2012. Kinocilia mediate mechanosensitivity in developing zebrafish hair cells. Developmental Cell 23 (2), 329-341.

Kozlowski, D.J., Whitfield, T.T., Hukriede, N.A., Lam, W.K., Weinberg, E.S., 2005. The zebrafish dog-eared mutation disrupts eya1, a gene required for cell survival and differentiation in the inner ear and lateral line. Developmental Biology 277 (1), 27-41.

Kwak, S.-J., Vemaraju, S., Moorman, S.J., Zeddies, D., Popper, A.N., Riley, B.B., 2006. Zebrafish pax5 regulates development of the utricular macula and vestibular function. Developmental Dynamics 235, 3026-3038.

Ladich, F., Fay, R.R., 2013. Auditory evoked potential audiometry in fish. Reviews in Fish Biology and Fisheries 23 (3), 317-364.

Lenzi, D., Runyeon, J.W., Crum, J., Ellisman, M.H., Roberts, W.M., 1999. Synaptic vesicle populations in saccular hair cells reconstructed by electron tomography. Journal of Neuroscience 19 (1), 119-132.

Liang, J., Wang, D., Renaud, G., Wolfsberg, T.G., Wilson, A.F., Burgess, S.M., 2012. The stat3/socs3a pathway is a key regulator of hair cell regeneration in zebrafish stat3/socs3a pathway: regulator of hair cell regeneration. Journal of

Neuroscience 32 (31), 10662-10673.

Lin, S.Y., Vollrath, M.A., Mangosing, S., Shen, J., Cardenas, E., Corey, D.P., 2016. The zebrafish pinball wizard gene encodes WRB, a tail-anchored-protein receptor essential for inner-ear hair cells and retinal photoreceptors. The Journal of Physiology 594 (4), 895-914.

Linbo, T.L., Stehr, C.M., Incardona, J.P., Scholz, N.L., 2006. Dissolved copper triggers cell death in the peripheral mechanosensory system of larval fish. Environmental Toxicology & Chemistry 25 (2), 597-603.

Liu, X., Lin, J., Zhang, Y., Guo, N., Li, Q., 2018. Sound shock response in larval zebrafish: a convenient and highthroughput assessment of auditory function. Neurotoxicology and Teratology 66, 1-7.

Lombarte, A., 1992. Changes in otolith area: sensory area ratio with body size and depth. Environmental Biology of Fishes 33 (4), 405-410.

Lu, Z., DeSmidt, A.A., 2013. Early development of hearing in zebrafish. The Association for Research in Otolaryngology 14 (4), 509-521.

Lundberg, Y.W., Xu, Y., Thiessen, K.D., Kramer, K.L., 2015. Mechanisms of otoconia and otolith development. Developmental Dynamics 244 (3), 239-253.

Lush, M.E., Diaz, D.C., Koenecke, N., Baek, S., Boldt, H., St Peter, M.K., Gaitan-Escudero, T., Romero-Carvaja, A., Busch-Nentwich, E.M., Perera, A.G., Hall, K.E., 2019. scRNA-Seq reveals distinct stem cell populations that drive hair cell regeneration after loss of Fgf and Notch signaling. eLife 8, e44431.

Lychakov, D.V., Rebane, Y.T., 2000. Otolith regularities. Hearing Research 143, 83-102.

Lychakov, D.V., Rebane, Y.T., 2005. Fish otolith mass asymmetry: morphometry and influence on acoustic functionality. Hearing Research 201, 55-69.

Ma, E.Y., Rubel, E.W., Raible, D.W., 2008. Notch signaling regulates the extent of hair cell regeneration in the zebrafish lateral line. Journal of Neuroscience 28 (9), 2261-2273.

Maeda, R., Pacentine, I.V., Erickson, T., Nicolson, T., 2017. Functional analysis of the transmembrane and cytoplasmic domains of Pcdh15a in zebrafish hair cells. Journal of Neuroscience 37 (12), 3231-3245.

Maier, E.C., Whitfield, T.T., 2014. RA and FGF signalling are required in the zebrafish otic vesicle to pattern and maintain ventral otic identities. PLoS Genetics 10 (12), e1004858.

Marsden, K.C., Jain, R.A., Wolman, M.A., Echeverry, F.A., Nelson, J.C., Hayer, K.E., Miltenberg, B., Pereda, A.E., Granato, M., 2018. A Cyfip2-dependent excitatory interneuron pathway establishes the innate startle threshold. Cell Reports 23 (3), 878-887.

Matthews, G., Fuchs, P., 2010. The diverse roles of ribbon synapses in sensory neurotransmission. Nature Reviews Neuroscience 11 (12), 812.

McCauley, R.D., Fewtrell, J., Popper, A.N., 2003. High intensity anthropogenic sound damages fish ears. Journal of the Acoustical Society of America 113 (1), 638-642.

Medan, V., Preuss, T., 2014. The Mauthner-cell circuit of fish as a model system for startle plasticity. The Journal of Physiology 108 (2e3), 129-140.

Mennerick, S., Matthews, G., 1996. Ultrafast exocytosis elicited by calcium current in synaptic terminals or retinal bipolar neurons. Neuron 17, 1241-1249.

Millimaki, B.B., Sweet, E.M., Dhason, M.S., Riley, B.B., 2007. Zebrafish atoh1 genes: classic proneural activity in the inner ear and regulation by Fgf and Notch. Development 134 (2), 295-305.

Moller, P., 2002. Multimodal sensory integration in weakly electric fish: a behavioral account. The Journal of Physiology 96 (5e6), 547-556.

Monroe, J.D., Manning, D.P., Uribe, P.M., Bhandiwad, A., Sisneros, J.A., Smith, M.E., Coffin, A.B., 2016. Hearing sensitivity differs between zebrafish lines used in auditory research. Hearing Research 341, 220-231.

Montgomery, J., Coombs, S., Janssen, J., 1994. Form and function relationships in lateral line systems: comparative data from six species of Antarctic notothenioid fish. Brain, Behavior and Evolution 44 (6), 299-306.

Montgomery, J.C., Baker, C.F., Carton, A.G., 1997. The lateral line can mediate rheotaxis in fish. Nature 389 (6654), 960.

Moser, T., Brandt, A., Lysakowski, A., 2006. Hair cell ribbon synapses. Cell and Tissue Research 326 (2), 347-359.

Münz, H., 1985. Single unit activity in the peripheral lateral line system of the cichlid fish *Sarotherodon niloticus* L. Journal of Comparative Physiology A 157 (5), 555-568.

Murayama, E., Okuno, A., Ohira, T., Takagi, Y., Nagasawa, H., 2000. Molecular cloning and expression of an otolith matrix protein cDNA from the rainbow trout, *Oncorhynchus mykiss*. Comparative Biochemistry and Physiology 126B, 511-520.

Murayama, E., Takagi, Y., Ohira, T., Davis, J.G., Green, M.I., Nagasawa, H., 2002. Fish otolith contains a unique structural protein, otolin-1. European Journal of Biochemistry 269, 688-696.
Murayama, E., Herbomel, P., Kawakami, A., Takeda, H., Nagasawa, H., 2005. Otolith matrix proteins OMP-1 and Otolin-1 are necessary for normal otolith growth and their correct anchoring onto the sensory maculae. Mechanisms of Development 122 (6), 791-803.
Nedelec, S.L., Campbell, J., Radford, A.N., Simpson, S.S., Merchant, N.D., 2016. Particle motion: the missing link in underwater acoustic ecology. Methods in Ecology and Evolution 7 (7), 836-842.
Nicolson, T., 2005. The genetics of hearing and balance in zebrafish. Annual Review of Genetics 39, 9-22.
Nicolson, T., 2015. Ribbon synapses in zebrafish hair cells. Hearing Research 330, 170-177.
Nicolson, T., 2017. The genetics of hair-cell function in zebrafish. Journal of Neurogenetics 31 (3), 102-112.
Nicolson, T., Rüsch, A., Friedrich, R.W., Granato, M., Ruppersberg, J.P., Nüsslein-Volhard, C., 1998. Genetic analysis of vertebrate sensory hair cell mechanosensation: the zebrafish circler mutants. Neuron 20 (2), 271-283.
Obholzer, N., Wolfson, S., Trapani, J.G., Mo, W., Nechiporuk, A., Busch-Nentwich, E., Seiler, C., Sidi, S., Söllner, C., Duncan, R.N., Boehland, A., 2008. Vesicular glutamate transporter 3 is required for synaptic transmission in zebrafish hair cells. Journal of Neuroscience 28 (9), 2110-2118.
Ogun, O., Zallocchi, M., 2014. Clarin-1 acts as a modulator of mechanotransduction activity and presynaptic ribbon assembly. The Journal of Cell Biology 207, 375-391.
Parvulescu, A., 1967. The acoustics of small tanks. Marine Bioacoustics 2, 7-13.
Pickett, S.B., Raible, D.W., 2019. Water waves to sound waves: using zebrafish to explore hair cell biology. The Association for Research in Otolaryngology 20 (1), 1-19.
Pickles, J.O., Corey, D.P., 1992. Mechanoelectrical transduction by hair cells. Trends in Neurosciences 15 (7), 254-259.
Pisam, M., Jammet, C., Laurent, D., 2002. First steps of otolith formation of the zebrafish: role of glycogen? Cell and Tissue Research 310, 163-168.
Platt, C., 1993. Zebrafish inner ear sensory surfaces are similar to those in goldfish. Hearing Research 65 (1-2), 133-140.
Popper, A.N., 1971. The effects of size on the auditory capabilities of the goldfish. Journal of Auditory Research 11, 239-247.
Popper, A.N., 1977. A scanning electron microscopy study of the sacculus and lagena in the ear of fifteen species of teleost fishes. Journal of Morphology 153, 397-418.
Popper, A.N., Coombs, S., 1982. The morphology and evolution of the ear in actinopterygian fishes. American Zoologist 22, 311-328.
Popper, A.N., Hastings, M.C., 2009. The effects of anthropogenic sources of sound on fishes. Journal of Fish Biology 75 (3), 455-489.
Popper, A.N., Fay, R.R., 1999. The auditory periphery in fishes. In: Fay, R.R., AN, P. (Eds.), Comparative Hearing: Fish and Amphibians. Springer-Verlag, New York, pp. 43-100.
Riley, B.B., Zhu, C., Janetopoulos, C., Aufderheide, K.J., 1997. A critical period of ear development controlled by distinct populations of ciliated cells in the zebrafish. Developmental Biology 191, 191-201.
Riley, B.B., Chiang, M.-Y., Farmer, L., Heck, R., 1999. The deltaA gene of zebrafish mediates lateral inhibition of hair cells in the inner ear and is regulated by pax2.1. Development 26, 5669-5678.
Rogers, P.H., Cox, M., 1988. Underwater sound as a biological stimulus. In: Atema, J., Fay, R.R., Popper, A.N., Tavolga, W.N. (Eds.), Sensory Biology of Aquatic Animals. Springer, New York, NY, pp. 131-149.
Rogers, P.H., Hawkins, A.D., Popper, A.N., Fay, R.R., Gray, M.D., 2016. Parvulescu revisited: small tank acoustics for bioacousticians. In: Popper, A., Hawkins, A. (Eds.), The Effects of Noise on Aquatic Life Ⅱ. Advances in Experimental Medicine and Biology. Springer, New York, pp. 933-941.
Rohmann, K.N., Tripp, J.A., Genova, R.M., Bass, A.H., 2014. Manipulation of BK channel expression is sufficient to alter auditory hair cell thresholds in larval zebrafish. Journal of Experimental Biology 217 (14), 2531-2539.
Romero-Carvajal, A., Navajas Acedo, J., Jiang, L., Kozlovskaja-Gumbriene, A., Alexander, R., Li, H., Piotrowski, T., 2015. Regeneration of sensory hair cells requires localized interactions between the notch and wnt pathways. Developmental Cell 34, 267-282.
Schibler, A., Malicki, J., 2007. A screen for genetic defects of the zebrafish ear. Mechanisms of Development 124 (7-8), 592-604.
Sewell, W.F., 1984. The relation between the endocochlear potential and spontaneous activity in auditory nerve fibres of the cat. The Journal of Physiology 347, 685-696.

Sheets, L., Trapani, J.G., Mo, W., Obholzer, N., Nicolson, T.T., 2011. Ribeye is required for presynaptic Ca(V)1.3a channel localization and afferent innervation of sensory hair cells. Development 138, 1309-1319.

Sheets, L., Kindt, K.S., Nicolson, T.T., 2012. Presynaptic CaV1.3 channels regulate synaptic ribbon size and are required for synaptic maintenance in sensory hair cells. Journal of Neuroscience 32, 17273-17286.

Shen, Y.C., Jeyabalan, A.K., Wu, K.L., Hunker, K.L., Kohrman, D.C., Thompson, D.L., Liu, D., Barald, K.F., 2008. The transmembrane inner ear (tmie) gene contributes to vestibular and lateral line development and function in the zebrafish (*Danio rerio*). Developmental Dynamics 237 (4), 941-952.

Sidi, S., Friedrich, R.W., Nicolson, T., 2003. NompC TRP channel required for vertebrate sensory hair cell mechanotransduction. Science 301 (5629), 96-99.

Sisneros, J.A., Popper, A.N., Hawkins, A.D., Fay, R.R., 2016. Auditory evoked potential audiograms compared with behavioral audiograms in aquatic animals. In: Popper, A., Hawkins, A. (Eds.), The Effects of Noise on Aquatic Life II. Springer, New York, NY, pp. 1049-1056.

Smith, M.E., Coffin, A.B., Miller, D.L., Popper, A.N., 2006. Anatomical and functional recovery of the goldfish (*Carassius auratus*) ear following noise exposure. Journal of Experimental Biology 209 (21), 4193-4202.

Smith, M.E., Monroe, J.D., 2016. Causes and consequences of sensory hair cell damage and recovery in fishes. In: Sisneros, J.A. (Ed.), Fish Hearing and Bioacoustics. Springer, New York, pp. 393-417.

Söllner, C., Burghammer, M., Busch-Nentwich, E., Berger, J., Schwarz, H., Riekel, C., Nicolson, T.T., 2003. Control of crystal size and lattice formation by starmaker in otolith biomineralization. Science 302, 282-286.

Söllner, C., Rauch, G.J., Siemens, J., Geisler, R., Schuster, S.C., Müller, U., Nicolson, T., Tübingen 2000 Screen Consortium, 2004. Mutations in cadherin 23 affect tip links in zebrafish sensory hair cells. Nature 428 (6986), 955.

Song, J., Yan, H.Y., Popper, A.N., 1995. Damage and recovery of hair cells in fish canal (but not superficial) neuromasts after gentamicin exposure. Hearing Research 91 (1-2), 63-71.

Steiner, A.B., Kim, T., Cabot, V., Hudspeth, A.J., 2014. Dynamic gene expression by putative hair-cell progenitors during regeneration in the zebrafish lateral line. Proceedings of the National Academy of Sciences of the United States of America 111 (14), E1393-E1401.

Stooke-Vaughan, G.A., Obholzer, N.D., Baxendale, S., Megason, S.G., Whitfield, T.T., 2015. Otolith tethering in the zebrafish otic vesicle requires Otogelin and a-Tectorin. Development 142 (6), 1137-1145.

Sumanas, S., Larson, J.D., Bever, M.M., 2003. Zebrafish chaperone protein GP96 is required for otolith formation during ear development. Developmental Biology 261, 443-455.

Sweet, E.M., Vemaraju, S., Riley, B.B., 2011. Sox2 and Fgf interact with Atoh1 to promote sensory competence throughout the zebrafish inner ear. Developmental Biology 358, 113-121.

Tanimoto, M., Ota, Y., Horikawa, K., Oda, Y., 2009. Auditory input to CNS is acquired coincidentally with development of inner ear after formation of functional afferent pathway in zebrafish. Journal of Neuroscience 29 (9), 2762-2767.

Tanimoto, M., Ota, Y., Inoue, M., Oda, Y., 2011. Origin of inner ear hair cells: morphological and functional differentiation from ciliary cells into hair cells in zebrafish inner ear. Journal of Neuroscience 31, 3784-3794.

Trapani, J.G., Obholzer, N., Mo, W., Brockerhoff, S.E., Nicolson, T., 2009. Synaptojanin1 is required for temporal fidelity of synaptic transmission in hair cells. PLoS Genetics 5 (5), e1000480.

Van Trum, W.J., Coombs, S., Duncan, K., McHenry, M.J., 2010. Gentamicin is ototoxic to all hair cells in the fish lateral line system. Hearing Research 261 (1-2), 42-50.

Vanwalleghem, G., Heap, L.A., Scott, E.K., 2017. A profile of auditory-responsive neurons in the larval zebrafish brain. The Journal of Comparative Neurology 525 (14), 3031-3043.

Viader-Llargués, O., Lupperger, V., Pola-Morell, L., Marr, C., López-Schier, H., 2018. Live cell-lineage tracing and machine learning reveal patterns of organ regeneration. eLife 7, e30823.

von, Frisch, K., 1938. The sense of hearing in fish. Nature 141, 8-11.

Wan, L., Almers, W., Chen, W., 2005. Two ribeye genes in teleosts: the role of Ribeye in ribbon formation and bipolar cell development. Journal of Neuroscience 25, 941-949.

Wang, J., Song, Q., Yu, D., Yang, G., Xia, L., Su, K., Yin, S., 2015. Ontogenetic development of the auditory sensory organ in zebrafish (*Danio rerio*): changes in hearing sensitivity and related morphology. Scientific Reports 5, 15943.

Webb, J.F., Shirey, J.E., 2003. Postembryonic development of the cranial lateral line canals and neuromasts in zebrafish. Developmental Dynamics 228 (3), 370-385.

Webb, J.F., Montgomery, J.C., Mogdans, J., 2008. Bioacoustics and the lateral line system of fishes. In: Fish Bioacoustics. Springer, New York, NY, pp. 145-182.

Weigele, J., Franz-Odendaal, T.A., Hilbig, R., 2016. Not all inner ears are the same: otolith matrix proteins in the inner ear of sub-adult cichlid fish, *Oreochromis mossambicus*, reveal insights into the biomineralization process. The Anatomical Record 299 (2), 234-245.

Westerfield, M., 1989. The Zebrafish Book. University of Oregon Press, Eugene Oregon. Whitfield, T., Granato, M., Van Eeden, F.J., Schach, U., Brand, M., Furutani-Seiki, M., Haffter, P., Hammerschmidt, M., Heisenberg, C.P., Jiang, Y.J., Kane, D.A., 1996. Mutations affecting development of the zebrafish inner ear and lateral line. Development 123 (1), 241-254.

Wiedenhoft, H., Hayashi, L., Coffin, A.B., 2017. PI3K and inhibitor of apoptosis proteins modulate gentamicin induced hair cell death in the zebrafish lateral line. Frontiers in Cellular Neuroscience 11, 326.

Xiong, W., Grillet, N., Elledge, H.M., Wagner, T.F., Zhao, B., Johnson, K.R., Kazmierczak, P., Müller, U., 2012. TMHS is an integral component of the mechanotransduction machinery of cochlear hair cells. Cell 151 (6), 1283-1295.

Yang, Q., Sun, P., Chen, S., Li, H., Chen, F., 2017. Behavioral methods for the functional assessment of hair cells in zebrafish. Frontiers of Medicine 11 (2), 178-190.

Yao, Q., DeSmidt, A.A., Tekin, M., Liu, X., Lu, Z., 2016. Hearing assessment in zebrafish during the first week postfertilization. Zebrafish 13 (2), 79-86.

Zeddies, D.G., Fay, R.R., 2005. Development of the acoustically evoked behavioral response in zebrafish to pure tones. Journal of Experimental Biology 208 (7), 1363-1372.

第六章

社会行为的形成发展

S. J. Stednitz[1], P. Washbourne[1]

6.1 物种间的社会发展

社会互动是一种在动物幼崽间快速发展的行为形式。大多数脊椎动物都表现出某种形式的社会行为，即使是那些只为了繁殖而彼此相遇的独居动物。社会互动可能是高度复杂的，其依赖复杂的多模式感觉输入与经验的整合来产生适应性反应（Chen and Hong, 2018）。

在脊椎动物中，控制社会行为的神经元回路在功能上是保守的，它们具有相同的分子特征和进化起源（Chen and Hong, 2018; O'Connell and Hofmann, 2012），还可以改变行为输出以适应独特的环境与物种特异性的选择压力。从根本上讲，社会行为是具有进化优势的，它既可以增加交配机会，也可以通过减少捕食与促进高级合作从而赋予动物无数的生存优势。

社会个体发育或成为社会个体的过程（Mason, 1979），在不同物种之间是高度可变的，并在动物的整个生存阶段受到基因和环境影响的控制。根据特定物种的生活历程，这些因素的相对贡献差异很大，这使得它们从社会性中获得了不同的进化优势（Alexander, 1974）。比如，与斑马鱼相比，未成熟期较长的哺乳动物可能更依赖于从父母 - 子女模式下的社会环境中学习，而斑马鱼在野外或实验室环境中都不太可能遇到自己的父母或后代。然而，即使在一生中没有父母互动的鱼类中，也观察到了依赖经验的可塑性，因此社会学习是通过与同种个体的相互作用发生的（Abril-de-Abreu et al., 2015; Brown and Laland, 2003; Webster and Laland, 2017）。

正如技能是在越来越复杂的行为储备中获得和实施的，社会行为在生长发育的过程中变得越来越复杂（Carpenter et al., 1998）。比如，人类婴儿表现出对人脸的直接偏好（Valenza et al., 1996），但是直到 6 ～ 9 个月大时才会开始关注他人的痛苦（Mundy and Jarrold, 2010; Valenza et al., 1996）。同样地，人类仿效他人行为的能力，即“模仿”，直到 12 个月后才完全建立起来（Woodward and Gerson, 2014）。重要的是，这些发育过程中

[1] Institute of Neuroscience, University of Oregon, Eugene, OR, United States.

的缺陷可以被诊断为神经发育障碍，如影响社会互动的自闭症谱系障碍（autism spectrum disorder, ASD）（Bhat et al., 2014）。ASD 早期形式的社会注意缺陷会损害随后的社会发展（Carpenter et al., 1998; Dawson et al., 2004），尽管这在斑马鱼自闭症的基因模型中是否成立还有待观察（Kozol, 2018）。

正如动物能向它们的亲代和不相关的同类学习如何理解社会线索并恰当地做出反应，社会行为的发展可能是具有经验依赖性的。当神经系统的某些方面成熟时，包括人类在内的动物在早期关键时期的社会剥夺，与生命后期的社会行为受损有关（Committee on Child Maltreatment Research, Policy, and Practice for the Next Decade: Phase Ⅱ, Board on Children, Youth, and Families, Committee on Law and Justice, Institute of Medicine, & National Research Council, 2014; Fone and Veronica Porkess, 2008）。幼年时代被忽视的长期影响包括认知和语言能力的缺陷，面部表情处理的困难，以及行为问题的增加（Spratt et al., 2012）。这些能力的获得依赖于对通过接触而获得的社会线索和对来自同类的反馈的学习。

早期的社会经历塑造了动物对熟悉的社会伙伴的偏好，这种偏好可以持续一生。例如，以无生命物体作为父母替代物的鸟类会发展出亲子关系，甚至在某种程度上，合适的真正的父母会被拒绝，而更喜欢熟悉的“父母”（Bolhuis, 1991），而且在哺乳动物中也观察到类似的对性伴侣的印记（Kendrick et al., 1998）。此外，关键时期的隔离会损害髓鞘形成，并破坏大脑各远处解剖区域有效沟通的能力，这种影响在对被忽视的人类受害者和受控的啮齿动物的研究中都出现过（Hanson et al., 2013; Makinodan et al., 2012）。

斑马鱼等已建立的动物模型为操纵研究发育轨迹提供了极好的平台，有利于更好地理解其在复杂疾病（如 ASD）的病因学中的作用（Kozol, 2018）。斑马鱼是群居动物（Suriyampola et al., 2016），由于在实验室环境中容易诱发社会行为，它们非常适合用于行为学实验。由于基因的遗传可追溯性和外部发育，我们可以通过操纵基因和外部环境条件，观察其对神经系统、生理和行为的影响。

斑马鱼被认为拥有 70% 人类基因的同源序列，而且，重要的是，在人类疾病中受损的基因中，有很大比例（80%）的基因在斑马鱼基因组中有同源序列（Howe et al., 2013）。分子遗传学和功能神经解剖学的最新研究进展揭示了斑马鱼和哺乳动物之间的神经元群体不仅进化保守同时也具有功能同源性（Ganz et al., 2014; Lal et al., 2018; Mueller et al., 2011; Nieuwenhuys, 2010; Stednitz et al., 2018）。此外，环境和基因对斑马鱼早期发育的干扰会损害其晚年的社会行为。在胎儿乙醇综合征模型中，胚胎期间的乙醇暴露足以损害鱼的群聚行为（Buske and Gerlai, 2011），而早期铅暴露可改变成鱼的攻击行为（Weber and Ghorai, 2013）。相似地，在斑马鱼中，与人类自闭症相关的基因发生突变，也会导致社会行为障碍（Liu et al., 2018）。尽管哺乳动物和硬骨鱼的社会环境存在差异，但这些因素增加了它们在模拟具有社会表型的人类神经发育障碍方面的潜在应用价值。

6.2 斑马鱼社会行为的定义

社会行为大致上是由两只或两只以上动物之间的各种互动组成的。斑马鱼也不例外，在文献中对其多种不同的社会行为进行了分类。为了进一步讨论，我们将关注发生在性成熟之前观察到的且不同于繁殖及支配攻击行为的亲密行为。重要的是，这些亲密行为在不同发育阶段具有独特的特性，可以对潜在神经回路的特定组成部分进行详细剖析。

文献中提到了四种相关行为：

① 社会吸引力，向同类靠拢的倾向；

② 群聚，集体游泳；

③ 社交暗示，利用来自同类的信号来引导社会互动；

④ 社会偏好，一种对同族特定外表特征的视觉偏好。

个别社会行为在实验室环境中可以通过使用不同的实验来诱发。在开放的水箱场景中，斑马鱼使用所有感官输入进行互动，这是最自然的实验。然而，强调行为的一个方面的还原性实验应该理解社会刺激的特定过程或显著特征。我们通过比较多个实验模式下的发现来讨论斑马鱼社会行为的个体发生。

6.3 发展阶段的分类

解释斑马鱼社会个体发育的一个主要障碍是：考虑按受精后天数（days postfertilization, dpf）以外的发育阶段来报道年龄的情况比较少，这导致了当斑马鱼因品系差异或饲养条件不同而导致发育速度不同时，实验室之间的可重复性将面临挑战。文献已经详尽描述了斑马鱼从受精到胚胎发生，再到自由游动的仔鱼等这些早期发育阶段的特征（Kimmel et al., 1995），同时也证实了不同年龄段的解剖学特征取决于温度、初始摄食年龄、食物的质量和数量以及品系等（Carvalho et al., 2006; Maack et al., 2003）。毋庸置疑的是，这些影响会持续到后期的发育阶段，并对单个动物产生相当大的影响。利用活鱼的形态特征进行胚胎后期分类是可行的，但很少在行为文献中使用（Parichy et al., 2009; Singleman and Holtzman, 2014）。虽然体形最能预测鱼鳔和鱼鳍的发育特征，也是一个有用的替代指标，但温度和种群密度也对这些特征产生重大影响（Parichy et al., 2009）。此外，由形态定义的给定表型的鳍在大小上存在显著的差异，诸如此类的物理特征可能出现在 7 ～ 15mm 标准长度的斑马鱼中（Singleman and Holtzman, 2014）。

许多行为研究的学者都是按照受精后天数来报道实际年龄，而不是解剖学上定义的发育年龄（Dreosti et al., 2015; Hinz and de Polavieja, 2017; Larsch and Baier, 2018）。在发育过程中营养和其他环境因素的差异可能是不同研究小组所报道的一些变异性以及下文讨论的品系之间差异的原因。鉴于本章的目的，除了我们自己的数据外，还通过总结许多其他来源的数据简化了分类标准（Kimmel et al., 1995; Parichy et al., 2009; Singleman and Holtzman, 2014; Westerfield, 1997），并应用了一套基于年龄、形态特征和标准长度范围的命名系统（表 6-1，图 6-1）。我们通过卵黄囊、脊索弯曲（脊索后部的背侧弯曲）（Parichy et al., 2009）和鳍的形态，以及在外观类似于小型成鱼的仔鱼阶段来描述这些阶段。与其他鱼类相比，斑马鱼从仔鱼（larvae）表型到成鱼表型的变形重塑在很大程度上是一个连续的过程，而其他鱼类在仔鱼和幼鱼之间的过渡阶段可能具有相当大的形态变化（如比目鱼）或生理特

表 6-1 基于年龄、形态特征和近似最小标准长度的发育分期

发育阶段	年龄 /dpf	标准长度 /mm	形态特征
卵黄囊仔鱼（yolk sac larvae）	3 ～ 6	3.7	鳍连续折叠，无鳍条，存在卵黄囊
前弯期仔鱼（preflexion）	7 ～ 13	4.5	鳍线出现，卵黄囊消失，脊索开始弯曲
弯曲期仔鱼（early flexion）	14 ～ 20	6.2	鳍线发展，但尚未分叉
后弯期仔鱼（postflexion）	21 ～ 29	7.8	弯曲完成，双叶鱼鳔
幼鱼期（juvenile）	30 ～ 89	10	几乎完全的成年鳍和色素
成鱼期（adult）	90+	14	长成成鱼

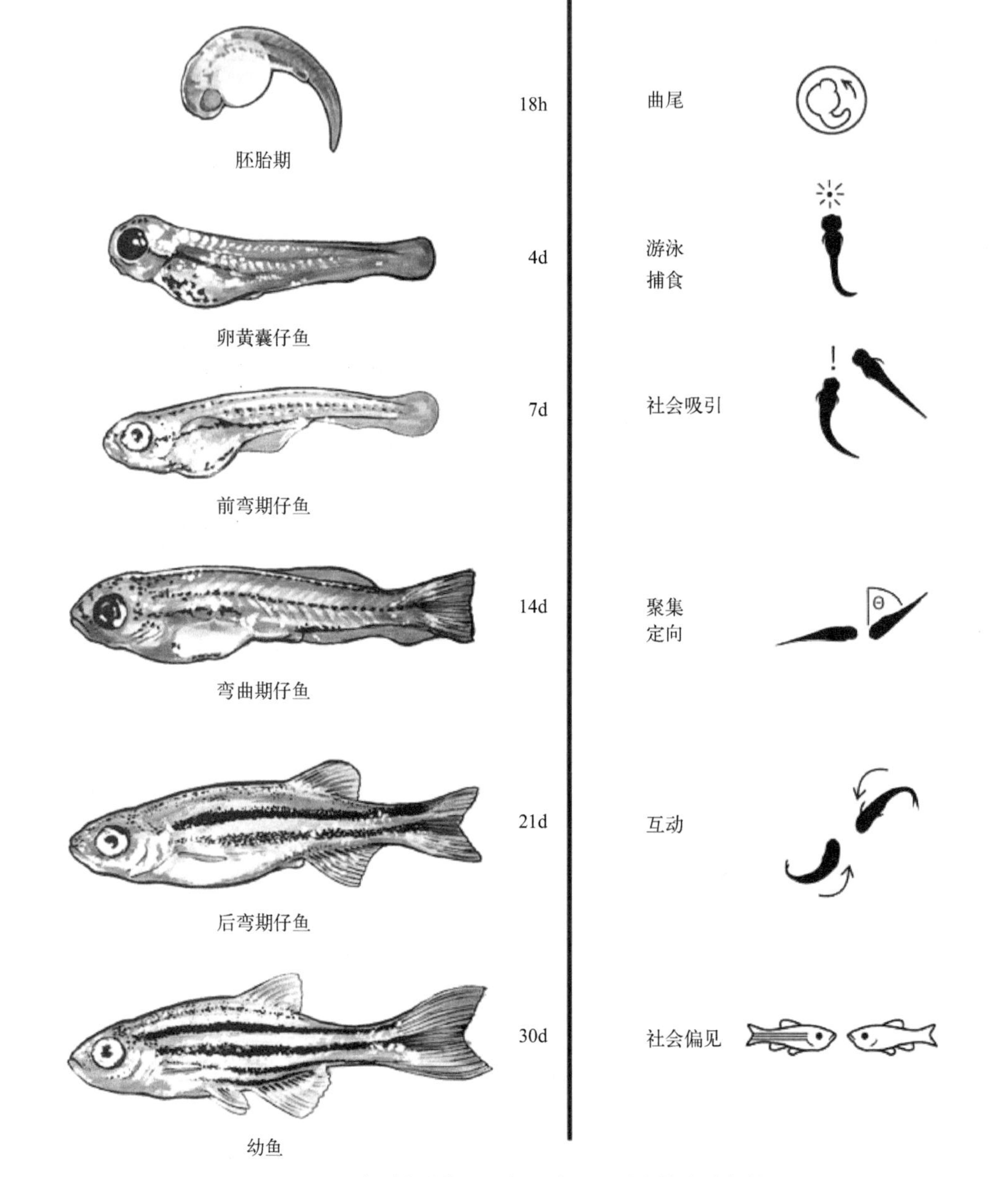

图 6-1 斑马鱼形态分期以及行为出现所对应的大致年龄

征变化（如鲑鱼）（McMenamin and Parichy, 2013; Parichy et al., 2009; Kendall et al., 1984）。

最后，我们认为，由于在发育过程中品系与条件的差异，绝对的按照时间计算的年龄不太可能成为一个高度准确预测社会行为存在与否的指标。我们的描述包括了大致的发育阶段和按时间计算的实际年龄，以帮助理解其社会行为，但我们认为在给定的时间窗口内社会行为可能具有变异性。

6.4 特定社会行为的开始

年龄为 1 周龄之前的斑马鱼仔鱼并不会对同类表现出吸引力，但值得注意的是，这并不是因为缺乏视觉敏感度或运动能力。早在受精后 18h 刻板行为开始出现，仔鱼期开始表现出自发的尾巴卷曲（Brustein et al., 2003; Saint-Amant and Drapeau, 1998），早在 3dph 时就能引起惊吓反应（Burgess and Granato, 2007）。到 4dph，鱼鳔可以完全充气，并观察到协调游泳，还包括避开环境中的潜在威胁（Lindsey et al., 2010），仔鱼开始出现对视觉要求很高的捕食动作（Fero et al., 2010; Marques et al., 2018）。

有证据表明，社会吸引行为早在 7dph 时就出现了，此时卵黄囊已不再存在，仔鱼已进入独立的取食阶段（Hinz and de Polavieja, 2017）。由于 7 ～ 13dph 时脊索弯曲过程已经开始，我们将此阶段称为前弯期仔鱼（preflexion）（Parichy et al., 2009）。在此之前，斑马鱼仔鱼可能会通过最大限度地增加它们之间的距离来避开同类（Creton, 2009）。仔鱼在很短的时间内实现由中性环境刺激物向同种生物作为引诱物的转换，其转换的潜在基础是我们对仔鱼神经回路理解中的一个有趣而悬而未决的问题。

相对于斑马鱼的总寿命来说，社会行为的开始是快速的，因此，幼鱼（30dph）与最大寿命为 66 个月的性成熟成年斑马鱼的社会亲密行为并无不同（Gerhard et al., 2002）。社会性的出现是与其他复杂行为如学习操作性任务的能力相关联的（Valente et al., 2012）。这种时间上的巧合表明，在生命的第一个月里，神经回路的功能迅速增加，使得这一时期成为特别适合探索更高阶的神经发育过程的阶段。

6.5 社会吸引力

社会吸引力或在游泳时优先转向同类的倾向是斑马鱼所有社会行为的基本组成部分。前弯期仔鱼在发生早弯曲期（7dpf）时开始优先转向对方，直到后弯曲期（21dpf）这些相互作用的频率和强度都在增加（Hinz and de Polavija, 2017）。即使被透明隔板隔开，7dpf 的仔鱼也会对其他斑马鱼群表现为弱吸引力（Dreosti et al., 2015）。在另一个使用分隔器的实验中，7dpf 的仔鱼并不会对单个的同类个体产生吸引力，这种现象表明，在 7dpf 这个阶段，群体可能会比个体发挥更强的吸引力。这些在物理分隔条件下的发现表明，社会吸引力的最初发展是由视觉介导的。与这个假设一致，我们组和其他组的研究表明，视觉刺激可以引发许多社会行为，包括吸引、群聚和社交提示（Dreosti et al., 2015; Gerlai, 2017; Larsch and Baier, 2018; Stednitz et al., 2018）。

随着时间的推移，前弯期仔鱼个体成熟，其社会吸引力发展成群聚，并导致鱼体间距离的接近。野生型前弯期仔鱼早在9dpf时就开始聚集，这种吸引力增加，以至于在12dpf时个体之间随机分布的平均距离显著减小（Hinz and de Polavija, 2017）。在此期间观察到一些品系差异，11dpf的自交系AB品系斑马鱼与上述非亲缘交配的野生品系相比，鱼间距离显著缩小（Hinz and de Polavija, 2017）。前面所讨论过的较小的发育阶段差异也可以解释这种差异，尽管它们也可能是品系间发育的真实差异造成的。

因此，社会吸引力在前弯期开始，并在第二至第三周增加。到14dpf，斑马鱼成群游动，社会吸引力增加，直到后弯期（21dpf）导致群体凝聚力增加。在斑马鱼仔鱼中，控制社会吸引行为的局部互动会一直持续到成鱼期，因此它们在早期就建立起来，并可能会指导以后更复杂的社会行为（Hinz and de Polavija, 2017），与人类的认知发展模式类似，这种模式依赖于技能的逐渐习得（Fischer, 1980）。

6.6 群聚与社会偏好性

群聚现象可以在野生斑马鱼中观察到，但也可以在简化的环境下引发，例如当鱼被一个透明的分隔物隔开时。在这些实验中，斑马鱼更有可能占用物理空间，在这些空间中，可以看到处于弯曲期（14pdf）的单个或成群的同种个体（Dreosti et al., 2015），这种空间偏好伴随着头部以大约45°的角度朝向同伴（刺激鱼）。到后弯期（21dpf），定向行为变得非常强烈，并会在动物的一生中持续存在（Stednitz et al., 2018）。

值得注意的是，这些发育阶段也存在个体差异，例如，一些斑马鱼表现出上述行为，倾向于避开同类，这种厌恶现象与吸引行为一起出现（Dreosti et al., 2015）。决定选择接近和回避行为的个体因素尚不清楚。我们的研究表明，成鱼在社交方式和取向方面也存在类似的个体差异，这可能与稳定的个体差异和短暂的环境线索（如营养状态或捕食者的存在）有关。众所周知，诸如速度等持续的个体差异会影响到刺鱼鱼群的凝聚力（Jolles et al., 2018）。影响斑马鱼回避性社会反应的因素仍然是一个有待研究的问题。

虚拟刺激能够驱动弯曲期（14dpf）斑马鱼的吸引行为（Larsch and Baier, 2018; Stowers et al., 2017）。无论刺激物的轨迹如何，当刺激物以模仿生物运动的非连续运动方式移动时，驱动吸引行为的效果是最好的。这也支持先前报道的前弯期时就出现吸引行为，直至27dpf的后弯期，这种吸引行为变得越来越强大（Dreosti et al., 2015）。

社交刺激的大小，无论是真实的同类还是虚拟的，对于推动社交互动是很重要的。卵黄囊仔鱼（7dpf）对出现的较大后弯期（21dpf）仔鱼的刺激，表现出轻微的厌恶反应（Dreosti et al., 2015）。相反，当后弯期仔鱼面临卵黄囊仔鱼刺激时，尽管其群聚行为减少了，但仍可观察到。在虚拟刺激中也会呈现回避反应，较动物大很多的点状物可引发回避反应而非接近反应（Larsch and Baier, 2018）。据推测，这种现象可以解释为：较小的鱼不会被对方视为潜在的威胁，而较大的鱼可能会被对方视为潜在的威胁。虚拟刺激的最佳直径随着年龄的增长而增加，这表明相对大小是幼小斑马鱼用来确定合适的社会伙伴的因素之一（Larsch and Baier, 2018）。这种偏好可能提供了一种斑马鱼会与年龄及发育程度相匹配的鱼群聚集在一起的机制。

6.7 社会暗示

斑马鱼利用来自同族的社会线索来指导其从属的社会行为，但是，关于这些线索在其哪个发育阶段变得更加突出方面的研究并不充分。在求偶期，斑马鱼会主动注意伴侣的身体动作和姿势（Darrow and Harris, 2004）。同样地，当隔离的鱼被允许观察成对的斑马鱼时，它们更有可能注意到有互动的成对斑马鱼而非无互动的成对斑马鱼（Abril-de-Abreu et al., 2015）。社会线索驱动社会从属行为的差异程度因实验模式的不同而不同。

在实验中观察斑马鱼的定向行为，斑马鱼被一个透明的分隔器分开，由视觉社会线索引导。在弯曲期（14dpf）时仔鱼就有关于同步定向的报道，我们发现成对幼鱼（30dpf）在 1s 内就能与其伴侣的角度相匹配（Dreosti et al., 2015; Stednitz et al., 2018）。在我们的二元分析实验中，测试鱼的社会定向的最大预测因素是刺激鱼的定向行为。当药物治疗或特定前脑区域的损伤导致正常的社会定向能力受损时，刺激鱼不再影响测试鱼（Stednitz et al., 2018）。然而，这些分析中较小的空间参数可能会引发其在自然环境中不经常发生的行为，这些互动的性质需要在开放环境中进一步研究。

尽管有上述说明，我们观察到社会定向行为开始于前弯期（早在 12dpf），这种行为在短时间内也得到提高，以至于到 16dpf 时与幼鱼期（1 个月）的定向行为并没有区别。同样，时滞互相关分析（time lag cross-correlation analysis）显示，在 12 ～ 16dpf 之间，反向转向事件迅速变得更加稳健，这反映了刺激鱼对测试鱼行为的影响增加（Stednitz and Washbourne, in press）。这种对社会定向行为的快速获得表明，线索依赖的定向行为在首次建立起来时，也经历了弯曲期和后弯期。

成对个体模式化定向行为的重要性很难被引入到自然环境中，因为在自然环境中斑马鱼并没有被限制在独立的容器中。考虑到同族个体的位置和行为会引导个体的社会行为（Hinz and de polavija, 2017），我们很容易推断斑马鱼在群聚期间可能会表现出一个相对于同族个体的偏好方向，以同时用两只眼睛巩固多个社会刺激。这可能是平行游泳的机制，平行游泳是在具有不同程度极化和内聚度的开放水箱中的主要行为。事实上，平行游泳受到环境习惯的强烈影响（Miller and Gerlai, 2012）。与上面提到的成对个体分析相类似，单一药物处理的斑马鱼的社会行为受损，足以破坏鱼群的凝聚性（Maaswinkel et al., 2013）。如果在新环境中存在更具有探索性的动物，斑马鱼也会动态地增加它们自身的探索行为（Guayasamin et al., 2017）。这些复杂的开放式水池群聚行为出现的发展阶段值得进一步探索研究。

虚拟刺激实验表明，虽然生物运动是推动社会互动的一个线索，但可能不需要互惠性来激发跟随（Larsch and Baier, 2018）。类似地，对于受测试鱼来说，照片级真实感虚拟社交伙伴的行为比一个被动刺激（其轨迹是预先确定的或完全由测试鱼决定的）更能有效地引发跟随（Stowers et al., 2017）。这种差异可能是由于虚拟刺激保持了与测试鱼足够近的距离而引发关注。相反，在二元分析实验中，当刺激鱼没有互动时，仅靠接近并不足以驱动测试鱼的空间偏好或定向，这表明不同的社会行为依赖于不同的线索。虚拟社会刺激的进一步发展可以解释所有简化的视觉特征，这些视觉特征等价于同类刺激鱼，可以驱动社会吸引力、群聚和互助，这是一个值得进一步研究的激动人心的课题。

6.8 早期经验

斑马鱼表现出视觉介导的社会偏好，这是由早期生活经历形成的，在这里被称为社会偏见（social bias）。在野生型成年斑马鱼中，条纹的粗度是导致社会偏见的一个有吸引力的线索，而与缺乏条纹的斑马鱼突变体相比，相关的条纹鲋属物种对野生型成年雄鱼来说是更有效的社会刺激（Engeszer et al., 2008）。使用模拟斑马鱼作为刺激物也观察到了类似的效果（Saverino and Gerlai, 2008）。用缺乏条纹的色素突变体斑马鱼做实验，可以颠覆人们认为的斑马鱼更喜好条纹鱼刺激的偏见，比如 *nacre* 突变体（Engeszer et al., 2007）。无论测试鱼自身的外观或遗传背景如何，都趋向于首选色素突变体。这种现象是由相似的同种鱼的出现所驱动的，而不是鱼类与生俱来的对条纹或类似基因型斑马鱼的偏好。有趣的是，社会偏见直到幼鱼期（30dpf）才出现，这表明经验依赖型偏见直到社会吸引、群聚和社会线索出现后才出现。这种效应在以后的发育中是稳定的，不具有可塑性，这表明存在一个关键时期，在这个时期可以建立一个理想的具有吸引力的同族外观（Engeszer et al., 2007）。

斑马鱼也容易受到早期隔离的不利影响。在 180 天之前不进行社会互动会扰乱成鱼期正常的社会偏好，从而强调了在生命早期加强社会反馈的重要性（Shams et al., 2018; Shams et al., 2015），而短期的社会隔离（7dpf）不足以驱动类似的效果。同样，将仔鱼隔离直至后弯期（21dpf）对隔离期后立即出现的虚拟点状的群聚伙伴的吸引力没有可检测到的影响（Larsch and Baier, 2018）。然而，长期隔离的动物显示出鱼群凝聚力受损，大脑中多巴胺和血清素的浓度降低，这是一种行为强化神经递质，已知这些物质参与斑马鱼和其他脊椎动物的社会互动（Gunaydin and Deisseroth, 2014; Kiser, 2012）。这些基于隔离的行为变化的神经机制值得进一步研究，以了解它们与哺乳动物隔离模型的关系。

有趣的是，在观察到基本的社会互动之前，较年长的仔鱼建立正常社会行为所需回路的关键时期可能就已经出现了。用丙戊酸（valproic acid, VPA）处理 4 ～ 5dpf 仔鱼足以降低后弯期仔鱼（21dpf）的社会偏好和相互作用（Dwivedi et al., 2019），其他研究显示，0 ～ 48hpf 仔鱼在接触 VPA 后，其长成成鱼（70dpf）后的社会偏好略有下降（Zimmermann et al., 2015）。然而，这种操纵的行为结果并不局限于社交缺陷，需要进一步的研究来缩小驱动社交行为的特定回路受到影响的发育时间窗。

6.9 结论

尽管上述研究在方法上存在差异，但其在关键结论上达成了一致。7dpf 龄前的卵黄囊期仔鱼基本不群居，对同族无偏好。在 7 ～ 13dpf 之间，早弯期仔鱼开始表现出社会吸引和群聚。在弯曲期（14dpf）时，仔鱼开始表现出群体游泳、定向和有限的社会线索。经过后弯曲期（21dpf），持续到成鱼期的强健的社会行为似乎完全建立起来，包括反向定向转身和协调的团体游泳。幼鱼期（30dpf）斑马鱼能够对特定类型的社会伙伴建立社会喜好，这种喜好一直持续到成鱼期。不同研究之间具有显著差异的现象提示我们，为待研究的社会行为选择适当的分析方法非常重要。同样，形态学定义的发育阶段可能比按照时间顺序排列的实际年龄更能预测社会行为的发展阶段，因此在实验设计中应加以考虑。

总的来说，研究表明，斑马鱼在社会互动方面表现出与其他动物一样的等级发展，例如，在现有的感觉运动和行为能力框架内，随着个体的发育，社会行为变得越来越复杂。

致谢

我们要感谢 Charles Kimmel、Judith Eisen 和 Johannes Larsch 的广泛反馈和在写作方面的帮助。

（李海芸翻译　李丽琴审校）

参考文献

Abril-de-Abreu, R., Cruz, J., Oliveira, R.F., 2015. Social eavesdropping in zebrafish: tuning of attention to social interactions. Scientific Reports 5, 12678.

Alexander, R.D., 1974. The evolution of social behavior. Annual Review of Ecology and Systematics 5 (1), 325-383.

Bhat, S., Rajendra Acharya, U., Adeli, H., Muralidhar Bairy, G., Adeli, A., 2014. Autism: cause factors, early diagnosis and therapies. Reviews in the Neurosciences 25 (6). https://doi.org/10.1515/revneuro-2014-0056.

Bolhuis, J.J., 1991. Mechanisms of avian imprinting: a review. Biological Reviews of the Cambridge Philosophical Society 66 (4), 303-345.

Brown, C., Laland, K.N., 2003. Social learning in fishes: a review. Fish and Fisheries 4 (3), 280-288.

Brustein, E., Saint-Amant, L., Buss, R.R., Chong, M., McDearmid, J.R., Drapeau, P., 2003. Steps during the development of the zebrafish locomotor network. Journal of Physiology Paris 97 (1), 77-86.

Burgess, H.A., Granato, M., 2007. Sensorimotor gating in larval zebrafish. Journal of Neuroscience: The Official Journal of the Society for Neuroscience 27 (18), 4984-4994.

Buske, C., Gerlai, R., 2011. Early embryonic ethanol exposure impairs shoaling and the dopaminergic and serotoninergic systems in adult zebrafish. Neurotoxicology and Teratology 33 (6), 698-707.

Carpenter, M., Nagell, K., Tomasello, M., 1998. Social cognition, joint attention, and communicative competence from 9 to 15 months of age. Monographs of the Society for Research in Child Development 63 (4), 1-143 i -vi.

Carvalho, A.P., Araujo, L., Santos, M.M., 2006. Rearing zebrafish (*Danio rerio*) larvae without live food: evaluation of a commercial, a practical and a purified starter diet on larval performance. Aquaculture Research 37 (11), 1107-1111.

Chen, P., Hong, W., 2018. Neural circuit mechanisms of social behavior. Neuron 98 (1), 16-30.

Committee on Child Maltreatment Research, Policy, and Practice for the Next Decade: Phase Ⅱ, Board on children, youth, and families, Committee on Law and Justice, Institute of Medicine, National Research Council, 2014. In: Petersen, A.C., Joseph, J., Feit, M. (Eds.), New Directions in Child Abuse and Neglect Research. National Academies Press (US), Washington (DC).

Creton, R., 2009. Automated analysis of behavior in zebrafish larvae. Behavioural Brain Research 203 (1), 127-136.

Darrow, K.O., Harris, W.A., 2004. Characterization and development of courtship in zebrafish, *Danio rerio*. Zebrafish 1 (1), 40-45.

Dawson, G., Toth, K., Abbott, R., Osterling, J., Munson, J., Estes, A., Liaw, J., 2004. Early social attention impairments in autism: social orienting, joint attention, and attention to distress. Developmental Psychology 40 (2), 271-283.

Dreosti, E., Lopes, G., Kampff, A.R., Wilson, S.W., 2015. Development of social behavior in young zebrafish. Frontiers in Neural Circuits 9, 39.

Dwivedi, S., Medishetti, R., Rani, R., Sevilimedu, A., Kulkarni, P., Yogeeswari, P., 2019. Larval zebrafish model for studying the effects of valproic acid on neurodevelopment: an approach towards modeling autism. Journal of Pharmacological and Toxicological Methods 95, 56-65.

Engeszer, R.E., Barbiano, L.A.D.A., Ryan, M.J., Parichy, D.M., 2007. Timing and plasticity of shoaling behaviour in the zebrafish, *Danio rerio*. Animal Behaviour 74 (5), 1269-1275.

Engeszer, R.E., Wang, G., Ryan, M.J., Parichy, D.M., 2008. Sex-specific perceptual spaces for a vertebrate basal social aggregative behavior. Proceedings of the National Academy of Sciences of the United States of America 105 (3), 929-933.

Fero, K., Yokogawa, T., Burgess, H.A., 2010. The behavioral repertoire of larval zebrafish. In: Neuromethods, pp. 249-291.

Fischer, K.W., 1980. A theory of cognitive development: the control and construction of hierarchies of skills. Psychological Review 87 (6), 477-531.

Fone, K.C.F., Veronica Porkess, M., 2008. Behavioural and neurochemical effects of post-weaning social isolation in rodentsdrelevance to developmental neuropsychiatric disorders. Neuroscience and Biobehavioral Reviews 32 (6), 1087-1102.

Ganz, J., Kroehne, V., Freudenreich, D., Machate, A., Geffarth, M., Braasch, I., et al., 2014. Subdivisions of the adult zebrafish pallium based on molecular marker analysis. F1000Research 3, 308.

Gerhard, G.S., Kauffman, E.J., Wang, X., Stewart, R., Moore, J.L., Kasales, C.J., et al., 2002. Life spans and senescent phenotypes in two strains of Zebrafish (*Danio rerio*). Experimental Gerontology 37 (8-9), 1055-1068.

Gerlai, R., 2017. Animated images in the analysis of zebrafish behavior. Current Zoology 63 (1), 35e44.

Guayasamin, O.L., Couzin, I.D., Miller, N.Y., 2017. Behavioural plasticity across social contexts is regulated by the directionality of inter-individual differences. Behavioural Processes 141 (Pt 2), 196-204.

Gunaydin, L.A., Deisseroth, K., 2014. Dopaminergic dynamics contributing to social behavior. Cold Spring Harbor Symposia on Quantitative Biology 79.

Hanson, J.L., Adluru, N., Chung, M.K., Alexander, A.L., Davidson, R.J., Pollak, S.D., 2013. Early neglect is associated with alterations in white matter integrity and cognitive functioning. Child Development 84 (5), 1566-1578.

Hinz, R.C., de Polavieja, G.G., 2017. Ontogeny of collective behavior reveals a simple attraction rule. Proceedings of the National Academy of Sciences of the United States of America 114 (9), 2295-2300.

Howe, K., Clark, M.D., Torroja, C.F., Torrance, J., Berthelot, C., Muffato, M., et al., 2013. The zebrafish reference genome sequence and its relationship to the human genome. Nature 496 (7446), 498-503.

Jolles, J.W., Laskowski, K.L., Boogert, N.J., Manica, A., 2018. Repeatable group differences in the collective behaviour of stickleback shoals across ecological contexts. Proceedings of the Royal Society B: Biological Sciences (1872), 285. https://doi.org/10.1098/rspb.2017.2629.

Kendall, A.W., Ahlstrom, E.H., Moser, H.G., 1984. Early Life History Stages of Fishes and Their Characters. Ontogeny and Systematics of Fishes. American Society of Ichthyologists and Herpetologists, Lawrence, Kansas. Kendrick, K.M., Hinton, M.R., Atkins, K., Haupt, M.A., Skinner, J.D., 1998. Mothers determine sexual preferences. Nature 395 (6699), 229-230.

Kimmel, C.B., Ballard, W.W., Kimmel, S.R., Ullmann, B., Schilling, T.F., 1995. Stages of embryonic development of the zebrafish. Developmental Dynamics: An Official Publication of the American Association of Anatomists 203 (3), 253-310.

Kiser, D., et al, 2012. The reciprocal interaction between serotonin and social behaviour. Neuroscience and Biobehavioral Reviews 36 (2).

Kozol, R.A., 2018. Prenatal neuropathologies in autism spectrum disorder and intellectual disability: the gestation of a comprehensive zebrafish model. Journal of Developmental Biology 6 (4). https://doi.org/10.3390/jdb6040029.

Lal, P., Tanabe, H., Suster, M.L., Ailani, D., Kotani, Y., Muto, A., et al., 2018. Identification of a neuronal population in the telencephalon essential for fear conditioning in zebrafish. BMC Biology 16 (1), 45.

Larsch, J., Baier, H., 2018. Biological motion as an innate perceptual mechanism driving social affiliation. Current Biology 28 (22), 3523-3532.e4.

Lindsey, B.W., Smith, F.M., Croll, R.P., 2010. From inflation to flotation: contribution of the swimbladder to wholebody density and swimming depth during development of the zebrafish (*Danio rerio*). Zebrafish 7 (1), 85-96.

Liu, C.-X., Li, C.-Y., Hu, C.-C., Wang, Y., Lin, J., Jiang, Y.-H., et al., 2018. CRISPR/Cas9-induced shank3b mutant zebrafish display autism-like behaviors. Molecular Autism 9. https://doi.org/10.1186/s13229-018-0204-x.

Maack, G., Segner, H., Tyler, C.R., 2003. Ontogeny of sexual differentiation in different strains of zebrafish (*Danio rerio*). Fish Physiology and Biochemistry 28 (1-4), 125-128.

Maaswinkel, H., Zhu, L., Weng, W., 2013. Assessing social engagement in heterogeneous groups of zebrafish: a new paradigm for autism-like behavioral responses. PLoS One 8 (10), e75955.

Makinodan, M., Rosen, K.M., Ito, S., Corfas, G., 2012. A critical period for social experience-dependent oligodendrocyte maturation and myelination. Science 337 (6100), 1357-1360.

Marques, J.C., Lackner, S., Félix, R., Orger, M.B., 2018. Structure of the zebrafish locomotor repertoire revealed with unsupervised behavioral clustering. Current Biology 28 (2), 181-195.e5.

Mason, W., 1979. Ontogeny of social behavior. In: Social Behavior and Communication, pp. 1-28.

McMenamin, S.K., Parichy, D.M., 2013. Metamorphosis in teleosts. In: Current Topics in Developmental Biology, pp. 127-165.

Miller, N., Gerlai, R., 2012. From schooling to shoaling: patterns of collective motion in zebrafish (*Danio rerio*). PLoS One 7, e48865. https://doi.org/10.1371/journal.pone.0048865.

Mueller, T., Dong, Z., Berberoglu, M.A., Guo, S., 2011. The dorsal pallium in zebrafish, *Danio rerio* (Cyprinidae, Teleostei). Brain Research 1381, 95-105.

Mundy, P., Jarrold, W., 2010. Infant joint attention, neural networks and social cognition. Neural Networks: The Official Journal of the International Neural Network Society 23 (8-9), 985-997.

Nieuwenhuys, R., 2010. The development and general morphology of the telencephalon of actinopterygian fishes: synopsis, documentation and commentary. Brain Structure and Function 215 (3-4), 141-157.

O'Connell, L.A., Hofmann, H.A., 2012. Evolution of a vertebrate social decision-making network. Science 336 (6085), 1154-1157.

Parichy, D.M., Elizondo, M.R., Mills, M.G., Gordon, T.N., Engeszer, R.E., 2009. Normal table of postembryonic zebrafish development: staging by externally visible anatomy of the living fish. Developmental Dynamics: An Official Publication of the American Association of Anatomists 238 (12), 2975-3015.

Saint-Amant, L., Drapeau, P., 1998. Time course of the development of motor behaviors in the zebrafish embryo. Journal of Neurobiology 37 (4), 622-632.

Saverino, C., Gerlai, R., 2008. The social zebrafish: behavioral responses to conspecific, heterospecific, and computer animated fish. Behavioural Brain Research 191 (1), 77-87.

Shams, S., Amlani, S., Buske, C., Chatterjee, D., Gerlai, R., 2018. Developmental social isolation affects adult behavior, social interaction, and dopamine metabolite levels in zebrafish. Developmental Psychobiology 60 (1), 43-56.

Shams, S., Chatterjee, D., Gerlai, R., 2015. Chronic social isolation affects thigmotaxis and whole-brain serotonin levels in adult zebrafish. Behavioural Brain Research 292, 283-287.

Singleman, C., Holtzman, N.G., 2014. Growth and maturation in the zebrafish, *Danio rerio*: a staging tool for teaching and research. Zebrafish 11 (4), 396-406.

Spratt, E.G., Friedenberg, S.L., Swenson, C.C., Larosa, A., De Bellis, M.D., Macias, M.M., et al., 2012. The effects of early neglect on cognitive, language, and behavioral functioning in childhood. Psychology 3 (2), 175-182.

Stednitz, S.J., McDermott, E.M., Ncube, D., Tallafuss, A., Eisen, J.S., Washbourne, P., 2018. Forebrain control of behaviorally driven social orienting in zebrafish. Current Biology 28 (15), 2445-2451.e3.

Stowers, J.R., Hofbauer, M., Bastien, R., Griessner, J., Higgins, P., Farooqui, S., et al., 2017. Virtual reality for freely moving animals. Nature Methods 14 (10), 995-1002.

Suriyampola, P.S., Shelton, D.S., Shukla, R., Roy, T., Bhat, A., Martins, E.P., 2016. Zebrafish social behavior in the wild. Zebrafish 13 (1), 1-8.

Valente, A., Huang, K.-H., Portugues, R., Engert, F., 2012. Ontogeny of classical and operant learning behaviors in zebrafish. Learning and Memory 19 (4), 170-177.

Valenza, E., Simion, F., Cassia, V.M., Umiltà, C., 1996. Face preference at birth. Journal of Experimental Psychology: Human Perception and Performance 22 (4), 892-903.

Weber, D.N., Ghorai, J.K., 2013. Experimental design affects social behavior outcomes in adult zebrafish developmentally exposed to lead. Zebrafish 10 (3), 294-302.

Webster, M.M., Laland, K.N., 2017. Social information use and social learning in non-grouping fishes. Behavioral Ecology: Official Journal of the International Society for Behavioral Ecology 28 (6), 1547-1552.

Westerfield, M., 1997. The Zebrafish Book: A Guide for the Laboratory Use of Zebrafish (Danio Rerio*).

Woodward, A.L., Gerson, S.A., 2014. Mirroring and the development of action understanding. Philosophical Transactions of the Royal Society B: Biological Sciences 369 (1644), 20130181.

Zimmermann, F.F., Gaspary, K.V., Leite, C.E., De Paula Cognato, G., Bonan, C.D., 2015. Embryological exposure to valproic acid induces social interaction deficits in zebrafish (*Danio rerio*): a developmental behavior analysis. Neurotoxicology and Teratology 52 (Pt A), 36-41.

第七章

斑马鱼仔鱼的经典和操作性条件反射

David Pritchett[1], Caroline H. Brennan[1]

7.1 引言

联想学习是动物对两种刺激之间的关联或者对某种特定动作与某种刺激之间的关联进行学习的过程。在习惯化（habituation）和敏感化（sensitization）这两种非关联式的学习形式中，动物应对刺激的反应强度会随着反复刺激而改变，而联想学习则与它们有所不同。自19世纪末以来，人们就开始研究联想学习、经典和操作性条件反射，研究对象包括狗、猫、大鼠、小鼠、兔子、鸟、昆虫、鱼、人类以及非人灵长类动物（见第28章和第29章）。

那么，为什么斑马鱼仔鱼（*Danio rerio*）是一种对研究联想学习具有重大意义的模式生物呢？是因为斑马鱼仔鱼的小巧大脑和透明头部使其非常适合于全脑单细胞分辨率钙成像，研究人员在采用斑马鱼进行特定行为测试时，能对斑马鱼整个大脑神经元的活动进行可视化观察（Ahrens et al., 2012; Feierstein et al., 2015; Orger and Portugues, 2016）。因此，研究人员可识别与联想学习回路相关的组成部分（不论其分布有多广），观察学习活动时回路中的神经元是如何变化的，并最终通过光遗传操作来探测它们的功能。

虽然单细胞分辨率全脑成像同样适用于秀丽线虫和普通果蝇等其他模式生物，但斑马鱼作为一种脊椎动物物种，与人类基因的同源性达到约70%，因此具有更大的研究潜力（Howe et al., 2013）。但是，作为模式生物的斑马鱼也有其局限性（Orger and Polavieja, 2017）。随着斑马鱼生长发育，头部变得又大又厚，对其大脑更深部位的观测受到限制；此外在实验中，体形越大的斑马鱼也越难操控；最后，许多用于成像的转基因品系在受精后6～7天（6～7dpf）其基因编码的GCaMP指示剂的表达减少，因为其表达受到早期神经发育期间最活跃的泛

[1] School of Biological and Chemical Sciences, Queen Mary University of London, London, United Kingdom.

神经元启动子 *elavl3* 的控制（Kim et al., 2014; Bergmann et al., 2018）。

正是基于以上原因，全脑单细胞分辨率钙成像研究几乎不适用于受精后超过 1 周的斑马鱼。因此，确定斑马鱼仔鱼是否能够在这个年龄范围内进行强有力的联想学习至关重要。本章综述了目前的研究现状，描述了仔鱼的经典和操作性条件反射的个体发育过程，并阐述了这些行为背后有关的神经生物学机制。

7.2 经典条件反射

在典型的经典条件反射案例中，当不引起动物内在反射的刺激（称为条件刺激，the conditioned stimulus，CS）早于第二个更有力的刺激［称为非条件刺激（the unconditioned stimulus，US），该刺激可引发非条件反射（the unconditioned response，UR）］出现，且两种刺激共同终止的情况下，CS 可触发原本由 US 引发的非条件反射。实验中，通过条件刺激和非条件刺激的重复配对出现，动物学会了两种刺激之间的关联，因此条件反射会触发非条件反射。若只反复单独暴露于条件刺激而不进行非条件刺激，那么条件刺激将不再引起条件反射，这种现象称为反射消失。

7.2.1 受约束斑马鱼仔鱼的经典条件反射

2011 年，Aizenberg 和 Schuman 首次证明了斑马鱼仔鱼的经典条件反射（Aizenberg and Schuman, 2011）。在这项开创性的研究中，将 6 ～ 8dpf 的斑马鱼暂时性麻醉后埋入低熔点琼脂糖中，小心地切开凝胶使其尾部自由活动。然后将斑马鱼放置在一个半圆形的发光二极管阵列中心，依次点亮这些发光二极管会产生一个移动白点，作为条件刺激（CS）；用压电马达驱动的不锈钢钢丝接触身体侧面作为非条件刺激（US）；由高速摄像机监控条件刺激诱发的尾部运动即条件反射（CR）。

在实验方案的学习阶段，一个对照组设置单一重复的条件刺激；另一个对照组条件刺激与非条件刺激以不固定方式配对出现；而实验组设置 10 组条件刺激与非条件刺激配对出现。结果显示实验组仔鱼条件刺激诱发的尾部运动次数逐渐增加，并在第 7 组时显著增加，而两个对照组均没有增加现象。因此，证明了条件反射的结果是由联想学习导致的而不是由条件刺激敏感化导致的。为了测试新形成的记忆能够持续保留的时间，在仔鱼学习阶段结束 30min 后进行了只出现条件刺激的三次实验。在第一次试验中条件反射现象很明显，但随后迅速消失。进而，不同保留时间间隔的研究表明，新形成的记忆能够持续的时间不超过 1h。上述实验结果清楚地表明，斑马鱼从受精后 1 周开始具有条件反射反应。但是，实验过程存在极高的损耗率，因为 40% 的仔鱼在实验前由于表现出低水平的自发活动或对条件刺激的强烈反应而被剔除。

近年来，研究人员又开发了两种经典的恐惧条件反射方法（Matsuda et al., 2017; Harmon et al., 2017）。第一种方案中，Matsuda 及其同事采用 LED 光源的漫射白光照亮试验场地，停止光照让整个场地黑暗持续 5s，作为条件刺激。通过靠近仔鱼尾巴的电极进行 1ms 的电击，作为非条件刺激。条件刺激诱发的心动过缓即为条件反射，心动过缓是一种由自主神经系统控制的恐惧反应，可由红外（IR）相机检测到。

整个实验分为适应阶段、学习阶段和测试阶段，分别由10～15、20和10个试验组成。在适应和测试阶段，单独呈现“突然黑暗”的条件刺激。大多数鱼类在第一次经历突然黑暗时会表现出心动过缓，但适应了一到两次后，这种反应逐渐消失。在学习阶段，心动过缓最初仅由电击才能触发，但在10～15次实验后，心动过缓由突然黑暗后就触发，这种反应一直持续到测试阶段结束。因此，相对于适应阶段，在测试阶段表现出应对突然黑暗时，心动过缓反应更为强烈的仔鱼，称为“学习者”。该实验过程中仔鱼总体表现不佳，但结果随着仔鱼年龄的增长而改善：5～9dpf的仔鱼中有0%的“学习者”；10～16dpf的仔鱼中有25%的“学习者”；17～25 dpf的仔鱼中有37.5%的“学习者”。

另外，在电击呈现在突然黑暗之前的对照组中，研究人员发现在测试阶段仔鱼处于黑暗时没有发生心动过缓。因此推断，突然黑暗诱发的心动过缓反应是因为仔鱼的联想记忆，而不是因为突然黑暗的敏感化。尽管这个结论比较准确，但因为条件刺激和非条件刺激之间仍然存在可预测的关系，一般认为倒置条件（backward conditioning）被视为标准经典调节的不适当刺激条件，可能导致动物对条件刺激的抑制性反应（Rescorla, 1967）。

Harmon及其同事采用另一种相似的经典恐惧条件反射方案。使用2s的蓝色LED光作为条件刺激；直接作用于尾巴尖端的5 ms电击作为非条件刺激（Harmon et al., 2017）。用针固定6～8dpf仔鱼的头部和尾部，通过腹根的电生理记录测定蓝光诱发的自主游动行为即为条件反射。训练开始前，将仔鱼单独暴露在蓝光中，逐渐增加背景光照，以降低与蓝光的对比度，直到蓝光不再引起自主游动行为。

实验组仅有45%的仔鱼可触发自主游动行为，剩下55%的仔鱼没有表现出任何学习能力，还有另一批仔鱼则因蓝光诱发过度活动而在学习阶段之前就已被剔除。在70次的学习阶段的最后30次实验中，蓝光触发约60%的仔鱼出现自主游动行为。在“非配对”对照组中，暴露于随机蓝光和电击下的仔鱼，自主游动行为反应没有增加。研究人员还证明，在试验阶段后，单独使用4～6次蓝光可使自主游动行为消失。因此。“学习者”的行为证明了联想学习而不是敏感化。

7.2.2 自由游动斑马鱼仔鱼的经典条件反射

除了被束缚的斑马鱼仔鱼，自由游动的仔鱼也存在相同的经典的条件恐惧效应（Valente et al., 2012）。仔鱼在墙壁不透明和底座透明的矩形水池中自由游动，水池下方的液晶屏幕作为视觉刺激，对水池进行70 ms的电击作为非条件刺激，并用红外相机追踪记录鱼的踪迹。与之前描述的实验相反，作者采用了一种不同的条件反射模式，采用两种不同的视觉刺激；两种视觉刺激分别使用黑白网格图案和灰色图案。其中黑色网格图案作为威胁提示与非条件刺激配对，而灰色图案作为安全提示。

整个方案包括基础（baseline）、学习（acquisition）和测试（test）三个阶段，分别持续30min、1.5min和32.5min。在基础和测试阶段，一半水池呈现黑白网格图案，另一半水池呈现灰色图案，均不实施电击。基础阶段的目的就是确保仔鱼对这两种图案没有天生的偏好。在学习阶段的9个实验中，两种图案依次呈现并覆盖整个水池底部。灰色图案先显示 8.5s，而后黑白网格图案和电击一起呈现 1.5s 后同时结束。随后，在测试阶段，通过观察仔鱼在水池中的空间位置以及远离黑白网格图案的倾向，考察仔鱼对黑白网格图案的厌恶程度。接下来，2min 后交换黑白网格图案和灰色图案的位置，以确保仔鱼的厌恶行为来自黑白网格图案，而不是因为偏爱另一半水池。

接下来，Valente 及其同事又重新优化了他们的试验方案，以表征斑马鱼不同发育阶段的经典条件反射。对 7 ～ 42dpf 的 17 条斑马鱼分组训练和测试，每天 6 次，每周 5 天。受精后 4 周的斑马鱼有显著的学习行为，在第 6 周时到达峰值。然而，正如作者自己所说，因为斑马鱼是社会性动物，从小就表现出聚集的倾向，无法确定是否一组内所有的斑马鱼都是“学习者”，因此进行分组训练和测试的实验方法存在争议。其次，在其他研究中，每个发育阶段均使用未经过实验的斑马鱼，而且一天内只进行一次实验。本实验中斑马鱼却经历连续几天、几周的反复训练，使得实验结果很难与其他的研究结果之间进行比较。

7.2.3 仔鱼经典条件反射的神经生物学机制

小脑回路在很多物种的经典条件反射中发挥至关重要的作用（Thompson and Steinmetz, 2009）。为了考察斑马鱼是否也是如此，Aizenberg 和 Schuman 使用可注射用的钙指示剂 OGB-1 AM 进行活体小脑钙成像（Aizenberg and Schuman, 2011）。条件反射和非条件反射激活大量的细胞，分别占成像小脑神经元的 22% 和 17%，前者主要位于小脑的内侧和前部，后者主要位于小脑的腹侧和前部。与哺乳动物的证据一致，条件刺激诱发的动物小脑活动在学习阶段明显增强，而在单独暴露于条件刺激的阶段却没有出现小脑活动明显增强的现象。研究人员还发现了一种潜在的联想学习神经基质，即在条件反射过程中形成双峰的神经元。一部分神经元作为非条件刺激细胞开始于学习阶段，但在训练过程中学习了显著的条件刺激反应。最后，研究表明，用激光切除双侧小脑可以阻止条件反射的学习。

Harmon 及其同事通过电生理技术开展进一步的研究（Harmon et al., 2017）。同 Aizenberg 和 Schuman 一样，他们发现小脑消融阻止了条件反射的发生，证明了经典恐惧条件反射反应是小脑依赖性的。随后，对小脑中的浦肯野细胞［小脑中的大型分支 GABA 能（即抑制性）神经元］进行了细胞内记录，并根据它们在条件反射期间产生的复合峰（complex spikes）的数量，将其分为三类：多复合峰（multiple complex spike, MCS）细胞、单复合峰（single complex spike, SCS）细胞和零复合峰（zero complex spike, ZCS）细胞。这三种细胞亚型在小脑内的位置也不同。

MCS 细胞产生多复合峰与条件刺激发生之间的一致性趋于增加，而 SCS 细胞产生单复合峰则与条件反射相关。在学习阶段，MCS 细胞表现出与学习相关的复合峰生成增加，而 ZCS 细胞显示与条件刺激相关的简单峰生成增加。通过光遗传学抑制条件刺激期间的简单峰可损害仔鱼条件反射的学习获得，但不影响已经习得条件反射的仔鱼的表现。这意味着简单峰（零复合峰）与小脑学习密切相关，但在良好学习行为维持方面却似乎不需要 ZCS 细胞产生的简单峰。

最近，Matsuda 及其同事利用表达基因编码钙指示剂 GCaMP7A 的转基因仔鱼，在经典恐惧反射实验中对小脑内的神经元活动进行成像（Matsuda et al., 2017）。他们在小脑体（corpus cerebelli, CCe）中发现了个别神经元，初步认定为颗粒细胞，在学习阶段其条件刺激诱发活动逐渐增加，然后在测试阶段下降。关键的是，在没有习得条件反射的仔鱼中没有观察到这种细胞。因此，根据其反应的时间动力学，这些神经元可分为两类：Ⅰ型神经元的激活与条件刺激的呈现具有一致性，而在条件刺激之后，Ⅱ型神经元的激活达到最大。

7.3 操作性条件反射

在经典条件反射范式中，动物对条件刺激没有反应，所以经典条件反射是一种非自发的、被动的反射反应。相比之下，操作性条件反射与自发行为有关，非条件刺激的呈现有赖于动物行为，因此操作性条件反射是一种自发的、主动的反射反应。在操作性条件反射中，发生特定自发行为的可能性大小取决于触发该行为的后果。自发行为发生的可能性大即称为强化，反之则称为惩罚。强化又可分为正强化和负强化：正强化即为期望的行为可导致出现满欲刺激（appetitive stimulus）；负强化即为期望的行为可导致终止厌恶性刺激（aversive stimulus）（逃避）或避免出现厌恶性刺激（主动回避）。同样，惩罚也可分为正惩罚和负惩罚：正惩罚即为不期望的行为可导致出现厌恶性刺激；负惩罚即不期望的行为可导致终止满欲刺激（见图 7-1）。

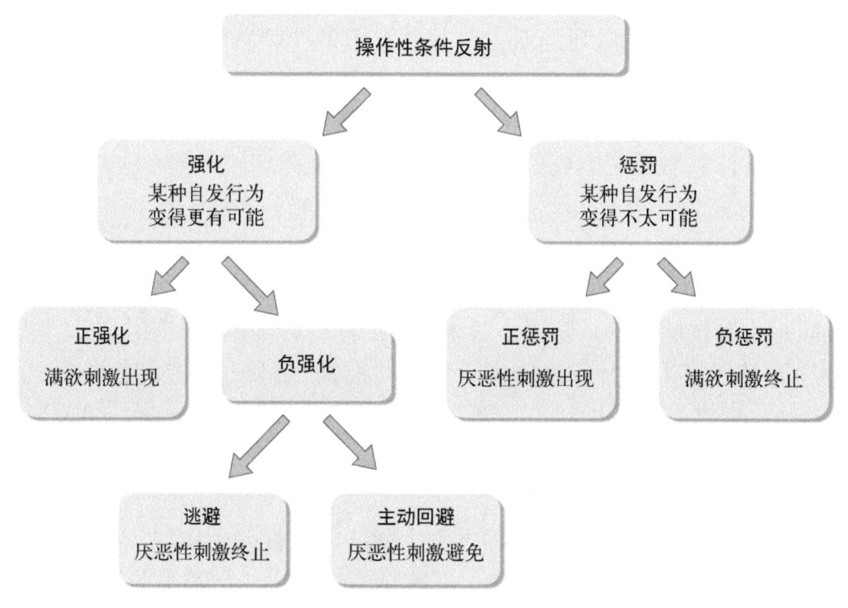

图 7-1 **操作性条件反射的分型**

强化使某种自发行为更有可能发生，而惩罚则使其不太可能发生。根据一个满欲刺激或者厌恶性刺激是否可以出现（introduced）、终止（terminated）或避免（prevented）而将操作性条件反射细分成不同的类型

迄今为止，针对仔鱼操作性条件反射的研究主要集中在主动回避方面。因为非条件刺激与特定的条件刺激共存，因此本文所描述的大多数方案可定义为辨别性回避分析

（discriminated avoidance assays）。严格地说，许多行为科学家将这些视为经典和操作性条件反射的混合模式，因为在动物由于不好的预期而开始回避条件刺激前，条件刺激和非条件刺激已经通过经典条件反射联系在一起，这通常称为主动回避学习的"双因素"理论（Mowrer, 1947）。

7.3.1　自由游动仔鱼的操作性调节

Valente 及其同事对他们原先的方案进行修改，将经典条件反射模式转化为操作性条件反射模式（Valente et al., 2012）。与之前实验设计的黑白网格图案和灰色图案依次呈现的方式不同，在目前的实验设计中，在学习阶段两种图案同时呈现，每一个图案覆盖一半水池，并且每当仔鱼游入黑白网格区域时就会进行电击，仔鱼可以通过避开黑白网格图案来躲避电击。实验包括基础阶段、学习阶段和测试阶段，依次为 30min、30min 和 60min。在这种情况下，对仔鱼进行单独的训练和测试，并在每个发育阶段使用全新的动物。实验结果表明，受精后 3 周的斑马鱼有显著的学习行为，大多数斑马鱼到第 6 周时均有良好的学习行为。

最近，研究人员使用改良的实验方案，证明了 7 ～ 10dpf 的斑马鱼的操作性学习（Yang et al., 2019）。改良的方案由 10min 的基础阶段、20min 的学习阶段、1min 的维持时间和 18min 的测试阶段组成。与原先的方案不同，在整个实验过程中，有规律地颠倒黑白网格图案和灰色图案的位置。在学习阶段，每次更换图案后，仔鱼有 7s 时间逃离黑白图案区，然后进行 100ms 电击。随后每隔 3s 进行一次电击，直到仔鱼完成逃逸。

在 Yang 及其同事的实验中，其中一个视觉刺激是均匀的灰色图案，另一个视觉刺激要么是黑白网格图案，要么是黑红网格图案，要么是均匀的黑色图案。通过基础和学习阶段，仔鱼对黑色图案的厌恶程度呈现指数级增加。当视觉刺激是黑白网格图案时，厌恶程度无明显变化，但是在黑红图案中观察到显著的增加，在黑色图案中观察到非常显著的增加。同样，以个体为基础对数据进行评估也显示相同的趋势：在三组中，分别有 6%、25% 和 50% 的仔鱼被归类为"学习者"。平均而言，对黑色图案和黑红网格图案的条件厌恶在测试阶段持续了大约 12min。这些结果反映了该行为属于联想学习而非敏感化，因为对照组条件刺激引发鱼的厌恶程度并没有明显增加。

尽管"非学习者"的数量占比很高，"学习者"的表现也不够出色，习得行为的持续时间也很短，但这些数据足以表明仔鱼出现操作性条件反射的时间可能比之前报道的时间要早 2 周。有一种解释是，视觉刺激图案的反复更换会增加电击次数以及使仔鱼学习完全不受空间位置限制，从而增加了学习的机会。另一种解释是，仔鱼表现取决于黑色和灰色图案之间的对比度。因此当使用黑白网格和灰色这两种亮度几乎相同的图案时，动物无法分辨和学习；然而，采用比灰色图案更暗的黑色图案时，动物的学习表现会更佳。另外，Valente 及其同事在之前的实验中也使用了对比不明显的两种图案，也可能使仔鱼在较弱的对比度下无法进行有效的分辨和学习。

在上述实验前，2010 年，研究人员首次发表了一个更简化的操作方法（Lee et al., 2010; Cheng and Jesuthasan, 2012; Lupton et al., 2017）。让仔鱼探索一个长方形的水箱，水箱两端都安装了红色的 LED 灯。每次试验都从仔鱼进入水箱一侧开始，同时水箱一侧的 LED 灯发光作为条件刺激；为了回避 100ms 或 200ms 电击，仔鱼有 5s 的时间游到另一半水箱，每天学习 10 次。在 3 到 4 周龄段内，随着年龄增长，仔鱼成功回避电击的比例显著上升，4 天后平均达到 85%（Cheng and Jesuthasan, 2012）。同样，单纯条件刺激组（只有 LED 灯发光刺激，无电击的非条件刺激）和未配对对照组（LED 灯发光的条件刺激和电击的非条件刺激两者不配对出现）的结果均未见改善。另外，将实验方案中条件刺激持续时间延长至 8s，并通过添加部分分隔器，在水箱创建两个不同的隔间（Lupton et al., 2017）。这种方法于 20 世纪 30 年代在啮齿动物中首次提出，是主动回避“穿梭箱”的经典示例（例如，Warner, 1932）。

若将狗或其他啮齿类动物预先暴露于强烈和无法逃避的刺激源（如电击或冷水）后，它们学习主动回避的能力将大受影响（Overmier and Seligman, 1967; Weiss and Glazer, 1975; Amat et al., 2001）。针对这一现象的解释是：若动物已经接受了自己无法控制自己的处境，它们便不再试图逃跑，这一说法被称为“习得性无助”假说。许多人认为这种行为类似于人类的抑郁，抑郁通常源于对个人环境缺乏控制。为了研究斑马鱼是否有与哺乳动物类似的反应，Lee 及其同事对 3 到 5 周的幼鱼进行了 10 次学习试验，如前所述，随后进行了单独测试试验，以评估它们的学习表现。实验结果显示，对照组 80% 的鱼成功避免了电击，而经历过无法逃避刺激源的动物却全部无法成功避免电击（Lee et al., 2010）。

7.3.2 被约束仔鱼的操作性条件反射

目前，只有 Li 研究组成功开发了一种用于被约束仔鱼的操作性条件反射模式（Li, 2012）。在此实验中，将 3 ～ 8dpf 仔鱼包埋在琼脂糖凝胶中，用摄像机跟踪其尾部运动。实验中，使用红外激光将 32℃的热脉冲传送到仔鱼的头部和身体，产生的热能作为仔鱼的厌恶性的非条件刺激。每当试验开始时，激光随之激活，一旦仔鱼朝着预定的方向转动一圈，激光就会消失。如果第一次转弯的方向跟特定方向一致，则计为得分；相反则不得分，激光会一直保持激活直到转向正确。为了最大化可能学习的动态范围，在仔鱼学习阶段设定正确的转弯方向总是与个体天生的转弯方向相反。这个实验是一个逃避而不是主动回避学习的例子（图 7-1），因为非条件刺激的出现是无法避免的，但可以通过实施特定的行为迅速终止。本实验中非条件刺激与条件刺激没有联系，这也是本章中唯一一个“单纯的”操作性条件反射的例子。

学习阶段包括 25 ～ 30 次试验，每次持续 2min。大多数 7 ～ 8dpf 的仔鱼能够学习所需的行为，70% 的仔鱼被归类为“学习者”，30% 的仔鱼被定义为“非学习者”。“学习者”经过大约 6 次试验建立起对特定转向的显著偏好，正确转向的平均比例峰值达到 75%。经过最初的几次试验后，首次正确转身的潜伏期也迅速减少。可以预见的是，对照组斑马鱼没有学习行为，激光照射的持续时间是随机的，激光热刺激的停止并不依赖于（取决于）动物的行为。在“学习者”中，条件性反射行为在 1h 后仍然存在。此外，仔鱼具有反转学习能力，如果在 25 ～ 30 次试验后改变特定转向方向，仔鱼也可以很容易学会新的转向偏好。操作性条件反射的个体发育的进一步研究表明，斑马鱼在 3dpf 就具备学习能力，并且在 3 ～ 7dpf 之间学习能力显著提高。

7.3.3 仔鱼操作性条件反射的神经生物学机制

缰核（habenula）与主动回避有关，缰核损伤会影响啮齿动物主动回避学习行为（Wilcox et al., 1986; Thornton and Bradbury, 1989）。前文描述的“穿梭箱”实验也验证了这个结论（Lee et al., 2010）。Lee 及其同事使用 KR11 转基因系斑马鱼，其缰核传入纤维中选择性表达光敏剂 KillerRed，绿光照射后神经元细胞产生光毒性导致膜损伤甚至死亡，因此影响机体主动回避学习。同样，若缰核中表达 GAL4^{s1019t}/UAS：TeTXlc-CFP 这一神经元沉默毒素基因，也会损害仔鱼学习过程。同时，也有研究人员进一步研究成鱼缰核在主动回避学习中的作用（Amo et al., 2014）。

由 *Kissi* 基因编码的肽神经调节剂 Kisspeptin1 在缰腹侧表达，可能对主动回避学习有重要作用。通过 CRISPR/Cas9 产生 *Kissi* 突变体仔鱼，在“穿梭箱”实验中，受精后 4 ～ 6 周的斑马鱼表现出行为受损，6 ～ 8dpf 的仔鱼遭受电击时，中缝核反应性降低（Lupton et al., 2017）。尽管以上作用机制尚不清楚，但可以看出不同浓度的 Kisspeptin1 可以使得缰核腹侧神经元去极化或超极化（Lupton et al., 2017）。

缰核活动也与逃避学习相关，因为仔鱼学会了通过鱼体左转或右转缓解热刺激。Li 及其同事使用 *elavl3*：*GCaMP5G* 仔鱼进行了全脑单细胞分辨率钙成像（Li, 2012），通过神经元位置及不同反应特征揭示了多种神经元亚型。首先，他们发现感觉神经元在大脑中广泛分布，在缰核中尤其丰富，其激活与热刺激开始或终止一致。其次，他们观察了前脑、中脑和后脑中与方向特异性运动相关的神经元，无论是自发的还是刺激诱发的运动，神经元放电都与它们有所关联。之后，他们找到了在缰核富集的运动选择性相关（action selection-related, ASR）神经元。虽然 ASR 具有方向特异性，但它们的放电只与刺激诱发的运动有关。与特异性运动相关神经元不同的是，ASR 放电可以在运动发生之前或之后几秒钟，说明它们参与了运动反应的“计划（planning）”，也暗示它们是“工作记忆（working memory）”的潜在性神经基础。因为只有在动作执行后才能形成强化，因此保留先前执行动作的记忆痕迹对于操作性学习是必不可少的。

在缰核内，Li 及其同事还发现了另外三种细胞类型。与感觉神经元一样，消除预测（relief prediction, RP+）和无消除预测（no-relief prediction, RP−）神经元的激活与热刺激出现相一致。然而，与感觉神经元不同的是，这种反应的幅度是由是否学习强化调节的：当仔鱼成功学习强化行为时，RP+ 神经元受热刺激后活性增强，而 RP− 神经元的活性减弱。第三类细胞是消除预测误差（relief prediction error, RPE）神经元，其对热刺激的开始和终止都有反应：前者受到学习强化行为的正向调节，后者则受负向调节。因此，在仔鱼中，RPE 神经元主要由热刺激结束而激活，但在成功学习强化行为后由热刺激开始而激活。

在斑马鱼大脑中 RP 和 RPE 信号的观察结果与哺乳动物的报告一致（Schultz, 2016）。也就是说，RP+ 和 RPE 神经元明显富集在右侧缰核，而 RP− 神经元则局限于左侧缰核，目前这种不对称分布在其他物种中未见报道。僵核中的预测环路被认为可以识别非自适应性的优势行为反应（例如，左转），通过调节 ASR 神经元的活性来驱动另一种行为反应（例如，右转）。与这一假说一致，当右转行为增强时，左侧特异性 ASR 神经元活性受到抑制，当左转行为增强时，左侧特异性 ASR 神经元活性增强。同样，对于右侧特异性

ASR 神经元活性也是如此。此外，“非学习者”的大脑相对于“学习者”的大脑中的 RP 和 ASR 神经元数量急剧减少，表明这些神经元信号在逃逸学习中非常重要。

7.4 满欲刺激的联想学习

通过厌恶性刺激的学习效果比用食物奖励刺激的学习效果好得多（Powell et al., 2005）。而且，对于仔鱼来说，较少的食量需求限制了其每天可进行实验的次数，因此食物本身就不是一种合适的奖励刺激。因此，在像活体成像这种要求在一次过程中学习强烈行为的实验中，厌恶性刺激学习得到了广泛应用。尽管如此，研究人员越来越意识到，除了食物以外的很多满欲刺激都可以用于学习分析。例如，对成鱼来说，视觉上接近同种属动物是一种奖赏行为（Sison and Gerlai, 2011）。目前看来，这也是唯一一种适用于仔鱼联想学习研究的满欲刺激（Hinz et al., 2013）。

在 Hinz 及其同事的实验中，用一个投影仪由下往上照亮一个长方形的水池，水池中一半的光照比另一半亮四倍，仔鱼可以同时进入这两个视觉上截然不同的环境。该方案包括 30min 基础阶段、3h 学习阶段、14h 维持时间和 30 min 测试阶段。在基础和测试阶段，水池的侧壁完全不透明。然而，在学习阶段，黑暗环境中的侧壁是透明的，使得仔鱼能够在两个相邻的水池中看到个别其他仔鱼。从基础阶段到测试阶段，仔鱼在黑暗环境中花费的时间增加。6 ～ 8dpf 仔鱼在预期方向上显示出显著的变化，并且这种效应可以持续 36h，而对照组中没有观察到这种变化。若在学习阶段用蛋白质合成抑制剂或 *N*- 甲基 -D- 天冬氨酸受体（*N*-methyl-D-aspartate receptor, NMDAR）拮抗剂 MK-801 处理仔鱼，它们的行为会受到严重影响，结果表明满欲刺激的联想行为与蛋白质合成和 NMDAR 依赖性密切相关。

7.5 综合讨论

7.5.1 对现有联想学习分析方法的评价

尽管仔鱼联想学习的研究已有十余年，但目前仍处于起步阶段，仅有少数已发表的研究。但已有研究证明，在独特的实验条件下，斑马鱼在受精后 1 周就可以同时进行经典条件反射和操作性条件反射。不过，目前看来，整体研究还未处于稳定阶段。总体来说，仔鱼的行为是非常不稳定的，其特点是：①个体间差异较大；②学习能力处于较低水平；③在一半以上的实验中，出现相当大比例的“非学习者”。

然而，这些实验探索是研究人员在新的领域上迈出的关键一步，实验结果与成败似乎也就不那么重要了。从长远来看，啮齿动物的联想学习研究持续了近一个世纪，研究人员可以从几十年的经验和优化中获益。在接下来的几年里，随着最佳实验参数的发现，仔鱼的学习能力可能会逐渐提高。事实上，目前已经有文献发表了相关方法，就是通过对原先方案进行了部分调整和改进而进行实验（Yang et al., 2019）。

另一个可能的解决办法就是根据仔鱼的长度而不是鱼龄来挑选仔鱼。因为长度是一个更准确的个体发展的阶段性指标，这样可能会减少一些个体间的差异（参考 Cheng and Jesuthasan, 2012）。未来，研究遗传背景对联想学习的影响也很重要。值得注意的是，遗传背景会影响 5dpf 仔鱼的惊吓习惯和光 / 暗诱导运动行为（van den Bos et al., 2017）。在本章所述的实验中使用了各种自交系（见表 7-1 和表 7-2），有助于通过对比说明实验结果。

表 7-1 仔鱼经典条件作用实验总结

作者	是否约束	品种	鱼龄	方案	因变量	关键发现
Aizenberg and Schuman (2011)	是	*nacre*	6～8dpf	白色光点运动（条件刺激）与尾部钢丝接触（非条件刺激）	条件刺激诱发尾部运动	小脑消融阻止了学习。体内钙成像显示学习相关的小脑神经元活动的变化
Valente et al. (2012)	不	AB	1～6 周	第一种：视觉模式（黑白图案条件刺激 CS+）伴电击（非条件刺激） 第二种：视觉模式（灰色图案条件刺激 CS−）不伴有电击	进入灰色图案区域多于黑白图案区域	直到受精后 4 周，学习活动才明显出现
Matsuda et al. (2017)	是	AB; gSA2AzGFF152B; *elavl*3: GAL4-VP16[nns6]; UAS: GCaMP7A[zf415]; UAS:BoTxBLC-GFP[icm21]	5～25dpf	白光偏移（条件刺激）伴电击（非条件刺激）	条件刺激诱发心动过缓	体内钙成像显示小脑有两类不同的学习相关颗粒细胞
Harmon et al. (2017)	是	*nacre*; Arch-tagRFP-T: car8: GCaMP5	6～8dpf	蓝光（条件刺激）伴电击（非条件刺激）	条件刺激诱发腹根活动	小脑消融阻止了学习。体内电生理学显示小脑有三种不同类型的学习相关浦肯野细胞

如果仔鱼行为分析的可靠性和可重复性得到提高，那么就有可能将其用作筛选工具，以研究遗传、药理学或环境操作对联想学习的影响。Lupton 及其同事的工作为这种方法提供了原理证明（Lupton et al., 2017），这是在操作试验中首次证明突变仔鱼的行为受损。另一方面，仔鱼的联想学习能力可能有一个与现有水平接近，但又不可逾越的上限值。有研究显示，相当一部分仔鱼可能因为大脑发育不足而无法学习。简单地说，仔鱼尚未建立需要交流的大脑区域之间的联系（Florian Engert 未发表的数据）。若是这样，那么再多的程序优化也不能使这些动物进行有效学习。

7.5.2 将联想学习分析与体内成像相结合

仔鱼作为模式生物的主要优势之一是其适合全脑单细胞分辨率钙成像。如引言所述，

表 7-2 **仔鱼操作性条件作用实验总结**

作者	是否约束	品种	鱼龄	方案	因变量	关键发现
Lee et al. (2010)；Cheng and Jesuthasan (2012)	不	AB; KR4 和 KR11 (AB); GAL4[s1019t]/ UAS: TeTXlc-CFP	3～5 周	红灯（条件刺激 CS）伴触电（非条件刺激 US）。通过逃离条件刺激而避免电击	成功回避反应的数量	由于预先暴露于无法逃避的压力源（即“习得性无助”）而行为受损
Valente et al. (2012)	不	AB	1～8 周	第一种视觉模式（黑白图案 CS+）伴电击（非条件刺激 US）。第二种视觉模式（灰色图案 CS−）不伴有电击。留在灰色图案区域避免电击	进入灰色图案区域（CS−）多于黑白图案区域（CS+）	直到受精后 3 周，学习活动才明显出现
Hinz et al. (2013)	不	AB	6～8dpf	黑暗环境（条件刺激 CS+）伴有视觉接近同种动物（非条件刺激 US）。明亮环境（条件刺激 CS−）不伴有视觉接近同种动物（非条件刺激 US）	进入黑暗区域（CS+）多于进入明亮区域（CS−）	使用满欲刺激首次证明仔鱼进行联想学习
Lupton et al. (2017)；	不	*Kiss1* 突变 (*elavl*3：GCaMP6F)	4～6 周	红灯（条件刺激 CS）伴电击（非条件刺激 US）。通过逃离条件刺激避免电击	成功回避反应的次数	肽神经调节剂 Kisspeptin1 参与主动回避学习
Yang et al. (2019)	不	AB；*elavl*3：H2B-GCaMP6F	7～10dpf	第一个视觉模式（条件刺激 CS+）伴电击（非条件刺激）。第二个视觉模式（条件刺激 CS −）不伴电击（非条件刺激）。逃离条件刺激 CS+ 可避免电击	进入灰色图案区域（CS−）多于进入黑白图案区域（CS+）	提高黑白图案和灰色图案之间的对比度有助于提高学习能力
Li (2012)	是	*elavl*3：GCaMP5G	3～8dpf	通过向预定方向转动终止热刺激（非条件刺激 US）	成功逃避反应次数；逃逸潜伏期	全脑单细胞分辨钙成像显示，不对称缰核回路编码的消除预测和消除预测错误

出于一系列技术考虑，全脑单细胞分辨率钙成像适用的仔鱼年龄范围较窄。然而，研究结果表明，1 周龄仔鱼即可进行联想学习。因此，将这些行为分析与全脑成像技术相结合，可以描述经典或操作性条件反射下的完整脑回路。虽然已经对这些过程背后的神经生物学机制有了一些了解，尤其是对小脑和缰核的研究，但目前只有 Li 及其同事在进行仔鱼的整个大脑水平上可视化联想学习回路的研究（Li, 2012）。未来的研究应该致力于发掘这种巨大的潜力。此外，可以利用光遗传学工具对已识别的脑回路进行有针对性的研究，从而使该领域研究不局限于相关性和因果关系推断。

本章描述的许多实验都是利用无约束、自由游动的仔鱼，虽然这种方案可能符合更高的生态有效性，但同时仔鱼的快速与不可预测的移动也使成像更加困难。因此，研究人员目前正在开发用于无约束仔鱼的活体成像工具（Cong et al., 2017; Kim et al., 2017）。

在现有的实验中，相当一部分仔鱼根本无法学习，可能会成为综合行为和成像研究的障碍。除非研究的目的是明确比较“学习者”和“非学习者”，否则可能需要放弃多达一半的实验组。当然，使用完全不同的物种也是一种冒险尝试。例如，*Danionella translucida* 与斑马鱼关系密切，成年后鱼体仍然小而透明，长约 1.2cm，是人类已知的最小脊椎动物之一。有趣的是，研究人员在 *neurod1* 启动子的控制下，在表达 GCaMP6F 的成鱼中进行了体内钙成像（Schulze et al., 2018），同时也在一个简单的实验中证明了关联学习（Penalva et al., 2018）。因此，利用 *Danionella* 可以对神经系统成熟、行为完整的成鱼进行全脑单细胞分辨率成像。

7.6 展望

操作性条件反射可分为几大类，如图 7-1 所示。在这个简单的分类之外，还有大量实验组合和早已确立的行为现象，这些都超出了本章的范围。经典条件反射同样如此。迄今为止，对仔鱼的联想学习的研究仍旧停留在表面阶段，在未来几年，随着空白的填补，深入研究的内容将会越来越丰富。

致谢

本章由 David Pritchett 编写，Caroline H. Brennan 编辑。David Pritchett 得到 Leverhulme 信托基金（RPG-2016-143）资助。Caroline H. Brennan 得到 Leverhulme 信托基金（RPG-2016-143）、人类前沿科学计划（RGP0008/2017）和美国国家卫生研究院（U01DA044400-02）资助。

（朱思庆翻译　李丽琴审校）

参考文献

Ahrens, M.B., Li, J.M., Orger, M.B., Robson, D.N., Schier, A.F., Engert, F., Portugues, R., 2012. Brain-wide neuronal dynamics during motor adaptation in zebrafish. Nature 485, 471-477. https://doi.org/10.1038/nature11057.

Aizenberg, M., Schuman, E.M., 2011. Cerebellar-dependent learning in larval zebrafish. Journal of Neuroscience 31, 8708. https://doi.org/10.1523/JNEUROSCI.6565-10.2011.

Amat, J., Sparks, P.D., Matcs-Amat, P., Griggs, J., Watkins, L.R., Maier, S.F., 2001. The role of the habenular complex in the elevation of dorsal raphe nuclecs serotonin and the changes in the behavioral responses produced by uncontrollable stress. Brain Research 917, 118-126. https://doi.org/10.1016/S0006-8993(01)02934-1.

Amo, R., Fredes, F., Kinoshita, M., Aoki, R., Aizawa, H., Agetsuma, M., Aoki, T., Shiraki, T., Kakinuma, H., Matsuda, M., Yamazaki, M., Takahoko, M., Tsuboi, T., Higashijima, S., Miyasaka, N., Koide, T., Yabuki, Y., Yoshihara, Y., Fukai, T., Okamoto, H., 2014. The habenulo-raphe serotonergic circuit encodes an aversive expectation value essential for adaptive active avoidance of danger. Neuron 84, 1034-1048. https://doi.org/10.1016/j.neuron.2014.10.035.

Bergmann, K., Meza Santoscoy, P., Lygdas, K., Nikolaeva, Y., MacDonald, R.B., Cunliffe, V.T., Nikolaev, A., 2018. Imaging neuronal activity in the optic tectum of late stage larval zebrafish. Journal of Developmental Biology 6, 6. https://doi.org/10.3390/jdb6010006.

Cheng, R.-K., Jesuthasan, S., 2012. Automated conditioning in larval zebrafish. In: Kalueff, A.V., Stewart, A.M. (Eds.), Zebrafish Protocols for Neurobehavioral Research. Humana Press, Totowa, NJ, pp. 107e120. https://doi.org/ 10.1007/978-1-61779-597-8_8.

Cong, L., Wang, Z., Chai, Y., Hang, W., Shang, C., Yang, W., Bai, L., Du, J., Wang, K., Wen, Q., 2017. Rapid whole brain imaging of neural activity in freely behaving larval zebrafish (*Danio rerio*). eLife 6, e28158. https://doi.org/ 10.7554/ eLife.28158.

Feierstein, C.E., Portugues, R., Orger, M.B., 2015. Seeing the whole picture: a comprehensive imaging approach to functional mapping of circuits in behaving zebrafish. Neuroscience 296, 26-38. https://doi.org/10.1016/ j.neuroscience.2014.11.046.

Harmon, T.C., Magaram, U., McLean, D.L., Raman, I.M., 2017. Distinct responses of Purkinje neurons and roles of simple spikes during associative motor learning in larval zebrafish. eLife 6, e22537. https://doi.org/10.7554/ eLife.22537.

Hinz, F.I., Aizenberg, M., Tcshev, G., Schuman, E.M., 2013. Protein synthesis-dependent associative long-term memory in larval zebrafish. Journal of Neuroscience 33, 15382. https://doi.org/10.1523/JNEUROSCI.0560-13.2013.

Howe, K., Clark, M.D., Torroja, C.F., Torrance, J., Berthelot, C., Muffato, M., Collins, J.E., Humphray, S., McLaren, K., Matthews, L., McLaren, S., Sealy, I., Caccamo, M., Churcher, C., Scott, C., Barrett, J.C., Koch, R., Rauch, G.-J., White, S., Chow, W., Kilian, B., Quintais, L.T., Guerra-Assunção, J.A., Zhou, Y., Gu, Y., Yen, J., Vogel, J.-H., Eyre, T., Redmond, S., Banerjee, R., Chi, J., Fu, B., Langley, E., Maguire, S.F., Laird, G.K., Lloyd, D., Kenyon, E., Donaldson, S., Sehra, H., Almeida-King, J., Loveland, J., Trevanion, S., Jones, M., Quail, M., Willey, D., Hunt, A., Burton, J., Sims, S., McLay, K., Plumb, B., Davis, J., Clee, C., Oliver, K., Clark, R., Riddle, C., Elliot, D., Threadgold, G., Harden, G., Ware, D., Begum, S., Mortimore, B., Kerry, G., Heath, P., Phillimore, B., Tracey, A., Corby, N., Dunn, M., Johnson, C., Wood, J., Clark, S., Pelan, S., Griffiths, G., Smith, M., Glithero, R., Howden, P., Barker, N., Lloyd, C., Stevens, C., Harley, J., Holt, K., Panagiotidis, G., Lovell, J., Beasley, H., Henderson, C., Gordon, D., Auger, K., Wright, D., Collins, J., Raisen, C., Dyer, L., Leung, K., Robertson, L., Ambridge, K., Leongamornlert, D., McGuire, S., Gilderthorp, R., Griffiths, C., Manthravadi, D., Nichol, S., Barker, G., Whitehead, S., Kay, M., Brown, J., Murnane, C., Gray, E., Humphries, M., Sycamore, N., Barker, D., Saunders, D., Wallis, J., Babbage, A., Hammond, S., MashreghiMohammadi, M., Barr, L., Martin, S., Wray, P., Ellington, A., Matthews, N., Ellwood, M., Woodmansey, R., Clark, G., Cooper, J.D., Tromans, A., Grafham, D., Skuce, C., Pandian, R., Andrews, R., Harrison, E., Kimberley, A., Garnett, J., Fosker, N., Hall, R., Garner, P., Kelly, D., Bird, C., Palmer, S., Gehring, I., Berger, A., Dooley, C.M., Ersan-Ürün, Z., Eser, C., Geiger, H., Geisler, M., Karotki, L., Kirn, A., Konantz, J., Konantz, M., Oberländer, M., Rudolph-Geiger, S., Teucke, M., Lanz, C., Raddatz, G., Osoegawa, K., Zhu, B., Rapp, A., Widaa, S., Langford, C., Yang, F., Schcster, S.C., Carter, N.P., Harrow, J., Ning, Z., Herrero, J., Searle, S.M.J., Enright, A., Geisler, R., Plasterk, R.H.A., Lee, C., Westerfield, M., de Jong, P.J., Zon, L.I., Postlethwait, J.H., Nüsslein-Volhard, C., Hubbard, T.J.P., Roest Crollics, H., Rogers, J., Stemple, D.L., 2013. The zebrafish reference genome sequence and its relationship to the human genome. Nature 496, 498-503. https://doi.org/10.1038/nature12111.

Kim, C.K., Miri, A., Leung, L.C., Berndt, A., Mourrain, P., Tank, D.W., Burdine, R.D., 2014. Prolonged, brain-wide expression of nuclear-localized GCaMP3 for functional circuit mapping. Frontiers in Neural Circuits 8, 138. https://doi.org/10.3389/ fncir.2014.00138.

Kim, D.H., Kim, J., Marques, J.C., Grama, A., Hildebrand, D.G.C., Gu, W., Li, J.M., Robson, D.N., 2017. Pan-neuronal calcium imaging with cellular resolution in freely swimming zebrafish. Nature Methods 14, 1107.

Lee, A., Mathuru, A.S., Teh, C., Kibat, C., Korzh, V., Penney, T.B., Jesuthasan, S., 2010. The habenula prevents helpless behavior in larval zebrafish. Current Biology 20, 2211-2216. https://doi.org/10.1016/j.cub.2010.11.025.

Li, J.M., 2012. Identification of an Operant Learning Circuit by Whole Brain Functional Imaging in Larval Zebrafish. Doctoral dissertation, Harvard University.

Lupton, C., Sengupta, M., Cheng, R.-K., Chia, J., Thirumalai, V., Jesuthasan, S., 2017. Loss of the habenula intrinsic neuromodulator kisspeptin1 affects learning in larval zebrafish. eNeuro 4. https://doi.org/10.1523/ ENEURO.0326-16.2017.

Matsuda, K., Yoshida, M., Kawakami, K., Hibi, M., Shimizu, T., 2017. Granule cells control recovery from classical conditioned fear responses in the zebrafish cerebellum. Scientific Reports 7, 11865. https://doi.org/10.1038/ s41598-017-10794-0.

Mowrer, O.H., 1947. On the dual nature of learningda re-interpretation of "conditioning" and "problem-solving. Harvard Educational Review 17, 102-148.

Orger, M.B., de Polavieja, G.G., 2017. Zebrafish behavior: opportunities and challenges. Annual Review of Neuroscience 40, 125-147. https://doi.org/10.1146/annurev-neuro-071714-033857.

Orger, M.B., Portugues, R., 2016. Correlating whole brain neural activity with behavior in head-fixed larval zebrafish. In: Kawakami, K., Patton, E.E., Orger, M. (Eds.), Zebrafish: Methods and Protocols. Springer New York, New York, NY, pp. 307-320. https://doi.org/10.1007/978-1-4939-3771-4_21.

Overmier, J.B., Seligman, M.E., 1967. Effects of inescapable shock upon subsequent escape and avoidance responding. Journal of Comparative and Physiological Psychology 63, 28-33. https://doi.org/10.1037/h0024166.

Penalva, A., Bedke, J., Cook, E.S.B., Barrios, J.P., Bertram, E.P.L., Douglass, A.D., 2018. Establishment of the miniature fish species *Danionella translucida* as a genetically and optically tractable neuroscience model. bioRxiv. https://doi.org/10.1101/444026.

Powell, R.A., Symbaluk, D.G., MacDonald, S.E., 2005. Introduction to Learning and Behavior. Thomson/Wadsworth, Belmont, CA.

Rescorla, R.A., 1967. Pavlovian conditioning and its proper control procedures. Psychological Review 74, 71-80. https://doi.org/10.1037/h0024109.

Schultz, W., 2016. Dopamine reward prediction error coding. Dialogues in Clinical Neuroscience 18, 23-32.

Schulze, L., Henninger, J., Kadobianskyi, M., Chaigne, T., Faustino, A.I., Hakiy, N., Albadri, S., Schuelke, M., Maler, L., Del Bene, F., Judkewitz, B., 2018. Transparent *Danionella translucida* as a genetically tractable vertebrate brain model. Nature Methods 15, 977e983. https://doi.org/10.1038/s41592-018-0144-6.

Sison, M., Gerlai, R., 2011. Associative learning performance is impaired in zebrafish (*Danio rerio*) by the NMDA-R antagonist MK-801. Neurobiology of Learning and Memory 96, 230-237. https://doi.org/10.1016/ j.nlm.2011.04.016.

Thompson, R.F., Steinmetz, J.E., 2009. The role of the cerebellum in classical conditioning of discrete behavioral responses. Neuroscience 162, 732-755. https://doi.org/10.1016/j.neuroscience.2009.01.041.

Thornton, E.W., Bradbury, G.E., 1989. Effort and stress influence the effect of lesion of the habenula complex in oneway active avoidance learning. Physiology and Behavior 45, 929-935. https://doi.org/10.1016/0031-9384(89) 90217-5.

Valente, A., Huang, K.-H., Portugues, R., Engert, F., 2012. Ontogeny of classical and operant learning behaviors in zebrafish. Learning and Memory (Cold Spring Harbor, N.Y.) 19, 170-177. https://doi.org/10.1101/ lm.025668.112.

van den Bos, R., Mes, W., Galligani, P., Heil, A., Zethof, J., Flik, G., Gorissen, M., 2017. Further characterisation of differences between TL and AB zebrafish (*Danio rerio*): gene expression, physiology and behaviour at day 5 of the larval stage. PLoS One 12, e0175420. https://doi.org/10.1371/journal.pone.0175420.

Warner, L.H., 1932. The association span of the white rat. The Pedagogical Seminary and Journal of Genetic Psychology 41, 57-90. https://doi.org/10.1080/08856559.1932.9944143.

Weiss, J.M., Glazer, H.I., 1975. Effects of acute exposure to stressors on subsequent avoidance-escape behavior. Psychosomatic Medicine 37.

Wilcox, K.S., Christoph, G.R., Double, B.A., Leonzio, R.J., 1986. Kainate and electrolytic lesions of the lateral habenula: effect on avoidance responses. Physiology and Behavior 36, 413-417. https://doi.org/10.1016/0031- 9384(86)90307-0.

Yang, W., Meng, Y., Li, D., Wen, Q., 2019. Visual contrast modulates operant learning responses in larval zebrafish. Frontiers in Behavioral Neuroscience 13, 4. https://doi.org/10.3389/fnbeh.2019.00004.

第 三 部 分

斑马鱼成鱼的行为

第八章

斑马鱼运动形式和游动路径特征的行为学研究

Anton M. Lakstygal[1,2], Konstantin A. Demin[1,3], Allan V. Kalueff[4,5,6,7,8,9,10]

8.1 引言

动物行为谱是动物表现特定行为的表现形式（Sakamoto et al., 2009），所有可定义的行为都可归纳在不同类别中，如攻击性、社交性、认知性、生殖行为、探索性以及情绪性方面等（Kalueff et al., 2013）。一些以图形方式展示的行为谱，不仅对行为进行了分类，还表明了行为发生的频率以及一种行为接着另一种行为发生的可能性（Sakamoto et al., 2009）。通过直接观察动物行为并制作行为谱是理解动物行为的基础（Marques et al., 2018）。然而，这种方法不适用于研究飞行或潜水的动物，因为超出了人类视觉的观察极限。另外，行为图谱由构成动物行为全貌的各种离散部分组成（Casarrubea et al., 2017），这些离散的部分间可能会重叠，而且啮齿类动物重叠性更高，因此很难明确地识别它们并将其分开。在这

[1] Institute of Translational Biomedicine, St. Petersburg State University, St. Petersburg, Russia.

[2] Laboratory of Preclinical Bioscreening, Granov Russian Research Center of Radiology and Surgical Technologies, Ministry of Healthcare of Russian Federation, Pesochny, Russia.

[3] Institute of Experimental Medicine, Almazov National Medical Research Centre, Ministry of Healthcare of Russian Federation, St. Petersburg, Russia.

[4] School of Pharmacy, Southwest University, Chongqing, China.

[5] Almazov National Medical Research Centre, St. Petersburg, Russia.

[6] Ural Federal University, Ekaterinburg, Russia.

[7] Granov Russian Research Center of Radiology and Surgical Technologies, St. Petersburg, Russia.

[8] Laboratory of Biological Psychiatry, Institute of Translational Biomedicine, St. Petersburg State University, St. Petersburg, Russia.

[9] Laboratory of Biopsychiatry, Scientific Research Institute of Physiology and Basic Medicine, Novosibirsk, Russia.

[10] Institute, Slidell, LA, United States.

种情况下，可以通过生物遥测这种手段，远程监测动物行为（Cooke et al., 2012）。此外，可携带数据传感技术的发展，使得即使动物处于移动状态，多个传感器也可以记录完整的生物行为信息（Cooke et al., 2012）。目前，研究人员已经制定了斑马鱼行为的详细目录（斑马鱼行为目录，ZBC，表 8.1），涵盖了仔鱼和成鱼模型（Kalueff et al., 2013）。虽然斑马鱼的行为没有啮齿动物复杂，但它们行为的复杂程度与哺乳动物较为接近。本章总结了当前研究阶段已知的仔鱼和成鱼的运动行为。

8.2 斑马鱼运动行为的表型分析

尽管有多种方法可以观测斑马鱼行为，但关键在于如何准确评估其行为并进行实验验证。实验中通常会涉及特定的评估参数，例如：持续时间、频率、速度、距离或潜伏期。有经验的实验人员可以手动评估部分参数（Cachat et al., 2011a），例如，通过观测斑马鱼进入陌生鱼缸顶部的次数和停留时间评价焦虑样行为。但是，如果没有特定的、精确的设备，就无法对速度、游动距离等参数进行评估。因此，目前斑马鱼实验研究中，录像是应用最广泛的方法。再通过特定软件对录像进行在线或离线分析，以获取无法直接观测到的行为信息（Stewart et al., 2015）（表 8-1）。

表 8-1 仔鱼和成鱼的常见运动行为 [根据斑马鱼行为目录，ZBC (Kalueff et al., 2013)]

仔鱼	成鱼	描述
警惕反应	警惕反应	一种适应性逃避反应，对引起恐惧刺激的一种反抗。特征是运动速度加快和方向迅速改变
回避	回避	远离某个物体或刺激物的运动或者时间增加（例如，捕食者、明亮的光线）
吸引力		接近或朝向某物体（视觉）或刺激物（例如食物）游动的时间增加
	向后游动	某些情况下的异常行为，例如暴露于某些致幻药物（例如麦角酸二乙胺，LSD）
拍打滑行（beat-and-glide）		一种间歇性的游动方式，其特点是先摆尾巴，然后滑行；在仔鱼 4dpf 时出现
弯曲游动（bending）		身体朝一侧弯曲游动的异常神经表型；可以作为癫痫发作的特征
猛冲和惯性滑行（burst-and-coast behavior）		仔鱼不能连续游动的猛冲模式。特征是向前猛冲后以一个缓慢的速度滑行，或者再次向前猛冲
突然游动（burst swim）		2dpf 仔鱼快速游动时，鱼体中部弯曲角度最大。缓慢游动时表现较大的弯曲幅度，更快的速度，更大的转向。通常与逃跑行为有关。此时，鱼胸鳍紧靠身体，不活跃
C 形启动（C 形弯曲，Mauthner 反射）		鱼体首先弯曲成 C 形，然后快速游动，以与先前位置成一定角度的方向推动身体离开
	转圈	以圆周方向重复游动（通常在癫痫发作、神经损伤和药物作用后出现）
	惯性滑行	无身体或鳍参与的被动游动。
卷体（coiling）		表现为全身收缩的胚胎运动，尾巴的尖端与头部卷曲重合；行为可以是自发的，也可以是触摸引发的

续表

仔鱼	成鱼	描述
	螺旋式游动	螺旋式游动，速度加快，方向不明确；通常作为某些疾病发作表型
	匍匐	非常缓慢地游动，只有胸鳍推动鱼前行
	猛冲	使用尾鳍在一个方向上的单一快速加速（例如逃逸行为）
	疾驰	一系列定向（推进）猛冲运动；通常被认为是一种逃避反应
	尾巴下垂	与神经功能缺损、运动功能减退有关。可由衰老、运动障碍、遗传和药物干预引起
	无规律运动	以方向或速度的急剧变化和重复的快速猛冲为特征的复杂行为
急转弯		仔鱼的逃逸状转向，特点是快速、大角度转向，以大角度弯曲整个身体
	“八”字形游动	某些药物干预（例如尼古丁或氯胺酮）后的一种特殊的游动模式
逃跑	逃跑	加速远离另一条鱼或刺激物的运动
浮游	浮游	在水面附近被动游动
	僵直	除鳃和眼睛，鱼在池底时完全停止运动
过度运动	过度运动	长时间的异常快速的游动。通常与精神刺激、痉挛或焦虑样行为有关
运动减退	运动减退	长时间不正常地缓慢游动。通常与镇静、神经障碍或运动障碍有关
静止不动	静止不动	除鳃和眼睛，在水箱底部完全停止移动；但又不同于僵直和休息
倾斜游动		相对于水面以一定的异常角度游动，通常由神经毒性物质引起
	“战战兢兢”游动	一种特殊的游动模式，特征是多次短暂的“急促”运动，游动轨迹的平滑度降低，常见于癫痫发作
环形（looping）		仔鱼在身体外的一个虚拟点周围有明显的环形游动行为
	蜿蜒	没有固定方向或路径的运动
O 形弯曲		仔鱼定向弯曲改变游动方向的运动
视动反应		由重复性运动刺激（例如旋转的滚筒）引起的运动，因为斑马鱼通常向与刺激物运动相同的方向游动
光电反应		仔鱼对光刺激的一系列运动行为，表现出运动兴奋
缓慢游动		仔鱼缓慢游动，特点是弯曲角度小，弯曲位置靠近尾巴
S 形启动或弯曲		惊吓逃跑反应，身体弯曲形成 S 形，头侧和尾侧同时活动
惊吓反应	惊吓反应	应对突然的刺激，通常是厌恶的刺激（例如振动、光、声音、触摸）表现出的保守的适应性反应
定向运动	定向运动	斑马鱼在某些条件下（例如用尼古丁和咖啡因等精神刺激剂或伊波加因等致幻剂干预）诱发的一种僵硬、重复的行为模式（例如从一个角落游到另一个角落）
游动	游动	简单的斑马鱼运动；按其持续时间可分为“延长”，可维持数分钟，或“持续”，可维持数小时
震颤	震颤	特殊颤抖样行为，由某些神经毒素或惊厥药物诱发
	起伏运动	与攻击行为有关。特征为波浪状或蛇形的运动
	垂直漂移	一种包括被动垂直漂浮的异常表现
	垂直游动	一种包括垂直游动的异常表现（通常在水面上抬头）
	迂回游动	异常震颤或颤抖样表现，通常由神经毒素或惊厥剂引起

三维表型是最新的斑马鱼行为评价方法，但这种方法更适用于成鱼研究，因为成鱼表现出复杂的三维运动行为，而仔鱼主要表现为二维运动。简单来说，三维表型是首先利用侧视和俯视两个摄像头记录斑马鱼的游动轨迹（图 8-1），然后通过软件重建游动路径，最后分析路径的空间坐标（Cachat et al., 2011b）。斑马鱼行为的三维分析显著改善了斑马鱼的神经表型分析技术，可用于评估成鱼的焦虑样行为（Stewart et al., 2015; Cachat et al., 2011b）。与二维摄像或多角度录像比较，三维数据表现出更高的可靠性（Cachat et al., 2011b）。事实上，三维分析最大的优势就是能分析不同组斑马鱼的游动方向（Stewart et al., 2015）。除了游动远近、处于鱼缸的上或下部等主观特征，如游动速度、游动转角、特定身体部位的运动距离等大多数精细行为都可以通过软件工具进行评估。这不仅有助于定义斑马鱼在测试过程中表现出的行为类型，还可以评估这种行为的可靠性。例如，特定身体部位（如头部）移动速度或距离的改变可以作为压力增加或减少的指标（该研究尚未发表）。

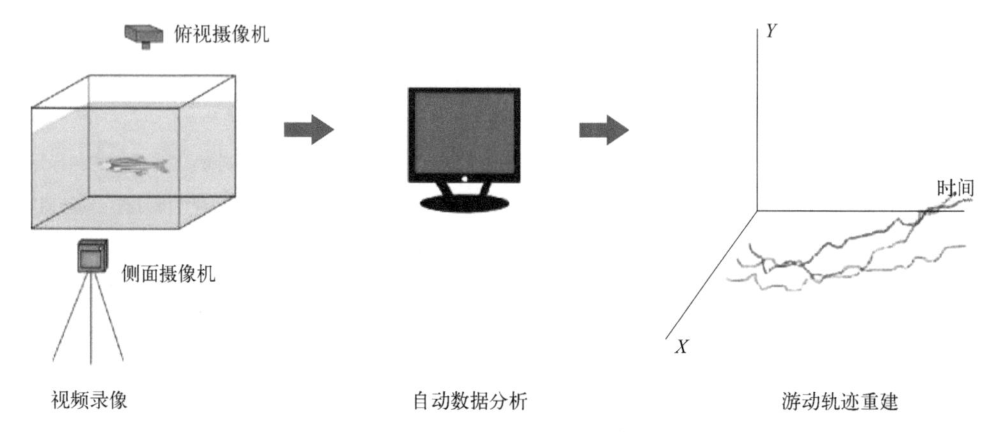

图 8-1 **斑马鱼行为的三维表型**

先进行录像，然后对获得的数据进行处理，并进一步重建游动轨迹，以验证在特定条件下表达的行为表型

在药物测试中，分层聚类分析也是一种有用的工具。在聚类图中，每个格子代表高于或低于对照组的平均相对标准偏差值（Cachat et al., 2011b）。该分析不仅再次确认了传统研究行为方法的有效性，而且还确定了一些计算机生成的新方法的可靠性。这些参数（图 8-2）可以反映斑马鱼焦虑水平的高低（Cachat et al., 2011b），有助于全面剖析和量化成鱼的行为，而且类似的方法也可以应用于仔鱼模型。

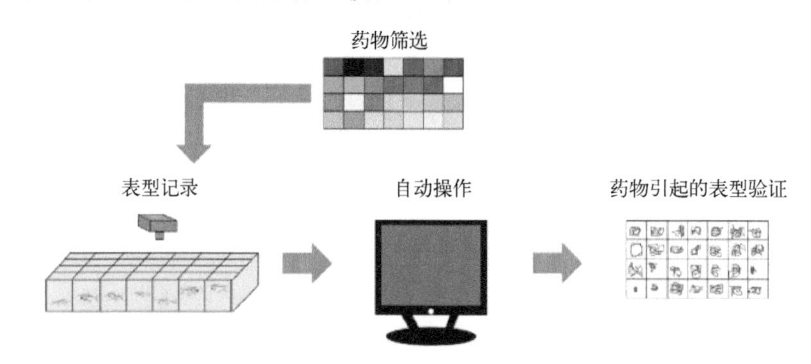

图 8-2 **斑马鱼高通量药物筛选**

斑马鱼体形小，可以同时筛选多种物质。斑马鱼是临床前药物检测的模式生物，对数百条斑马鱼进行视频记录，并通过专门软件进行自动化处理，这一过程更加快速和节约成本

8.3 斑马鱼的运动形式

作为一种用来观测运动行为的简单透明的动物模型（Fetcho et al., 2008），斑马鱼主要运用于研究生物力学、姿势平衡、聚集成群和神经肌肉形成等方面（Stewart et al., 2010）。通过运动学分析，研究人员已经建立了仔鱼体轴和胸鳍的不同协调模式（Budick and O'Malley, 2000）。这些运动形式可能由分散的相互作用环路产生，而来自大脑的信息输入可以选择性激活这些环路。事实上，研究人员已经开始探索仔鱼不同类型中间神经元在逃避和游动过程中的作用，以及被下行神经元激活的过程（Gahtan et al., 2002; Liao and Fetcho, 2008）。斑马鱼通过大约 300 个神经元从后脑和中脑进入脊髓来控制运动行为。这些神经元大部分已经根据胞体的背腹侧和头端尾轴位置以及树突和轴突树状结构进行了形态学分类（Thiele et al., 2014; Lee and Eaton, 1991）。

阐明感觉运动回路的神经结构是理解神经行为的基础（Roh et al., 2011; Briggman and Kristan, 2008）。对于斑马鱼来说，不同的神经元群驱动特定的运动行为，例如与逃逸相关的 C 形弯曲的速度（O'Malley et al., 1996; Liu and Fetcho, 1999; Prugh et al., 1982），自发和刺激诱发的转向（Orger et al., 2008; Huang et al., 2013），向前游动（Kimura et al., 2013）以及眼球共轭运动（ballistic eye movements）（Schoonheim et al., 2010）。斑马鱼在发育早期就开始形成运动行为及其神经基础。具体来说，斑马鱼受精后第一天（dpf）的特征就是快速的运动行为（Saint-Amant and Drapeau, 1998）。斑马鱼透明胚胎和仔鱼为研究早期胚胎阶段的发育提供了独特的模型，通过使用基因操作、光学成像和常规电生理技术，很容易观察到产生鱼类行为（例如游泳）节律模式的机制（Masino and Fetcho, 2005）。

在斑马鱼受精后 17h 的胚胎内就可观察到其首次的活动（Saint-Amant and Drapeau, 1998），这种活动包括尾部像节拍器般的对侧收缩以及尾部和头部缓慢环绕的行为。实际上，在完整的受精卵中就可以观察到这种活动，将胚胎从绒毛膜中取出后更加容易观察（Saint-Amant and Drapeau, 1998）。另外，1dpf 仔鱼中也可以观察到诱发活动，其头部或尾部被触碰后会快速地自发地进行对侧收缩（Saint-Amant and Drapeau, 1998）。

上述 1dpf 仔鱼表现出的活动似乎是反射性活动，真正意义上的自发活动才刚刚开始形成。2dpf 时，仔鱼没有表现出活跃的运动行为，即使被束缚，也没有表现出强烈和快速的逃避或挣扎反应。5dpf 时，仔鱼尾部开始轻微且缓慢地弯曲，以便探索环境和寻找食物（Wyart and Thirumalai, 2019）。2dpf 时的仔鱼被触碰后的回避向前游动是斑马鱼的首次逃逸反应（Saint-Amant and Drapeau, 1998）。这时的游动幅度较低，频率较高。相比较而言，17dpf 时自发游动的幅度较大，频率较低（Saint-Amant and Drapeau, 1998）。当第一次游动发生后，接下来的游动的频率就会不断增加（Saint Amant and Drapeau, 1998）。与其他鱼类行为一样，斑马鱼首次游动的突然出现，表明其潜在神经回路可能是遗传的，而不是后天学习的（Saint-Amant and Drapeau, 1998）。

仔鱼表现出缓慢游动和加速游动两种向前游动形式（表 8-1），两者具有不同的游动速度、摆尾频率，以及身体的弯曲角度或位置（Day, 2008）。5dpf 时，仔鱼就表现出向前游动、转身、捕获猎物、逃避反应和挣扎等多种运动形式。虽然这些运动形式间有细微的差别，但可以通过实验进行调节，从而分析其潜在的神经回路。尽管这种分析不能精确地回答控制运动的神经回路，例如回路的解剖细节，但也提供了一个分析神经输出的结构框架（Day, 2008）。

斑马鱼可能通过改变摆尾频率、弯曲幅度等运动学变量，以调整游动速度和位置。在不同的导航转向行为中也使用了保守的运动模式，包括光运动和暗诱导的转向。所有这些运动形式似乎都采用了“慢”运动系统，但这种“慢”与仔鱼在逃逸时C形启动所采用的高频、快速运动有所不同。根据仔鱼的惊吓程度，其身体弯曲的程度也分为C形、S形和中间型。一般来说，仔鱼存在两种不同的运动形式：第一种是以C形启动为特征的沿身体一侧同时进行的活动；第二种是以S形启动为特征的在头、尾两侧同时进行的活动。头部刺激通常引起C形启动，而尾部刺激可引起C形启动和S形启动。S形启动在早期发育阶段发生频率较高，可能是鱼类的一种基本逃逸行为，通过与C形启动的对比来理解简单的神经回路（Liu et al., 2012）。同时，像游动捕猎这种运动形式，摆尾频率的调节和躯干弯曲更加复杂，自然也就要求软件具有更高的计算能力（Liu et al., 2012）。

在斑马鱼胚胎和仔鱼中，运动神经元遵循尺寸大小原则进行募集，最终形成中枢神经系统和肌肉组织之间的分界面（Gabriel et al., 2011）。在较低游动频率下，小的运动神经元运动活跃，但随着游动速度的增加，更大的运动神经元开始募集，活跃细胞的数量随之增加（Gabriel et al., 2011）。兴奋性中间神经元也通过设置运动网络的兴奋性结构来改变游动速度（Eklöf-Ljunggren et al., 2012）。后脑中的V2a神经元群与脊髓中的神经元相连（Kimura et al., 2013; Kinkhabwala et al., 2011），可以调控斑马鱼的运动行为。大量实验表明，后脑中发育的V2a中间神经元是实施精密行为的关键结构（Pujala and Koyama, 2019）。早期生成的V2a神经元是逃避和挣扎反射行为的基础。进一步的实验表明，V2a神经元细胞的生成顺序决定了它们的生物特性以及它们如何连接到脊髓神经元，从而导致不同时间生成的神经元支持不同的行为（Wyart and Thirumalai, 2019）。在仔鱼中，旧的中间神经元位于背侧，而新的中间神经元位于腹侧（McLean and Fetcho, 2009）。腹侧新的运动神经元可能在发育后期生成，此时动物开始缓慢地游动（Gabriel et al., 2011）。这些新发育的运动神经元的募集受到突触驱动和内在特性的控制，但与它们的募集阻力无关（Gabriel et al., 2011）。在孵化后2周，斑马鱼会出现红色肌肉纤维（Van Raamsdonk et al., 1982）。

8.4　成鱼的运动行为

斑马鱼生长过程中，它们的身体大小和形状会发生变化，运动模式和速度也会相应地改变。事实上，鱼体的大小直接影响运动的生物力学，通常随着身体尺寸的增大，游动频率会降低。在仔鱼阶段，成对的胸鳍、臀鳍以及未配对的背鳍和尾鳍取代了鱼体正中的鳍褶（Cubbage and Mabee, 1996; Van Eeden et al., 1996）。因此，游动中推进力和阻力的分布和作用都发生了改变。同样，斑马鱼成长过程中，控制运动神经元募集的机制也在发生变化（Ausborn et al., 2012），这与脊髓区域下行输入的变化一致（Gabriel et al., 2011; Menelaou and McLean, 2012），有助于提高其游动速度和游动距离。

仔鱼和成鱼的游动行为都涉及红色肌纤维（缓慢游动时）和白色肌纤维（快速游动时）（Müller and van Leeuwen, 2004）。因此，随着斑马鱼的生长，白色肌肉成熟（Müller and van Leeuwen, 2004）、鱼体流线型化和神经系统发育的综合作用可能增加持续爆发式游动时间。有趣的是，仔鱼阶段鱼体应对阻力产生的形态变化早于流线型行为的变化（McHenry and Lauder, 2005）。在发育过程中，不同游动形式之间能够快速转换，这表明驱动运动的神经回路的形成优先于特定行为的出现。

成鱼控制缓慢游动的运动神经元具有特定的生物膜特性，以便接受更大的突触驱动，但此时神经元的募集顺序与仔鱼时输入阻力（input resistance）之间没有联系（Gabriel et al., 2011）。运动神经元以不同的频率逐步募集，并形成四个特定的形态区域（Gabriel et al., 2011）。运动神经元的募集顺序由特定的生物特性和突触电流强度共同决定，并不遵循输入阻力规律。募集遵循由外侧到内侧的激活顺序，并且不依赖于运动神经元的输入阻力（Gabriel et al., 2011）。位于腹外侧的次级运动神经元在低游动频率下募集，并在运动发作期间一直保持活跃（Gabriel et al., 2011）。它们接受大的突触电流，并且在抑制后反弹爆发（Gabriel et al., 2011）。募集的腹内侧运动神经元的数量随着游动频率的增加而递增，这与它们处于腹背侧的位置有关（Gabriel et al., 2011）。背侧运动神经元表现出强烈的适应性，终止放电后需要大电流刺激才能达到阈值（Gabriel et al., 2011）。因此，不同的运动神经元区域似乎具有特定的突触和内在特性，这些特性的有序组合用于限制游动行为中的神经元的逐步聚集。在发育过程中，运动柱（运动神经元前体区域）的组织结构和运动神经元募集的原则似乎是受调控的，以适应仔鱼从快速爆发游动到其成体缓慢、稳定游动的变化过程。总体看来，观察成鱼的运动回路和游动行为的方法众多，例如兴奋性中间神经元释放的谷氨酸是脊髓运动网络内重要的兴奋性递质，因此在药理学上，可以激活谷氨酸能 *N*- 甲基 -D- 天冬氨酸（NMDA）受体，实现斑马鱼的兴奋活动（Kyriakatos et al., 2011）。然而，NMDA 诱导的兴奋活动表现出较少的频率变化，因此不适用于在很大的频率范围内检测单个脊髓网络神经元的活动。此外，NMDA 受体的持续激活会引起神经元膜电位的强去极化，显著影响其膜电位振荡的波形。因此，与药理学工具相比，使用电刺激激活脊髓运动回路具有许多优势，可以在更大的频率范围内反复激发并募集运动神经元（Kyriakatos et al., 2011）。

研究特定行为的神经基础，最简单的评估方法就是损伤对应的大脑区域。在斑马鱼中，后脑和脊髓的损伤对它们的行为有不同的影响。刚开始，后脑损伤对自发收缩无影响，然而，随着损伤沿着脊髓尾端蔓延，斑马鱼行为逐渐受到抑制，最后当损伤到达第六个体节外，其行为完全停止（Saint-Amant and Drapeau, 1998）。后脑损伤确实影响斑马鱼的触觉反应和游动。后脑损伤可分为三类：①在前端，耳石完整，但与中脑交界处受损；②在中间，耳石附近区域受损；③在末端，耳石受损，但与脊髓的边界保持完整（Saint-Amant and Drapeau, 1998）。后脑不同损伤对斑马鱼的运动行为有不同的影响，完全切除后脑就会消除这些影响（Saint-Amant and Drapeau, 1998）。因此，除了脊髓，完整的后脑对于斑马鱼产生触觉反应是必不可少的。有趣的是，Rohon-Beard 神经元与网织脊髓神经元相互接触，并与三叉神经同时投射到后脑并在那里呈束状排列（Kimmel et al., 1990）。

值得注意的是，不同品系斑马鱼的运动行为也有显著差异，例如成年 ABstrg、TU 和 WIK 品系的斑马鱼的游动距离低于 AB、*casper* 或 EK 品系（Lange et al., 2013）。同时，ABstrg、 *casper*、EK、TU 和 WIK 品系斑马鱼的游动时间比 AB 品系鱼短（Lange et al., 2013）。成年 AB 和 *casper* 品系斑马鱼始终表现出比所有其他品系更长的游动距离（Lange et al., 2013）。最后，一些品系的鱼体大小与其平均游动速度之间也存在相关性，例如 AB、ABstrg、*casper* 和 EK 鱼体大小与速度呈负相关（Lange et al., 2013）。总之，斑马鱼运动的个体发育代表了神经系统和肌肉组织发育之间复杂的相互作用。仔鱼和成鱼的特征性运动形式构成每个生命阶段的特定表现形式（Lange et al., 2013）。

8.5 遗传控制与运动突变体

斑马鱼可以用于制备具有运动缺陷的突变动物，使其成为研究运动神经控制的良好模式生物（Granato et al., 1996）。斑马鱼突变体有多种表型，例如运动丧失或运动降低突变体、机械感觉缺陷突变体、“痉挛”突变体、旋转突变体和运动回路缺陷突变体（Granato et al., 1996）。斑马鱼行为缺陷分析表明，不同的基因参与斑马鱼不同的运动形式，例如它们的触觉反应、有节奏的尾巴运动、平衡控制或一般运动。在一些突变体中，神经系统发育中的特定缺陷也可以被检测到，例如 *nevermind* (*nev*) 和 *macho* (*mao*) 两个基因的突变可影响视顶盖的轴突投射，而 *diwanka* (*diw*) 和 *unplugged* (*unp*) 两个基因的突变则破坏了运动神经元的轴突形成和延伸（Granato et al., 1996）。这些突变体的行为表型是不同的：*accordion*、*bandoneon*、*diwanka* 突变可能会导致头尾两侧同时收缩；*macho* 和 *space cadet* 突变会导致触摸反应丧失；*twice once* 和 *shocked* 突变会加速疲劳；*space cadet* 和 *techno trousers* 突变会导致鱼体形过宽和无规则运动（Fetcho et al., 2008）。

其中一些突变与肌肉功能有关，例如乙酰胆碱受体或二氢吡啶受体突变，可产生以下移动障碍突变体 *Nic1*、*sofa patato* 和 *relaxed*（Ono et al., 2001; Schredelseker et al., 2005; Sepich et al., 1998; Zhou et al., 2006），支架蛋白（rapsyn）基因 *twitch once* 突变体可使乙酰胆碱受体全身扩散，导致鱼体无法连续游动（Ono et al., 2002）。*accordion* 型突变是异质性的，可由以下几种突变引起：*accordion* 突变体中，钙转运蛋白突变可导致肌肉收缩后钙超载而引发肌肉松弛障碍（Gleason et al., 2004; Hirata et al., 2004; Olson et al., 2010）；*zieharmonika* 突变体中，乙酰胆碱酯酶突变可导致高乙酰胆碱突触水平（Downes and Granato, 2004）；*bandoneon* 突变体中，甘氨酸受体突变可导致弯曲时对侧抑制性传递中断（Hirata et al., 2005）。例如，斑马鱼基因组编码五个甘氨酸受体 α 亚单位（*glra1*、*glra2*、*glra3*、*glra4a*、*glra4b*），为研究每个亚型对一般运动行为的独立作用提供了一个可选择的模型（Samarut et al., 2019）。最近研究发现，*glra2*、*a3*、*a4a* 和 *a4b* 不同亚基之间存在很强的功能重叠，只有 *glra1*–/–（*hic*）斑马鱼表现出低活性，表明 *a1* 亚基是参与运动功能的甘氨酸抑制性神经传递中唯一特别需要的亚基（Samarut et al., 2019）。

斑马鱼其他的基因突变涉及发育和神经缺陷。例如 *unplugged* 和 *diwanka* 基因分别参与神经肌肉突触形成（Zhang et al., 2004）和生长锥迁移（Schneider and Granato, 2006）。两种突变都会损害运动轴突的功能，*unplugged* 突变可导致鱼体不能移动，但后来还可以游动，而 *diwanka* 突变会导致手风琴样的收缩（Hutson and Chien, 2002）。这两种突变都会损害运动轴突功能。*diwanka* 基因突变，鱼的吻侧初级运动神经元在脊髓出口处停止，中部、尾侧初级运动神经元在到达水平肌间隔之前停止（Zeller and Granato, 1999），所有 *unplugged* 基因突变神经元都能正确投射于水平肌间隔处，但尾侧和头侧初级神经元分支出现异常（Granato et al., 1996; Zhang et al., 2001）。

一些与基因突变相关的 notch 信号通路，如 *mindbomb* 和 *deadly seven*，改变了与运动相关的神经元的发育（Schier et al., 1996; Gray et al., 2001）。例如两种突变都增加了 Mauthner 细胞的数量，*deadly seven* 突变会增加原代运动神经元以及 RoL2 和 MiD3cm 两种后脑网状脊髓神经元的数量（Schier et al., 1996; Gray et al., 2001）。同时，虽然 *deadly seven* 突变可导致突触靶点在额外神经元之间的均匀再分配（Liu et al., 2003），但会导致 *mindbomb* 胚胎中最外侧的 Mauthner 细胞异常投射（Schier et al., 1996）。另一种 *twitch twice* 突变，在惊吓反应后引发连续的单侧 C 形弯曲，与 Mauthner 细胞的轴突不能穿过中线并向尾部投射有关（Burgess et al., 2009）。类似地，*space cadet* 突变体由于螺旋纤维轴突缺陷会引发连续的单侧 C 形弯曲（Lorent et al., 2001）。

总的来说，由于斑马鱼与人类有很高的遗传同源性，研究斑马鱼运动行为的遗传基础可以成为探索人类中枢神经系统的有力工具。例如，暴露于帕金森病相关毒素或帕金森病易感基因敲除可导致仔鱼阶段游动减少（Flinn et al., 2008）。触摸后鱼的运动减少被用于评估与人类自闭症相关的基因 *shank3* 的功能（图 8-3）（Gauthier et al., 2010）。另外，通过预先的非启动刺激减少惊吓反应也被作为斑马鱼精神分裂症的内在表型（Morris, 2009），因此可以很容易地应用于精神病样状态的遗传模型表征。

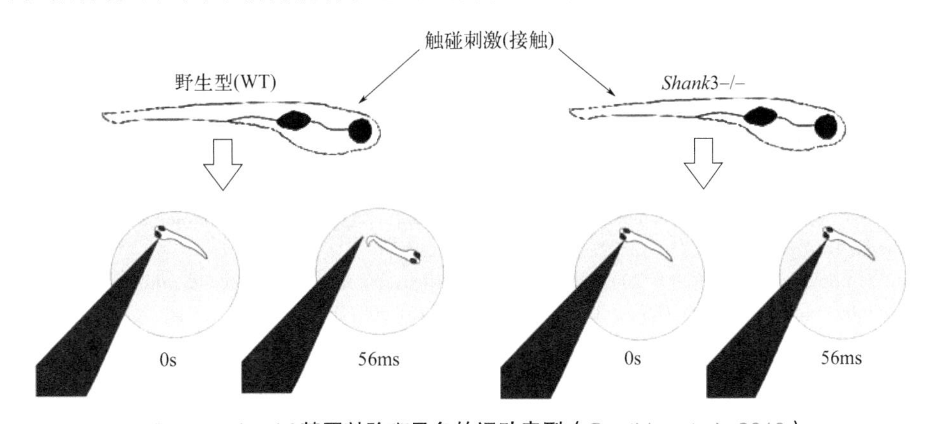

图 8-3 ***shank3*基因敲除斑马鱼的运动表型**（Gauthier et al., 2010）

shank3 基因与自闭症的发病机制有关。对斑马鱼的研究表明 *shank3* 基因敲除导致仔鱼在触觉刺激后运动逃避反应受到抑制

8.6 结论

总之，我们概述了斑马鱼运动行为的基本特征。斑马鱼有特定的神经回路，负责控制从仔鱼到成鱼的整个生命周期中的运动行为（Fetcho et al., 2008; Muto et al., 2011; Arber, 2012）。不同的神经群调节斑马鱼运动行为的发展，而体重、大小和形态的变化影响斑马鱼不同生

长阶段的运动行为。例如改变红、白肌肉比例（Muller and van Leeuwen, 2004），鳍的发育，增加鱼体长度和宽度都有助于斑马鱼运动行为的发育和表达。

从生理学和遗传学角度来看，斑马鱼与哺乳动物有高度同源性（Lakstygal et al., 2018a）。因此，它们具有很高的转化生物医学研究潜力。运动行为的改变与各种神经系统疾病有关，因此可以通过斑马鱼评估这些疾病的潜在生物标志物。斑马鱼的遗传易处理性和与人类的高度同源性使其成为研究运动行为遗传基础的重要模型，在研究中枢神经系统疾病的标记物或发病机制方面做出突出贡献。另外，参与中枢疾病发病机制的一些基因的异常活动（如 *shank3* 基因，图 8-3）也会影响斑马鱼的运动（Gauthier et al., 2010）。另一方面，由于斑马鱼具有体外受精和卵子透明等特征，斑马鱼比其他啮齿动物更容易进行转基因操作，利用这一优势产生了各种具有运动异常性的转基因品系。因此，对斑马鱼运动行为的神经遗传学研究可以为防治人类中枢疾病做出巨大贡献。

最后，斑马鱼也是进行药物干预比较方便的模型。由于斑马鱼可以通过鳃和皮肤快速吸收水中的物质，因此不需要通过鱼体注射药物来评估其运动效果（Lakstygal et al., 2018b）。可以将成鱼或仔鱼浸入药物溶液中，克服了方法学方面的困难（Irons et al., 2010; Gupta et al., 2014），因此斑马鱼比啮齿动物更适合进行高通量药物筛选（图 8-2）。综上所述，斑马鱼是研究运动活动模式、遗传学基础、潜在的神经基质以及药物干预等强有力的模式生物。

（朱思庆翻译　李丽琴审校）

参考文献

Arber, S., 2012. Motor circuits in action: specification, connectivity, and function. Neuron 74 (6), 975-989.

Ausborn, J., Mahmood, R., El Manira, A., 2012. Decoding the rules of recruitment of excitatory interneurons in the adult zebrafish locomotor network. Proceedings of the National Academy of Sciences of the United States of America 109 (52), E3631-E3639.

Briggman, K.L., Kristan Jr., W., 2008. Multifunctional pattern-generating circuits. Annual Review of Neuroscience 31, 271-294.

Budick, S.A., O'Malley, D.M., 2000. Locomotor repertoire of the larval zebrafish: swimming, turning and prey capture. Journal of Experimental Biology 203 (17), 2565-2579.

Burgess, H.A., Johnson, S.L., Granato, M., 2009. Unidirectional startle responses and disrupted lefteright coordination of motor behaviors in robo3 mutant zebrafish. Genes, Brain and Behavior 8 (5), 500-511.

Cachat, J.M., et al., 2011. Video-aided analysis of zebrafish locomotion and anxiety-related behavioral responses. In: Zebrafish Neurobehavioral Protocols. Springer, pp. 1-14.

Cachat, J., et al., 2011. Three dimensional neurophenotyping of adult zebrafish behavior. PLoS One 6 (3), e17597

Casarrubea, M., Santangelo, A., Crescimanno, G., 2017. Multivariate approaches to behavioral physiology. Oncotarget 8 (21), 34022.

Cooke, S.J., et al., 2012. Biotelemetry and Biologging. Fisheries Techniques, third ed. American Fisheries Society, Bethesda, Maryland, pp. 819-860.

Cubbage, C.C., Mabee, P.M., 1996. Development of the cranium and paired fins in the zebrafish *Danio rerio* (Ostariophysi, Cyprinidae). Journal of Morphology 229 (2), 121-160.

Day, L.J., 2008. The Kinematics and Conservation of Motor Patterns in Larval Zebrafish, *Danio rerio*. Northeastern University.

Downes, G.B., Granato, M., 2004. Acetylcholinesterase function is dispensable for sensory neurite growth but is critical for neuromuscular synapse stability. Developmental Biology 270 (1), 232-245.

Eklöf-Ljunggren, E., et al., 2012. Origin of excitation underlying locomotion in the spinal circuit of zebrafish. Proceedings of the National Academy of Sciences of the United States of America 109 (14), 5511-5516.

Fetcho, J.R., Higashijima, S.-i., McLean, D.L., 2008. Zebrafish and motor control over the last decade. Brain Research Reviews 57 (1), 86-93.

Flinn, L., et al., 2008. Zebrafish as a new animal model for movement disorders. Journal of Neurochemistry 106 (5), 1991-1997.

Gabriel, J.P., et al., 2011. Principles governing recruitment of motoneurons during swimming in zebrafish. Nature Neuroscience 14 (1), 93.

Gahtan, E., et al., 2002. Evidence for a widespread brain stem escape network in larval zebrafish. Journal of Neurophysiology 87 (1), 608-614.

Gauthier, J., et al., 2010. De novo mutations in the gene encoding the synaptic scaffolding protein SHANK3 in patients ascertained for schizophrenia. Proceedings of the National Academy of Sciences of the United States of America 107 (17), 7863-7868.

Gleason, M.R., et al., 2004. A mutation in serca underlies motility dysfunction in accordion zebrafish. Developmental Biology 276 (2), 441-451.

Granato, M., et al., 1996. Genes controlling and mediating locomotion behavior of the zebrafish embryo and larva. Development 123, 399-413.

Gray, M., et al., 2001. Zebrafish deadly seven functions in neurogenesis. Developmental Biology 237 (2), 306-323.

Gupta, P., et al., 2014. Assessment of locomotion behavior in adult zebrafish after acute exposure to different pharmacological reference compounds. Drug Development and Therapeutics 5 (2), 127.

Hirata, H., et al., 2004. accordion, a zebrafish behavioral mutant, has a muscle relaxation defect due to a mutation in the ATPase Ca2þ pump SERCA1. Development 131 (21), 5457-5468.

Hirata, H., et al., 2005. Zebrafish bandoneon mutants display behavioral defects due to a mutation in the glycine receptor b-subunit. Proceedings of the National Academy of Sciences of the United States of America 102 (23), 8345-8350.

Huang, K.-H., et al., 2013. Spinal projection neurons control turning behaviors in zebrafish. Current Biology 23 (16), 1566-1573.

Hutson, L.D., Chien, C.-B., 2002. Wiring the zebrafish: axon guidance and synaptogenesis. Current Opinion in Neurobiology 12 (1), 87-92.

Irons, T., et al., 2010. Acute neuroactive drug exposures alter locomotor activity in larval zebrafish. Neurotoxicology and Teratology 32 (1), 84-90.

Kalueff, A.V., et al., 2013. Towards a comprehensive catalog of zebrafish behavior 1.0 and beyond. Zebrafish 10 (1), 70-86.

Kimmel, C.B., Hatta, K., Metcalfe, W.K., 1990. Early axonal contacts during development of an identified dendrite in the brain of the zebrafish. Neuron 4 (4), 535-545.

Kimura, Y., et al., 2013. Hindbrain V2a neurons in the excitation of spinal locomotor circuits during zebrafish swimming. Current Biology 23 (10), 843-849.

Kinkhabwala, A., et al., 2011. A structural and functional ground plan for neurons in the hindbrain of zebrafish. Proceedings of the National Academy of Sciences of the United States of America 108 (3), 1164-1169.

Kyriakatos, A., et al., 2011. Initiation of locomotion in adult zebrafish. Journal of Neuroscience 31 (23), 8422-8431.

Lakstygal, A.M., de Abreu, M.S., Kalueff, A.V., 2018a. Zebrafish models of epigenetic regulation of CNS functions. Brain Research Bulletin 142, 344-351.

Lakstygal, A.M., et al., 2018b. Zebrafish models of diabetes related CNS pathogenesis. Progress in Neuro-Psychopharmacology and Biological Psychiatry 92, 48-58.

Lange, M., et al., 2013. Inter-individual and inter-strain variations in zebrafish locomotor ontogeny. PLoS One 8 (8), e70172.

Lee, R.K., Eaton, R.C., 1991. Identifiable reticulospinal neurons of the adult zebrafish, *Brachydanio rerio*. Journal of Comparative Neurology 304 (1), 34-52.

Liao, J.C., Fetcho, J.R., 2008. Shared versus specialized glycinergic spinal interneurons in axial motor circuits of larval zebrafish. Journal of Neuroscience 28 (48), 12982-12992.

Liu, K.S., Fetcho, J.R., 1999. Laser ablations reveal functional relationships of segmental hindbrain neurons in zebrafish. Neuron 23 (2), 325-335.

Liu, K.S., et al., 2003. Mutations in deadly seven/notch1a reveal developmental plasticity in the escape response circuit. Journal of Neuroscience 23 (22), 8159-8166.

Liu, Y.C., Bailey, I., Hale, M.E., 2012. Alternative startle motor patterns and behaviors in the larval zebrafish (*Danio rerio*). Journal of Comparative Physiology A Neuroethology, Sensory, Neural, and Behavioral Physiology 198 (1), 11-24.

Lorent, K., et al., 2001. The zebrafish space cadet gene controls axonal pathfinding of neurons that modulate fast turning movements. Development 128 (11), 2131-2142.

Marques, J.C., et al., 2018. Structure of the zebrafish locomotor repertoire revealed with unsupervised behavioral clustering. Current Biology 28 (2), 181-195. e5.

Masino, M.A., Fetcho, J.R., 2005. Fictive swimming motor patterns in wild type and mutant larval zebrafish. Journal of Neurophysiology 93 (6), 3177-3188.

McHenry, M.J., Lauder, G.V., 2005. The mechanical scaling of coasting in zebrafish (*Danio rerio*). Journal of Experimental Biology 208 (12), 2289-2301.

McLean, D.L., Fetcho, J.R., 2009. Spinal interneurons differentiate sequentially from those driving the fastest swimming movements in larval zebrafish to those driving the slowest ones. Journal of Neuroscience 29 (43), 13566-13577.

Menelaou, E., McLean, D.L., 2012. A gradient in endogenous rhythmicity and oscillatory drive matches recruitment order in an axial motor pool. Journal of Neuroscience 32 (32), 10925-10939.

Morris, J.A., 2009. Zebrafish: a model system to examine the neurodevelopmental basis of schizophrenia. In: Progress in Brain Research. Elsevier, pp. 97-106.

Müller, U.K., van Leeuwen, J.L., 2004. Swimming of larval zebrafish: ontogeny of body waves and implications for locomotory development. Journal of Experimental Biology 207 (5), 853-868.

Muto, A., et al., 2011. Genetic visualization with an improved GCaMP calcium indicator reveals spatiotemporal activation of the spinal motor neurons in zebrafish. Proceedings of the National Academy of Sciences of the United States of America 108 (13), 5425-5430.

O'Malley, D.M., Kao, Y.-H., Fetcho, J.R., 1996. Imaging the functional organization of zebrafish hindbrain segments during escape behaviors. Neuron 17 (6), 1145-1155.

Olson, B.D., Sgourdou, P., Downes, G.B., 2010. Analysis of a zebrafish behavioral mutant reveals a dominant mutation in atp2a1/SERCA1. Genesis 48 (6), 354-361.

Ono, F., et al., 2001. Paralytic zebrafish lacking acetylcholine receptors fail to localize rapsyn clusters to the synapse. Journal of Neuroscience 21 (15), 5439-5448.

Ono, F., et al., 2002. The zebrafish motility mutant twitch once reveals new roles for rapsyn in synaptic function. Journal of Neuroscience 22 (15), 6491-6498.

Orger, M.B., et al., 2008. Control of visually guided behavior by distinct populations of spinal projection neurons. Nature Neuroscience 11 (3), 327.

Prugh, J., Kimmel, C.B., Metcalfe, W.K., 1982. Noninvasive recording of the Mauthner neurone action potential in larval zebrafish. Journal of Experimental Biology 101 (1), 83-92.

Pujala, A., Koyama, M., 2019. Chronology-based architecture of descending circuits that underlie the development of locomotor repertoire after birth. elife 8, e42135.

Roh, J., Cheung, V.C., Bizzi, E., 2011. Modules in the brain stem and spinal cord underlying motor behaviors. Journal of Neurophysiology 106 (3), 1363-1378.

Saint-Amant, L., Drapeau, P., 1998. Time course of the development of motor behaviors in the zebrafish embryo. Journal of Neurobiology 37 (4), 622-632.

Sakamoto, K.Q., et al., 2009. Can ethograms be automatically generated using body acceleration data from free-ranging birds? PLoS One 4 (4), e5379.

Samarut, E., et al., 2019. Individual knock out of glycine receptor alpha subunits identifies a specific requirement of glra1 for motor function in zebrafish. PLoS One 14 (5), e0216159.

Schier, A.F., et al., 1996. Mutations affecting the development of the embryonic zebrafish brain. Development 123 (1), 165.

Schneider, V.A., Granato, M., 2006. The myotomal diwanka (lh3) glycosyltransferase and type XVIII collagen are critical for motor growth cone migration. Neuron 50 (5), 683-695.

Schoonheim, P.J., et al., 2010. Optogenetic localization and genetic perturbation of saccade-generating neurons in zebrafish. Journal of Neuroscience 30 (20), 7111-7120.

Schredelseker, J., et al., 2005. The b1a subunit is essential for the assembly of dihydropyridine-receptor arrays in skeletal muscle. Proceedings of the National Academy of Sciences of the United States of America 102 (47), 17219e17224.

Sepich, D.S., et al., 1998. An altered intron inhibits synthesis of the acetylcholine receptor a-subunit in the paralyzed zebrafish mutant nic1. Genetics 148 (1), 361-372.

Stewart, A., et al., 2010. The developing utility of zebrafish in modeling neurobehavioral disorders. International Journal of Comparative Psychology 23 (1).

Stewart, A.M., et al., 2015. A novel 3D method of locomotor analysis in adult zebrafish: implications for automated detection of CNS drug-evoked phenotypes. Journal of Neuroscience Methods 255, 66-74.

Thiele, T.R., Donovan, J.C., Baier, H., 2014. Descending control of swim posture by a midbrain nucleus in zebrafish. Neuron 83 (3), 679-691.

Van Eeden, F., et al., 1996. Genetic analysis of fin formation in the zebrafish, *Danio rerio*. Development 123 (1), 255-262.

Van Raamsdonk, W., et al., 1982. Differentiation of muscle fiber types in the teleost *Brachydanio rerio*, the zebrafish. Anatomy and Embryology 164 (1), 51-62.

Wyart, C., Thirumalai, V., 2019. Locomotion: building behaviors, one layer at a time. eLife 8, e46375.

Zeller, J., Granato, M., 1999. The zebrafish diwanka gene controls an early step of motor growth cone migration. Development 126 (15), 3461-3472.

Zhang, J., et al., 2001. A dual role for the zebrafish unplugged gene in motor axon pathfinding and pharyngeal development. Developmental Biology 240 (2), 560-573.

Zhang, J., et al., 2004. Zebrafish unplugged reveals a role for muscle-specific kinase homologs in axonal pathway choice. Nature Neuroscience 7 (12), 1303-1309.

Zhou, W., et al., 2006. Non-sense mutations in the dihydropyridine receptor b1 gene, CACNB1, paralyze zebrafish relaxed mutants. Cell Calcium 39 (3), 227-236.

第九章

斑马鱼睡眠的行为标准和研究技术

David Zada[1], Lior Appelbaum[1]

9.1 引言

睡眠是一种进化保守的行为，对所有动物包括从水母到人类来说都是必不可少的（Campbell and Tobler, 1984; Nath et al., 2017）。睡眠障碍是一个普遍存在的危险因素，与衰老和诸如癌症及神经退行性病变等各种疾病有关（Stankiewicz et al., 2017; Taillard et al., 2019）。睡眠的典型量化方法包括采用整体动物行为学标准（Toth and Bhargava, 2013）、脑电图（electroencephalogram, EEG）、肌电图（electromyogram, EMG）和眼电图（electrooculogram, EOG）（Brown et al., 2012）以及神经元网络的活动（Vyazovskiy and Harris, 2013）进行表征。最近的研究表明，可采用分子标记物在单细胞分辨率水平监测睡眠情况（Zada et al., 2019）。睡眠对各种生理过程都很重要，比如，能量守恒（Berger and Phillips, 1995）、大分子合成（Mackiewicz et al., 2008）、代谢清除（Xie et al., 2013）、突触可塑性（Tononi and Cirelli, 2019）、记忆和学习（Maquet, 2001）、氧化应激恢复（Hill et al., 2018; Villafuerte et al., 2015）和细胞核稳态等（Bellesi et al., 2016; Cheung et al., 2019; Zada et al., 2019）。

对人类而言，睡眠可由一系列行为学标准和大量神经元的电生理指标记录来定义。EEG 可记录皮质神经元的平均活动，EMG 可监测骨骼肌的活动，EOG 可量化眼球的水平运动。这些技术可用于定义三种主要的大脑状态：清醒状态、非快速眼动（non-rapid-eye-movement, NREM）睡眠和快速眼动（rapid-eye-movement, REM）睡眠（Brown et al., 2012）。呈现低振幅高频 EEG 波和高肌张力可定义为清醒状态；呈现高振幅低频 EEG 波和低肌张力可定义为 NREM 睡眠；而 EEG 波与清醒状态相似，同时伴有肌张力完全丧失和特别的快速眼球运动则定义为 REM 睡眠（Brown et al., 2012）。值得注意的是，NREM 和 REM

[1] The Faculty of Life Sciences and the Multidisciplinary Brain Research Center, Bar-Ilan University, Ramat Gan, Israel.

睡眠主要在哺乳动物中发现，但也在几种鸟类和爬行动物类中发现（Campbell and Tobler, 1984; Zimmerman et al., 2008; Rattenborg et al., 2016; Shein-Idelson et al., 2016）。到目前为止，在低等脊椎动物如鱼类和无脊椎动物中，还没有发现像哺乳动物一样的EEG，其原因可能在于其前脑和皮层发育不全，或者缺乏合适的工具和实验装置来监测其EEG。

睡眠和清醒过程构成了全部的大脑状态。然而，已有报道表明，在某些动物中具有单个大脑半球睡眠和特定脑区局部睡眠的特征，比如海豚和候鸟（Krueger et al., 2019; Rattenborg, 2017; Mascetti, 2016）。因此，阐明不同物种白天和晚上所有时间内睡眠与清醒时整个大脑的神经元活动和神经回路的特性是极其重要的（Weber and Dan, 2016）。随着新的遗传学和影像学工具的开发，例如基因编码的钙成像和电压敏感指示剂的出现，使得通过单细胞分辨率记录神经元活性成为可能（Ahrens et al., 2012; Akerboom et al., 2012）。特别是在哺乳动物研究中，采用这些工具和传统的电生理实验来鉴定睡眠和清醒活动区域，包括促进睡眠的大脑区域——腹外侧视前区（ventrolateral preoptic area, VLPO），以及促进清醒的核团——蓝斑（locus coeruleus, LC）（SternNaidoo, 2015）。清醒和睡眠状态之间的转换是由几种神经递质进行调节的。例如，下丘脑泌素 / 食欲素（hypocretin/orexin, Hcrt）神经元诱导从睡眠到觉醒的转变，这些神经元的丧失会导致嗜睡症（Sakurai, 2007; Sutcliffe and de Lecea, 2002）。已经有大量文献报道了关于调节睡眠和清醒期间大脑活动的神经元回路方面的研究（Eban-Rothschildet et al., 2017; Saper and Fuller, 2017; Weber and Dan, 2016）。

斑马鱼是一种简单的脊椎动物，在进化树中处于重要位置，仅次于无脊椎动物。它是一种白天活动的动物，神经系统相对简单，和人类一样都是晚上睡觉（Elbaz et al., 2013）。此外，转基因斑马鱼非常适合于基因操作、分子和药理筛选，也适用于活体动物的单分子和细胞成像。考虑到这些优势，在过去的20年里，斑马鱼已经成为研究睡眠的最受欢迎的动物模型。此外，对突变斑马鱼和转基因斑马鱼进行了睡眠相关的药理学和遗传学操作。药理实验表明，药物对斑马鱼的唤醒和睡眠具有与对哺乳动物一样的进化上的保守作用（Rihel et al., 2010b）。例如，褪黑素可促进斑马鱼睡眠（Zhdanova et al., 2001），而莫达非尼可诱导斑马鱼觉醒（Nishimura et al., 2015; Sigurgeirsson et al., 2011）。候选基因过表达和突变以及特定神经元群体的消融、沉默和激活均可用于鉴定和表征参与调节睡眠和觉醒的基因和神经回路（Bringmann, 2019; Levitas-Djerbi Appelbaum, 2017）。例如，对Hcrt系统的操作表明，与哺乳动物类似，Hcrt可维持斑马鱼的觉醒状态（Elbaz et al., 2017）。此外，下丘脑神经肽神经降压素的突变可导致斑马鱼夜间睡眠增加（Levitas-Djerbi et al., 2019），缺乏下丘脑神经肽QRFP（一种由136个氨基酸组成的广泛存在于脑内的神经内分泌肽）的斑马鱼睡眠时间明显减少（Chen et al., 2016）。已经有广泛的研究报道了可调控斑马鱼睡眠的基因和神经元回路。因此，本章将主要讨论关于斑马鱼仔鱼和成鱼睡眠的行为学标准，并推荐一些其他的睡眠标志物。

9.2 表征睡眠的行为学标准

睡眠的定义通常是指在行为学上静止不动的一段时间，睡眠受以下因素调节：①生物钟，生物钟设定了睡眠时间，即夜间睡眠或日间睡眠；②内稳态，可以通过睡眠剥夺和睡眠反弹来研究。睡眠依赖性的行为学上的静止不动还可能与以下因素有关：③对外部刺激的觉醒阈值增加；④物种的特定体位；⑤快速反转（quick reversibility）（图 9-1）。

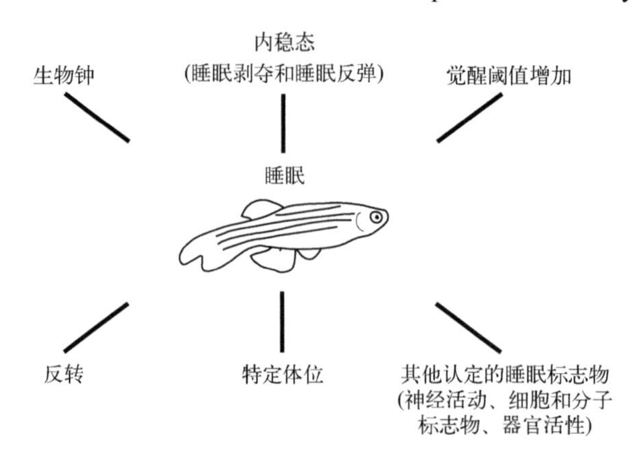

图 9-1 **定义斑马鱼仔鱼和成鱼睡眠的工具和方法**

与其他动物的标准类似，斑马鱼的睡眠可以通过实验来定义，这些实验显示睡眠由生物钟和体内平衡过程调节，睡眠时觉醒阈值增加、保持特定的体位、快速反转。其他睡眠标志物，如神经活动、细胞和分子标志物，以及器官活性（心率、眼睛或尾巴运动），也可能与斑马鱼清醒和睡眠有关

在大多数研究中，实验动物的睡眠都是通过其行为学实验来定义的。*Cassiopea* 水母的睡眠状态定义为脉冲率降低（Nath et al., 2017）；秀丽隐杆线虫（*Caenorhabditis elegans*）在昏睡期表现为静止不动（Iwanir et al., 2013; Raizen et al., 2008）；美国加州海兔（*Aplysia californica*）的睡眠状态表现为至少 1min 静止不动（Vorster et al., 2014）；黑腹果蝇（*Drosophila melanogaster*）（Hendricks et al., 2000）和蜜蜂（Eban-Rothschild and Bloch, 2008; Sauer et al., 2003）的睡眠状态表现为至少 5min 静止不动。青蛙的睡眠行为定义为全身僵直，表现为一种紧张性或放松性的肌肉紧张状态，并且闭眼（Kulikov et al., 1994; Popova et al., 1984）。小鼠的睡眠行为通过 EEG/EMG 进行监测和定义，与睡眠高度相关的行为表现为持续不动的时间超过 40s（Pack et al., 2007）。对猴而言，NREM 睡眠表现为闭眼和不运动，而 REM 睡眠在行为学上表现为闭眼，头部下垂，眼睛持续运动，面部和四肢抽搐，以及外耳突然出现快速运动等（Balzamo et al., 1998）。与无脊椎动物和脊椎动物相似，斑马鱼的仔鱼和成鱼睡眠分别定义为持续不动的时间分别为 1min 和 6s（Elbaz et al., 2012; Prober et al., 2006; Yokogawa et al., 2007）。值得注意的是，总睡眠时间并不一定反映自发活动的水平（Elbaz et al., 2012）。用来定义斑马鱼睡眠的行为学标准和实验是多种多样的。本章将讨论每个行为学标准的利与弊。

9.3 清醒 / 睡眠周期的生物钟调节

在昼夜循环的过程中，动物行为的许多方面都会发生显著变化。为了适应环境因素的节律性变化，如食物的可获得性、捕食风险、温度和光线，动物已经进化出一种内在的节律器，即生物钟，它可以预测即将发生的变化并控制节律性的生理和行为（Pittendrigh, 1993; Takahashi, 2017）。这种生物钟可能通过环境信号（主要是光线）产生，并使内源性节律与 24h 的太阳日同步（Ben-Moshe et al., 2014）。睡眠 / 觉醒周期是生物钟标志性的行为输出。然而，每天的睡眠时间因物种而异。例如，小鼠主要在白天睡觉（Pack et al., 2007），而蝇主要在晚上睡觉（Hendricks et al., 2000; Shaw et al., 2000）。蠕虫主要在昏睡期（lethargus period）睡眠，其睡眠可能发生在白天或晚上（Raizen et al., 2008）。和人类一样，斑马鱼也是白天活动晚上睡觉的动物，斑马鱼的睡眠 / 觉醒周期是由生物钟调节的（Elbaz et al., 2013）。有节律的睡眠 / 觉醒周期通常在不同的光照 / 黑暗（light/dark, LD）条件下监测，使用红外线摄像机可以检测动物在黑暗中的行为。为了区分光照和生物钟效应，测定了包括持续光照（continuous light, LL）、黑暗（dark, DD），或昏暗光照（dim light, dLL）等恒定条件下的自发活动。

对于成年斑马鱼而言，光照 / 黑暗（LD）周期下自发活动和清醒状态在白天达到高峰（Cruz et al., 2017; Hurd et al., 1998; Sigurgeirsson et al., 2013; Singh et al., 2013; Yoko-gawa et al., 2007）。值得注意的是，尽管 LD 下的节律是稳定的，但在恒定的条件下，其振幅会降低。在 DD 下，在实验观察日，自发活动和清醒时间增加，但节律幅度降低（Hurd et al., 1998; Sigurgeirsson et al., 2013; Singh et al., 2013; Yokogawa et al., 2007）。在 LL 下，由于光驱动了斑马鱼的持续活动和觉醒效应，其自发活动和睡眠 / 觉醒周期可能不具有节律性（Hurd et al., 1998; Sigurgeirsson et al., 2013; Singh et al., 2013; Yokogawa et al., 2007）。无论如何，成年斑马鱼在 LL 环境下超过一周，最终表现为夜间活动减少，表现为睡眠行为（Yokogawa et al., 2007）。为了减少光线效应，可以监测 dLL 条件下鱼的活动情况。实际上，在这种情况下，成鱼表现出周期性的自发活动，而这种自发活动在夜间减少（Zhdanova et al., 2008）。值得注意的是，在 3 岁和 4 岁的斑马鱼中，衰老与其在白天和夜间活动水平的显著减少有关（Zhdanova et al., 2008）。此外，与年轻成鱼相比，3 岁斑马鱼在夜间睡眠减少，而在白天没有观察到两者的睡眠情况有明显的差异（Stankiewicz et al., 2017）。

4dpf 的斑马鱼仔鱼表现出稳定的自发活动昼夜节律（Hurd and Cahill, 2002）。与成鱼相似，在 LD 条件下，仔鱼节律性自发活动在白天达到高峰（Cahill et al., 1998; Grone Cahill et al., 2017; Rihel Cahill et al., 2010a）。在DD条件下，仔鱼也表现出有节律性自发活动，但是日间和夜间的差异性要比 LD 条件下的低（Hurdand Cahill, 2002）。在 LL 条件下，节律性自发活动被光线效应所彻底破坏。为了克服这个问题，可以在 dLL 条件下测试，光强的降低能使斑马鱼产生与其在 DD 条件下相同的节律性自发活动，但是，斑马鱼在 dLL 条件下较其在 DD 条件下的自发活动具有更高的振幅和较低的变异性（Appelbaum et al., 2009, 2010; Elbaz et al., 2013; Smadja Storz et al., 2013; Tovin et al., 2012）。值得注意的是，无论是白天还是晚上，仔鱼的睡眠都呈碎片状，睡眠的持续时间为一至几十分钟不等。睡眠的持续时间和进入睡眠状态的次数在夜间增加。此外，与白天相比，夜间清醒期的运动活动减少（Zada et al., 2019）。

为了进一步了解光线和内源性生物钟的作用，一种简洁的方法就是敲除生物钟基因或使其突变，并检测基因突变的仔鱼在各种不同的光照条件下的行为（Ben-Moshe Livne et al., 2016; Tan et al., 2008）。然而，监测这些不同光照条件下的节律性自发活动并不能将睡眠 /

觉醒周期与节律性自发活动区分开，因为这两种节律都是由光线和生物钟控制的。因此，在睡眠研究中，除了经典的昼夜节律实验外，最好在白天和夜间都进行内稳态过程的影响的测试。

9.4 睡眠剥夺和内稳态

睡眠的内稳态系统与设定睡眠时间的生物钟共同调节睡眠机制。因此，人们预测在夜间睡眠较少也即被剥夺睡眠的斑马鱼会在第二天以更多的睡眠时间来弥补这种睡眠不足的情况，也即睡眠反弹。在所有睡眠剥夺研究方法中，长时间睡眠剥夺会使受试者产生应激反应（stress effect）。啮齿类动物的睡眠剥夺方案及其对应激反应的影响研究表明，温和的处理方式更有利于减少应激反应（Colavito et al., 2013）。因此，所有的睡眠剥夺方案都试图减少斑马鱼的应激反应。

在鱼体内几乎所有的细胞中都含有光受体，所以光是一种高效的睡眠抑制剂（Pagano et al., 2017; Vatine et al., 2009）。例如，在整个晚上或晚上的部分时间使用恒定的光或者光脉冲会抑制仔鱼或成鱼的睡眠（Lv et al., 2019; Pinheiro-da-Silva et al., 2017; Sigurgeirsson et al., 2013; Singh et al., 2013; Yoko-gawa et al., 2007）。然而，因为光线是控制生物钟的关键因素，所以仅仅利用光线进行睡眠剥夺实验是不够的。因为独立于睡眠内稳态压力的昼夜循环会重新调整，所以夜间光照处理可能会导致第二天的睡眠反弹。因此，开展夜间鱼的睡眠剥夺实验时，建议监测第二天黑暗条件下的睡眠反弹情况。为了剥夺成年斑马鱼的睡眠，在 6h 的夜晚期间，只要斑马鱼休息时间达到 6 s 就给予电击脉冲，正如预期的那样，斑马鱼第二天的睡眠时间相应增加，并且与其对外部机械刺激的敏感性降低有关（Sigurgeirsson et al., 2013; Yokogawa et al., 2007）。使用中度电击脉冲可能会对斑马鱼的大脑产生直接的神经生理影响（Liu et al., 2018），在理解睡眠依赖的神经元连接和分子机制的研究中应该考虑到这一点。

在仔鱼睡眠剥夺实验中，在夜晚持续 6h 采用恒定的振动刺激，这些睡眠剥夺的仔鱼在次日均出现了睡眠反弹（Zhdanova et al., 2001）。我们应该考虑到持续的和同步的刺激会导致产生行为习惯，因此，刺激的强度和频率应该是随机的，建议重复刺激之间的间隔最长为 15s（Burgess and Granato, 2007; Woods et al., 2014）。事实上，在夜晚最初的 4 ～ 6h 内给予随机的轻柔的机械拍打会延长鱼的活动时间，并且随后也会出现睡眠反弹（Appelbaum et al., 2010; Yelin-Bekerman et al., 2015; Zada et al., 2019）。然而，就像其他睡眠剥夺实验方案中的情况一样，不能排除应激反应。为了尽量减少应激反应，也发展了另外的实验方案。水流可以用来阻止睡眠，在用水流对鱼进行 6h 的睡眠剥夺后，睡眠不足的仔鱼在接下来的黑暗期和光亮开始期表现出来的应激反应都减少，表明鱼出现睡眠反弹现象（Aho et al., 2017）。促觉醒药物如莫达非尼（Nishimura et al., 2015; Sigur-geirsson et al., 2011）和咖啡因（Aho et al., 2017）也可以用来剥夺斑马鱼的睡眠。综上所述，使用斑马鱼进行的睡眠剥夺实验可以持续 4 ～ 12h，甚至几天，排除非特异性应激反应和避免习惯化是很重要的。斑马鱼和其他动物模型睡眠剥夺实验的一大挑战就是将人为操作导致的应激反应与缺乏睡眠引起的应激反应区分开。因此，为了使斑马鱼仔鱼和成鱼的睡眠剥夺实验标准化，应该制定其他的实验方案并进行测试。

9.5 唤醒阈值

多种动物实验研究表明，在睡眠期间，唤醒阈值增加。为了区分静止不动的斑马鱼是处于清醒状态还是熟睡状态，有必要证明唤醒静止不动的处于睡眠状态的斑马鱼需要更强的刺激条件。为了确定唤醒阈值，需要使用不同强度的外界刺激。对于成年斑马鱼，在24h的LD周期内每30min给予10ms的电刺激，电刺激增加的量在0～2V。在白天和夜间，对非常低的电压刺激能做出反应的处于活动（清醒）状态的鱼与那些不活动（睡眠）状态的鱼相比，无论处于何种活动（清醒和睡眠）状态，所有鱼类均对较高的电压刺激产生反应（Yokogawa et al., 2007）。同样，年龄较大的鱼（1～4岁）表现出唤醒阈值增加，与白天相比，夜间的鱼对于轻度电击的反应性降低（Zhdanov et al., 2008）。

人们也对仔鱼的唤醒阈值和睡眠进行了研究。采用机械刺激测定仔鱼的唤醒阈值。在白天，给予单次刺激可以引起行为反应，而在晚上，需要给予三次重复的类似刺激才能引起同样的行为反应（Zhdanova et al., 2001）。与之相似，在仔鱼的研究中也使用了多种不同强度的机械式声音刺激，在夜间每分钟给予一次刺激，仔鱼的唤醒阈值由每一次刺激发生反应的仔鱼比例来测定（Singh et al., 2015）。除了声音和机械刺激之外，光或暗脉冲也可以用来确定斑马鱼在睡眠期间的唤醒阈值。一种突然的暗脉冲可以急剧增加成鱼和仔鱼的自发活动，其原因可能是这种刺激与捕食者阴影相类似（Kimmel et al., 1974; Easter and Nicola, 1996）。因此，在白天清醒状态时，所有活动的仔鱼都会在15s内对暗脉冲做出反应，表现为其活动性均增加了。然而，对于那些在暗刺激前就已经入睡达到1min或更长时间的仔鱼来说，其对暗刺激的反应性相应减弱（即唤醒阈值增加）（Prober et al., 2006）。同样的，在夜晚期间的光脉冲也会增加仔鱼的自发活动。尽管所有处于清醒状态的活动的仔鱼对两种强度（相对高一点的光强度大于25lx，低的在14～18lx之间）的光刺激均能在1min内作出反应，但是，在光刺激前静止不动时间超过1min的大多数仔鱼对低强度的光刺激是没有反应的（Elbaz et al., 2012）。惊吓反应也可以作为量化斑马鱼对外界刺激的反应的一个行为指标。在白天和夜间及睡眠剥夺之后，用5V（10 ms）的温和、稳定的时序性电刺激处理仔鱼，仔鱼在夜晚期间和白天开始的睡眠剥夺后对惊吓反应的反应性减弱（即唤醒阈值增加）（Aho et al., 2017）。此外，在白天采用相对较低的声音强度足以让仔鱼的逃逸反应达到最大，然而，在夜间诱发类似反应却需要更高的声音强度（Pantoja et al., 2016; Xiao et al., 2018）。虽然光和黑暗刺激已被成功地用于证明唤醒阈值是睡眠依赖性增加的，但是，鉴于机械、声音和电刺激不太可能影响昼夜节律，因此，应该在需要记录几天行为的长期实验中采用该类刺激。总之，不同的研究结果都表明，仔鱼和成鱼在睡眠过程中唤醒阈值都会增加。

9.6 睡眠体位

人们已经对自然环境或实验室中各种鱼类的睡眠都进行了研究。在睡眠期间，鱼的心跳和呼吸频率降低，并表现出典型的睡眠体位（Campbell and Tobler, 1984）。当鱼类睡觉时，它们为了保护自己可能会到水体中特定的区域睡觉，或者将自己部分或全部埋在沙子中（McNamara et al., 2010）。成年斑马鱼晚上会减少自发活动，有约 60% 到 80% 的时间都在水面附近或水箱底部睡觉（Kacprzak et al., 2017; Singh et al., 2013）。事实上，成年斑马鱼至少 6s 的静止不动时间与偏好处于以上那些位置是相关的（Yokogawa et al., 2007）。与清醒期间观察到的常见活动不同，睡眠体位以静止不动和尾鳍下垂为其特征（Chiu and Prober, 2013; Yokogawa et al., 2007）。与哺乳动物不同的是，在成年斑马鱼上检测不到 REM 睡眠。然而，与哺乳动物类似的是，斑马鱼在睡眠期间的呼吸频率也降低（Árnason et al., 2015）。仔鱼睡眠通常有两种主要体位，一种是在睡眠时低头不动或者位于靠近水箱底部的水平位置躺着（Zhdanova et al., 2001）。显然，为了更好地描述成鱼和仔鱼的睡眠体位，需要开展更多的研究。由于斑马鱼睡眠的大多数行为学研究是通过视频追踪系统同时跟踪几十条仔鱼进行的，因此，获得的数据较为有限。对单条斑马鱼采用高放大倍数的视频追踪系统将有助于更好地描述其睡觉时表现出的精细行为方面的细节。

9.7 定义睡眠的未来工具和标准

在斑马鱼和其他简单脊椎动物中，传统上广泛使用行为学标准来定义睡眠。在斑马鱼仔鱼中，采用高通量方法为睡眠的定义提供了基础。然而，这些行为学标准要求动物能自由移动。此外，一些微视行为，例如，静止鱼体的短期睡眠、运动的鳍和尾巴这些行为学特征都被忽略了。正如在蝇体内进行的研究结果一样，这些附加的行为学特征可以提供更多关于斑马鱼睡眠行为的细节（Geissmann et al., 2019）。

行为学标准的一个主要不足之处在于目前还缺乏在显微镜下定义斑马鱼仔鱼睡眠的手段，这一点至关重要，因为该模型的一个主要优点是能够对透明活体脊椎动物中的单个粒子和细胞成像。在未来的研究中，将相对简单的睡眠生理指标，如眼球运动、心跳、肌肉收缩和尾巴运动等组合起来可以用于定义斑马鱼睡眠。有趣的是，使用基因编码的钙离子指示剂和电压指示剂的全脑成像（Akerboom et al., 2012; Yang and St-Pierre, 2016）或集成有电极的微流控系统设备可用于检测癫痫发作（Hong et al., 2016; Hunyadi et al., 2017），同样可以用来识别清醒期间和睡眠状态的神经活动变化。支持这一观点的是，蝇前脑区域的局部场电位记录和基于 GCaMP 的蘑菇体（昆虫脑内重要的感觉整合中枢）成像均可显示睡眠状态下的神经元活动减少，并且对感觉刺激的神经反应变得迟钝（Bushey et al., 2015; Nitz et al., 2002）。如果在无脊椎动物体内可发现这种睡眠依赖性的脑区变化，那么在斑马鱼体内也可能发现。事实上，斑马鱼仔鱼在休息期间视顶盖的平均神经元活动略有减少（Lee et al., 2017）。即早基因 *c-fos* 表达在仔鱼睡眠剥夺后增加而在夜间减少，也支持这一观察现象（Appelbaum et al., 2010; Elbaz et al., 2015, 2012）。此外，GCaMP 标记的仔鱼后脑神经元单细胞成像结果显示神经元活动在夜间减弱（Zada et al., 2019）。因此，未来对不同脑区神经元活动的表征预计可以用作斑马鱼的睡眠和觉醒标记。

在细胞水平上定义睡眠的能力可以促进所有动物模型和人类睡眠的研究。利用仔鱼在染色体标志物的活体成像研究发现睡眠会增加染色体动力学。在夜间睡眠期间这种染色质的快速移动能够有效地减少 DNA 的损伤，而在清醒期间 DNA 的损伤是积聚的（Zada et al., 2019）。因此，有望在单细胞分辨率水平，基于细胞过程和核过程来定义睡眠。总之，斑马鱼仔鱼和成鱼的睡眠可以应用行为标准来定义。先进的遗传学和成像工具的发展有望进一步阐明细胞和神经网络水平的更为复杂的睡眠结构。这些新的实验方法有望揭示斑马鱼睡眠机制的重要方面，并可以用于其他动物模型，最终用于人体。

致谢

作者感谢 Yael Laure 女士对稿件编辑提供的帮助。这项工作得到了以下单位的支持，包括以色列科学基金会（批准号：961/19）、美国以色列两国科学基金会（BSF，批准号：2017105）、美国国家科学基金会（NSF）-BSF 联合资金 EDGE 计划（NSF-BSF，资助号：2019604）以及美国国家卫生研究院（NIH, R01 MH116470）。

（张瑞华翻译　张毅审校）

参考文献

Aho, V., Vainikka, M., Puttonen, H.A.J., Ikonen, H.M.K., Salminen, T., Panula, P., Porkka-Heiskanen, T., Wigren, .-K., 2017. Homeostatic response to sleep/rest deprivation by constant water flow in larval zebrafish in both dark and light conditions. Journal of Sleep Research 26, 394-400. https://doi.org/10.1111/jsr.12508.

Ahrens, M.B., Li, J.M., Orger, M.B., Robson, D.N., Schier, A.F., Engert, F., Portugues, R., 2012. Brain-wide neuronal dynamics during motor adaptation in zebrafish. Nature 485, 471-477. https://doi.org/10.1038/nature11057.

Akerboom, J., Chen, T.-W., Wardill, T.J., Tian, L., Marvin, J.S., Mutlu, S., Calderón, N.C., Esposti, F., Borghuis, B.G.,Sun, X.R., Gordus, A., Orger, M.B., Portugues, R., Engert, F., Macklin, J.J., Filosa, A., Aggarwal, A., Kerr, R.A., Takagi, R., Kracun, S., Shigetomi, E., Khakh, B.S., Baier, H., Lagnado, L., Wang, S.S.-H., Bargmann, C.I.,Kimmel, B.E., Jayaraman, V., Svoboda, K., Kim, D.S., Schreiter, E.R., Looger, L.L., 2012. Optimization of a GCaMP calcium indicator for neural activity imaging. The Journal of Neuroscience 32, 13819e13840. https://doi.org/10.1523/JNEUROSCI.2601-12.2012.

Appelbaum, L., Wang, G., Yokogawa, T., Skariah, G.M., Smith, S.J., Mourrain, P., Mignot, E., 2010. Circadian and homeostatic regulation of structural synaptic plasticity in hypocretin neurons. Neuron 68, 87e98. https://doi.org/10.1016/j.neuron.2010.09.006.

Appelbaum, L., Wang, G.X., Maro, G.S., Mori, R., Tovin, A., Marin, W., Yokogawa, T., Kawakami, K., Smith, S.J.,Gothilf, Y., Mignot, E., Mourrain, P., 2009. Sleep-wake regulation and hypocretin-melatonin interaction in zebrafish. Proceedings of the National Academy of Sciences of the United States of America 106, 21942-21947. https://doi.org/10.1073/pnas.906637106.

Árnason, B.B., Þorsteinsson, H., Karlsson, K.Æ., 2015. Absence of rapid eye movements during sleep in adult zebrafish. Behavioural Brain Research 291, 189-194. https://doi.org/10.1016/j.bbr.2015.05.017.

Balzamo, E., Van Beers, P., Lagarde, D., 1998. Scoring of sleep and wakefulness by behavioral analysis from video recordings in rhesus monkeys: comparison with conventional EEG analysis. Electroencephalography and Clinical Neurophysiology 106, 206-212.

Bellesi, M., Bushey, D., Chini, M., Tononi, G., Cirelli, C., 2016. Contribution of sleep to the repair of neuronal DNA double-strand breaks: evidence from flies and mice. Scientific Reports 6, 36804. https://doi.org/10.1038/srep36804.

Ben-Moshe, Z., Foulkes, N.S., Gothilf, Y., 2014. Functional development of the circadian clock in the zebrafish pinealgland. BioMed Research International 2014, 235781. https://doi.org/10.1155/2014/235781.

Ben-Moshe Livne, Z., Alon, S., Vallone, D., Bayleyen, Y., Tovin, A., Shainer, I., Nisembaum, L.G., Aviram, I., Smadja-Storz, S., Fuentes, M., Falcón, J., Eisenberg, E., Klein, D.C., Burgess, H.A., Foulkes, N.S., Gothilf, Y., 2016. Genetically blocking the zebrafish pineal clock affects circadian behavior. PLoS Genetics 12, e1006445. https://doi.org/10.1371/journal.pgen.1006445.

Berger, R.J., Phillips, N.H., 1995. Energy conservation and sleep. Behavioural Brain Research 69, 65-73.

Bringmann, H., 2019. Genetic sleep deprivation: using sleep mutants to study sleep functions. EMBO Reports 20. https://doi.org/10.15252/embr.201846807.

Brown, R.E., Basheer, R., McKenna, J.T., Strecker, R.E., McCarley, R.W., 2012. Control of sleep and wakefulness. Physiological Reviews 92, 1087e1187. https://doi.org/10.1152/physrev.00032.2011.

Burgess, H.A., Granato, M., 2007. Modulation of locomotor activity in larval zebrafish during light adaptation. Journal of Experimental Biology 210, 2526e2539. https://doi.org/10.1242/jeb.003939.

Bushey, D., Tononi, G., Cirelli, C., 2015. Sleep- and wake-dependent changes in neuronal activity and reactivity demonstrated in fly neurons using in vivo calcium imaging. Proceedings of the National Academy of Sciencesof the United States of America 112, 4785-4790. https://doi.org/10.1073/pnas.1419603112.

Cahill, G.M., Hurd, M.W., Batchelor, M.M., 1998. Circadian rhythmicity in the locomotor activity of larval zebrafish. NeuroReport 9, 3445-3449.

Campbell, S.S., Tobler, I., 1984. Animal sleep: a review of sleep duration across phylogeny. Neuroscience and Biobehavioral Reviews 8, 269-300.

Chen, A., Chiu, C.N., Mosser, E.A., Kahn, S., Spence, R., Prober, D.A., 2016. QRFP and its receptors regulate locomotor activity and sleep in zebrafish. The Journal of Neuroscience 36, 1823-1840. https://doi.org/10.1523/JNEUR-OSCI.2579-15.2016.

Cheung, V., Yuen, V.M., Wong, G.T.C., Choi, S.W., 2019. The effect of sleep deprivation and disruption on DNA damage and health of doctors. Anaesthesia 74, 434-440. https://doi.org/10.1111/anae.14533.

Chiu, C.N., Prober, D.A., 2013. Regulation of zebrafish sleep and arousal states: current and prospective approaches.Frontiers in Neural Circuits 7, 58. https://doi.org/10.3389/fncir.2013.00058.

Colavito, V., Fabene, P.F., Grassi-Zucconi, G., Pifferi, F., Lamberty, Y., Bentivoglio, M., Bertini, G., 2013. Experimental sleep deprivation as a tool to test memory deficits in rodents. Frontiers in Systems Neuroscience 7, 106. https://doi.org/10.3389/fnsys.2013.00106.

Cruz, B.P., Brongar, L.F., Popiolek, P., Gonçalvez, B.S.B., Figueiredo, M.A., Amaral, I.P.G., Da Rosa, V.S.,Nery, L.E.M., Marins, L.F., 2017. Clock genes expression and locomotor activity are altered along the light-dark cycle in transgenic zebrafish overexpressing growth hormone. Transgenic Research 26, 739e752. https://doi.org/10.1007/s11248-017-0039-9.

Easter, S.S., Nicola, G.N., 1996. The development of vision in the zebrafish (*Danio rerio*). Developmental Biology 180,646-663.

Eban-Rothschild, A., Appelbaum, L., de Lecea, L., 2017. Neuronal mechanisms for sleep/wake regulation and modulatory drive. Neuropsychopharmacology, The official publication of the American College of Neuropsychopharmacology. https://doi.org/10.1038/npp.2017.294.

Eban-Rothschild, A.D., Bloch, G., 2008. Differences in the sleep architecture of forager and young honeybees (*Apis mellifera*). Journal of Experimental Biology 211, 2408-2416. https://doi.org/10.1242/jeb.016915.

Elbaz, I., Foulkes, N.S., Gothilf, Y., Appelbaum, L., 2013. Circadian clocks, rhythmic synaptic plasticity and the sleep-wake cycle in zebrafish. Frontiers in Neural Circuits 7, 9. https://doi.org/10.3389/fncir.2013.00009.

Elbaz, I., Lerer-Goldshtein, T., Okamoto, H., Appelbaum, L., 2015. Reduced synaptic density and deficient locomotor response in neuronal activity-regulated pentraxin 2a mutant zebrafish. Federation of American Societies for Experimental Biology 29, 1220-1234. https://doi.org/10.1096/fj.14-258350.

Elbaz, I., Levitas-Djerbi, T., Appelbaum, L., 2017. The hypocretin/orexin neuronal networks in zebrafish. Current Topics in Behavioral Neurosciences 33, 75-92. https://doi.org/10.1007/7854_2016_59.

Elbaz, I., Yelin-Bekerman, L., Nicenboim, J., Vatine, G., Appelbaum, L., 2012. Genetic ablation of hypocretin neurons alters behavioral state transitions in zebrafish. The Journal of Neuroscience 32, 12961-12972. https://doi.org/10.1523/JNEUROSCI.1284-12.2012.

Geissmann, Q., Beckwith, E.J., Gilestro, G.F., 2019. Most sleep does not serve a vital function: evidence from *Drosophila melanogaster*. Science Advances 5. https://doi.org/10.1126/sciadv.aau9253.

Grone, B.P., Qu, T., Baraban, S.C., 2017. Behavioral comorbidities and drug treatments in a zebrafish scn1lab model of Dravet syndrome. eNeuro 4. https://doi.org/10.1523/ENEURO.0066-17.2017.

Hendricks, J.C., Finn, S.M., Panckeri, K.A., Chavkin, J., Williams, J.A., Sehgal, A., Pack, A.I., 2000. Rest in *Drosophila* is a sleep-like state. Neuron 25, 129-138.

Hill, V.M., O'Connor, R.M., Sissoko, G.B., Irobunda, I.S., Leong, S., Canman, J.C., Stavropoulos, N., Shirasu-Hiza, M.,2018. A bidirectional relationship between sleep and oxidative stress in *Drosophila*. PLoS Biology 16, e2005206. https://doi.org/10.1371/journal.pbio.2005206.

Hong, S., Lee, P., Baraban, S.C., Lee, L.P., 2016. A novel long-term, multi-channel and non-invasive electrophysiology platform for zebrafish. Scientific Reports 6, 28248. https://doi.org/10.1038/srep28248.

Hunyadi, B., Siekierska, A., Sourbron, J., Copmans, D., de Witte, P.A.M., 2017. Automated analysis of brain activity for seizure detection in zebrafish models of epilepsy. Journal of Neuroscience Methods 287, 13e24. https://doi.org/10.1016/j.jneumeth.2017.05.024.

Hurd, M.W., Cahill, G.M., 2002. Entraining signals initiate behavioral circadian rhythmicity in larval zebrafish. Journal of Biological Rhythms 17, 307-314. https://doi.org/10.1177/074873002129002618.

Hurd, M.W., Debruyne, J., Straume, M., Cahill, G.M., 1998. Circadian rhythms of locomotor activity in zebrafish. Physiology and Behavior 65, 465-472.

Iwanir, S., Tramm, N., Nagy, S., Wright, C., Ish, D., Biron, D., 2013. The microarchitecture of *C. elegans* behavior during lethargus: homeostatic bout dynamics, a typical body posture, and regulation by a central neuron. Sleep 36,385e395. https://doi.org/10.5665/sleep.2456.

Kacprzak, V., Patel, N.A., Riley, E., Yu, L., Yeh, J.-R.J., Zhdanova, I.V., 2017. Dopaminergic control of anxiety in young and aged zebrafish. Pharmacology Biochemistry and Behavior 157, 1-8. https://doi.org/10.1016/j.pbb.2017.01.005.

Kimmel, C.B., Patterson, J., Kimmel, R.O., 1974. The development and behavioral characteristics of thestartle response in the zebra fish. Developmental Psychobiology 7, 47-60. https://doi.org/10.1002/dev.420070109.

Krueger, J.M., Nguyen, J.T., Dykstra-Aiello, C.J., Taishi, P., 2019. Local sleep. Sleep Medicine Reviews 43, 14-21.https://doi.org/10.1016/j.smrv.2018.10.001.

Kulikov, A.V., Karmanova, I.G., Kozlachkova, E.Y., Voronova, I.P., Popova, N.K., 1994. The brain tryptophan hydroxylase activity in the sleep-like states in frog. Pharmacology Biochemistry and Behavior 49, 277-279.

Lee, D.A., Andreev, A., Truong, T.V., Chen, A., Hill, A.J., Oikonomou, G., Pham, U., Hong, Y.K., Tran, S., Glass, L.,Sapin, V., Engle, J., Fraser, S.E., Prober, D.A., 2017. Genetic and neuronal regulation of sleep by neuropeptide VF.eLife 6. https://doi.org/10.7554/eLife.25727.

Levitas-Djerbi, T., Appelbaum, L., 2017. Modeling sleep and neuropsychiatric disorders in zebrafish. Current Opinionin Neurobiology 44, 89-93. https://doi.org/10.1016/j.conb.2017.02.017.

Levitas-Djerbi, T., Sagi, D., Lebenthal-Loinger, I., Lerer-Goldshtein, T., Appelbaum, L., 2019. Neurotensin enhances locomotor activity and arousal, and inhibits melanin-concentrating hormone signalings. Neuroendocrinology. https://doi.org/10.1159/000500590.

Liu, A., Vöröslakos, M., Kronberg, G., Henin, S., Krause, M.R., Huang, Y., Opitz, A., Mehta, A., Pack, C.C.,Krekelberg, B., Berényi, A., Parra, L.C., Melloni, L., Devinsky, O., Buzsáki, G., 2018. Immediate neurophysiological effects of transcranial electrical stimulation. Nature Communications 9, 5092. https://doi.org/10.1038/s41467-018-07233-7.

Lv, D.-J., Li, L.-X., Chen, J., Wei, S.-Z., Wang, F., Hu, H., Xie, A.-M., Liu, C.-F., 2019. Sleep deprivation caused a memory defects and emotional changes in a rotenone-based zebrafish model of Parkinson's disease. Behavioural Brain Research 372, 112031. https://doi.org/10.1016/j.bbr.2019.112031.

Mackiewicz, M., Naidoo, N., Zimmerman, J.E., Pack, A.I., 2008. Molecular mechanisms of sleep and wakefulness. Annals of the New York Academy of Sciences 1129, 335-349. https://doi.org/10.1196/annals.1417.030.

Maquet, P., 2001. The role of sleep in learning and memory. Science 294, 1048-1052. https://doi.org/10.1126/science.1062856.

Mascetti, G.G., 2016. Unihemispheric sleep and asymmetrical sleep: behavioral, neurophysiological, and functional perspectives. Nature and Science of Sleep 8, 221-238. https://doi.org/10.2147/NSS.S71970.

McNamara, P., Barton, R.A., Nunn, C.L. (Eds.), 2010. Evolution of Sleep: Phylogenetic and Functional Perspectives.Cambridge University Press, Cambridge, UK; New York.

Nath, R.D., Bedbrook, C.N., Abrams, M.J., Basinger, T., Bois, J.S., Prober, D.A., Sternberg, P.W., Gradinaru, V.,Goentoro, L., 2017. The jellyfish *Cassiopea* exhibits a sleep-like state. Current biology: CB 27. https://doi.org/10.1016/j.cub.2017.08.014, 2984-2990.e3.

Nishimura, Y., Okabe, S., Sasagawa, S., Murakami, S., Ashikawa, Y., Yuge, M., Kawaguchi, K., Kawase, R.,Tanaka, T., 2015. Pharmacological profiling of zebrafish behavior using chemical and genetic classification of sleep-wake modifiers. Frontiers in Pharmacology 6, 257. https://doi.org/10.3389/fphar.2015.00257.

Nitz, D.A., van Swinderen, B., Tononi, G., Greenspan, R.J., 2002. Electrophysiological correlates of rest and activity in *Drosophila melanogaster*. Current biology: CB 12, 1934-1940.

Pack, A.I., Galante, R.J., Maislin, G., Cater, J., Metaxas, D., Lu, S., Zhang, L., Von Smith, R., Kay, T., Lian, J.,Svenson, K.,

Peters, L.L., 2007. Novel method for high-throughput phenotyping of sleep in mice. Physiological Genomics 28, 232-238. https://doi.org/10.1152/physiolgenomics.00139.2006.

Pagano, C., Ceinos, R.M., Vallone, D., Foulkes, N.S., 2017. The fish circadian timing system: the illuminating case of light-responsive peripheral clocks. In: Kumar, V. (Ed.), Biological Timekeeping: Clocks, Rhythms and Behaviour. Springer India, New Delhi, pp. 177-192.

Pantoja, C., Hoagland, A., Carroll, E.C., Karalis, V., Conner, A., Isacoff, E.Y., 2016. Neuromodulatory regulation of behavioral individuality in zebrafish. Neuron 91, 587-601. https://doi.org/10.1016/j.neuron.2016.06.016.

Pinheiro-da-Silva, J., Silva, P.F., Nogueira, M.B., Luchiari, A.C., 2017. Sleep deprivation effects on object discrimination task in zebrafish (*Danio rerio*). Animal Cognition 20, 159-169. https://doi.org/10.1007/s10071-016-1034-x.

Pittendrigh, C.S., 1993. Temporal organization: reflections of a Darwinian clock-watcher. Annual Review of Physiology 55, 16-54. https://doi.org/10.1146/annurev.ph.55.030193.000313.

Popova, N.K., Lobacheva, I.I., Karmanova, I.G., Shilling, N.V., 1984. Serotonin in the control of the sleep-like states in frogs. Pharmacology Biochemistry and Behavior 20, 653-657.

Prober, D.A., Rihel, J., Onah, A.A., Sung, R.-J., Schier, A.F., 2006. Hypocretin/orexin overexpression induces an insomnia-like phenotype in zebrafish. The Journal of Neuroscience 26, 13400-13410. https://doi.org/10.1523/JNEUROSCI.4332-06.2006.

Raizen, D.M., Zimmerman, J.E., Maycock, M.H., Ta, U.D., You, Y., Sundaram, M.V., Pack, A.I., 2008. Lethargus is a *Caenorhabditis elegans* sleep-like state. Nature 451, 569-572. https://doi.org/10.1038/nature06535.

Rattenborg, N.C., 2017. Sleeping on the wing. Interface Focus 7, 20160082. https://doi.org/10.1098/rsfs.2016.0082.

Rattenborg, N.C., Voirin, B., Cruz, S.M., Tisdale, R., Dell'Omo, G., Lipp, H.-P., Wikelski, M., Vyssotski, A.L., 2016.Evidence that birds sleep in mid-flight. Nature Communications 7, 12468. https://doi.org/10.1038/ncomms12468.

Rihel, J., Prober, D.A., Arvanites, A., Lam, K., Zimmerman, S., Jang, S., Haggarty, S.J., Kokel, D., Rubin, L.L.,Peterson, R.T., Schier, A.F., 2010a. Zebrafish behavioral profiling links drugs to biological targets and rest/wake regulation. Science 327, 348-351. https://doi.org/10.1126/science.1183090.

Rihel, J., Prober, D.A., Schier, A.F., 2010b. Monitoring sleep and arousal in zebrafish. Methods in Cell Biology 100,281-294. https://doi.org/10.1016/B978-0-12-384892-5.00011-6.

Sakurai, T., 2007. The neural circuit of orexin (hypocretin): maintaining sleep and wakefulness. Nature Reviews Neuroscience 8, 171-181. https://doi.org/10.1038/nrn2092.

Saper, C.B., Fuller, P.M., 2017. Wake-sleep circuitry: an overview. Current Opinion in Neurobiology 44, 186-192. https://doi.org/10.1016/j.conb.2017.03.021.

Sauer, S., Kinkelin, M., Herrmann, E., Kaiser, W., 2003. The dynamics of sleep-like behaviour in honey bees. Journal of Comparative Physiology A: Neuroethology, Sensory, Neural, and Behavioral Physiology 189, 599-607. https://doi.org/10.1007/s00359-003-0436-9.

Shaw, P.J., Cirelli, C., Greenspan, R.J., Tononi, G., 2000. Correlates of sleep and waking in *Drosophila melanogaster*. Science 287, 1834-1837.

Shein-Idelson, M., Ondracek, J.M., Liaw, H.-P., Reiter, S., Laurent, G., 2016. Slow waves, sharp waves, ripples, and REM in sleeping dragons. Science 352, 590-595. https://doi.org/10.1126/science.aaf3621.

Sigurgeirsson, B., Thornorsteinsson, H., Sigmundsdóttir, S., Lieder, R., Sveinsdóttir, H.S., Sigurjónsson, Ó.E.,Halldórsson, B., Karlsson, K., 2013. Sleep-wake dynamics under extended light and extended dark conditions in adult zebrafish. Behavioural Brain Research 256, 377-390. https://doi.org/10.1016/j.bbr.2013.08.032.

Sigurgeirsson, B., Thorsteinsson, H., Arnardóttir, H., Jóhannesdóttir, I.T., Karlsson, K.A., 2011. Effects of modafinil on sleep-wake cycles in larval zebrafish. Zebrafish 8, 133-140. https://doi.org/10.1089/zeb.2011.0708.

Singh, A., Subhashini, N., Sharma, S., Mallick, B.N., 2013. Involvement of the a1-adrenoceptor in sleep-waking and sleep loss-induced anxiety behavior in zebrafish. Neuroscience 245, 136-147. https://doi.org/10.1016/j.neuroscience.2013.04.026.

Singh, C., Oikonomou, G., Prober, D.A., 2015. Norepinephrine is required to promote wakefulness and for hypocretin-induced arousal in zebrafish. eLife 4, e07000. https://doi.org/10.7554/eLife.07000.

Smadja Storz, S., Tovin, A., Mracek, P., Alon, S., Foulkes, N.S., Gothilf, Y., 2013. Casein kinase 1d activity: a key element in the zebrafish circadian timing system. PLoS One 8, e54189. https://doi.org/10.1371/journal.pone.0054189.

Stankiewicz, A.J., McGowan, E.M., Yu, L., Zhdanova, I.V., 2017. Impaired sleep, circadian rhythms and neurogenesis in diet-induced premature aging. International Journal of Molecular Sciences 18. https://doi.org/10.3390/ijms18112243.

Stern, A.L., Naidoo, N., 2015. Wake-active neurons across aging and neurodegeneration: a potential role for sleep disturbances in promoting disease. SpringerPlus 4, 25. https://doi.org/10.1186/s40064-014-0777-6.

Sutcliffe, J.G., de Lecea, L., 2002. The hypocretins: setting the arousal threshold. Nature Reviews Neuroscience 3,339-349.

https://doi.org/10.1038/nrn808.

Taillard, J., Sagaspe, P., Berthomier, C., Brandewinder, M., Amieva, H., Dartigues, J.-F., Rainfray, M., Harston, S., Micoulaud-Franchi, J.-A., Philip, P., 2019. Non-REM sleep characteristics predict early cognitive impairment in an aging population. Frontiers in Neurology 10, 197. https://doi.org/10.3389/fneur.2019.00197.

Takahashi, J.S., 2017. Transcriptional architecture of the mammalian circadian clock. Nature Reviews Genetics 18, 164e179. https://doi.org/10.1038/nrg.2016.150.

Tan, Y., DeBruyne, J., Cahill, G.M., Wells, D.E., 2008. Identification of a mutation in the Clock1 gene affecting zebrafish circadian rhythms. Journal of Neurogenetics 22, 149-166. https://doi.org/10.1080/01677060802049738.

Tononi, G., Cirelli, C., 2019. Sleep and synaptic down-selection. European Journal of Neuroscience. https://doi.org/10.1111/ejn.14335.

Toth, L.A., Bhargava, P., 2013. Animal models of sleep disorders. Comparative Medicine 63, 91-104.

Tovin, A., Alon, S., Ben-Moshe, Z., Mracek, P., Vatine, G., Foulkes, N.S., Jacob-Hirsch, J., Rechavi, G., Toyama, R.,Coon, S.L., Klein, D.C., Eisenberg, E., Gothilf, Y., 2012. Systematic identification of rhythmic genes reveals camk1gb as a new element in the circadian clockwork. PLoS Genetics 8, e1003116. https://doi.org/10.1371/journal.pgen.1003116.

Vatine, G., Vallone, D., Appelbaum, L., Mracek, P., Ben-Moshe, Z., Lahiri, K., Gothilf, Y., Foulkes, N.S., 2009. Lightdirects zebrafish period2 expression via conserved D and E boxes. PLoS Biology 7, e1000223. https://doi.org/10.1371/journal.pbio.1000223.

Villafuerte, G., Miguel-Puga, A., Rodríguez, E.M., Machado, S., Manjarrez, E., Arias-Carrión, O., 2015. Sleep deprivation and oxidative stress in animal models: a systematic review. Oxidative Medicine and Cellular Longevity2015, 234952. https://doi.org/10.1155/2015/234952.

Vorster, A.P.A., Krishnan, H.C., Cirelli, C., Lyons, L.C., 2014. Characterization of sleep in *Aplysia californica*. Sleep 37,1453-1463. https://doi.org/10.5665/sleep.3992.

Vyazovskiy, V.V., Harris, K.D., 2013. Sleep and the single neuron: the role of global slow oscillations in individual cellrest. Nature Reviews Neuroscience 14, 443-451. https://doi.org/10.1038/nrn3494.

Weber, F., Dan, Y., 2016. Circuit-based interrogation of sleep control. Nature 538, 51e59. https://doi.org/10.1038/nature19773.

Woods, I.G., Schoppik, D., Shi, V.J., Zimmerman, S., Coleman, H.A., Greenwood, J., Soucy, E.R., Schier, A.F., 2014. Neuropeptidergic signaling partitions arousal behaviors in zebrafish. The Journal of Neuroscience 34, 3142-3160.https://doi.org/10.1523/JNEUROSCI.3529-13.2014.

Xiao, T., Ackerman, C.M., Carroll, E.C., Jia, S., Hoagland, A., Chan, J., Thai, B., Liu, C.S., Isacoff, E.Y., Chang, C.J.,2018. Copper regulates rest-activity cycles through the locus coeruleus-norepinephrine system. Nature ChemicalBiology 14, 655-663. https://doi.org/10.1038/s41589-018-0062-z.

Xie, L., Kang, H., Xu, Q., Chen, M.J., Liao, Y., Thiyagarajan, M., O'Donnell, J., Christensen, D.J., Nicholson, C.,Iliff, J.J., Takano, T., Deane, R., Nedergaard, M., 2013. Sleep drives metabolite clearance from the adult brain. Science 342, 373-377. https://doi.org/10.1126/science.1241224.

Yang, H.H., St-Pierre, F., 2016. Genetically encoded voltage indicators: opportunities and challenges. The Journal of Neuroscience 36, 9977-9989. https://doi.org/10.1523/JNEUROSCI.1095-16.2016.

Yelin-Bekerman, L., Elbaz, I., Diber, A., Dahary, D., Gibbs-Bar, L., Alon, S., Lerer-Goldshtein, T., Appelbaum, L., 2015. Hypocretin neuron-specific transcriptome profiling identifies the sleep modulator Kcnh4a. eLife 4, e08638. https://doi.org/10.7554/eLife.08638.

Yokogawa, T., Marin, W., Faraco, J., Pézeron, G., Appelbaum, L., Zhang, J., Rosa, F., Mourrain, P., Mignot, E., 2007. Characterization of sleep in zebrafish and insomnia in hypocretin receptor mutants. PLoS Biology 5, e277. https://doi.org/10.1371/journal.pbio.0050277.

Zada, D., Bronshtein, I., Lerer-Goldshtein, T., Garini, Y., Appelbaum, L., 2019. Sleep increases chromosome dynamics to enable reduction of accumulating DNA damage in single neurons. Nature Communications 10, 895. https://doi.org/10.1038/s41467-019-08806-w.

Zhdanova, I.V., Wang, S.Y., Leclair, O.U., Danilova, N.P., 2001. Melatonin promotes sleep-like state in zebrafish. Brain Research 903, 263-268.

Zhdanova, I.V., Yu, L., Lopez-Patino, M., Shang, E., Kishi, S., Guelin, E., 2008. Aging of the circadian system in zebrafish and the effects of melatonin on sleep and cognitive performance. Brain Research Bulletin 75, 433-441.https://doi.org/10.1016/j.brainresbull.2007.10.053.

Zimmerman, J.E., Naidoo, N., Raizen, D.M., Pack, A.I., 2008. Conservation of sleep: insights from non-mammalian model systems. Trends in Neurosciences 31, 371-376. https://doi.org/10.1016/j.tins.2008.05.001.

第 十 章

斑马鱼的恐惧反应和抗捕食行为——一种良好的实验模型

Robert T. Gerlai[1]

10.1 引言

在过去的20年里，斑马鱼在神经科学研究领域越来越受欢迎，因为该物种既具有操作方面的简单性又具有系统的复杂性，使神经科学家能够开展脊椎动物大脑功能方面的研究，并有效地模拟人类器官的疾病（Gerlai, 2014, 2012, 2011）。焦虑是人类最普遍的神经精神疾病之一（Remes et al., 2016）。虽然临床上已经有几种治疗焦虑症的药物，但是焦虑症的医疗需求在很大程度上未能得到满足。因为相当大比例的病人对现有的药物不敏感，且这些药物也具有副作用，并且/或者在长时间服用后须停止治疗（Davidson et al., 2010）。焦虑症并不代表一种疾病，而可能是具有不同潜在神经生物学机制且由遗传和/或环境原因引起的疾病的统称，这一情况也阻碍了焦虑症治疗方法的开发。为了了解这一系列疾病的发病机制，或者为解决这一复杂问题而开展精神药理学治疗，人们可能需要进行临床前研究，即使用动物模型（Green and Hodges, 1991）。解决此类复杂问题的一种方法是进行系统的和彻底的筛选，即非假设驱动的分析，可以在受控的实验条件下对形成焦虑的许多生物学因素进行研究（Gerlai, 2010a）。我们可以进行两种完全不同的筛选：突变筛选和药物（小分子）筛选。有人建议将斑马鱼作为传统实验模式生物的一个很好的替代品用于两种筛选（Gerlai, 2010a; Stewart et al., 2015）。人们推荐将斑马鱼作为传统实验模式生物的替代品具有多种原因，本书将分章节讨论其中的部分原因。简单地说，斑马鱼是一种小型脊椎动物，可以大量饲养并且成本较低，可用于很多研究中。此外，它又足够复杂，具有

[1] Department of Psychology, University of Toronto Mississauga, Mississauga, ON, Canada.

许多进化上的保守特征，通过对斑马鱼进行研究获得的结果可以很好地类推到哺乳动物和人类。例如，在突变筛选中，人们可能鉴定出大量潜在的突变体，并且这些携带突变的基因最终会改变焦虑，或者改变抗焦虑药物或致焦虑药物的作用。一旦发现具有这样的作用，后续研究就可以阐明这些基因的蛋白质产物在作用机制、神经生物学过程、生化途径中发挥的作用。同样，大规模药物筛选也有助于我们理解焦虑的潜在机制，因为识别有效的小分子意味着识别与特定靶标结合并改变其功能的化学物质。因此，对这种结合相互作用及随之产生的功能变化方面的研究将有助于阐明机制问题。然而，对于突变和药物筛选，人们需要有适当的筛选范式，即可以量化研究生物体焦虑的测试。本章重点讨论量化斑马鱼焦虑的行为学测试（Gerlai, 2010b, 2013）。主要以在焦虑研究领域得到很好发展的成年斑马鱼的行为学测试方面的讨论为主，另外，也将扩展到年龄更小的斑马鱼研究中。然而，在深入讨论此类测试的细节之前，需首先对焦虑这个术语进行定义，并讨论其与恐惧的差异性。

10.2　焦虑还是恐惧

临床上，人们认为焦虑是一种病理状态，即患者在未接收到令人厌恶的刺激时也会展现出恐惧表现，这种表现可能反应在行为上，也可能反应在生理上，包括血压升高、心跳加快和皮质醇水平升高。大多数行为学研究能够将恐惧与焦虑区分开来。我们认为后者是病态的，而前者是适应性的，恐惧是一种具有有益功能的自然的行为或状态。从实验的角度分析，焦虑和恐惧也是有区别的。恐惧是机体做出的对明显的厌恶性刺激的一系列反应，也就是说，恐惧是对环境信号的直接反应，而且是急性的。从机体表型方面来看，焦虑可能与恐惧相同，但是，焦虑是机体在未接收到令人厌恶的刺激情况下也会出现的一系列反应，并且它可能是慢性的，其症状会持续很长时间。关于焦虑和恐惧是否属于生物学上的两种不同现象目前仍然是有争议的，但大多数人都同意这两种现象的行为表现有很多重叠之处。在斑马鱼的研究中，恐惧和焦虑的定义取决于反应的诱发方式，而非其展现的行为表现。斑马鱼的恐惧可能会通过多种厌恶性刺激而引起，而焦虑则是以更加分散的方式引起的，例如，将鱼放到一个新的环境中可能会引发焦虑。从机制角度来看，对斑马鱼的这些操作性定义是否正确仍有待研究，但大多数人同意斑马鱼的行为筛查范式应同时使用恐惧和焦虑诱导方法，以涵盖所有可能的潜在机制。

10.3　恐惧：对捕食者的一种自然反应

在详细介绍可能引起斑马鱼恐惧和/或焦虑的方法之前，我们需要强调一个重要的观点，即在人类和斑马鱼以及其他生物中，应将恐惧认为是一种适应性的自然反应，其功能是帮助独立的个体避开自然界中的危险。自然界中存在的危险情形通常为：捕食者出现和/或捕食者可能比猎物更占优势的情况下出现。猎物往往对捕食者的特征刺激或者对其容易受到捕食的环境表现出物种特定的恐惧反应。人们认为这些恐惧反应可以提高猎物避开捕食者的能力。斑马鱼有几种天然的捕食者，它们的捕食方式各不相同，因此斑马鱼可能会出现不同的反捕食反应（Engeszer et al., 2007）。本书的第一章（由 Parichy 和 Postlethwait 编写）也介绍了这类捕食者的例子，并详细介绍了有关它们出现的栖息地。以下只是简要地对它们进行讨论。了解关于斑马鱼天敌的知识可能有助于设计引发恐惧和 / 或焦虑的范式，因此，这种讨论也是必要的（Gerlai, 2013, 2012；另见 Gerlai and Clayton, 1999a, 1999b 对本主题的一般性讨论）。

斑马鱼在其自然栖息地遇到的两种截然不同的捕食者分别为捕鱼鸟和肉食鱼类。斑马鱼占据水体的中上层，其大多数栖息地的水并不深，因此，斑马鱼很容易受到空中捕食者的攻击，比如鸟类。因此，位于实验水箱上方的大型移动物体可引起斑马鱼强烈的恐惧反应就不足为奇了（Miller and Gerlai, 2007）。还有几种与斑马鱼同域的肉食鱼（Engeszer et al., 2007）。这些肉食鱼在捕食猎物的方式上有所不同。布氏鳠（例如，*Mystus bleekeri*）习惯待在水体的底部，并以相对恒定的速度移动，主要是通过触觉或嗅觉而不是视觉来发现猎物。南鲈（*Nandus nandus*）是一种伏击捕食者，这种鱼通常隐藏在水体的水生植物中或其他生物体中，直到斑马鱼或其他小型待捕获的鱼游到距其足够近的地方时，它们就会迅速跳跃并袭击猎物。一些斑鳢（例如，*Channa maculata*）以及针鱼或尖嘴扁颌针鱼（*Xenentodon cancila*）的仔鱼通常待在靠近水面的地方并在那里伏击小型猎物。研究已经发现，包括叉尾斗鱼（*Macropodus opercularis*）在内的几种鱼在事先没有接触过它们的捕食者的情况下，也能够对它们的天敌（同域）做出特定反应，即“识别”（Gerlai, 1993），因此，斑马鱼也能以猎物特有的方式对上述同域捕食者做出反应（Bass and Gerlai, 2008）。然而，这种捕食者特异性反应的系统行为学分析尚未在斑马鱼中进行。但是，目前活的捕食者物种，以及模拟捕食者某些方面的计算机动画刺激已成功应用（Gerlai, 2010b; Gerlai et al., 2009）。除了这些视觉刺激之外，人们也成功利用了另一种自然的可诱发恐惧反应的刺激——嗅觉线索（Gerlai, 2010b; Speedie and Gerlai, 2008）。从斑马鱼的皮肤中提取的报警物质或报警信息素实际上不是单一的化学成分，而是一组分子的集合，研究已表明，该类物质可使斑马鱼产生强烈的恐惧反应（Speedie and Gerlai, 2008）。

在以下部分，我们首先回顾一下用于诱导斑马鱼产生恐惧的不同视觉刺激，随后将列举一个同样成功应用的嗅觉线索的例子。此外，简要回顾一下预期会诱发或量化“焦虑”的范式。关注的重点是成年斑马鱼，因为大多数与恐惧和焦虑相关的文献都是关于成年斑马鱼的。尽管如此，也会简要地提及一个用于使斑马鱼仔鱼产生厌恶性反应的实例。

10.4 视觉刺激和成年斑马鱼传递恐惧的方法

斑马鱼是一种典型的具有较好视觉的白天活动的鱼类。使用视觉较好的实验物种有许多优点（Gerlai, 2017）。首先，由于视觉也是人类首选的获取信息的模式，因此市场上有许多价格合理的消费产品（例如，相机、电视显示器等）。其次，与嗅觉或听觉线索不同，视觉刺激易于转换，其精确位置以及启动和终止都可以很容易地控制。目前已经使用许多简单的视觉刺激来诱导斑马鱼的恐惧反应。

虽然没有对斑马鱼进行系统分析，但已经发现其他鱼类对同域捕食者表现出天生的反捕食反应。例如，叉尾斗鱼（*M. opercularis*）只对同域的捕食者表现出物种特有的恐惧反应，而斑鳢（*Channa*）对异域捕食者（如马拉维湖淡水鱼 *Nimbochromis compressiceps*）和同域、异域非食肉鱼类，包括三星攀鲈（*Trichopodus trichopterus*）和鲫鱼（*Carassius auratus*），均不产生恐惧反应（Gerlai, 1993）。研究也表明，斑马鱼对其天然（同域）捕食者南鲈（*N. nandus*）（Bass and Gerlai, 2008）以及铠甲弓背鱼（*Chitala chitala*）（Ahmed et al., 2012）则表现出强烈的恐惧反应。虽然目前尚不清楚斑马鱼关注同域捕食者的哪些特征，即“识别”先天捕食者的基础特征可能是什么，但是计算机模拟这些捕食者图像（移动）的出现可在斑马鱼中引起强烈的恐惧反应（Ahmed et al., 2012）。如前所述，空中捕食者也对斑马鱼构成重大威胁。与此相一致的是，放置在实验水箱上方的电脑屏幕上出现移动的类似猛禽的黑色轮廓时，斑马鱼也表现出了强烈的反捕食反应（Luca and Gerlai, 2012a）。最后，在实验水箱的玻璃墙一侧，用在电脑屏幕上显示一个黑点不断扩大的图像以模拟鱼类捕食者快速游向猎物的场景，即显示猎物物种可能看到的东西，这时斑马鱼也表现出强烈的恐惧反应（Luca and Gerlai, 2012b）。利用斑马鱼进行视觉（和嗅觉）的刺激的一些例子如图 10-1 所示。

值得注意的是，上述刺激引起恐惧反应的程度或类型并不完全相同。人们已经发现，斑马鱼的反捕食反应具有刺激特异性（例如，Luca and Gerlai, 2012b）。这种情况也不足为奇，因为在自然界中根据捕食者的行为 / 状态、位置 / 距离以及种类，不同的反捕食策略可能是与环境相适应的结果。

10.5　视觉刺激特异性的反捕食反应

由上述例子中所述刺激引起的斑马鱼反捕食（恐惧）反应的分析表明，斑马鱼具有可以最大限度减少捕食威胁的复杂行为库。Bass 和 Gerlai（2008）研究了斑马鱼对不同种类活鱼的反应。这些研究人员利用同域捕食性鱼类（*N. nandus*）和异域捕食性鱼类（*N. compressiceps*）以及同域的（*Devario aequipinnatus*，波条鲘鱼）和异域的无害鱼类（*Xiophhorus heleri*，剑尾鱼）作为刺激鱼。斑马鱼对同域捕食者表现出与对所有其他鱼类包括异域捕食者不同的行为学反应方式，即实验用斑马鱼的跳跃次数增加，不规则运动的持续时间增加，不规则运动也称为“之”字形运动。Gerlai 等人（2009）利用这些结果开发了一种自动化产生厌恶（诱导恐惧）性刺激的范式，在任务中向实验斑马鱼展示计算机模拟的同域捕食者南鲈的动画（移动）图像。在该测试中，采用两种不同的方法测量斑马鱼的行为反应。一种是采用观察和事件记录的方法记录其运动和体位模式；另一种是采用视频跟踪方法。事件记录数据表明，当电脑屏幕上出现南鲈时，斑马鱼的不规则运动持续

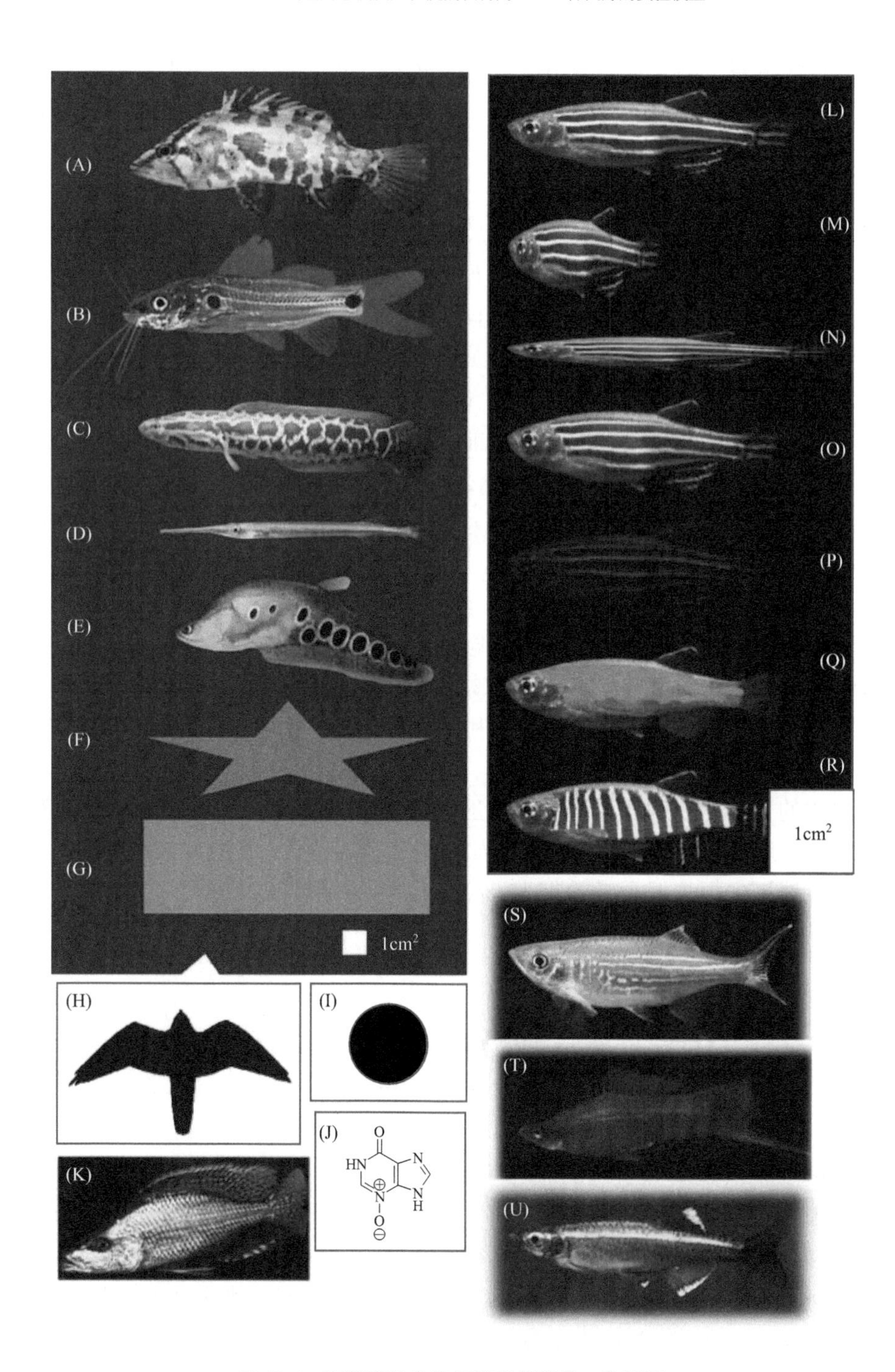

图 10-1 **利用斑马鱼进行视觉刺激的一些例子**

目前还没有系统分析哪些刺激可能是厌恶性的（引起恐惧或焦虑），哪些刺激是愉悦性的（奖励、诱导，例如群聚行为），我们也不知道斑马鱼是如何区分这些刺激的。尽管如此，已经成功应用了许多刺激来引发这些行为。本文列举了一些利用刺激成功地引发斑马鱼的抗捕食反应（A）～（J）或社会行为反应（L）～（U）的例子。注意：其中一些刺激物以活体刺激鱼的形式呈现，而另一些则以计算机屏幕上的动画图像形式呈现。图中也显示了嗅觉刺激，列举了一个嗅觉刺激的例子。（A）南鲈（*Nandus nandus*）；（B）鲶鱼（*Mystus bleekeri*）；（C）黑鱼（*Channa maculata*）；（D）针鱼（或尖嘴扁颌针鱼，*Xenentodon cancila*）；（E）小丑刀鱼（*Chitala chitala*）；（F）移动的星形物体；（G）一个移动的矩形；（H）一个移动的黑色鸟的剪影；（I）一个不断扩大的黑点；（J）次黄嘌呤 3-*N*- 氧化物或 H_3NO 的结构。右侧：（L）～（U）列举了一些预期为中性（不同鱼类的动画图像，或此类物种的活的刺激鱼）或诱发聚群（形成群体）行为反应的刺激物。(L) 野生型斑马鱼（*Danio rerio*）；(M) 短鱼；(N) 细长鱼；(O) 黄色鱼；(P) 红色鱼；(Q) 无条纹鱼；(R) 斑马鱼的垂直条纹动画图案；(S) 巨型斑马鱼（*Devario aequipinnatus*）；剑尾鱼（*Xiphophorus helleri*）；(U) 白云样鱼（*Tanichthys albonubes*）。值得注意的是，研究发现活的南鲈和它的电脑动画图像都会引起斑马鱼强烈的恐惧反应。此外，还发现其他同域捕食者（B）～（E）的动画图像也会引起强烈的恐惧反应。抽象的形状（F）、（G）以及鸟的轮廓（H）和不断膨胀的点（I）也会引起强烈的恐惧反应。然而，当我们进行比较时，发现活的异域捕食者（K）在诱导恐惧反应方面远不如活的同域捕食者（A），而活的同域无害鱼（S）也没有引起恐惧反应，这表明也许运动 / 行为以及形状 / 图案 / 颜色在斑马鱼决定该物种是危险还是无害方面都可能发挥作用。斑马鱼动画图像［(L)、(M)、(O)、(P)、(Q) 和 (R)］都引起了聚群行为，除了细长鱼的图像（N），但（N）它引起了显著的回避反应，可能是由于它与同域的斑马鱼捕食者针鱼（D）相类似。另请注意，当斑马鱼与活的无害（非捕食性）刺激鱼一起游动时，斑马鱼趋向于与其同种的鱼聚集成群，而不与非群聚物种鱼（T）或异域物种鱼（U）聚集成群。综上所述，使用上述刺激物的研究结果表明，斑马鱼可以依靠形状、颜色、图案、运动特征、出现的位置和所显示刺激物的大小来决定是否需要以及需要何种类型的反捕食反应。更多详细信息请参阅文献。修改自 Bass,S.L., Gerlai, R., 2008. Zebrafish (Danio rerio) responds differentially to stimulus fish: the effects of sympatric and allopatric predators and harmless fish. Behavioural Brain Research 186, 107e117; Saverino, C., Gerlai, R., 2008. The social zebrafish: behavioral responses to conspecific, heterospecific, and computer animated fish. Behavioural Brain Research 191, 77e87; Ahmed, T.S., Fernandes, Y., Gerlai, R., 2012. Effects of animated images of sympatric predators and abstract shapes on fear responses in zebrafish. Behaviour 149, 1125e1153

时间增加，跳跃频率增加（Gerlai et al., 2009）。最重要的是，作者发现视频跟踪系统的测量结果与基于人观察的行为结果具有高度相关性。例如，Gerlai（2009）等人报道，在屏幕上出现南鲈期间，受试斑马鱼个体的游泳速度（暂时的）的变异性增加、转弯角度增加以及转弯角度的个体内变异性增加。不规则运动和跳跃行为主要包括快速加速、快速转弯和突然停止等行为，上述视频跟踪测试系统可以很好地捕捉和量化这些行为特征，该系统不仅可以呈现引起恐惧的刺激，还可以记录由计算机自动模拟的刺激引起的恐惧行为反应（另请参见图 10-2）。

在随后的研究中，Ahmed 等人（2012）使用了相同的厌恶刺激（南鲈的动画图像），但研究了不同的行为反应。到目前为止，有几篇文献中的斑马鱼焦虑和恐惧反应使用所谓的“潜水实验”进行评价。这种“实验方法”是基于这样一种假设和观察，即自然界中的斑马鱼可能会遇到掠食性鸟类，而逃离这些鸟类的最好办法是潜到水底并躲藏起来。因此，Ahmed 等人（2012）利用高精度的视频跟踪系统测量了实验鱼与鱼缸底部之间的距离，以及斑马鱼游泳路径的其他参数。然而，他们发现这个距离不受南鲈出现的影响。有趣的是，其他行为学参数明显受到这种同域捕食者动画图像的影响。例如，当同域捕食者的动画图像出现时，斑马鱼与呈现刺激的电脑屏幕之间的距离显著增加。同样值得注意的是，当捕食者的图像出现时，斑马鱼与鱼缸底部的距离的个体间变异性显著降低。换句话说，作为对捕食者图像的反应，实验用斑马鱼并没有向底部移动得更近，而是停止了上下移动，也就是说，它们对鱼缸的垂直探索减少了。同样值得注意的是，在本次测试中观察到斑马鱼到鱼缸底部的距离显著缩短，但这不是因为斑马鱼对捕食者图像的反应，而是在记录过程开始时斑马鱼对人工操作和 / 或对实验鱼缸新颖性的反应。因此，这些结果表明，斑马鱼可能对不同的威胁做出不同的反应。例如，鱼类捕食者的出现引起不规则的运动，然而，在没有明显的恐惧诱导刺激的情况下，一种新颖的环境可使实验斑马鱼更接近底部。

为了研究特定捕食者引发的不同恐惧效应，Ahmed 等人（2012）使用了各种计算机动画图像，包括南鲈的图像以及与斑马鱼同域的其他四种鱼类［布氏鳠（*M. bleekeri*）、铠甲弓背鱼（*C. chitala*）、尖嘴扁颌针鱼（*X. cancila*）和斑鳢（*C. maculata*）］的图像，这

(A)

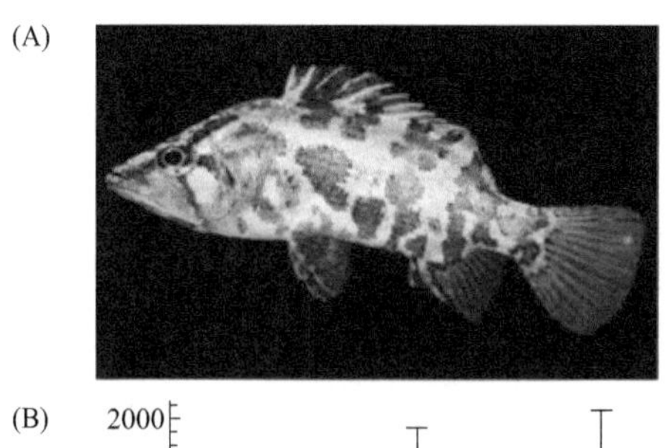

(B)

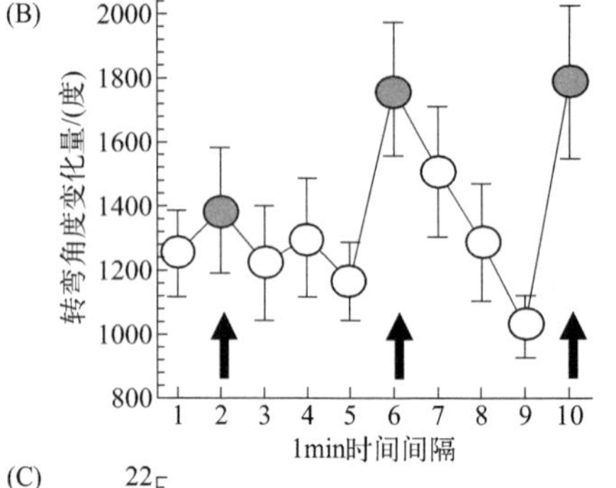

(C)

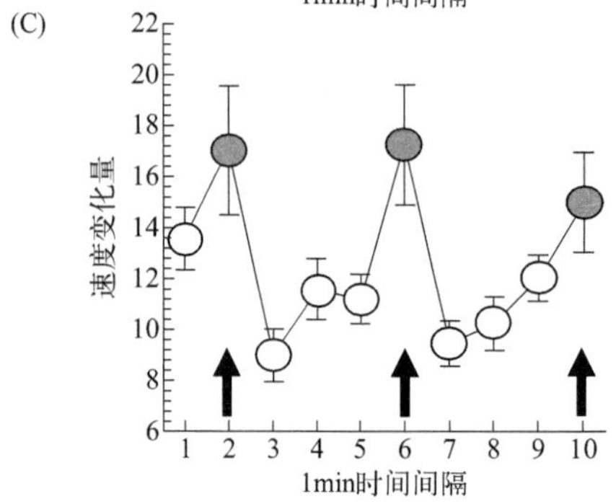

(D)

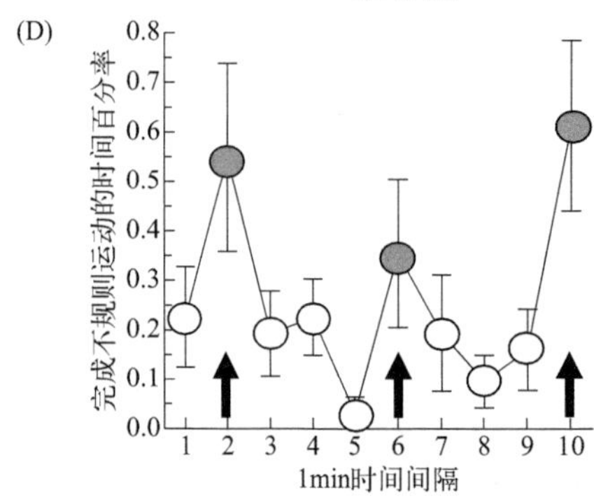

检测恐惧和／或焦虑的行为学例子

基于观察（或视频追踪）的测量

- 不规律运动或“之”字形运动（转弯的角度或／和速度变化幅度加大）
- 蹦或跳跃（速度变化的幅度加大）
- 僵化或静止不动（运动的频率或者持续时间不超过0.1cm/s）
- 待在底部或潜水时间增加（到底部距离减少）
- 群聚行为增加（每条鱼或者鱼群之间的距离减小）
- 回避反应（远离厌恶刺激）
- 趋触性或靠近壁（离鱼缸壁的距离减小）
- 垂直探索减少（到底部的距离变化幅度减小）

图10-2　**对厌恶刺激的行为反应示例**

（A）同域捕食者，南鲈（*Nandus nandus*）的移动（计算机动画）图像在指定的1min内呈现［由图（B）～（D）上的箭头所示］。通过两种方式，即对实验斑马鱼游泳路径的各种视频跟踪测量以及对运动和体位模式的基于观察的事件记录对行为反应进行量化。（B）转弯角度的变异性。（C）速度的变异性，采用视频跟踪测量方法。请注意，在这种情况下，变异性代表个体的行为随时间的变异性，这是衡量每条实验鱼的速度和转向角度方面的不一致程度。（D）不规则的运动，一种基于观察的测量指标。灰色实心圆圈表示在捕食者的动画图像出现时记录的行为，而白色空心圆圈表示在没有捕食者图像时的行为反应（测定值为平均值＋标准误差）。需要指出的是：视频跟踪测量方法和基于观测的测量方法之间具有较强的相关性。不规则运动（或“之”字形运动）与游泳速度（加速和减速）以及游泳方向（转弯）的快速变化有关。在图的右侧，列出了一些关于恐惧和／或焦虑的行为测量的例子，首先是基于观察的测量方法，括号中提到了可能对应的视频跟踪测量。要了解更多细节，请参见以下文献。修改自：Gerlai, R., Fernandes, Y., Pereira, T.,2009. Zebrafish (Danio rerio) responds to the animated image of a predator: towards the development of an automated aversive task. Behavioural Brain Research 201, 318e324

些鱼在自然界中可能捕食斑马鱼，还包括抽象形状的图像，例如星形和矩形，其线性尺寸和整体像素色彩组成整体看起来符合捕食者图像的特征。这些结果证实了之前对南鲈研究的发现，但也表明斑马鱼的所有反应取决于图像显示的捕食鱼的种类。同样重要的是，研究发现，与南鲈引起的反应相比，斑马鱼对刀鱼和斑鳢的反捕食反应更强，后一个结果值得注意，因为与其他捕食者图像相比的唯一共同视觉线索是矩形的大小和移动速度。

为了寻找最佳（最有效的）的恐惧诱发图像呈现的方法，Luca 和 Gerlai（2012）比较了一组不同动画图像的效果，包括南鲈、针鱼以及在水箱侧面或上方移动的鸟类剪影。此外，还测试了模拟食鱼动物从水箱侧面和上方快速接近的膨胀点扩展的效果。再次，发现斑马鱼的反捕食反应取决于所显示刺激的类型和位置。例如，鸟的轮廓和从上方呈现出的扩展点引起了最强烈的潜水反应，并且从上方呈现的扩展点引起了最强的不规律运动和最高的跳跃频率，这证实了自然界中斑马鱼最危险的捕食者可能是快速接近它的从上方攻击的鸟类。

10.6 嗅觉线索引起的抗捕获反应

嗅觉线索也可能在探测和对捕食者做出反应方面发挥重要作用。众所周知，骨鳔鱼（ostariophysan，硬骨淡水鱼）通常对被称为警报信息素的报警物质做出反应（Maximino et al., 2019）。这种物质实际上是大量分子的混合物，当鱼的表皮破损时，这些报警物质会从位于表皮中的特殊棒状细胞中释放到水中。有趣的是，从某种意义上说，该物质不是物种特异性的，即使是远亲的骨鳔鱼类间也会发生交叉反应。从生态或进化适应性的角度来看，这种现象是合理的。如果猎物鱼受到攻击，其皮肤受损，任何物种的邻近鱼类都会因为捕食者的存在而处于危险之中。

报警物质也作为实验条件开始应用。例如，可以将刚刚麻醉好的斑马鱼侧面切开，清洗切割区域并收集清洗液，就可以很容易地提取到这种物质。目前已发现将几滴这种液体滴入测试水箱中会使斑马鱼产生强烈的警报（恐惧）反应，包括不规则运动、静止不动和待在底部的持续时间增加以及群聚量增加（Speedie and Gerlai, 2008）。天然报警物质虽然易于收集，但主要问题是其准确浓度无法确定。在实验中这可能并不是主要问题，因为这种物质通过系列稀释后，实验人员就可建立其相对浓度差。然而，在不同的实验中其浓度会有所不同，取决于在切割提取过程中获得的报警物质的量，即使采用完全相同的方式也会导致提取到的报警物质的产量不同。为了避免这个问题出现，已经开始对警报物质的活性成分进行鉴定。此外，在已鉴定分子的基础上，已经开始合成其类似物。其中的一种物质是次黄嘌呤 3-*N*- 氧化物或 H_3NO。

H_3NO 已成功应用于斑马鱼实验（Parra et al., 2009）。研究已经发现它会引起斑马鱼跳跃次数和不规则运动的大幅增加。尽管随着时间的推移，这种合成化合物容易分解和失效，但它的主要优点是具有已知的分子结构和分子量，这使得实验人员可以精确地控制实验中及实验间该物质的浓度。

10.7 引起焦虑的范式

与恐惧不同，在没有明显的厌恶性刺激出现时也会产生焦虑。在未给新鱼缸中的鱼提供具有明显危险特征的特定视觉或嗅觉信号时，斑马鱼也会出现焦虑症状。然而，即使没有这种明确的厌恶性刺激信号，新环境也可能会造成危险。鱼类和其他脊椎动物在新环境中成功躲避捕食者的能力较低，这可能是因为在这种环境中，它们还不知道躲藏的地方和逃跑路线（Gerlai, 1993; Gerlai et al., 1990）。不管是什么原因，研究已经发现新型水箱会引起前面讨论过的大多数恐惧反应。同样值得注意的是，这些焦虑 / 恐惧反应会随着时间的推移而减弱，并且在大约 5 ～ 10min 后，它们表现出的焦虑症状明显减少，环境适应速度可以有效地用于衡量抗焦虑或致焦虑化合物的疗效（Stewart et al., 2013）。

用斑马鱼量化“焦虑”的另一项厌恶性任务是采用哺乳动物的明 - 暗模式（Hascoët et al., 2001）。在这项任务中，斑马鱼要在实验水箱中黑暗的一面和明亮的一面之间做出选择。已经发现夜间活动的啮齿动物表现出对黑暗隔间的强烈偏好（Hascoët et al., 2001）。研究已经证明抗焦虑药物可以减少动物对黑暗的偏好（Hascoët et al., 2001）。这项任务对啮齿类动物的生态学意义在于，夜间活动的啮齿类动物在黑暗中更安全，不会受到捕食者的攻击，因此具有避光性。然而，斑马鱼是一种生活在光线充足的浅水区的白天活动的昼行性动物。与此相一致的是，以斑马鱼为对象采用明 - 暗模式的第一项研究（Gerlai et al., 2000）发现了斑马鱼会避开黑暗的地方，即对光线的偏好。然而，自从这项开创性工作发表以来，出现了一些矛盾的结果，即一些研究表明斑马鱼对光具有偏好性，即趋光性（例如，Blaser and Penalosa, 2011），而另一些研究则发现斑马鱼对黑暗具有偏好性，即趋暗性（Maximino et al., 2010）。基于经验证据，我们认为这些争议出现的主要原因可能是实验人员如何定义“光明与黑暗”的问题。似乎有几项研究混淆了光照水平和背景阴影，即假设光与白色是同义词，暗与黑色是同义词（Facciol et al., 2019）。在斑马鱼的明 - 暗模式中，另一个潜在的重要因素是，隔间的开放性与封闭 / 覆盖性也可能发挥作用。目前已经开始系统地分析这些因素对斑马鱼穿梭箱一侧偏好的贡献（Blaser and Penalosa, 2011; Facciol et al., 2019, 2017）。虽然数据不够完整，但到目前为止的结果表明，斑马鱼更喜欢照明良好（明亮）的、背景为深色（黑色）的、开放的（透明玻璃）隔间，而非没有照明的、上面被覆盖（洞穴状）的隔间，该环境条件类似于斑马鱼自然栖息地的特征。

10.8 仔鱼的恐惧反应

在恐惧和焦虑研究中应用成年斑马鱼的优势在于，在这个阶段斑马鱼体形更大，更容易处理，因此它们的行为更容易量化。但也许更重要的是，在发育到成年阶段，斑马鱼更为稳定，也就是说，它们的大脑已经发育完全，不会由于个体发育而迅速改变其结构和功能。然而，斑马鱼鱼苗，在文献也有人称为“仔鱼（larval）”的斑马鱼则更难以处理。它们的小尺寸给视频跟踪或基于观察的行为记录带来了技术问题，受到与时间相关的快速个体发育变化的混杂影响。尽管如此，使用这样的仔鱼开展工作也有好处。因为它们是几天大的斑马鱼，所以可以使用多孔板，例如标准的96孔板，同时监测大量受试者的行为反应，也就是说，极大地增加了行为筛选的通量。较小的鱼类具有较大的体表面积，因此化合物可以被更快地吸收并更快地到达它们的靶器官。就中枢神经系统而言，使用斑马鱼仔鱼有一个另外的好处，那就是在受精后10天之前，血脑屏障还没有完全发育好（Fleming et al., 2013）。对于斑马鱼仔鱼来说，确实存在恐惧/焦虑范式成功的例子。其中一种试验称为弹跳球试验（bouncing ball assay）（Pelkowski et al., 2011）。在这项任务中，将仔鱼放入一个培养皿中，实验者在培养皿下方放置了一个电脑屏幕、平板电脑或iPad，从下方提供照明，便于更好地进行视频跟踪，在上方呈现一个大的动画（移动）彩色实心圆圈，即“球”。仔鱼表现为尽量避开这个移动的大物体，可量化这种逃避反应并将其定义为一种恐惧反应。

10.9　斑马鱼恐惧和焦虑的精神药理学

在抗焦虑化合物的精神药理学分析中，斑马鱼是相对较新的模式动物。然而，到目前为止，使用这一物种证明这些化合物效能的几项研究已经发表。对这些文献的详细介绍并不属于本章的描述范围，这些文献已由其他作者公开发表（Stewart et al., 2011; Kalueff et al., 2014）。下面，我们仅仅综述该快速发展的工作体系中的一些具有代表性的、概念上有趣的例子。

其中一项最早的研究由Bencan等人（2009）采用斑马鱼证明了药物的抗焦虑特性。作者在所谓的“潜水试验”中研究了氯氮卓、地西泮和丁螺环酮对斑马鱼行为的影响。如前所述，这种分析研究是基于斑马鱼的自然活动趋势，在令其厌恶的条件下包括将其置于一个新奇的环境中时，斑马鱼趋向于靠近底部。然而，作者发现在这项任务中氯氮卓并没有改变斑马鱼的潜水反应，而地西泮却是有效的，但详细的研究结果也发现，剂量反应曲线呈“W”形，低、中、高浓度间没有显著性差异，只有在低浓度和中浓度之间以及中浓度和高浓度之间的部分浓度间有显著性差异。这种不寻常的剂量反应曲线表明可能存在随机变化，且可能是由于对1型误差处理不当造成的（例如，在进行重复组间比较时，作者并没有提到他们是否使用了Bonferroni或Bonferronie-Holm校正检验）。然而，他们的确发现丁螺环酮在潜水反应中具有线性剂量效应关系，这种药物对降低斑马鱼的潜水反应具有剂量依赖性。丁螺环酮是一种$5HT_{1A}$受体激动剂，在较高浓度时也与多巴胺D2受体具有一定的结合亲和力。重要的是，众所周知，控制鱼体内浮力的鱼鳔可被5-羟色胺能神经元所支配，该神经元也存在于其他神经递质系统中（Zaccone et al., 2012）。根据我们自己的观察（Gerlai et al., 未发表），丁螺环酮确实改变了底部停留时间，但可能是通过影响鱼鳔功能，因为用这个药物处理的鱼会被动地浮到水面，并且它们似乎试图挣扎着从水面返回到水底，但未能成功。换句话说，在上述研究中，以降低潜水反应而评价的丁螺环酮的“抗焦虑”效应可能是鱼鳔功能异常的假象。这种可能与外周运动功能相关的异常情况表明，通过将分析范围扩展到目标表型检测之外的功能 / 行为领域方面的研究，对于彻底调查和描述药物效应是非常重要的。为了实现这一目标，建议可以采用两种基本不同但也不相互排斥的方法。首先，人们最好使用多个任务来测量多个行为终点，这些任务都具有类似的恐惧 / 焦虑相关的行为状态 / 反应。如果这些任务与任务特异性行为表现（知觉和运动功能相关方面）相关，那么人们就可以将这些方面与受试药物引起的焦虑和 / 或恐惧相关的效应分离开来。其次，除了恐惧 / 焦虑测试之外，人们还可以使用一系列能够直接量化表现特征行为的任务，包括独立于焦虑和恐惧反应之外的感知和运动功能。

Gebauer 等人（2011）基于上述思想进行了研究，并分析了包括氯硝西泮、溴西泮、地西泮、丁螺环酮、普萘洛尔和乙醇在内的多种抗焦虑药物。他们研究了白黑偏好、身体颜色变化以及群聚行为，并且发现抗焦虑药物具有明显的行为效应。这些变化包括浅滩内群聚减少，潜水反应降低，白色回避减少，身体颜色亮度降低。除了普萘洛尔，所有其他受试药物都有这些效应。

Maximino 等人（2014）进行了一项类似的研究，他们在“光 - 暗”（实际上是白 - 黑）模式下用斑马鱼测试了各种精神活性化合物，包括咖啡因、乙醇、安非他酮、维拉帕米、丁螺环酮和苯二氮䓬类药物如地西泮。作者在这项任务中量化的一些行为参数包括：在白色隔室停留的时间、进入白色隔室的潜伏期。量化的一些活动性参数包括：跨格（squares crossed）、进入白色隔室的次数、不规则运动的次数、僵直（静止不动）、趋向性、风险评估尝试次数（快速进入并快速离开白色隔室）。作者根据这些行为测量结果对受试药物进行了分类。他们将行为测量结果分为 4 个类别，分别称为“回避”“自发活动”“风险评估”和“恐惧”，并推断根据这些测量结果可以正确地区分具有特定效应和作用模式的药物。

这种药物效应的行为指纹图谱的概念并不新鲜（Gerlai, 2002）。例如，PsychoGenics 公司（Paramus, NJ, USA）开发了一种称为“智能立方体（smart cube）”的专有行为指纹测试组合（test battery），该公司通过一系列行为测量手段对越来越多的药物进行了表征。这种方法的价值在于可以分析大量化合物引起的多种行为。这样的数据库随着获得的药物和化合物的结果而扩展，研究人员以更高的分辨率和精度来区分化合物效应谱的能力得到了提高。这也使得药物可以根据其作用机制进行聚类，有助于研究人员预测新化合物的作用机制和潜在的临床疗效，并提高可靠性。不幸的是，在斑马鱼上进行焦虑调节药物测试的化合物数量以及使用的行为测量和范例数量是非常有限的，但 Gebauer 等人（2011）和 Maximino 等人（2014）获得的结果清楚地证明了这种方法具有可行性。

在精神药理学方面，我们讨论的最后一个案例集中在人类常用和滥用药物的效应方面的研究。两种最常见的滥用药物是尼古丁和乙醇。研究已经证明这些药物对人类和非人哺乳动物具有抗焦虑和致焦虑的特性，主要取决于给药剂量和给药方案（Picciotto et al., 2002; Steven et al., 2008; Gatch and Lal, 2001)。研究发现，这两种药物也会影响斑马鱼的焦虑 / 恐惧相关反应。例如，在明 - 暗和新型鱼缸测试中，发现尼古丁可以消除斑马鱼的焦虑反应（Duarte et al., 2019; Levin t al., 2007）。此外，还发现尼古丁会改变斑马鱼的群聚行为。当急性给药时，它显著降低了鱼群的内聚性，但对鱼群的极化（同步游动方向，即鱼的运动矢量）影响不大（Miller et al., 2012）。有趣的是，研究发现急性乙醇给药具有相反的效果：它显著地破坏了群聚的极化，但只适度地降低了群聚的内聚性（Miller et al., 2012）。鱼类群聚与焦虑之间具有相关性已经得到证实。也就是说，在厌恶性刺激条件下，斑马鱼的鱼群会变得更紧密，鱼群成员之间游得更近（Miller and Gerlai, 2007; Speedie and Gerlai, 2008）。因此，急性给予尼古丁和乙醇导致的鱼群凝聚力降低可解释为这些药物具有抗焦虑作用。在其他多种行为范式中也发现乙醇具有改变斑马鱼焦虑的作用。例如，在新鱼缸试验中，急性给予 1% 乙醇可以显著减少底部停留时间（至底部的距离增加）和减少转弯角度（不规则运动），尤其是在人工处理引起的压力之后，出现这样的降低焦虑性的反应更为明显（Tran et al., 2016）。

10.10 斑马鱼恐惧和焦虑的遗传分析

尽管斑马鱼是分子生物学中最受青睐的模式生物之一，并且针对该物种已经发展了许多重组 DNA 技术，但分析这一物种的恐惧和焦虑产生的生物机制时，可用的遗传学方法仍然很少见。仅有几个例子，包括携带瘦素（leptin）突变基因的斑马鱼突变体，该突变体是采用转录激活因子样效应物核酸酶（Transcription Activator-Like Effector Nuclease, TALEN）方法产生的。Audira 等人（2018）使用该方法敲除了瘦素 A（Leptin_A）基因编码区内 4 个碱基对序列，产生了一个终止密码子，导致突变等位基因的转录提前终止，从而在突变鱼中缺乏该功能性蛋白表达。瘦素除了在调节食欲和觅食方面发挥作用外，推测其在大脑的其他功能中也发挥作用。为了揭示这种可能的作用，作者使用一系列行为测试方法对 Leptin_A 基因敲除斑马鱼的特征进行了描述。他们将基因突变鱼与正常对照鱼进行比较，即对两种鱼在新的鱼缸中、在叮镜子任务中、在躲避捕食者任务中、在社会互动中以及在群聚测试中的行为进行了比较。他们还研究了突变鱼的昼夜节律、颜色偏好和记忆表现。作者发现，Leptin_A 基因敲除的突变型斑马鱼表现出行为反应的改变，包括游泳速度增加，快速移动量增加，角速度提高（转动的速度），待在实验水箱顶层的时间增加，静止不动量减少，对捕食者的恐惧反应减少，社交互动减少，聚群行为减少，昼夜节律失调和颜色偏好减弱，但短期记忆没有变化（Audira et al., 2018）。基于这些行为分析和一些后续分子水平的分析，作者得出结论，瘦素不仅参与食欲的调节，还参与焦虑、恐惧、攻击性和昼夜节律的调节。对焦虑和恐惧方面这些结果的解释特别有趣。作者认为，他们在 Leptin_A 基因敲除鱼身上发现的变化证明了突变鱼在新型鱼缸中的焦虑增加，以及对捕食者出现的恐惧反应减少。这也意味着焦虑和恐惧之间隐含的区别是值得注意的。然而，就焦虑而言，对这些结果的解释是有争议的。尽管游泳速度增加、角速度增加、实验水箱顶层的快速运动增加和对这些反应的习惯性减弱似乎与焦虑增加的行为相一致，但发现突变鱼在水箱的顶层花费的时间更多，进入水箱上层区域的频率增加并且进入水箱的这个区域也更快，这与焦虑增加的说法相矛盾。

为了研究脆性X综合征（Fragile X syndrome, FXS），Kim等人（2014）对*fmr1*基因敲除（$fmr1^{hu2787}$）斑马鱼的行为进行了分析。这些突变鱼可从斑马鱼国际资源中心（ZIRC; https://zebrafish.org/home/guide.php）购买。作者在一个开放式水箱中将突变鱼与野生鱼进行了比较，开放式水箱的两个侧壁是透明的，另外两个侧壁是白色的。作者报告表明，与野生鱼相比，突变鱼在这种陌生环境中的前10min发生了许多行为改变，包括僵直行为时间缩短、游泳距离增加、与底部的距离增加以及在水箱白壁附近停留的时间增加。突变鱼的这些变化可解释为新奇恐惧减少，以及新环境引起的焦虑也减少（Kim et al., 2014）。

与焦虑表型相关的基因操纵的最后一个案例采用了多学科研究，该研究举例说明了在这种情况下如何在生物组织的多个层次（即跨学科）进行分析，并使用多个物种，即斑马鱼、小鼠和人类，这种多学科研究可能增强我们发现和解释研究结果的能力。这项由Cheol-Hee Kim带头的大型研究包括发现一个编码趋化因子样因子的新基因家族（samdori），敲除编码该家族一个成员的基因（*sam2*），鉴定*sam2*体外和体内功能，确定该基因的内源性表达脑区及其工作方式，在斑马鱼和小鼠中出现了相同的研究结果，最后分别分析了*sam2*在人中枢神经系统相关疾病中可能发挥的作用（Choi et al., 2018）。简而言之，这些作者发现*sam2*主要在斑马鱼大脑的背侧缰核、端脑和下丘脑中表达。在新型鱼缸测试、明-暗测试和群聚试验中，研究人员发现携带*sam2*沉默等位基因的基因敲除斑马鱼表现出较高的焦虑样反应，这些结果在作者制作的*sam2*基因敲除小鼠中也能重现出来。最后，由于斑马鱼和人类基因组的核苷酸序列高度保守，作者能够鉴定出斑马鱼*sam2*基因的人类同源物，找到包含*sam2*序列以及另外21个注释基因的候选区域，并且确认该区域与12q14.1缺失综合征中的智力障碍和自闭症谱系障碍有关（Choi et al., 2018）。

10.11 结论

在焦虑和恐惧的分析，以及与这些复杂行为现象相关的人类中枢神经系统疾病建模方面，对斑马鱼的研究是一个相对较新的领域。尽管如此，到目前为止，研究人员已经开发了几种方法，采用这些方法研究者能够诱导和量化斑马鱼的恐惧和焦虑反应。其中一些范式可以自动化，因此具有高度可扩展性，这意味着它们可以用于高通量突变和药物筛选。恐惧和焦虑的精神药理学特性已经开始在斑马鱼模式生物中进行研究，尽管也有争议，但许多结果表明，改变哺乳动物焦虑的经典药物在斑马鱼模式生物中以预期的方式发挥作用，这意味着进化过程的保守性以及物种之间转化具有相关性。在用斑马鱼进行焦虑方面的研究时很少用到遗传学方法。然而，为数不多的使用遗传学方法进行研究的例子以及为斑马鱼开发出来的令人印象深刻的重组DNA方法表明这些方法也具有光明的前景。因此，无论是在学术界，还是在生物技术和生物制药研究行业，斑马鱼都可能在发现焦虑潜在的机制和为人类临床开发治疗方案方面发挥作用。

（张瑞华翻译　李丽琴审校）

参考文献

Ahmed, T.S., Fernandes, Y., Gerlai, R., 2012. Effects of animated images of sympatric predators and abstract shapes on fear responses in zebrafish. Behaviour 149, 1125-1153.

Audira, G., Sarasamma, S., Chen, J.R., Juniardi, S., Sampurna, B.P., et al., 2018. Zebrafish mutants carrying leptin a (lepa) gene deficiency display obesity, anxiety, less aggression and fear, and circadian rhythm and color preference dysregulation. International Journal of Molecular Sciences 19 (12) pii: E4038.

Bass, S.L., Gerlai, R., 2008. Zebrafish (*Danio rerio*) responds differentially to stimulus fish: the effects of sympatric and allopatric predators and harmless fish. Behavioural Brain Research 186, 107-117.

Bencan, Z., Sledge, D., Levin, E.D., 2009. Buspirone, chlordiazepoxide and diazepam effects in a zebrafish model of anxiety. Pharmacology Biochemistry and Behavior 94, 75-80.

Blaser, R.E., Penalosa, Y.M., 2011. Stimuli affecting zebrafish (*Danio rerio*) behavior in the light/dark preference test. Physiology and Behavior 104, 831-837.

Choi, J.-H., Jeong, Y.-M., Kim, S., Lee, B., Ariyasiri, K., Kim, H.-T., Jung, S.-H., et al., 2018. Targeted knockout of anovel chemokine-like gene increases anxiety and fear responses. Proceedings of the National Academy of Sciences of the United States of America 115, E1041-E1050.

Davidson, J.R., Feltner, D.E., Dugar, A., 2010. Management of generalized anxiety disorder in primary care: identifying the challenges and unmet needs. Primary Care Companion to the Journal of Clinical Psychiatry 12 (2) pii: PCC.09r00772. https://doi.org/10.4088/PCC.09r00772blu.

Duarte, T., Fontana, B.D., Müller, T.E., Bertoncello, K.T., Canzian, J., Rosemberg, D.B., 2019. Nicotine prevents anxiety-like behavioral responses in zebrafish. Progress in Neuro-Psychopharmacology and Biological Psychiatry 94, 109655. https://doi.org/10.1016/j.pnpbp.2019.109655.

Engeszer, R.E., Patterson, L.B., Rao, A.A., Parichy, D.M., 2007. Zebrafish in the wild: a review of natural history andnew notes from the field. Zebrafish 4, 21-40.

Facciol, A., Iqbal, M., Eada, A., Tran, S., Gerlai, R., 2019. The light-dark task in zebrafish confuses two distinct factors: interaction between background shade and illumination level preference. Pharmacology Biochemistry and Behavior 179, 9-21.

Facciol, A., Tran, S., Gerlai, R., 2017. Re-examining the factors affecting choice in the light-dark preference test in zebrafish. Behavioural Brain Research 327, 21-28.

Fleming, A., Diekmann, H., Goldsmith, P., 2013. Functional characterisation of the maturation of the blood-brain barrier in larval zebrafish. PLoS One 8 (10), e77548. https://doi.org/10.1371/journal.pone.0077548.

Gebauer, D.L., Pagnussat, N., Piato, A.L., et al., 2011. Effects of anxiolytics in zebrafish: similarities and differences between benzodiazepines, buspirone and ethanol. Pharmacology Biochemistry and Behavior 99, 480-486.

Gatch, M.B., Lal, H., 2001. Animal models of the anxiogenic effects of ethanol withdrawal. Drug Development Research 54, 95-115.

Gerlai, R., Fernandes, Y., Pereira, T., 2009. Zebrafish (*Danio rerio*) responds to the animated image of a predator: towards the development of an automated aversive task. Behavioural Brain Research 201, 318-324.

Gerlai, R., Lahav, M., Guo, S., Rosenthal, A., 2000. Drinks like a fish: zebra fish (*Danio rerio*) as a behavior genetic model to study alcohol effects. Pharmacology Biochemistry and Behavior 67, 773e782.

Gerlai, R., Clayton, N.S., 1999a. Analysing hippocampal function in transgenic mice: an ethological perspective. Trends in Neurosciences 22, 47-51, 161.

Gerlai, R., Clayton, N.S., 1999b. Tapping artificially into natural talents. Trends in Neurosciences 22, 301-302.

Gerlai, R., Crusio, W.E., Csányi, V., 1990. Inheritance of species specific behaviors in the paradise fish (*Macropodus opercularis*): a diallel study. Behavior Genetics 20, 487-498.

Gerlai, R., 2017. Animated Images in the analysis of zebrafish behaviour. Current Zoology 63, 35-44.

Gerlai, R., 2014. Fish in behavior research: unique tools with a great promise! Journal of Neuroscience Methods 234,54-58.

Gerlai, R., 2013. Antipredatory behavior of zebrafish: adaptive function and a tool for translational research. Evolutionary Psychology 11, 1-15.

Gerlai, R., 2012. Using zebrafish to unravel the genetics of complex brain disorders. Current Topics in Behavioral Neurosciences 12, 3-24.

Gerlai, R., 2011. A small fish with a big future: zebrafish in behavioral neuroscience. Reviews in the Neurosciences 22,3-4.

Gerlai, R., 2010a. High-throughput behavioral screens: the first step towards finding genes involved in vertebrate brain function using zebrafish. Molecules 15, 2609-2622.

Gerlai, R., 2010b. Zebrafish antipredatory responses: a future for translational research? Behavioural Brain Research 207, 223-231.

Gerlai, R., 2002. Phenomics: fiction or the future? Trends in Neurosciences 25, 506-509.

Gerlai, R., 1993. Can paradise fish (*Macropodus opercularis*) recognize its natural predator? An ethological analysis. Ethology 94, 127-136.

Green, S., Hodges, H., 1991. Animal models of anxiety. In: Willner, P. (Ed.), Behavioural Models in Psychopharmacology: Theoretical, Industrial and Clinical Perspectives. Cambridge University Press, New York, NY, US, pp. 21-49.

Hascoët, M., Bourin, M., Nic Dhonnchadha, B.A., 2001. The mouse light-dark paradigm: a review. Progress in Neuro-Psychopharmacology and Biological Psychiatry 25, 141-166.

Kalueff, A.V., Stewart, A.M., Gerlai, R., 2014. Zebrafish as an emerging model for studying complex brain disorders.Trends in Pharmacological Sciences 35, 63-75.

Kim, L., He, L., Maaswinkel, H., Zhu, L., Sirotkin, H., Weng, W., 2014. Anxiety, hyperactivity and stereotypy in a zebrafish model of fragile X syndrome and autism spectrum disorder. Progress in Neuro-Psychopharmacology and Biological Psychiatry 55, 40-49.

Levin, E.D., Bencan, Z., Cerutti, D.T., 2007. Anxiolytic effects of nicotine in zebrafish. Physiology and Behavior 90, 54-58.

Luca, R., Gerlai, R., 2012a. Animated bird silhouette above the tank: acute alcohol diminishes fear responses in zebrafish. Behavioural Brain Research 229, 194-201.

Luca, R., Gerlai, R., 2012b. In search of optimal fear inducing stimuli: differential behavioral responses to computer animated images in zebrafish. Behavioural Brain Research 226, 66-76.

Maximino, C., do Carmo Silva, R.X., Dos Santos Campos, K., de Oliveira, J.S., Rocha, S.P., et al., 2019. Sensory ecology of ostariophysan alarm substances. Journal of Fish Biology 95, 274-286.

Maximino, C., da Silva, A.W., Araújo, J., Lima, M.G., Miranda, V., et al., 2014. Fingerprinting of psychoactive drugs inzebrafish anxiety-like behaviors. PLoS One 9 (7), e103943. https://doi.org/10.1371/journal.pone.0103943.

Maximino, C., de Brito, T.M., Colmanetti, R., Pontes, A.A., de Castro, H.M., et al., 2010. Parametric analyses of anxiety in zebrafish scototaxis. Behavioural Brain Research 210, 1-7.

Miller, N., Greene, K., Dydynski, A., Gerlai, R., 2012. Effects of nicotine and alcohol on zebrafish (*Danio rerio*) shoaling. Behavioural Brain Research 240, 192-196.

Miller, N., Gerlai, R., 2007. Quantification of shoaling behaviour in zebrafish (*Danio rerio*). Behavioural Brain Research184, 157-166.

Parra, K.V., Adrian Jr., J.C., Gerlai, R., 2009. The synthetic substance hypoxanthine 3-N-oxide elicits alarm reactions inzebrafish (*Danio rerio*). Behavioural Brain Research 205, 336-341.

Pelkowski, S.D., Kapoor, M., Richendrfer, H.A., Wang, X., Colwill, R.M., Creton, R., 2011. A novel high-throughput imaging system for automated analyses of avoidance behavior in zebrafish larvae. Behavioural Brain Research 223, 135-144.

Picciotto, M.R., Brunzell, D.H., Caldarone, B.J., 2002. Effect of nicotine and nicotinic receptors on anxiety and depression. NeuroReport 13, 1097-1106.

Remes, O., Brayne, C., van der Linde, R., Lafortune, L., 2016. A systematic review of reviews on the prevalence ofanxiety disorders in adult populations. Brain and Behavior 6 (7), e00497. https://doi.org/10.1002/brb3.497.

Saverino, C., Gerlai, R., 2008. The social zebrafish: behavioral responses to conspecific, heterospecific, and computer animated fish. Behavioural Brain Research 191, 77-87.

Speedie, N., Gerlai, R., 2008. Alarm substance induced behavioral responses in zebrafish (*Danio rerio*). Behavioural Brain Research 188, 168-177.

Steven, S., Rist, F., Gerlach, A.L., 2008. A review of experimental findings concerning the effect of alcohol on clinically relevant anxiety. Zeitschrift für Klinische Psychologie und Psychotherapie 37, 95-102.

Stewart, A.M., Gerlai, R., Kalueff, A.V., 2015. Developing highER-throughput zebrafish screens for in-vivo CNS drug discovery. Frontiers in Behavioral Neuroscience 9, 14. https://doi.org/10.3389/fnbeh.2015.00014.

Stewart, A.M., Cachat, J., Green, J., Gaikwad, S., Kyzar, E., Roth, A., Davis, A., et al., 2013. Constructing the habituome for phenotype-driven zebrafish research. Behavioural Brain Research 236, 110-117.

Stewart, A., Wu, N., Cachat, J., Hart, P., Gaikwad, S., Wong, K., Utterback, E., Gilder, T., et al., 2011. Pharmacological modulation of anxiety-like phenotypes in adult zebrafish behavioral models. Progress in Neuro-Psychopharmacology and Biological Psychiatry 35, 1421-1431.

Tran, S., Nowicki, M., Fulcher, N., Chatterjee, D., Gerlai, R., 2016. Interaction between handling induced stress and anxiolytic effects of ethanol in zebrafish: a behavioral and neurochemical analysis. Behavioural Brain Research298, 278-285.

Zaccone, D., Sengar, M., Lauriano, E.R., Pergolizzi, S., Macri, F., et al., 2012. Morphology and innervation of the teleost physostome swim bladders and their functional evolution in non-teleostean lineages. Acta Histochemica 114, 763-772.

第十一章

斑马鱼的社会行为及其精神药理学和遗传学分析

Noam Miller[1]

11.1 引言

社会行为是行为生态学与比较心理学研究的交叉领域。不同领域的研究人员对于社会行为有着不同类型的解释：生态学家关注适应性功能；而心理学家则关注认知机制（Zentall, 2013）。比如，行为生态学教科书通常根据受试者的得与失对社会行为进行分类（Rubinstein and Alcock, 2019），而心理学教科书则更加关注可能导致特定社会行为的认知功能（Shettleworth, 2010）。社会行为遗传学的研究正处于这两种研究的中间区域，某些等位基因被保留的适应性优势和基因影响行为的机制都是研究的重点。本章结合了这两种领域的手段，对斑马鱼社会行为的行为遗传学进行粗略解读。

由于行为的发生受基因和环境的相互作用影响并通过认知进行调节［图 11-1（A）］，社会行为的遗传学研究存在一些特殊的挑战。尤其对于无语言功能的动物，其认知内容是无法直接观察到的，那么，关于基因对认知影响的假说必然是基于一种双重推理，即从基因到行为，然后从行为再回到认知机制。社会行为中个体间的互动可能对彼此的行为决策有不同程度的干扰，例如亲属个体间不同程度重叠的基因或者个体居住环境的共同经历，都会让这种双重推理变得更加复杂和困难［图 11-1（B）］。所有动物的社会行为都是复杂的、多变的、环境敏感的。这种多基因的和定性的特征，给研究带来了重重挑战（Wright, 2011）。同时，有研究人员认为脊椎动物的社会行为（也称为社会决策网络），由众多脑区调控，不仅仅由某一个脑区控制（O’Connell and Hofmann, 2012; Bshary et al., 2014）。基

[1] Departments of Psychology & Biology, Wilfrid Laurier University, Waterloo, ON, Canada.

于以上原因，社会行为的遗传学研究仍处于起步阶段。

本章主要关注斑马鱼的社会行为以及遗传和环境因素如何影响其社会行为。在一些疾病模型中，研究人员利用一系列基因操作和药理学实验来揭示斑马鱼社会行为的机制（Fontana et al., 2018）。然而，斑马鱼社会行为是很复杂的，仅仅把它们直接与一个或两个基因产物的作用联系起来是不太可能的。因此，虽然斑马鱼在这一领域的许多研究比较成熟，但本章将补充其他鱼类的数据，让大家对这一复杂主题有较新的认识。

通过集体决策、合作和社会促进现象，社会行为中从学习到交流的一系列过程可细分为多种方式（Shettleworth, 2010; Oliveria, 2013）。本章主要关注斑马鱼群体活动和社会选择这两种社会行为。

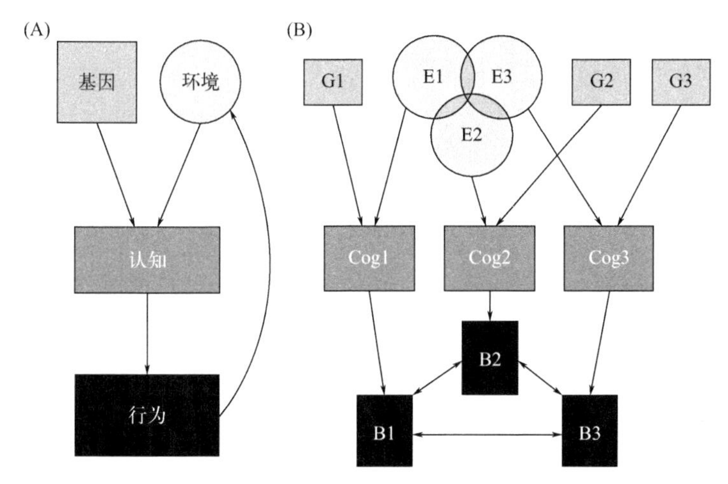

图 11-1 **遗传和环境对行为的影响**

（A）遗传效应和环境效应共同驱动认知，认知产生行为，通过行为改变所处的环境；

（B）就社会行为而言，具有相同环境和遗传背景的个体间也会受彼此的影响

G1：基因 1；G2：基因 2；G3：基因 3；E1：环境 1；E2：环境 2；E3：环境 3；Cog1：认知 1；Cog2：认知 2；Cog3：认知 3；B1：行为 1；B2：行为 2；B3：行为 3

11.2 群体活动：集群和群聚

迄今为止，群体活动是有关鱼类研究最深入、最成熟的社会现象（Mirabet et al., 2007; Sumpter, 2010）。斑马鱼几乎不单独活动，总是以大约 10 个个体的小群体进行移动和觅食，称为群聚（鱼群聚集，shoals）或集群（schools）（Pritchard et al., 2001; Suriyampola et al., 2016; Pitcher, 1983; Miller and Gerlai, 2012a）。若它们栖息在野外平缓的水域中，群体数目甚至会多达 300 个（Suriyanpola et al., 2016）。

11.2.1 斑马鱼群聚的特征

大量实验（Katz et al., 2011）和理论研究（Couzin et al., 2002）都集中于既能调节群

体活动以避免鱼群分裂，又同时保持个体间的足够距离以避免争夺资源方面（Miller and Gerlai, 2008）。研究人员对鱼类群聚的利与弊做了深入探讨（Krause and Ruxton, 2002）。一方面，为什么处于群体中的个体被捕食概率较低？第一，由于“多眼睛效应”，一个庞大群体会更容易发现捕食者（Godin et al., 1988）；第二，“稀释效应”，鱼群中的独立个体被捕食的风险大大降低（Pitcher and Parrish, 1993）；第三，“捕食者混淆效应”，捕食者容易被紧密排列的相似体型的个体所迷惑，导致捕食成功率降低（Landeau and Terborgh, 1986）。以上三种效应可能有助于确定个体特征以及群体行为，例如，斑马鱼鱼体的水平条纹可能有助于迷惑捕食者（Hogan et al., 2016）。另一方面，斑马鱼群体活动也会给其自身的觅食带来优势，例如，捕获比单独个体大或危险猎物能力有所提高（Schmitt and Strand, 1982），以及群体间可共享食物资源（Reebs and Gallant, 1997）。但是，鱼群活动也有弊端，寄生虫感染的风险增加或个体间可能发生有限资源的争夺（Krause and Ruxton, 2002）。

研究人员同时在实验室（Miller and Gerlai 2011, 2012a）和野外环境下（Suriyanpola et al., 2016）对斑马鱼的群聚进行了研究。研究人员通过以下几个指标来描述聚集的鱼群：第一，鱼群的大小或数量，但是这一标准对于特定群体是否合适仍存在争议（Miller and Gerlai, 2011; Quera et al., 2013）；第二，鱼群个体间的距离，这个距离可以是“核心鱼”与其相邻个体之间的距离（Clark and Evans, 1954）也可以是所有个体之间距离的平均值（Miller and Gerlai, 2012b）。另外，如果所有的鱼朝向同一个方向运动，则称为极化（polarization）。高度极化的群聚称为集群（Miller and Gerlai, 2012a）。实验研究和理论研究都表明，群体中个体的活动主要受相邻个体活动的影响（Couzin et al., 2002; Partridge, 1981）。

斑马鱼鱼群有以下几种鲜明的特征。将一小群斑马鱼放在一个大的无遮盖的鱼缸中，刚开始熟悉环境时，鱼聚成极化鱼群（鱼个体之间距离紧密）的时间是聚成相对松散鱼群的时间的 1.5 倍，但随着其在鱼缸中的时间延长，斑马鱼的聚集时间会更多些（Miller and Gerlai, 2012a）。鱼群每隔几秒就可以从一种模式切换到另一种模式，反映了个体决策间复杂的相互作用。鱼缸中的鱼群极化是相对固定的，但伴随着对环境的适应，其极化会逐渐减少（Miller and Gerlai, 2012a）。鱼缸集群中的鱼游动速度比群聚中的鱼更快，个体间距离更远（Miller and Gerlai, 2012a; Parrish et al., 2002）。随着反复暴露在水箱中，鱼群逐渐分散成单个个体，并远离鱼群整体（Miller and Gerlai, 2011, 2012b）。在实验室环境下，每 5 ～ 15s 周期，群体间就会出现大约 3 ～ 4cm 的最近距离（Miller 和 Gerlai, 2008），也正好是鱼身长度（Dlugos and Rabin, 2003; Miller et al., 2013）。这一结果与在野外静水环境中观察到的斑马鱼鱼群的数值接近（Suriyanpola et al., 2016）。但实际上，野外环境的复杂性更大，比如水中的植被根系可能会分散鱼群，流动的水可能会增加群聚能力（Suriyanpola et al., 2016），斑马鱼逆流而上的能力也可能取决于鱼群的特征（Wiwchar et al., 2018）。因此，实验室条件几乎不能模拟野外的环境变化以及对鱼类群体活动的影响。

11.2.2　鱼群聚集的环境效应

斑马鱼的主要群体活动发生在 12 ～ 40dpf（Engeszer et al., 2007; Buske and Gerlai,

2011），对同种属相关位置的偏好可能早在孵化后1周就已经出现（Hinz et al., 2013）。因为斑马鱼会倾向于通过单侧眼睛以一定角度观察同龄个体（Dreosti et al., 2015），所以在接下来的两周内，这种偏好会有所增强。鱼群早期形成的聚集行为可以被认为是最终形成成鱼社会行为的组成部分。在发育过程中，群聚得越来越紧密（Buske and Gerlai, 2012; Mahabir et al., 2013），鱼类偏向在共同饲养的群体中寻找配偶（Engeszer et al., 2007）。然而，成年期暴露于由多种表型混合组成的社会环境也会影响群聚偏好性（Moretz et al., 2007）。

越来越多的证据表明，斑马鱼大脑中多巴胺和血清素水平驱动着群聚行为的变化。在鱼群聚期间，两种神经递质的数量会急剧增加（Buske and Gerlai, 2012）。而且，通过研究不同近交斑马鱼品系，证明了群聚和多巴胺水平之间存在明显相关性（Mahabir et al., 2013）。此外，也有研究证明，多巴胺拮抗剂会扰乱成年斑马鱼的群聚（Scerbina et al., 2012）。相反，若暴露于同类鱼图像时，斑马鱼成鱼大脑中的多巴胺水平会升高（Saif et al., 2013）。隔离饲养的斑马鱼在群聚时其个体间的距离会增加，并且大脑中多巴胺代谢物DOPAC的水平降低，但血清素的水平正常（Shams et al., 2018）。然而，长期隔离会降低斑马鱼成鱼大脑中5-羟色胺的水平，但不会降低多巴胺的水平（Shams et al., 2015）。不同品系之间神经递质受体的表达水平也存在差异（Pan et al., 2012），可能导致不同近交系斑马鱼群聚的能力也存在差异（Seguret et al., 2016）。综上所述，斑马鱼大脑中多巴胺能和5-羟色胺能系统复杂的相互作用影响其社会行为。同样，其他没有提及的神经递质和主要激素也可能发挥重要作用（Oliveira and Gongalves, 2008）。

群聚的个体间距离也反映了斑马鱼在竞争食物和避免被捕食之间作出的权衡（Ruzzante, 1994）。与不易被捕食的鱼类相比，网纹鳉（*Poecilia reticulata*）和真鲹（*Phoxinus phoxinus*）这些容易被捕食的鱼类，在实验室中遭受到捕食时，鱼群聚集得更加紧密（Seghers, 1974a, 1974b; Magurran and Seghers, 1991; Magurra and Pitcher, 1987; Huizinga et al., 2009）。然而，这种群体遗传效应是由经验（Magurran, 1990）和饲养条件共同调节的，例如，易被捕食的网纹鳉与不易被捕食的伴侣鱼一起饲养，鱼群聚集会更紧密，反之亦然（Song et al., 2011）。换句话说，网纹鳉聚集的遗传倾向可以通过与非紧密聚焦的鱼群同类一起饲养来改变，反之亦然。目前来说，在斑马鱼身上还没有直接观察到这种效应，但也观察到一些间接的现象。例如，斑马鱼会在水中的植被中躲避捕食者，这就部分解释了为什么植被中的斑马鱼鱼群更加分散（Suriyanpola et al., 2016）。另外，当鸟类从头顶掠过后，斑马鱼聚集时个体在一段时间内会保持较近的距离（Miller and Gerlai, 2007）。

当食物集中在可守护位置时，具有“统治力”的斑马鱼会试图垄断资源（Hamilton and Dill, 2002）。与食物分布在鱼缸表面进行饲养的鱼类对比，食物分布在可守护区域进行饲养的鱼类感受到捕食者信号时，鱼群聚集更加紧密（Ruzzante and Doyle, 1993）。如果一条鱼愿意与同种属鱼保持较近距离，那么它的后代成长速度也更快（Ruzzante and Doyle, 1993），这表明“社会容忍度”的可遗传性，可能会影响面对捕食者信号作出的反应等行为。同时，也有人认为，对存在同种属的“社会容忍度”增加是鱼类复杂社会结构进化的先决条件（White et al., 2017），也可能是支持人类出现累积文化的重要因素之一（Heyes, 2018）。

11.2.3 药理与遗传效应

斑马鱼药理学实验研究了大量滥用物质如乙醇是如何改变斑马鱼的聚集行为的（Echevarria et al., 2008）。此外，也有越来越多的研究探索神经递质激动剂和拮抗剂的药理作用。

急性乙醇暴露不仅以剂量依赖的方式扰乱鱼群聚集（Gerlai, 2000），也会对其他行为产生影响，这一机制由多巴胺D1受体介导（Tran et al., 2015）。虽然几天内反复短暂的乙醇暴露会稍微增加鱼群的聚集能力，但不会减少个体间的距离（Muller et al., 2017），这可能与改变有关神经递质生成与功能的基因的表达有关（Pan et al., 2011）。有趣的是，像斑马鱼一样，哺乳动物长期接触乙醇会产生耐受性，当移除乙醇后会出现戒断效应（Gerlai et al., 2009）。麦角酸二乙胺（lysergic acid diethylamide, LSD）、3,4-亚甲二氧基甲基苯丙胺（3,4-methylenedioxymethamphetamine, MDMA）和MK801——一种*N*甲基-D-天冬氨酸（*N*-methyl-D-aspartate, NMDA）受体拮抗剂均被证明能使斑马鱼鱼群变得散乱（Maaswinkel et al., 2013; Green et al., 2012）。其他实验也证明了药物对鱼群聚集有特殊的影响。例如，低浓度尼古丁会增加鱼类与其相邻个体之间的距离，但不会改变聚集或集群的极化。相反，乙醇暴露对鱼群个体间距离的影响较小，会更多地破坏极化（Miller et al., 2013）。这些结果表明，先对鱼群采取一些措施，而后可以尝试将群体活动的不同行为与特定的认知能力联系起来进行研究。

已有研究证明，某些基因对聚集行为有影响。例如，敲除唐氏综合征（Down syndrome）和自闭症相关基因 *dyrkla* 的斑马鱼，其聚集鱼群比野生型斑马鱼的聚集鱼群要小，并且靠近不影响鱼群视线的障碍物的时间也更少。值得注意的是，在社会行为分析中，这些鱼还表现出压力降低（Kim et al., 2017）。敲除自闭症相关基因 *shank3b* 的斑马鱼也显示出鱼群个体间的距离增加、社会偏好减少以及普遍存在的运动障碍（Liu et al., 2018）。

这些社会行为的药理学和遗传效应，很多可能是非特异性应激反应的结果。压力增加通常会增加鱼群的聚集能力，而一系列抗焦虑药物（乙醇、丁螺环酮或东莨菪碱）（Hamilton et al., 2017）可能以不同的方式导致鱼群的破坏或解散（Gebauer et al., 2011）。

群体活动不是一类社会认知，而是一种特殊的行为。然而，它是应对社会刺激的一类特殊反应。换言之，聚集的鱼类以及其他成群活动的动物可能使用一套“交通规则”（Parrish and Turchin, 1997），这就是它们群体活动的认知机制。这种机制通常不存在于社会认知的心理学范畴，但可以被认为是社会发展的一种特殊形式（Zajonc, 1965; Oliveira, 2013; Rosenthal et al., 2015）。

社会选择是一种有助于聚集的行为，而且已经被证明与群体活动无关，下一节将进行具体讨论。

11.3 社会选择

在网纹鳉中，所有鱼群的成员大约每10s改变一次（Croft et al., 2003）。与网纹鳉相似，斑马鱼群也在很短的时间内解散和重组。因此，鱼类必须反复决定加入附近几个鱼群中的哪一个鱼群。虽然选择一个群体加入（社会选择）和聚集成群（松散聚焦）看起来是同一回事，但至少在刺鱼（*Gasterosteus aculeatus*）身上，这两种行为在功能上是相互独立的，并且由不同的遗传组件所介导（Greenwood et al., 2015）。因此，研究人员试图探究那些更加具有吸引力的鱼群的特征。通常在一个透明的屏障后面设置两个鱼群，这两个鱼群在某些兴趣特征方面存在差异，然后让一条试验鱼选择接近两个鱼群中的一个。与这种行为相关的认知机制被称为“社会偏好”（Oliveira, 2013）。

结果显示，即使鱼群中的鱼与试验鱼的表型不同，试验鱼也更喜欢有鱼群的鱼缸而非空鱼缸。（Sneckser et al., 2006），另外，与不同表型的斑马鱼群相比较，斑马鱼更喜欢相似表型的鱼群（Rosenthal and Ryan, 2005; Sneckser et al., 2010）。在混合鱼鱼群中，试验鱼仍然更接近其同种属而不是异种属的鱼（Saverino and Gerlai, 2008）。许多鱼类更喜欢与自己大小相似的同种鱼类，以便保持相同的游泳速度，维持鱼群聚集（Krause et al., 2000; Croft et al., 2003）。不同颜色的鱼群（McRobert and Bradner, 1998）增加了鱼群的视觉同质性，也让捕食者产生视觉混淆。斑马鱼喜欢加入鱼数目多的鱼群，不过也受鱼群活动水平（Pritchard et al., 2001）以及鱼群个体性别的影响（Ruhl and McRobert, 2005）。

与群体活动一样，已有证明药物能改变斑马鱼的社会选择。MK801 是一种 NMDA 受体拮抗剂，可以干扰 3 周龄斑马鱼的社会偏好，乙醇也是如此（Dreosti et al., 2015）。如上所述，敲除与唐氏综合征和自闭症相关的 *dyrkla* 基因也会损害社会偏好（Kim et al., 2017）。如果斑马鱼在胚胎发育过程中暴露于低浓度乙醇，社会选择会剂量依赖性减少，这一效应类似于胎儿乙醇综合征（Fernandes and Gerlai, 2009）。

社会行为的复杂性和研究者如何衡量社会选择的巨大差异导致了一些相互矛盾的结果。例如，已有研究证明，催产素和精氨酸加压素以及它们的鱼类特异性同系物鱼神经叶激素和血管催产素，能以剂量依赖的方式使斑马鱼增加对具有相似表型的同种属的偏好，并且这种效应能被相应受体的拮抗剂所抑制（Braida et al., 2012）。然而，也有研究证明，不同剂量的血管催产素能减少对鱼群的偏好，其受体拮抗剂也有同样的效果（Lindeyer et al., 2015）。这些结果指出了社会选择行为的一些局限性，在这些行为中，试验鱼与社会刺激物在物理上是分离的，唯一的衡量标准是它们在一定距离内的时间比例。对以上方法存在的不足，也有研究人员给出了建议和改进方案（Agrillo et al., 2017）。

在通常的三室实验中，与空笼以及较熟悉的动物比较，啮齿动物经常更喜欢接近陌生的同种属个体（Yang et al., 2011）。斑马鱼实验结果也是如此（Barba Escobedo&Gould, 2012）。斑马鱼暴露于 $5HT_{1A}$ 受体激动剂丁螺环酮和大麻素受体激动剂 WIN 55212 时，会以不同的方式增加社会偏好（Barba Escobedo and Gould, 2012）。值得注意的是，鉴于丁螺环酮在斑马鱼中作为一种抗焦虑药物发挥作用（Gebauer et al., 2011），并且可能会减少社会偏好，因此该结果在某些方面与上文预期相反。与对照组不同的是，鱼类暴露于具有抗焦虑作用的乙醇后，虽然表达了对陌生同种属的偏好，但这种偏好程度没有超过更熟悉的同种属（Ariyasiri et al., 2019）。有趣的是，在 *sam2* 基因敲除斑马鱼身上也发现了类似的结果模式，此前已有研究表明，这些斑马鱼会表现出较高的焦虑水平（Choi et al., 2018）。总之，这些结果表明斑马鱼的社会选择是一个复杂的现象，受焦虑水平变化的影响，并受几种不同的神经递质系统的调节。

11.4 其他形式

斑马鱼还有一些其他形式的社会行为，但因为这些形式的研究数量远远少于鱼群聚集和社会选择，因此在上文中未提及（Oliveira, 2013）。然而，也有研究人员对斑马鱼其他的社会机制感兴趣。有关社会学习的研究越来越多，社会学习即动物通过观察对方的行为来了解自身环境的过程（Nunes et al., 2017）。在其他鱼类物种研究中，研究人员也越来越关注社会学习这一行为（Brown and Laland, 2003）。例如，在社会传递方面，斑马鱼通过习得可能呈现捕食者气味、光的条件反射，当面临捕食者的时候能释放一种预警信号——从受损皮肤释放出的一种报警信息素——该信号同样会在未接触过预警信号的鱼群中传递（Hall and Suboski, 1995）。

最近，研究人员比较关注斑马鱼能否识别同种属个体。众所周知，斑马鱼可能通过嗅觉线索识别近亲（Mann et al., 2003），并且能够区分熟悉的同种属个体（Madeira and Oliveira, 2017）。这表明，在野生环境中，斑马鱼的聚集可能不是完全随机的，近亲鱼群间聚集可能性更大（Krause and Ruxton, 2002）。个体识别也可能在斑马鱼优势等级的形成和维持中发挥重要作用（Paull et al., 2010）。

相较非社会性刺激和无相互作用的鱼，斑马鱼更倾向于关注有相互作用的鱼，并且已有研究表明，某些基因介导了这种效应（Lopes et al., 2015）。同时，对社会刺激和对附近个体的容忍度增加则是另一种机制，都对社会结构和文化发展有重要作用（Heyes, 2018）。

最后，研究人员对斑马鱼个体差异的研究与日俱增。除了早期针对其他鱼类的研究（Sih et al., 2004），现有研究大量关注斑马鱼个体的差异（Khan and Echevarria, 2017）以及如何更好地测量这种差异（Toms et al., 2010; Teles and Oliveira, 2016）。不仅群体中每个个体应对刺激的反应影响群体行为，而且在许多情况下，社会环境也会改变个体的反应行为（Guayasamin et al., 2017），所以可以说，个体差异是理解社会行为的关键（Dingemanse et al., 2010）。个体差异导致繁殖，形成了鱼类聚集成群的遗传背景（Vargas et al., 2018）。品系差异也属于斑马鱼个体差异（Roy and Bhat, 2018），也有研究初步阐述了品系差异的分子和遗传学基础（Laine and van Oers, 2017）。

11.5 结论

本章中斑马鱼的很多社会行为与本书其他章节所述的行为有共同之处。例如，对于几种鱼类来说，研究表明，攻击性和集群行为与遗传密切相关（Ruzzante, 1994），另外上文已经指出压力水平如何影响社会互动，反过来又受到社会刺激的影响。社会因素也可以调节其他行为。例如，斑马鱼受预警信号的影响，会停止运动、不稳定运动或出现其他的应激反应（Speedie and Gerlai, 2009; Green et al., 2012），但是当看到未暴露于预警信号的同种属时，这些行为则会减弱（Faustino et al., 2017）。阐明这些行为的遗传基础具有较大的挑战性，因为许多社会行为都依赖于参与社会行为的基因调控和受到多种认知机制的影响。

在上文提到的众多研究中，多巴胺能和5-羟色胺能系统在塑造社会行为方面起着重要作用，部分原因可能是大家对这些系统的研究比较完善。同时，也有研究人员认为是两个系统的协同作用为社会行为的出现创造了必要的条件（Van den Bos, 2015），多巴胺的激活增加了鱼类聚集的趋势，可能是通过接近同种可产生有益结果介导的（Saif et al., 2013），而5-羟色胺的激活降低了攻击性水平，增加了对周围个体的容忍度。随着对两个中枢系统的了解日益深入，以及越来越多的其他调节因素的发现，最终将阐释基因是如何影响斑马鱼和其他脊椎动物的社会认知和行为的。

致谢

本章的撰写得到了加拿大国家科学与工程研究委员会（National Science and Engineering Research Council of Canada, NSERC）的资助（RGPIN-2016-06138）。

（朱思庆翻译　李丽琴审校）

参考文献

Agrillo, C., Petrazzini, M.E., Bisazza, A., 2017. Numerical abilities in fish: a methodological review. Behavioural Processes 141, 161-171.

Ariyasiri, K., Choi, T.-I., Kim, O.-H., Hong, T.I., Gerlai, R., Kim, C.-H., 2019. Pharmacological (ethanol) and mutation (sam2 KO) induced impairment of novelty preference in zebrafish quantified using a new three-chamber social choice task. Progress in Neuropsychopharmacology and Biological Psychiatry 88, 53-65.

Barba-Escobedo, P.A., Gould, G.G., 2012. Visual social preferences of lone zebrafish in a novel environment: strain and anxiolytic effects. Genes, Brain and Behavior 11, 366-373.

Braida, D., Donzelli, A., Martucci, R., Capurro, V., Busnelli, M., Chini, B., Sala, M., 2012. Neurohypophyseal hormones manipulation modulate social and anxiety-related behavior in zebrafish. Psychopharmacology 220, 319-330.

Brown, C., Laland, K.N., 2003. Social learning in fishes: a review. Fish and Fisheries 4, 280e288.

Bshary, R., Gingins, S., Vail, A.L., 2014. Social cognition in fishes. Trends in Cognitive Sciences 18, 465-471.

Buske, C., Gerlai, R., 2011. Shoaling develops with age in Zebrafish (*Danio rerio*). Progress in Neuropsychopharmacology and Biological Psychiatry 35, 1409-1415.

Buske, C., Gerlai, R., 2012. Maturation of shoaling behavior is accompanied by changes in the dopaminergic and serotoninergic systems in zebrafish. Developmental Psychobiology 54, 28-35.

Choi, J.-H., Jeong, Y.-M., Kim, S., Lee, B., Ariyasiri, K., Kim, H.-T., et al., 2018. Targeted knockout of a chemokine-like gene increases anxiety and fear responses. Proceedings of the National Academy of Sciences of the USA 115, E1041-E1050.

Clark, P.J., Evans, F.C., 1954. Distance to nearest neighbor as a measure of spatial relationships in populations. Ecology 35, 445-453.

Couzin, I.D., Krause, J., James, R., Ruxton, G.D., Franks, N.R., 2002. Collective memory and spatial sorting in animal groups. Journal of Theoretical Biology 218, 1-11.

Croft, D.P., Arrowsmith, B.J., Bielby, J., Skinner, K., White, E., Couzin, I.D., Magurran, A.E., Ramnarine, I., Krause, J.,2003. Mechanisms underlying shoal composition in the Trinidadian guppy, Poecilia reticulata. Oikos 100, 429-438.

Dingemanse, N.J., Kazem, A.J.N., Réale, D., Wright, J., 2010. Behavioural reaction norms: animal personality meets individual plasticity. Trends in Ecology and Evolution 25, 81-89.

Dlugos, C.A., Rabin, R.A., 2003. Ethanol effects on three strains of zebrafish: model system for genetic investigations. Pharmacology, Biolchemistry and Behavior 74, 471-480.

Dreosti, E., Lopes, G., Kampff, A.R., Wilson, S.W., 2015. Development of social behavior in young zebrafish. Frontiers in Neural Circuits 9, 39.

Echevarria, D.J., Hammack, C.M., Pratt, D.W., Hosemann, J.D., 2008. A novel behavioral test battery to assess global drug effects using the zebrafish. International Journal of Comparative Psychology 21, 19-34.

Engeszer, R.E., da Barbiano, L.A., Ryan, M.J., Parichy, D.M., 2007. Timing and plasticity of shoaling behavior on the zebrafish (*Danio rerio*). Animal Behaviour 74, 1269-1275.

Faustino, A.I., Tacao-Monteiro, A., Oliveira, R.F., 2017. Mechanisms of social buffering of fear in zebrafish. Scientific Reports 7, 44329.

Fernandes, Y., Gerlai, R., 2009. Long-term behavioral changes in response to early develop-mental exposure to ethanol in zebrafish. Alcoholism: Clinical and Experimental Research 33, 601-609.

Fontana, B.D., Mezzomo, N.J., Kalueff, A.V., Rosemberg, D.B., 2018. The developing utility of zebrafish models of neurological and neuropsychiatric disorders: a critical review. Experimental Neurology 299, 157-171.

Gebauer, D.L., Pagnussat, N., Piato, A.L., Schaefer, I.C., Bonan, C.D., Lara, D.R., 2011. Effects of anxiolytics in zebrafish: similarities and differences between benzodiazepines, buspirone and ethanol. Pharmacology, Biochemistry and Behavior 99, 480-486.

Gerlai, R., 2000. Drinks like a fish: zebra fish (*Danio rerio*) as a behavior genetic model to study alcohol effects. Pharmacology, Biochemistry and Behavior 67, 773-782.

Gerlai, R., Chatterjee, D., Pereira, T., Sawashima, T., Krishnannair, R., 2009. Acute and chronic alcohol dose: population differences in behavior and neurochemistry of zebrafish. Genes, Brain and Behavior 8, 586-599.

Godin, J.-G.J., Classon, L.J., Abrahams, M.V., 1988. Group vigilance and shoal size in a small characin fish. Behaviour 104, 29-40.

Green, J., Collins, C., Kyzar, E.J., Pham, M., Roth, A., Gaikwad, S., et al., 2012. Automated high-throughput neurophenotyping of zebrafish social behavior. Journal of Neuroscience Methods 210, 266-271.

Greenwood, A.K., Wark, A.R., Yoshida, K., Peichel, C.L., 2015. Genetic and neural modularity underlie the evolution of schooling behavior in threespine sticklebacks. Current Biology 23, 1884-1888.

Guayasamin, O.L., Couzin, I.D., Miller, N.Y., 2017. Behavioural plasticity across social contexts is regulated by the directionality of inter-individual differences. Behavioural Processes 141, 196-204.

Hall, D., Suboski, M.D., 1995. Visual and olfactory stimuli in learned release of alarm reactions by zebra danio fish (*Brachydanio rerio*). Neurobiology of Learning and Memory 63, 229-240.

Hamilton, I.M., Dill, L.M., 2002. Monopolization of food by zebrafish (*Danio rerio*) increases in risky habitats. Canadian Journal of Zoology 80, 2164-2169.

Hamilton, T.J., Morrill, A., Lucas, K., Gallup, J., Harris, M., Healey, M., et al., 2017. Establishing zebrafish as a model to study the anxiolytic effects of scopolamine. Scientific Reports 7, 15081.

Heyes, C.M., 2018. Cognitive Gadgets: The Cultural Evolution of Thinking. Harvard University Press.

Hinz, F.I., Aizenberg, M., Tushev, G., Schuman, E.M., 2013. Protein synthesis-dependent associative long-term memory in larval zebrafish. Journal of Neuroscience 33, 15382-15387.

Hogan, B.G., Cuthill, I.C., Scott-Samuel, N.E., 2016. Dazzle camouflage, target tracking, and the confusion effect. Behavioral Ecology 27, 1547-1551.

Huizinga, M., Ghalambor, C.K., Reznick, D.N., 2009. The genetic and environmental basis of adaptive differences in shoaling behavior among populations of Trinidadian guppies, *Poecilia reticulata*. Journal of Evolutionary Biology 22, 1860-1866.

Katz, Y., Tunstrøm, K., Ioannou, C.C., Huepe, C., Couzin, I.D., 2011. Inferring the structure and dynamics of interactions in schooling fish. Proceedings of the National Academy of Sciences of the United States of America 108, 18720-18725.

Khan, K.M., Echevarria, D.J., 2017. Feeling fishy: trait differences in zebrafish (*Danio rerio*). In: Vonk, J., Weiss, A., Kuczaj, S.A. (Eds.), Personality in Nonhuman Animals. Springer, pp. 111-127.

Kim, O.-H., Cho, H.-J., Han, E., Hong, T.I., Ariyasiri, K., Choi, J.-H., et al., 2017. Zebrafish knockout of Down syndrome gene, DYRK1A, shows social impairments relevant to autism. Molecular Autism 8, 50.

Krause, J., Hoare, D.J., Croft, D., Lawrence, J., Ward, A., Ruxton, G.D., Godin, J.-G.J., James, R., 2000. Fish shoal composition: mechanisms and constraints. Proceedings of the Royal Society 267, 2011-2017.

Krause, J., Ruxton, G.D., 2002. Living in Groups. Oxford University Press.

Laine, V.N., van Oers, K., 2017. The quantitative and molecular genetics of individual differences in animal personality. In: Vonk, J., Weiss, A., Kuczaj, S.A. (Eds.), Personality in Nonhuman Animals. Springer, pp. 55-72.

Landeau, L., Terborgh, J., 1986. Oddity and the 'confusion effect' in predation. Animal Behaviour 34, 1372-1380.

Lindeyer, C.M., Langen, E., Swaney, W.T., Reader, S.M., 2015. Nonapeptide influences on social behaviour: effects of vasotocin and isotocin on shoaling and interaction in zebrafish. Behaviour 152, 897-915.

Liu, C.-X., Li, C.-Y., Hu, C.-C., Wang, Y., Lin, J., Jiang, Y.-H., et al., 2018. CRISPR/Cas9-induced shank3b mutant zebrafish display autism-like behaviors. Molecular Autism 9, 23.

Lopes, J.S., Abril-de-Abreu, R., Oliveira, R., 2015. Brain transcriptomic response to social eavesdropping in zebrafish (*Danio rerio*). PLoS One 10 (12), e0145801.

Maaswinkel, H., Zhu, L., Weng, W., 2013. Assessing social engagement in heterogeneous groups of zebrafish: a new paradigm for autism-like behavioral responses. PLoS One 8 (10), e75955.

Madeira, N., Oliveira, R.F., 2017. Long-term social recognition memory in zebrafish. Zebrafish 14, 305-310.

Magurran, A.E., 1990. The inheritance and development of minnow anti-predator behaviour. Animal Behaviour 39, 834-842.

Magurran, A.E., Pitcher, T.J., 1987. Provenance, shoal size and the sociobiology of predator evasion behaviour in minnow shoals. Proceedings of the Royal Society 229, 439-465.

Magurran, A.E., Seghers, B.H., 1991. Variation in schooling and aggression amongst guppy (*Poecilia reticulata*) populations in Trinidad. Behaviour 118, 214-234.

Mahabir, S., Chatterjee, D., Buske, C., Gerlai, R., 2013. Maturation of shoaling in two zebrafish strains: a behavioral and neurochemical analysis. Behavioural Brain Research 247, 1-8.

Mann, K.D., Turnell, E.R., Atema, J., Gerlach, G., 2003. Kin recognition in juvenile zebrafish (*Danio rerio*) based on olfactory cues. The Biological Bulletin 205, 224-225.

McRobert, S.P., Bradner, J., 1998. The influence of body coloration on shoaling preferences in fish. Animal Behaviour 56, 611-615.

Miller, N., Gerlai, R., 2007. Quantification of shoaling behavior in zebrafish (*Danio rerio*). Behavioural Brain Research 184, 157-166.

Miller, N.Y., Gerlai, R., 2008. Oscillations in shoal cohesion in zebrafish (*Danio rerio*). Behavioural Brain Research 193, 148-151.

Miller, N., Gerlai, R., 2011. Redefining membership in animal groups. Behavior Research Methods 43, 964-970.

Miller, N., Gerlai, R., 2012a. From schooling to shoaling: patterns of collective motion in zebrafish (*Danio rerio*). PLoS One 7 (11), e48865.

Miller, N., Gerlai, R., 2012b. Automated tracking of zebrafish shoals and the analysis of shoaling behavior. In:Kalueff, A.V., Stewart, A.M. (Eds.), Zebrafish Protocols for Neurobehavioral Research. Humana Press, NewYork, pp. 217-230.

Miller, N., Greene, K., Dydinski, A., Gerlai, R., 2013. Effects of nicotine and alcohol on zebrafish (*Danio rerio*) shoaling. Behavioural Brain Research 240, 192e196.

Mirabet, V., Auger, V., Lett, C., 2007. Spatial structures in simulations of animal grouping. Ecological Modelling 201, 468-476.

Moretz, J.A., Martins, E.P., Robison, B.D., 2007. The effects of early and adult social environment on zebrafish (*Danio rerio*) behavior. Environmental Biology of Fishes 80, 91-101.

Müller, T.E., Nune, S.Z., Silveira, A., Loro, V.L., Rosemberg, D.B., 2017. Repeated ethanol exposure alters social behavior and oxidative stress parameters of zebrafish. Progress in Neuropsychopharmacology and Biological Psychiatry 79, 105-111.

Nunes, A.R., Ruhl, N., Winberg, S., Oliveira, R.F., 2017. Social phenotypes in zebrafish. In: Kalueff, A.V. (Ed.), The Rights and Wrongs of Zebrafish: Behavioral Phenotyping of Zebrafish. Springer, pp. 95-130.

O'Connell, L.A., Hofmann, H.A., 2012. Evolution of a vertebrate social decision-making network. Science 336, 1154-1157.

Oliveira, R.F., 2013. Mind the fish: zebrafish as a model in cognitive social neuroscience. Frontiers in Neural Circuits 7, 131.

Oliveira, R.F., Gonçalves, D.M., 2008. Hormones and social behavior of teleost fish. In: Magnhagen, C., Braithwaite, V.A., Forsgren, E., Kapoor, B.G. (Eds.), Fish Behaviour. Science Publishers, pp. 61-150.

Pan, Y., Kaiguo, M., Razak, Z., Westwood, J.T., Gerlai, R., 2011. Chronic alcohol exposure induced gene expression changes in the zebrafish brain. Behavioural Brain Research 216, 66-76.

Pan, Y., Chatterjee, D., Gerlai, R., 2012. Strain dependent gene expression and neurochemical levels in the brain of zebrafish: focus on a few alcohol related targets. Physiology and Behavior 107, 773-780.

Parrish, J.K., Viscido, S.V., Grünbaum, D., 2002. Self-organized fish schools: an examination of emergent properties. The Biological Bulletin 202, 296-305.

Parrish, J., Turchin, P., 1997. Individual decisions, traffic rules, and emergent pattern in schooling fish. In: Parrish, J., Hamner, W. (Eds.), Animal Groups in Three Dimensions: How Species Aggregate. Cambridge University Press, Cambridge, pp. 126-142.

Partridge, B.L., 1981. Internal dynamics and the interrelations of fish in schools. Journal of Comparative Physiology 144, 313-325.

Paull, G.C., Filby, A.L., Giddins, H.G., Coe, T.S., Hamilton, P.B., Tyler, C.R., 2010. Dominance hierarchies in zebrafish (*Danio rerio*) and their relationship with reproductive success. Zebrafish 7, 109-117.

Pitcher, T.J., 1983. Heuristic definitions of fish shoaling behaviour. Animal Behaviour 31, 611-613.

Pitcher, T.J., Parrish, J.K., 1993. Functions of shoaling behavior in teleosts. In: Pitcher, T.J. (Ed.), Behavior of Teleost Fishes. Chapman & Hall, London, pp. 363-439.

Pritchard, V.L., Lawrence, J., Butlin, R.K., Krause, J., 2001. Shoal choice in zebrafish, *Danio rerio*: the influence of shoal size and activity. Animal Behaviour 62, 1085-1088.

Quera, V., Beltran, F.S., Givoni, I.E., Dolado, R., 2013. Determining shoal membership using affinity propagation. Behavioural Brain Research 241, 38-49.

Reebs, S.G., Gallant, B.Y., 1997. Food-anticipatory activity as a cue for local enhancement in golden shiners. Ethology 103, 1060-1069.

Rosenthal, S.B., Twomey, C.R., Hartnett, A.T., Wu, H.S., Couzin, I.D., 2015. Revealing the hidden networks of interaction in mobile animal groups allows prediction of complex behavioral contagion. Proceedings of the National Academy of Sciences of the United States of America 112, 4690-4695.

Rosenthal, G.G., Ryan, M.J., 2005. Assortative preferences for stripes in danios. Animal Behaviour 70, 1063-1066.

Roy, T., Bhat, A., 2018. Repeatability in boldness and aggression among wild zebrafish (*Danio rerio*) from two differing predation and flow regimes. Journal of Comparative Psychology 132, 349-360.

Rubinstein, D.R., Alcock, J., 2019. Animal Behavior, eleventh ed. Oxford University Press, Sunderland, MA.

Ruhl, N., McRobert, S.P., 2005. The effect of sex and shoal size on shoaling behaviour in *Danio rerio*. Journal of Fish Biology 67, 1318-1326.

Ruzzante, D.E., 1994. Domestication effects on aggressive and schooling behavior in fish. Aquaculture 120, 1-24.

Ruzzante, D.E., Doyle, R.W., 1993. Evolution of social behavior in a resource-rich, structured environment: selection experiments with medaka (*Oryzias latipes*). Evolution 47, 456-470.

Saif, M., Chatterjee, D., Buske, C., Gerlai, R., 2013. Sight of conspecific images induces changes in neurochemistry in zebrafish. Behavioural Brain Research 243, 294-299.

Saverino, C., Gerlai, R., 2008. The social zebrafish: behavioral responses to conspecific, heterospecifics, and computer animated fish. Behavioural Brain Research 191, 77-87.

Scerbina, T., Chatterjee, D., Gerlai, R., 2012. Dopamine receptor antagonism disrupts social preference in zebrafish: a strain comparison study. Amino Acids 43, 2059-2072.

Schmitt, R.J., Strand, S.W., 1982. Cooperative foraging by yellowtail, *Seriola lalandei* (*Carangidae*) on two species of fish prey. Copeia 1982, 714-717.

Seghers, B.H., 1974a. Schooling behavior on the Guppy (*Poecilia reticulata*): an evolutionary response to predation. Evolution 28, 486-489.

Seghers, B.H., 1974b. Geographic variation in the responses of Guppies (*Poecilia reticulata*) to aerial predators. Oecologia 14, 93-98.

Séguret, A., Collignon, B., Halloy, J., 2016. Strain differences in the collective behaviour of zebrafish (*Danio rerio*) in heterogeneous environment. Royal Society Open Science 3, 160451.

Shams, S., Chatterjee, D., Gerlai, R., 2015. Chronic social isolation affects thigmotaxis and whole-brain serotonin levels in adult zebrafish. Behavioural Brain Research 292, 283-287.

Shams, S., Amlani, S., Buske, C., Chatterjee, D., Gerlai, R., 2018. Developmental social isolation affects adult behavior, social interaction, and dopamine metabolite levels in zebrafish. Developmental Psychobiology 60, 43-56.

Shettleworth, S.J., 2010. Cognition, Evolution, and Behavior, second ed. Oxford University Press.

Sih, A., Bell, A., Johnson, J.C., 2004. Behavioral syndromes: an ecological and evolutionary overview. Trends in Ecology and Evolution 19, 372-378.

Sneckser, J.L., McRobert, S.P., Murphy, C.E., Clotfelter, E.D., 2006. Aggregation behavior in wildtype and transgenic zebrafish. Ethology 112, 181-187.

Sneckser, J.L., Ruhl, N., Bauer, K., McRobert, S.P., 2010. The influence of sex and phenotype on shoaling decisions in zebrafish. International Journal of Comparative Psychology 23, 70-81.

Song, Z., Boenke, M.C., Rodd, F.H., 2011. Interpopulation differences in shoaling behaviour in guppies (*Poecilia reticulata*): roles of social environment and population origin. Ethology 117, 1009-1018.

Speedie, N., Gerlai, R., 2009. Alarm substance induced behavioral responses in zebrafish (*Danio rerio*). Behavioural Brain Research 188, 168-177.

Sumpter, D.J.T., 2010. Collective Animal Behavior. Princeton University Press.

Suriyampola, P.S., Shelton, D.S., Shukla, R., Roy, T., Bhat, A., Martins, E.P., 2016. Zebrafish social behavior in the wild. Zebrafish 13, 1-8.

Teles, M.C., Oliveira, R.F., 2016. Quantifying aggressive behavior in zebrafish. In: Kawakami, K., et al. (Eds.), Zebrafish: Methods and Protocols. Springer, pp. 293-305.

Toms, C.N., Echevarria, D.J., Jouandot, D.J., 2010. A methodological review of personality-related studies in fish: focus on the shy-bold axis of behavior. International Journal of Comparative Psychology 23, 1-25.

Tran, S., Nowicki, M., Muraleetharan, A., Chatterjee, D., Gerlai, R., 2015. Differential effects of acute administration of SCH-23390, a D1 receptor antagonist, and of ethanol on swimming activity, anxiety-related responses, and neurochemistry of zebrafish. Psychopharmacology 232, 3709-3718.

van den Bos, R., 2015. The dorsal striatum and ventral striatum play different roles in the programming of social behaviour: a tribute to Lex Cools. Behavioural Pharmacology 26, 6-17.

Vargas, R., Mackenzie, S., Rey, S., 2018. 'Love at first sight': the effect of personality and colouration patterns in the reproductive success of zebrafish (*Danio rerio*). PLoS One 13 (9), e0203320.

Wiwchar, L.D., Gilbert, M.J.H., Kasurak, A.V., Tierney, K.B., 2018. Schooling improves critical swimming performance in zebrafish (*Danio rerio*). Canadian Journal of Fisheries and Aquatic Sciences 75, 653-661.

White, S.L., Wagner, T., Gowan, C., Braithwaite, V.A., 2017. Can personality predict individual differences in brook trout spatial learning ability? Behavioural Processes 141, 220-228.

Wright, D., 2011. QTL mapping of behavior in the zebrafish. In: Kalueff, A.V., Cahcat, J.M. (Eds.), Zebrafish Models in Neurobehavioral Research. Humana Press, Totowa, NJ, pp. 101-141.

Yang, M., Silverman, J.L., Crawley, J.N., 2011. Automated three-chambered social approach task for mice. Current Protocols in Neuroscience 8, 26.

Zajonc, R.B., 1965. Social facilitation. Science 149, 269-274.

Zentall, T.R., 2013. Observational learning in animals. In: Clark, K.B. (Ed.), Social Learning Theory. Nova Science, New York, pp. 3-33.

第十二章

成年斑马鱼的联想学习和非联想学习

Justin W. Kenney[1]

12.1 引言

学习和记忆对生物体生存来说是不可或缺的，为塑造生物体对环境的行为响应提供了经验平台。从线虫到昆虫，到脊椎动物的各种物种，记忆过程中涉及的许多基本分子和突触机制在不同物种中都是非常保守的（Kandel, 2012）。然而，为进一步理解人类相关的大脑 - 行为关系，我们在研究中需要使用具有复杂行为表型，且中枢神经系统所有组织与人具有相似特征的生物。因此，在学习和记忆研究中，啮齿动物一直以来都是主要的模式动物［图 12-1（A）］，其他模式生物占比相对较小，但目前与其他模式生物相关的文献量不断增长［图 12-1（B）］。在啮齿类动物的常用替代物种中，斑马鱼因其中枢神经系统的组织结构与哺乳动物的相似度更高而闻名（Panula et al., 2010; Wullimann and Mueller, 2004）。此外，有关成年斑马鱼学习的研究在过去的 20 年中已有大量报道，研究发现斑马鱼具有多种联想学习和非联想学习能力。斑马鱼作为动物模型的使用不仅仅限于神经科学，在免疫学（Meeker and Trede, 2008）和心血管研究（Gut et al., 2017）等其他领域也占有一席之地。斑马鱼的广泛应用促进了大量分子遗传学工具的开发，并且获得了几乎能与啮齿动物相匹敌的转基因动物资源。结合它们对高通量研究的适应性，斑马鱼作为脊椎动物模型的一个必要组成部分，有望继续成为理解学习和记忆的分子和神经基础的完整补充。

[1] Department of Biological Sciences, Wayne State University, Detroit, MI, United States.

12.2 斑马鱼在学习和记忆研究中的应用

神经科学在利用来自动物种属的各种动物模型方面有着丰富的历史（Marder, 2002）。然而，关于学习和记忆的研究主要采用两种密切相关的啮齿动物，即大鼠和小鼠，它们相关的论文约占最近发表论文的95%［图12-1（A）］。仅仅关注一两个高度相关的物种可能会带来一些风险，即研究发现的结果是这些物种所独有的，而不是更可能与人类相关的保守脊椎动物所具有的普遍现象（Marder, 2002; Brenowitz and Zakon, 2015; Yartsev, 2017）。因此，更多地考虑一种亲缘关系较远的模型，比如斑马鱼，可对理解中枢神经系统功能的一般原理提供一种重要的互补观点。事实上，斑马鱼和人类拥有相同的主要神经递质系统（Panula et al., 2010），并具有相同的中枢神经系统总体结构（Wullimann and Muel-ler, 2004）。此外，如下所述，研究显示斑马鱼具有各种各样的联想和非联想学习能力，能完成许多类似于哺乳动物能完成的任务。

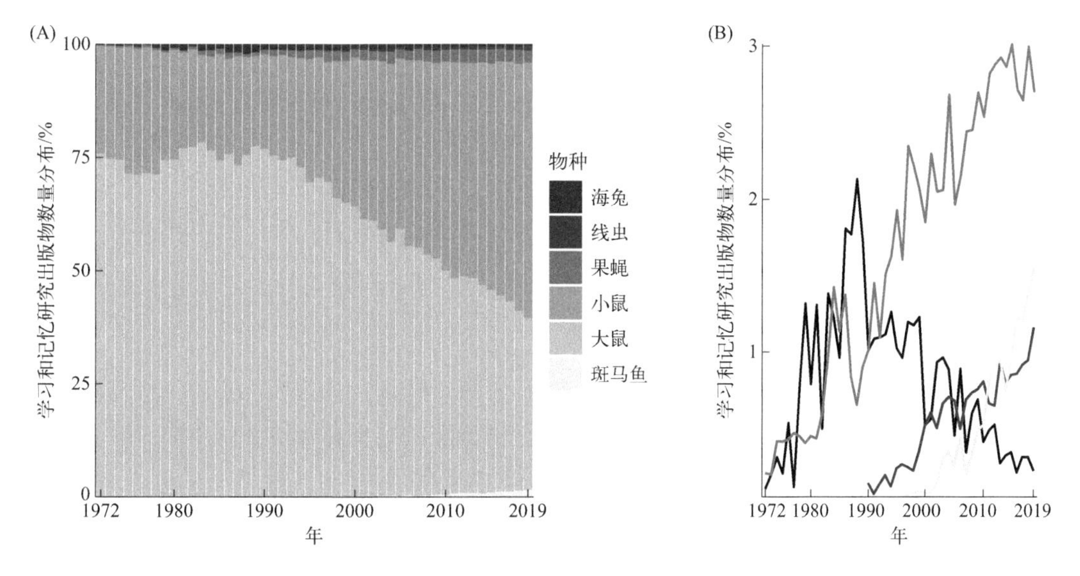

图12-1 **与常见动物模型学习和记忆研究相关出版物的数量分布**

（A）包括啮齿动物；（B）不包括啮齿动物。这一发现来自2019年12月27日在Pubmed进行的搜索，使用搜索词：（learning）或（memory）和（species），其中物种名称在图例中表示。使用R软件（R Development Core Team, 2016）（版本3.6.1）进行数据分析，使用ggplot2（Wickham, 2015）（版本3.2.1）和viridis（Garnier, 2018）（版本0.5.1）软件包进行数据可视化分析

成年斑马鱼用于研究的最大潜在优势是高繁殖力和高通量分析的适宜性。斑马鱼可以以更高的密度饲养，饲养成本约为啮齿动物的1/5（Kalueffetal., 2014）。尽管小鼠和斑马鱼性成熟的时间相似［斑马鱼受精后2～3个月（Castranova et al., 2011; Lawrence et al., 2012），小鼠妊娠3周后6～8周］，但是雌性斑马鱼每周可产下100多个卵。这种快速繁殖能力和低成本规模化饲养特性，使斑马鱼成为利用大量数据的最新数据挖掘和机器学习技术的理想选择。

用斑马鱼作研究的另一个主要优点是它们易于进行遗传操作。它们的卵子是体外受精的，因此大大简化了突变动物和转基因动物的繁殖过程。随着一些先进技术的迅速使用，操作

斑马鱼基因组的手段越来越多，CRISPR/Cas9 是最新的手段之一（Li et al., 2016）。由于该技术可应用于各领域，能为突变动物和转基因动物的培育提供足够的筛选样本（Varshney and Burgess, 2014）。神经科学研究中最令人兴奋的是存在几个大型鱼类资源库，利用模块化的 GAL4/UAS 和 Cre/loxP 系统操纵中枢神经系统中选定的细胞亚群的基因表达（Marquart et al., 2015; Asakawa et al., 2008; Jungke et al., 2013）。这些技术的集合，加上在单个实验室的水平上相对容易地产生独特的突变体，使斑马鱼成为探索学习和记忆的神经和分子基础的强大模型。

12.3 非联想学习

非联想学习是指在反复接触单一刺激后引起行为变化的一种学习模式，而不是由感觉或运动功能引起行为变化的学习模式，如疲劳（Rankin et al., 2009）。这种学习方式是高度保守的，并且已经在各种各样的生物体中得到了验证，包括海参、蝇和线虫等无脊椎动物以及斑马鱼和啮齿动物等常用脊椎动物的模型（Ceruttiand Levin, 2006）。非联想学习可以分为两类：习惯化和敏感化。其中习惯化是指对反复施加刺激的反应减弱，而敏感化是指对反复施加刺激的反应增强。对成年斑马鱼已经进行过“习惯化学习”测试，而斑马鱼的“敏感化学习”还没有得到证实。尽管这些学习方式相对简单，但我们对与其相关的分子和神经机制还缺乏全面了解（Rankin et al., 2009; Glanzman, 2009）。

12.3.1 习惯化学习

习惯化学习是一种有用的学习方式，可使动物在其环境中学会忽略重复刺激，而将其注意力集中于潜在的更相关的刺激。惊吓实验是测试习惯化学习的常用手段，在这个测试中，针对反复出现的有害的刺激，快速启动反应随着时间的推移而减弱。对于斑马鱼而言，惊吓反应可以通过用力敲击水箱或制造一个听觉刺激来实现。在这种刺激下，斑马鱼表现为突然游动，该反应是由一对 Mauthner 神经元及其脑干周围神经网络介导的（Eaton et al., 2001）。对于随后出现的重复刺激，鱼的反应幅度降低，表明已经形成了习惯（Eddinsetal., 2010; Parketal., 2018; Spagnolietal., 2015）。研究表明这些反应逐渐减弱是习惯化的典型特征（Rankinetal., 2009），随着刺激出现频率的增加，形成习惯化的概率也会增加（Park et al., 2018），长期没有刺激出现时，反应也会自行恢复（Spagnoli et al., 2015）。此外，对金鱼和斑马鱼仔鱼的研究也证明了其他与习惯化学习相关的现象，如给予不同的刺激（去习惯化）后反应就会恢复，并且也可以类推到其他相似的刺激中（Roberts et al., 2013; Rooney and Laming, 1988; Wolman et al., 2011）。

已经发现有许多方法可以干扰成年斑马鱼的习惯化学习，例如，将发育过程中的斑马鱼胚胎暴露于各种毒素（Bailey and Levin, 2015; Sledge et al., 2011; Glazer et al., 2018）以及用假单胞菌感染神经等（Spagnoli et al., 2015）。人们还发现社会等级也可以调节习惯：在刺激出现频率高（但不低）的情况下，占主导地位的公鱼比从属鱼更容易形成习惯（Park et al., 2018)。后续的计算模型表明，与主导鱼相比，从属鱼的 Mauthner 细胞可能接收的兴奋性输入较多，抑制性输入较少。尽管在金鱼和暹罗斗鱼的研究中发现端脑的各个部分都起着调节作用，但有关成年斑马鱼习惯化学习相关的分子、细胞或神经递质方面的研究较少（Rooney and Laming, 1986, 1988; Marino-Neto and Sabbatini, 1983）。尽管如此，研究人员已经使用斑马鱼仔鱼进行了大量机制研究，结果表明，5- 羟色胺和多巴胺对习惯化具有反向调节作用（Pantoja et al., 2016），而常见的遗忘剂通过靶向谷氨酸信号转导、基因转录和蛋白质合成可以阻断长期学习。（Wolman et al., 2011; Roberts et al., 2011, 2016）。最近的这些发现尤其重要，因为已知这些分子机制与哺乳动物和其他物种的长期记忆形成有关（Kandel et al., 2014），同时支持以下观点：许多学习的基本机制在斑马鱼体内也是保守的。

用于研究斑马鱼习惯化学习的另一项试验是新水箱试验。在这项试验中，将斑马鱼从它们原先所在的水箱里拿出来，放到一个新的水箱里，通常持续一小段时间（5 ～ 10min）。暴露于新的环境初期，鱼通常会长时间待在水箱的底部，这种行为被认为是焦虑的一种反应，因为鸟类是斑马鱼的天然捕食者，以靠近水体上层的鱼为食。随着时间的推移，无论是在特定时段内还是在间隔数天的多个时段之间，鱼会逐渐在水箱顶部停留更多时间，这被认为是对环境适应的一个表现（Wong et al., 2010; Levin et al., 2007; Gerlai et al., 2000）。这些发现类似于啮齿动物暴露于开阔场地时表现出的探索行为的变化（Crusio, 2001）。虽然新水箱试验主要用于研究焦虑，但最近的研究表明在这项任务中焦虑和习惯化是有区别的，这表明它可以用来同时检查这两种行为（Stewart et al., 2013）。类似惊吓反应，新水箱习惯化试验在用于评估斑马鱼暴露于某些毒素的发育毒性（Sledge et al., 2011; Glazer et al., 2018; Crosby et al., 2015; Oliveri et al., 2015）以及用于评估成鱼尼古丁慢性暴露毒性研究中是较为敏感的（Stewart et al., 2015）。然而，这一学习行为的神经和分子机制在很大程度上仍然是未知的。

习惯化测试具有相当大的应用前景，因为该测试方法易于实现，易于掌握和推广。此外，对斑马鱼习惯化学习相关的细胞、神经和分子机制的了解仍不完全（Rankin et al., 2009; Glanzman, 2009）。综上所述，鉴于斑马鱼适合高通量分析，它可能是唯一适合解析习惯化学习的神经和分子机制的脊椎动物模型。尽管如此，为了了解成年斑马鱼养成的习惯化学习范式在多大程度上表现出诸如非习惯化和普遍化等关键特征，还需要进行其他更为详细的工作（Rankin et al., 2009）。

12.4 联想学习

有两种类型的联想条件反射：经典（或巴甫洛夫）条件反射和操作性条件反射。经典条件反射涉及独立于动物行为发生的刺激的再现，而操作性条件反射涉及反应强化，其中行为反应影响非条件刺激的管理。虽然操作性条件反射在成年斑马鱼的行为范式中得到了广泛应用，但经典条件反射却得到相当少的关注。尽管如此，研究者已经使用多种刺激条件证明存在两种学习形式。

12.4.1 经典条件反射

在经典条件反射中，将最初不相关的条件刺激与产生某种行为的非条件刺激配对，会导致动物对随后出现的条件刺激表现出条件反射。条件反射的强弱被认为是已经形成的联想记忆强度的一种指标。经典条件反射可以根据条件刺激（扩散或离散线索）和非条件刺激（食欲刺激或厌恶刺激）的性质进行细分。许多任务是利用多种刺激来完成的，和人类一样，其中许多刺激是基于斑马鱼视觉是最重要的感觉形态之一这一事实来完成的（Fleisch and Neuhauss, 2006）。

（1）厌恶（恐惧）条件反射

在当前的学习和记忆研究中，恐惧条件反射是最常用的任务之一。在这项任务中，需要用到产生非条件刺激的恐惧条件（如电击），并将恐惧条件与一个新地方（情境）或一个无关的信号（典型的听觉或视觉刺激）进行配对使用。再次暴露于条件刺激下会引起恐惧相关的行为增加，即为条件反射。这项任务在啮齿动物中特别常用，因为它功能强大且可以快速获取，可以直接操纵记忆形成的获取、巩固和回忆阶段。信号和情境性恐惧条件作用范式也在成年斑马鱼中得到了阐述，因此对学习和记忆的比较分析具有相当大的潜力。

信号恐惧条件首次在成年斑马鱼研究中使用，是使用报警信息素（Schreckstoff）作为非条件刺激，使用视觉或嗅觉信号作为条件刺激（HallandSuboski, 1995a, 1995b）。报警信息素作为一个行为学相关刺激具有一定的优势，因为它是通过斑马鱼表皮细胞的机械破坏而产生的。人们认为报警信息素对于传递环境中存在捕食者或其他有害刺激是非常重要的（Stensmyr and Maderspacher, 2012）。然而，事实证明，报警信息素作为非条件刺激的作用是有限的，因为只有 40% 的鱼对这一物质表现出明确的非条件反射（Hall and Suboski, 1995a）。报警信息素引起的最一致的反射是不规律游泳行为增加（Hall and Suboski, 1995a, 1995b; Egan et al, 2009），而关于静止不动和停留在底部的时间增加方面的报道不一（Ogawa et al, 2014; Speedie and Gerlai, 2008）。此外，预期出现的条件反射尚不清楚，正如已经报道的不规则运动和底部停留时间即使在同一研究中也会出现一些不一致的情况（HallandSuboski, 1995a; Ruhletal., 2017）。尽管如此，最近已经从斑马鱼和其他鱼类中鉴定出了一些报警信息素的成分（Stensmyr and Maderspacher, 2012; Mathuruetal., 2012; Parraetal., 2009），为给予一定范围内的刺激提供了更大的潜在控制。

最近，使用电击作为非条件性刺激已经证明了存在信号和情境性恐惧学习（Agetsuma et al., 2010; Valente et al., 2012; Kenney et al., 2017）。电击的好处是在空间和时间上都易于控制，以及可以使动物产生相同的非条件反应（例如，突然活动）。通过将离散的视觉信号与电击配对使用证明了存在信号性恐惧条件反射，条件反射的反应表现为行为增加或者回避视觉信号。然而，目前尚不清楚信号引发的恐惧记忆能持续多长时间，因为还没有检测到经这种学习后的记忆超过 30min（Agetsuma et al., 2010; Valente et al., 2012）。在情境恐惧条件反射中，斑马鱼在新型水箱中受到给予的电击作为条件刺激，条件反射以相对于基线期游泳减少的量来衡量（Kenney et al., 2017）。斑马鱼的情境性恐惧条件反射有许多特征与啮齿动物相类似，即能够迅速获得（5min 内），对电击强度敏感，且在没有电击的情况下反复暴露于水箱后该条件反射会消失。重要的是，这种记忆是针对受到电击的水箱，而且可以通过给予 NMDA 型谷氨酸受体拮抗剂而阻断，两者都表明自发活动的变化不是由受到伤害引起的。然而，与啮齿动物不同的是，斑马鱼的情境性恐惧记忆仅仅持续约 2 周（Kenney et al., 2017），而啮齿动物的恐惧记忆通常持续很长时间（Houston et al., 1999）。

有关斑马鱼巴甫洛夫恐惧条件反射的分子和神经机制研究还没有广泛开展。然而，和啮齿动物一样（Bast et al., 2003），研究发现 NMDA 型谷氨酸受体对斑马鱼情境式恐惧记忆的巩固非常重要（Kenney et al., 2017）。在神经递质研究方面，Agetsuma 等人（2010）利用 GAL4/UAS 系统研究了外侧背缰核（lateral dorsal habenula, dHbL）在线索恐惧记忆中的作用。首先，他们构建了仅表达 dHbL 的 GAL4 载体。然后将该载体与表达破伤风毒素轻链的 UAS：TeTxLC 载体杂交，破伤风毒素轻链是一种可阻止突触传递的蛋白质，从而阻止 dHbL 神经元的输出。有趣的是，转基因鱼似乎有学习能力，但它们的条件反射发生了改变，它们对条件刺激的反应变为僵直（freezing）行为增加，而对照鱼的转弯频率升高。这些发现表明，与在啮齿动物中的结果相同（Yamaguchi et al., 2013; Okamoto and Aizawa, 2013），dHbL 在控制形成条件反射的特定行为中发挥着重要作用。

（2）食欲条件反射

斑马鱼的经典条件反射可以通过食欲条件刺激来实现，如食物或精神活性药物。食物是一种具有挑战性的非条件刺激，因为斑马鱼很快就吃饱了。尽管如此，气味（Braubach et al., 2009; Miklavc and Valentincic, 2012; Namekawa et al., 2018）、环境（Lau et al., 2006）、视觉和听觉刺激（Cerutti et al., 2013; Doyle et al., 2017）都能成功地与食物奖励配对使用。虽然具体的条件反应各不相同，但都是鱼对条件刺激的典型行为，表明鱼正在关注条件刺激，比如缩短与奖励地点的距离。其中两项研究值得关注，在这些研究中，任务是完全自动化的，鱼在整个实验过程中都留在自己的水箱（Doyle et al., 2017）或单独的水箱（Namekawa et al., 2018）。在一天当中，所有获取食物的途径都与条件刺激相匹配，在整个实验过程中鱼都处在可以给予条件刺激的水箱中，这样就避免了饱腹问题。在这些范例中鱼的学习也是相当快的，在第一天训练的过程中就会有明显的反应。因此，这些实验装置在鱼类学习的高通量分析方面具有相当大的潜力。

在食欲性经典条件反射过程中，人们认为能产生愉悦效果的精神活性药物也可用作非条件刺激物。鉴于斑马鱼与哺乳动物的神经递质系统存在大量重叠，斑马鱼对于研究药物效应是非常有用的（Panula et al., 2010）。通常用于啮齿动物（Tzschentke, 2007）的条件性位置偏好（conditioned place preference, CPP）实验也已成功应用于斑马鱼的研究（Collier and Echevarria, 2013）。在这种范式中，鱼最初暴露于一个分成两到三个隔间的水箱中，每个隔间都包含不同的视觉信号。首先，测定鱼对不同隔间偏好的基线。条件作用是通过施加非条件刺激（如药物）或者对照刺激，同时将动物限制在水箱的一个隔间内，使得在隔间内的刺激信号和药物之间形成关联。然后，通过将动物放回水箱，让其自由进入所有隔间，并测量它在与药物配对的一侧隔间内相对于基线所花费的时间，来测试这种关联的强度。尽管它看起来很简单，但在设计 CPP 实验时仍有许多潜在的难点需要考虑（Collier and Echevarria, 2013; Blaser and Vira, 2014）。

目前，已经使用多种药物成功开展了斑马鱼的 CPP 研究，包括可卡因（Darland and Dowling, 2001; Darland et al., 2012; Mersereau et al., 2016; Wold et al., 2017），苯丙胺（Ninkovic et al., 2006; Webb et al., 2009），吗啡（Lau et al., 2006），丹酚 A（Braida et al., 2007），尼古丁（Kily et al., 2008; Parker et al., 2013; Kedikian et al., 2013; Ponzoni et al., 2014; Faillace et al., 2018），

乙醇（Kily et al., 2008; Mathur et al., 2011; Chacon and Luchiari, 2014; Parker et al., 2016）和甲基苯丙胺（Ponzoni et al., 2016; Zhu et al., 2017）。有趣的是，即使面临惩罚，斑马鱼的CPP也会持续存在（Kily et al., 2008）。斑马鱼还被用于探索性研究胚胎暴露于药物和毒素对其成年后CPP学习的影响：研究发现，发育过程中暴露于镉或可卡因会剂量依赖性地干扰成年斑马鱼的可卡因CPP（Mersereau et al., 2016; Wold et al., 2017），胚胎时期暴露于乙醇会增强成年斑马鱼的乙醇CPP（Parker et al., 2016）。最后，在一项令人印象深刻的深入研究中，Brock等人（2017）利用斑马鱼具有高通量筛选的潜力对27种不同的化合物是否可以诱导CPP进行了研究，且每种化合物都有多个剂量。与之前的研究结果一致，他们证实尼古丁、乙醇、安非他明、可卡因和吗啡可以诱发偏好。此外，他们首次发现芬太尼、羟考酮、丁卡因、苯环己哌啶和氯苯那敏也可以作为食欲的非条件刺激。通过比较斑马鱼的CPP与啮齿动物和非人灵长类动物的相关范式，他们发现斑马鱼的CPP假阳性率低，但假阴性率高，因此在预测一般人类滥用药物方面有些受限。尽管如此，这项研究表明，斑马鱼对于理解阿片类药物和精神兴奋剂等药物亚类的奖励效应是非常有用的。

成年斑马鱼也被用来阐明CPP的分子机制，研究中发现了与哺乳动物模型相同的变化，并揭示了潜在的新机制。例如，斑马鱼在学习CPP过程中的转录组变化不仅与啮齿动物中涉及的基因和途径有大量的重叠（Kily et al., 2008），而且还指向新的机制，如通常与发育相关的转录因子具有新用途（Webb et al., 2009）。此外，已经发现乙酰胆碱酯酶对斑马鱼（Ninkovic et al., 2006）和啮齿动物（Hikida et al., 2003）建立苯异丙胺的CPP非常重要。最后，一些研究利用了斑马鱼模型的优势之一，即正向遗传筛选，并确定了没有可卡因（Darland and Dowling, 2001）和苯异丙胺（Webb et al., 2009）CPP的突变鱼系，尽管其潜在的基因尚未测定。尽管如此，这种方法在鉴定新的候选基因方面，尤其是随着资源的不断开发，在构建具有已知突变基因的大型动物文库，如斑马鱼突变计划项目等方面，将具有相当大的应用前景（Kettleborough et al., 2013）。

12.4.2 操作性条件反射

已通过使用多种条件刺激和非条件刺激证明了斑马鱼具有操作性条件反射。这种类型的学习可以通过增加（正强化）或去除（负强化）不同强度的非条件刺激来增加（强化）或减少（惩罚）特定行为的可能性。各种食欲性和厌恶性刺激已被用于证明斑马鱼的正强化和负强化以及正惩罚。迄今为止，采用负强化（主动回避）的研究为斑马鱼的大脑行为关系提供了最详细的解析。

（1）正强化

正强化学习包括使用食欲性非条件刺激，即在对刺激作出正确反应后给予食物，这导致向条件刺激趋向行为增加。食物和接近同类都被成功地用作非条件刺激，食物会引发更强烈的反应，但社交刺激不太容易产生饱腹感（Daggett et al., 2018）。多个实验室报道已经将视觉信号与食物出现成功配对使用（Sison and Gerlai, 2011a; Bilotta et al., 2005; Mueller and Neuhauss, 2012; Parker et al., 2012）。这一实验通常是在水箱或迷宫（十字迷宫或T形迷宫）中使用各种不同颜色的条件刺激来完成的（Colwill et al., 2005）。许多操作会改变食物信号引起的学习和记忆，并受到应激干扰（Gaikwad et al., 2011）和发育过程中铅或乙醇暴露（Chen et al., 2012; Fernandes et al., 2014）的干扰，而运动（Luchiari and Chacon, 2013）或给予药物吡拉西坦（普遍存在的抑制性神经递质 GABA 的衍生物）（Grossman et al., 2011）则可增强学习和记忆。人们也对大麻素的作用进行了研究，不同激动剂的作用各不相同，拮抗剂的长期处理导致记忆力增强（Ruhl et al., 2014, 2015）。最后，Roy 和 Bhat（2018）进行了一项独特的研究，他们发现从印度本土更复杂的栖息地捕获的野生斑马鱼记忆能力更强，这支持了认知生态学的观点，即环境背景会影响行为（Hutchins, 2010）。

社交非条件刺激，如接近同种动物或鱼游动的视频，通过与视觉信号配对也成功地支持了正强化学习（Al-Imari and Gerlai, 2008; Pather and Gerlai, 2009; Karnik and Gerlai, 2012; Fernandes et al., 2016）。社交刺激不仅不会像食物一样迅速地使动物厌烦，而且在训练前不需要剥夺食物。NMDA 型谷氨酸受体拮抗剂（Sison and Gerlai, 2011b）和睡眠剥夺（Pinheiro-da-Silva et al., 2017）能破坏对社交信号的学习。此外，研究发现多巴胺在接近同类（Scerbina et al., 2012）和联想学习（Naderi et al., 2016, 2018）中发挥重要作用，所有的发现都与之前在啮齿动物中的研究结果相一致。

（2）负强化

在负强化学习中，动物必须在施加条件刺激后作出反应以阻止或停止即将施加的厌恶性的非条件刺激。这种类型的学习在斑马鱼中已经通过主动回避范式得到证实。这些任务通常包括在穿梭箱的一个隔间中出现离散或扩散的视觉刺激，随后是厌恶性刺激，典型的是电击刺激，直到动物移动到水箱的其他部位（Pradel et al., 1999; Xu et al., 2007）。虽然斑马鱼获得主动回避通常需要几次试验，但强有力的证据表明，通常在训练的第一天就开始学习。与主动回避的获得或巩固有关的许多分子机制已明确，例如细胞黏附分子（Pradel et al., 1999, 2000）和一氧化氮，但令人惊讶的是，没有 NMDA 型谷氨酸受体的参与（Xu et al., 2007）。最近发现睡眠剥夺（Pinheiro-da-Silva et al., 2018）和衰老（Yang et al., 2018）也会干扰主动回避学习，并且成鱼或胚胎暴露于有毒物质，如甲基汞和铅，可能影响到暴露鱼的下一代（Xu et al., 2012a, 2012b, 2016a, 2016b）。然而，需要注意的是，胚胎暴露于 DMSO（一种常用的有机溶剂）也会干扰成年期的记忆（Truong et al., 2014），建议在设计药理学实验时必须谨慎考虑。

主动回避条件反射相关的神经机制方面的研究工作为理解成年斑马鱼与学习相关的各种大脑区域和通路的功能作用提供了一些详细信息。在接受主动回避训练后，Aoki 等人（2013）对清醒但固定不动的斑马鱼神经元活动进行了实时成像。他们发现，给予学会将视觉信号 - 电击关联起来的斑马鱼视觉刺激信号时，会导致外侧和中央背侧端脑的某个子区域的活动增加。他们鉴定出的神经元是谷氨酸能神经元，主要投射到腹侧端脑，该区域的消融导致长期而非短期记忆恢复的缺失。在后续研究中，Amo 等人（2014）通过采用一种令人印象深刻的实验组合，包括体内和体外电生理记录、自由游动鱼的光遗传学操作，以及结合 GAL4-UAS 和 theCre-loxP 系统功能的巧妙交叉遗传实验，发现主动回避学习的获得涉及谷氨酸能腹侧缰核到中缝核通路。作者还发现，腹侧缰核接收来自大脑皮层腹侧内足核的输入，该核也受可能来自中缝 5- 羟色胺能投射的支配，形成一个被认为在脊椎动物中高度保守的回路（Okamoto and Aizawa, 2013; Turner et al., 2016; Shabel et al., 2012; Stephenson-Jones et al., 2012）。最后，Laletal（2018）利用 Gal4-UAS 这个强大系统筛选了大量转基因鱼以确定内侧端脑背侧的神经元群体，这些神经元对主动回避学习是必需的。这些神经元是谷氨酸能神经元，并投射到几个区域，例如下丘脑和腹侧端脑，包括脚内核（entopeduncular nucleus），尽管目前尚未发现参与学习的具体投射区域。

（3）正惩罚

在成年斑马鱼中，抑制性回避是一种常见的学习任务，它利用了一种积极惩罚来操控学习的偶然性。在正惩罚中，施加厌恶性的非条件刺激会导致特定行为的减少。通常情况下，将鱼放置在一个穿梭箱中，其中一边的条件要比另一边更好，有一侧光线昏暗或更深。当鱼一开始被放置在水箱中其不喜欢的一侧时，它会迅速游到更喜欢的一侧。然而，如果当动物移动到喜好的一边时，施加了厌恶性非条件刺激，比如电击，当再次暴露在水箱中时，它们需要花更多的时间游到喜好的这一边，证明对先前受到的厌恶性刺激的经历有记忆（Blank et al., 2009）。正如预期的那样，电击的强度越大，记忆就越深刻（Manuel et al., 2014a; Ng et al., 2012）；然而，遗传背景的差异和饲养条件对于学习任务可产生较大影响。例如，Gorissen 等人（2015）发现 TL 品系的鱼有学习任务的能力，而 AB 品系的鱼则没有，Reolon 等人（2018）发现鱼缸中饲养的是同一性别的鱼时，鱼有学习任务的能力，而不同性别鱼混合饲养时，鱼则没有学习任务的能力。此外，年龄较大的鱼类、暴露在不可预测的慢性压力下的鱼类以及在丰富环境中饲养的鱼的学习能力也会受到损害（Manuel et al., 2014b, 2015）。总的来说，这些研究表明，研究人员在选择和报告行为实验中使用的鱼的背景系、饲养条件、年龄和性别时应该谨慎。

斑马鱼的抑制性回避记忆与几种神经递质受体有关，例如 NMDA 型谷氨酸受体（Blank et al., 2009）以及烟碱乙酰胆碱受体和毒蕈碱乙酰胆碱受体（Richetti et al., 2011; de Castro et al., 2009）。研究也发现，腺苷受体具有减轻大麻二酚（Nazario et al., 2015）和毒蕈碱乙酰胆碱受体拮抗剂（Bortolotto et al., 2015）遗忘效应的作用。任务学习也对胚胎阶段（de Castro et al., 2009; Haab Lutte et al., 2018）或成年期（Vuaden et al., 2012; Da Silva Acosta et al., 2016; Altenhofen et al., 2017; Bridi et al., 2017）接触各种有毒物质很敏感。虽然考虑到抑制性和主动回避有很多相似之处，两者之间可能有相当大的重叠，但正惩罚潜在的神经机制还尚未研究。

12.5 结论

过去几十年积累的大量研究表明，成年斑马鱼能够进行多种形式的联想和非联想学习。许多实验设计无疑是受到啮齿动物研究结果的启发，通过这些实验设计可以直接比较不同物种之间学习能力的差异性。特别令人兴奋的是，最近的一些研究利用了斑马鱼的一些关键优势，比如它们易于进行分子基因操纵（Agetsuma et al., 2010; Aoki et al., 2013; Amo et al., 2014）和高通量分析（Brock et al., 2017）或将这些实验设计组合使用（Lal et al., 2018）。随着 21 世纪的到来，鉴于斑马鱼具有高繁殖力、低成本、易于基因操作和具有复杂行为谱的优点，斑马鱼将继续成为学习和记忆研究中的首要模式生物。

（张瑞华翻译　李丽琴审校）

参考文献

Agetsuma, M., Aizawa, H., Aoki, T., et al., 2010. The habenula is crucial for experience-dependent modification of fear responses in zebrafish. Nature Neuroscience 13 (11), 1354-1356.
https://doi.org/10.1038/nn.2654.

Al-Imari, L., Gerlai, R., 2008. Sight of conspecifics as reward in associative learning in zebrafish (*Danio rerio*). Behavioural Brain Research 189 (1), 216-219. https://doi.org/10.1016/j.bbr.2007.12.007.

Altenhofen, S., Wiprich, M.T., Nery, L.R., Leite, C.E., Vianna, M.R.M.R., Bonan, C.D., 2017. Manganese(II) chloridealters behavioral and neurochemical parameters in larvae and adult zebrafish. Aquatic Toxicology 182,172e183. https://doi.org/10.1016/j.aquatox.2016.11.013.

Amo, R., Fredes, F., Kinoshita, M., et al., 2014. The habenulo-raphe serotonergic circuit encodes an aversive expectation value essential for adaptive active avoidance of danger. Neuron 84 (5), 1034-1048. https://doi.org/10.1016/j.neuron.2014.10.035.

Aoki, T., Kinoshita, M., Aoki, R., et al., 2013. Imaging of neural ensemble for the retrieval of a learned
behavioral program. Neuron 78 (5), 881-894. https://doi.org/10.1016/j.neuron.2013.04.009.

Asakawa, K., Suster, M.L., Mizusawa, K., et al., 2008. Genetic dissection of neural circuits by Tol2 transposon-mediated Gal4 gene and enhancer trapping in zebrafish. Proceedings of the National Academy of Sciences of the United States of America 105 (4), 1255-1260. https://doi.org/10.1073/pnas.0704963105.

Bailey, J.M., Levin, E.D., 2015. Neurotoxicity of FireMaster 550? in zebrafish (*Danio rerio*): chronic developmental and acute adolescent exposures. Neurotoxicology and Teratology 52, 210-219. https://doi.org/10.1016/j.ntt.2015.07.001.

Bast, T., Zhang, W.N., Feldon, J., 2003. Dorsal hippocampus and classical fear conditioning to tone and context in rats:effects of local NMDA-receptor blockade and stimulation. Hippocampus 13 (6), 657-675. http://www.ncbi.nlm. nih.gov/entrez/query.fcgi?cmd¼Retrieve&db¼PubMed&dopt¼Citation&list_uids¼12962312.

Bilotta, J., Risner, M.L., Davis, E.C., Haggbloom, S.J., 2005. Assessing appetitive choice discrimination learning in zebrafish. Zebrafish 2 (4), 259-268. https://doi.org/10.1089/zeb.2005.2.259.

Blank, M., Guerim, L.D., Cordeiro, R.F., Vianna, M.R.M.M., 2009. A one-trial inhibitory avoidance task to zebrafish: rapid acquisition of an NMDA-dependent long-term memory. Neurobiology of Learning and Memory 92 (4),529-534. https://doi.org/10.1016/j.nlm.2009.07.001.

Blaser, R.E., Vira, D.G., 2014. Experiments on learning in zebrafish (*Danio rerio*): a promising model of neurocognitive function. Neuroscience and Biobehavioral Reviews 42, 224-231. https://doi.org/10.1016/j.neubiorev.2014.03.003.

Bortolotto, J.W., de Melo, G.M., de Cognato, G.P., Vianna, M.R.M., Bonan, C.D., 2015. Modulation of adenosine signaling prevents scopolamine-induced cognitive impairment in zebrafish. Neurobiology of Learning and Memory 118, 113-119. https://doi.org/10.1016/j.nlm.2014.11.016.

Braida, D., Limonta, V., Pegorini, S., et al., 2007. Hallucinatory and rewarding effect of salvinorin A in zebrafish: k-opioid and CB1-cannabinoid receptor involvement. Psychopharmacology 190 (4), 441-448. https://doi.org/10.1007/s00213-006-0639-1.

Braubach, O.R., Wood, H.D., Gadbois, S., Fine, A., Croll, R.P., 2009. Olfactory conditioning in the zebrafish (*Danio rerio*). Behavioural Brain Research 198 (1), 190-198. https://doi.org/10.1016/j.bbr.2008.10.044.

Brenowitz, E.A., Zakon, H.H., 2015. Emerging from the bottleneck: benefits of the comparative approach to modern neuroscience. Trends in Neurosciences 38 (5), 273-278. https://doi.org/10.1016/j.tins.2015.02.008.

Bridi, D., Altenhofen, S., Gonzalez, J.B., Reolon, G.K., Bonan, C.D., 2017. Glyphosate and Roundup?alter morphology and behavior in zebrafish. Toxicology 392, 32-39. https://doi.org/10.1016/j.tox.2017.10.007.

Brock, A.J., Goody, S.M.G., Mead, A.N., Sudwarts, A., Parker, M.O., Brennan, C.H., 2017. Assessing the value of the zebrafish conditioned place preference model for predicting human abuse potential. Journal of Pharmacology and Experimental Therapeutics 363 (1), 66e79. https://doi.org/10.1124/jpet.117.242628.

Castranova, D., Lawton, A., Lawrence, C., et al., 2011. The effect of stocking densities on reproductive performance in laboratory zebrafish (*Danio rerio*). Zebrafish 8 (3), 141-146. https://doi.org/10.1089/zeb.2011.0688.

Cerutti, D.T., Levin, E.D., 2006. Cognitive impairment models using complementary species. In: Levin, E.D., Buccafusco, J.J. (Eds.), Animal Models of Cognitive Impairment. Taylor & Francis Group, LLC. http://www.ncbi.nlm.nih.gov/pubmed/21204361.

Cerutti, D.T., Jozefowiez, J., Staddon, J.E., 2013. Rapid, accurate time estimation in zebrafish (*Danio rerio*). Behavioural Processes 99, 21-25. https://doi.org/10.1016/j.beproc.2013.06.007.

Chacon, D.M., Luchiari, A.C., 2014. A dose for the wiser is enough: the alcohol benefits for associative learning in zebrafish.

Progress in Neuro-Psychopharmacology and Biological Psychiatry 53, 109-115. https://doi.org/10.1016/j.pnpbp.2014.03.009.
Chen, J., Chen, Y., Liu, W., et al., 2012. Developmental lead acetate exposure induces embryonic toxicity and memory deficit in adult zebrafish. Neurotoxicology and Teratology 34 (6), 581-586. https://doi.org/10.1016/j.ntt.2012.09.001.
Collier, A.D., Echevarria, D.J., 2013. The utility of the zebrafish model in conditioned place preference to assess the rewarding effects of drugs. Behavioural Pharmacology 24 (5e6), 375-383. https://doi.org/10.1097/FBP.0b013e328363d14a.
Colwill, R.M., Raymond, M.P., Ferreira, L., Escudero, H., 2005. Visual discrimination learning in zebrafish (*Danio rerio*). Behavioural Processes 70 (1), 19-31. https://doi.org/10.1016/j.beproc.2005.03.001.
Crosby, E.B., Bailey, J.M., Oliveri, A.N., Levin, E.D., 2015. Neurobehavioral impairments caused by developmental imidacloprid exposure in zebrafish. Neurotoxicology and Teratology 49, 81e90. https://doi.org/10.1016/j.ntt.2015.04.006.
Crusio, W.E., 2001. Genetic dissection of mouse exploratory behaviour. Behavioural Brain Research 125 (1e2),127-132. https://doi.org/10.1016/S0166-4328(01)00280-7.
Da Silva Acosta, D., Danielle, N.M., Altenhofen, S., et al., 2016. Copper at low levels impairs memory of adult zebrafish (*Danio rerio*) and affects swimming performance of larvae. Comparative Biochemistry and Physiology e PartC: Toxicology and Pharmacology 185e186, 122-130. https://doi.org/10.1016/j.cbpc.2016.03.008.
Daggett, J.M., Brown, V.J., Brennan, C.H., 2018. Food or friends? What motivates zebrafish (*Danio Rerio*) performing a visual discrimination. Behavioural Brain Research 359, 190-196. https://doi.org/10.1016/J.BBR.2018.11.002.
Darland, T., Dowling, J.E., 2001. Behavioral screening for cocaine sensitivity in mutagenized zebrafish. Proceedings of the National Academy of Sciences of the United States of America 98 (20), 11691-11696. https://doi.org/10.1073/pnas.191380698.
Darland, T., Mauch, J.T., Meier, E.M., Hagan, S.J., Dowling, J.E., Darland, D.C., 2012. Sulpiride, but not SCH23390, modifies cocaine-induced conditioned place preference and expression of tyrosine hydroxylase and elongation factor 1a in zebrafish. Pharmacology Biochemistry and Behavior 103 (2), 157-167. https://doi.org/10.1016/j.pbb.2012.07.017.
de Castro, M.R., Lima, J.V., Salomão de Freitas, D.P., et al., 2009. Behavioral and neurotoxic effects of arsenic exposure in zebrafish (*Danio rerio*, Teleostei: Cyprinidae). Comparative Biochemistry and Physiology e Part C: Toxicology and Pharmacology 150 (3), 337-342. https://doi.org/10.1016/j.cbpc.2009.05.017.
Doyle, J.M., Merovitch, N., Wyeth, R.C., et al., 2017. A simple automated system for appetitive conditioning of zebrafish in their home tanks. Behavioural Brain Research 317, 444-452. https://doi.org/10.1016/j.bbr.2016.09.044.
Eaton, R.C., Lee, R.K., Foreman, M.B., 2001. The Mauthner cell and other identified neurons of the brainstem escape network of fish. Progress in Neurobiology 63 (4), 467-485. https://doi.org/10.1016/S0301-0082(00)00047-2.
Eddins, D., Cerutti, D., Williams, P., Linney, E., Levin, E.D., 2010. Zebrafish provide a sensitive model of persisting neurobehavioral effects of developmental chlorpyrifos exposure: comparison with nicotine and pilocarpine effects and relationship to dopamine deficits. Neurotoxicology and Teratology 32 (1), 99-108. https://doi.org/10.1016/j.ntt.2009.02.005.
Egan, R.J., Bergner, C.L., Hart, P.C., et al., 2009. Understanding behavioral and physiological phenotypes of stress and anxiety in zebrafish. Behavioural Brain Research 205 (1), 38-44. https://doi.org/10.1016/j.bbr.2009.06.022.
Faillace, M.P., Pisera-Fuster, A., Bernabeu, R., 2018. Evaluation of the rewarding properties of nicotine and caffeine by implementation of a five-choice conditioned place preference task in zebrafish. Progress In Neuro-Psychopharmacology and Biological Psychiatry 84 (Pt A), 160-172. https://doi.org/10.1016/j.pnpbp.2018.02.001.
Fernandes, Y., Tran, S., Abraham, E., Gerlai, R., 2014. Embryonic alcohol exposure impairs associative learning performance in adult zebrafish. Behavioural Brain Research 265, 181-187. https://doi.org/10.1016/j.bbr.2014.02.035.
Fernandes, Y.M., Rampersad, M., Luchiari, A.C., Gerlai, R., 2016. Associative learning in the multichamber tank: a new learning paradigm for zebrafish. Behavioural Brain Research 312, 279-284. https://doi.org/10.1016/j.bbr.2016.06.038.
Fleisch, V.C., Neuhauss, S.C., 2006. Visual behavior in zebrafish. Zebrafish 3 (2), 191-201. https://doi.org/10.1089/zeb.2006.3.191.
Gaikwad, S., Stewart, A., Hart, P., et al., 2011. Acute stress disrupts performance of zebrafish in the cued and spatial memory tests: the utility of fish models to study stress-memory interplay. Behavioural Processes 87 (2), 224-230.https://doi.org/10.1016/j.beproc.2011.04.004.
Garnier, S., 2018. Viridis: Default Color Maps from "Matplotlib". R Package Version 0.5.1. Available from: https://CRAN.R-project.org/package=viridis https://cran.r-project.org/package=viridis.
Gerlai, R., Lahav, M., Guo, S., Rosenthal, A., 2000. Drinks like a fish: zebra fish (*Danio rerio*) as a behavior genetic model to study alcohol effects. Pharmacology Biochemistry and Behavior 67 (4), 773-782. https://doi.org/10.1016/S0091-3057(00)00422-6.

Glanzman, D.L., 2009. Habituation in Aplysia: the Cheshire Cat of neurobiology. Neurobiology of Learning and Memory 92 (2), 147-154. https://doi.org/10.1016/j.nlm.2009.03.005.

Glazer, L., Wells, C.N., Drastal, M., et al., 2018. Developmental exposure to low concentrations of two brominated flame retardants, BDE-47 and BDE-99, causes life-long behavioral alterations in zebrafish. Neurotoxicology 66,221-232. https://doi.org/10.1016/j.neuro.2017.09.007.

Gorissen, M., Manuel, R., Pelgrim, T.N.M., et al., 2015. Differences in inhibitory avoidance, cortisol and brain gene expression in TL and AB zebrafish. Genes, Brain and Behavior 14 (5), 428-438. https://doi.org/10.1111/gbb.12220.

Grossman, L., Stewart, A., Gaikwad, S., et al., 2011. Effects of piracetam on behavior and memory in adult zebrafish.Brain Research Bulletin 85 (1e2), 58-63. https://doi.org/10.1016/j.brainresbull.2011.02.008.

Gut, P., Reischauer, S., Stainier, D.Y.R., Arnaout, R., 2017. Little fish, big data: zebrafish as a model for cardiovascular and metabolic disease. Physiological Reviews 97 (3), 889-938. https://doi.org/10.1152/physrev.00038.2016.

Haab Lutte, A., HuppesMajolo, J., Reali Nazario, L., Da Silva, R.S., 2018. Early exposure to ethanol is able to affect the memory of adult zebrafish: possible role of adenosine. Neurotoxicology 69, 17-22. https://doi.org/10.1016/j.neuro.2018.08.012.

Hall, D., Suboski, M.D., 1995. Visual and olfactory stimuli in learned release of alarm reactions by zebra danio fish (*Brachydanio rerio*). Neurobiology of Learning and Memory 63 (3), 229-240. https://doi.org/10.1006/nlme.1995.1027.

Hall, D., Suboski, M.D., 1995. Sensory preconditioning and second-order conditioning of alarm reactions in zebra danio fish (*Brachydanio rerio*). Journal of Comparative Psychology 109 (1), 76-84. https://doi.org/10.1037/0735-7036.109.1.76.

Hikida, T., Kitabatake, Y., Pastan, I., Nakanishi, S., 2003. Acetylcholine enhancement in the nucleus accumbens prevents addictive behaviors of cocaine and morphine. Proceedings of the National Academy of Sciences of the United States of America 100 (10), 6169-6173. https://doi.org/10.1073/pnas.0631749100.

Houston, F.P., Stevenson, G.D., McNaughton, B.L., Barnes, C.A., 1999. Effects of age on the generalization and incubation of memory in the F344 rat. Learning and Memory 6 (2), 111-119. https://doi.org/10.1101/LM.6.2.111.

Hutchins, E., 2010. Cognitive ecology. Topics in Cognitive Science 2 (4), 705-715. https://doi.org/10.1111/j.1756-8765.2010.01089.x.

Jungke, P., Hans, S., Brand, M., 2013. The zebrafish CreZoo: an easy-to-handle database for novel CreER^{T2}-driver lines. Zebrafish 10 (3), 259-263. https://doi.org/10.1089/zeb.2012.0834.

Kalueff, A.V., Stewart, A.M., Gerlai, R., 2014. Zebrafish as an emerging model for studying complex brain disorders. Trends in Pharmacological Sciences 35 (2), 63-75. https://doi.org/10.1016/j.tips.2013.12.002.

Kandel, E.R., 2012. The molecular biology of memory: CAMP, PKA, CRE, CREB-1, CREB-2, and CPEB. Molecular Brain 5 (1), 14. https://doi.org/10.1186/1756-6606-5-14.

Kandel, E.R., Dudai, Y., Mayford, M.R., 2014. The molecular and systems biology of memory. Cell 157 (1), 163e186. https://doi.org/10.1016/j.cell.2014.03.001.

Karnik, I., Gerlai, R., 2012. Can zebrafish learn spatial tasks? An empirical analysis of place and single CS-US associative learning. Behavioural Brain Research 233 (2), 415-421. https://doi.org/10.1016/j.bbr.2012.05.024.

Kedikian, X., Faillace, M.P., Bernabeu, R., 2013. Behavioral and molecular analysis of nicotine-conditioned place preference in zebrafish. In: Kalueff, A.V. (Ed.), PLoS One 8 (7), e69453. https://doi.org/10.1371/journal.pone.0069453.

Kenney, J.W., Scott, I.C., Josselyn, S.A., Frankland, P.W., 2017. Contextual fear conditioning in zebrafish. Learning & Memory 24 (10), 516e523. https://doi.org/10.1101/lm.045690.117.

Kettleborough, R.N., Busch-Nentwich, E.M., Harvey, S.A., et al., 2013. A systematic genome-wide analysis of zebrafish protein-coding gene function. Nature 496 (7446), 494-497. https://doi.org/10.1038/nature11992.

Kily, L.J., Cowe, Y.C., Hussain, O., et al., 2008. Gene expression changes in a zebrafish model of drug dependency suggest conservation of neuro-adaptation pathways. Journal of Experimental Biology 211 (10), 1623-1634. https://doi.org/10.1242/jeb.014399.

Lal, P., Tanabe, H., Suster, M.L., et al., 2018. Identification of a neuronal population in the telencephalon essential for fear conditioning in zebrafish. BMC Biology 16 (1). https://doi.org/10.1186/s12915-018-0502-y.

Lau, B., Bretaud, S., Huang, Y., Lin, E., Guo, S., 2006. Dissociation of food and opiate preference by a genetic mutation in zebrafish. Genes, Brain and Behavior 5 (7), 497-505. https://doi.org/10.1111/j.1601-183X.2005.00185.x.

Lawrence, C., Adatto, I., Best, J., James, A., Maloney, K., 2012. Generation time of zebrafish (*Danio rerio*) and medakas (oryziaslatipes) housed in the same aquaculture facility. Lab Animal 41 (6). https://doi.org/10.1038/laban.092.

Levin, E.D., Bencan, Z., Cerutti, D.T., 2007. Anxiolytic effects of nicotine in zebrafish. Physiology & Behavior 90 (1),54-58.

https://doi.org/10.1016/j.physbeh.2006.08.026.

Li, M., Zhao, L., Page-McCaw, P.S., Chen, W., 2016. Zebrafish genome engineering using the CRISPR-Cas9 system. Trends in Genetics 32 (12), 815-827. https://doi.org/10.1016/j.tig.2016.10.005.

Luchiari, A.C., Chacon, D.M., 2013. Physical exercise improves learning in zebrafish, *Danio rerio*. Behavioural Processes 100, 44-47. https://doi.org/10.1016/j.beproc.2013.07.020.

Manuel, R., Gorissen, M., Piza Roca, C., et al., 2014. Inhibitory avoidance learning in zebrafish (*Danio rerio*): effects of shock intensity and unraveling differences in task performance. Zebrafish 11 (4), 341-352. https://doi.org/10.1089/zeb.2013.0970.

Manuel, R., Gorissen, M., Zethof, J., et al., 2014. Unpredictable chronic stress decreases inhibitory avoidance learning in Tuebingen long-fin zebrafish: stronger effects in the resting phase than in the active phase. Journal of Experimental Biology 217 (21), 3919-3928. https://doi.org/10.1242/jeb.109736.

Manuel, R., Gorissen, M., Stokkermans, M., et al., 2015. The effects of environmental enrichment and age-related differences on inhibitory avoidance in zebrafish (*Danio rerio* Hamilton). Zebrafish 12 (2), 152-165. https://doi.org/10.1089/zeb.2014.1045.

Marder, E., 2002. Non-mammalian models for studying neural development and function. Nature 417 (6886), 318-321. https://doi.org/10.1038/417318a.

Marino-Neto, J., Sabbatini, R.M., 1983. Discrete telencephalic lesions accelerate the habituation rate of behavioralarousal responses in Siamese fighting fish (Betta splendens). Brazilian Journal of Medical and Biological Research16 (3), 271-278. http://www.ncbi.nlm.nih.gov/pubmed/6360269.

Marquart, G.D., Tabor, K.M., Brown, M., et al., 2015. A 3D searchable database of transgenic zebrafish Gal4 and Cre lines for functional neuroanatomy studies. Frontiers in Neural Circuits 9, 78. https://doi.org/10.3389/fncir.2015.00078.

Mathur, P., Berberoglu, M.A., Guo, S., 2011. Preference for ethanol in zebrafish following a single exposure. Behavioural Brain Research 217 (1), 128-133. https://doi.org/10.1016/j.bbr.2010.10.015.

Mathuru, A.S., Kibat, C., Cheong, W.F., et al., 2012. Chondroitin fragments are odorants that trigger fear behavior in fish. Current Biology 22 (6), 538-544. https://doi.org/10.1016/j.cub.2012.01.061.

Meeker, N.D., Trede, N.S., 2008. Immunology and zebrafish: spawning new models of human disease. Developmental & Comparative Immunology 32 (7), 745-757. https://doi.org/10.1080/00223131.2008.10875772.

Mersereau, E.J., Boyle, C.A., Poitra, S., et al., 2016. Longitudinal effects of embryonic exposure to cocaine on morphology, cardiovascular physiology, and behavior in zebrafish. International Journal of Molecular Sciences 17 (6), 847. https://doi.org/10.3390/ijms17060847.

Miklavc, P., Valentinčič, T., 2012. Chemotopy of amino acids on the olfactory bulb predicts olfactory discrimination capabilities of zebrafish *Danio rerio*. Chemical Senses 37 (1), 65-75. https://doi.org/10.1093/chemse/bjr066.

Mueller, K.P., Neuhauss, S.C., 2012. Automated visual choice discrimination learning in zebrafish (*Danio rerio*). Journal of Integrative Neuroscience 11 (01), 73-85. https://doi.org/10.1142/S0219635212500057.

Naderi, M., Jamwal, A., Chivers, D.P., Niyogi, S., 2016. Modulatory effects of dopamine receptors on associative learning performance in zebrafish (*Danio rerio*). Behavioural Brain Research 303, 109-119. https://doi.org/10.1016/j.bbr.2016.01.034.

Naderi, M., Salahinejad, A., Ferrari, M.C.O., Niyogi, S., Chivers, D.P., 2018. Dopaminergic dysregulation and impaired associative learning behavior in zebrafish during chronic dietary exposure to selenium. Environmental Pollution 237, 174-185. https://doi.org/10.1016/j.envpol.2018.02.033.

Namekawa, I., Moenig, N.R., Friedrich, R.W., 2018. Rapid olfactory discrimination learning in adult zebrafish. Experimental Brain Research 1e11. https://doi.org/10.1007/s00221-018-5352-x.

Nazario, L.R., Antonioli, R., Capiotti, K.M., et al., 2015. Reprint of "caffeine protects against memory loss induced by high and non-anxiolytic dose of cannabidiol in adult zebrafish (*Danio rerio*). Pharmacology Biochemistry and Behavior 139, 134-140. https://doi.org/10.1016/j.pbb.2015.11.002.

Ng, M.C., Hsu, C.P., Wu, Y.J., Wu, S.Y., Yang, Y.L., Lu, K.T., 2012. Effect of MK-801-induced impairment of inhibitory avoidance learning in zebrafish via inactivation of extracellular signal-regulated kinase (ERK) in telencephalon. Fish Physiology and Biochemistry 38 (4), 1099-1106. https://doi.org/10.1007/s10695-011-9595-8.

Ninkovic, J., Folchert, A., Makhankov, Y.V., et al., 2006. Genetic identification of AChE as a positive modulator of addiction to the psychostimulant D-amphetamine in zebrafish. Journal of Neurobiology 66 (5), 463-475. https://doi.org/10.1002/neu.20231.

Ogawa, S., Nathan, F.M., Parhar, I.S., 2014. Habenular kisspeptin modulates fear in the zebrafish. Proceedings of the National Academy of Sciences of the United States of America 111 (10), 3841-3846. https://doi.org/10.1073/pnas.1314184111.

Okamoto, H., Aizawa, H., 2013. Fear and anxiety regulation by conserved affective circuits. Neuron 78 (3), 411-413. https://doi.org/10.1016/j.neuron.2013.04.031.

Oliveri, A.N., Bailey, J.M., Levin, E.D., 2015. Developmental exposure to organophosphate flame retardants causes behavioral effects in larval and adult zebrafish. Neurotoxicology and Teratology 52, 220-227. https://doi.org/10.1016/j.ntt.2015.08.008.

Pantoja, C., Hoagland, A., Carroll, E.C., Karalis, V., Conner, A., Isacoff, E.Y., 2016. Neuromodulatory regulation of behavioral individuality in zebrafish. Neuron 91 (3), 587-601. https://doi.org/10.1016/j.neuron.2016.06.016.

Panula, P., Chen, Y.C., Priyadarshini, M., et al., 2010. The comparative neuroanatomy and neurochemistry of zebrafish CNS systems of relevance to human neuropsychiatric diseases. Neurobiology of Disease 40 (1), 46-57. https://doi.org/10.1016/j.nbd.2010.05.010.

Park, C., Clements, K.N., Issa, F.A., Ahn, S., 2018. Effects of social experience on the habituation rate of zebrafish startle escape response: empirical and computational analyses. Frontiers in Neural Circuits 12, 7. https://doi.org/10.3389/fncir.2018.00007.

Parker, M.O., Gaviria, J., Haigh, A., et al., 2012. Discrimination reversal and attentional sets in zebrafish (*Danio rerio*). Behavioural Brain Research 232 (1), 264e268. https://doi.org/10.1016/j.bbr.2012.04.035.

Parker, M.O., Brock, A.J., Millington, M.E., Brennan, C.H., 2013. Behavioral phenotyping of Casper mutant and 1-pheny-2-thiourea treated adult zebrafish. Zebrafish 10 (4), 466-471. https://doi.org/10.1089/zeb.2013.0878.

Parker, M.O., Evans, A.M., Brock, A.J., Combe, F.J., Teh, M.T., Brennan, C.H., 2016. Moderate alcohol exposure during early brain development increases stimulus-response habits in adulthood. Addiction Biology 21 (1), 49-60. https://doi.org/10.1111/adb.12176.

Parra, K.V., Adrian, J.C., Gerlai, R., 2009. The synthetic substance hypoxanthine 3-N-oxide elicits alarm reactions in zebrafish (*Danio rerio*). Behavioural Brain Research 205 (2), 336-341. https://doi.org/10.1016/j.bbr.2009.06.037.

Pather, S., Gerlai, R., 2009. Shuttle box learning in zebrafish (*Danio rerio*). Behavioural Brain Research 196 (2), 323-327. https://doi.org/10.1016/j.bbr.2008.09.013.

Pinheiro-da-Silva, J., Tran, S., Silva, P.F., Luchiari, A.C., 2017. Good night, sleep tight: the effects of sleep deprivation on spatial associative learning in zebrafish. Pharmacology Biochemistry and Behavior 159, 36-47. https://doi.org/10.1016/j.pbb.2017.06.011.

Pinheiro-da-Silva, J., Tran, S., Luchiari, A.C., 2018. Sleep deprivation impairs cognitive performance in zebrafish: a matter of fact? Behavioural Processes. https://doi.org/10.1016/j.beproc.2018.04.004.

Ponzoni, L., Braida, D., Pucci, L., et al., 2014. The cytisine derivatives, CC4 and CC26, reduce nicotine-induced conditioned place preference in zebrafish by acting on heteromeric neuronal nicotinic acetylcholine receptors. Psychopharmacology 231 (24), 4681-4693. https://doi.org/10.1007/s00213-014-3619-x.

Ponzoni, L., Daniela, B., Sala, M., 2016. Abuse potential of methylenedioxymethamphetamine (MDMA) and its derivatives in zebrafish: role of serotonin 5HT2-type receptors. Psychopharmacology 233 (15-16), 3031e3039. https://doi.org/10.1007/s00213-016-4352-4.

Pradel, G., Schachner, M., Schmidt, R., 1999. Inhibition of memory consolidation by antibodies against cell adhesion molecules after active avoidance conditioning in zebrafish. Journal of Neurobiology 39 (2), 197-206. http://doi.wiley.com/10.1002/%28SICI%291097-4695%28199905%2939%3A2%3C197%3A%3AAID-NEU4%3E3.0.CO%3B2-9.

Pradel, G., Schmidt, R., Schachner, M., 2000. Involvement of L1.1 in memory consolidation after active avoidance conditioning in zebrafish. Journal of Neurobiology 43 (4), 389-403.

R Development Core Team, 2016. R: a language and environment for statistical computing. R Foundation for Statistical Computing 1 (2.11.1), 409. https://doi.org/10.1007/978-3-540-74686-7.

Rankin, C.H., Abrams, T., Barry, R.J., et al., 2009. Habituation revisited: an updated and revised description of thebehavioral characteristics of habituation. Neurobiology of Learning and Memory 92 (2), 135-138. https://doi.org/10.1016/j.nlm.2008.09.012.

Reolon, G.K., de Melo, G.M., da Rosa, J.G.D.S., Barcellos, L.J.G., Bonan, C.D., 2018. Sex and the housing: effects onbehavior, cortisol levels and weight in zebrafish. Behavioural Brain Research 336, 85-92. https://doi.org/10.1016/j.bbr.2017.08.006.

Richetti, S.K., Blank, M., Capiotti, K.M., et al., 2011. Quercetin and rutin prevent scopolamine-induced memoryimpairment in zebrafish. Behavioural Brain Research 217 (1), 10-15. https://doi.org/10.1016/j.bbr.2010.09.027.

Roberts, A.C., Reichl, J., Song, M.Y., et al., 2011. Habituation of the C-start response in larval zebrafish exhibits severaldistinct phases and sensitivity to NMDA receptor blockade. PLoS One 6 (12), e29132. https://doi.org/10.1371/journal.pone.0029132.

Roberts, A.C., Bill, B.R., Glanzman, D.L., 2013. Learning and memory in zebrafish larvae. Frontiers in Neural Circuits7, 126. https://doi.org/10.3389/fncir.2013.00126.

Roberts, A.C., Pearce, K.C., Choe, R.C., et al., 2016. Long-term habituation of the C-start escape response in zebrafish larvae. Neurobiology of Learning and Memory 134, 360-368. https://doi.org/10.1016/j.nlm.2016.08.014.

Rooney, D.J., Laming, P.R., 1986. Localization of telencephalic regions concerned with habituation of cardiac andventilatory responses associated with arousal in the goldfish (*Carassius auratus*). Behavioral Neuroscience 100(1), 45-50. https://doi.org/10.1037/0735-7044.100.1.45.

Rooney, D.J., Laming, P.R., 1988. Effects of telencephalic ablation on habituation of arousal responses, within and between daily training sessions in goldfish. Behavioral and Neural Biology 49 (1), 83-96. https://doi.org/10.1016/S0163-1047(88)91267-8.

Roy, T., Bhat, A., 2018. Divergences in learning and memory among wild zebrafish: do sex and body size play a role?Learning and Behavior 46 (2), 124-133. https://doi.org/10.3758/s13420-017-0296-8.

Ruhl, T., Prinz, N., Oellers, N., et al., 2014. Acute administration of THC impairs spatial but not associative memory function in zebrafish. Psychopharmacology 231 (19), 3829-3842. https://doi.org/10.1007/s00213-014-3522-5.

Ruhl, T., Moesbauer, K., Oellers, N., von der Emde, G., 2015. The endocannabinoid system and associative learning and memory in zebrafish. Behavioural Brain Research 290, 61-69. https://doi.org/10.1016/j.bbr.2015.04.046.

Ruhl, T., Zeymer, M., von der Emde, G., 2017. Cannabinoid modulation of zebrafish fear learning and its functional analysis investigated by c-Fos expression. Pharmacology Biochemistry and Behavior 153, 18-31. https://doi.org/10.1016/j.pbb.2016.12.005.

Scerbina, T., Chatterjee, D., Gerlai, R., 2012. Dopamine receptor antagonism disrupts social preference in zebrafish: astrain comparison study. Amino Acids 43 (5), 2059-2072. https://doi.org/10.1007/s00726-012-1284-0.

Shabel, S.J., Proulx, C.D., Trias, A., Murphy, R.T., Malinow, R., 2012. Input to the lateral habenula from the basalganglia is excitatory, aversive, and suppressed by serotonin. Neuron 74 (3), 475-481. https://doi.org/10.1016/j.neuron.2012.02.037.

Sison, M., Gerlai, R., 2011. Associative learning in zebrafish (*Danio rerio*) in the plus maze. Behavioural Brain Research207 (1), 1-16. https://doi.org/10.1016/j.bbr.2009.09.043 (Associative).

Sison, M., Gerlai, R., 2011. Associative learning performance is impaired in zebrafish (*Danio rerio*) by the NMDA-Rantagonist MK-801. Neurobiology of Learning and Memory 96 (2), 230-237. https://doi.org/10.1016/j.nlm.2011.04.016.

Sledge, D., Yen, J., Morton, T., et al., 2011. Critical duration of exposure for developmental chlorpyrifos-induced neurobehavioral toxicity. Neurotoxicology and Teratology 33 (6), 742-751. https://doi.org/10.1016/j.ntt.2011.06.005.

Spagnoli, S., Xue, L., Kent, M.L., 2015. The common neural parasite *Pseudoloma neurophilia* is associated with altered startle response habituation in adult zebrafish (*Danio rerio*): implications for the zebrafish as a model organism.Behavioural Brain Research 291, 351-360. https://doi.org/10.1016/j.bbr.2015.05.046.

Speedie, N., Gerlai, R., 2008. Alarm substance induced behavioral responses in zebrafish (*Danio rerio*). Behavioural Brain Research 188 (1), 168-177. https://doi.org/10.1016/j.bbr.2007.10.031.

Stensmyr, M.C., Maderspacher, F., 2012. Pheromones: fish fear factor. Current Biology 22 (6), R183-R186. https://doi.org/10.1016/j.cub.2012.02.025.

Stephenson-Jones, M., Floros, O., Robertson, B., Grillner, S., 2012. Evolutionary conservation of the habenular nuclei and their circuitry controlling the dopamine and 5-hydroxytryptophan (5-HT) systems. Proceedings of the National Academy of Sciences of the United States of America 109 (3), E164-E173. https://doi.org/10.1073/pnas.1119348109.

Stewart, A.M., Cachat, J., Green, J., et al., 2013. Constructing the habituome for phenotype-driven zebrafish research. Behavioural Brain Research 236 (1), 110-117. https://doi.org/10.1016/j.bbr.2012.08.026.

Stewart, A.M., Grossman, L., Collier, A.D., Echevarria, D.J., Kalueff, A.V., 2015. Anxiogenic-like effects of chronic nicotine exposure in zebrafish. Pharmacology Biochemistry and Behavior 139, 112-120. https://doi.org/10.1016/j.pbb.2015.01.016.

Truong, L., Mandrell, D., Mandrell, R., Simonich, M., Tanguay, R.L., 2014. A rapid throughput approach identifies cognitive deficits in adult zebrafish from developmental exposure to polybrominated flame retardants. Neurotoxicology 43, 134-142. https://doi.org/10.1016/j.neuro.2014.03.005.

Turner, K.J., Hawkins, T.A., Yáñez, J., Anadón, R., Wilson, S.W., Folgueira, M., 2016. Afferent connectivity of the zebrafish habenulae. Frontiers in Neural Circuits 10, 30. https://doi.org/10.3389/fncir.2016.00030.

Tzschentke, T.M., 2007. Measuring reward with the conditioned place preference (CPP) paradigm: update of the last decade. Addiction Biology 12 (3e4), 227-462. https://doi.org/10.1111/j.1369-1600.2007.00070.x.

Valente, A., Huang, K.H., Portugues, R., Engert, F., 2012. Ontogeny of classical and operant learning behaviors in zebrafish. Learning and Memory 19 (4), 170-177. https://doi.org/10.1101/lm.025668.112.

Varshney, G.K., Burgess, S.M., 2014. Mutagenesis and phenotyping resources in zebrafish for studying development and human disease. Briefings in Functional Genomics 13 (2), 82-94. https://doi.org/10.1093/bfgp/elt042.

Vuaden, F.C., Savio, L.E., Piato, A.L., et al., 2012. Long-term methionine exposure induces memory impairment on inhibitory avoidance task and alters acetylcholinesterase activity and expression in Zebrafish (*Danio rerio*). Neuro-chemical Research 37 (7), 1545-1553. https://doi.org/10.1007/s11064-012-0749-6.

Webb, K.J., Norton, W.H., Trümbach, D., et al., 2009. Zebrafish reward mutants reveal novel transcripts mediating the behavioral effects of amphetamine. Genome Biology 10 (7), R81. https://doi.org/10.1186/gb-2009-10-7-r81.

Wickham, H., 2015. Elegant Graphics for Data Analysis, vol. 35. Springer. https://doi.org/10.1007/978-0-387-98141-3.

Wold, M., Beckmann, M., Poitra, S., et al., 2017. The longitudinal effects of early developmental cadmium exposure on conditioned place preference and cardiovascular physiology in zebrafish. Aquatic Toxicology 191, 73-84.https://doi.org/10.1016/j.aquatox.2017.07.017.

Wolman, M.A., Jain, R.A., Liss, L., Granato, M., 2011. Chemical modulation of memory formation in larval zebrafish. Proceedings of the National Academy of Sciences of the United States of America 108 (37), 15468-15473. https://doi.org/10.1073/pnas.1107156108.

Wong, K., Elegante, M., Bartels, B., et al., 2010. Analyzing habituation responses to novelty in zebrafish (*Danio rerio*). Behavioural Brain Research 208 (2), 450-457. https://doi.org/10.1016/j.bbr.2009.12.023.

Wullimann, M.F., Mueller, T., 2004. Teleostean and mammalian forebrains contrasted: evidence from genes to behavior. The Journal of Comparative Neurology 475 (2), 143-162. https://doi.org/10.1002/cne.20183.

Xu, X., Scott-Scheiern, T., Kempker, L., Simons, K., 2007. Active avoidance conditioning in zebrafish (*Danio rerio*). Neurobiology of Learning and Memory 87 (1), 72-77. https://doi.org/10.1016/j.nlm.2006.06.002.

Xu, X., Lamb, C., Smith, M., Schaefer, L., Carvan III, M.J., Weber, D.N., 2012. Developmental methylmercury exposure affects avoidance learning outcomes in adult zebrafish. Journal of Toxicology and Environmental Health Sciences 4 (5), 85-91. https://doi.org/10.5897/JTEHS12.004.

Xu, X., Weber, D., Carvan, M.J., et al., 2012. Comparison of neurobehavioral effects of methylmercury exposure in older and younger adult zebrafish (*Danio rerio*). Neurotoxicology 33 (5), 1212-1218. https://doi.org/10.1016/j.neuro.2012.06.011.

Xu, X., Weber, D., Martin, A., Lone, D., 2016. Trans-generational transmission of neurobehavioral impairments produced by developmental methylmercury exposure in zebrafish (*Danio rerio*). Neurotoxicology and Teratology 53,19-23. https://doi.org/10.1016/j.ntt.2015.11.003.

Xu, X., Weber, D., Burge, R., VanAmberg, K., 2016. Neurobehavioral impairments produced by developmental lead exposure persisted for generations in zebrafish (*Danio rerio*). Neurotoxicology 52, 176-185. https://doi.org/10.1016/j.neuro.2015.12.009.

Yamaguchi, T., Danjo, T., Pastan, I., Hikida, T., Nakanishi, S., 2013. Distinct roles of segregated transmission of the septo-habenular pathway in anxiety and fear. Neuron 78 (3), 537-544. https://doi.org/10.1016/j.neuron.2013.02.035.

Yang, P., Kajiwara, R., Tonoki, A., Itoh, M., 2018. Successive and discrete spaced conditioning in active avoidance learning in young and aged zebrafish. Neuroscience Research 130, 1-7. https://doi.org/10.1016/j.neures.2017.10.005.

Yartsev, M.M., 2017. The emperor's new wardrobe: rebalancing diversity of animal models in neuroscience research. Science 358 (6362), 466-469. https://doi.org/10.1126/science.aan8865.

Zhu, C., Liu, W., Luo, C., et al., 2017. Inhibiting effects of rhynchophylline on methamphetamine-dependent zebrafish are related with the expression of tyrosine hydroxylase (TH). Fitoterapia 117, 47-51. https://doi.org/10.1016/j.fitote.2017.01.001.

第十三章

斑马鱼中的关系型学习：人类陈述性记忆的模型？

Robert T. Gerlai[1]

13.1 引言

自从在 Henry Gustav Molaison 先生身上（以患者 H. M. 的名字被科学界所熟知）发现双侧海马切除术的破坏性影响以来（Scoville and Liner, 1957），科学家越来越意识到记忆是多方面的，它具有多种形式和多个阶段（Squire and Zola-Morgan, 1991）。特别有说服力的是发现 H. M. 的短期记忆完好无损，且可以获得长期程序记忆（long-term procedural memory），而他对所接受的训练没有情景记忆（episodic memory）。当他进行长期程序记忆测试时，没有表现出任何有缺陷的迹象。下面是这种测试的一个示例：他的任务是当看到一幅复杂图画时练习描绘它，且需要通过一面镜子采用镜像绘画的方式用铅笔描绘它的轮廓。在训练过程中，H. M. 的表现稳步提高，清晰地表明他对任务的掌握以及记忆的调取。然而，在训练结束后，有人问他怎么这么擅长这个任务，他却不记得曾经练习过。H. M. 无法说明或有意识地回忆起他曾经完成过这项任务，尽管如此，H. M. 的短期记忆是完整的，因为他可以继续进行日常对话，并且可以记住几分钟内的谈话主题。但他不记得十几岁时术后的生活发生了什么，因此每次刮胡子和照镜子时都会看到一个老人回头看他，即他患了顺行性健忘症。术后，H. M. 发展为严重的顺行性健忘症［无法记住新学到的情景信息（episodic information）］和时间分级的逆行性遗忘症（回忆术前的记忆能力降低）。这些现象催生了整个认知研究领域，即对情景记忆或陈述性记忆（declarative memory）的研究。

"情景记忆"（episodic）这个术语通常用于描述连接关于"在哪里""什么"和"何时"发生的记忆片段（Tulving and Markowitsch, 1998）这种类型的记忆。本章对不同形式的记

[1] Department of Psychology, University of Toronto Mississauga, Mississauga, ON, Canada.

忆和/或学习类型不做系统描述。简而言之，情景记忆也常被称为“陈述性记忆”，以强调人可以有意识地回忆和陈述这一事实，即谈论一个人记得的情景。这种类型记忆的一个显著特征是它由多个记忆轨迹组成，即它有几个组成部分，形成我们所说的“关系集（relational set）”。Eichenbaum 提倡使用“关系学习和记忆”（relational learning and memory）这个术语来强调这种类型记忆最重要的方面（Howard and Eichenbaum, 2015）：事实上，零星相关的空间和时间信息必须由大脑编织在一起，以便一个人能够回忆起某种情景（episode）。尽管关系学习从根本上属于联想学习（associative learning），即与联想学习类似，它基于条件刺激（conditioned stimulus, CS）和非条件刺激（unconditioned stimulus, US）之间的关联之上，但联想学习的简单（基本）形式与关系学习（relational learning）之间的区别在于后者有多个条件刺激，它们的时间连续性（时间重叠或接近）以及它们的连贯性（与非条件刺激以及彼此之间的时间顺序）不是很强。

特别是在过去的20年里，在关系学习和记忆的神经解剖学方面以及神经生物学和分子机制方面的认知有了巨大提高，本章对其中的细节不展开描述。简而言之，海马是 H. M. 大脑中被切除的主要结构之一，已经证实海马在关系学习和记忆中非常重要。大量研究表明，海马充当标记模块，即根据它们是否属于一个关系集来“标记”以分布式方式存储在皮层中的记忆痕迹（Shapiro and Eichenbaum, 1999; Squire, 1992; Rolls, 2018）。简而言之就是，例如当一个人感知到一种特别的气味，或回忆起某种视觉线索时，与这种感觉相对应的神经回路活动便会触发编码记忆痕迹的回路/神经元的活动，其中也包括这种特定的嗅觉或视觉线索。因此，当人们回忆起一个情景时，不仅记得“哦，我见过这朵玫瑰”或“闻到过这朵花的香味”，而且还记得整个情景，比如“1998 年大学毕业后，当我和朋友们在巴黎的香榭丽舍大街上散步时，我看到（或闻到）这朵美丽的玫瑰”，即一组空间的、自传式的和时间的信息。

很明显，关系学习和记忆是至关重要的认知功能。没有它们，人们可能无法理解世界是如何构成的，由什么构成，以及自己过去记忆的哪些方面彼此独立或属于一个整体。从医学的角度来看，这种记忆形式也很重要。神经退行性疾病，包括几种形式的痴呆症，与海马功能和关系记忆缺陷有关（Gallagher and Koh, 2011; Dickerson and Eichenbaum, 2010）。例如，阿尔茨海默病最明显的行为表现通常是患者不记得他/她把车钥匙放在哪里，或者汽车停在停车场的哪里，甚至他/她的家在哪里。最终，神经退行性过程甚至可发展到患者可能不记得他/她最亲近的亲属。因此，了解那些负责关系记忆形成的神经元损害而导致神经退行性过程的机制具有非常重要的意义。许多实验室正在使用动物（主要是包括小鼠在内的啮齿动物）对关系记忆过程进行建模，以分析形成关联记忆或此类功能障碍的潜在机制（Morris, 2001）。最近，已经采用斑马鱼开展这方面的研究（Gerlai, 2017, 2011）。下面进一步讨论采用斑马鱼开展相关研究的优势。

13.2 斑马鱼应用的优点

许多疾病和生物学功能的大量代表性研究表明，斑马鱼与哺乳动物具有疾病和生物学功能的相关性（Pickart&Klee, 2014; Brennan, 2011）。这种相关性在基因水平上尤为明显。已发现斑马鱼基因的核苷酸序列与相应哺乳动物基因核苷酸序列具有平均约 70% 的相似度，并且大约有相似比例的人类基因在斑马鱼中被鉴定为同源基因（Howe et al, 2013）。如果人们希望研究人类疾病的分子机制，那么斑马鱼可提供一种合理的遗传解决方案。换句话说，如果人们确定了一种参与特定功能的斑马鱼基因，那么很可能会有一个涉及类似功能的人类或哺乳动物同源基因，如果首先关注人类或哺乳动物的疾病基因，那么很可能会在斑马鱼上找到相应的同源基因，并且可以在斑马鱼上开展研究。但是，当其他与人类关系更密切的模式生物（包括小鼠和大鼠）可用时，为什么要使用该物种进行此类基因操作呢？原因是斑马鱼在操作简单性和生物复杂性之间提供了很好的折中。换句话说，斑马鱼本身具有足够的复杂性以及与人类具有高度的相似性，其已成为一个相当好的模型，但同时又足够简单，可以为开展研究工作和疾病建模提供廉价而有效的解决方案。最重要的是，正如本书和其他资料所介绍的（Gerlai, 2014），这种鱼在进化上的简单性使得研究者可以采用一种简化论方法，进而专注于开展脊椎动物常见疾病的发生和发展机制研究。

为什么斑马鱼很简单？实际上它是一种小型脊椎动物，完全长大后长 4cm。它可以大量饲养在小鱼缸中，因为该物种具有高度的社会性，喜欢成群结队地游泳（也称为集群）。例如，我们自己的设施占地约 100m^2，可容纳多达 60000 条斑马鱼，如果饲养小鼠或大鼠，实验对象数则需要整个动物饲养室。斑马鱼发育非常快，从受精卵发育成功能齐全的自由游泳小鱼，例如，可对各种形式的刺激做出反应，捕捉猎物并学习，只需 5 天。同样重要的是，在这 5 天的时间里，鱼几乎保持完全透明，可以观察其器官，并通过荧光或其他染色方法量化许多可视化的标志物。斑马鱼的寿命很长，喂养时间大约为 5 年，即比小鼠或大鼠活得更长，因此可能适合衰老研究（Van Houcke et al, 2015），包括与年龄相关的认知障碍和记忆力下降。针对斑马鱼，已开发出有效的高密度和其他工业规模的维护和饲养方法 / 设备，并形成商业化。最后，正在开发高通量和大规模筛选方法，甚至可以用于斑马鱼学习和记忆等复杂功能方面的研究。总而言之，斑马鱼是啮齿类动物一种廉价而有效的替代品，它可以帮助研究人员确定进化上保守的人类疾病（包括中枢神经系统疾病）的普遍机制。

13.3 为什么要研究学习和记忆，为什么要研究关系记忆？

人们从多个角度对学习和记忆进行了深入研究。心理学家一直在绘制许多不同形式和阶段的记忆图，分析动物和人类可以学习和记忆的方式和内容。神经解剖学家已经阐明了涉及学习和记忆不同方面的大脑区域。神经生理学家一直在研究学习和记忆背后的神经元可塑性的突触机制。分子生物学家、生物化学家一直在研究大脑认知和记忆功能的生化机制。有关这些主题的出版物已有数千种。尽管学习和记忆有多种形式，但这些出版物大多数只关注其中之一：与学习和记忆相关。这本书包含关于斑马鱼非联想学习和联想学习（nonassociative and associative learning）的章节。本章重点介绍关系学习和记忆，这是一种复杂的联想学习形式。这种关注实际上是有道理的，因为这可能是最重要的学习和记忆形式之一，是中枢神经系统的一项奇妙的功能，没有它，就像 H. M. 一样，我们将无法有意识地回忆我们一生发生的事情，我们将无法说出我们记得什么。神经退行性疾病严重损害了这一功能，这一事实也证明了这一点，因此从医学角度来看，这也是一个重要问题。

在接下来的部分，我们将探讨为什么空间学习已成为关系学习研究的焦点，以及斑马鱼是否可以完成空间学习记忆任务。

13.4　空间学习：关系学习的一种形式

普通动物（包括斑马鱼）不能说话，因此无法在它们身上测试陈述性 - 情景记忆。事实上，有些人认为普通动物不存在人类定义的情景记忆（Tulving, 2001）。还有一些人持不同观点，认为情景 / 陈述性记忆的特征也存在于不能言语的脊椎动物中（Salwiczek et al, 2010）。为了将这种形式的记忆与人类情景/陈述性记忆区分开，他们称之为类情景式记忆（episodic-like memory）。对我们来说，这场辩论似乎有点学术性。多项研究表明，动物（包括哺乳动物和鸟类）可以学习和记忆发生在它们身上的“何时何地”，这是情景记忆的基本定义。这种记忆是否与人类相似，即动物（在这种情况下是鸟类、啮齿动物、狗、灵长类动物）是否可以像人类一样有意识地在心理时间中来回思考并在他们的脑海中回忆起这样的情景，这一点是无关紧要的。至少对我们而言，相关的是，在关系记忆研究的背景下，哺乳动物和人类甚至鸟类之间几乎所有生物组织水平上都发现了大量的进化相似性。

鱼有什么不同吗？一些人说没有（Woodruff, 2017）。从我们自己的角度来看，这是一个必须通过实验来回答的经验问题（Gerlai, 2017b）。但是如何研究鱼的情景记忆或关系学习呢？也许最简单的方法是检查鱼是否可以完成空间学习任务。

空间学习是关系学习的一种形式，其必须记住和回忆空间线索之间的动态关系，以便动物（或人类）在空间中找到一个由非条件刺激（通常是食物或其他类型的奖励）强化的位置。空间记忆任务在啮齿动物研究中已经普遍应用，因为它可以让研究人员使用相对简单的测试方法和设备来研究复杂的关系学习。例如，最常用的啮齿动物空间记忆测试之一是 Morris 水迷宫，其中，将啮齿动物（大鼠或小鼠）放进一个大型圆形水箱中，水中加入奶粉或无毒的白色水彩（最近使用）使其变得不透明（Morris, 1984）。Morris 水迷宫的任务就是让动物找到一个隐藏在水面下的平台。爬上平台是奖励（从水中逃生）。条件刺激是围绕在水箱外的视觉线索。另一项常用的空间任务称为场景和线索依赖性恐惧条件反射（the context and cue-dependent fear conditioning）（Curzon et al, 2009）。在这项任务中，将啮齿动物放在一个以特定视觉、嗅觉和触觉提示为特征的电击室（背景）中。动物在几分钟内（通常为 6min）接受几次（通常为 3 次）电击，并配合预测性听觉信号（在轻度电击到来之前出现的声音信号），即任务的训练阶段。第二天，将测试对象放置在同一个电击室中，该室没有声音信号提示或者电击提示（测试空间记忆或者场景记忆任务），或者将测试对象放入可提供声音信号的另一个不同的电击室中［测试预测性显著联想线索（the predictive salient associative cue）的记忆，条件刺激（CS）］。通过对僵直（freezing，一种对疼痛和恐惧的自然反应）时间的测试来判定记忆强度。人们已经发现啮齿动物在解决此类空间 / 场景任务方面表现出色，但前提是它们的海马体功能正常。目前已发现海马体功能的药理学、解剖学或遗传学破坏会导致其对空间或场景信息的反应性显著受损（空间记忆受损），而其动机（在水迷宫中寻找平台，或对痛苦的电击做出反应）方面则没有改变，啮齿动物的基本（简单的条件刺激 - 非条件刺激）联想学习（例如，对先前与电击配对的声音的记忆，或之前在水迷宫中标记逃生平台的视觉线索的记忆）也没有改变。因此，Morris 水迷宫、场景和线索依赖性恐惧条件反射实验已经成为测试动物海马体功能的常用方法。斑马鱼能解决类似的任务吗？回想一下，人们经常认为斑马鱼的大脑没有皮层或海马体，而这些皮层或海马体的大脑结构在关系学习和记忆中起着至关重要的作用。

13.5 斑马鱼与哺乳动物物种的空间学习和记忆

我们开始讨论下一个问题，即斑马鱼大脑的相关结构。尽管斑马鱼的大脑看起来确实与哺乳动物大脑有很大的不同，但研究者认为，它确实包含与哺乳动物大脑皮层和海马结构同源的区域。例如，Mueller（2012）认为，斑马鱼的背侧大脑皮层对应于哺乳动物的皮层，而内侧大脑皮层对应于海马体，该结论是基于细胞结构和连通性特征以及发育过程得出的。尤其是大脑发育过程值得注意，因为鱼类和哺乳动物的大脑发育完全不同。众所周知，鱼的大脑通过向外翻转（eversion）过程发育，而哺乳动物的大脑通过向外膨出（evagination）过程发育。这种高度差别的发育过程造成的结果是，鱼类和哺乳动物之间的同源大脑结构最终位于成年大脑中显著不同的解剖位置，但也许鱼类和哺乳动物相应的大脑区域所承担的功能差异可能并不大（Salas et al, 2006）。

但是，当我们在空间记忆任务中测试实验对象时，我们真的知道其确实获得了空间信息吗？这个问题并不像人们想象的那么简单。大多数研究空间学习的研究者认为，当实验对象在迷宫中找到指定的（奖励）空间位置时，证明该对象获得了空间记忆。然而，这可能并不总是正确的结论。良好的空间记忆表现可能是基本学习记忆的结果，如果实验对象从空间线索中挑出一个线索并将该线索（CS）与无条件刺激（US）相关联，则这种 CS-US 联想学习策略不需要获取大量空间线索之间的动态关系信息，从而绕过海马体。研究人员试图准备建立程序和空间线索，以最大限度地减少空间任务中的这种可能性，但在此类研究中通常无法获得实际空间学习的明确行为证据。

当然，空间学习任务的问题在于我们无法口头询问主体对象实际学到了什么，只能通过实验解决这个问题。但是这样的实验非常困难，例如，它需要操控每一个空间线索，并确定空间学习发生在它们复杂的动态空间关系的背景下，这几乎是不可能实现的，因为我们通常甚至不知道动物关注哪些空间线索。另一种建议是让动物根据空间的几何形状寻找空间位置。动物进行的基于几何线索的探索已经在简单的矩形迷宫中进行了测试，但这正是这些研究的问题所在：它们的几何结构的简单性不允许测试复杂的关系记忆。也许另一种方法可以确定或研究受试对象是否在迷宫中寻找奖励位置中利用了空间学习和记忆，比如水迷宫，就是要对受试对象的运动路径进行详细的分析。这种分析可能会揭示动物的运动轨迹是基于单个视觉（一种简单的条件刺激）还是基于复杂的空间线索关系。然而，在这一点上，尚不清楚运动路径的哪些方面会清楚地区分基于地标（基于简单的条件刺激）

和基于多空间线索的搜索策略。总之，在大多数空间学习研究中，我们无法确定动物是否真的学习并记住了空间信息。

也许有一个例外，那就是使用场景和线索依赖的恐惧条件任务实验对啮齿动物（大鼠和小鼠）进行的少数研究（Kim and Fanselow, 1992; Phillips and Le Doux, 1992; Gerlai, 1998）。我们对这个范式及其方法论概念进行了更仔细地研究，因为这个主题也与我们对鱼类空间学习的文献综述直接相关。正如我们上面提到的，在这项任务中，啮齿动物需要学习和记住与强化（电击）相关的两种不同类型的线索。一种是简单的基本线索，声音提示用作条件刺激，另一种是描述电击训练发生的房间和背景的一组空间线索。具有完整海马体的啮齿动物可以获得两种类型的记忆（基本记忆和空间或场景记忆）。那么，海马体功能受损的啮齿动物会如何表现呢？三项独立研究回答了这个问题，在这些研究中使用的啮齿动物的海马体要么受损，要么由基因修饰导致动物的（例如 DBA/2 小鼠）海马体功能失调（Kim and Fanselow, 1992; Phillips and Le Doux, 1992; Gerlai, 1998）。这些研究令人惊奇地发现，海马体功能障碍的啮齿动物可以通过条件反射对场景做出反应，即当动物被放置在接受电击训练的房间中时，第二天它对该房间做出僵直反应。这怎么可能？到进行这些研究为止，我们有大量证据表明海马体功能障碍会导致空间 / 关系学习和记忆受损（Morris, 1990; Barnes, 1988）。研究者认为，海马体受损的啮齿动物“识别”电击室的唯一可能方式是，如果它从场景中挑选出一个线索并将该单一线索与 US 联系起来，实际上将复杂的空间任务变成了一个简单的基本任务（Gerlai, 1998）。我们怎样才能防止这种情况发生？场景和线索依赖的恐惧条件任务实验中提供了一个解决方案。如果实验者提供了一个极其显著的提示，如声音提示，它与电击具有良好的时间连续性（在电击之前发生），并且它具有良好的时序性（总是在电击之前出现，可预测随后电击的发生），那么海马体功能失调的动物将无法从场景中分辨出另外的线索。因此，接受过声音提示的海马体功能障碍的动物在第二天并不记得电击室，只有海马体完好的动物才能记得。换句话说，健康对照受试对象在第二天进行测试时对声音提示和场景（电击室）都有反应，然而海马体功能障碍的动物只能对声音提示做出反应，这与我们所预期的完全一样，这是基于已知海马体参与空间学习和记忆，而不是基础联想学习和记忆（elemental associative learning and memory）。

上述逻辑如何适用于鱼类研究呢？一些研究已经证明鲤科鱼类具有出色的空间学习和记忆能力，斑马鱼也属于鲤科。例如，只有在大脑皮层没有破坏的情况下，金鱼才能表现出空间学习能力（Broglio et al, 2010; Dalas et al, 1996）。目前发现斑马鱼在空间学习任务中也表现良好（Williams et al, 2002; Spence et al, 2011）。然而，就像在啮齿动物研究中一样，其中一些研究具有这样的可能性，即斑马鱼的空间记忆能力是基础（例如地标）的学习和记忆。然而，其中也有例外（Karnik and Gerlai, 2012）。

在后一项研究中，作者表明，就像场景和线索依赖性恐惧条件实验中的啮齿动物一样，当基础联想线索（elemental associative cue）CS 和空间位置这两种线索同时呈现和同时强化时，斑马鱼也能够习得这些线索。本研究主要开展了如下工作。将斑马鱼单独放置在一个大水箱中，在那里它们可以自由游泳。单条斑马鱼非常有动力寻找并加入同种鱼群（Al-Imari and Gerlai, 2008）。在这个实验水箱中，可以看到放置在其大型测试水箱内刺激水箱中一组同种类鱼。刺激水箱的位置在 5 天的训练期间保持不变，并且还标有显著的、极易观察的视觉提示，即在刺激水箱透明壁后面放置红色塑料提示卡（图 13-1）。因此，可以

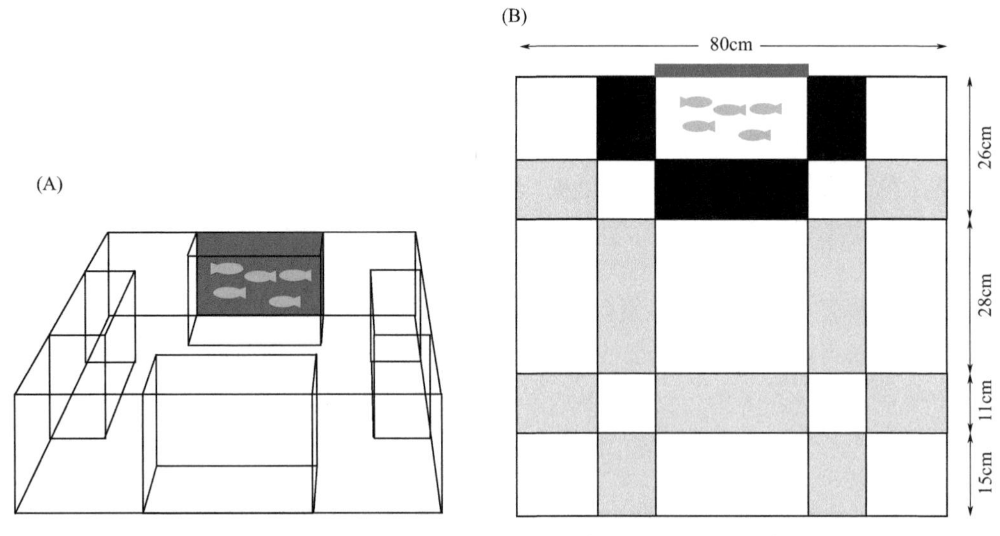

图 13-1 **斑马鱼元素、空间学习和记忆的训练箱和测试箱**

图（A）显示了水箱的3D视图，其中在训练水箱内放置四个刺激水箱。其中一个刺激水箱墙后放置了一张红色提示卡，并装有活的刺激鱼（与测试鱼大小相同的同种鱼）。其他三个刺激水箱的大小和形状与含有鱼的刺激水箱相同，但没有刺激鱼。图（B）从上方显示了带有四个刺激箱的相同测试箱，追踪摄像机的视图记录了测试鱼的行为。黑色矩形代表测试水箱的邻近区域，测试鱼可以从该区域清楚地观察到刺激鱼以及红色提示卡，而灰色矩形代表靠近其他刺激水箱的等效区域。图中也标明了水箱的尺寸、刺激鱼存在与否以及红色提示卡。配对组中的鱼接受了多次训练试验，训练试验中的刺激鱼和红色提示卡总是一起出现并位于同一位置；而未配对组的鱼除了红色提示卡和刺激鱼在整个试验中随机出现外，其他部分均是接受相同数量的训练试验和相同的训练过程。在训练结束时，50%的鱼接受了只有红色提示卡（红色提示卡放置于测试鱼的不同位置）而没有刺激鱼的探测试验。其他50%的鱼接受了没有红色提示卡和刺激鱼的探测试验［引自文献并作适当修改：Karnik I, Gerlai R (2012). Can zebrafish learn spatial tasks? An empirical analysis of place and single CS-US associative learning.Behavioural Brain Research 233: 415-421］

根据两个不同的线索，即红色提示卡和由许多箱外视觉空间线索定义的刺激水箱的位置来找到刺激鱼（作为奖励或US）。经过5天的训练期后（每天2次试验，即总共20次试验），对鱼进行了两种测试实验。50%的鱼在随机位置带有红色提示卡和无刺激鱼的测试箱中进行测试，这一探测任务测试CS-US关联的获得性，即基础学习（elemental learning）和记忆功能；另外50%的鱼在没有红色提示卡和无刺激鱼的测试箱中进行测试，这一探测任务测试了强化物（刺激鱼）与其空间位置之间的关联，即关系记忆的获得。Karnik和Gerlai（2012）还训练了另一组鱼，这些鱼接受了与前面描述组（配对组）相同的过程和训练，但接受了刺激鱼信号和在随机位置红色提示卡信号（称为未配对组）。这两个探测试验的结果显示在图13-2。首先要注意的是，在探测试验期间，未配对组的鱼对红色提示卡或迷宫的任何位置都没有表现出偏好。这一结果表明，本研究中使用的斑马鱼对条件刺激没有先天的偏好，并且对迷宫中的任何区域也没有空间偏好。结果中第二个值得注意的是，在单独使用红色提示卡的探测试验期间，尽管将红色测试卡放置在每条测试鱼的不同位置，测试配对组的鱼仍然表现出对红色提示卡的显著偏好并保持靠近它。这一结果表明，斑马鱼可以获得、记忆和回忆起CS-US关联，即具有基础学习能力和长期记忆。但或许这项研究最重要的发现是：配对组的斑马鱼表现出显著的空间偏好，即在探测试验期间对刺激鱼以前所在的确切空间位置表现出高度偏好（图13-2）。鉴于实验鱼是随机选择进行试探试验1（基本记忆测试）和试探试验2（空间记忆测试），我们可以得出结论，就像具有完整海马功能的啮齿动物一样，斑马鱼可以同时学习CS-US基本关联和空间位置-US关系关联。

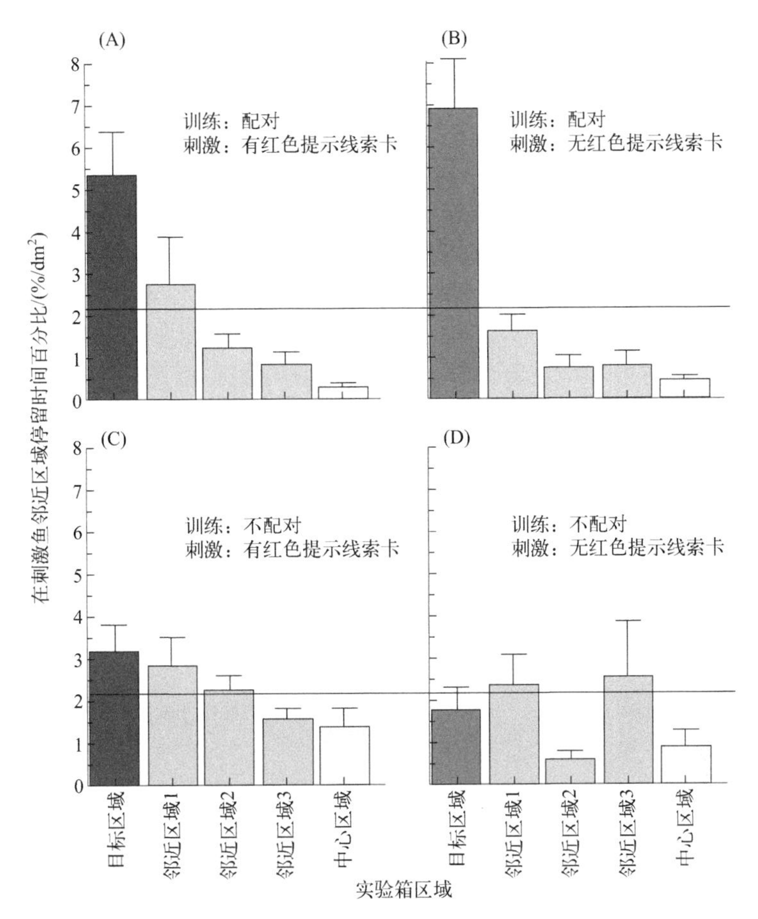

图13-2 **在探测试验期间实验斑马鱼停留在邻近区域的时间表明它们可以同时学习联想线索（CS）以及训练期间形成的强化位置**

图（A）和（B）显示了配对组鱼的结果，图（C）和（D）显示了未配对组鱼的结果。图（A）显示，即使没有刺激鱼存在，斑马鱼在探测实验期间仍停留在红色提示卡附近，证明了红色提示卡与刺激鱼存在（基础学习和记忆）之间的关联记忆。图（B）表明，即使没有红色提示卡和刺激鱼存在，斑马鱼停留在刺激鱼的先前位置附近，表明对刺激鱼先前位置的记忆（空间学习）。图（C）和（D）显示，当红色提示卡和刺激鱼出现在随机位置时，未配对组的鱼在水箱区域停留的时间与时序性没有显著性差异，即这些鱼没有表现出对红色提示卡的偏好性或者对在配对组中鱼表现的对刺激鱼的位置偏好（或回避）性。[引自文献并作适当修改：Karnik I, Gerlai R (2012). Canzebrafish learn spatial tasks? Anempirical analysis of place and single CS-US associative learning.Behavioural Brain Research233: 415-421.]

13.6 斑马鱼空间学习的预测和建构效度

在使用动物疾病模型时，我们经常提到可接受的模型的三个必要标准：表面效度（face validity）、预测效度（predictive validity）和建构效度（construct validity）。表面效度是指动物模型与人类状况之间的表观相似性。上述提及表面效度的一个例子是斑马鱼在空间和基础学习任务中表现得与啮齿动物相似。斑马鱼的学习方式似乎与海马功能完好的啮齿动物相似。但是，这怎么可能呢？斑马鱼没有类似于哺乳动物海马体典型的三突触回路的任何结构，在哺乳动物的大脑中，皮层下区域的突触投射到齿状回颗粒细胞上，这些颗粒细胞通过苔状纤维通路连接到 CA3 区锥体细胞，CA3 区锥体细胞通过谢氏侧支和穿通通路（the Schaffer collateral/commissural pathway）与 CA1 区锥体细胞（众所周知的“定位”细胞）的树突形成突触，然后这些神经元将它们的投射物发送回皮层下和其他大脑区域。这种映射良好的海马皮层回路已被证明在哺乳动物的空间学习和记忆中起着至关重要的作用，但在鱼的大脑中没有发现。然而，人们已经证明鱼类大脑与哺乳动物海马（大脑皮层的一部分）相对应的结构受损会导致鲤科鱼类的空间学习缺陷（Salas et al, 1996; Broglio et al, 2010）。是否存在以下可能，即哺乳动物海马或非哺乳动物物种的与海马相对应的结构对空间 / 关联学习很重要？而非神经元回路对空间 / 关联学习很重要？就如本章前文所说的，为什么斑马鱼可能是一种极好的利用还原论研究关联学习的模型，以及它是如何使研究人员关注到在进化上更为保守因而更为重要的物种间共同的关系学习机制以及相关方面。

其中一种想法认为，与哺乳动物不同，斑马鱼没有不必要的复杂回路。因此，它应该使我们能够研究关系记忆的最重要、更基本的机制方面，例如突触生理学和生物化学（Gerlai, 2014）。尽管这些领域研究还处于初步阶段，但已经收集到一些证明斑马鱼模型建构效度的证据。建构效度是指动物模型与人类特征或状况之间的机制相似性。对斑马鱼关系学习分析模型的建构效度进行系统综述并不属于本章的范围。因此，我们在此将只关注几个我们认为可能最重要的例子。第一个是关于鱼的长时程增强（long-term potentiation, LTP）作用。第二个是 *N*- 甲基 -D- 天冬氨酸（*N*-methyl-D-aspartate, NMDA）受体参与鱼的学习和记忆过程。

LTP 是 1973 年 Bliss 和 Lomo 发现的一种突触水平的现象（Bliss and Lomo, 1973）。目前，在解释行为层面上的记忆表现方面，LTP 与长时程抑制作用（long-term depression, LTD）一起被认为是最佳突触机制（Teyler, 1988; Morris et al, 1990; Lisman et al, 2018）。也就是说，关系记忆模式可以解释为跨多个神经元的突触的强化（LTP）和抑制（LTD）。LTP 和 LTD 不仅多方面模仿行为水平的记忆，而且大量研究也表明两者之间有因果关系。LTP 已在鱼类研究中得到证实（Satou et al, 2006; Lewis and Teyler, 1986），包括在斑马鱼端脑中发现 NMDA 受体依赖性的 LTP（Nam et al, 2004），这表明突触水平的神经元可塑性在机制上与哺乳动物相似。

NMDA 受体是突触可塑性的分子机制研究方面研究得最深入的分子之一。NMDA 受体是一种突触后神经递质受体，一种由谷氨酸（一种神经递质）和电压（突触后去极化）共同控制的通道（Morris et al, 1990; Collingridge et al, 1991）。就分子水平而言，它是中枢神经系统研究得最充分和最重要的符合检测器（coincidence detector）之一。要了解这种符合检测器的工作原理，我们需要考虑以下情况。当一个刺激，比如说一种动作电位 S1，从神经元 1 到达突触前末梢时，它会触发谷氨酸释放到突触间隙中。谷氨酸与位于靶神经元突触后膜的 NMDA 受体结合。NMDA 受体是一种阳离子通道，可以让钙离子通过，但单独的谷氨酸不能打开 NMDA 受体通道，因为通道受到镁离子的阻塞。另一个刺激 S2，一种来自神经元 2 的动作电位，可使靶神经元的膜去极化，这种去极化传导到突触（突触前膜的动作电位 S1 诱发释放谷氨酸）的突触后膜。只有当谷氨酸已经与 NMDA 受体结合时，突触后膜的去极化才会除去镁离子的阻塞。也就是说，当 S1 和 S2 都发生时，NMDA 受体才能打开并允许钙离子流入突触后末端。钙离子是第二信使，然后触发一系列分子事件，这些事件可以导致突触的加强（或减弱），这是作为学习和记忆基础的突触可塑性（LTP 或 LTD）。不仅如此，S1 必须在 S2 之前，这与 CS 必须在 US 之前获取有效记忆的方式完全相同。

在哺乳动物中人们已经对 NMDA 受体进行了详细研究，有趣的是，斑马鱼中这种复杂的符合检测器的多亚基通道也具有进化上的高度保守性，该通道的某些亚基与人类同源物具有 90% 的一致性（Cox et al, 2005）。这种保守进化的结果之一是发现可有效调节哺乳动物（含人类）NMDA 受体功能的药物对斑马鱼也起作用（Sison & Gerlai, 2011a）。MK801 就是一种非竞争性 NMDA 受体拮抗剂药物。因为它与 NMDA 受体的结合区域与谷氨酸（NMDA 受体的天然配体）结合区域不同，它不与天然配体竞争，所以称之为非竞争性拮抗剂。当 MK801 与 NMDA 受体结合时，拮抗剂会阻止 NMDA 受体的开放以及钙离子的流入。目前已有研究证明 MK801 会损害斑马鱼的学习和记忆（Sison & Gerlai, 2011b）。

通过将斑马鱼浸入 MK801 药物溶液中来给药，从而影响学习和记忆的不同时间过程（Sison& Gerlai, 2011b）。例如，在训练前给药是为了研究药物对记忆获得的潜在影响，在训练后给药是为了研究药物是否改变了记忆巩固。众所周知，记忆巩固会在获得记忆后 90 min 的时间窗口内立即发生。最后，在训练后、探索试验之前立即将斑马鱼浸入药物溶液中来测试该药物对记忆唤醒的潜在影响。斑马鱼接受不同时间药物处理后，在复杂的十字迷宫中单独训练（图 13-3）。在训练期间，试验斑马鱼在迷宫的一个臂上遇到了一群

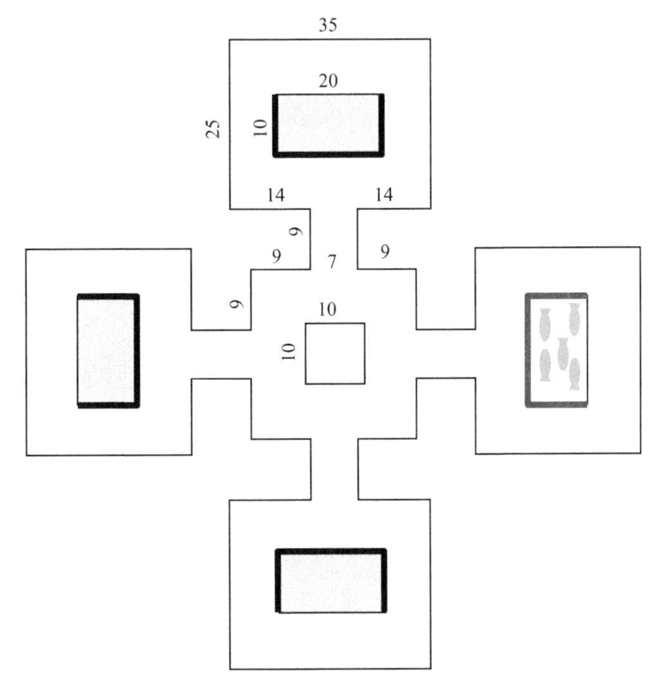

图 13-3 **在十字迷宫中量化评价关联学习及 MK801 给药的影响**

迷宫有四个臂，每个臂都包含一个刺激水箱。在目标臂中，刺激水箱包含与测试鱼相同大小的鱼群，而其他刺激水箱中没有鱼群。而且，有同种鱼群的刺激水箱三边被红墙包围，只有远离迷宫中心的一侧是透明的。这种装置确保了测试鱼只有在围绕刺激水箱游来游去时才能看到刺激水箱的东西。没有刺激鱼群的刺激水箱在它们的三边周围也有类似的不透明墙，但是墙是黑色的。因此，该装置确保：当测试鱼在其中游泳时，它首先看到条件刺激（例如红色提示卡，CS），然后才看到同种刺激鱼群（US）。因此，仅通过迷宫的空间布置就可以满足关联学习所需的 CS-US 连续性（contiguity）和时序性（contingency）。图中显示了迷宫的大小、红色或黑色墙壁以及刺激鱼群的分布情况［引用自文献并作修改：Sison M, Gerlai R. (2011b). Associative learning performance is impaired in zebrafish (Danio rerio) by the NMDA-R antagonist MK-801. Neurobiology of Learning and Memory. 96: 230-237.］

留在迷宫内刺激水箱中的同种刺激鱼群（US）。使用明显的视觉线索（CS）标记刺激鱼的位置，红色的墙壁将刺激水箱三侧包围，这样只有在测试鱼游过红色墙壁后，才能看到刺激鱼群（图 13-3）。图 13-4 结果表明，哺乳动物 NMDA 受体拮抗剂 MK801 药物显著破坏了斑马鱼的记忆巩固和唤醒。实验人员还训练了另一组鱼，在随机位置呈现刺激鱼和红色提示墙。图 13-4 显示，这些未配对训练的鱼的表现是偶然性的，且与巩固期和唤醒前给予 MK801 的配对训练鱼相比，两者的表现没有区别。也就是说，即 MK801 完全消除了斑马鱼的记忆巩固和唤醒。

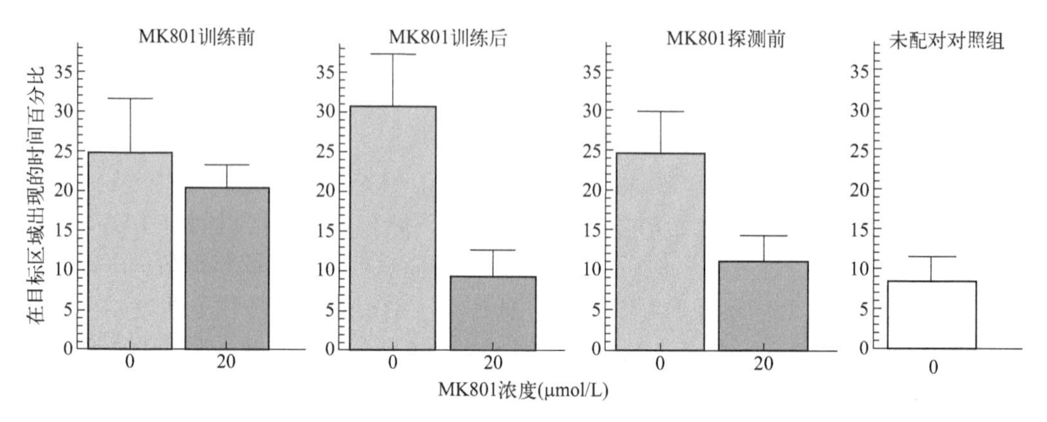

图 13-4 **在探索实验期间实验斑马鱼在目标隔室中花费的时间百分比**

注意，此实验是在图 13-3 所示的鱼在十字迷宫中训练后测试的。另需注意的是，在探索实验期间，任何刺激水箱中均不存在刺激鱼群。停留在刺激水箱周围有红色墙壁的目标臂上，表明该墙壁颜色与刺激鱼群的存在两者之间的关联记忆。MK801 给药时间显示在每个图的上方，剂量在 *x* 轴下方（对照组：0μmol/L；MK801 给药组：20μmol/L）。注意：训练前给药 MK801，表示训练目标记忆的获得；训练后给药，表示巩固记忆；探索实验前给药，表示对记忆的回忆。图中还显示了在红墙和刺激鱼群的随机位置（未配对对照）出现鱼的情况。注意，当训练（配对）鱼在训练后或探索实验之前立即给药 MK801，与未配对对照鱼相比，没有出现显著性损害（损害表现为：减少在目标臂上花费的时间）［修改自文献：Sison M, Gerlai R. (2011b).Associative learning performance is impaired in zebrafish (*Danio rerio*) by the NMDA-R antagonist MK-801. Neurobiology of Learning and Memory. 96: 230-237］

这些结果证明了斑马鱼作为人类 / 哺乳动物关联学习和记忆模型的建构和预测效度。在建构效度方面，它证明了斑马鱼中枢神经系统（CNS）功能涉及的分子机制，而该分子机制同样与哺乳动物的相应 CNS 功能有关。在预测效度方面，它在斑马鱼中证明了与哺乳动物和人类相同 / 相似的药物作用。

13.7 高通量应用

人们认为斑马鱼可作为生物医学研究领域中一种有应用前景的动物模型，其原因之一是它可能为药物和突变效应提供高效、廉价和高通量的筛选（Gerlai, 2010）。然而，一些人可能会认为需要进行 1 周的学习训练，并没有实现高通量。那为什么上述结果与生物医学研究相关呢？答案就是尽管上面提到的学习训练的确需要一周时间才能完成，而且在此期间只测试了少量研究对象（几十个），但这些实验是可扩展的。训练和探索试验的箱子相对较小，大量实验对象可以在小型测试空间进行。训练试验不需要对鱼的行为进行太多监测，并且使用视频跟踪系统可自动量化探索试验结果。因此，试验可以平行进行大量训练和随后的探索试验。例如，一个单独 25m^2 的行为测试室可能容纳 20 个测试箱，因此在一周内，一个人可以测量 20×25，即 500 条鱼，这将允许对大型药物库进行高通量筛选，或在突变筛选中对大量突变鱼进行全面筛选。

同样值得关注的是，也可利用其他潜在的高通量学习训练方法。例如，最近引入了一种基于潜在学习的实验方法，它允许在空间任务中使用单个实验箱训练一群鱼，这一系统将极大提高实验通量（Gómez-Laplaza and Gerlai, 2010）。

13.8 结论

尽管对斑马鱼认知（例如学习）和记忆相关特征的表征相对新颖，但专门为斑马鱼开发或对其适用的新型学习方法的数量正在迅速增加。本书中讨论关联和非关联学习的章节也很好地说明了斑马鱼在学习和记忆分析中的作用。关联学习可能是基础科学和医学研究中最复杂和最重要的认知现象之一。空间学习是关联学习的一个例子，已经在不同的物种中对其进行了深入的研究，但关注这种现象的一些研究未考虑空间任务的表现可能依赖于简单基础学习而不是复杂的关联学习和记忆。尽管旨在寻找鱼类空间学习能力的明确证据很困难，但是，现在一些研究已经证明斑马鱼中存在这种大脑功能。其中一些研究者也开始揭示斑马鱼空间学习的机制。迄今为止，这些研究的结果表明，斑马鱼在模拟哺乳动物（和人类）学习和记忆过程建模方面表现出良好的表型、建构和预测效度，因此它可用于分析研究此类过程的功能障碍，包括人类神经退行性疾病。

致谢

由 NSERC 研究基金支持（#311637）。

（靳倩翻译　王陈审校）

参考文献

Al-Imari, L., Gerlai, R., 2008. Sight of conspecifics as reward in associative learning tasks for zebrafish (*Danio rerio*). Behavioural Brain Research 189, 216-219.

Barnes, C.A., 1988. Spatial learning and memory processes: the search for their neurobiological mechanisms in the rat. Trends in Neurosciences 11, 163-169.

Bliss, T.V., Lomo, T., 1973. Long-lasting potentiation of synaptic transmission in the dentate area of the anaesthetized rabbit following stimulation of the perforant path. Journal of Physiology 232, 331-356.

Brennan, C.H., 2011. Zebrafish behavioural assays of translational relevance for the study of psychiatric disease. Reviews in the Neurosciences 22, 37-48.

Broglio, C., Rodríguez, F., Gómez, A., Arias, J.L., Salas, C., 2010. Selective involvement of the goldfish lateral pallium in spatial memory. Behavioural Brain Research 210, 191-201.

Collingridge, G.L., Blake, J.F., Brown, M.W., Bashir, Z.I., Ryan, E., 1991. Involvement of excitatory amino acid receptors in long-term potentiation in the Schaffer collateral-commissural pathway of rat hippocampal slices. Canadian Journal of Physiology and Pharmacology 69, 1084-1090.

Cox, J.A., Kucenas, S., Voigt, M.M., 2005. Molecular characterization and embryonic expression of the family of *N*-methyl-D-aspartate receptor subunit genes in the zebrafish. Developmental Dynamics 234, 756-766.

Curzon, P., Rustay, N.R., Browman, K.E., 2009. Cued and contextual fear conditioning for rodents. Chapter 2. In: Buccafusco, J.J. (Ed.), Methods of Behavior Analysis in Neuroscience, second ed. CRC Press/Taylor & Francis, Boca Raton (FL).

Dickerson, B.C., Eichenbaum, H., 2010. The episodic memory system: neurocircuitry and disorders. Neuropsychopharmacology 35, 86-104.

Gallagher, M., Koh, M.T., 2011. Episodic memory on the path to Alzheimer's disease. Current Opinion in Neurobiology 21, 929-934.

Gerlai, R., 2017. Zebrafish and relational memory: could a simple fish be useful for the analysis of biological mechanisms of complex vertebrate learning? Behavioural Processes 141, 242-250.

Gerlai, R., 2017b. Learning, memory, cognition and the question of sentience in fish. Invited Commentary Animal Sentience 13 (8), 1-4.

Gerlai, R., 2014. Fish in behavior research: unique tools with a great promise! Journal of Neuroscience Methods 234, 54-58.

Gerlai, R., 2011. Associative learning in zebrafish (*Danio rerio*). Methods in Cell Biology 101, 249-270.

Gerlai, R., 2010. High-throughput behavioral screens: the first step towards finding genes involved in vertebrate brain function using zebrafish. Molecules 15, 2609-2622.

Gerlai, R., 1998. Contextual learning and cue association in fear conditioning in mice: a strain comparison and a lesion study. Behavioral Brain Research 95, 191-203.

Gómez-Laplaza, L.M., Gerlai, R., 2010. Latent learning in zebrafish (*Danio rerio*). Behavioural Brain Research 208, 509-515.

Howard, M.W., Eichenbaum, H., 2015. Time and space in the hippocampus. Brain Research 1621, 345-354.

Howe, K., Clark, M.D., Torroja, C.F., Torrance, J., Berthelot, C., Muffato, M., Collins, J.E., Humphray, S., et al., 2013.The zebrafish reference genome sequence and its relationship to the human genome. Nature 496, 498-503.

Karnik, I., Gerlai, R., 2012. Can zebrafish learn spatial tasks? An empirical analysis of place and single CS-US associative learning. Behavioural Brain Research 233, 415-421.

Kim, J.J., Fanselow, M.S., 1992. Modality-specific retrograde amnesia of fear. Science 256, 675-657.

Lisman, J., Cooper, K., Sehgal, M., Silva, A.J., 2018. Memory formation depends on both synapse-specific modifications of synaptic strength and cell-specific increases in excitability. Nature Neuroscience 21, 309-314.

Lewis, D., Teyler, T.J., 1986. Long-term potentiation in the goldfish optic tectum. Brain Research 375, 246-250.

Morris, R.G., 2001. Episodic-like memory in animals: psychological criteria, neural mechanisms and the value of episodic-like tasks to investigate animal models of neurodegenerative disease. Philosophical Transactions of the Royal Society of London B Biological Sciences 356, 1453-1465.

Morris, R.G., 1990. Toward a representational hypothesis of the role of hippocampal synaptic plasticity in spatial and other forms of learning. Cold Spring Harbor Symposia on Quantitative Biology 55, 161-173.

Morris, R., 1984. Developments of a water-maze procedure for studying spatial learning in the rat. Journal of Neuroscience Methods 11, 47-60.

Morris, R.G., Davis, S., Butcher, S.P., 1990. Hippocampal synaptic plasticity and NMDA receptors: a role in information storage? Philosophical Transactions of the Royal Society of London B Biological Sciences 329, 187-204.

Mueller, T., 2012. What is the thalamus in zebrafish? Frontiers in Neuroscience. https://doi.org/10.3389/fnins.2012.00064.

Nam, R.H., Kim, W., Lee, C.J., 2004. NMDA receptor-dependent long-term potentiation in the telencephalon of the zebrafish. Neuroscience Letters 370, 248e-251.

Phillips, R.G., LeDoux, J.E., 1992. Differential contribution of amygdala and hippocampus to cued and contextual fear conditioning. Behavioral Neuroscience 106, 274-285.

Pickart, M.A., Klee, E.W., 2014. Zebrafish approaches enhance the translational research tackle box. Translational Research 163, 65-78.

Rolls, E.T., 2018. The storage and recall of memories in the hippocampo-cortical system. Cell and Tissue Research 373, 577-604.

Salas, C., Broglio, C., Durán, E., Gómez, A., Ocaña, F.M., Jiménez-Moya, F., Rodríguez, F., 2006. Neuropsychology of learning and memory in teleost fish. Zebrafish 3, 157-171.

Salas, C., Rodríguez, F., Vargas, J.P., Durán, E., Torres, B., 1996. Spatial learning and memory deficits after telencephalic ablation in goldfish trained in place and turn maze procedures. Behavioral Neuroscience 110, 965-980.

Salwiczek, L.H., Watanabe, A., Clayton, N.S., 2010. Ten years of research into avian models of episodic-like memory and its implications for developmental and comparative cognition. Behavioural Brain Research 215, 221-234.

Satou, M., Hoshikawa, R., Sato, Y., Okawa, K., 2006. An in vitro study of long-term potentiation in the carp (*Cyprinus carpio* L.) olfactory bulb. Journal of Comparative Physiology A, Neuroethology, Sensory, Neural and Behavioral Physiology 192, 135-150.

Scoville, W.B., Milner, B., 1957. Loss of recent memory after bilateral hippocampal lesions. Journal of Neurology, Neurosurgery, and Psychiatry 20, 11-21.

Shapiro, M.L., Eichenbaum, H., 1999. Hippocampus as a memory map: synaptic plasticity and memory encoding by hippocampal neurons. Hippocampus 9, 365-384.

Sison, M., Gerlai, R., 2011a. Behavioral performance altering effects of MK-801 in zebrafish (*Danio rerio*). Behavioural Brain Research 220, 331-337.

Sison, M., Gerlai, R., 2011b. Associative learning performance is impaired in zebrafish (*Danio rerio*) by the NMDA-R antagonist MK-801. Neurobiology of Learning and Memory 96, 230-237.

Spence, R., Magurran, A.E., Smith, C., 2011. Spatial cognition in zebrafish: the role of strain and rearing environment. Animal Cognition 14, 607-612.

Squire, L.R., Zola-Morgan, S., 1991. The medial temporal lobe memory system. Science 253, 1380-1386.

Squire, L.R., 1992. Memory and the hippocampus: a synthesis from findings with rats, monkeys, and humans. Psychological Review 99, 195-231.

Teyler, T.J., 1988. Long-term potentiation and memory. International Journal of Neurology 22, 163-171.

Tulving, E., 2001. Episodic memory and common sense: how far apart? Philosophical Transactions of the Royal Society of London B Biological Sciences 356, 1505-1515.

Tulving, E., Markowitsch, H.J., 1998. Episodic and declarative memory: role of the hippocampus. Hippocampus 8, 198-204.

Van Houcke, J., De Groef, L., Dekeyster, E., Moons, L., 2015. The zebrafish as a gerontology model in nervous system aging, disease, and repair. Ageing Research Reviews 24, 358-368.

Williams, F.E., White, D., Messer, W.S., 2002. A simple spatial alternation task for assessing memory function in zebrafish. Behavioural Processes 58, 125-132.

Woodruff, M.L., 2017. Consciousness in teleosts: there is something it feels like to be a fish. Animal Sentience 13,010.

第四部分

斑马鱼遗传学方法

第十四章

斑马鱼复杂行为的分子遗传学研究

Jiale Xu[1], Su Guo[1]

14.1 引言

自 1951 年霍尔（Hall's）早期总结“情绪的主要维度，即情绪、积极性、攻击性、反应性都有胚质（germplasm）来源”以来，剖析复杂行为特征背后的遗传结构成为一个长期课题。特别是从神经科学家的角度来看，行为遗传基础的破译不仅对于理解基本的分子和细胞机制非常重要，而且对于开发人类神经精神疾病的新治疗方法也至关重要。

一种用于行为遗传学研究的理想模式生物，应该适合在良好控制条件下以定量方式对自然行为进行大规模评估（Orger and de Polavieja, 2017），而且，该模型也应适用于在基因组和转录组水平对行为模式进行高通量遗传表征。模式生物的理想状态还包括在进化上应该高度保守，从而可以较为容易识别人类同源基因和细胞类型（Gerlai, 2014）。研究者虽然认为斑马鱼是模式生物领域中较新的物种，但最近人们已经认识到斑马鱼可满足上述所有要求，进而也促使其在神经科学领域的应用迅速增加（Gerlai, 2014; Guo, 2004; Kalueff et al., 2014; Shams et al., 2018; Sison et al., 2006）。

目前人们已经在斑马鱼中建立了行为范式，它可以进行有效的表型筛选。遗传和基因组资源的开发也同步推进。随着生物技术和计算机科学的进步，斑马鱼正成为探索复杂行为特征的生物学基础的日益突出的模式生物。

在本章中，我们将首先描述为评估斑马鱼行为而建立的几种范式，作为与人类疾病相关的潜在衡量标准。随后将针对每个范式，对斑马鱼仔鱼和成鱼的性能、基本测试仪器和量化方法以及药理作用进行阐述。同时，我们也将讨论如何结合遗传学和基因组方法进一步分析，重点是全基因组关联分析（genome-wide association study, GWAS）和单细胞分辨

[1] Department of Bioengineering and Therapeutic Sciences, Programs in Biological Sciences , Institute of Human Genetics, University of California, San Francisco, CA, United States.

率的RNA测序（sequencing of RNA at single-cell resolution, scRNA-seq），以发现潜在的遗传网络。

14.2 斑马鱼仔鱼和成鱼的行为谱

斑马鱼从早期发育阶段开始就表现出丰富的行为能力，许多行为是受神经系统控制的，而这些神经系统异常也可导致人类疾病。一些行为表现为对不熟悉的环境或新事物有喜好或厌恶（Hascoët and Bourin, 2009）。还有其他更为复杂的刺激行为，例如社交暗示（Spence et al., 2008）。世界各地的实验室正在积极研究这些行为的潜在分子和细胞基础。鉴于脊椎动物系统中保守的遗传结构和类似的神经解剖学特征，预计将获得斑马鱼在进化上保守的一些信息，进而促使其迅速转化到人类相关的研究中。

14.2.1 光/暗偏好行为

光/暗偏好行为是对简单视觉刺激的复杂反应，这种行为允许动物在半明半暗的平台上自由穿梭数十分钟。平台中光/暗面的光强和对比度可以精确控制，在这样一个特定的行为环境中，每个个体面临着自由探索整个平台的动力，表现出对黑暗或光亮一面的偏好（图14-1）。

我们的地球有光/暗循环。生物体适应它们的生活环境，在这些环境中，黑暗或光亮有不同的意义。例如，类似捕食者影子一样的黑暗对于小鱼来说是一种危险信号，但如果觉察自身很容易被捕食者发现，个体则会避免自身暴露在明亮环境中。事实上，光/暗偏好在进化上是保守的，并且可以在果蝇（Gong et al., 2010）和小鼠（Bourin and Hascoët, 2003）等不同物种中观察到。人们认为哺乳动物中的光/暗偏好是一种类似焦虑的特质，可用于评估抗焦虑药物（Bilkei-Gorzo et al., 1998）。

在斑马鱼中，光/暗偏好行为在仔鱼和成鱼阶段都有其特征（Bai et al., 2016; Chen et al., 2015; Magno et al., 2015; Maximino et al., 2011; Steenbergen et al., 2011）。成年斑马鱼通常表现出对由墙壁颜色处理导致的黑暗有先天的偏好（厌恶亮光）（Blaser et al., 2010; Lau et al., 2011; Maximino et al., 2011; Serra et al., 1999）。然而，仔鱼在相同的刺激下普遍表现出对光的偏好（厌恶黑暗）（Bai et al., 2016; Dahlén et al., 2019; Lau et al., 2011; Steen-bergen et al., 2011; Wagle et al., 2017），这种受发育调节的选择性改变在一些研究中被归因于色素沉着（pigmentations）。研究者使用良好的表面效度（即隐藏，避免捕食者）解释了成年斑马鱼的黑暗偏好，仔鱼的反向行为可以从行为学角度来解释，即仔鱼将黑暗认为是四处游荡的掠食者的阴影，或是缺乏温暖的信号（Bai et al., 2016）。

用于斑马鱼光/暗选择性测试的一般实验方法如下（图14-1）。一种方法为：在起始时间将个体放入半黑半白室的中心，然后进行几十分钟的视频跟踪（Lau et al., 2011; Maximino et al., 2011; Serra et al., 1999）。利用顶端红外摄像机探测跟踪鱼的探索位置，生成包含多个输出参数的数字视频文件，也包括可作为偏好衡量标准的选择指数。选择指数可以列举为总体的均值（Bai et al., 2016），或者列举为个体水平上多次试验的均值（Dahlén et al., 2019; Wagle et al., 2017）。另一种方法为：揭示了具有不同遗传特征的实验室品系AB（Wagle et al., 2017）[图 14-1（C）] 和野生品系 EK-WT（Dahlén et al., 2019）的鱼的避暗谱。野生 EK 品系具有广泛的遗传多样性和可遗传表型变异，它将是未来全基因组关联研究揭示光 / 暗偏好基因控制的理想模型。

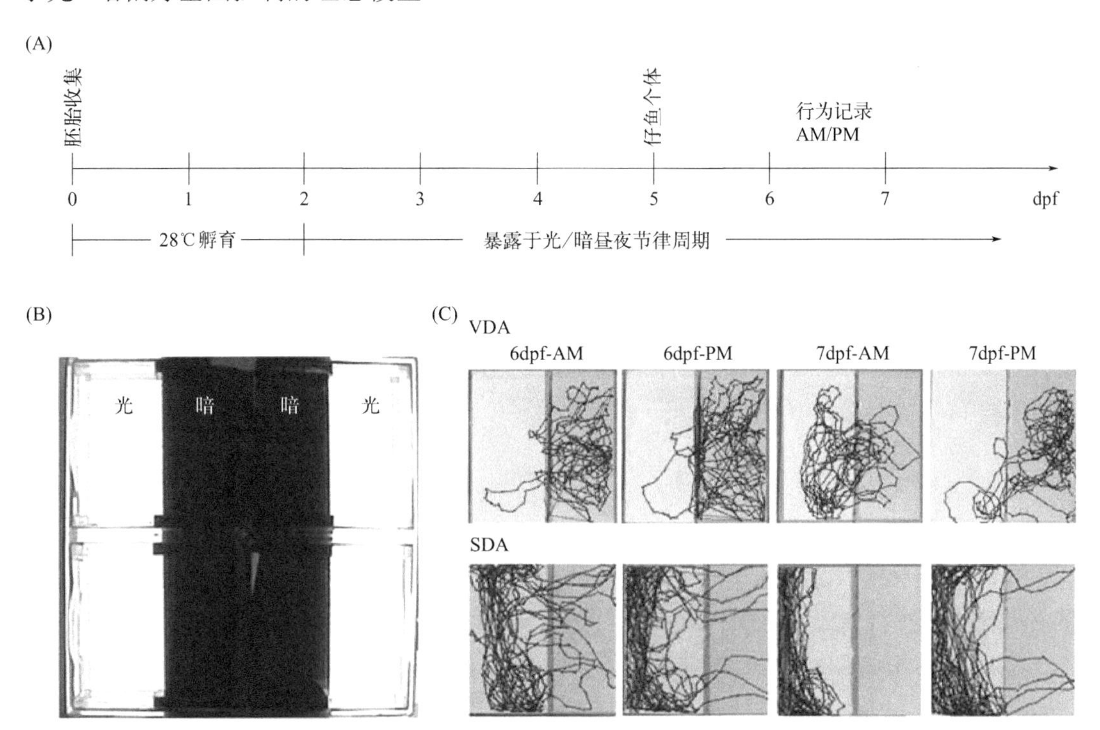

图 14-1 由 Wagle 等人（2017）设计的明暗偏好测试方案

（A）饲养和行为测试的时间轴；（B）测试所用行为室的照片；（C）具有不同避暗和强烈避暗行为的仔鱼的代表性轨迹

研究者们使用已知的抗焦虑药物，已经证实了仔鱼（Bai et al., 2016; Chen et al., 2015; Steenbergen et al., 2011）和成鱼（Gebauer et al., 2011; Lau et al., 2011; Magno et al., 2015; Maximino et al., 2011; Sackerman et al., 2010）的光 / 暗选择是一种类焦虑样行为反应。具有厌恶性的非药物刺激也能产生类似于抗焦虑药物的效果（Bai et al., 2016）。例如，乙醇急性暴露可剂量依赖性改善成鱼的避光行为，而停止长期乙醇暴露可增强避光行为（Mathur and Guo, 2011）。通过以皮质醇等内源性激素作为焦虑生理学指标的检测进一步证实了光 / 暗偏好行为与焦虑相关（Colwill and Creton, 2011）。这些结果表明，斑马鱼的光 / 暗偏好行为是筛选抗焦虑药物的一种实用、灵敏且经济的方法（Gebauer et al., 2011）。

值得注意的是，在使用不同背景色调（黑色与白色）而照明水平保持不变的情况下，已经观察到成年斑马鱼具有暗偏好。白色背景具有反光性，因此白色背景部分会比黑色背景部分更亮。这种黑暗偏好也称为趋暗性（Maximino et al., 2010），也可转换为：在封闭区域或在完全黑暗的房间中应用渐变照明时，对更亮照明区域的偏好（Blaser and Peñalosa, 2011; Champagne et al., 2010; Gerlai et al., 2000）。然而，在最近的一项研究中（Facciol et al., 2017），这种被称为趋光性的对比偏好似乎已经削弱，在该研究中采用了顶部未覆盖的暗室，从而表明另一个因素“开放性”也对光暗偏好性产生影响。对有盖实验水箱的厌恶可能是由于其有洞穴状结构，斑马鱼在其栖息地很少遇到这种结构（Engeszer et al., 2007b），从而导致恐惧症。进一步研究发现背景色调和照明水平之间存在联合效应，且可以通过急性乙醇暴露来改变（Facciol et al., 2019）。总体而言，这些不一致的研究表明，我们应根据刺激的复杂性仔细解释光 / 暗偏好测试的结果。

14.2.2 趋触性

趋触性是指与空旷区域相比，个体对于边缘（图 14-2）的偏好性，是另一种类型的焦虑样行为反应，这种行为反应从斑马鱼（Schnörr et al., 2012）到哺乳动物（Komada et al., 2008; Prut and Belzung, 2003; Treit and Fundytus, 1988）在进化上都是保守的。当将成年斑马鱼放入圆形水箱时，其倾向于沿着外围游动并避开水箱的中心（Buske and Gerlai, 2012; Champagne et al., 2010; Colwill and Creton, 2011; Grossman et al., 2010; López-Patiño et al., 2008）。与光 / 暗偏好不同，这种行为没有表现出年龄依赖性的变化，因为斑马鱼仔鱼的趋触性特征与在斑马鱼成鱼中观察到的特征相类似（Best and Vijayan, 2018; Buske and Gerlai, 2012; Schnörr et al., 2012）。

Schnörr 等人（Schnörr et al., 2012）使用标准化的 24 孔板设计了斑马鱼仔鱼趋触性测试装置，其中每个孔划分为具有等效空间面积的外区域和内区域［图 14-2（A）］。将 5dpf 的仔鱼放入每个孔中，实验包括以下 2 个步骤：一是在光照下适应环境（0 ～ 6min）；二是在黑暗下进行视觉运动（7 ～ 10min）［图 14-2（B）］。采用自动录像系统追踪记录这些鱼在第二阶段的游泳行为。趋触性指数表示为在外区域总移动距离的百分比或在外区域花费时间的百分比（Schnörr et al., 2012）。由于黑暗被认为是一种引发焦虑的刺激信号，能够有效地促进斑马鱼仔鱼的趋触性，因此该实验是在从明到暗的过渡中进行的。

(A) 测试装置

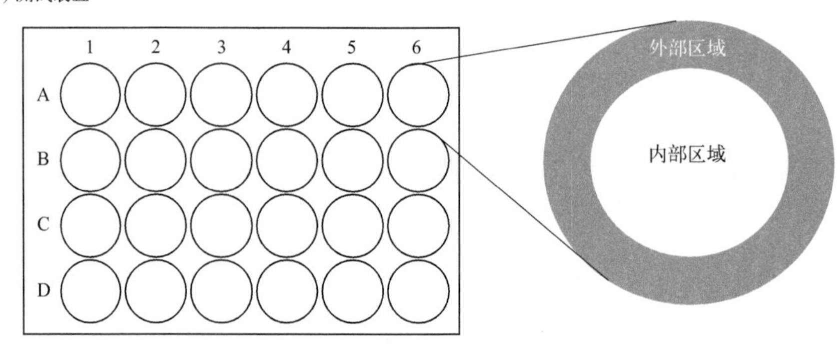

(B) 实验过程

图 14-2 由 Schnörr 等人设计的趋触性实验方案（Schnörr, 2012）

（A）测试装置由 24 孔板组成。划定每个孔的内部区域和外部区域；（B）实验包括 2 个步骤，包括在光照下适应环境（0 ～ 6min）和黑暗条件下的视觉运动（7 ～ 10min）

药物的趋触性与其对其他焦虑行为的影响一致。常用的抗焦虑药物地西泮可显著减弱趋触性，而产生焦虑的药物如咖啡因可增强趋触性（Schnörr et al., 2012）。注射 α- 氟甲基组氨酸（α-fluoromethylhistidine, α-FMH）可改变行为状态，使其在中央区域花费的时间增加（Peitsaro et al., 2004）。α-FMH 这种作用可能是由于其抑制了组胺（histamine, HA）的释放，组胺是一种神经递质，涉及多种大脑机制，包括警觉性和睡眠、癫痫阈值、激素分泌和疼痛等（Brown et al., 2001; Fernández-Novoa and Cacabelos, 2001; Haas and Panula, 2003; Schwartz et al., 1991）。另外，从 100% 光亮过渡到 100% 黑暗的突然转换会引发斑马鱼对边缘区域的偏好，因为这种情况代表了一种引起焦虑的情况，而如果向 90% 和 80% 的黑暗过渡的话，这种程度就会减轻（Schnörr et al., 2012）。

14.2.3 栖底性

栖底性（图 14-3）是一种在初始时间表现为对新型水箱底部偏好的行为反应，之后逐渐增加对水箱的更高水平位置的探索（Bencan et al., 2009; Levin et al., 2007; Mathur et al., 2011a）。这种新型水箱潜水反应伴随着随后对水箱上部的探索，这种探索类似于啮齿动物的趋触性，这通常被解释为对压力的焦虑样反应（Schnörr et al., 2012; Treit and Fundytus, 1988）。栖底性和趋触性两者都可以被视为试图靠近固体物体，该物体提供了躲避掠食者的藏身之处（Gerlai, 2010）。

新型水箱潜水试验首先由 Levin 等人提出（Levin et al., 2007）。水箱分为三个虚拟隔间，并由侧面的数字摄像机监控（图 14-3）。研究者可以通过跟踪和表征 5min 内的鱼的游泳行为，以选择位置（底部 / 上层的比值）作为焦虑指数。在测试期间已使用多个参数来量化其轨迹，包括实验对象与水箱底部的平均距离（Levin et al., 2007）、在水箱的三个虚拟隔间中的每个隔间所花费的时间（Bencan et al., 2009; Mathur and Guo, 2011; Sackerman et al., 2010; Wong et al., 2010），或进入水箱顶部三分之一的延迟时间（Egan et al., 2009）等。研究者认为这些指标都可以正确量化斑马鱼的焦虑程度（Tran and Gerlai, 2016）。

已有大量实例证明上述行为是通过药理学调节的，证实了新的水箱潜水反应与焦虑

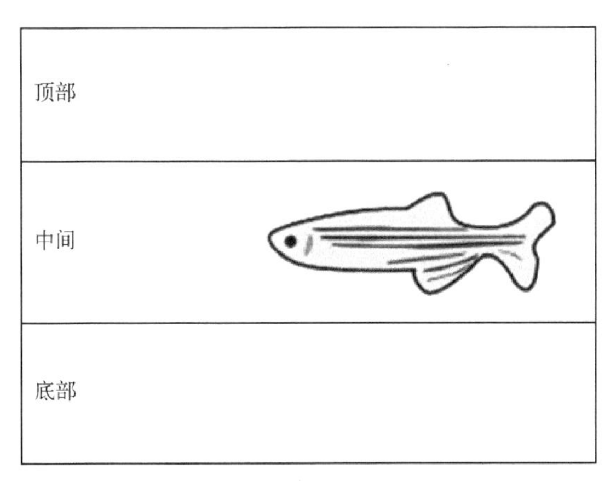

图 14-3 **新型水箱潜水试验方案**

潜水测试在一个分为 3 个虚拟隔间的水箱中进行。刚开始斑马鱼经过 2min 的适应期，之后采用摄像机追踪记录其 5min 内的活动

相关。丁螺环酮等抗焦虑药物可显著减弱斑马鱼的栖底行为，并增加鱼类探索水箱上部的时间；然而，几种苯二氮䓬类药物（氯硝西泮、溴西泮和地西泮）不会产生任何明显的影响，这可能是由于它们对运动的影响会损害向上游泳的能力（Bencan et al., 2009）。通过接触单胺再摄取抑制剂（例如地昔帕明和西酞普兰）和尼古丁也可以降低这种潜水反应，后者在与其拮抗剂一起给药时这种反应可被抵消（Levin et al., 2007）。相比之下，报警物质（Quadros et al., 2016）、慢性乙醇暴露戒断（Mathur and Guo, 2011）和可卡因中毒（López-Patiño et al., 2008）均会产生焦虑反应。另外，据报道新的水箱潜水反应是经验依赖的，反复暴露新型水箱可以适应并减弱栖底行为（Wong et al., 2010）。相反，减少适应会导致栖底行为加剧（Tran et al., 2016）。这些不同的调节作用共同表明，斑马鱼新型水箱潜水测试在区分药物类型和揭示其生物学神经回路方面具有一定的功效。

14.2.4　条件位置偏好行为

动物对某些奖赏物质（例如食物）的自然和强化行为，可能是由神经系统所介导的，而这些神经系统对厌食症、肥胖症和成瘾等人类疾病具有重要的病理研究意义（Saper et al., 2002）。为了研究相关的神经回路，之前在啮齿动物中建立的条件位置偏好行为（the conditioned place preference, CPP）（Tzschentke, 2007）实验已经在斑马鱼中得到应用，证明了自然奖励或成瘾物质包括可卡因（Darland and Dowling, 2001）、吗啡（Lau et al., 2006）、安非他明（Ninkovic and Bally-Cuif, 2006）、迷幻鼠尾草（Braida et al., 2007）、尼古丁（Kedikian et al., 2013; Kily et al., 2008）、乙醇（Mathur et al., 2011b; Parmar et al., 2011）和其他药物（Collier et al., 2014）等具有强大的强化作用。

Mathur 等人（Mathur et al., 2011b）描述了斑马鱼 CPP 测试的具体过程。CPP 测试过程经历连续 2 天时间，分 3 个步骤进行。测试水箱由 2 个视觉上不同的隔间组成，中间隔开（图 14-4）。第 1 步进行初始偏好测试，排除具有强烈仪器位置偏好的个体；第 2 步将鱼限制在每个隔间中一段时间，在此期间进行试验组或对照组处理；第 3 步允许鱼对所有隔间进行自由探索，以确定其最终偏好程度。步骤 1 和步骤 3 均持续 5min，建议步骤 2 中的药物暴露持续时间为 20min（Mathur et al., 2011a）。为了量化 CPP 的程度，我们从最终偏好减去初始偏好来计算偏好百分比（条件化后在隔间中的时间百分比减去条件化前在同一隔间中的时间百分比）。

前面提到的那些药物暴露会诱导显著的 CPP，其中有些具有浓度依赖性。例如，在吗啡诱导的 CPP 测试中，低剂量（1.5μmol/L）或高剂量（15μmol/L）吗啡低于中剂量（Lau et al., 2006）产生的 CPP。此外，吗啡诱导的 CPP 可以通过使用多巴胺受体拮抗剂预处理来阻断，这表明斑马鱼中多巴胺参与吗啡诱导产生的 CPP。急性乙醇暴露诱导的 CPP 中也有多巴胺的参与，而慢性乙醇暴露则导致这种偏好性减弱（Collier et al., 2014）。采用传统检测仪器检测咖啡因诱导的 CPP 相对较弱。然而，在新设计的测试 CPP 方法中，在测试 CPP 过程中引入新的刺激，咖啡因的奖励特性都可以得到加强，从而产生更强的 CPP（Faillace et al., 2018）。无论这些药物有什么不同的强化作用，其背后的斑马鱼神经系统中都存在着类似哺乳动物的中脑边缘通路（例如多巴胺能通路）（Gould, 2011）。因此，斑马鱼的 CPP 方法可应用于药物筛选和对神经回路的理解。

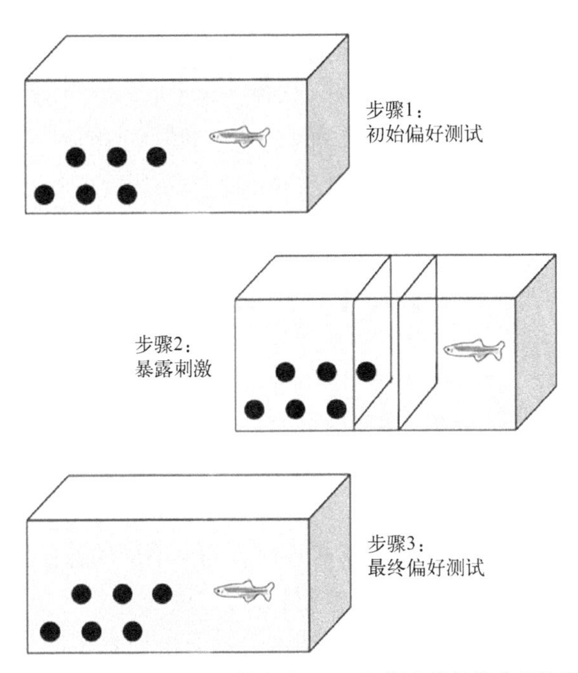

图 14-4 Guo（2004）和Mathur等人（2011b）**提出的条件位置偏好分析方法**

该过程包括 3 个步骤。第 1 步，进行初始偏好测试，排除具有强烈仪器位置偏好的个体；第 2 步，将鱼限制在每个隔间中一段时间，在此期间进行试验组或对照组处理；第 3 步，允许鱼对所有隔间进行自由探索，以确定其最终偏好程度

14.3 前脉冲抑制

自然环境中丰富的感官刺激要求动物能够消除多余的信息（Braff et al., 2001）。这是通过选择性门控运动系统来实现的。研究发现，如果提供微弱的初始刺激，则强刺激引起的惊吓反应可能会减弱（Swerdlow et al., 2001）。这种现象称为前脉冲抑制（prepulse inhibition, PPI），已经作为一种测量感觉运动门控的强有力方法。精神分裂症（Braff et al., 2001; Grillon et al., 1992; Swerdlow et al., 2006）和其他几种神经系统精神疾病（Kohl et al., 2013; Nestler and Hyman, 2010）均会导致感觉运动门控受损，引发 PPI 功能障碍。

Burgess 和 Granato（2007）首次在斑马鱼仔鱼中引入了 PPI 检测方法，它们表现出强烈的惊吓反应，包括身体的“C 形弯曲”，然后是较弱反向弯曲（smaller counter bend）和游泳（Kimmel et al., 1974）。一般来说，仔鱼仅接受声音惊吓刺激，或者前脉冲和声音惊吓刺激的混合型刺激，对两种处理中的任一种（仅惊吓刺激、前脉冲 + 惊吓刺激）的反应延迟和反应率参数进行测试，获得两种处理之间的差异量化值，该值可作为 PPI 的检测值（Bhandiwad et al., 2013; Burgess and Granato, 2007）。

Burgess 和 Granato（2007）报告了两种不同调节方式下的声学惊吓反应模式，只有一种模式会因 Mauthner 细胞的消融而改变。Mauthner 细胞介导斑马鱼的快速逃逸反应（Liu and Fetcho, 1999）。通过多巴胺能药物的药理调节作用研究已将斑马鱼的 PPI 缺陷与其多巴胺机制联系起来，在其他高等脊椎动物中也是如此（Nestler and Hyman, 2010;

Schall et al., 1999），这种现象表明斑马鱼的 PPI 调节非常适用于转化研究。此外，基因筛选已经确定了一种 PPI 值降低的 *Ophelia* 突变体（Burgess and Granato, 2007）。CRISPR-Cas9 介导的突变体群体的深入鉴定进一步确定了两个 PPI 降低的突变体 *atxn7* 和 *akt3*（Thyme et al., 2019）。

14.4 社交行为

斑马鱼成鱼是具有社交互动特征的社交动物（Oliveira, 2013），其包括松散聚集和集群（Green et al., 2012; Krause et al., 2000; McCann et al., 1971; Miller and Gerlai, 2012）、同种定向攻击（Gerlai et al., 2000; Jones and Norton, 2015）、交配（Engeszer et al., 2008）和其他社交行为（Arganda et al., 2012）。尽管这些社交行为由不同的通信系统介导，但在某种程度上，许多社交行为可归因于对同一物种成员的视觉接近。

针对松散聚集社交行为的一项研究表明，与具有不同表型的个体相比，鱼体具有相似图案的个体间更有可能出现松散聚集式互动（Engeszer et al., 2004）。松散聚集和称为集群的更极化行为，在动物行为学上更加有利于发现捕食者、识别食物块以及获得配偶（Miller and Gerlai, 2007; Spence et al., 2008）。虽然这些行为都是先天性的，但如果斑马鱼小于 7 dpf（受精后），则不存在松散聚集行为，这表明该行为的获得有赖于发育阶段（Buske and Gerlai, 2012; Engeszer et al., 2007a; Shams et al., 2018）。对斑马鱼仔鱼的进一步研究发现，斑马鱼仔鱼对同一物种的视觉接触偏好是逐渐增强的（Dreosti et al., 2015）。这种偏好与仔鱼个体及其社交伙伴之间的清晰互动相结合，形成了一种同步运动（Dreosti et al., 2015）。然而，这种视觉接触偏好对形成松散聚集行为或其他社交互动（如集群行为或攻击性）的影响，与感官因素（如嗅觉和触觉）具有不可割舍的联系。未来的研究需要对多模态刺激的复杂性进行行为学分析。

除了松散聚集行为，斑马鱼行为实验研究还包括更复杂的特征，如攻击性和决策能力（Orger and de Polavieja, 2017）。许多社会行为偏好是针对某些可能导致其恐惧或焦虑的厌恶性刺激（例如，捕食者）而形成的。因此，人们对鉴定可能影响社会行为的非社会行为的兴趣与日俱增（Al-Imari and Gerlai, 2008; McRobert and Bradner, 1998; Seguin et al., 2016）。例如，在没有光的情况下，人们可能不会形成早期的社会偏好（Dreosti et al., 2015），这表明与暗 / 明偏好一样，这是一种视觉引导的偏好行为。相反，在松散聚集和趋触性这一对社会和非社会行为之间，存在较少的相关性，因为它们受到年龄增加的不同影响（Buske and Gerlai, 2012）。栖底行为是另一种与焦虑相关的非社会行为，它也受到一种神经递质系统的调节，这种神经递质系统与松散聚集行为相关的神经递质系统是不同的（Gebauer et al., 2011）。社会和非社会行为之间的联系需要进一步探索研究。

药理学操作能以多种方式改变斑马鱼的社交行为。*N*- 甲基 -D- 天冬氨酸（NMDA）受体信号通路参与社交学习（Maaswinkel et al., 2013）。NMDA 受体拮抗剂 MK-801 可破坏 3 周龄鱼的社会偏好（Dreosti et al., 2015）。乙醇治疗对社会偏好的影响呈现剂量依赖性关系；

高剂量通常产生抑制作用，而低剂量则产生适当的促进作用（Fernandes et al., 2014; Gerlai et al., 2000; Ladu et al., 2014）。然而，胚胎暴露于低剂量乙醇会导致松散聚集行为受损，个体距离增加（Fernandes and Gerlai, 2009; Fernandes et al., 2015）。使用抗焦虑药物（氟西汀、丁螺环酮或地西泮）处理也可以观察到类似的行为改变（Gebauer et al., 2011; Giacomini et al., 2016）。另一种抗焦虑药物尼古丁可显著增强成年斑马鱼的学习能力（Levin et al., 2006）。总而言之，这些研究结果表明社交行为对大量的药物都高度敏感（Shams et al., 2018）。

14.5 采用正向和反向遗传策略进行行为筛查

斑马鱼的正向遗传筛查加深了对神经回路构成和遗传调控功能的理解（Guo, 2004）。通过研究遗传缺陷和行为改变之间的关联，人们已经发现了大量对多巴胺能（DA）神经元分化、习惯化学习、人类自闭症谱系障碍等有影响的基因（Melgoza and Guo et al., 2018）。一种新型的基因组编辑技术 CRISPR（规律成簇间隔短回文重复序列，clustered regularly interspaced short palindromic repeats）具有对目的基因突变的能力（Cong et al., 2013; Jinek et al., 2012; Mali et al., 2013），该技术精确度和效率更高，使得反向遗传筛查成为可能（Shah et al., 2015）。最近一项针对候选基因中 Crispr-Cas9 突变体的研究优先考虑了 30 多个精神分裂症相关基因，有待进一步研究（Thyme et al., 2019）。有关斑马鱼系统遗传筛查的深入研究，可以参阅 Melgoza 和 Guo 的书籍（Melgoza and Guo, 2018）。

14.6 用于行为研究的斑马鱼遗传种群

斑马鱼（*Danio rerio*）所属的 *Danio* 属，由 20 个经过分类验证的物种组成（Parichy, 2015），这些物种广泛分布于南亚和东南亚，栖息地和当地条件具有高度多样性（Arunachalam et al., 2013; Engeszer et al., 2007b; Spence et al., 2008）。对当地广泛变化的环境的适应性导致物种间巨大的遗传多样性，其中一部分已被收集并保存在实验室中。斑马鱼国际资源中心（http://zfin.org/）称，目前全世界 400 多个实验室驯化了大约 30 种公认的野生型斑马鱼品系（Parichy, 2015; Spence et al., 2008）。

基于斑马鱼形态和生理特征的多样化反映了其遗传多样性（Long et al., 2013; Patterson et al., 2014; Spence et al., 2008; Ulloa et al., 2015）。相比之下，控制斑马鱼行为变异的遗传机制仍未得到充分研究（Orger and de Polavieja, 2017; Parichy, 2015）。目前对其行为学遗传剖析较少的主要原因是许多行为表现难以定义和量化（Sokolowski, 2001; Spence et al., 2008），特别是那些由个体行为组成的社会互动的行为表现。复杂行为特征的遗传变异不仅源于等位基因加性效应，还源于等位基因内和等位基因间的相互作用（Sokolowski, 2001; Wright et al., 2006a, 2006b）。此外，许多基因以动态空间模式表达，并受自然条件和社会环境的影响（Oliveira et al., 2016）。

14.7 实验室养殖和野生斑马鱼品系的遗传多样性

高通量行为方法的建立促使人们发现了同一个群体内或不同群体之间存在着巨大的行为差异。Sackerman 等人（2010）报道了 4 种品系的斑马鱼在新型水箱和十字迷宫中的焦虑样反应的差异。Quadros 等人（2016）研究确定了报警物质在野生型和豹纹型斑马鱼之间的对比效应：报警物质暴露对豹纹型斑马鱼有抗焦虑作用，但对野生型斑马鱼却有致焦虑作用。据报道，乙醇效应也具有品系依赖性（Dlugos and Rabin, 2003; Gerlai et al., 2008）。对三种斑马鱼品系的聚集行为的评估表明，不同品系的斑马鱼对乙醇的敏感性不同，并且对急性与慢性暴露都有耐受性（Dlugos, 2011）。对其他四个不同群体的观察结果也表明了乙醇处理效应的品系依赖性（Gerlai et al., 2008）。值得注意的是，AB 品系与其他品系斑马鱼对中等剂量乙醇的反应性不同，AB 品系鱼在新型水箱中距离底部的位置并没有增加（Egan et al., 2009; Gerlai et al., 2000; Sackerman et al., 2010）。然而，在另一项同一品系的独立研究中，这种反应却是可以观察到的（Mathur and Guo, 2011）。这种差异的一个解释是个别实验室的品系可能发生了遗传漂变（LaFave et al., 2014; Mathur and Guo, 2011; Trevarrow and Robison, 2004）。最近的两项研究中发现了更多的品系内行为变异，其中分别在 AB 品系和 EK-WT 品系（野生型品系）的仔鱼中揭示了避暗光谱，并且两个实验中对交配后代的评估结果表明了不同行为的遗传性（Dahlén et al., 2019; Wagle et al., 2017）。上述提到的对斑马鱼仔鱼的社会偏好性研究中，少数 3 周大的仔鱼个体对于同物种表现出厌恶性反应，而大多数仔鱼对于同物种则表现出明显的偏好性（Dreosti et al., 2015）。综上所述，这些品系效应的发现表明导致行为差异的相关遗传结构存在多样性。

14.8 斑马鱼作为全基因组关联分析的候选模型

过去十年来，全基因组关联分析（genome-wide association study, GWAS）在识别单核苷酸多态性（single nucleotide polymorphisms, SNP）方面得到了广泛应用，其中等位基因频率随着表型特征值变化而发生系统的变化（Marees et al., 2018）。迄今为止，研究者们已在人类基因组针对疾病相关性状进行了 2000 多项 GWAS 研究，阐明了成百上千的独特表型与数百万基因之间的关联（Leslie et al., 2014）。因此需要利用动物模型建立基因型和表型之间的因果关系，广泛应用的模型是啮齿动物（Flint and Eskin, 2012; Lawson and Cheverud, 2010; Parker and Palmer, 2011; Waterston et al., 2002）。相反，斑马鱼更常用于 GWAS 后分析（post-GWAS analysis）（Made-laine et al., 2018）而不是关联检测。然而，斑马鱼几个显著特征弥补了许多哺乳动物模型的缺陷，这表明它是一种潜在的强大的 GWAS 模型，可以与人类和啮齿动物模型互补。

GWAS 的目的是检测与形成目标复杂性状的非基因型致病变异体相关的基因型位点。这两个基因座之间的关联是基于它们的连锁不平衡（linkage disequilibrium, LD）。LD 的强度取决于相关基因座的等位基因频率，因此在群体中频率较低（<0.01）的罕见变异将处于低 LD 且附近有 GWAS 信号（Visscher et al., 2012）。由于缺乏 SNP 表型关联的能力，GWAS 的线性回归模型通常可排除罕见变异（Marees et al., 2018）。换句话说，GWAS 在设计上能够检测与人群中相对常见的因果变异的关联。经验数据分析也表明罕见变异对疾病风险非常重要（Anderson et al., 2011），因此，研究者们认为，能够排除遗传方差比例较低的罕见变异是 GWAS 模型可用的原因之一（Manolio et al., 2009）。实验室培育的斑马鱼品系可能没有这个问题，因为它们中的大多数来自斑马鱼总遗传多样性的一小部分（Brown et al., 2012; Coe et al., 2009; Patowary et al., 2013; Whiteley et al., 2011），这表明罕见等位基因的比例很低（Parker and Palmer, 2011）。另一方面，正如小鼠模型中报道的，有限的遗传多样性则意味着不能使用一个单一群体来分析可能存在于不同品系之间的所有变异的问题（Yalcin et al., 2010）。我们必须研究多个群体以获得全部谱系的遗传变异（Roberts et al., 2007）。

除了遗传多样性之外，使用 GWAS 成功检测因果变异还需要通过分解 LD 来实现高的定位分辨率。每次减数分裂的重组数量相对较少（Moens, 2006），因此需要具有多代异交（多代远亲繁殖）的大种群来分解 LD，使检测到的 SNP 与真正的因果变异之间的物理距离值最小化。对 2000 多个 GWAS 位点回顾可知，其检测能力与种群规模之间存在着很强的相关性（Parichy, 2015）。与小鼠或其他哺乳动物模式生物相比，体形较小的斑马鱼由于其多产的特性（Guo, 2004）将更适合于庞大的种群繁殖发展和维持。此外，大样本量还提供了对基因间相互作用（上位性）影响的评估能力，尽管行为特征比形态特征具有更多的上位方差，但很少对其进行分析（Wright et al., 2006a）。

虽然与 GWAS 相关的 SNP 通常可以解释不到 20% 复杂性状的遗传变异，但考虑到所有 GWAS 的 SNP（包括那些没有统计学意义的 SNP），则与 GWAS 相关的 SNP 可以代表群体中一半的加性遗传变异（Allen et al., 2010; Anderson et al., 2011; Yang et al., 2011）。在许多 GWAS 中，由于控制了基因组多重检测的假阳性，SNP 无法与目标表型相关联，随着所研究的 SNP 集的扩大（Fernando and Garrick, 2013），这种误差控制变得越来越严格，导致检测能力大大降低。为了避免检测能力与测试数量之间的矛盾，需要将 GWAS 与贝叶斯回归方法组合应用，将所有阳性结果中的假阳性结果的比例限制在某个值来控制假阳性（Fernando and Garrick, 2013; Meuwissen et al., 2001; Yoshida et al., 2017）。在 GWAS 中也经常应用另一种基于置换检验的方法（a permutation-based approach）控制假阳性（Abney, 2015; Parker et al., 2016）。

检测到的基因座效应的集合虽然相对较小，但是其对阐明具有生物学意义的途径却是非常重要的。多基因遗传的经典理论表明，对于大多数复杂性状，一组影响较小的遗传变异能比相对罕见的单基因突变发挥更大的作用（Fuchsberger et al., 2016; Golan et al., 2014）。为了挖掘 GWAS 的更多实际效用，来自精神病学领域的研究人员研发了一种专注于多基因风险评分（polygenic risk score, PRS）的方法，它允许将多个 SNP 的效应大小集合为一个可用于预测疾病风险的单一评分（Dudbridge, 2013; Khera et al., 2018; Purcell et al., 2009）。随着种群的增加和算法的改进，PRS 方法可进一步用于基于另一种表型的 GWAS 预测某种表型的风险（Howe et al., 2013; Power et al., 2015; Purcell et al., 2009; Stringer et al., 2014）。

目前，已经完成了斑马鱼参考基因组的测序和注释（Howe et al., 2013），其中约 70%

的基因具有至少一种人类同源基因。人们已发现超过 80% 的人类疾病相关基因与至少一种斑马鱼同源基因有关。进一步对斑马鱼和人类之间基因组对比分析发现在人类染色体两边都有双重保守模块（double-conserved synteny blocks）（Howe et al., 2013）。此外，参考基因组和测序技术可以检测不同品系的全基因组多态性，从而解决斑马鱼缺乏高通量基因分型阵列的问题。总之，与人类高度进化同源的参考基因组序列的使用、统计工具和方法的发展，与基因的精确操作等手段相结合，可对 GWAS 关联进行有效的实验测试，并随后将斑马鱼的发现转化为在人体中的应用。

14.9 行为特征的转录组学研究

鉴于行为特征的复杂性，仅依靠与基因组多态性的关联不太可能完全理解整个生物回路。基因表达谱的综合分析可弥补对基因组全景的研究，以进一步鉴定致病性变异基因以及揭示其生物学途径。自 20 世纪 90 年代以来，基于原位杂交的 DNA 微阵列一直是大规模基因表达研究中最受欢迎的技术之一（Schena et al., 1995）。与这种杂交方法相辅相成的是 PCR 检测，即定量 PCR（quantitative PCR, qPCR），它具有定量 RNA 转录的能力（Higuchi et al., 1993）。该技术与高通量测序方法的最新进展相结合，产生了一种基于序列的转录组分析方法，将其称为转录组测序技术（RNA-seq），与基于 DNA 的微阵列方法相比具有多种优势（Wang et al., 2009; Zhao et al., 2014）。

对斑马鱼基因表达的研究已经发现了与行为驱动神经元活动有关的基因。自 1980 年以来，早期转录因子编码基因 *c-fos* 是被鉴定出来的最好的活性依赖性基因之一（Kaczmarek, 1993; Kovács, 2008），目前已被广泛用作神经元活性定位的功能性分子标记。基于 *c-fos* 基因的组织特异性杂交，Lau 等人（2011）将成年斑马鱼光暗选择的神经元活动映射到两个活跃的大脑区域，包括背侧端脑区（dorsal telecephalic region, Dm）和腹侧端脑区的背核（the dorsal nucleus of the ventral telencephalic area, Vd）。采用 qPCR 技术表明，另一个编码 *ERβ* 受体的基因 *esr2b* 也与同样的神经元活性有关（Chen et al., 2015）。进一步的药物调节表明，*ERβ* 信号的激活可减弱应激反应。Kedikian 等人（2013）在尼古丁配对 CPP 测定中采用 qPCR 技术，已经检测到四种编码烟碱受体亚基基因的不同表达模式。结合 RNA-seq 方法，最近的一项关于小鼠 GWAS 的研究，绘制了 eQTL（QTLs 解释了基因表达表型遗传变异的一小部分），进一步细化了相关的多位点区域（Parker et al., 2016）。这些发现共同说明了基因组和转录组数据综合分析在剖析复杂行为特征的生物网络方面是非常有用的。

14.10 单细胞分辨率转录组学分析

Tang 等人（2009）首先提出了单细胞分辨率下更灵敏的 mRNA 测序（mRNA-Seq）

检测。这项技术也称为单细胞转录组测序（scRNA-seq）技术，极大地改变了转录组分析，特别是对于早期胚胎发育，RNA 数量不足以进行早已存在的基因表达分析（Tang et al., 2009）。此后，为了提高检测效率和准确度对方法进行了修改，产生了一系列方法，包括 STRT（Islam et al., 2011）、Smart-Seq（Ramsköld et al., 2012）、CEL-Seq（Hashimshony et al., 2016）和 inDrop（Klein et al., 2015; Zilionis et al., 2016）。目前已有将 sgRNA-seq 应用于斑马鱼胚质的一些案例，从而进一步将斑马鱼确立为生物学和医学的重要遗传模式生物。

scRNA-seq 技术可根据基因表达水平和转录物变异体来区分细胞类型（Hashimshony et al., 2016; Islam et al., 2011; Ramsköld et al., 2012）。计算方法的并行开发进一步实现了具有高时间分辨率的单细胞转录组的动态分析。结合基于图形方法的一项研究，对 90000 多个胚胎细胞（从 4hpf 到 24hpf）进行转录组分析，建立了斑马鱼胚胎发育的完整发育图谱（Wagner et al., 2018）。一项类似的研究以更高的时间分辨率分析了数万个胚胎细胞的转录组（Farrell et al., 2018）。这两项研究都通过构建分支树（branching tree）的方法对细胞状态进行跟踪。此外，Wagner 等人（2018）指出，因为一些细胞状态是通过树外互连观察到的，所以需要更复杂的拓扑图来完全破译细胞状态转换的轨迹。两项研究使用了不同的突变体，但是都得出结论认为突变并没有引起明显的转录状态的改变。相比之下，第三项针对另一种脊椎动物模型热带爪蟾（*Xenopus tropicalis*）的研究已经发现了胚胎发育过程中细胞状态的动态变化既有多样化特征，也有保守性特征（Briggs et al., 2018）。在 CRISPR 产生的突变群体中应用 sgRNA-seq 进行综合研究，揭示了转录因子 *znf536* 在与社会行为和应激反应相关的前脑神经元发育中具有重要作用。总之，以上研究均为行为特征方面的研究奠定了基础，这些行为特征很可能是由具有不同转录组动态特征的细胞类型之间的协同相互作用引起的。

14.11 结论

探索行为变异的遗传基础是一项极具挑战性的工作。正如本章所述，目前研究者们已经建立的应用于高通量基因分型和动力学表型分析的斑马鱼模型与先进的分析方法相结合，将成为可以揭示行为特征的潜在遗传机制的有效方法。同时，斑马鱼对药理调节的敏感性及其与人类的生理和遗传方面的同源性，使其非常适合应用于快速发展的转化生物医学方面的研究。

致谢

我们感谢 Michael Munchua 和 Vivian Yuan 对鱼类精心的养护，以及与 Guo 实验室成员进行的有益讨论。作者实验室的工作得到了美国国家卫生研究院（R01NS095734 和 R01GM132500）的资助。

（靳倩翻译　王陈审校）

参考文献

Abney, M., 2015. Permutation testing in the presence of polygenic variation. Genetic Epidemiology 39, 249-258.

Al-Imari, L., Gerlai, R., 2008. Sight of conspecifics as a reward in associative learning in zebrafish (*Danio rerio*). Behavioural Brain Research 189, 219.

Allen, H.L., Estrada, K., Lettre, G., Berndt, S.I., Weedon, M.N., Rivadeneira, F., Willer, C.J., Jackson, A.U.,Vedantam, S., Raychaudhuri, S., et al., 2010. Hundreds of variants clustered in genomic loci and biological pathways affect human height. Nature 467, 832-838.

Anderson, C.A., Soranzo, N., Zeggini, E., Barrett, J.C., 2011. Synthetic associations are unlikely to account for many common disease genome-wide association signals. PLoS Biology 9, e1000580.

Arganda, S., Pérez-Escudero, A., de Polavieja, G.G., 2012. A common rule for decision making in animal collective sacross species. Proceedings of the National Academy of Sciences United States of America 109, 20508-20513.

Arunachalam, M., Raja, M., Vijayakumar, C., Malaiammal, P., Mayden, R.L., 2013. Natural history of zebrafish (*Danio rerio*) in India. Zebrafish 10, 1-14.

Bai, Y., Liu, H., Huang, B., Wagle, M., Guo, S., 2016. Identification of environmental stressors and validation of light preference as a measure of anxiety in larval zebrafish. BMC Neuroscience 17, 63-0298.

Bencan, Z., Sledge, D., Levin, E.D., 2009. Buspirone, chlordiazepoxide and diazepam effects in a zebrafish model of anxiety. Pharmacology Biochemistry and Behavior 94, 75-80.

Best, C., Vijayan, M.M., 2018. Cortisol elevation post-hatch affects behavioural performance in zebrafish larvae. General and Comparative Endocrinology 257, 220-226.

Bhandiwad, A.A., Zeddies, D.G., Raible, D.W., Rubel, E.W., Sisneros, J.A., 2013. Auditory sensitivity of larval zebrafish (*Danio rerio*) measured using a behavioral prepulse inhibition assay. Journal of Experimental Biology 216,3504-3513.

Bilkei-Gorzo, A., Gyertyan, I., Levay, G., 1998. mCPP-induced anxiety in the light-dark box in rats—a new method forscreening anxiolytic activity. Psychopharmacology 136, 291-298.

Blaser, R.E., Peñalosa, Y.M., 2011. Stimuli affecting zebrafish (*Danio rerio*) behavior in the light/dark preference test. Physiology and Behavior 104, 831-837.

Blaser, R.E., Chadwick, L., McGinnis, G.C., 2010. Behavioral measures of anxiety in zebrafish (*Danio rerio*). Behavioural Brain Research 208, 56-62.

Bourin, M., Hascoët, M., 2003. The mouse light/dark box test. European Journal of Pharmacology 463 (1-3), 55-65.

Braff, D.L., Geyer, M.A., Light, G.A., Sprock, J., Perry, W., Cadenhead, K.S., Swerdlow, N.R., 2001. Impact of prepulse characteristics on the detection of sensorimotor gating deficits in schizophrenia. Schizophrenia Research 49,171-178.

Braida, D., Limonta, V., Pegorini, S., Zani, A., Guerini-Rocco, C., Gori, E., Sala, M., 2007. Hallucinatory and rewarding effect of salvinorin A in zebrafish: k-opioid and CB1-cannabinoid receptor involvement. Psychopharmacology190, 441-448.

Briggs, J.A., Weinreb, C., Wagner, D.E., Megason, S., Peshkin, L., Kirschner, M.W., Klein, A.M., 2018. The dynamics of gene expression in vertebrate embryogenesis at single-cell resolution. Science 360.

Brown, K.H., Dobrinski, K.P., Lee, A.S., Gokcumen, O., Mills, R.E., Shi, X., Chong, W.W.S., Chen, J.Y.H., Yoo, P.,David, S., et al., 2012. Extensive genetic diversity and substructuring among zebrafish strains revealed through copy number variant analysis. Proceedings of the National Academy of Sciences of the United States of America109, 529-534.

Brown, R.E., Stevens, D.R., Haas, H.L., 2001. The physiology of brain histamine. Progress in Neurobiology 63 (6),637-672.

Burgess, H.A., Granato, M., 2007. Sensorimotor gating in larval zebrafish. Journal of Neuroscience 27, 4984-4994.

Buske, C., Gerlai, R., 2012. Maturation of shoaling behavior is accompanied by changes in the dopaminergic and serotoninergic systems in zebrafish. Developmental Psychobiology 54, 28-35.

Champagne, D.L., Hoefnagels, C.C.M., de Kloet, R.E., Richardson, M.K., 2010. Translating rodent behavioral repertoire to zebrafish (*Danio rerio*): relevance for stress research. Behavioural Brain Research 214, 332-342.

Chen, F., Chen, S., Liu, S., Zhang, C., Peng, G., 2015. Effects of lorazepam and WAY-200070 in larval zebrafish light/dark choice test. Neuropharmacology 95, 226-233.

Coe, T.S., Hamilton, P.B., Griffiths, A.M., Hodgson, D.J., Wahab, M.A., Tyler, C.R., 2009. Genetic variation in strains of zebrafish (*Danio rerio*) and the implications for ecotoxicology studies. Ecotoxicology 18, 144-150.

Collier, A.D., Khan, K.M., Caramillo, E.M., Mohn, R.S., Echevarria, D.J., 2014. Zebrafish and conditioned place preference: a translational model of drug reward. Progress In Neuro-Psychopharmacology and Biological Psychiatry 55, 16-25.

Colwill, R.M., Creton, R., 2011. Imaging escape and avoidance behavior in zebrafish larvae. Reviews in the Neurosciences 22, 63-73.

Cong, L., Ran, F.A., Cox, D., Lin, S., Barretto, R., Habib, N., Hsu, P.D., Wu, X., Jiang, W., Marraffini, L.A., et al., 2013. Multiplex genome engineering using CRISPR/Cas systems. Science 339, 819-823.

Dahlén, A., Wagle, M., Zarei, M., Guo, S., 2019. Heritable natural variation of light/dark preference in an outbred zebrafish population. Journal of Neurogenetics 1-10.

Darland, T., Dowling, J.E., 2001. Behavioral screening for cocaine sensitivity in mutagenized zebrafish. Proceedings ofthe National Academy of Sciences of the United States of America 98, 11691-11696.

Dlugos, C.A., 2011. Genetics of ethanol-related behaviors. NeuroMethods 52, 143-161.

Dlugos, C.A., Rabin, R.A., 2003. Ethanol effects on three strains of zebrafish: model system for genetic investigations. Pharmacology Biochemistry and Behavior 74, 471-480.

Dreosti, E., Lopes, G., Kampff, A., Wilson, S., 2015. Development of social behavior in young zebrafish. Frontiers in Neural Circuits 9.

Dudbridge, F., 2013. Power and predictive accuracy of polygenic risk scores. PLoS Genetics 9.

Egan, R.J., Bergner, C.L., Hart, P.C., Cachat, J.M., Canavello, P.R., Elegante, M.F., Elkhayat, S.I., Bartels, B.K.,Tien, A.K., Tien, D.H., et al., 2009. Understanding behavioral and physiological phenotypes of stress and anxiety in zebrafish. Behavioural Brain Research 205, 38-44.

Engeszer, R.E., Ryan, M.J., Parichy, D.M., 2004. Learned social preference in zebrafish. Current Biology 14, 881-884.

Engeszer, R.E., Da Barbiano, L.A., Ryan, M.J., Parichy, D.M., 2007a. Timing and plasticity of shoaling behaviour inthe zebrafish, *Danio rerio*. Animal Behaviour 74, 1269-1275.

Engeszer, R.E., Patterson, L.B., Rao, A.A., Parichy, D.M., 2007b. Zebrafish in the wild: a review of natural history andnew notes from the field. Zebrafish 4, 21-40.

Engeszer, R.E., Wang, G., Ryan, M.J., Parichy, D.M., 2008. Sex-specific perceptual spaces for a vertebrate basal social aggregative behavior. Proceedings of the National Academy of Sciences United States of America 105, 929-933.

Facciol, A., Tran, S., Gerlai, R., 2017. Re-examining the factors affecting choice in the light-dark preference test in zebrafish. Behavioural Brain Research 327, 21-28.

Facciol, A., Iqbal, M., Eada, A., Tran, S., Gerlai, R., 2019. The light-dark task in zebrafish confuses two distinct factors:interaction between background shade and illumination level preference. Pharmacology Biochemistry andBehavior 179, 9-21.

Faillace, M.P., Pisera-Fuster, A., Bernabeu, R., 2018. Evaluation of the rewarding properties of nicotine and caffeine by implementation of a five-choice conditioned place preference task in zebrafish. Progress In Neuro-Psychopharmacology and Biological Psychiatry 84, 160-172.

Farrell, J.A., Wang, Y., Riesenfeld, S.J., Shekhar, K., Regev, A., Schier, A.F., 2018. Single-cell reconstruction of developmental trajectories during zebrafish embryogenesis. Science 360.

Fernandes, Y., Gerlai, R., 2009. Long-term behavioral changes in response to early developmental exposure to ethanol in zebrafish. Alcoholism: Clinical and Experimental Research 33, 601-609.

Fernandes, Y., Tran, S., Abraham, E., Gerlai, R., 2014. Embryonic alcohol exposure impairs associative learning performance in adult zebrafish. Behavioural Brain Research 265, 181-187.

Fernandes, Y., Rampersad, M., Gerlai, R., 2015. Embryonic alcohol exposure impairs the dopaminergic system and social behavioral responses in adult zebrafish. International Journal of Neuropsychopharmacology 18, 1-8.

Fernández-Novoa, L., Cacabelos, R., 2001. Histamine function in brain disorders. Behavioural Brain Research 124,213-233.

Fernando, R.L., Garrick, D., 2013. Bayesian methods applied to GWAS. Methods in Molecular Biology 1019, 237-274.

Flint, J., Eskin, E., 2012. Genome-wide association studies in mice. Nature Reviews Genetics 13, 807-817.

Fuchsberger, C., Flannick, J., Teslovich, T.M., Mahajan, A., Agarwala, V., Gaulton, K.J., Ma, C., Fontanillas, P., Moutsianas, L., McCarthy, D.J., et al., 2016. The genetic architecture of type 2 diabetes. Nature 536, 41-47.

Gebauer, D.L., Pagnussat, N., Piato, Â.L., Schaefer, I.C., Bonan, C.D., Lara, D.R., 2011. Effects of anxiolytics in zebrafish: similarities and differences between benzodiazepines, buspirone and ethanol. Pharmacology Biochemistry and Behavior 99, 480-486.

Gerlai, R., 2010. Zebrafish antipredatory responses: a future for translational research? Behavioural Brain Research 207, 223-231.

Gerlai, R., 2014. Fish in behavior research: unique tools with a great promise! Journal of Neuroscience Methods 234,54-58.

Gerlai, R., Lahav, M., Guo, S., Rosenthal, A., 2000. Drinks like a fish: zebra fish (*Danio rerio*) as a behavior genetic model to study alcohol effects. Pharmacology Biochemistry and Behavior 67, 773-782.

Gerlai, R., Ahmad, F., Prajapati, S., 2008. Differences in acute alcohol-induced behavioral responses among zebrafish populations. Alcoholism: Clinical and Experimental Research 32, 1763-1773.

Giacomini, A.C.V.V., Abreu, M.S., Giacomini, L.V., Siebel, A.M., Zimerman, F.F., Rambo, C.L., Mocelin, R.,Bonan, C.D., Piato, A.L., Barcellos, L.J.G., 2016. Fluoxetine and diazepam acutely modulate stress induced behavior. Behavioural Brain Research 296, 301-310.

Golan, D., Lander, E.S., Rosset, S., 2014. Measuring missing heritability: inferring the contribution of common variants. Proceedings of the National Academy of Sciences of the United States of America 111, E5272-E5281.

Gong, Z., Liu, J., Guo, C., Zhou, Y., Teng, Y., Liu, L., 2010. Two pairs of neurons in the central brain control *Drosophila* innate light preference. Science 330, 499-502.

Gould, G.G., 2011. Modified associative learning T-maze test for zebrafish (*Danio rerio*) and other small teleost fish. In:Kalueff, A.V., Cachat, J.M. (Eds.), Neuromethods. Humana Press, Totowa, NJ, pp. 61-73.

Green, J., Collins, C., Kyzar, E.J., Pham, M., Roth, A., Gaikwad, S., Cachat, J., Stewart, A.M., Landsman, S., Grieco, F.,2012. Automated high-throughput neurophenotyping of zebrafish social behavior. Journal of Neuroscience Methods 210, 266-271.

Grillon, C., Ameli, R., Charney, D.S., Krystal, J., Braff, D., 1992. Startle gating deficits occur across prepulse intensitiesin schizophrenic patients. Biological Psychiatry 32, 939-943.

Grossman, L., Utterback, E., Stewart, A., Gaikwad, S., Chung, K.M., Suciu, C., Wong, K., Elegante, M., Elkhayat, S.,Tan, J., et al., 2010. Characterization of behavioral and endocrine effects of LSD on zebrafish. Behavioural BrainResearch 214, 277-284.

Guo, S., 2004. Linking genes to brain, behavior and neurological diseases: what can we learn from zebrafish? Genes, Brain and Behavior 3, 63-74.

Haas, H., Panula, P., 2003. The role of histamine and the tuberomamillary nucleus in the nervous system. Nature Reviews Neuroscience 4, 121-130.

Hall, C.S., 1951. The genetics of behavior. In: Stevens, S.S. (Ed.), Handbook of Experimental Psychology. Wiley, NewYork.

Hascoët, M., Bourin, M., 2009. The mouse light-dark box test. Neuro Methods.

Hashimshony, T., Senderovich, N., Avital, G., Klochendler, A., de Leeuw, Y., Anavy, L., Gennert, D., Li, S.,Livak, K.J., Rozenblatt-Rosen, O., et al., 2016. CEL-Seq2: sensitive highly-multiplexed single-cell RNA-Seq.Genome Biology 17, 77.

Higuchi, R., Fockler, C., Dollinger, G., Watson, R., 1993. Kinetic PCR analysis: real-time monitoring of DNA amplification reactions. Bio-Technology 11, 1026-1030.

Howe, K., Clark, M.D., Torroja, C.F., Torrance, J., Berthelot, C., Muffato, M., Collins, J.E., Humphray, S., McLaren, K., Matthews, L., et al., 2013. The zebrafish reference genome sequence and its relationship to the human genome.Nature 496 (7446), 498-503.

Islam, S., Kjallquist, U., Moliner, A., Zajac, P., Fan, J.-B., Lonnerberg, P., Linnarsson, S., 2011. Characterization of thesingle-cell transcriptional landscape by highly multiplex RNA-seq. Genome Research 21, 1160-1167.

Jinek, M., Chylinski, K., Fonfara, I., Hauer, M., Doudna, J.A., Charpentier, E., 2012. A Programmable dual-RNA-guided DNA endonuclease in adaptive bacterial immunity. Science 337, 816-821.

Jones, L.J., Norton, W.H.J., 2015. Using zebrafish to uncover the genetic and neural basis of aggression, a frequent comorbid symptom of psychiatric disorders. Behavioural Brain Research 276, 171-180.

Kaczmarek, L., 1993. Molecular biology of vertebrate learning: is c-fos a new beginning? Journal of Neuroscience Research 34, 377-381.

Kalueff, A.V., Stewart, A.M., Gerlai, R., 2014. Zebrafish as an emerging model for studying complex brain disorders.Trends in Pharmacological Sciences 35, 63-75.

Kedikian, X., Faillace, M.P., Bernabeu, R., 2013. Behavioral and molecular analysis of nicotine-conditioned place preference in zebrafish. PLoS One 8, e69453.

Khera, A.V., Chaffin, M., Aragam, K.G., Haas, M.E., Roselli, C., Choi, S.H., Natarajan, P., Lander, E.S., Lubitz, S.A.,Ellinor, P.T., et al., 2018. Genome-wide polygenic scores for common diseases identify individuals with risk equivalent to monogenic mutations. Nature Genetics 50, 1219-1224.

Kily, L.J.M., Cowe, Y.C.M., Hussain, O., Patel, S., McElwaine, S., Cotter, F.E., Brennan, C.H., 2008. Gene expression changes in a zebrafish model of drug dependency suggest conservation of neuro-adaptation pathways. Journal ofExperimental Biology 211, 1623-1634.

Kimmel, C.B., Patterson, J., Kimmel, R.O., 1974. The development and behavioral characteristics of the startle response in the zebra fish. Developmental Psychobiology 7, 47-60.

Klein, A.M., Mazutis, L., Akartuna, I., Tallapragada, N., Veres, A., Li, V., Peshkin, L., Weitz, D.A., Kirschner, M.W.,2015. Droplet barcoding for single-cell transcriptomics applied to embryonic stem cells. Cell 161, 1187-1201.

Kohl, S., Heekeren, K., Klosterkötter, J., Kuhn, J., 2013. Prepulse inhibition in psychiatric disorders-apart from schizophrenia. Journal of Psychiatric Research 47, 445-452.

Komada, M., Takao, K., Miyakawa, T., 2008. Elevated plus maze for mice. Journal of Visualized Experiments e1088.

Kovács, K.J., 2008. Measurement of immediate-early gene activation-c-fos and beyond. Journal of Neuroendocrinology 20, 665-672.

Krause, J., Butlin, R.K., Peuhkuri, N., Pritchard, V.L., 2000. The social organization of fish shoals: a test of the predictive power of laboratory experiments for the field. Biological Reviews 75, 477-501.

Ladu, F., Butail, S., Macrí, S., Porfiri, M., 2014. Sociality modulates the effects of ethanol in zebra fish. Alcoholism:Clinical and Experimental Research 38, 2096-2104.

LaFave, M.C., Varshney, G.K., Vemulapalli, M., Mullikin, J.C., Burgess, S.M., 2014. A defined zebrafish line for high-throughput genetics and genomics: NHGRI-1. Genetics 198, 167-170.

Lau, B., Bretaud, S., Huang, Y., Lin, E., Guo, S., 2006. Dissociation of food and opiate preference by a genetic mutationin zebrafish. Genes, Brain and Behavior 5, 497-505.

Lau, B.Y.B., Mathur, P., Gould, G.G., Guo, S., 2011. Identification of a brain center whose activity discriminates achoice behavior in zebrafish. Proceedings of the National Academy of Sciences United States of America 108,2581-2586.

Lawson, H.A., Cheverud, J.M., 2010. Metabolic syndrome components in murine models. Endocrine, Metabolic andImmune Disorders e Drug Targets 10, 25-40.

Leslie, R., O'Donnell, C.J., Johnson, A.D., 2014. GRASP: analysis of genotype-phenotype results from 1390 genome-wide association studies and corresponding open access database. Bioinformatics 30, i185-i194.

Levin, E.D., Limpuangthip, J., Rachakonda, T., Peterson, M., 2006. Timing of nicotine effects on learning in zebrafish. Psychopharmacology 184, 547-552.

Levin, E.D., Bencan, Z., Cerutti, D.T., 2007. Anxiolytic effects of nicotine in zebrafish. Physiology and Behavior 90,54-58.

Liu, K.S., Fetcho, J.R., 1999. Laser ablations reveal functional relationships of segmental hindbrain neurons in zebrafish. Neuron 23, 325-335.

Long, Y., Song, G., Yan, J., He, X., Li, Q., Cui, Z., 2013. Transcriptomic characterization of cold acclimation in larval zebrafish. BMC Genomics 14, 612.

López-Patiño, M.A., Yu, L., Cabral, H., Zhdanova, I.V., 2008. Anxiogenic effects of cocaine with drawal in zebrafish. Physiology and Behavior 93, 160-171.

Maaswinkel, H., Zhu, L., Weng, W., 2013. Assessing social engagement in heterogeneous groups of zebrafish: a new paradigm for autism-like behavioral responses. PLoS One 8, e75955.

Madelaine, R., Notwell, J.H., Skariah, G., Halluin, C., Chen, C.C., Bejerano, G., Mourrain, P., 2018. A screen for deeply conserved non-coding GWAS SNPs uncovers a MIR-9-2 functional mutation associated to retinal vasculature defects in human. Nucleic Acids Research 46, 3517-3531.

Magno, L.D.P., Fontes, A., Gonçalves, B.M.N., Gouveia Jr., A., 2015. Pharmacological study of the light/dark preference test in zebrafish (*Danio rerio*): waterborne administration. Pharmacology Biochemistry and Behavior 135,169-176.

Mali, P., Yang, L., Esvelt, K.M., Aach, J., Guell, M., DiCarlo, J.E., Norville, J.E., Church, G.M., 2013. RNA-guided human genome engineering via Cas9. Science 339, 823-826.

Manolio, T.A., Collins, F.S., Cox, N.J., Goldstein, D.B., Hindorff, L.A., Hunter, D.J., McCarthy, M.I., Ramos, E.M.,Cardon, L.R., Chakravarti, A., et al., 2009. Finding the missing heritability of complex diseases. Nature 461,747-753.

Marees, A.T., de Kluiver, H., Stringer, S., Vorspan, F., Curis, E., Marie-Claire, C., Derks, E.M., 2018. A tutorial on conducting genome-wide association studies: quality control and statistical analysis. International Journal of Methodsin Psychiatric Research 27, e1608.

Mathur, P., Guo, S., 2011. Differences of acute versus chronic ethanol exposure on anxiety-like behavioral responses in zebrafish. Behavioural Brain Research 219, 234-239.

Mathur, P., Lau, B., Guo, S., 2011a. Conditioned place preference behavior in zebrafish. Nature Protocols 6, 338-345.

Mathur, P., Berberoglu, M.A., Guo, S., 2011b. Preference for ethanol in zebrafish following a single exposure. Behavioural Brain Research 217, 128-133.

Maximino, C., Marques de Brito, T., Dias, C.A., G. de M., Gouveia, A., Morato, S., 2010. Scototaxis as anxiety-like behavior in fish. Nature Protocols 5, 209-216.

Maximino, C., da Silva, A.W.B., Gouveia, A., Herculano, A.M., 2011. Pharmacological analysis of zebrafish (*Danio rerio*) scototaxis. Progress In Neuro-Psychopharmacology & Biological Psychiatry 35, 624-631.

McCann, L.I., Koehn, D.J., Kline, N.J., 1971. The effects of body size and body markings on nonpolarized schooling behavior of zebra fish (*Brachydanio rerio*). Journal of Psychology 79, 71-75.

McRobert, S.P., Bradner, J., 1998. The influence of body coloration on shoaling preferences in fish. Animal Behaviour 56, 611-615.

Melgoza, A., Guo, S., 2018. Systematic screens in zebrafish shed light on cellular and molecular mechanisms of complex brain phenotypes. In: Molecular-Genetic and Statistical Techniques for Behavioral and Neural Research. Elsevier, pp. 385-400.

Meuwissen, T.H., Hayes, B.J., Goddard, M.E., 2001. Prediction of total genetic value using genome-wide dense marker maps. Genetics 157, 1819-1829.

Miklósi, Á., Andrew, R.J., 2006. The zebrafish as a model for behavioral studies. Zebrafish 3, 227-234.

Miller, N., Gerlai, R., 2007. Quantification of shoaling behaviour in zebrafish (*Danio rerio*). Behavioural Brain Research184, 157-166.

Miller, N., Gerlai, R., 2012. From schooling to shoaling: patterns of collective motion in zebrafish (*Danio rerio*). PLoSOne 7, e48865.

Moens, P.B., 2006. Zebrafish: chiasmata and interference. Genome 49, 205-208.

Nestler, E.J., Hyman, S.E., 2010. Animal models of neuropsychiatric disorders. Nature Neuroscience 13, 1161-1169.

Ninkovic, J., Bally-Cuif, L., 2006. The zebrafish as a model system for assessing the reinforcing properties of drugs ofabuse. Methods 39, 262-274.

Oliveira, R., 2013. Mind the fish: zebrafish as a model in cognitive social neuroscience. Frontiers in Neural Circuits 7, 131.

Oliveira, R.F., Simões, J.M., Teles, M.C., Oliveira, C.R., Becker, J.D., Lopes, J.S., 2016. Assessment of fight outcome is needed to activate socially driven transcriptional changes in the zebrafish brain. Proceedings of the National Academy of Sciences United States of America 113, E654-E661.

Orger, M.B., de Polavieja, G.G., 2017. Zebrafish behavior: opportunities and challenges. Annual Review of Neuroscience 40, 125-147.

Parichy, D.M., 2015. Advancing biology through a deeper understanding of zebrafish ecology and evolution. eLife 4,e05635.

Parker, C.C., Palmer, A.A., 2011. Dark matter: are mice the solution to missing heritability? Frontiers in Genetics 2, 32.

Parker, C.C., Gopalakrishnan, S., Carbonetto, P., Gonzales, N.M., Leung, E., Park, Y.J., Aryee, E., Davis, J.,Blizard, D.A., Ackert-Bicknell, C.L., et al., 2016. Genome-wide association study of behavioral, physiological and gene expression traits in outbred CFW mice. Nature Genetics 48, 919-926.

Parmar, A., Parmar, M., Brennan, C.H., 2011. Zebrafish conditioned place preference models of drug reinforcement and relapse to drug seeking. In: Kalueff, A.V., Cachat, J.M. (Eds.), Zebrafish Neurobehavior Protocols. Humana Press, Totowa, NJ, pp. 75-84.

Patowary, A., Purkanti, R., Singh, M., Chauhan, R., Singh, A.R., Swarnkar, M., Singh, N., Pandey, V., Torroja, C.,Clark, M.D., et al., 2013. A sequence-based variation map of zebrafish. Zebrafish 10, 15-20.

Patterson, L.B., Bain, E.J., Parichy, D.M., 2014. Pigment cell interactions and differential xanthophore recruitment underlying zebrafish stripe reiteration and Danio pattern evolution. Nature Communications 5, 5299.

Peitsaro, N., Kaslin, J., Anichtchik, O.V., Panula, P., 2004. Modulation of the histaminergic system and behaviour by α-fluoromethyl histidine in zebrafish. Journal of Neurochemistry 86, 432-441.

Power, R.A., Steinberg, S., Bjornsdottir, G., Rietveld, C.A., Abdellaoui, A., Nivard, M.M., Johannesson, M.,Galesloot, T.E., Hottenga, J.J., Willemsen, G., et al., 2015. Polygenic risk scores for schizophrenia and bipolar disorder predict creativity. Nature Neuroscience 18, 953-955.

Prut, L., Belzung, C., 2003. The open field as a paradigm to measure the effects of drugs on anxiety-like behaviors: a review. European Journal of Pharmacology 463, 3-33.

Purcell, S.M., Wray, N.R., Stone, J.L., Visscher, P.M., O'Donovan, M.C., Sullivan, P.F., Ruderfer, D.M., McQuillin, A.,Morris, D.W., Oĝdushlaine, C.T., et al., 2009. Common polygenic variation contributes to risk of schizophrenia and bipolar disorder. Nature 460, 748-752.

Quadros, V.A., Silveira, A., Giuliani, G.S., Didonet, F., Silveira, A.S., Nunes, M.E., Silva, T.O., Loro, V.L.,Rosemberg, D.B., 2016. Strain- and context-dependent behavioural responses of acute alarm substance exposurein zebrafish. Behavioural Processes 122, 1-11.

Ramsköld, D., Luo, S., Wang, Y.-C., Li, R., Deng, Q., Faridani, O.R., Daniels, G.A., Khrebtukova, I., Loring, J.F.,Laurent, L.C., et al., 2012. Full-length mRNA-Seq from single-cell levels of RNA and individual circulating tumor cells. Nature

Biotechnology 30, 777-782.

Roberts, A., Pardo-Manuel de Villena, F., Wang, W., McMillan, L., Threadgill, D.W., 2007. The polymorphism architecture of mouse genetic resources elucidated using genome-wide resequencing data: implications for QTL discovery and systems genetics. Mammalian Genome 18, 473-481.

Sackerman, J., Donegan, J.J., Cunningham, C.S., Nguyen, N.N., Lawless, K., Long, A., Benno, R.H., Gould, G.G., 2010. Zebrafish behavior in novel environments: effects of acute exposure to anxiolytic compounds and choice of *Danio rerio* line. International Journal of Comparative Psychology 23, 43-61.

Saper, C.B., Chou, T.C., Elmquist, J.K., 2002. The need to feed. Neuron 36, 199-211.

Schall, U., Keysers, C., Kast, B., 1999. Pharmacology of sensory gating in the ascending auditory system of the pigeon(*Columba livia*). Psychopharmacology 145, 273-282.

Schena, M., Shalon, D., Davis, R.W., Brown, P.O., 1995. Quantitative monitoring of gene expression patterns with a complementary DNA microarray. Science 270, 467-470.

Schnörr, S.J., Steenbergen, P.J., Richardson, M.K., Champagne, D.L., 2012. Measuring thigmotaxis in larval zebrafish. Behavioural Brain Research 228, 367-374.

Schwartz, J.C., Arrang, J.M., Garbarg, M., Pollard, H., Ruat, M., 1991. Histaminergic transmission in the mammalian brain. Physiological Reviews 71, 1-51.

Seguin, D., Shams, S., Gerlai, R., 2016. Behavioral responses to novelty or to a predator stimulus are not altered in adult zebrafish by early embryonic alcohol exposure. Alcoholism: Clinical and Experimental Research 40, 2667-2675.

Serra, E.L., Medalha, C.C., Mattioli, R., 1999. Natural preference of zebrafish (*Danio rerio*) for a dark environment. Brazilian Journal of Medical and Biological Research 32, 1551-1553.

Shah, A.N., Davey, C.F., Whitebirch, A.C., Miller, A.C., Moens, C.B., 2015. Rapid reverse genetic screening using CRISPR in zebrafish. Nature Methods 12, 535-540.

Shams, S., Rihel, J., Ortiz, J.G., Gerlai, R., 2018. The zebrafish as a promising tool for modeling human brain disorders:a review based upon an IBNS Symposium. Neuroscience and Biobehavioral Reviews 85, 176-190.

Sison, M., Cawker, J., Buske, C., Gerlai, R., 2006. Fishing for genes influencing vertebrate behavior: zebrafish making headway. Laboratory Animals 35, 33-39.

Sokolowski, M.B., 2001. Drosophila: genetics meets behaviour. Nature Reviews Genetics 2, 879-890.

Spence, R., Gerlach, G., Lawrence, C., Smith, C., 2008. The behaviour and ecology of the zebrafish, *Danio rerio*. Biological Reviews 83, 13-34.

Steenbergen, P.J., Richardson, M.K., Champagne, D.L., 2011. Patterns of avoidance behaviours in the light/dark preference test in young juvenile zebrafish: a pharmacological study. Behavioural Brain Research 222, 15-25.

Stringer, S., Kahn, R.S., de Witte, L.D., Ophoff, R.A., Derks, E.M., 2014. Genetic liability for schizophrenia predict srisk of immune disorders. Schizophrenia Research 159, 347-352.

Swerdlow, N., Geyer, M., Braff, D., 2001. Neural circuit regulation of prepulse inhibition of startle in the rat: current knowledge and future challenges. Psychopharmacology 156, 194-215.

Swerdlow, N.R., Light, G.A., Cadenhead, K.S., Sprock, J., Hsieh, M.H., Braff, D.L., 2006. Startle gating deficits in a large cohort of patients with schizophrenia: relationship to medications, symptoms, neurocognition, and level of function. Archives of General Psychiatry 63, 1325-1335.

Tang, F., Barbacioru, C., Wang, Y., Nordman, E., Lee, C., Xu, N., Wang, X., Bodeau, J., Tuch, B.B., Siddiqui, A., et al.,2009. mRNA-Seq whole-transcriptome analysis of a single cell. Nature Methods 6, 377-382.

Thyme, S.B., Pieper, L.M., Li, E.H., Pandey, S., Wang, Y., Morris, N.S., Sha, C., Choi, J.W., Herrera, K.J., Soucy, E.R.,et al., 2019. Phenotypic landscape of schizophrenia-associated genes defines candidates and their shared functions. Cell 177, 478-491.e20.

Tran, S., Gerlai, R., 2016. The novel tank test: handling stress and the context specific psychopharmacology of anxiety. Current Psychopharmacology 5, 169-179.

Tran, S., Nowicki, M., Fulcher, N., Chatterjee, D., Gerlai, R., 2016. Interaction between handling induced stress and anxiolytic effects of ethanol in zebrafish: a behavioral and neurochemical analysis. Behavioural Brain Research 298, 278-285.

Treit, D., Fundytus, M., 1988. Thigmotaxis as a test for anxiolytic activity in rats. Pharmacology Biochemistry and Behavior 31, 959-962.

Trevarrow, B., Robison, B., 2004. Genetic backgrounds, standard lines, and husbandry of zebrafish. In: Methods in Cell Biology. Academic Press, pp. 599-616.

Tzschentke, T.M., 2007. Measuring reward with the conditioned place preference (CPP) paradigm: update of the last decade. Addiction Biology 12, 227-462.

Ulloa, P.E., Rincón, G., Islas-Trejo, A., Araneda, C., Iturra, P., Neira, R., Medrano, J.F., 2015. RNA sequencing to study gene expression and SNP variations associated with growth in zebrafish fed a plant protein-based diet. Marine Biotechnology 17, 353-363.

Visscher, P.M., Brown, M.A., McCarthy, M.I., Yang, J., 2012. Five years of GWAS discovery. The American Journal of Human Genetics 90 (1), 7-24.

Wagle, M., Nguyen, J., Lee, S., Zaitlen, N., Guo, S., 2017. Heritable natural variation of an anxiety-like behavior in larval zebrafish. Journal of Neurogenetics 31, 138-148.

Wagner, D.E., Weinreb, C., Collins, Z.M., Briggs, J.A., Megason, S.G., Klein, A.M., 2018. Single-cell mapping of gene expression landscapes and lineage in the zebrafish embryo. Science 360, 981-987.

Wang, Z., Gerstein, M., Snyder, M., 2009. RNA-Seq: a revolutionary tool for transcriptomics. Nature Reviews Genetics 10, 57-63.

Waterston, R.H., Lindblad-Toh, K., Birney, E., Rogers, J., Abril, J.F., Agarwal, P., Agarwala, R., Ainscough, R.,Alexandersson, M., An, P., et al., 2002. Initial sequencing and comparative analysis of the mouse genome. Nature420, 520-562.

Whiteley, A.R., Bhat, A., Martins, E.P., Mayden, R.L., Arunachalam, M., Uusi-HeikkilÎ, S., Ahmed, A.T.A.,Shrestha, J., Clark, M., Stemple, D., et al., 2011. Population genomics of wild and laboratory zebrafish (*Danio rerio*). Molecular Ecology 20, 4259-4276.

Wong, K., Elegante, M., Bartels, B., Elkhayat, S., Tien, D., Roy, S., Goodspeed, J., Suciu, C., Tan, J., Grimes, C., 2010. Analyzing habituation responses to novelty in zebrafish (*Danio rerio*). Behavioural Brain Research 208, 450-457.

Wright, D., Butlin, R.K., Carlborg, Ö., 2006a. Epistatic regulation of behavioural and morphological traits in the zebrafish (*Danio rerio*). Behavior Genetics 36, 914-922.

Wright, D., Nakamichi, R., Krause, J., Butlin, R.K., 2006b. QTL analysis of behavioral and morphological differentiation between wild and laboratory zebrafish (*Danio rerio*). Behavior Genetics 36, 271.

Yalcin, B., Nicod, J., Bhomra, A., Davidson, S., Cleak, J., Farinelli, L., Østerås, M., Whitley, A., Yuan, W., Gan, X.,et al., 2010. Commercially available outbred mice for genome-wide association studies. PLoS Genetics 6 (9),e1001085.

Yang, J., Manolio, T.A., Pasquale, L.R., Boerwinkle, E., Caporaso, N., Cunningham, J.M., de Andrade, M., Feenstra, B.,Feingold, E., Hayes, M.G., et al., 2011. Genome partitioning of genetic variation for complex traits using common SNPs. Nature Genetics 43, 519-525.

Yoshida, G.M., Lhorente, J.P., Carvalheiro, R., Yáñez, J.M., 2017. Bayesian genome-wide association analysis for body weight in farmed Atlantic salmon (*Salmo salar* L.). Animal Genetics 48, 698-703.

Zhao, S., Fung-Leung, W.-P., Bittner, A., Ngo, K., Liu, X., 2014. Comparison of RNA-Seq and microarray in transcriptome profiling of activated T cells. PLoS One 9, e78644.

Zilionis, R., Nainys, J., Veres, A., Savova, V., Zemmour, D., Klein, A.M., Mazutis, L., 2016. Single-cell barcoding and sequencing using droplet microfluidics. Nature Protocols 12, 44.

第十五章

斑马鱼的行为研究：压力致变异

Ruud van den Bos[1], Gert Flik[1], Marnix Gorissen[1]

15.1 引言

斑马鱼（*Danio rerio*）已成为神经和精神病学等生物医学研究中越来越受欢迎的模型（Norton, 2013; Stewart et al., 2014）。在这些研究领域，无论是在成鱼还是仔鱼阶段，研究者们通常利用斑马鱼行为学来筛选化合物或研究神经或精神疾病的潜在机制。一般来说，行为学分析对遗传背景、饲养条件和测试环境等因素很敏感（Chesler et al., 2002; de Visser, 2008; Lange et al., 2013; Wahlsten et al., 2006; 最近的相关综述和讨论，请参阅 Kafkafi et al., 2018）。

一些研究已经表明，在监测化合物镇痛活性的试验中，引起小鼠缩尾反射结果变异的原因可以分为基因型（小鼠品系）、实验室环境（包括人类处理）以及它们之间的相互作用（Chesler et al., 2002）。此外，即便通过使用遗传背景明确的近交系小鼠和标准化测试环境来控制基因型，实验室之间的焦虑测试结果也有所不同，这可能是由实验室的舍饲差异（de Visser, 2008）引起的。最终，有人提出一些比行为测试对环境条件更敏感的其他测试，例如，焦虑相关测试似乎比小鼠运动或感官相关测试更易受到试验中各类因素的影响（de Visser, 2008; Wahlsten et al., 2006）。

因此，如果我们对试验中影响因素控制不当或缺乏相关知识，那么将会降低实验室内和各个实验室之间试验结果的可重复性，这反过来会降低科学的总体可信度（Kafkafi et al., 2018; Stevens, 2017）。目前大家已经观察到斑马鱼品系之间的显著行为学差异（见下文），所以我们将品系（基因型）作为斑马鱼行为学研究的一个重要因素。

[1] Department of Animal Ecology and Physiology, Institute of Water and Wetland Research, Faculty of Science, Radboud University, Nijmegen, the Netherlands.

接下来将讨论与行为学相关的问题：实验结果变异来源（第 2 节）、品系和遗传变异简史（第 3 节），以及仔鱼和成鱼行为学中品系差异的案例，重点关注 AB 和 TL 品系之间的差异（第 4 节），最后是结论（第 5 节）。

15.2 实验数据的变异来源

图 15-1 指出了三种不同的变异来源，它们最终会影响科学研究结果的可重复性。这些来源从后到前依次为：①测试环境，②饲养条件或生活环境，以及③基因型。

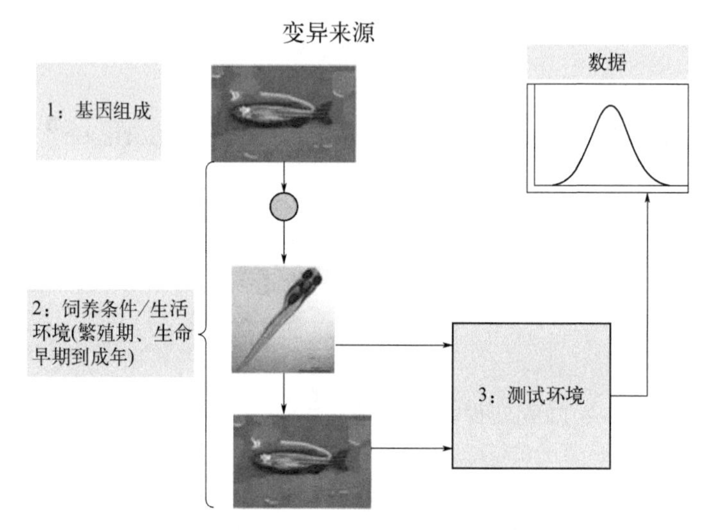

图 15-1 **行为学实验中的变异来源**

例如，就测试环境而言，目前有关明暗测试研究结果表明斑马鱼仔鱼（Steenbergen et al., 2011）和成鱼（Stephenson et al., 2011）对明暗对比很敏感。因此，测试环境标准化程度越高，数据就越可靠。同样，这种自动化分析的结果应始终与现实生活中行为相一致，行为的自动化分析越多，人类观察者的即时主观评估越少，数据就越可靠。用于仔鱼研究的 Daniovision（Noldus Information Technology, Wageningen, the Netherlands）和 ZebraBox（Viewpoint Behavior Technology, Lyon, France）两种设备提供了可控的封闭测试微环境，以及标准化行为测试和自动化行为分析的方法，这是斑马鱼研究向前迈出的重要一步。目前已经开发出对斑马鱼成鱼测试的标准化设备，这种设备拥有更受控制的封闭测试微环境，类似于目前已经开发的啮齿动物装置［几家公司可提供监测啮齿动物活动行为的笼养（home-cage）分析系统；de Visser, 2008; Koot et al., 2009］，这是一个显著的进步。

关于生活环境，许多不同因素可导致上述结果变化，或者明显地发现它们可能会产生影响。通常，这些因素与下丘脑 - 垂体 - 肾上腺轴（HPI 轴）的活性相关，而 HPI 轴是研究对象与环境之间的关键调节因子。例如，卵母细胞中母体皮质醇的含量会调节胚胎发育中皮质醇的基线水平，并可能产生终生影响（Hartig et al., 2016; Nesan and Vijayan, 2016; van den Bos et al., 2019）。例如，早期生活的明暗条件（Ahmad, 2014; van den Bos et al., 2017b;

Villamizar et al., 2014）、养殖系统的背景颜色（Pavlidis et al., 2013）和水箱密度（Pavlidis et al., 2013; Ramsay et al., 2006; Ribas et al., 2017），这些对 HPI 轴功能都有影响，从而对健康行为可能产生直接或间接影响（Ramsay et al., 2009）。

本章主题是斑马鱼品系，目前已使用许多品系，例如 AB、Tübingen（TU）、Tüpfel Long Fin（TL）和 Wild India Karyotype（或 Kolkata; WIK）。此外，当地宠物店斑马鱼的变异仍在研究中。一些研究已经表明，这些品系和变体在许多方面都不同：基因型（Butler et al., 2015; Coe et al., 2009; Guryev et al., 2006; Holden and Brown, 2018）和表型（行为和生理学 Gorissen et al., 2015; Lange et al., 2013; van den Bos et al., 2017a, 2017b, 2018, 2019; Vignet et al., 2013; 见第 4 部分）两者不同。

15.3 品系和遗传变异

15.3.1 概述

作为实验室物种的斑马鱼起源于尼泊尔、印度和孟加拉国（图 15-2）。正如线粒体 DNA 分析所揭示的一样，斑马鱼种群远非同质的，在遗传上存在差异性（Whiteley et al., 2011）。此外，第一代后代的新异物体识别实验和集群测试结果显示，已经发现种群新异物体识别实验中存在显著差异，但在集群趋势上没有差异（Wright et al., 2003）。因此，当使用不同的种群作为实验室品系的研究对象时，这种显著差异可能会对实验室种群产生影响（图 15-2）。

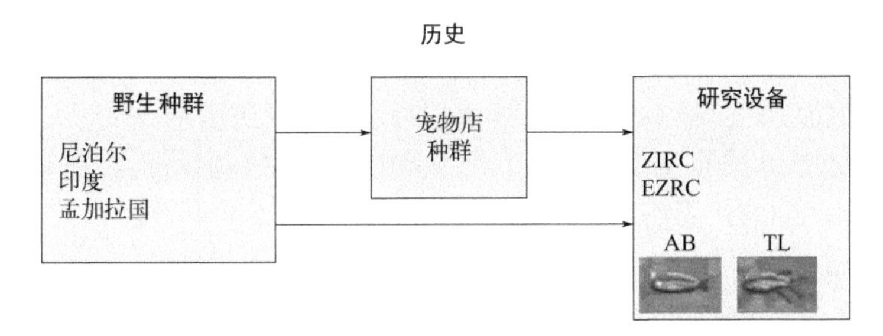

图 15-2 **斑马鱼品系来源流程图**

ZIRC 和 EZRC 保留了许多不同野生型品系和突变体 / 转基因品系的种群。此外，新的品系可以提交到斑马鱼基因数据库（https://zfin.org/）。ZIRC（https://zebrafish.org/home/guide.php）；EZRC（http://www.ezrc.kit.edu/）

如上所述，实验室保存有许多斑马鱼品系，例如 AB、TU、TL 和 WIK，这些品系可以从 ZIRC 或 EZRC 等公司购买，因为这些公司保存有这些种群（见图 15-2）。这些斑马鱼品系是从宠物店或野外获得的。例如，ZIRC 网站揭示了这些品系的“自然历史”，即目前的种群是如何产生的。从这些“自然历史”中可以清楚地看出，这些品系背景存在相当大的差异：虽然 TU、AB 和 WIK 都被认为是常见的野生型实验室品系，但它们仍有明显不同的时间线和“地理”起源。TU 起源于 1997 年 Tübingen（德国）一家宠物店的几个种群。早在 1970 年，AB 源于奥尔巴尼（美国俄勒冈州）两家宠物店种群的杂交品种。

线粒体DNA分析表明，AB品系源于印度加尔各答附近的恒河/雅鲁藏布江地区（Whiteley et al., 2011）。1997年，WIK起源于印度加尔各答的野生捕捞鱼的单对交配品种（Holden and Brown, 2018; Rauch et al., 1997）。此外，AB经历了两轮雌核生殖，其中六对是随后AB品系斑马鱼的祖先。TU品系斑马鱼具有胚胎致死突变基因，用于基因组测序前敲除该致死突变基因。最后，TL品系包含两个突变，使其在外观上有所不同：lof^{dt1}（导致成年鱼出现斑点的隐性突变）和lof^{dt2}（导致长鳍的显性突变）是纯合子。

15.3.2 野生种群和实验室品系之间的差异

实验室品系的遗传变异性总体而言低于野生种群的遗传变异性（Coe et al., 2008; Whiteley et al., 2011）。然而，野生种群拷贝数变异（copy number variants, CNV）数量和平均CNV大小低于实验室品系（Brown et al., 2012），CNV通常是有害的，在实验室中可能并不存在野生种群中所存在的选择压力，因此可以解释这一点（Brown et al., 2012）。

实验室品系与野生种群相比可能被认为是更有冒险性的（bolder）（Drew et al., 2012; Oswald and Robison, 2008; Wright et al., 2006）。脑转录组分析表明，野生种群和实验室品系之间的突触可塑性和细胞信号通路存在差异（Drew et al., 2012）。野生种群和AB品系进行的数量性状基因座（Quantitative Trait Locus, QTL）定位分析表明，大胆/逃避捕食者行为的QTL位于9号染色体上，16号和21号染色体上的QTL较弱。

在性别决定/性别分化方面，野生种群以及实验室品系WIK在4号染色体上有一个基因座，该基因决定了性别（纯合子：雄性；杂合子：主要是雌性），而在实验室品系AB和TU中该基因座是缺失的（Holtzman et al., 2016; Wilson et al., 2014）。在斑马鱼品系AB和TU中，性别决定/性别分化似乎是通过环境中另一组基因和因素的相互作用而发生的（Holtzman et al., 2016; Liew and Orban, 2013）。目前已有研究证明高密度、高温、低食物供应和缺氧等因素会导致性别比例偏向雄性（Holtzman et al., 2016）。因此，在性别决定/性别分化方面，AB和TU可能比野生鱼对这些因素更敏感。

最后，实验室品系具有生长速度快、体重和脂肪含量高的特点，在其23号染色体上发现了控制该性状的QTL位点（Oswald and Robison, 2008; Wright et al., 2006）。

15.3.3 实验室品系的差异

在实验室品系中，可能会发现遗传变异：WIK与AB、TU或TL相比，具有更多的遗传变异（Butler et al., 2015; Coe et al., 2009; Rauch et al., 1997）。然而，TU中CNV的数量和长度高于WIK或AB（Brown et al., 2012）。这可能是由于TU相对于其他两个品系具有较大的起源规模和复合种群来源（见上文；Brown et al., 2012）。肝脏mRNA表达谱显示，TU、AB和WIK之间也存在差异（Holden and Brown, 2018）。例如，虽然AB和WIK中存在广泛变化的性别特异性表达基因，但不会发生在TU品系中（Holden and Brown, 2018）。另外，也存在一些基因在一种品系或其他品系中的表达更强或更弱（Holden and Brown, 2018; Whiteley, 2011）。此外，同一品系的实验室种群之间也存在大量遗传变异（Brown et al., 2012; Butler et al., 2015; Coe et al., 2009）。

与小鼠（https://phenome.jax.org/）和青鳉鱼（https://www.ebi.ac.uk/birney-srv/medaka-ref-panel/index.html#）等实验室动物相比，斑马鱼中不存在近交系。近交系被定义为由一对祖先建立并且已经近交（兄弟 × 姐妹）连续 20 代或更多代的种群。经历 20 多代后纯合率超过 98%。目前还未发现满足这些标准的斑马鱼品系，但研究者们认为 C32（91% 纯合）、SJD（90% 纯合）和 IM（95% 纯合）是最“同基因”或“近交”的斑马鱼品系（Nechiporuk et al., 1999; Shinya and Sakai, 2011）。因此，通常来说，斑马鱼品系是具有不同程度多样性的远交品系（Guryev et al., 2006）。例如，当使用转基因鱼进行实验时可能会有问题，因为实验组和对照组之间最好是只有目的基因不同。

15.3.4 遗传差异：功能后果

这些遗传变异研究的普遍结论认为：无论是实验室内野生品系和实验室品系之间，还是实验室之间关于同一品系的差异，都可能会对斑马鱼作为生态毒理学（Coe et al., 2009）和疾病建模（Brown et al., 2012; Holden and Brown, 2018）的模型物种产生影响，尤其是一般意义上的外推和可复制性方面（Brown et al., 2012; Guryev et al., 2006; Holden and Brown, 2018）。目前尚不清楚实验室品系遗传谱差异的功能后果是什么。例如，这是否意味着某些品系可能更适合作为特定疾病的模型物种。

15.4 品系之间的行为差异

15.4.1 TL 与 AB

15.4.1.1 概述

我们最近开始了一系列实验来研究 TL 和 AB 品系斑马鱼，观察仔鱼和成鱼阶段两种品系在生理学（HPI 轴）和行为（包括其他声学/振动惊吓反应和社会行为）方面的潜在差异，发现 TL 品系斑马鱼成鱼表现出抑制性回避学习（Gorissen et al., 2015; Manuel et al., 2016）。如上所述，TL 有两种不同的突变，使得它在外观上与 AB、TU 和 WIK 不同：它是 leo^{t1} 的纯合子，leo^{t1} 为豹纹基因的隐性突变，在成鱼中引起斑点，而 lof^{dt2} 是导致长鳍的显性突变基因。

迄今为止，关于这些突变行为后果的系统研究很少。因此，我们开始比较 TL 品系和 AB 品系仔鱼和成鱼在几种检测方法中的行为。鉴于它们不同的背景和繁殖历史，很明显，无论这些特定突变如何变化，AB 品系和 TL 品系在整个基因组的许多基因座上都会有所不同。所以，在一些实验中，我们使用了同样携带 *leo* 基因突变的豹纹斑马鱼品系来评估结果的特异性。此外，将我们的数据与文献中的数据进行比较，评估结果是否特定属于我们实验室，还是具有普遍的性质。到目前为止，数据似乎支撑我们的假设，即这些突变会影响行为实验的结果。在以下部分中，我们将讨论结果并进一步检验我们的假设。

15.4.1.2 连接蛋白41.8突变

leo^{t1} 突变是连接蛋白41.8（Cx41.8）基因中的一个无义突变，人们已经广泛将其用作研究皮肤模型（Watanabe et al., 2006；综述见 Watanabe, 2017）。Cx41.8 蛋白有四个跨膜区、两个细胞外环和一个胞质环，在细胞质位点上有N端和C端（Watanabe et al., 2006）。*leo*^{t1} 突变位于第一个细胞外环，将蛋白质从42 kDa截短到8 kDa。这种突变会导致纯合子的皮肤出现斑点（“tüpfel”），但在杂合子中不会出现，因此，它是一种隐性突变（Watanabe et al., 2006）。其他区域（第一个和最后一个跨膜结构区）存在另外两个错义突变，它们都是显性负突变（Watanabe et al., 2006）。N端区域包含一个ExxxE基序，一个推测的多胺结合位点（Watanabe et al., 2012）。研究者们认为此突变在调节间隙连接功能方面发挥作用（Watanabe et al., 2012, 2016; Watanabe, 2017）。Cx41.8通常编码的亚基（异聚和同聚）组成缝隙连接，这些缝隙连接涉及离子（如 Ca^{2+}）、二级信使（如肌醇-1,4,5-三磷酸）和小代谢物的细胞间运输（Irion et al., 2014; Watanabeet al., 2006, 2016）。目前认为Cx41.8与Cx39.4都用于调节钙敏感跨膜电流（Watanabe et al., 2016）。

虽然大多数研究都是针对Cx41.8突变对皮肤模型影响进行的（Irion et al., 2014; Watanabe et al., 2006, 2012, 2016; Watanabe, 2017），但人们已经证明Cx41.8存在于心脏、眼睛和大脑中。哺乳动物同源基因Cx40（间隙连接5a）也存在于大脑和心脏（神经组织）中（Bai, 2014; De Bock et al., 2014a; 更多信息请参见网站 https://www.proteinatlas.org/ENSG00000265107-GJA5/tissue）。由于它存在于与行为相关的多个组织中，可以假设其影响是在行为分析中发现的。

15.4.2 视觉运动反应

在哺乳动物中，Cx40存在于视网膜屏障的内皮细胞中，在视网膜发育中发挥作用（De Bock et al., 2014b）。研究表明，在仔鱼阶段，从黑暗到光明的突然变化反应取决于眼睛，而从亮到暗的反应取决于眼睛和脑深部光感受器（Fernandes et al., 2012, 2013）。因此，可以预测，当Cx41.8突变对视网膜功能有影响时，这种影响就能通过亮-暗转换的视觉运动反应来测定。现有数据表明TL和AB品系仔鱼在视觉运动反应中存在差异（De Esch et al., 2012; Gao et al., 2016; Liu et al., 2015; van den Bos et al., 2017a）。最近的数据表明，TL和AB品系之间视网膜功能的差异很小（Dona et al., 2018）。因此我们目前需要更多的研究来探索Cx41.8在视网膜功能中的作用，从而研究其在视觉运动行为中的作用。

15.4.3 声学/振动惊吓反应

在斑马鱼行为轨迹跟踪系统和斑马鱼仔鱼行为观察箱中，斑马鱼仔鱼声学/振动惊吓反应可以很容易标准化，并且具有界限分明的神经元回路（Medan and Preuss, 2014），非常适合进行斑马鱼品系间分析。我们使用斑马鱼行为轨迹跟踪系统（Noldus Information Technology, Wageningen, the Netherlands）开发了一种声学/振动惊吓方法，其中包含两个系列的刺激：一个系列每20s提供一次刺激（总共10次），另一个系列我们每秒提供一次刺激（共30次）（van den Bos et al., 2017a）。第二个系列比第一个系列产生更强的习惯（van den Bos et al., 2017a, 2018, 2019）。我们使用最大速度（mm/s）作为惊吓反应的参数（van den Bos et al., 2017a）。

我们通常观察到TL和AB品系的5dpf仔鱼在响应第一个声学/振动刺激时没有差异（van den Bos et al., 2017a, 2018, 2019）。TL品系成鱼的听觉阈值（频率范围100～3200Hz）高于AB品系成鱼（Monroe et al., 2016）。同样在仔鱼阶段，TL品系的听觉灵敏度似乎低于AB品系（频率范围90～310Hz）（Monroe et al., 2016）。实验数据表明两种品系中振动/声学刺激都高于听觉阈值。

我们在研究中最大的发现是，无论使用10次刺激系列还是30次刺激系列，在适应重复系列的声学/振动刺激方面，TL仔鱼都比AB仔鱼慢（图15-3 A; van den Bos et al., 2017a, 2018, 2019）。将我们的数据与文献中报道数据进行比较，发现了一个相似的现象：AB仔鱼比TL仔鱼适应得更快（TL: Wolman et al., 2011; AB: Best et al., 2008）。这表明这是AB和TL之间真正存在的差异，而不是我们当地实验室条件引起的结果。

为了研究这种习惯差异是否与*leo*基因的突变有关，我们将AB和TL仔鱼的习惯与携带*leo*基因突变但在TU背景下的豹纹品系仔鱼的习惯进行比较（详见图15-3）。从表型上看，这种豹纹品系成鱼在斑点图案方面类似于TL成鱼。虽然TL、豹纹和AB品系的仔鱼对第一次刺激的反应没有差异，但与AB仔鱼相比，TL与豹纹品系仔鱼对一系列重复的声学/振动刺激表现出更弱的习惯化［van den Bos et al., 2018，图15-3（A）］。这一发现支撑了*leo*基因突变在造成AB和TL之间习惯化差异中的作用。

N-甲基-D-天冬氨酸（NMDA）受体的长期激活或阻断对Cx40（小鼠运动神经元：Personius et al., 2008; 大鼠心脏：Shi et al., 2017）的表达有影响。在哺乳动物中，Cx40存在于神经胶质血管单元（由神经元、星形胶质细胞、小胶质细胞、内皮细胞、血管周细胞、平滑肌细胞和循环血细胞组成）的一些结构中，控制突触连接、血脑屏障功能、局部血液供应、神经元发育和监测/免疫功能（Ca^{2+}作用）（De Bock et al., 2014a; Takeuchi and Suzumura, 2014）。惊吓习惯（不是对第一次刺激的反应）取决于NMDA/甘氨酸受体复合物的激活与Ca^{2+}信号功能（Best et al., 2008; Marsden and Granato, 2015; Roberts et al., 2011; Wolman et al., 2011）。因此，可以假设TL和AB之间的习惯化差异也与NMDA/甘氨酸受体复合物和Ca^{2+}信号转导功能差异有关。为了验证这个假设，我们在TL和AB仔鱼中进行了一系列实验，将斑马鱼暴露于一种阻断Ca^{2+}通道和NMDA的非竞争性NMDA拮抗剂MK801中。已有数据证明MK801可有效阻止习惯化（Marsden and Granato, 2015; Wolman et al., 2011），而NMDA会增加习惯化（Best et al., 2008）。

我们观察到，总体上MK801可减缓习惯化，对第一次刺激的反应没有影响，这与其他人的数据一致（Marsden and Granato, 2015; Wolman et al., 2011）。不同剂量的MK801对习惯化的影响，30次刺激系列较10次刺激系列更强，这是可以预知的，因为1s间隔触发较20s间隔的习惯化更强［图15-3（A）］。此外，不同剂量的MK801在AB中的作用强于TL品系［图15-3(B)］，而另一方面，不同剂量NMDA在TL中的作用强于AB品系［图15-3（C）］。通过对10次刺激中20s间隔中最大速度不规则模式的描述，发现MK801似乎增加了对刺激的敏感性（数据未发表；AB和TL实验常规方法van den Bos et al., 2017a; Wolman et al., 2011）。然而，MK801没有系统地影响对第一次刺激无反应的受试者数量

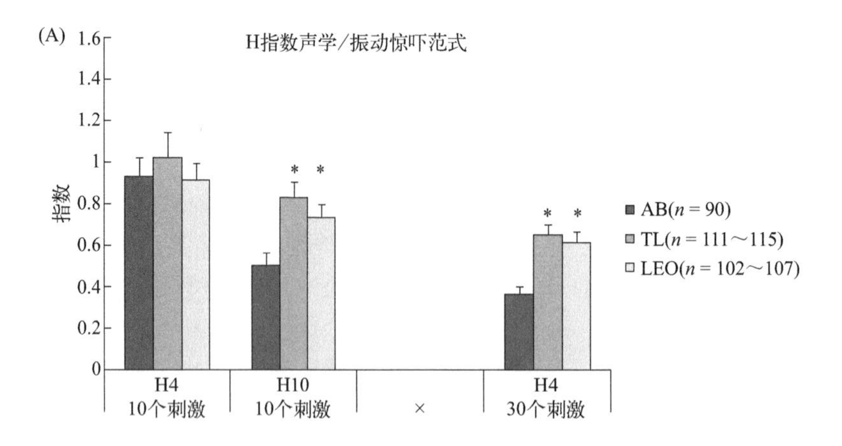

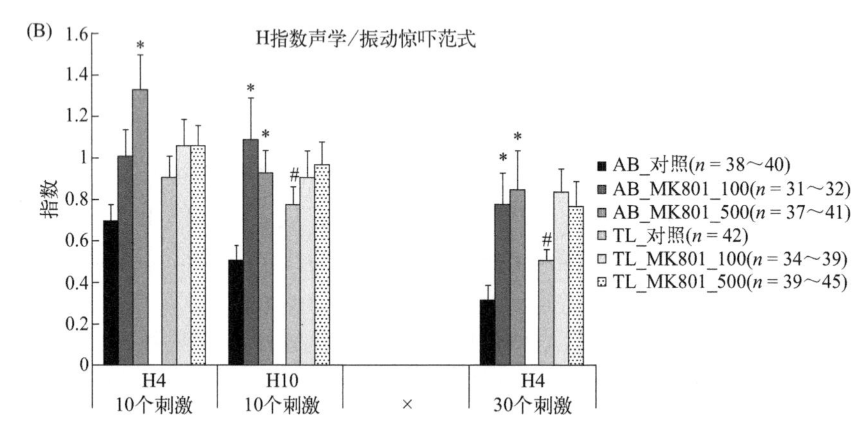

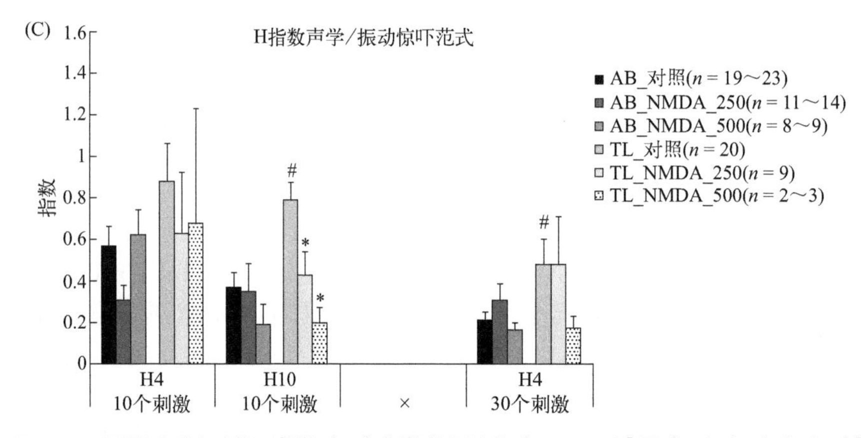

图 15-3 **不同系列实验的习惯化（H）指数的平均值（±SEM）[图（A）、（B）和（C）]**

所有图中，H4 是第四次刺激的惊吓反应值（最大速度，mm/s）除以第一次刺激的值（最大速度，mm/s），而 H10 是第 10 次刺激的惊吓反应值（最大速度，mm/s）除以第一次刺激的惊吓反应值（最大速度，mm/s）。每个图的左侧部分（10 个刺激）描绘了 10 个重复刺激的系列，而每个图的右侧部分（30 个刺激）描绘了 30 个重复刺激的系列。惊吓反应 <15mm/s 的受试者从分析中排除（受试者未达到标准）；因此，10 个刺激和 30 个刺激系列之间的受试者数量可能不同。H 指数是一系列重复刺激后习惯的良好近似值（van den Bos et al., 2018, 2019）

图（A）：AB、TL 和豹纹品系之间 H 指数的差异。第一次刺激的惊吓反应没有发现显著性差异。*：$P \leqslant 0.05$（双尾；Dunnet's-T 检验与 AB 具有显著品系效应）。AB 品系来自 ZIRC（#ZIRC：ZL1；ZFIN ID：ZDB-GENO-960809-7）；TL 品系来自 ZIRC（#ZIRC：ZL86；ZFIN ID：ZDB-GENO-990623-2）；豹纹品系来自 EZRC（编号 #20467；提供者：Christiane Nüsslein-Volhard，马克斯·普朗克发育生物学研究所，德国图宾根 Tübingen）

图（B）：MK801（100μmol/L 或 500μmol/L）处理后 AB 或 TL 的 H 指数。在惊吓刺激开始前 10 ～ 20min 使用药物（Sigma Aldrich, Zwijndrecht，荷兰）（van den Bos et al., 2017a, 2018, 2019）。第一次刺激的惊吓反应没有发现显著差异。数值低于 15mm/s 的受试者数量如下。10 个刺激系列：AB 对照组（16.7%），TL 对照组（12.5%）；AB 100μmol/L 组（32.6%），TL 100μmol/L 组（18.8%）；AB 500μmol/L 组（12.8%），TL 500μmol/L 组（6.3%）。仅 AB 组的卡方值有显著性。AB：卡方 =6.28，$P<0.04$（双尾）；TL：卡方 =3.34，ns。30 个刺激系列：AB 对照组（20.8%），TL 对照组（12%，5%）；AB 100μmol/L 组（28.3%），TL 100μmol/L 组（29.2%）；AB 500μmol/L 组（21.3%），TL 500μmol/L 组（18.8%）。AB 和 TL 组卡方值都不显著。AB：卡方 = 0.91，ns；TL：卡方 =4.23，ns。*：$P \leqslant 0.05$（双尾 Dunnet's-T 检验，AB 空白对照组与处理组或者 TL 空白对照组与处理组间具有显著差异性）；#：$P \leqslant 0.05$（TL 组与 AB 组间具有显著差异性）

图（C）：NMDA（250μmol/L 或 500μmol/L）处理后，AB 或 TL 的 H 指数。在惊吓刺激开始前 10 ～ 20min 使用药物（Sigma Aldrich, Zwijndrecht，荷兰）（van den Bos et al., 2017a, 2018, 2019）。仅在 AB 处理的仔鱼中，发现 10 个刺激系列中的第一个刺激有显著影响（250μmol/L 显著低于对照组）；在其他情况下，均未发现显著性影响。数值低于 15 mm/s 的受试者数量如下。10 个刺激系列：AB 对照组（14.8%），TL 对照组（13.0%）；AB 250μmol/L 组（41.7%），TL 250μmol/L 组（62.5%）；AB 500μmol/L 组（60%），TL 500μmol/L 组（87%）；AB 和 TL 卡方值均有显著性。AB：卡方 =10.5，$P<0.01$（双尾）；TL：卡方 = 26.31，$P<0.001$（双尾）。30 个刺激系列：AB 对照组（25.9%），TL 对照组（13.0%）；AB 250μmol/L 组（54.2%），TL 250μmol/L 组（62.5%）；AB 500μmol/L 组（55%），TL 500μmol/L 组（91.3%）。仅 TL 组的卡方值有显著性。AB：卡方 =5.57，ns, TL：卡方 =29.23，$P<0.001$（双尾）；*：$P \leqslant 0.05$（双尾 Dunnet's-T 检验，AB 空白对照组与处理组或者 TL 空白对照组与处理组间具有显著差异性）；#：$P \leqslant 0.05$（TL 组与 AB 组间具有显著差异性）

[图 15-3（B）]。相比之下，对第一次刺激没有反应的受试者数量随着 NMDA 剂量的增加而显著增加，尤其是在 TL 品系中 [图 15-3（C）]。我们观察到暴露于 NMDA 不同剂量的仔鱼出现抽搐 / 震颤样行为。当这些剂量的 NMDA 被视为产生神经毒性时，这些数据表明 NMDA 对 TL 比对 AB 具有更强的神经毒性作用。

总体而言，这些数据表明，涉及 Cx41.8 间隙连接的 NMDA/ 甘氨酸受体 Ca^{2+} 通道复合物（可能）的激活在 AB 品系比在 TL 品系更强，引起更快的习惯化。金鱼 Mauthner 细胞上第八对脑神经（听觉前庭）的大型有髓鞘末端显示为混合突触：在重复刺激后，Ca^{2+} 通过 NMDA 受体刺激进入细胞，激活 Ca^{2+}- 钙调蛋白依赖性激酶Ⅱ（CaMK Ⅱ），进而磷酸化谷氨酸受体和连接蛋白或调节分子（Pereda et al., 2013）。相关的连接蛋白是形成电突触的 Cx35 和 C34.7。目前的数据使得研究 NMDA 受体和 Cx41.8 间隙连接在惊吓习惯中的关联有了进一步的发展。

15.4.4 焦虑相关和恐惧相关的行为

AB 和 TL 品系成鱼在移入亮暗箱中的暗室时表现出相似的潜伏期（Gorissen et al., 2015），这表明它们在焦虑上没有区别。然而，TL 品系斑马鱼成鱼在受到电击时很容易学会避开暗室，而 AB 品系斑马鱼成鱼则不能（Gorissen et al., 2015；综述：Manuel et al., 2016）。我们将这种差异归因于 AB 品系斑马鱼的全身皮质醇基线水平高于 TL 斑马鱼，这可能会对海马驱动的情境学习产生负面影响（Gorissen et al., 2015; Manuel et al., 2016）。随后的一项研究表明，AB 品系斑马鱼更容易在类似于提示学习（依赖杏仁核）任务中学会避免冲击，而不是情境学习（依赖海马体）任务中（Kenney et al., 2016）。已有研究证明 AB 品系斑马鱼可以轻松掌握与提示相关的任务（Vignet et al., 2013）。

在仔鱼阶段，TL 品系斑马鱼全身皮质醇水平（HPI 轴活动）比 AB 品系低（van den Bos et al., 2017a, 2019）。我们观察到与 AB 品系相比，TL 斑马鱼类固醇激素合成急性调节蛋白（steroid acute regulatory protein, StAR）水平极低，StAR 参与了类固醇合成的第一步限速步骤（van den Bos et al., 2019）。迄今为止，我们实验室（van den Bos et al., 2017a,

2017b, 2019）和其他实验室（TL: Bestet al., 2017; AB: Nesan and Vijayan, 2016）的数据都表明，AB 和 TL 品系之间的 HPI 轴功能存在差异。

在哺乳动物的肾上腺皮质细胞中发现了 Cx40，这表明它在类固醇合成中起作用（https://www.proteinatlas.org/ENSG00000265107-GJA5/tissue）。很明显，这个发现值得进一步研究，因为皮质醇基线水平的一般差异对新陈代谢、免疫功能和行为表现等有影响（Nesan and Vijayan, 2013; van den Bos et al., 2019）。

15.4.5 心脏功能

在人类和啮齿动物中，Cx40 存在于心房心肌细胞和心室传导系统中（Bai, 2014）。Cx40 突变与房性心律失常、房室传导阻滞和束支传导阻滞有关（Bai, 2014; Simon et al., 1998）。目前 Cx41.8 突变对斑马鱼心脏功能影响的数据仍待研究。在 Cx40 基因敲除小鼠中，与野生型（每分钟约 16 次）相比，基因敲除小鼠心率略低。我们还观察到受精后三天，TL 品系（157±8bpm; n=10）心率略低于 AB 品系（166±4bpm; n=17）。到目前为止还没有测量过其他参数。然而，当血液循环需要达到最佳时（另见下文），心房功能的变化可能会影响行为，尤其是能量需求行为。值得注意的是，目前已有研究表明，不同品系之间的心脏功能存在显著差异（TE, SUN, WIK 和 AB）（Wang et al., 2017）。

15.4.6 小结

虽然 AB 和 TL 品系的研究对象在整个基因组的许多基因座上都有所不同，但此处提供的数据表明，Cx41.8 突变引起的影响可能比单独的皮肤模型更广泛。例如可以进一步研究，使用当前可用的遗传工具可以在 AB 品系中进行 Cx41.8 功能（Musa and Veenstra, 2003）影响实验或控制 AB 品系 Cx41.8 表达水平的实验。根据可用的数据，可以对这些实验结果做出明确的预测。

斑马鱼在白天比在夜间活动得更多，即它们是白天活动的鱼（图 15-4）。我们观察到 TL 品系成鱼比 AB 品系成鱼移动距离少，这是白天在鱼最活跃时最容易观察到的现象［图 15-4（A）］。虽然我们没有观察到鱼移动时游泳速度的差异（未发表的数据），但我们观察到 TL 品系鱼表现出长时间的不活动，即躺在水箱底部并依靠鳍休息；相反，AB 几乎连续不断地游［图 15-4（B）］。我们没有使用进 / 出水系统更新换水，否则会产生水流迫使鱼移动。当我们将 TL 和 AB 品系成鱼放在一个 60L 的大型水箱中，氧气泵的入口在水箱的顶部，产生强烈的局部水流，AB 鱼在水流中大力游动，而 TL 鱼漂浮在水箱中间或靠近底部，这也就证实了小水箱的观察结果。

这些数据表明 TL 鱼在不被强迫时不会移动，这样似乎可以节约体力。对于这些发现，可以提出几种解释。除了可能存在心房颤动或（相对）低水平的皮质醇，它们对能量需求的游泳行为有负面影响，长鳍突变可能是造成这种情况的原因，因为长鳍对游泳行为产生负面影响。与我们的研究一致，Plaut（2000）的研究表明长鳍鱼比短鳍鱼表现出较少的活动，尤其是在白天。此外在游泳隧道中，长鳍斑马鱼的临界游泳速度（即鱼在一段时间内可以保持的最大速度）比短鳍斑马鱼的游泳速度要低。这可能由于长尾鳍产生的推动动物

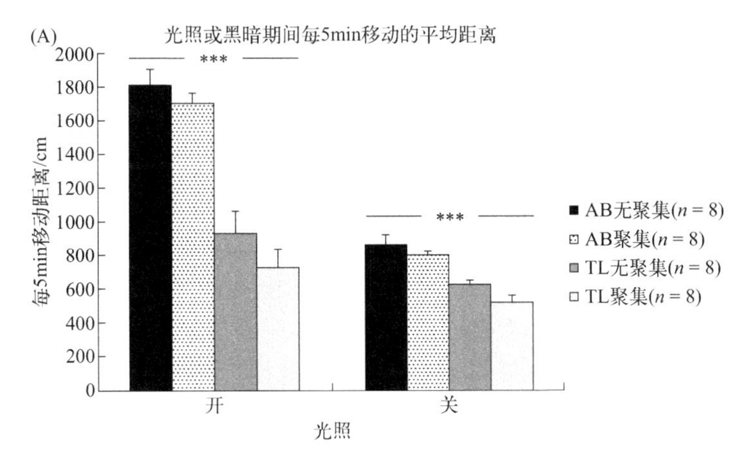

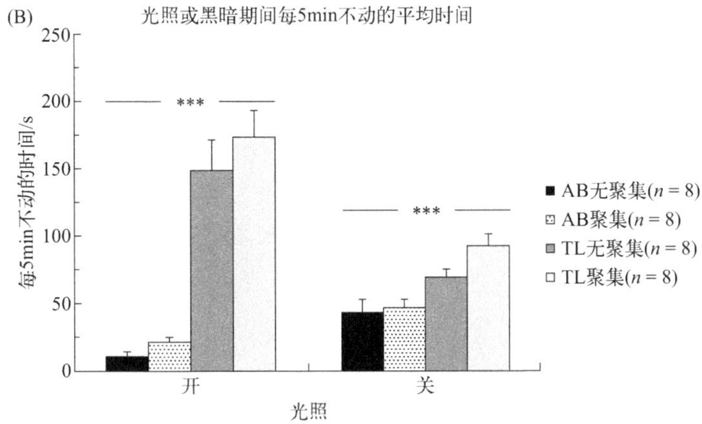

图 15-4 **TL和AB斑马鱼成鱼在光照或黑暗时间段内每5min移动距离（图A）或每5min不动的时间（图B）的平均值（±SEM）**

将四条 AB 或 TL 品系斑马鱼放置在一个 41cm×24.5cm×15cm 大小的水箱（水深 8cm；体积 8L），14h 光照（335 ~ 380Lux）-10h 黑暗环境下的气候控制室中（22:17 时关灯；08:17 时亮灯）。水箱放置在红外背光单元（Noldus Information Technology, Wageningen, the Netherlands），并使用顶部的红外敏感相机进行记录。使用 Ethovision 软件（Noldus Information Technology, Wageningen, the Netherlands）提取相关参数。每种条件运行八次重复［AB 或 TL 组，有或无聚集（中间放有塑料植物）］。数据表示在光明或黑暗期间平均 5min 一个周期的八次重复（每次运动为 1 分）的平均值。统计（三因素方差分析；品系和聚集作为独立因素；昼夜重复测量），移动距离：昼 - 夜，$F(1,28)=194.560$，$P<0.001$；昼 - 夜 * 品系，$F(1,28)=60.469$，$P<0.001$；品系，$F(1,28)=90.150$，$P<0.001$；聚集，$F(1,28)=4.088$，$P\leqslant 0.05$。不动时间：昼 - 夜，$F(1,28)=5.453$，$P\leqslant 0.03$；昼 - 夜 * 品系，$F(1,28)=48.425$，$P<0.001$；品系，$F(1,28)=98.721$，$P<0.001$。***：$P<0.001$ 品系效应遵循双向方差分析（品系和聚集作为独立因素）

前进的力比短尾鳍小（Plaut, 2000）。这也解释了在新的水箱潜水测试中，长鳍鱼比短鳍鱼移动到顶部的潜伏期更长（Egan et al., 2009）。

在最近的一项研究中，在具有固定或浮动地标的大型场地（1m×1m; 1.2m×1.2m）中进行 1h 观察，研究 AB 和 TL 品系成鱼的鱼群凝聚力和集群行为（Séguret et al., 2016）。观察结果显示：① AB 品系鱼之间的个体间距离大于 TL 品系鱼之间的距离；②当 10 条鱼一组时，AB 品系鱼似乎较 TL 品系鱼使用地标更强烈，但在 5 条一组时则不然；③ AB 品系鱼在地标之间显示出更多的“一对一过渡”，而 TL 品系鱼则进行了更多的集体 U 形转弯。此外，沿墙游泳没有显著差异。当 AB 和 TL 鱼探索环境时，在放入水箱后的第一个小时进行记录分析，图 15-4 结果确认了对个体间距离的观察：AB 品系斑马鱼之间的距离［聚集：（11.1±0.6）cm（*n*=8）；非聚集：（12.3±0.5）cm（*n*=8）］高于 TL 品系斑马鱼之间的距离［聚集：（9.2±0.5）cm（*n*=8）；非聚集：（8.9±0.5）cm（*n*=8）］；品系 $F(1,28)=26.808$，$P<0.001$（双因素方差分析：品系和聚集作为独立因素）。在研究聚集时，我们使用了“记号”，我们观察到 TL 品系在聚集地附近花费的时间［（257.1±10.8）s, *n*=8］略多于 AB 品系［（227.3±9.7）s, *n*=8；$F(1,14)=4.291$，$P\leqslant0.06$］。最后，AB 品系［聚集：（129.7±6.2）s，（*n*=8）；非聚集：（159.8±13.5）s，（*n*=8）］在水箱壁上花费的时间比 TL 品系多［聚集：（78.1±16.0）s，（*n*=8）；非聚集：（39.2±22.7）s，（*n*=8）；品系 $F(1,28)=4.421$，$P<0.001$（双因素方差分析：品系和聚集作为独立因素）］。这些数据总体上表明 AB 和 TL 品系在空间鱼群凝聚力和集体行为方面存在差异。还指出群体大小和水箱大小可能有影响。

Séguret 等人（2016）对鱼群凝聚力差异提供了两种解释。第一种解释为：与 AB 品系鱼的短鳍相比，TL 品系鱼长鳍会向侧向感知系统产生更强大的信号以感知同种个体，从而导致 TL 品系中的距离比 AB 品系短。另一种解释为：是由于斑马鱼使用视觉系统进行吸引和侧向感知系统进行排斥，从而调节接近度，Cx41.8 突变可能对 TL 中的侧向系统产生影响，导致信号传输的敏感度较低，排斥力较小，因此 TL 品系中的距离比 AB 品系的短。我们研究了 AB 品系和购买的豹纹斑马鱼品系之间的差异（见上文），发现豹纹斑马鱼品系在第一个小时内个体间距离也比 AB 低（未发表的数据），这表明第二种解释可能比第一种解释更可信。

15.5 结论

从偶然观察到的 AB 和 TL 品系成鱼在抑制性回避学习上的差异开始（Gorissen et al., 2015），我们开始更详细地比较这些品系。TL 品系斑马鱼数据表明，*leo* 和 *lof* 突变对行为有影响。尽管对 Cx41.8 的研究主要集中在皮肤状态方面，但很明显 Cx41.8 突变需要在不同器官进行研究，以了解连接蛋白 Cx41.8 基因突变是如何影响它们功能的。这反过来也表明了生物医学研究（如疾病建模）中使用 TL 品系的局限性，另一方面为探索 Cx41.8 在细胞中的作用提供了机会，例如与心脏（Bai, 2014）和 / 或中枢神经系统相关的疾病建模（De Bock et al., 2014a）。

我们从对抑制性回避学习研究开始，就以 AB 作为参考品系（Gorissen et al., 2015），但很明显，AB 并不是所有野生型品系的代表。事实上，大家都已经观察到 AB 和 TU 在惊吓行为（Van den Bos et al., 2018）、焦虑（Vignet et al., 2013），包括潜在神经生物学在内的社会行为发展（Mahabir et al., 2013）以及对 MK801 的药理敏感性（Liu et al., 2014）等方面存在实质性差异。

在最近的一系列实验中，在 AB 和 TL 品系斑马鱼受精后的前 6h（0～6hpf），我们（van den Bos et al., 2019）将胚胎暴露于皮质醇来模拟母体应激对仔鱼行为、免疫功能和 HPI 轴功能的影响。与受精后 5 天（5dpf）AB 品系仔鱼对照相比，HPI 轴活性上调，但 TL 品系仔鱼的 HPI 轴活性与文献中的数据不一致（AB: Nesan and Vijayan, 2016; TL: Best et al.,

2017）。与对照组相比，AB 品系仔鱼的趋触性增强，该表型表明其更焦虑，但 TL 品系仔鱼趋触性减少，该表型表明其更大胆（van den Bos et al., 2019）。其他实验室也发现 TL 品系仔鱼的相同表型（Best et al., 2017）。相反，AB 和 TL 品系仔鱼的免疫相关基因上调并且惊吓习惯化较慢（van den Bos et al., 2019）。因此，根据测试的特定参数，相同的处理导致两种品系产生相似或不同的效果。这表明研究的初始阶段使用更多的品系（最好也结合其他实验室的平行研究）是有效的，可以揭示哪些品系（和 / 或当地畜牧业或实验室条件）更稳定，哪些对品系（和 / 或当地畜牧业或实验室条件）更敏感，这将提高结果的外推有效性（van den Bos, 2018; Kafkafi et al., 2018）。

实验室内和实验室之间实验结果差异本身并不是问题，除非明确差异来源是什么。当这些基因被识别出来时，就变成了“可控的变异”。特别是在啮齿动物研究领域，人们已经就如何优化与基因型（使用特定的近交系或近交系）、环境（生活条件和测试环境）相关研究的可重复性以及它们之间的相互作用进行了许多讨论，并且在实验设计和统计数据应用方面取得了实质性进展（参见 Kafkafi 等人 2018 年发表的优秀论文和与本文相关的讲座）。斑马鱼研究领域应该从这些知识中获得经验，尤其在目前缺乏近交系研究的情况下。在生物医学研究中使用斑马鱼作为实验模型来替代啮齿动物模型，关键取决于在实验室之间和实验室内的数据可重复性。从长远来看，在所有影响因素中，只有当使用斑马鱼作为模型动物的结果可能会受到严重影响时，才会仔细考虑斑马鱼品系的差异性。

致谢

我们要感谢 Luuk Meulman、Niels Kraemer、Thomas Heuberger 和 Gema Tuckwood 在实习期间获得的数据，感谢我们的同事 Jan Zethof 和 Tom spanings 分别提供出色的分析和动物护理。我们也要感谢两位匿名的审稿人，他们对手稿提出了建设性的意见，这有助于提高质量。

（靳倩翻译　王陈审校）

参考文献

Ahmad, F., 2014. Zebrafish Embryos and Larvae as a Complementary Model for Behavioural Research. University of Leiden, The Netherlands, pp. 1e238. Ph.D. Thesis.

Bai, D., 2014. Atrial fibrillation-linked GJA5/connexin40 mutants impaired gap junctions via different mechanisms. FEBS Letters 588, 1238-1243.

Best, J.D., Berghmans, S., Hunt, J.J.F.G., Clarke, S.C., Fleming, A., Goldsmith, P., Roach, A.G., 2008. Non-associative learning in larval zebrafish. Neuropsychopharmacology 33, 1206-1215. https://doi.org/10.1038/sj.npp.1301489.

Best, C., Kurrasch, D.M., Vijayan, M.M., 2017. Maternal cortisol stimulates neurogenesis and affects larval behaviour in zebrafish. Scientific Reports 7, 40905.

Brown, K.H., Dobrinskia, K.P., Lee, A.S., Gokcumen, O., Mills, R.E., Shia, X., Chong, W.W.S., Chen, J.Y.H., Yoo, P., Davida, S., Peterson, S.M., Rajb, T., Choy, K.W., Stranger, B.E., Williamson, R.E., Zon, L.I., Freeman, J.L., Lee, C., 2012. Extensive genetic diversity and substructuring among zebraish strains revealed through copy number variant analysis. Proceedings of the National Academy of Sciences USA 109, 529-534 (2012).

Butler, M.G., Iben, J.R., Marsden, K.C., Epstein, J.A., Grenato, M., Weinstein, B.M., 2015. SNPfisher: tools for probing genetic variation in laboratory-reared zebrafish. Development 142, 1542e1552. https://doi.org/10.1242/dev.118786.

Chesler, E.J., Wilson, S.G., Lariviere, W.R., Rodriguez-Zas, S.L., Mogil, J.S., 2002. Influences of laboratory environment on behavior. Nature Neuroscience 5, 1101-1102.

Coe, T.S., Hamilton, P.B., Griffiths, A.M., Hodgson, D.J., Wahab, M.A., Tyler, C.R., 2009. Genetic variation in strains of zebrafish (*Danio rerio*) and the implications for ecotoxicology studies. 9. Ecotoxicology 18, 144e150. https://doi.org/10.1007/s10646-008-0267-0.

De Bock, M., Decrock, E., Wang, N., Bol, M., Vinken, M., Bultynck, G., Leybaert, L., 2014a. The dual face of connexin-based astroglial Ca2þ communication: a key player in brain physiology and a prime target in pathology. Biochimica et Biophysica Acta 1843, 2211-2232.

De Bock, M., Vandenbroucke, R.E., Decrock, E., Culot, M., Cecchelli, R., Leybaert, L., 2014b. A new angle on bloode CNS interfaces: a role for connexins? FEBS Letters 588, 1259-1270.

De Esch, C., van der Linde, H., Slieker, R., Willemsen, R., Wolterbeek, A., Woutersen, R., De Groot, D., 2012. Locomotor activity assay in zebrafish larvae: influence of age, strain and ethanol. Neurotoxicology and Teratology 34,425-433. https://doi.org/10.1016/j.ntt.2012.03.002.

De Visser, L., 2008. Home Sweet Home; Home Cage Testing for Behavioural Phenotyping of Mice. PhD dissertation Utrecht University. NPN Drukkers Breda.

Dona, M., Spijkerman, R., Lerner, K., Broekman, S., Wegner, J., Howat, T., Peters, T., Hetterschijt, L., Boon, N., De Vrieze, E., Sorusch, N., Wolfrum, U., Kremer, H., Neuhauss, S., Zang, J., Kamermans, M., Westerfield, M., Philips, J., van Wijk, E., 2018. Usherin defects lead to early-onset retinal dysfunction in zebrafish. Experimental Eye Research 173, 148-159.

Drew, R.E., Settles, M.L., Churchill, E.J., Williams, S.M., Balli, S., Robison, B.D., 2012. Brain transcriptome variation among behaviorally distinct strains of zebrafish (*Danio rerio*). BMC Genomics 13, 323.

Egan, R.J., Bergner, C.L., Hart, P.C., Cachat, J.M., Canavello, P.R., Elegante, M.F., Elkhayat, S.I., Bartels, B.K., Tien, A.K., Tien, D.H., Mohnot, S., Beeson, E., Glasgow, E., Amri, H., Zukowska, Z., Kalueff, A.V., 2009. Understanding behavioral and physiological phenotypes of stress and anxiety in zebrafish. Behavioural Brain Research 205, 38-44.

Fernandes, A.M., Fero, K., Arrenberg, A.B., Bergeron, S.A., Driever, W., Burgess, H.A., 2012. Deep brain photo receptors control light-seeking behavior in zebrafish larvae. Current Biology 22, 2042-2047.

Fernandes, A.M., Fero, K., Driever, W., Burgess, H.A., 2013. Enlightening the brain: linking deep brain photoreception with behavior and physiology. BioEssays 35, 775-779.

Gao, Y., Zhang, G., Jelfs, B., Carmer, R., Venkatraman, P., Ghadami, M., Brown, S.A., Pang, C.P., Leung, Y.F., Chan, R.H.M., et al., 2016. Computational classification of different wild-type zebrafish strains based on their variation in light-induced locomotor response. Computers in Biology and Medicine 69, 1-9.

Gorissen, M., Manuel, R., Pelgrim, T.N.M., Mes, W., de Wolf, M.J.S., Zethof, J., Flik, G., van den Bos, R., 2015. Differences in inhibitory avoidance, cortisol and brain gene expression in TL and AB zebrafish. Genes, Brain and Behavior 14, 428-438.

Guryev, V., Koudijs, M.J., Berezikov, E., Johnson, S.L., Plasterk, R.H.A., van Eeden, F.J.M., Cuppen, E., 2006. Genetic variation in the zebrafish. Genome Research 16, 491-497.

Hartig, E.I., Zhu, S., King, B.L., Coffman, J.A., 2016. Cortisol-treated zebrafish embryos develop into proinflammatory adults with aberrant immune gene regulation. Biology Open bio-020065.

Holden, L.A., Brown, K.H., 2018. Baseline mRNA expression differs widely between common laboratory strains of zebrafish. Scientific Reports 8, 4780.

Holtzman, N.G., Iovine, M.K., Liang, J.O., Morris, J., 2016. Learning to fish with genetics: a primer on the vertebrate model *Danio rerio*. Genetics 203, 1069-1089.

Irion, U., Frohnhoefer, H.G., Krauss, J., Colak Champollion, T., Maischein, H.M., Geiger-Rudolph, S., Weiler, C., Nuesslein-Volhard, C., 2014. Gap junctions composed of connexins 41.8 and 39.4 are essential for colour pattern formation in zebrafish. eLife 3, e05125.

Kafkafi, N., Agassia, J., Chesler, E.J., Crabbe, J.C., Crusio, W.E., Eilam, D., et al., 2018. Reproducibility and replicability of rodent phenotyping in preclinical studies. Neuroscience & Biobehavioral Reviews 87, 218-232. https://doi.org/10.1016/j.neubiorev.2018.01.003.

Kenney, J.W., Scott, I.C., Josselyn, S.A., Frankland, P.W., 2016. Contextual Fear Conditioning in Zebrafish. https://doi.org/10.1101/068833 doi: bioRxiv preprint first posted online August 10, 2016.

Koot, S., Adriani, W., Saso, L., van den Bos, R., Laviola, G., 2009. Home cage testing of delay discounting in rats. Behavior Research Methods 41, 1169-1176.

Lange, M., Neuzeret, F., Fabreges, B., Froc, C., Bedu, S., Bally-Cuif, L., Norton, W.H.J., 2013. Inter-individual andinter-strain variations in zebrafish locomotor ontogeny. PLoS One 8, e70172. https://doi.org/10.1371/journal.pone.0070172.

Liew, W.C., Orban, L., 2013. Zebrafish sex: a complicated affair. Briefings in Functional Genomics 13, 172-187.

Liu, X., Guo, N., Lin, J., Zhang, Y., Chen, X.Q., Li, S., He, L., Li, Q., 2014. Strain-dependent differential behavioral responses of zebrafish larvae to acute MK-801 treatment. Pharmacology, Biochemistry and Behavior 127, 82-89.

Liu, Y., Carmer, R., Zhang, G., Venkatraman, P., Brown, S.A., Pang, C.P., Zhang, M., Ma, P., Leung, Y.F., 2015. Statistical analysis of zebrafish locomotor response. PLoS One 10. https://doi.org/10.1371/journal.pone.0139521.

Mahabir, S., Chatterjee, D., Buske, C., Gerlai, R., 2013. Maturation of shoaling in two zebrafish strains: a behavioral and neurochemical analysis. Behavioural Brain Research 247, 1-8. https://doi.org/10.1016/j.bbr.2013.03.013.

Manuel, R., Gorissen, M., van den Bos, R., 2016. Relevance of test- and subject-related factors on inhibitory avoidance (performance) of zebrafish for psychopharmacology studies. Current Psychopharmacology 5, 152e168. https://doi.org/10.2174/2211556005666160526111427.

Marsden, K.C., Granato, M., 2015. In vivo Ca2þ imaging reveals that decreased dendritic excitability drives startle habituation. Cell Reports 13 (9), 1733-1740.

Medan, V., Preuss, T., 2014. The Mauthner-cell circuit of fish as a model system for startle plasticity. Journal of Physiology Paris 108, 129e140. https://doi.org/10.1016/j.jphysparis.2014.07.006.

Monroe, J.D., Manning, D.P., Urib, P.M., Bhandiwad, A., Sisneros, J.A., Smith, M.E., Coffin, A.B., 2016. Hearing sensitivity differs between zebrafish lines used in auditory research. Hearing Research 341, 220-231.

Musa, H., Veenstra, R.D., 2003. Voltage-dependent blockade of Connexin40 gap junctions by spermine. Biophysical Journal 84, 205-219.

Nechiporuk, A., Finney, J.E., Keating, M.T., Johnson, S.L., 1999. Assessment of polymorphism in zebrafish mapping strains. Genome Research 9, 1231-1238.

Nesan, D., Vijayan, M.M., 2013. Role of glucocorticoid in developmental programming: evidence from zebrafish. General and Comparative Endocrinology 181, 35-44.

Nesan, D., Vijayan, M.M., 2016. Maternal cortisol mediates hypothalamus-pituitary-interrenal axis development in zebrafish. Scientific Reports 6, 22582.

Norton, W.H.J., 2013. Toward developmental models of psychiatric disorders in zebrafish. Frontiers in Neural Circuits 7, 79. https://doi.org/10.3389/fncir.2013.00079.

Oswald, M., Robison, B.D., 2008. Strain-specific alteration of zebrafish feeding behavior in response to aversive stimuli. Canadian Journal of Zoology 86 (10), 1085e1094. https://doi.org/10.1139/Z08-085.

Pavlidis, M., Digka, N., Theodoridi, A., Campo, A., Barsakis, K., Skouradakis, G., et al., 2013. Husbandry of zebrafish, *Danio rerio*, and the cortisol stress response. Zebrafish 10 (4), 524-531.

Pereda, A.E., Curti, S., Hoge, G., Cachope, R., Flores, C.E., Rash, J.E., 2013. Gap junction-mediated electrical transmission: regulatory mechanisms and plasticity. Biochimica et Biophysica Acta 1828 (1), 134e146. https://doi.org/10.1016/j.bbamem.2012.05.026.

Personius, K.E., Karnes, J.L., Parker, S.D., 2008. NMDA receptor blockade maintains correlated motor neuron firing and delays synapse competition at developing neuromuscular junctions. Journal of Neuroscience 28 (36),8983-8992.

Plaut, I., 2000. Effects of fin size on swimming performance, swimming behaviour and routine activity of zebrafish *Danio rerio*. Journal of Experimental Biology 203, 813-820.

Ramsay, J.M., Feist, G.W., Varga, Z.M., Westerfield, M., Kent, M.L., Schreck, C.B., 2006. Whole-body cortisol is an indicator of crowding stress in adult zebrafish, *Danio rerio*. Aquaculture 258, 565-574.

Ramsay, J.M., Watral, V., Schreck, C.B., Kent, M.L., 2009. Husbandry stress exacerbates mycobacterial infections inadult zebrafish, *Danio rerio* (Hamilton). Journal of Fish Disease 32 (11), 931e941. https://doi.org/10.1111/j.1365-2761.2009.01074.x.

Rauch, G.J., Granato, M., Haffter, P., 1997. A polymorphic zebrafish line for genetic mapping using SSLPs on high-percentage agarose gels. Technical Tips Online T 01208 2, 148-150.

Ribas, L., Valdivieso, A., Dıaz, N., Piferrer, F., 2017. Appropriate rearing density in domesticated zebrafish to avoid masculinization: links with the stress response. Journal of Experimental Biology 220, 1056e1064. https://doi.org/10.1242/jeb.144980.

Roberts, A.C., Reichl, J., Song, M.Y., Dearinger, A.D., Moridzadeh, N., Lu, E.D., Pearce, K., Esdin, J., Glanzman, D.L., 2011. Habituation of the C-start response in larval zebrafish exhibits several distinct phases and sensitivity toNMDA receptor blockade. PLoS One 6, e29132. https://doi.org/10.1371/journal.pone.0029132.

Séguret, A., Collignon, B., Halloy, J., 2016. Strain Differences in the Collective Behaviour of Zebrafish (*Danio rerio*) in Heterogeneous Environment, vol. 3. Royal Society Open science, p. 160451.

Shi, S., Liu, T., Wang, D., Zhang, Y., Liang, J., Yang, B., Hu, D., 2017. Activation of *N*-methyl-d-aspartate receptors reduces heart rate variability and facilitates atrial fibrillation in rats. Europace 19, 1237-1243.

Shinya, M., Sakai, N., 2011. Generation of highly homogeneous strains of zebrafish through full sib-pair mating. G3(Genes Genomes Genetics) 1, 377-386.

Simon, A.M., Goodenough, D.A., Paul, D.L., 1998. Mice lacking connexin40 have cardiac conduction abnormalities

characteristic of atrioventricular block and bundle branch block. Current Biology 8, 295-298.
Steenbergen, P.J., Richardson, M.K., Champagne, D.L., 2011. Patterns of avoidance behaviours in the light/dark preference test in young juvenile zebrafish: a pharmacological study. Behavioural Brain Research 222, 15-25.
Stephenson, J.F., Whitlock, K.E., Partridge, J.C., 2011. Zebrafish preference for light or dark is dependent on ambient light levels and olfactory stimulation. Zebrafish 8, 17-22.
Stevens, J.R., 2017. Replicability and reproducibility in comparative psychology. Frontiers in Psychology 8, 862.https://doi.org/10.3389/fpsyg.2017.00862.
Stewart, A.M., Braubach, O., Spitsbergen, J., Gerlai, R., Kalueff, A.V., 2014. Zebrafish models for translational neuroscience research: from tank to bedside. Trends in Neurosciences 37, 264-278.
Takeuchi, H., Suzumura, A., 2014. Gap junctions and hemichannels composed of connexins: potential therapeutic targets for neurodegenerative diseases. Frontiers in Cellular Neuroscience 8, 189.
Vignet, C., Bégout, M.-L., Péan, S., Lyphout, L., Leguay, D., Cousin, X., 2013. Systematic screening of behavioral responses in two zebrafish strains. Zebrafish 10, 365-375.
Villamizar, N., Maria Vera, L., Foulkes, N.S., Javier Sanchez-Vazquez, F., 2014. Effect of lighting conditions on zebrafish growth and development. Zebrafish 11, 173-181.
van den Bos, R., 2018. Ecological validity. In: Bornstein, M.H. (Ed.), The SAGE Encyclopedia of Lifespan Human Development. SAGE Publications, Inc., Thousand Oaks (CL) USA, pp. 700-701. https://doi.org/10.4135/9781506307633.n262.
van den Bos, R., Mes, W., Galligani, P., Heil, A., Zethof, J., Flik, G., et al., 2017a. Further characterisation of differences between TL and AB zebrafish (*Danio rerio*): gene expression, physiology and behaviour at day 5 of the larval stage.PLoS One 12 (4), e0175420.
van den Bos, R., Zethof, J., Flik, G., Gorissen, M., 2017b. Light regimes differentially affect baseline transcript abundance of stress-axis and (neuro) development-related genes in zebrafish (*Danio rerio*, Hamilton 1822) AB and TLlarvae. Biology Open 6, 1692e1697. https://doi.org/10.1242/bio.028969.
van den Bos, R., Flik, G., Gorissen, M., 2018. Strain differences in acoustic/vibrational startle response in zebrafish larvae. In: Grant, R.A., Spink, A., Sullivan, M. (Eds.), Proceedings Measuring Behavior 2018; 5e8th June 2018; Manchester, UK, pp. 443-446.
van den Bos, R., Althuizen, J., Tschigg, K., Bomert, M., Zethof, J., Filk, G., Gorissen, M., 2019. Early life exposure to cortisol in zebrafish (*Danio rerio*): similarities and differences in behaviour and physiology between larvae of the AB and TL strains. Behavioural Pharmacology 30, 260-271. https://doi.org/10.1097/FBP.0000000000000470.
Wahlsten, D., Bachmanov, A., Finn, D.A., Crabbe, J.C., 2006. Stability of inbred mouse strain differences in behaviour and brain size between laboratories and across decades. Proceedings of the National Academy of Science United States of America 103, 16364-16369.
Wang, L.W., Huttner, I.G., Santiago, C.F., Kesteven, S.H., Yu, Z.-Y., Feneley, M.P., Fatkin, D., 2017. Standardized echocardiographic assessment of cardiac function in normal adult zebrafish and heart disease models. Disease Models & Mechanisms 10, 63-76.
Watanabe, M., Iwashita, M., Ishii, M., Kurachi, Y., Kawakami, A., Kondo, S., Okada, N., 2006. Spot pattern of leopard *Danio* is caused by mutation in the zebrafish connexin41.8 gene. EMBO Reports 7, 893-897.
Watanabe, M., Watanabe, D., Kondo, S., 2012. Polyamine sensitivity of gap junctions is required for skin pattern formation in zebrafish. Scientific Reports 2, 473.
Watanabe, M., Sawada, R., Aramaki, T., Skerrett, I.M., Kondo, S., 2016. The physiological characterization of Connexin41.8 and Connexin39.4, which are involved in the striped pattern formation of zebrafish. Journal of Biological Chemistry 291, 1053-1063.
Watanabe, M., 2017. Gap junction in the teleost fish lineage: duplicated connexins may contribute to skin pattern formation and body shape determination. Frontiers in Cellular and Developmental Biology 5, 13.
Whiteley, A.R., Bhat, A., Martins, E.P., Mayden, R.L., Arunachalam, M., Uusi-Heikkilä, Ahmed, A.T.A., Shrestha, J., Clark, M., Stemple, D., et al., 2011. Population genomics of wild and laboratory zebrafish (*Danio rerio*). Molecular Ecology 20 (20), 4259-4276.
Wilson, C.A., High, S.K., McCluskey, B.M., Amores, A., Yan, Y.L., Titus, T.A., Anderson, J.L., Batzel, P.,Carvan III, M.J., Schartl, M., et al., 2014. Wild sex in zebrafish: loss of the natural sex determinant in domesticated strains. Genetics 198, 1291-1308.
Wolman, M.A., Jain, R.A., Liss, L., Granato, M., 2011. Chemical modulation of memory formation in larval zebrafish. Proceedings National Academy of Sciences United States of America 108, 15468-15473.
Wright, D., Rimmer, L.B., Pritchard, V.L., Krause, J., Butlin, R.K., 2003. Inter and intrapopulation variation in shoaling and boldness in the zebrafish (*Danio rerio*). Naturwissenschaften 90 (8), 374-377.
Wright, D., Nakamichi, R., Krause, J., Butlin, R.K., 2006. QTL analysis of behavioral and morphological differentiation between wild and laboratory zebrafish (*Danio rerio*). Behavior Genetics 36 (2), 271-284.

第十六章

用于行为遗传学的突变体设计

Han B. Lee[1], Rodsy Modhurima[1], Amanda A. Heeren[1], Karl J. Clark[1]

16.1 用于靶向基因组修饰的可编程核酸酶

ZFN。锌指核酸酶（zinc finger nucleases, ZFN）的开发和其在靶向基因组编辑（Carroll 2011; Kim et al., 1996）方面的应用开创了“可编程”核酸酶的潜力，靶向基因组编辑应用也包括斑马鱼基因组的修饰（Doyon et al., 2008; Foley et al., 2009; Meng et al., 2008）。ZFN 是通过将 *Fok* Ⅰ限制酶的酶切域与锌指蛋白（zinc finger proteins, ZFP）融合而成。ZFP 由锌指串联阵列组成，每个锌指作为一个模块识别基因组中 3 ～ 4 个碱基的短序列（Kim et al., 1996）。在应用中，ZFN 是成对设计的，正向和反向 ZFP 分别融合一个 *Fok* Ⅰ核酸酶结构域并分别结合于 5 ～ 7 个碱基间隔区的两侧。两个 ZFNs 在靶位的结合使得 *Fok* Ⅰ核酸酶产生二聚化和激活，从而使间隔区内的 DNA 产生双链断裂（double-strand break, DSB）。ZFN 的基因编辑难点在于其碱基识别的模块化特性，因为每个锌指结构域的结合受到其相邻结构域的影响并且无法准确判定。尽管如此，ZFN 作为第一代可靶向核酸内切酶，已为该领域带来了多项显著的创新，当确定了一个好的核酸对时，ZFN 仍然可作为有效的靶向核酸酶。

TALEN。天然转录激活因子样效应子（transcription activator-like effectors, TALE），源自植物病原菌（例如，黄单胞菌属 *Xanthomonas* spp.）。在植物宿主细胞中诱导有利于病原菌感染活性的基因表达（参见综述，Doyle et al., 2013）。DNA 结合域由 33 ～ 35 个氨基酸重复序列组成。与 ZFN 不同的是，这些氨基酸重复模块每个都与特定的 DNA 碱基结合，这就使得优化 DNA 结合变得更加容易，可以在单个 DNA 结合模块和单个 DNA 碱基之间实现一对一的靶向结合。重复可变的双氨基酸残基（repeat variable di-residues, RVD;

[1] Department of Biochemistry and Molecular Biology, Mayo Clinic, Rochester, MN, United States.

在 33 ～ 35 个氨基酸重复序列中间的第 12 个和第 13 个氨基酸）决定了该模块结合的特定 DNA 碱基：HD（胞嘧啶）、NG（鸟嘌呤）、NI（腺嘌呤）和 NN（胸腺嘧啶）。转录激活因子样效应物核酸酶（transcription activator-like effector nucleases, TALEN）是通过将 *Fok* Ⅰ核酸酶的酶切区与 TALE 的 C 端 DNA 结合区融合而产生的。在实验室设计的 TAL 重复阵列，包括斑马鱼基因（Clark et al., 2011; Huang et al., 2011; Sander et al., 2011; Lee et al., 2016b），对目的基因具有高亲和力，同时较易设计并且目标广泛（Miller et al., 2011）。一对位于 14 ～ 20bp 间隔区两侧的 TALEN，可在其靶位点产生 DSB。ZFN 的大部分创新都适用于 TALEN，包括专性异源二聚体 *Fok* Ⅰ和切口酶。在实验室中设计产生 TALEN 有几个简单的方案，其中，生成 TALEN 的标准方法是从编码 TALE 重复区的质粒库中克隆包含所需重复序列的质粒。“Golden Gate”克隆方法可在单管中 3 天内组装出在斑马鱼体内具有高活性的 TALENs（Cermak et al., 2011; Ma et al., 2016）。

CRISPR/Cas。成簇规律间隔短回文重复序列及其相关系统（CRISPR/Cas）来自细菌细胞的适应性免疫反应（Wiedenheft et al., 2012）。CRISPR 利用 RNA 引导靶向技术，依赖于引导 RNA 和目标序列之间 Watson-Crick 碱基配对（Deltcheva et al., 2011; Wiedenheft et al., 2012），这使得它在非专业实验室的设计和应用比 TALEN 更加简单。单链引导 RNAs（single guide RNAs, sgRNA）和 Cas9 蛋白生产的相对便利性及其在应用上的简单性是该技术广泛应用的主要驱动力（Jinek et al., 2012）。第一代 CRISPR/Cas 是通过设计 sgRNA 和使用化脓性链球菌 Cas9（*Streptococcus pyogenes* Cas9，spCas9）从天然 CRISPR/Cas 系统的最简单形式（Ⅱ型）改造而来（Jinek et al., 2012）。当被引入细胞时，sgRNA：Cas9 复合物扫描整个基因组，直到 Cas9 蛋白发现一个原始间隔区相邻基序（protospacer adjacent motif, PAM）序列和互补序列，促使 sgRNA 与基因组上的 PAM 序列在反义链上形成配对的碱基对。一个 PAM 序列（对于 SpCas9 来说是 5′-NGG-3′）需要与 Cas9 蛋白的目标位点结合才能成功地切割靶基因 DNA 并形成 DSB（Jinek et al., 2012）。优化后的 CRISPR/Cas 系统可以更方便地用于斑马鱼的基因打靶（Chang et al., 2013; Hwang et al., 2013），包括大规模筛选（Varshney et al., 2015）。

然而，sgRNA 和基因组 DNA 之间碱基配对的特异性不如两个 TALEN 蛋白提供的特异性强，这导致最初研究者们担心 CRISPR/Cas 介导的基因组编辑中的脱靶效应会增加。目前已经发展出许多机制和实用手段来限制 CRISPR 和其他核酸酶的脱靶切割，包括靶点选择的特异性（Cho et al., 2013; Frock et al., 2015; Fu et al., 2014; Guilinger et al., 2014; Kleinstiver et al., 2016; Ran et al., 2013; Shen et al., 2014; Slaymaker et al., 2016; Tsai et al., 2014; Wyvekens et al., 2015）。为此，通过各种 sgRNA 设计工具来寻找特异的目标位点。CHOPCHOP 是一个网络工具，用来搜索 TALEN 和 CRISPR/Cas9 sgRNA 设计的目标位点，根据特异性和脱靶效应的可能性来表示目标位点的“质量”（Montague et al., 2014）。CHOPCHOP 还可以指出目标位点上的酶切位点，识别可能扩增目标区域的引物，并显示在基因或基因组上的酶切位点。CRISPOR 通过综合 8 个效率评分指南，使用算法对脱靶和在靶 DSB 的可能性进行评分（Haeussler et al., 2016）。Mojo-Hand 为 TALEN、CRISPR/Cas9 核酸酶诱导 DSB 和 CRISPR/Cas9 限制酶诱导单链断裂提供最佳设计策略，包括可用于限制性片段长度多态性（restriction fragment length polymorphism, RFLP）分析的限制性内切酶位点和微同源介导的末端连接（microhomology-mediated end joining, MMEJ）可能性的信息（www.talendesign.org/betatest; Neff et al., 2013）。

为了扩展潜在的目标位点，人们已经开发出目前最常用的 SpCas9 的替代版本。在自然界中，已经鉴定出来了六类 Cas 蛋白变种（Ⅰ类到Ⅵ类），每一类都包括几种不同的 Cas 蛋白，可以识别不同的 PAM 序列（Jiang et al., 2015; Makarova et al., 2018; Murugan et al., 2017; Shui et al., 2016; Wiedenheft et al., 2012; Wright et al., 2016）。在不同的 Cas 蛋白中，由于Ⅱ类系统所需的组件最少，Ⅱ类变异体的设计应用最为广泛（Makarova et al., 2018）。例如，作为 Cas9（Ⅱ类）的替代品，Cas12a（cpf1；Ⅱ类）具有能够通过单个活性位点诱导 DSB 的能力（Zetsche et al., 2015）。Cas12a 与 SpCas9 的互补效用在于它进行交错切割而不是 Cas9 进行的钝切割，并且它有一个富含 5′T 的 PAM 序列，可以靶向富含 AT 的基因组区域。从直肠真杆菌中分离出的一种 Cas12a（ErCas12a）已被证明在斑马鱼和人类细胞中具有活性（Wierson et al., 2019）。随着我们不断发现和检测新的 Cas 变异体，Cas 蛋白可靶向的基因组序列也在不断增加。

16.2 DNA 修复和 DNA 修饰

大多数情况下，靶向突变是通过可编程内切酶使用设计来识别 DNA 靶序列和内切酶的机制诱导基因组靶位点产生 DSB。这引发了三种主要的 DNA 修复机制：非同源末端连接（nonhomologous end joining, NHEJ）[图 16-1（A）]、微同源介导的末端连接（microhomology-mediated end joining, MMEJ）[图 16-1（B）] 和同源定向修复（homology-directed repair, HDR）[图 16-1（C）]。尽管进行了数十年的研究，这些修复途径仍未被完全定义，在本章中，我们主要基于实际应用及其结果对它们进行讨论。

自从人们设计出靶向核酸内切酶以来，非 NHEJ 修复一直是研究中最常用的方法，因为 NHEJ 修复过程容易出错而产生短的核苷酸缺失和 / 或插入，从而导致移码突变（Daley et al., 2005）。尽管可能通过选择性启动子、外显子剪接或翻译起始位点产生其他蛋白异构体，但经过这种移码突变的目的基因产物将不再以其天然形式产生（图 16-2）。因此，根据产生目的基因产物的代偿机制的激活程度，NHEJ 可以有效地敲除基因或根据预期的基因敲除产生代偿表型（Heidenreich et al., 2003）。经典的 NHEJ 是一种灵活的修复机制，可应用于任何类型和周期的细胞，这意味着分裂活化期、静止或终末分化的细胞可以利用这一途径（Lieber, 2008）。

MMEJ 是另一种末端连接途径，它使用 DNA 断裂两侧存在的短顺式同源序列（>3nt）。在切除两个断裂 DNA 片段的 3′ 末端后，正反义链中的两段短同源序列的退火导致插入序列和一个同源臂的缺失（Deng et al., 2014; Sinha et al., 2016）。这使得 MMEJ 修复途径非常适合诱导具有可预测结果的移码突变以产生基因敲除（Ata et al., 2018; Sabharwal et al., 2019; Shen et al., 2018）。MMEJ 修复途径在细胞周期的 M 期和早期 S 期发生得最明显（Lieber, 2008），似乎利用了与经典同源重组（homologous recombination, HR）和 NHEJ 途径不同的分子成分（尽管有一些重叠）（Sharma et al., 2015）。相关的算法可以定位可能使用

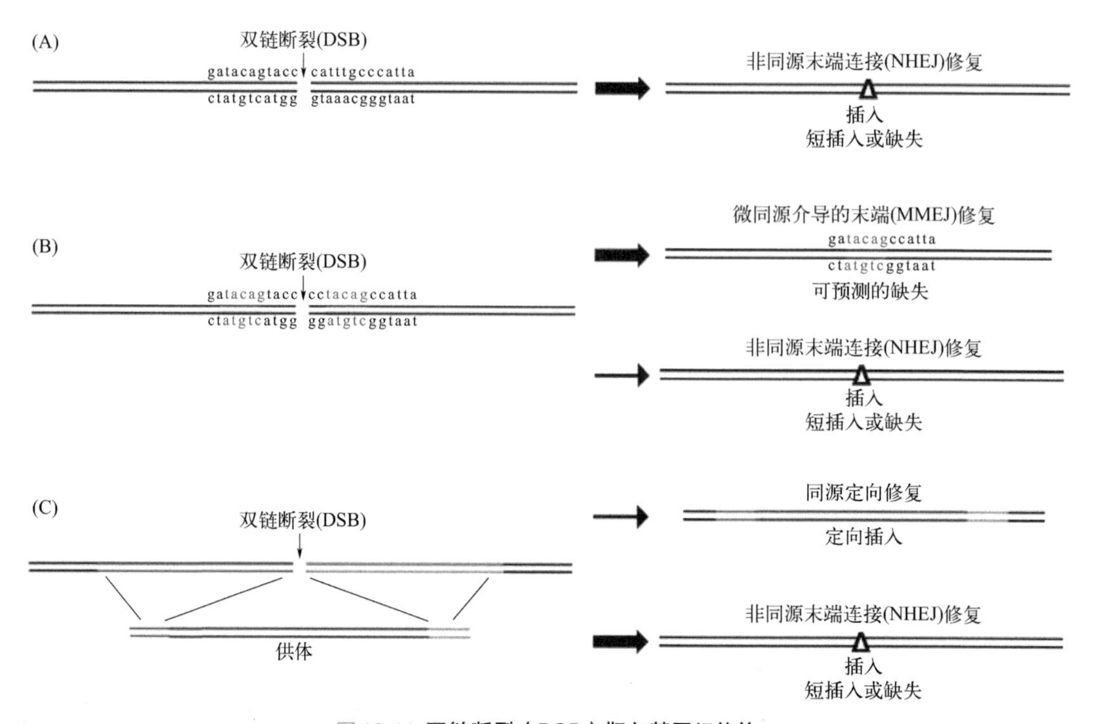

图 16-1 **双链断裂（DSB）靶向基因组修饰**

（A）当 DSB 由自定义核酸酶形成时，断裂通常会通过非同源末端连接（nonhomologous end joining, NHEJ）修复。NHEJ 是一种容易出错的机制，通常会导致各种短插入或缺失。（B）如果在 DSB 两侧或附近有短同源区域，≥ 3nt（由红色文本 / 线显示），细胞可以通过微同源介导的末端连接（MMEJ）修复此断裂。关闭中断会导致丢失重复序列和任何中间序列。这些结果在某种程度上是可预测的，并且在 DSB 位点附近具有微同源结构域的位点上比 NHEJ 的发生率更高。（C）如果供体与基因组断裂点末端（蓝色和绿色）具有同源性，则该供体可用于修复 DSB 并合并插入序列（红色）。该过程通常比通过 NHEJ 引起关闭的效率低

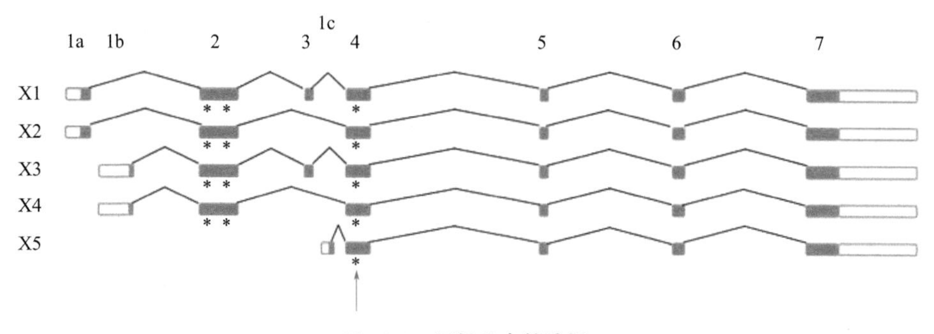

图 16-2 **目标位点的选择**

在这里，我们展示了具有五种替代转录本（X1 ～ X5）基因的通用转录图。在这些转录本中有三个替代启动子（外显子 1a、1b 和 1c）、一个被跳过的外显子（外显子 3）和几个较好的内部 AUG 起始位点（*）。所有转录本的第一个通用靶点将在外显子 4 中，并且建议在内部 AUG 之后进行靶向

MMEJ修复途径的目的基因位点，以实现精确和可预测的目标序列缺失（Ata et al., 2018; Mann et al., 2019; Shen et al., 2018）。

我们在本书中广泛使用HDR，包括利用同源供体DNA序列的修复途径。HDR用于通过使用短或长同源臂引入不同长度的外源DNA。这类外源性DNA包括具有数十个碱基对同源结构域的单链寡聚脱氧核糖核酸（single-strand oligo deoxyribonucleotide, ssODN）和具有可达数千个碱基的短或长同源结构域的双链DNA分子（Albadri et al., 2017; Bedell et al., 2012; Wierson et al., 2018; Zhang et al., 2017）。尽管HDR促进的敲入可能是有用的，但敲入结果可能效率较低或整合不精确。随着敲入效率的提高，我们可以设想，在靶位点敲入荧光报告基因或表位标记可用于研究基因表达模式，或提供一种以无创方式对动物进行基因鉴定的方法，而不是基于DNA分离和PCR。

16.3 目标选择：选择性启动子、外显子和起始密码子的潜力

使用任何基于网络的软件来靶标一个基因，都可以通过TALENs和CRISPR/Cas系统在外显子和内含子中提供大量目标位点。无论你是想生成简单的移码敲除等位基因还是敲入外源基因序列，都有几个需要考虑的注意事项。许多基因有多个近端启动子，启动转录并产生不同的第一外显子，以及由选择性剪接形成的多个转录异构体。虽然在斑马鱼中通常没有完全注释，但是检查人类同源基因可以提供潜在的选择性转录本的证据。如果你的目标是第一外显子中的一个位点，当存在其他启动子，你可以靶向一个单转录本。同样，如果你以一个特定的外显子为目标，而这个外显子并不包含在多个转录本中，那么这些选择性转录本也不会受到影响（图16-2）。此外，每个转录本在序列中间可以有多个翻译起始位点（称为替代AUG或下游翻译起始位点），这可能会产生具有完全或部分功能的蛋白质。虽然一些下游的AUG未被高度利用，但部分用于生成组织特异性蛋白亚型。因此，在试图敲除所有转录和翻译的亚型时，应考虑将这些变量的下游作为靶点。

16.4 减少脱靶突变或非目标突变影响的育种应用

16.4.1 遗传背景

有了所有可用于对基因组进行定向编辑的工具，研究单个基因在一种或多种特定行为中的作用比以往任何时候都要容易。虽然测量的行为通常看起来很简单（例如运动），但影响这些行为的诱因和神经回路往往比较复杂。因此，许多基因可以引起或改变观察到的行为。我们在单个基因中创建靶向突变或变异的目的是评估该基因如何导致一种或多种简单或复杂的行为。因此，重要的是通过实验将目的基因的影响与其他有助于或改变行

为的基因隔离开来。在啮齿动物系统中，如小鼠，通常使用近交系小鼠进行行为分析。根据定义，近交系小鼠是至少 20 代纯同胞交配的结果，近交系数（coefficient of inbreeding, COI）为 98.6% 或更高（Green, 1981; Carter et al., 1952）。这限制了非目的基因的变异，从而更容易衡量单个基因变化的影响。尽管如此，这并没有改变基因组中的其他基因可能有参与或改变观察到的表型或行为的事实。实际上，对不同近交系的基因敲除或变异进行测试的结果表明，表型可能发生巨大变化（Bothe et al., 2005; Crawley et al., 1997）。

在培育斑马鱼近交系方面进行了多次尝试，但取得的成功都是有限的（Shinya and Sakai, 2011）。然而，斑马鱼和其他生物一样，对近交衰退很敏感，多代近交开始导致斑马鱼的健康状况和繁殖能力明显下降，这通常不是积极的研究型实验室所希望的（Monson and Sadler, 2010）。因此，大多数常用的斑马鱼品种（如 AB、TU、TL）都是为了保持遗传多样性而进行的封闭种群繁殖，最常用的方法是循环交配。尽管这些品系种群的性质是封闭的，但仍可能存在大量的变异。例如，Burgess 实验室在选择单对交配以产生其定义的 NHGRI 品系之前，将源自 TU 和 AB 品系的杂交品种 TAB-5 鱼的封闭种群维持了大约 5 年。对单个育种对的基因组进行了测序，鉴定出超过 1700 万个序列变异（LaFave et al., 2014）。如果今天的研究型实验室可以轻松获得许多近交系的斑马鱼，为了充分了解单个基因或位点的作用，对每个近交系进行相似的编辑是很重要的，这样才能充分了解基因或位点在不同遗传修饰背景下的作用。值得庆幸的是，在斑马鱼中，由于一次交配可获得大量后代，使用封闭的远交鱼种群是非常可行的（通常单次交配产生大于 100 个后代，且至少每 2 周可重复一次）。这些数字使定量研究成为可能，即使在表型因遗传影响而表现出大范围的变异时，也可检测到“真实”信号。因此，对同基因种群的绝对需求很低，使用远交种属的好处是它允许在遗传修饰不断变化的背景下观察单个基因的影响。

根据斑马鱼养殖设施的规模和品系维持的监督，严格遵守的循环繁殖方案可能不实际或在实践中不可行。由于这些种群的封闭性质，在大多数鱼房里可能存在某种程度的背景 COI。如果没有适当的照顾，同胞交配很快会导致斑马鱼种群的遗传瓶颈。只需要几代同胞交配就可以产生高达 50% 的 COI（F）（图 16-3），其中包括越来越稳定的遗传区域。第一次同胞交配的后代（X1）的 COI 为 25%，这意味着任何特定的变异或染色体区域从单个祖先遗传的概率为 25%（图 16-3）。在 X1 阶段，比较同胞之间的变异或染色体区域遗传时，25% 的遗传随机分布在整个基因组中。然而，由 X1 同胞交配产生的第二代后代（X2）的 COI 增加至 37.5%。与 X1 一样，从共同祖先继承的大部分共享变异体或染色体区域（约 31.25%）随机分布在整个基因组中。然而，6.25% 的共享变异体或染色体区域将由任何特定一对（X1 对）的所有后代遗传。到 X3 代，计算出的总 COI 为 50%，其中同胞之间的随机性约为 35.94%，族系中共享的有 7.81%，这一代所有鱼类的稳定 COI 约为 6.25%。尽管通常直到 6 代或 6 代以上的同胞杂交才会发现有明显的近亲繁殖抑制，但高水平的近亲繁殖，特别是单个家族或整个世代的共同遗传，可能会影响行为研究。特别是，不同鱼类种群之间的比较，无论是两个研究人员在同一设施中养殖自己的鱼，还是与其他机构的已发表文献进行比较，都可能导致表型差异源于背景变异，而不是预期的遗传干扰。锁定在鱼类遗传背景中的基因变异可以改变背景行为和对刺激的反应。

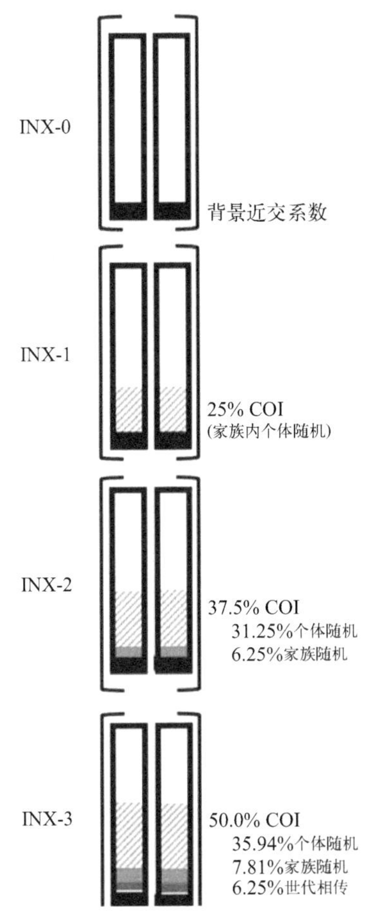

图 16-3 **完全同胞交配的影响**

对于每一代 INX-0、INX-1、INX-2 和 INX-3，单个连锁群（linkage group, LG）的配对染色体显示在括号中，代表动物的所有连锁群。以斑马鱼为例，有 25 个连锁群。假设我们从非亲缘代 INX-0 开始，在背景 COI 之外几乎没有遗传重叠（如 LG 底部的黑条所示）。在单次全同胞杂交后，尽管实际共享的遗传区域和实际数量是随机的，且因同胞而异，但产生的一代 INX-1 的 COI 为 25%。如果随后 INX-1 代的两个同胞交配，则最终 INX-2 代的 COI 为 37.5%，其中约 6.25% 的 COI 在一个家族的所有同胞共享，其余的是随机的。不同的配对将导致基因组的不同区域（约 6.25%）在所有同胞之间共享。当 INX-2 家族的任何一对同胞杂交时，产生的 COI 约为 50%。无论 INX-2 亲本如何，约 6.25% 将在所有 INX-3 后代之间共享，约 7.6% 将在单个 INX-2 对的所有 INX-3 同胞之间共享，其余 35.9% 将在 INX-3 同胞中随机遗传

当观察新的目的基因编辑的影响时，背景 COI 可以影响潜在的观察表型。因此，有必要通过规范的育种方案来保持野生型种群的遗传多样性。如果这在特定的鱼类实验室中不可行，则可以从饲养中心购买鱼，例如斑马鱼国际资源中心，在这里通常遵循维持遗传多样性的育种方案。

16.4.2 为行为分析创建自定义突变体的最佳应用

长期目标是创建新的功能缺失突变体或变异体，以评估这些变化对行为的影响，在此过程中可以采取几个重要步骤，以最好地评估预期变化的影响。首先，要么遵循育种方案，要么从 ZIRC 等机构购买远交鱼，此类机构在繁殖鱼类时会考虑这些品系的遗传多样

性。使用远交系繁殖、具有遗传多样性的鱼类将保持整体种群的背景COI相对较低，并防止潜在的修饰因子以独特的方式影响行为表型的准确性。尽管背景突变可能会改变下游检测中观察到的表型，但可能无法避免实验室种群中固有的COI。例如，一组有助于适应水族馆生活和后来的提高实验室生产力的变异将不断被选择。对于现有的AB品系，早期选择的瓶颈是基于对AB品系进行早期施压，选择21只雌性用于制造雌性二倍体子代（http://zfin.org/action/genotype/view/ZDB-GENO-960809-7）。21只雌性用于生产雌性生殖二倍体和随后的循环交配的选择标准都需要通过挤压雌性和雄性获得卵子和精子。因此，选择AB品系的部分原因是它适合体外受精，而不需要自然交配。

当使用基因编辑工具创建新的靶向编辑时，理想情况下，将不相关的斑马鱼杂交获得用于微注射的单细胞胚胎（图16-4）。有很多方法可以评估靶向编辑是否成功。在大多数情况下，对目标区域的PCR扩增产物进行进一步的分析，如限制性片段长度多态性（Ma et al., 2013; Narayanan, 1991）、错配核酸酶分析（Qiu et al., 2004; Yang et al., 2000），或在测序后手动或用算法辅助分析混合种群（即使在单个胚胎中也存在）的基因的插入/缺失分解（Brinkman et al., 2014; Etard et al., 2017）。注射后的鱼培养至性成熟时，可以通过杂交来观察注射后鱼的种系中是否有预期的变化。

有几个因素有助于提高基因编辑的整体效率，基因变化经常以镶嵌的方式在发育的胚胎中发生。因此，许多经过注射的鱼不会将新变异遗传给F1胚胎（图16-4，非携带者）。对于那些确实传递新等位基因突变体的F0代（图16-4，携带者），将很可能以亚孟德尔遗传数量将突变体保留下来。筛选时，通常会检测一小部分胚胎的分子变化情况。如果有证据表明这种变异在这个小样本中传播，则饲养其余的同胞。当这些F1鱼大到足以进行鳍活检时，将对它们进行单独评估，并将期望得到的基因编辑鱼进行整合。其他未产生预期编辑的同胞鱼则被安乐死。重要的是，由于潜在的嵌合、延迟编辑，在单个F0的生殖细胞中可能存在多个独立的等位基因，因此，在进行完整的分子表征之前应认为来自单个携带者的每个F1可能是唯一的。对于每个新的靶向编辑，应识别并提出两个或

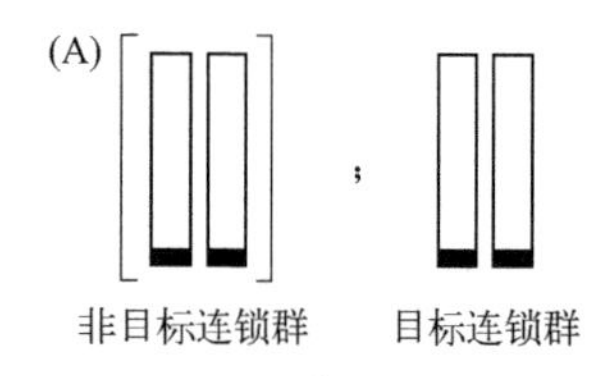

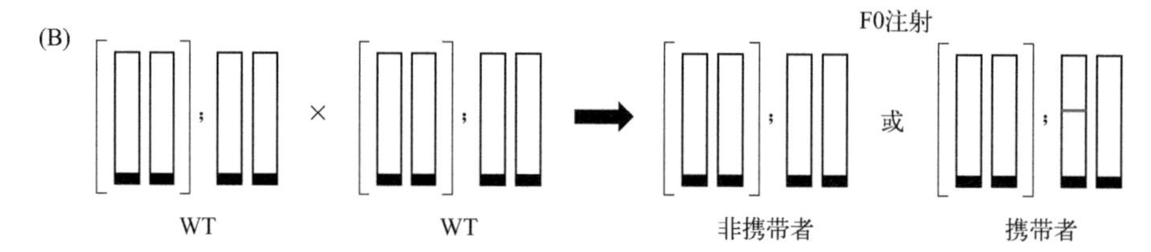

图16-4　**显微注射创建携带者**

（A）对于每条鱼，我们都有一组非靶向连锁群（LGs）。在斑马鱼中，存在24个非靶向LG和一个单一靶向LG。目标LG会根据目的基因或靶标而改变。（B）这里我们描述了两种不相关的野生型鱼的杂交。野生型（WT）杂交的后代F0，通过显微注射药物靶向基因组编辑或突变。许多被注射的后代的基因组不会有任何变化。这些胚胎是非携带者，基本上是野生型，除非发生脱靶效应。有些胚胎会在目标位点出现变化，则这些胚胎是携带者

多个等位基因。至少，这两个等位基因应该来自不同的F0母本（图16-5）。理想情况下，这两个等位基因应来自两个不同的胚胎簇（家族），以使新靶向改变关联的染色体多样化。如果两个等位基因来自同一个家族的不同F0s，则目的基因有25%的概率在同一个亲本连锁群上。因此，两个等位基因都与可能影响表型变化的相同旁观者突变体有关。如果从单个F0携带者中鉴定出两个分子上不同的等位基因（即4-bp缺失与7-bp缺失），则共享连锁遗传的概率增加到50%，因为这种变化可能发生在F0中存在的连锁群的两个副本之一。

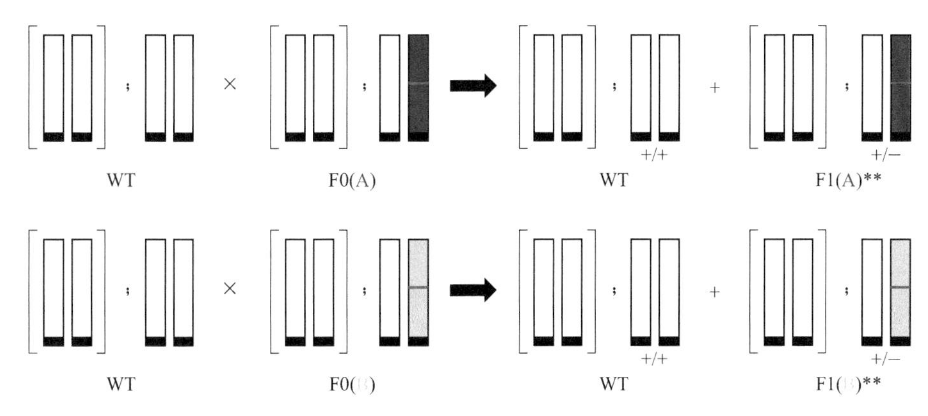

图16-5 **建立携带者鱼系**

在这里，我们展示了一个远缘或非亲缘的野生型鱼与注射的F0携带者A和B杂交。F0(A)和F0(B)分别显示为蓝色或绿色，表明每个染色体现在都与新的基因编辑紧密相连。现在与F0（A）和F0（B）编辑相关联的染色体很可能在整个连锁群的非编码区和编码区都存在大量的差异。不同的连锁群表型可能发生在遗传多样性养殖的斑马鱼，特别是当F0鱼来自不相关的胚胎。F1后代将是WT非携带者或携带者。** 由于注入单个F0可能是嵌合体，因为插入缺失和其他修饰经常发生在单细胞阶段之后，编辑的遗传可能是非孟德尔遗传，每个F1都应被视为潜在的不同等位基因

为了繁殖通过远交注入F0成鱼的两个或多个等位基因（图16-5），可以将F1鱼与不相关的野生型鱼远交［图16-6（A）］。远交主要有两个目的。首先，具有独特编辑的F1的数量可能很少，需要增加整体数目以进行行为分析。其次，除了与目标染色体直接相关的编辑，远交会稀释编辑过程中任何潜在的脱靶效应。假设一个特定的脱靶效应的发生率明显低于预期的在靶效应，那么来自F0代的两个不同个体的等位基因发生相同脱靶的概率就会大大降低。为了防止携带者鱼与野生种群的意外混淆，我们在这些杂交中采用了显性豹纹等位基因（$leo^{tq270+/-}$）（Watanabe et al., 2006）。携带显性豹纹等位基因的鱼已经与野生型种群繁殖了近十年，所以除了$leo^{tq270+/-}$等位基因和与其紧密相连的区域外，它们在基因上与野生型种群非常相似。通过让携带者与不相关的鱼类交配，COI的水平不会高于本底水平。当这些F2代鱼成年后，需要再次鉴定杂合子携带者，通常是通过鳍活检和分子检测。然而，在这一点上，遗传将遵循孟德尔规律，大约一半的鱼具有编辑过的等位基因。

为了寻找功能缺失的表型，需要分析纯合子后代，并使用任何先前描述的方法或等位基因特异性定量PCR确认后代的基因型（Lee et al., 2016a; Simone et al., 2018）。当携带相同等位基因的F2同胞进行杂交时，约25%的子代为纯合子［图16-6（B）］。这些胚胎的COI为25%。然而，每个个体的共享基因在相互比较时是随机的，因此，任何潜在的非目标突变或脱靶突变都同样与期望的等位基因及其野生型对应基因分离。因此，其影响可以

从种群效应中有效地减去。唯一不容易做到这一点的情况是，与目标等位基因在同一连锁群上的任何非目标或脱靶突变，它们仍然共享大部分原始目标染色体，仅因减数分裂而发生微小变化。以相同的方式从非亲缘 F0 中建立第二个等位基因，可在评估 F2 同胞杂交产生的 F3 后代中观察到表型时增加可信度。最终，这两个不同的等位基因可以杂交得到复合杂合子动物，而不存在相关同胞杂交中观察到的背景 COI［图 16-6（C）］。对于研究人员来说，对个体鱼类进行基因分型是一个巨大的负担，对于某些行为分析来说，很容易扩大至数百甚至数千条仔鱼，需要通过 DNA 分离和 PCR 进行基因分型。因此，当鱼可以作为成年纯合鱼存活时，人们希望简单地将纯合子的同胞进行杂交，并将它们的行为表型与不相关的野生型鱼进行比较［图 16-7（A）］。

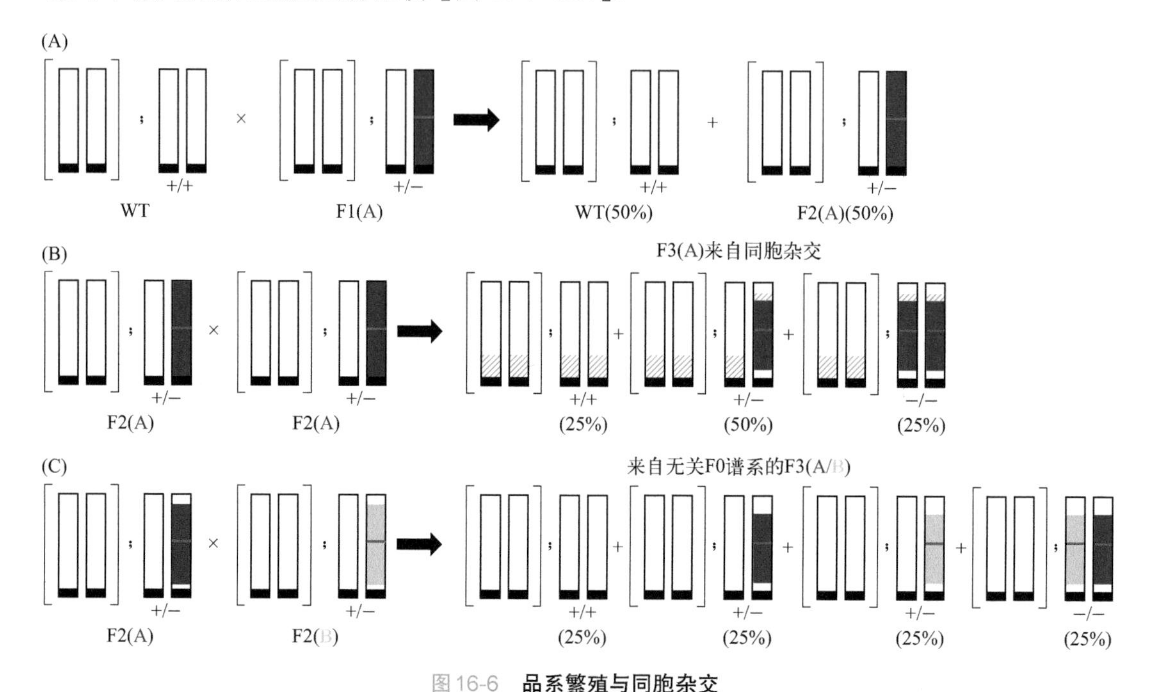

图 16-6 **品系繁殖与同胞杂交**

（A）每个 F1 代等位基因可能是唯一的，因此通常将它们与远缘或不相关的野生型鱼杂交以繁殖该系。在每次远交中，一些相连的染色体在减数分裂期间交换，这表现为染色体的蓝色或绿色覆盖率下降。（B）为了在纯合胚胎中寻找功能丧失表型，F2（A）鱼的同胞杂交可以产生野生型（+/+）、杂合子（+/−）和纯合子（−/−）的孟德尔比率，用于编辑目的基因。这些后代的近亲繁殖系数（COI）为 25%，但通过在早期杂交中使用远亲或非亲缘鱼，共享的遗传信息在兄弟姐妹之间随机遗传。每个纯合同胞间也仅存在目标连锁群（linkage group, LG）的紧密连锁区域。同样的过程可以用 F2（B）鱼（未显示）完成，它在目标 LG 上应该具有不同的连锁基因。（C）作为替代方案，为了减少所有的 COI，可以将远缘或非亲缘的 F2（A）和 F2（B）鱼进行杂交产生复合杂合子，这些杂合子是 −/− 编辑所需的

然而，通常用于这些类型分析的连续两轮同胞杂交后，纯合子种群中的任何单个育种对的所有后代都有显著的 COI 和一部分基因组共享。基因组的这一共享部分很容易影响观察到的表型，与不相关的野生型鱼进行比较并不能消除共享基因组中非靶旁观者突变的影响。因此，如果目标突变的鱼是纯合子，那么，在两个不相关等位基因的纯合子成年鱼之间精确繁殖［图 16-7（B）］可以缓解基因组上的 COI 压力。例如，同时培养纯合子目标等位基因的野生型同胞，并将这两种鱼杂交，可以降低 COI 系数增加的风险，并能更好地解释仅目标位点的作用效果，而不会出现旁观者突变体带来的干扰。

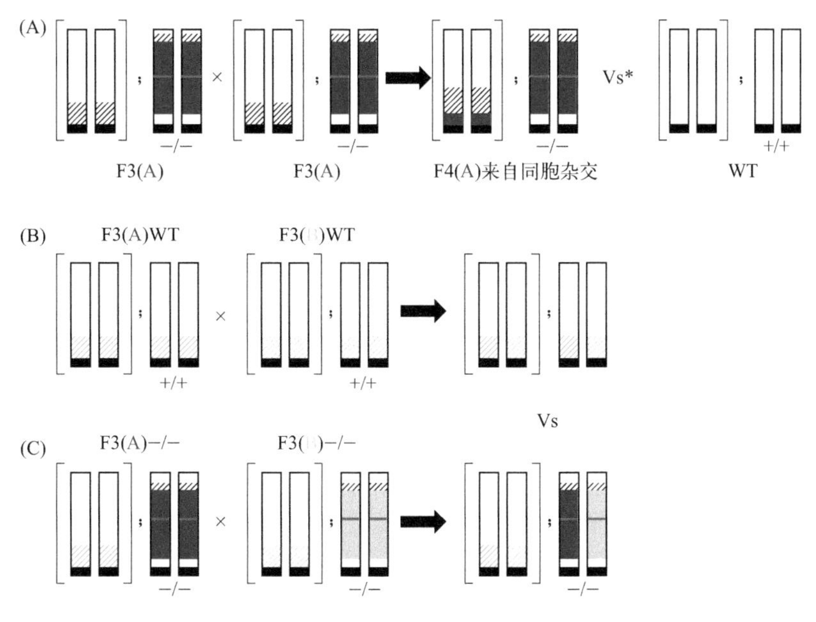

图 16-7　**绕过分子基因分型**

（A）为了避免对单个斑马鱼进行分子基因分型，在 F2 阶段甚至更早的 F1 阶段，如图 16-6 所示，对单个等位基因（例如 A 等位基因）进行同胞杂交的情况并不罕见。由此产生的一代具有纯合（−/−）个体。如果 −/− 个体是成年可存活的，则可以将它们杂交以产生全部为 −/− 个体的后代。这些 F4（或 F3）后代通常是两代同胞杂交的结果，其 COI 约为 37.5%，其中约有 6.5% 的共享基因在任何一个家族中都是常见的。然后将这些鱼（Vs*）与非亲缘关系的野生型后代进行比较。然而，在任何单个家族中，旁观者基因都有巨大的潜力影响所测量的行为。（B）当目标基因组变化的纯合子代可行时，避免分子基因分型的更好选择是利用两个等位基因（例如，A 和 B 等位基因），它们已经繁殖产生野生型（+/+）、杂合（+/−）或纯合（−/−）同胞，如图 16-6（B）所示。然后，可以培养来自每个等位基因（A 和 B）的野生型和纯合兄弟姐妹。这些鱼成年后可以杂交，基本上是远亲或无亲缘关系。然而，纯合子家族和野生型家族之间的比较将比示例（A）中所做的比较更接近

致谢

我们感谢梅奥诊所（the Mayo Clinic）的斑马鱼核心实验室及其工作人员（Casey Phillips、Julie Arkells 和 Camryn Hayft），感谢他们出色的斑马鱼养殖技术。感谢 Tanya Schwab、MS、Grace Boyum 和 Bradley Bowles 对手稿的批评指正。本章涉及的研究计划是由美国国家卫生研究院 R01（GM134732）和 R24（OD20166）基金向 KJC 资助的。

（石晶晶翻译　李丽琴审校）

参考文献

Albadri, S., Bene, F., Revenu, C., 2017. Genome editing using CRISPR/Cas9-based knock-in approaches in zebrafish. Methods 121, 77-85.

Ata, H., Ekstrom, T.L., Martínez-Gálvez, G., Mann, C.M., Dvornikov, A.V., Schaefbauer, K.J., Ma, A.C., Dobbs, D., Clark, K.J., Ekker, S.C., 2018. Robust activation of microhomology-mediated end joining for precision gene editing applications. PLoS Genetics 14, e1007652.

Bedell, V.M., Wang, Y., Campbell, J.M., Poshusta, T.L., Starker, C.G., Randall, K.G., Tan, W., Penheiter, S.G., Ma, A.C., Leung, A.Y., et al., 2012. In vivo genome editing using a high-efficiency TALEN system. Nature 491, 114-118.

Bothe, G.W., Bolivar, V.J., Vedder, M.J., Geistfeld, J.G., 2005. Behavioral differences among fourteen inbred mouse strains commonly used as disease models. Comparative Medicine 55, 326-334.

Brinkman, E.K., Chen, T., Amendola, M., van Steensel, B., 2014. Easy quantitative assessment of genome editing by sequence trace decomposition. Nucleic Acids Research 42, e168.

Carroll, D., 2011. Genome engineering with zinc-finger nucleases. Genetics 188, 773-782.

Carter, T.C., Dunn, L.C., Falconer, D.S., Grüneberg, H., Heston, W.E., Snell, G.D., 1952. Prepared by The Committee on Standardized Nomenclature for Inbred Strains of Mice. Cancer Research 12, 602-613.

Cermak, T., Doyle, E.L., Christian, M., Wang, L., Zhang, Y., Schmidt, C., Baller, J.A., Somia, N.V., Bogdanove, A.J., Voytas, D.F., 2011. Efficient design and assembly of custom TALEN and other TAL effector-based constructs for DNA targeting. Nucleic Acids Research 39, e82.

Chang, N., Sun, C., Gao, L., Zhu, D., Xu, X., Zhu, X., Xiong, J.-W., Xi, J., 2013. Genome editing with RNA-guided Cas9 nuclease in zebrafish embryos. Cell Research 23, 465-472.

Cho, S., Kim, S., Kim, J., Kim, J.-S., 2013. Targeted genome engineering in human cells with the Cas9 RNA-guided endonuclease. Nature Biotechnology 31, 230-232.

Clark, K.J., Voytas, D.F., Ekker, S.C., 2011. A TALE of two nucleases: gene targeting for the masses? Zebrafish 8, 147-149.

Crawley, J., Belknap, J.K., Collins, A., Crabbe, J.C., Frankel, W., Henderson, N., Hitzemann, R.J., Maxson, S.C., Miner, L.L., Silva, A.J., et al., 1997. Behavioral phenotypes of inbred mouse strains: implications and recommendations for molecular studies. Psychopharmacology 132, 107-124.

Daley, J.M., Palmbos, P.L., Wu, D., Wilson, T.E., 2005. Nonhomologous end joining in yeast. Annual Review of Genetics 39, 431-451.

Deltcheva, E., Chylinski, K., Sharma, C.M., Gonzales, K., Chao, Y., Pirzada, Z.A., Eckert, M.R., Vogel, J., Charpentier, E., 2011. CRISPR RNA maturation by trans-encoded small RNA and host factor RNase III. Nature 471, 602-607.

Deng, S.K., Gibb, B., de Almeida, M., Greene, E.C., Symington, L.S., 2014. RPA antagonizes microhomology mediated repair of DNA double-strand breaks. Nature Structural & Molecular Biology 21, 405-412.

Doyle, E.L., Stoddard, B.L., Voytas, D.F., Bogdanove, A.J., 2013. TAL effectors: highly adaptable phytobacterial virulence factors and readily engineered DNA-targeting proteins. Trends in Cell Biology 23, 390-398.

Doyon, Y., McCammon, J.M., Miller, J.C., Faraji, F., Ngo, C., Katibah, G.E., Amora, R., Hocking, T.D., Zhang, L., Rebar, E.J., et al., 2008. Heritable targeted gene disruption in zebrafish using designed zinc-finger nucleases. Nature Biotechnology 26, 702-708.

Etard, C., Joshi, S., Stegmaier, J., Mikut, R., Strähle, U., 2017. Tracking of indels by DEcomposition is a simple and effective method to assess efficiency of guide RNAs in zebrafish. Zebrafish 14, 586-588.

Foley, J.E., Yeh, J.-R.J., Maeder, M.L., Reyon, D., Sander, J.D., Peterson, R.T., Joung, K.J., 2009. Rapid mutation of endogenous zebrafish genes using zinc finger nucleases made by Oligomerized Pool ENgineering (OPEN). PLoS One 4, e4348.

Frock, R.L., Hu, J., Meyers, R.M., Ho, Y.-J., Kii, E., Alt, F.W., 2015. Genome-wide detection of DNA double-stranded breaks induced by engineered nucleases. Nature Biotechnology 33, 179-186.

Fu, Y., Sander, J.D., Reyon, D., Cascio, V.M., Joung, K.J., 2014. Improving CRISPR-Cas nuclease specificity using truncated guide RNAs. Nature Biotechnology 32, 279-284.

Green, E.L., 1981. Genetics and Probabliity in Animal Breeding Experiments. Guilinger, J.P., Thompson, D.B., Liu, D.R., 2014. Fusion of catalytically inactive Cas9 to Fok Ⅰ nuclease improves the specificity of genome modification. Nature Biotechnology 32, 577-582.

Haeussler, M., Schönig, K., Eckert, H., Eschstruth, A., Mianné, J., Renaud, J.-B., Schneider-Maunoury, S., Shkumatava, A., Teboul, L., Kent, J., et al., 2016. Evaluation of off-target and on-target scoring algorithms and integration into the guide RNA selection tool CRISPOR. Genome Biology 17, 148.

Heidenreich, E., Novotny, R., Kneidinger, B., Holzmann, V., Wintersberger, U., 2003. Non-homologous end joining as an important mutagenic process in cell cycle-arrested cells. The EMBO Journal 22, 2274e2283.

Huang, P., Xiao, A., Zhou, M., Zhu, Z., Lin, S., Zhang, B., 2011. Heritable gene targeting in zebrafish using customized TALENs. Nature Biotechnology 29, 699-700.

Hwang, W.Y., Fu, Y., Reyon, D., Maeder, M.L., Tsai, S.Q., Sander, J.D., Peterson, R.T., Yeh, J.-R., Joung, K.J., 2013. Efficient genome editing in zebrafish using a CRISPR-Cas system. Nature Biotechnology 31, 227-229.

Jiang, Y., Yin, S., Dudley, E.G., Cutter, C.N., 2015. Diversity of CRISPR loci and virulence genes in pathogenic *Escherichia coli* isolates from various sources. International Journal of Food Microbiology 204, 41-46.

Jinek, M., Chylinski, K., Fonfara, I., Hauer, M., Doudna, J.A., Charpentier, E., 2012. A programmable dual-RNA guided DNA endonuclease in adaptive bacterial immunity. Science 337, 816-821.

Kim, Y., Cha, J., Chandrasegaran, S., 1996. Hybrid restriction enzymes: zinc finger fusions to Fok Ⅰ cleavage domain. Proceedings of the National Academy of Sciences of the United States of America 93, 1156-1160.

Kleinstiver, B.P., Pattanayak, V., Prew, M.S., Tsai, S.Q., Nguyen, N.T., Zheng, Z., Joung, K.J., 2016. High-fidelity CRISPR-Cas9 nucleases with no detectable genome-wide off-target effects. Nature 529, 490-495.

LaFave, M., Varshney, G., Vemulapalli, M., Mullikin, J., Burgess, S., 2014. A defined zebrafish line for high throughput genetics and genomics: NHGRI-1. Genetics 198 (1), 167e170. https://doi.org/10.1534/ genetics.114.166769.

Lee, H.B., Sundberg, B.N., Sigafoos, A.N., Clark, K.J., 2016a. Genome engineering with TALE and CRISPR systems in neuroscience. Frontiers in Genetics 7, 47.

Lee, H.B., Schwab, T.L., Koleilat, A., Ata, H., Daby, C.L., Cervera, R., McNulty, M.S., Bostwick, H.S., Clark, K.J., 2016b. Allele-specific quantitative PCR for accurate, rapid, and cost-effective genotyping. Human Gene Therapy 27, 425-435.

Lieber, M.R., 2008. The mechanism of human nonhomologous DNA end joining. Journal of Biological Chemistry 283, 1-5.

Ma, A.C., Lee, H.B., Clark, K.J., Ekker, S.C., 2013. High efficiency in Vivo genome engineering with a simplified 15- RVD GoldyTALEN design. PLoS One 8, e65259.

Ma, A., McNulty, Poshusta, T., Campbell, J., Martinez-Galvez, G., Argue, D., Lee, H., Urban, Bullard, C., Blackburn, P., et al., 2016. FusX: a rapid one-step transcription activator-like effector assembly system for genome science. Human Gene Therapy 27, 451-463.

Makarova, K.S., Wolf, Y.I., Koonin, E.V., 2018. Classification and nomenclature of CRISPR-Cas systems: where from here? The CRISPR Journal 1, 325-336.

Mann, C.M., Martínez-Gálvez, G., Welker, J.M., Wierson, W.A., Ata, H., Almeida, M.P., Clark, K.J., Essner, J.J., McGrail, M., Ekker, S.C., et al., 2019. The Gene Sculpt Suite: a set of tools for genome editing. Nucleic Acids Research 47, W175-W182.

Meng, X., Noyes, M.B., Zhu, L.J., Lawson, N.D., Wolfe, S.A., 2008. Targeted gene inactivation in zebrafish using engineered zinc-finger nucleases. Nature Biotechnology 26, 695-701.

Miller, J.C., Tan, S., Qiao, G., Barlow, K.A., Wang, J., Xia, D.F., Meng, X., Paschon, D.E., Leung, E., Hinkley, S.J., et al., 2011. A TALE nuclease architecture for efficient genome editing. Nature Biotechnology 29, 143-148.

Monson, C.A., Sadler, K.C., 2010. Inbreeding depression and outbreeding depression are evident in wild-type zebrafish lines. Zebrafish 7 (2), 189e197. https://doi.org/10.1089/zeb.2009.0648.

Montague, T.G., Cruz, J.M., Gagnon, J.A., Church, G.M., Valen, E., 2014. CHOPCHOP: a CRISPR/Cas9 and TALEN web tool for genome editing. Nucleic Acids Research 42, W401-W407.

Murugan, K., Babu, K., Sundaresan, R., Rajan, R., Sashital, D.G., 2017. The revolution continues: newly discovered systems expand the CRISPR-Cas toolkit. Molecular Cell 68, 15-25.

Narayanan, S., 1991. Applications of restriction fragment length polymorphism. Annals of Clinical and Laboratory Science 21, 291-296.

Neff, K.L., Argue, D.P., Ma, A.C., Lee, H.B., Clark, K.J., Ekker, S.C., 2013. Mojo Hand, a TALEN design tool for genome editing applications. BMC Bioinformatics 14 (1), 1.

Qiu, P., Shandilya, H., D'Alessio, J., O'Connor, K., Durocher, J., Gerard, G., 2004. Mutation detection using Surveyor nuclease. BioTechniques 36 (4), 702e707. https://doi.org/10.2144/04364pf01.

Ran, A.F., Hsu, P.D., Lin, C.-Y., Gootenberg, J.S., Konermann, S., Trevino, A.E., Scott, D.A., Inoue, A., Matoba, S., Zhang, Y., et al., 2013. Double nicking by RNA-guided CRISPR Cas9 for enhanced genome editing specificity. Cell 154, 1380-1389.

Sabharwal, A., Campbell, J.M., WareJoncas, Z., Wishman, M., Ata, H., Liu, W., Ichino, N., Bergren, J.D., Urban, M.D., Urban, R., et al., 2019. A primer genetic toolkit for exploring mitochondrial biology and disease using zebrafish. bioRxiv 542084.

Sander, J.D., Cade, L., Khayter, C., Reyon, D., Peterson, R.T., Joung, K.J., Yeh, J.-R.J., 2011. Targeted gene disruption in somatic zebrafish cells using engineered TALENs. Nature Biotechnology 29, 697-698.

Sharma, S., Javadekar, S., Pandey, M., ivastava, Kumari, R., Raghavan, S., 2015. Homology and enzymatic requirements of microhomology-dependent alternative end joining. Cell Death and Disease 6, e1697.

Shen, B., Zhang, W., Zhang, J., Zhou, J., Wang, J., Chen, L., Wang, L., Hodgkins, A., Iyer, V., Huang, X., et al., 2014. Efficient genome modification by CRISPR-Cas9 nickase with minimal off-target effects. Nature Methods 11, 399-402.

Shen, M.W., Arbab, M., Hsu, J.Y., Worstell, D., Culbertson, S.J., Krabbe, O., Cassa, C.A., Liu, D.R., Gifford, D.K., Sherwood, R.I., 2018. Predictable and precise template-free CRISPR editing of pathogenic variants. Nature 563, 646-651.

Shinya, M., Sakai, N., 2011. Generation of highly homogeneous strains of zebrafish through full sib-pair mating. G3: genes genomes genetics 1 (5), 377-386.

Shui, B., Matias, L., Guo, Y., Peng, Y., 2016. The rise of CRISPR/Cas for genome editing in stem cells. Stem Cells International 2016, 1-17.

Simone, B.W., Martínez-Gálvez, G., WareJoncas, Z., Ekker, S.C., 2018. Fishing for understanding: unlocking the zebrafish gene editor's toolbox. Methods 150, 3-10.

Sinha, S., Villarreal, D., Shim, E., Lee, S., 2016. Risky business: microhomology-mediated end joining. Mutation Research: Fundamental and Molecular Mechanisms of Mutagenesis 788, 17-24.

Slaymaker, I.M., Gao, L., Zetsche, B., Scott, D.A., Yan, W.X., Zhang, F., 2016. Rationally engineered Cas9 nucleases with improved specificity. Science 351, 84-88.

Tsai, S.Q., Wyvekens, N., Khayter, C., Foden, J.A., Thapar, V., Reyon, D., Goodwin, M.J., Aryee, M.J., Joung, K.J., 2014. Dimeric CRISPR RNA-guided FokⅠ nucleases for highly specific genome editing. Nature Biotechnology 32, 569-576.

Varshney, G.K., Pei, W., LaFave, M.C., Idol, J., Xu, L., Gallardo, V., Carrington, B., Bishop, K., Jones, M., Li, M., et al., 2015. High-throughput gene targeting and phenotyping in zebrafish using CRISPR/Cas9. Genome Research 25, 1030-1042.

Watanabe, M., Iwashita, M., Ishii, M., Kurachi, Y., Kawakami, A., Kondo, S., Okada, N., 2006. Spot pattern of leopard *Danio* is caused by mutation in the zebrafish connexin41.8 gene. EMBO Reports 7, 893-897.

Wiedenheft, B., Sternberg, S.H., Doudna, J.A., 2012. RNA-guided genetic silencing systems in bacteria and archaea. Nature 482, 331-338.

Wierson, W.A., Welker, J.M., Almeida, M.P., Mann, C.M., Webster, D.A., Torrie, M.E., Weiss, T.J., Vollbrecht, M.K., Lan, M., McKeighan, K.C., et al., 2018. GeneWeld: a method for efficient targeted integration directed by short homology. bioRxiv 431627.

Wierson, W.A., Simone, B.W., WareJoncas, Z., Mann, C., Welker, J.M., Kar, B., Gendron, W.A., Barry, M.A., Clark, K.J., Dobbs, D.L., et al., 2019. Expanding the CRISPR toolbox with ErCas12a in zebrafish and human cells. biorxiv 650515.

Wright, A.V., Nuñez, J.K., Doudna, J.A., 2016. Biology and applications of CRISPR systems: harnessing nature's toolbox for genome engineering. Cell 164, 29-44.

Wyvekens, N., Topkar, V.V., Khayter, C., Joung, K.J., Tsai, S.Q., 2015. Dimeric CRISPR RNA-guided FokⅠ-dCas9 nucleases directed by truncated gRNAs for highly specific genome editing. Human Gene Therapy 26, 425-431.

Yang, B., Wen, X., Kodali, N., Oleykowski, C., Miller, C., Kulinski, J., Besack, D., Yeung, J., Kowalski, D., Yeung, A., 2000. Purification, Cloning, and Characterization of the CEL I Nuclease y. Biochemistry 39 (13), 3533e3541. https://doi.org/10.1021/bi992376z.

Zetsche, B., Gootenberg, J.S., Abudayyeh, O.O., Slaymaker, I.M., Makarova, K.S., Essletzbichler, P., Volz, S.E., Joung, J., van der Oost, J., Regev, A., et al., 2015. Cpf1 is a single RNA-guided endonuclease of a class 2 CRISPR-Cas system. Cell 163, 759-771.

Zhang, J.-P., Li, X.-L., Li, G.-H., Chen, W., Arakaki, C., Botimer, G.D., Baylink, D., Zhang, L., Wen, W., Fu, Y.-W., et al., 2017. Efficient precise knockin with a double cut HDR donor after CRISPR/Cas9-mediated double-stranded DNA cleavage. Genome Biology 18, 35.

第十七章

光遗传学

Sachiko Tsuda[1]

17.1 引言

神经科学的中心目标是了解大脑在各种行为条件下是如何工作的。为此，揭示大脑中负责信息处理的多种神经元和神经元回路的功能是至关重要的。电生理技术提供了关于神经元和回路功能的信息集合。这些技术在时间分辨率和灵敏度方面具有优势；然而，它们在检测神经元回路功能方面存在局限性。例如，使用电极进行细胞外刺激时，很难刺激特定类型和群体的神经元。对于细胞内的电刺激，只能刺激有限数量的神经元。在电生理记录方面，尽管它们具有出色的时间分辨率和高灵敏度，但它们在提供空间信息方面存在局限性，只能测量有限数量的神经元活动。考虑到大脑中神经元的多样性，这些局限性使得通过电生理技术来确定每种神经元类型和神经元回路的功能是非常困难的。

光学技术的最新进展提供了强大的工具来克服上述在研究神经元功能方面的局限性。一个突出的例子是光遗传学，它使神经元活动的光学测量和控制成为可能（Yizhar et al., 2011）。光遗传学是指光学和遗传学的结合，使光敏蛋白在遗传定义的神经元群体中表达。有两种类型的光遗传探针：光遗传执行器（actuators）和光遗传传感器（sensors）。光遗传执行器可以控制神经元活动，而光遗传传感器可以检测神经元活动（图 17-1）。通过一系列研究确定了以下微生物视蛋白可以用作光遗传执行器的物质：可以激活神经元的通道视紫红质（channelrhodopsin）和可以抑制神经元的氯视紫红质（halorhodopsin）。在光遗传传感器方面，已经开发了多种基因编码的离子指示剂，例如 GCaMPs（钙离子敏感探针）、Clomeleon（氯离子敏感探针）和 pHluorin（质子敏感探针）。此外，在开发遗传编码电压指示器（genetically encoded voltage indicators, GEVI）方面取得的重大进展促进了对膜电位的快速和直接测量（Nakajima et al., 2016）。

在本章中，我们重点介绍光遗传执行器和传感器及其在大脑回路功能分析中的应用，

[1] Graduate School of Science and Engineering, Saitama University, Saitama, Japan.

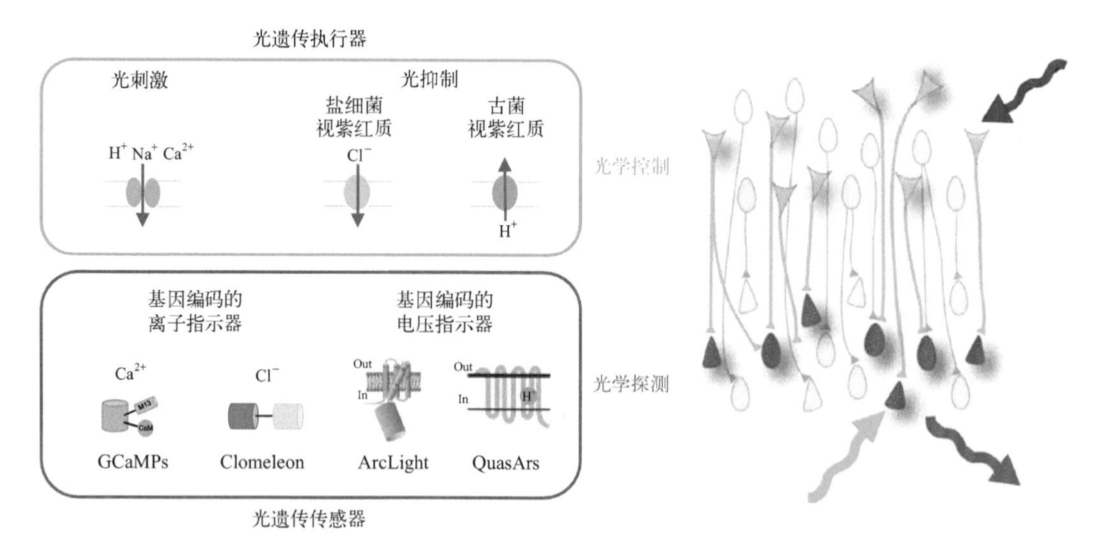

图 17-1 **使用光遗传执行器和传感器对神经元回路进行光遗传释疑**
根据 Strace 等人（2016）发表资料的修改

尤其是其在斑马鱼中的应用。斑马鱼在光学技术方面的优势使光遗传学能够应用于其各种行为研究，从而提供神经元和神经元回路在这些行为中的功能相关的基本信息。

17.2 神经元活动的光遗传学控制

17.2.1 光遗传学激活

通过对古菌、细菌、真菌和藻类中光敏泵的研究，在光学控制神经元活性方面取得了进展，其中包括光驱动的外向质子泵（Oesterhelt and Stoeckenius, 1973）和光驱动的内向氯离子泵（Lanyi and Weber, 1980; Matsuno-Yagi and Mukohata, 1977）。21 世纪初，通过将视紫红质和其他成分应用于果蝇的视觉系统，开发了遗传靶向光学控制的藻类衍生光门控离子通道的第一个版本（Zemelman et al., 2003）。在这项研究之后，表达 P2X2 受体的转基因果蝇被通过光遗传学诱导了行为改变（Lima and Miesenbock, 2005）。2005 年，将一种微生物视蛋白基因——通道视紫红质引入哺乳动物神经元，并诱发了毫秒级的动作电位（Boyden et al., 2005）。通道视紫红质在光遗传学方面具有显著优势，因为它不需要任何额外的分子就能对神经元活动进行光学控制。

光敏感通道（channelrhodopsin-2, ChR2）是一种广泛使用的通道视紫红质，它是一种在蓝光激活时打开的阳离子通道（Nagel et al., 2003）。通过在神经元中表达 ChR2，光可以在这些细胞中产生足够的光电流，从而触发动作电位（Boyden et al., 2005; Zhang et al., 2006）。这种光刺激可以以极快的速度（毫秒）进行，从而快速控制神经元活动。

通道视紫红质的分子修饰提供了一系列光遗传学应用（Yizhar et al., 2011; Deisseroth and Hegemann et al., 2017）。其中包括具有不同激发光谱的通道视紫红质，如红移通道视紫红质、ReaChR、Chronos 和 Chrimson（Inagaki et al., 2014; Klapoetke et al., 2014）。通过将这些探针与 ChR2 一起使用，可以分别控制不同类型的神经元，从而能够识别每种类型神经元的功能。此外，它还可以通过光同时控制和检测神经元活动（全光遗传学方法）。

17.2.2 光遗传学抑制

（1）氯视紫红质

盐细菌视紫红质（*Natronomonas* halorhodopsin, NpHR）的运用实现了神经元活性的光遗传光抑制（Zhang et al., 2007），这是一种光驱动的氯离子泵，可导致细胞超极化（Asrican et al., 2013）。这使得分析神经元回路的功能丧失成为可能（Yizhar et al., 2011）。由于 NpHR 的激发光谱在黄绿色范围内（约 580nm），与 ChR2 相比发生了红移，因此可以通过同时表达 NpHR 和 ChR2 来进行神经元活动的双向光遗传控制（Zhang et al., 2007; Arrenberg et al., 2009）。然而，NpHR 在细胞内转运至内质网（endoplasmic reticulum, ER），而不是外质膜，所以 NpHR 导致其光电流大小相对较小（Zhao et al., 2008; Gradinaru et al., 2008）。为了克服这一限制，通过在 NpHR 中添加 ER 输出基序设计了第二代增强型 NpHR（second-generation enhanced NpHR, eNpHR2.0）（Zhao et al., 2008; Gradinaru et al., 2008）。eNpHR2.0 改进了光敏性，具有更快的动力学，并已应用于体内和体外分析（Asrican et al., 2013; Tonnesen et al., 2009; Sohal et al., 2009）。此外，第三代 NpHR（third-generation NpHR, eNpHR3.0）也被开发出来，其通过添加来自 Kir2.1 钾通道的轴突运输序列来增强光抑制（Yizhar et al., 2011; Gradinaru et al., 2010）。eNpHR3.0 的表达强度足以使红光 / 远红光（约 680nm）产生光抑制，从而使光线更深入地穿透脑组织，该技术已在体内成功应用（Tye et al., 2011; Tonnesen et al., 2011）。

（2）古菌视紫红质

古菌视紫红质 -3（archaerhodopsin-3, Arch）也可以对神经元进行光抑制。Arch 是一种光驱动的外向质子泵，最初在盐细菌中发现（Mukohata et al., 1988）。在黄绿色光（约 570nm）的照射下，Arch 泵出质子，导致表达 Arch 的神经元受到光抑制（Chow et al., 2010）。尽管由于氯离子的积累，NpHR 逐渐改变 GABA（A）受体的逆转电位，从而导致抑制性突触变得兴奋，但 Arch 没有这个问题（Raimondo et al., 2012）。当存在抑制性突触输入时，这种差异使 Arch 成为光抑制的合适选择。然而，Arch 也有一个缺点，即由于质子外流，它会暂时增加细胞内的 pH 值（Chow et al., 2010）。

除了这些用于光抑制的光遗传执行器外，最近的研究还通过使用光门控钾通道提供了基于钾通道的光遗传学沉默工具（Cosentino et al., 2015; Bernal Sierra et al., 2018; Alberio et al., 2018）。

（3）其他类型的光遗传执行器

受体和分子信号的光遗传转换为研究大脑活动和行为过程中发生的分子现象提供了一种有效的方法。这是通过开发新的光遗传探针实现的，如 AMPA 受体（Kakegawa et al., 2018）、cAMP（Gasser et al., 2014）、Nodal（Sako et al., 2016）、Frizzled 7（Capek et al., 2019）和趋化因子（Sarris et al., 2016）。例如，PhotonSABER 能够对 AMPA 型谷氨酸受体内吞作用进行时间、空间和细胞类型特异性控制，开辟了一种通过光控制长时程抑制的方法，并揭示了其在小鼠运动学习中的作用（Kakegawa et al., 2018）。据报道，这些光遗传探针大多数对斑马鱼也适用（Gasser et al., 2014; Sako et al., 2016; Capek et al., 2019; Sarris et al., 2016），从而为神经元功能的分析提供了更广泛的选择。

17.3 神经元活动的光遗传学检测

17.3.1 钙成像

神经元放电通常与通过电压门控 Ca^{2+} 通道增加 Ca^{2+} 水平有关。因此，最早通过使用钙染料，钙成像已被广泛用于检测神经元活动（Homma et al., 2009; Yuste et al., 1992）。1997 年，开发了 Cameleon，这是一种基于钙调蛋白基因编码的钙指示剂（genetically encoded calcium indicator, GECI）（Miyawaki et al., 1997, 1999）。此后一系列研究发现了 GECIs 的各种改进版本，包括 GCaMP，它利用循环排列的绿色荧光蛋白和钙调蛋白，导致更高的 Ca^{2+} 亲和力和信噪比（signal-to-noise ratio, SNR）（Nakai et al., 2001）。在这一成就之后，GCaMPs 的新版本和改进版本已经陆续开发出来，包括红移 GECIs，它们有助于检测不同神经元群体的功能（Dana et al., 2016; Looger and Griesbeck, 2012）。然而，钙成像的局限性在于钙信号可以通过独立于神经元放电的细胞信号转导过程而改变（Charpak et al., 2001; Helmchen et al., 1999），这可能会高估神经元活动。尽管如此，GECIs 被广泛用于作为峰值活动的指标，这主要归因于 GECIs 的高信噪比和钙动力学的缓慢动力学，这有助于使成像更加明亮。

目前认为斑马鱼小而透明的大脑是一个进行神经元群体钙成像的合适系统。一系列研究使用了全脑钙成像技术，通常将该技术与行为分析和光遗传控制神经元活动相结合（Keller and Ahrens, 2015; Ahrens and Engert, 2015）。

17.3.2 氯成像

抑制信号在神经网络的信息处理中起着至关重要的作用，通常依赖于使用 GABA 和甘氨酸受体的 Cl^- 跨膜通量。为了观测突触抑制，一种基于 FRET 的基因编码氯化物指示剂 Clomeleon 被开发出来（Kuner and Augustine, 2000）。Clomeleon 由黄色荧光蛋白受体组成，其荧光被 Cl^- 和其他与青色荧光蛋白供体相关的卤化物猝灭（Wachter and Remington, 1999）。Cl^- 浓度的增加会降低 FRET，从而将 Clomeleon 的颜色从黄色变为青色。Clomeleon 已经在转基因小鼠的神经元中成功表达，从而能够分析多个大脑区域（例如海马体、小脑和杏仁核）中抑制回路的时空动态（Berglund et al., 2006）。但是，Clomeleon 也有一些缺点，如信噪比较低、动态范围狭窄。这些缺点主要归因于其对 Cl^- 的低亲和力和相对较高的细胞内 Cl^- 背景水平。利用体外蛋白工程技术，开发了具有更好信噪比、更高结合亲和力和更宽动态范围的 Clomeleon 变异体 SuperClomeleon（Grimley et al., 2013）。此外，已开发出另一种类型的传感器 ClopHensor，可以同时测量氯离子和 pH 值（Arosio et al., 2010）。

17.3.3 电压成像

通过电压成像对膜电位进行光学测量是一种非常有吸引力的方法，可以快速、直接和同时检测神经元群体的膜电位（Cohen and Salzberg, 1978; Loew et al., 1979; Djurisic et al., 2003; Storace et al., 2016; Knopfel and Song, 2019; Tsuda et al., 2013）。其高速和直接性能够检测突触和动作电位以及超极化。最近，遗传编码的基于 GPCR 活化的多巴胺（GPCR-activation-based DA, GRABDA）传感器的开发取得了重大进展，可以对来自不同神经元群体的神经元活动进行细胞类型特异性检测。越来越多的 GEVIs 具有更快的动力学和更亮的荧光，从而克服了先前在速度和信噪比方面的弱点（Knopfel and Song, 2019; Perron et al., 2009; St-Pierre et al., 2014; Kralj et al., 2012; Jin et al., 2012; Hochbaum et al., 2014）。关于 GEVIs 的结构，有几类：一类是一系列嵌合蛋白，它们使用来自海鞘和其他脊椎动物的电压敏感磷酸酶的电压敏感域［例如，ArcLight（Jin et al., 2012），VSFP-Butterfly 1.2.（Akemann et al., 2012）和 ASAP1（St-Pierre et al., 2014）］；另一类 GEVIs 利用微生物视紫红质的电压敏感性，例如，QuasArs（Hochbaum et al., 2014）、Ace2N-4AAmNeon（Gong et al., 2015）、Archon（Piatkevich et al., 2018）和 Voltron（Abdelfattah et al., 2019）。最近开发的 GEVIs 的比较评估可在已发表的综述中找到（Storace et al., 2016; Bando et al., 2019）。关于 GEVIs 在斑马鱼中的应用，部分 GEVIs 已通过建立转基因鱼品系成功应用于斑马鱼大脑的电压成像（Piatkevich et al., 2018; Abdelfattah et al., 2019; Miyazawa et al., 2018）。

17.3.4 其他的光遗传传感器

除了上述的光遗传传感器，研究者们开发了越来越多的蛋白传感器检测神经递质，包括谷氨酸和多巴胺（dopamine, DA）。iGluSnFR 是一种基于强度的谷氨酸敏感荧光报告基因，可对神经元和星形胶质细胞释放的谷氨酸进行体内成像。（Marvin et al., 2013）。已有文献报道了 iGluSnFR 在斑马鱼视顶盖中的应用（Marvin et al., 2013）。最近，研究人员开发了遗传编码电压指示剂（genetically encoded voltage indicator, GEVIs），能够以高时空精度测量活体动物（包括斑马鱼）的 DA 变化（Sun et al., 2018）。

17.4 光遗传学回路分析

17.4.1 体内光遗传学控制

光遗传学的强大之处在于它能在体内调控遗传定义的神经元簇。在大脑中，许多类型的神经元具有不同的功能，但通常在空间上彼此靠近。随着光遗传执行器在特定细胞类型上的基因靶向表达，现在可以在不干扰邻近细胞的情况下选择性激活或抑制特定细胞类型。将这些方法与行为测试相结合，可以进行特定神经元类型以及体内神经元回路的功能丧失和获得实验。使用这些方法得到的结果可以建立特定神经元类型和行为输出之间的因果关系。在第五章中解释了一些使用斑马鱼的例子。

17.4.2 回路连通性的光遗传学探测

基于光遗传中光刺激的另一个应用是鉴定体内和体外不同位置神经元之间的功能连接（图 17-2）。通过一个小光点照射表达光遗传执行器的神经元可以光激活这些神经元。通过将这种光刺激与突触后神经元的电生理或光学记录结合起来，就有可能识别这些神经元之间的突触连接。根据光点的位置，可以得到不同回路连通性的信息。通过光点靶向细胞体，可以识别长程投射，而轴突的光刺激可以识别目标结构内的局部突触连接。斑马鱼脑神经回路图谱的绘制也应用了类似的方法（Helmbrecht et al., 2018）。

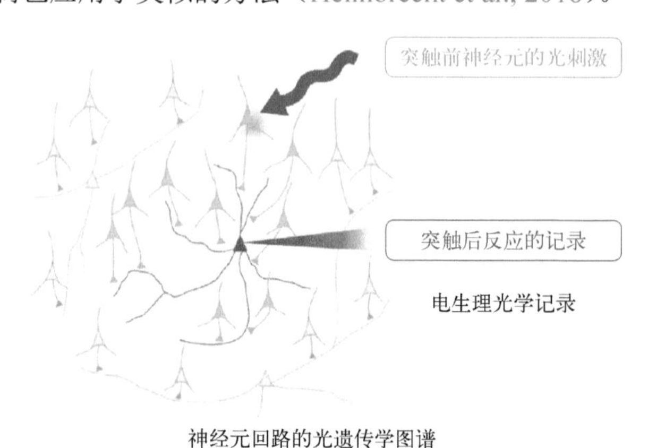

图 17-2 **神经回路的光遗传学图谱绘制**
修改自：Mancuso et al., 2011

上述方法通过小光点扫描脑组织，同时记录突触后神经元的活动，实现大脑的功能定位。它为分析大脑回路的功能和空间结构提供了一种有效的方法（Asrican et al., 2013; Kim et al., 2014; Wang et al., 2007）。

17.5 光遗传学在斑马鱼行为研究中的应用

斑马鱼透明且相对较小的大脑为分析不同行为模式下的大脑功能结构提供了一个极好的系统（Friedrich et al., 2010; Del Bene and Wyart, 2012; Baier and Scott, 2009; Portugues et al., 2013）。此外，建立透明的成年斑马鱼品系，即色素沉着突变体，可以精确观察和控制斑马鱼的神经元活动（White et al., 2008）。使用光遗传学的研究越来越多。本节介绍了光遗传学在斑马鱼几种行为中的一些应用，主要侧重于利用光遗传执行器的研究。最近使用包括钙成像的光遗传传感器的研究综述，可在其他地方查阅（Ahrens and Engert, 2015; Portugues et al., 2013）。

17.5.1 运动系统

在运动系统首次实现了通道视紫红质在斑马鱼大脑中的应用（Arrenberg et al., 2009;

Wyart et al., 2009)。利用在 Gal4/UAS 系统的控制下表达增强型 NpHR 和 ChR2 的转基因斑马鱼仔鱼，鉴定并表征了在后脑区的游泳指令回路（Arrenberg et al., 2009）。在脊髓中，斑马鱼仔鱼自发运动是通过正向驱动脊髓中枢模式发生器实现的，而这是利用光门控通道 LiGluR 鉴定出来的（Wyart et al., 2009）。自此，基于光遗传学的研究迅速增加，从而提供了调节后脑（Kimura et al., 2013; Lacoste et al., 2015; Song et al., 2016）、小脑（Matsui et al., 2014; Knogler et al., 2019）和三叉神经节（Itoh et al., 2014）运动行为的神经元回路和运动学习（Kawashima et al., 2016）的基本信息。

17.5.2 感觉系统

斑马鱼有多种与感觉系统有关的行为（Friedrich et al., 2010）。其中，视觉系统的解剖、发育、功能等方面的研究较为深入。斑马鱼具有广泛且保守的视觉行为，如视动性和视动反应（Beck et al., 2004; Neuhauss et al., 1999）。作为光遗传学在视觉系统上应用的早期尝试，人们利用 ChR2 光刺激确定了假定的后脑扫视发生器的位置（Schoonheim et al., 2010）。此后，利用 ChR2、NpHR 和 CalipHluorin（遗传编码的突触 pH 值报告分子）等光遗传执行器进行的研究，为斑马鱼仔鱼如何处理视觉输入并将信息整合到运动输出中提供了重要的见解，如视网膜计算（Wang et al., 2014）、动眼肌整合器（Goncalves et al., 2014）、光流响应（Kubo et al., 2014）和方向选择性（Antinucci et al., 2016）。此外，顶盖的钙成像表明，神经元回路能够产生视觉感知和视觉运动感知（Muto and Kawakami, 2013; Perez-Schuster et al., 2016）。关于其他感觉系统，越来越多的光遗传学方法描述了斑马鱼嗅觉系统中神经元回路的功能（Blumhagen et al., 2011; Bundschuh et al., 2012）。

17.5.3 恐惧、睡眠和癫痫发作

斑马鱼还有其他各种各样的行为。在这里，我们回顾了有关恐惧反应、睡眠和癫痫发作的研究。恐惧反应的一种类型是通过学习适应策略来主动回避潜在的危险环境。光遗传学揭示了背侧缰核（dorsal habenular nuclei, dHb）回路的功能不对称性，这相当于哺乳动物的内侧缰核（Hong et al., 2013）。成年斑马鱼体内的光遗传激活实验表明，腹侧缰核 - 中缝回路对于位置回避行为至关重要（Amo et al., 2014）。还有一些研究利用光遗传学揭示了应激反应的神经元基础（De Marco et al., 2013, 2016）。

光遗传学方法也阐明了睡眠的机制。通过对下丘脑分泌素表达神经元的光遗传激活以及对突变斑马鱼的分析表明，去甲肾上腺素对于促进斑马鱼仔鱼的觉醒和下丘脑分泌素诱导的觉醒至关重要（Singh et al., 2015）。最近一项使用光遗传学方法的研究表明，5- 羟色胺能中缝神经元促进斑马鱼仔鱼的睡眠（Oikonomou et al., 2019）。

关于癫痫，研究导致癫痫发作的状态转换的科研工作一直在增加。例如，最近在斑马鱼仔鱼中使用 ChR2 的研究描述了胶质细胞 - 神经元相互作用对癫痫发作状态转变的重要性（Diaz Verdugo et al., 2019）。

在上述工作中，光遗传学主要是通过转基因的方式应用于斑马鱼，将编码光遗传探针的基因整合到基因组中。这一方法的实现是由于在斑马鱼靶向基因表达技术上的重大

进展。突出的例子是利用 Tol2 转座子系统的高效转基因方法和基因捕获和增强子捕获方法，以及使用 Gal4-UAS 系统在神经回路功能研究中的应用（Asakawa et al., 2008）。通过筛选数以千计的增强子诱捕品系，鉴定出一组在特定细胞中表达 Gal4 的转基因斑马鱼品系，并通过将它们与 UAS 系杂交应用于光遗传学方法，从而触发细胞中所需光遗传学探针的表达（Asakawa et al., 2008）。虽然以病毒为载体感染是将光遗传学探针引入大脑的另一种主要方法，但转基因方法因其便捷性和同质性探针表达动物的可靠性而更为有用（Yizhar et al., 2011）。

17.6　光遗传学的未来展望

光遗传学的最新进展通过分析神经元活动和行为之间的因果关系，使神经科学领域发生了革命性的变化。光遗传执行器和传感器列表的扩展，以及表达这些执行器的转基因品系，为了解神经元回路的功能提供了强大的工具。此外，显微技术的改进也推进了大脑功能的研究，特别是通过全脑光学方法。这些技术的进步促进了光遗传学应用于各种行为条件，包括自由活动的斑马鱼和使用透明斑马鱼品系的成年斑马鱼（White et al., 2008; Kim et al., 2017; Naumann et al., 2010）。然而，光遗传学方法仍然存在局限性。例如，光遗传学探针表达模式的特异性通常依赖于合适的启动子 / 增强子序列的适用性。随着各种基因调控序列信息的积累、增强子诱捕品系的收集以及利用光学来精确激活特定的神经元群体知识的增加，将进一步促进光遗传控制和检测遗传定义的神经元群体中神经元的活动（Asakawa et al., 2008）。

大脑由多个紧密相连的神经元组成的神经回路组成。为了更深入地了解大脑功能，在观察行为结果的同时分析这些神经元的活动是至关重要的。光遗传学现在能够通过活体全光方法将光遗传学检测突触后神经元的反应与使用光遗传学光刺激突触前神经元相结合来完成这些任务。改进光学系统来高时空分辨率地光刺激和检测神经元群体的活动，以及增加有关斑马鱼脑回路神经解剖学的详细信息，对于更精确地检查神经元功能也至关重要，这将有助于更深入地理解大脑的功能结构。

致谢

本研究由日本埼玉大学 JSPS KAKENHI（科研资助）和 JST 终身教职跟踪计划（SUTT）资助。

（石晶晶翻译　李丽琴审校）

参考文献

Ahrens, M.B., Engert, F., 2015. Large-scale imaging in small brains. Current Opinion in Neurobiology 32C, 78-86.

Abdelfattah, A.S., Kawashima, T., Singh, A., Novak, O., Liu, H., Shuai, Y., et al., 2019. Bright and photostable chemigenetic indicators for extended in vivo voltage imaging. Science 365 (6454), 699-704.

Akemann, W., Mutoh, H., Perron, A., Park, Y.K., Iwamoto, Y., Knopfel, T., 2012. Imaging neural circuit dynamics with a voltage-sensitive fluorescent protein. Journal of Neurophysiology 108 (8), 2323-2337.

Alberio, L., Locarno, A., Saponaro, A., Romano, E., Bercier, V., Albadri, S., et al., 2018. A light-gated potassium channel for sustained neuronal inhibition. Nature Methods 15 (11), 969-976.

Amo, R., Fredes, F., Kinoshita, M., Aoki, R., Aizawa, H., Agetsuma, M., et al., 2014. The habenulo-raphe serotonergic circuit encodes an aversive expectation value essential for adaptive active avoidance of danger. Neuron 84 (5), 1034-1048.

Antinucci, P., Suleyman, O., Monfries, C., Hindges, R., 2016. Neural mechanisms generating orientation selectivity in the retina. Current Biology 26 (14), 1802-1815.

Arosio, D., Ricci, F., Marchetti, L., Gualdani, R., Albertazzi, L., Beltram, F., 2010. Simultaneous intracellular chloride and pH measurements using a GFP-based sensor. Nature Methods 7 (7), 516-518.

Arrenberg, A.B., Del Bene, F., Baier, H., 2009. Optical control of zebrafish behavior with halorhodopsin. Proceedings of the National Academy of Sciences of the United States of America 106 (42), 17968-17973.

Asakawa, K., Suster, M.L., Mizusawa, K., Nagayoshi, S., Kotani, T., Urasaki, A., et al., 2008. Genetic dissection of neural circuits by Tol2 transposon-mediated Gal4 gene and enhancer trapping in zebrafish. Proceedings of the National Academy of Sciences of the United States of America 105 (4), 1255-1260.

Asrican, B., Augustine, G.J., Berglund, K., Chen, S., Chow, N., Deisseroth, K., et al., 2013. Next-generation transgenic mice for optogenetic analysis of neural circuits. Frontiers in Neural Circuits 7, 160.

Baier, H., Scott, E.K., 2009. Genetic and optical targeting of neural circuits and behavior—zebrafish in the spotlight. Current Opinion in Neurobiology 19 (5), 553-560.

Bando, Y., Sakamoto, M., Kim, S., Ayzenshtat, I., Yuste, R., 2019. Comparative evaluation of genetically encoded voltage indicators. Cell Reports 26 (3), 802-813 e4.

Beck, J.C., Gilland, E., Tank, D.W., Baker, R., 2004. Quantifying the ontogeny of optokinetic and vestibuloocular behaviors in zebrafish, medaka, and goldfish. Journal of Neurophysiology 92 (6), 3546-3561.

Berglund, K., Schleich, W., Krieger, P., Loo, L.S., Wang, D., Cant, N.B., et al., 2006. Imaging synaptic inhibition in transgenic mice expressing the chloride indicator, Clomeleon. Brain Cell Biology 35 (4e6), 207-228.

Bernal Sierra, Y.A., Rost, B.R., Pofahl, M., Fernandes, A.M., Kopton, R.A., Moser, S., et al., 2018. Potassium channel-based optogenetic silencing. Nature Communications 9 (1), 4611.

Blumhagen, F., Zhu, P., Shum, J., Scharer, Y.P., Yaksi, E., Deisseroth, K., et al., 2011. Neuronal filtering of multiplexed odour representations. Nature 479 (7374), 493-498.

Boyden, E.S., Zhang, F., Bamberg, E., Nagel, G., Deisseroth, K., 2005. Millisecond-timescale, genetically targeted optical control of neural activity. Nature Neuroscience 8 (9), 1263-1268.

Bundschuh, S.T., Zhu, P., Scharer, Y.P., Friedrich, R.W., 2012. Dopaminergic modulation of mitral cells and odor responses in the zebrafish olfactory bulb. Journal of Neuroscience 32 (20), 6830-6840.

Capek, D., Smutny, M., Tichy, A.M., Morri, M., Janovjak, H., Heisenberg, C.P., 2019. Light-activated Frizzled7 reveals a permissive role of non-canonical wnt signaling in mesendoderm cell migration. eLife 8.

Charpak, S., Mertz, J., Beaurepaire, E., Moreaux, L., Delaney, K., 2001. Odor-evoked calcium signals in dendrites of rat mitral cells. Proceedings of the National Academy of Sciences of the United States of America 98 (3), 1230-1234.

Chow, B.Y., Han, X., Dobry, A.S., Qian, X., Chuong, A.S., Li, M., et al., 2010. High-performance genetically targetable optical neural silencing by light-driven proton pumps. Nature 463 (7277), 98-102.

Cohen, L.B., Salzberg, B.M., 1978. Optical measurement of membrane potential. Reviews of Physiology, Biochemistry & Pharmacology 83, 35-88.

Cosentino, C., Alberio, L., Gazzarrini, S., Aquila, M., Romano, E., Cermenati, S., et al., 2015. Optogenetics. Engineering of a light-gated potassium channel. Science 348 (6235), 707-710.

Dana, H., Mohar, B., Sun, Y., Narayan, S., Gordus, A., Hasseman, J.P., et al., 2016. Sensitive red protein calcium indicators for imaging neural activity. eLife 5.

De Marco, R.J., Groneberg, A.H., Yeh, C.M., Castillo Ramirez, L.A., Ryu, S., 2013. Optogenetic elevation of endogenous glucocorticoid level in larval zebrafish. Frontiers in Neural Circuits 7, 82.

De Marco, R.J., Thiemann, T., Groneberg, A.H., Herget, U., Ryu, S., 2016. Optogenetically enhanced pituitary corticotroph cell activity post-stress onset causes rapid organizing effects on behaviour. Nature Communications 7, 12620.

Deisseroth, K., Hegemann, P., 2017. The form and function of channelrhodopsin. Science 357 (6356).

Del Bene, F., Wyart, C., 2012. Optogenetics: a new enlightenment age for zebrafish neurobiology. Developmental Neurobiology 72 (3), 404-414.

Diaz Verdugo, C., Myren-Svelstad, S., Aydin, E., Van Hoeymissen, E., Deneubourg, C., Vanderhaeghe, S., et al., 2019. Glia-neuron interactions underlie state transitions to generalized seizures. Nature Communications 10 (1), 3830.

Djurisic, M., Zochowski, M., Wachowiak, M., Falk, C.X., Cohen, L.B., Zecevic, D., 2003. Optical monitoring of neural activity using voltage-sensitive dyes. Methods in Enzymology 361, 423-451.

Friedrich, R.W., Jacobson, G.A., Zhu, P., 2010. Circuit neuroscience in zebrafish. Current Biology 20 (8), R371-R381.

Gasser, C., Taiber, S., Yeh, C.M., Wittig, C.H., Hegemann, P., Ryu, S., et al., 2014. Engineering of a red-ligh activated human cAMP/cGMP-specific phosphodiesterase. Proceedings of the National Academy of Sciences of the United States of America 111 (24), 8803-8808.

Goncalves, P.J., Arrenberg, A.B., Hablitzel, B., Baier, H., Machens, C.K., 2014. Optogenetic perturbations reveal the dynamics of an oculomotor integrator. Frontiers in Neural Circuits 8, 10.

Gong, Y., Huang, C., Li, J.Z., Grewe, B.F., Zhang, Y., Eismann, S., et al., 2015. High-speed recording of neural spikes in awake mice and flies with a fluorescent voltage sensor. Science 350 (6266), 1361-1366.

Gradinaru, V., Thompson, K.R., Deisseroth, K., 2008. eNpHR: a Natronomonas halorhodopsin enhanced for optogenetic applications. Brain Cell Biology 36 (1-4), 129-139.

Gradinaru, V., Zhang, F., Ramakrishnan, C., Mattis, J., Prakash, R., Diester, I., et al., 2010. Molecular and cellular approaches for diversifying and extending optogenetics. Cell 141 (1), 154-165.

Grimley, J.S., Li, L., Wang, W., Wen, L., Beese, L.S., Hellinga, H.W., et al., 2013. Visualization of synaptic inhibition with an optogenetic sensor developed by cell-free protein engineering automation. Journal of Neuroscience 33 (41), 16297-16309.

Helmbrecht, T.O., Dal Maschio, M., Donovan, J.C., Koutsouli, S., Baier, H., 2018. Topography of a visuomotor transformation. Neuron 100 (6), 1429-14245 e4.

Helmchen, F., Svoboda, K., Denk, W., Tank, D.W., 1999. In vivo dendritic calcium dynamics in deep-layer cortical pyramidal neurons. Nature Neuroscience 2 (11), 989-996.

Hochbaum, D.R., Zhao, Y., Farhi, S.L., Klapoetke, N., Werley, C.A., Kapoor, V., et al., 2014. All-optical electrophysiology in mammalian neurons using engineered microbial rhodopsins. Nature Methods 11 (8), 825-833.

Homma, R., Baker, B.J., Jin, L., Garaschuk, O., Konnerth, A., Cohen, L.B., et al., 2009. Wide-field and two-photon imaging of brain activity with voltage- and calcium-sensitive dyes. Philosophical Transactions of the Royal Society of London B Biological Sciences 364 (1529), 2453-2467.

Hong, E., Santhakumar, K., Akitake, C.A., Ahn, S.J., Thisse, C., Thisse, B., et al., 2013. Cholinergic left-right asymmetry in the habenulo-interpeduncular pathway. Proceedings of the National Academy of Sciences of the United States of America 110 (52), 21171-21176.

Inagaki, H.K., Jung, Y., Hoopfer, E.D., Wong, A.M., Mishra, N., Lin, J.Y., et al., 2014. Optogenetic control of *Drosophila* using a red-shifted channelrhodopsin reveals experience-dependent influences on courtship. Nature Methods 11 (3), 325-332.

Itoh, M., Yamamoto, T., Nakajima, Y., Hatta, K., 2014. Multistepped optogenetics connects neurons and behavior. Current Biology 24 (24), R1155-R1156.

Jin, L., Han, Z., Platisa, J., Wooltorton, J.R., Cohen, L.B., Pieribone, V.A., 2012. Single action potentials and subthreshold electrical events imaged in neurons with a fluorescent protein voltage probe. Neuron 75 (5), 779-785.

Kakegawa, W., Katoh, A., Narumi, S., Miura, E., Motohashi, J., Takahashi, A., et al., 2018. Optogenetic control of synaptic AMPA receptor endocytosis reveals roles of LTD in motor learning. Neuron 99 (5), 985-998.e6.

Kawashima, T., Zwart, M.F., Yang, C.T., Mensh, B.D., Ahrens, M.B., 2016. The serotonergic system tracks the outcomes of actions to mediate short-term motor learning. Cell 167 (4), 933-946.e20.

Keller, P.J., Ahrens, M.B., 2015. Visualizing whole-brain activity and development at the single-cell level using lightsheet microscopy. Neuron 85 (3), 462-483.

Kim, J., Lee, S., Tsuda, S., Zhang, X., Asrican, B., Gloss, B., et al., 2014. Optogenetic mapping of cerebellar inhibitory circuitry reveals spatially biased coordination of interneurons via electrical synapses. Cell Reports 7 (5), 1601-1613.

Kim, D.H., Kim, J., Marques, J.C., Grama, A., Hildebrand, D.G.C., Gu, W., et al., 2017. Pan-neuronal calcium imaging with cellular resolution in freely swimming zebrafish. Nature Methods 14 (11), 1107-1114.

Kimura, Y., Satou, C., Fujioka, S., Shoji, W., Umeda, K., Ishizuka, T., et al., 2013. Hindbrain V2a neurons in the excitation of spinal locomotor circuits during zebrafish swimming. Current Biology 23 (10), 843-849.

Klapoetke, N.C., Murata, Y., Kim, S.S., Pulver, S.R., Birdsey-Benson, A., Cho, Y.K., et al., 2014. Independent optical excitation of distinct neural populations. Nature Methods 11 (3), 338-346.

Knogler, L.D., Kist, A.M., Portugues, R., 2019. Motor context dominates output from purkinje cell functional regions during reflexive visuomotor behaviours. eLife 8.

Knopfel, T., Song, C., 2019. Optical voltage imaging in neurons: moving from technology development to practical tool. Nature Reviews Neuroscience 20 (12), 719-727.

Kralj, J.M., Douglass, A.D., Hochbaum, D.R., Maclaurin, D., Cohen, A.E., 2012. Optical recording of action potentials in mammalian neurons using a microbial rhodopsin. Nature Methods 9 (1), 90-95.

Kubo, F., Hablitzel, B., Dal Maschio, M., Driever, W., Baier, H., Arrenberg, A.B., 2014. Functional architecture of an optic flow-responsive area that drives horizontal eye movements in zebrafish. Neuron 81 (6), 1344-1359.

Kuner, T., Augustine, G.J., 2000. A genetically encoded ratiometric indicator for chloride: capturing chloride transients in cultured hippocampal neurons. Neuron 27 (3), 447-459.

Lacoste, A.M., Schoppik, D., Robson, D.N., Haesemeyer, M., Portugues, R., Li, J.M., et al., 2015. A convergent and essential interneuron pathway for Mauthner-cell-mediated escapes. Current Biology 25 (11), 1526-1534.

Lanyi, J.K., Weber, H.J., 1980. Spectrophotometric identification of the pigment associated with light-driven primary sodium translocation in *Halobacterium halobium*. Journal of Biological Chemistry 255 (1), 243-250.

Lima, S.Q., Miesenbock, G., 2005. Remote control of behavior through genetically targeted photostimulation of neurons. Cell 121 (1), 141-152.

Loew, L.M., Scully, S., Simpson, L., Waggoner, A.S., 1979. Evidence for a charge-shift electrochromic mechanism in a probe of membrane potential. Nature 281 (5731), 497-499.

Looger, L.L., Griesbeck, O., 2012. Genetically encoded neural activity indicators. Current Opinion in Neurobiology 22 (1), 18-23.

Mancuso, J.J., Kim, J., Lee, S., Tsuda, S., Chow, N.B., Augustine, G.J., 2011. Optogenetic probing of functional brain circuitry. Experimental Physiology 96 (1), 26-33.

Marvin, J.S., Borghuis, B.G., Tian, L., Cichon, J., Harnett, M.T., Akerboom, J., et al., 2013. An optimized fluorescent probe for visualizing glutamate neurotransmission. Nature Methods 10 (2), 162-170.

Matsui, H., Namikawa, K., Babaryka, A., Koster, R.W., 2014. Functional regionalization of the teleost cerebellum analyzed in vivo. Proceedings of the National Academy of Sciences of the United States of America 111 (32), 11846-11851.

Matsuno-Yagi, A., Mukohata, Y., 1977. Two possible roles of bacteriorhodopsin; a comparative study of strains of *Halobacterium halobium* differing in pigmentation. Biochemical and Biophysical Research Communications 78 (1), 237-243.

Miyawaki, A., Llopis, J., Heim, R., McCaffery, J.M., Adams, J.A., Ikura, M., et al., 1997. Fluorescent indicators for Ca2t based on green fluorescent proteins and calmodulin. Nature 388 (6645), 882-887.

Miyawaki, A., Griesbeck, O., Heim, R., Tsien, R.Y., 1999. Dynamic and quantitative Ca2t measurements using improved cameleons. Proceedings of the National Academy of Sciences of the United States of America 96 (5), 2135-2140.

Miyazawa, H., Okumura, K., Hiyoshi, K., Maruyama, K., Kakinuma, H., Amo, R., et al., 2018. Optical interrogation of neuronal circuitry in zebrafish using genetically encoded voltage indicators. Scientific Reports 8 (1), 6048.

Mukohata, Y., Sugiyama, Y., Ihara, K., Yoshida, M., 1988. An Australian Halobacterium contains a novel proton pump retinal protein: archaerhodopsin. Biochemical and Biophysical Research Communications 151 (3), 1339-1345.

Muto, A., Kawakami, K., 2013. Prey capture in zebrafish larvae serves as a model to study cognitive functions Frontiers in Neural Circuits 7, 110.

Nagel, G., Szellas, T., Huhn, W., Kateriya, S., Adeishvili, N., Berthold, P., et al., 2003. Channelrhodopsin-2, a directly light-gated cation-selective membrane channel. Proceedings of the National Academy of Sciences of the United States of America 100 (24), 13940-13945.

Nakai, J., Ohkura, M., Imoto, K., 2001. A high signal-to-noise Ca2t probe composed of a single green fluorescent protein. Nature Biotechnology 19 (2), 137-141.

Nakajima, R., Tsuda, S., Kim, J., Augustine, G., 2016. Optogenetics Enables Selective Control of Cellular Electrical Activity, Molecular Neuroendocrinology: From Genome to Physiology. Wiley-Blackwell, pp. 275-300.

Naumann, E.A., Kampff, A.R., Prober, D.A., Schier, A.F., Engert, F., 2010. Monitoring neural activity with bioluminescence during natural behavior. Nature Neuroscience 13 (4), 513-520.

Neuhauss, S.C., Biehlmaier, O., Seeliger, M.W., Das, T., Kohler, K., Harris, W.A., et al., 1999. Genetic disorders of vision revealed by a behavioral screen of 400 essential loci in zebrafish. Journal of Neuroscience 19 (19), 8603-8615.

Oesterhelt, D., Stoeckenius, W., 1973. Functions of a new photoreceptor membrane. Proceedings of the National Academy of Sciences of the United States of America 70 (10), 2853-2857.

Oikonomou, G., Altermatt, M., Zhang, R.W., Coughlin, G.M., Montz, C., Gradinaru, V., et al., 2019. The serotonergic raphe promote sleep in zebrafish and mice. Neuron 103 (4), 686-701.

Perez-Schuster, V., Kulkarni, A., Nouvian, M., Romano, S.A., Lygdas, K., Jouary, A., et al., 2016. Sustained rhythmic brain activity underlies visual motion perception in zebrafish. Cell Reports 17 (4), 1098-1112.

Perron, A., Mutoh, H., Akemann, W., Gautam, S.G., Dimitrov, D., Iwamoto, Y., et al., 2009. Second and third generation voltage-sensitive fluorescent proteins for monitoring membrane potential. Frontiers in Molecular Neuroscience 2, 5.

Piatkevich, K.D., Jung, E.E., Straub, C., Linghu, C., Park, D., Suk, H.J., et al., 2018. A robotic multidimensional directed evolution approach applied to fluorescent voltage reporters. Nature Chemical Biology 14 (4), 352-360.

Portugues, R., Severi, K.E., Wyart, C., Ahrens, M.B., 2013. Optogenetics in a transparent animal: circuit function in the larval

zebrafish. Current Opinion in Neurobiology 23 (1), 119-126.
Raimondo, J.V., Kay, L., Ellender, T.J., Akerman, C.J., 2012. Optogenetic silencing strategies differ in their effects on inhibitory synaptic transmission. Nature Neuroscience 15 (8), 1102-1104.
Sako, K., Pradhan, S.J., Barone, V., Ingles-Prieto, A., Muller, P., Ruprecht, V., et al., 2016. Optogenetic control of nodal signaling reveals a temporal pattern of nodal signaling regulating cell fate specification during gastrulation. Cell Reports 16 (3), 866-877.
Sarris, M., Olekhnovitch, R., Bousso, P., 2016. Manipulating leukocyte interactions in vivo through optogenetic chemokine release. Blood 127 (23) e35-e41.
Schoonheim, P.J., Arrenberg, A.B., Del Bene, F., Baier, H., 2010. Optogenetic localization and genetic perturbation of saccade-generating neurons in zebrafish. Journal of Neuroscience 30 (20), 7111-7120.
Singh, C., Oikonomou, G., Prober, D.A., 2015. Norepinephrine is required to promote wakefulness and for hypocretin-induced arousal in zebrafish. eLife 4, e07000.
Sohal, V.S., Zhang, F., Yizhar, O., Deisseroth, K., 2009. Parvalbumin neurons and gamma rhythms enhance cortical circuit performance. Nature 459 (7247), 698-702.
Song, J., Ampatzis, K., Bjornfors, E.R., El Manira, A., 2016. Motor neurons control locomotor circuit function retrogradely via gap junctions. Nature 529 (7586), 399-402.
St-Pierre, F., Marshall, J.D., Yang, Y., Gong, Y., Schnitzer, M.J., Lin, M.Z., 2014. High-fidelity optical reporting of neuronal electrical activity with an ultrafast fluorescent voltage sensor. Nature Neuroscience 17 (6), 884-889.
Storace, D., Sepehri Rad, M., Kang, B., Cohen, L.B., Hughes, T., Baker, B.J., 2016. Toward better geneticall encoded sensors of membrane potential. Trends in Neurosciences 39 (5), 277-289.
Sun, F., Zeng, J., Jing, M., Zhou, J., Feng, J., Owen, S.F., et al., 2018. A genetically encoded fluorescent sensor enables rapid and specific detection of dopamine in flies, fish, and mice. Cell 174 (2), 481-496.e19.
Tonnesen, J., Sorensen, A.T., Deisseroth, K., Lundberg, C., Kokaia, M., 2009. Optogenetic control of epileptiform activity. Proceedings of the National Academy of Sciences of the United States of America 106 (29), 12162-12167.
Tonnesen, J., Parish, C.L., Sorensen, A.T., Andersson, A., Lundberg, C., Deisseroth, K., et al., 2011. Functional integration of grafted neural stem cell-derived dopaminergic neurons monitored by optogenetics in an in vitro Parkinson model. PLoS One 6 (3), e17560.
Tsuda, S., Kee, M.Z., Cunha, C., Kim, J., Yan, P., Loew, L.M., et al., 2013. Probing the function of neuronal populations: combining micromirror-based optogenetic photostimulation with voltage-sensitive dye imaging. Neuroscience Research 75 (1), 76-81.
Tye, K.M., Prakash, R., Kim, S.Y., Fenno, L.E., Grosenick, L., Zarabi, H., et al., 2011. Amygdala circuitry mediating reversible and bidirectional control of anxiety. Nature 471 (7338), 358-362.
Wachter, R.M., Remington, S.J., 1999. Sensitivity of the yellow variant of green fluorescent protein to halides and nitrate. Current Biology 9 (17), R628-R629.
Wang, H., Peca, J., Matsuzaki, M., Matsuzaki, K., Noguchi, J., Qiu, L., et al., 2007. High-speed mapping of synaptic connectivity using photostimulation in Channelrhodopsin-2 transgenic mice. Proceedings of the National Academy of Sciences of the United States of America 104 (19), 8143-8148.
Wang, T.M., Holzhausen, L.C., Kramer, R.H., 2014. Imaging an optogenetic pH sensor reveals that protons mediate lateral inhibition in the retina. Nature Neuroscience 17 (2), 262-268.
White, R.M., Sessa, A., Burke, C., Bowman, T., LeBlanc, J., Ceol, C., et al., 2008. Transparent adult zebrafish as a tool for in vivo transplantation analysis. Cell Stem Cell 2 (2), 183-189.
Wyart, C., Del Bene, F., Warp, E., Scott, E.K., Trauner, D., Baier, H., et al., 2009. Optogenetic dissection of a behavioural module in the vertebrate spinal cord. Nature 461 (7262), 407-410.
Yizhar, O., Fenno, L.E., Davidson, T.J., Mogri, M., Deisseroth, K., 2011. Optogenetics in neural systems. Neuron 71 (1), 9-34.
Yuste, R., Peinado, A., Katz, L.C., 1992. Neuronal domains in developing neocortex. Science 257 (5070), 665-669.
Zemelman, B.V., Nesnas, N., Lee, G.A., Miesenbock, G., 2003. Photochemical gating of heterologous ion channels: remote control over genetically designated populations of neurons. Proceedings of the National Academy of Sciences of the United States of America 100 (3), 1352-1357.
Zhang, F., Wang, L.P., Boyden, E.S., Deisseroth, K., 2006. Channelrhodopsin-2 and optical control of excitable cells. Nature Methods 3 (10), 785-792.
Zhang, F., Wang, L.P., Brauner, M., Liewald, J.F., Kay, K., Watzke, N., et al., 2007. Multimodal fast optical interrogation of neural circuitry. Nature 446 (7136), 633-639.
Zhao, S., Cunha, C., Zhang, F., Liu, Q., Gloss, B., Deisseroth, K., et al., 2008. Improved expression of halorhodopsin for light-induced silencing of neuronal activity. Brain Cell Biology 36 (1e4), 141-154.

第十八章

斑马鱼的 CRISPR/Cas 系统

Flavia De Santis[1], Javier Terriente[1], Vincenzo Di Donato[1]

18.1 引言

规律成簇间隔短回文重复序列及其相关（CRISPR/Cas）系统是原核生物天然适应性免疫系统的重要组成部分，负责降解来自噬菌体或质粒的入侵遗传物质（Marraffini and Sontheimer, 2008）。

在过去的几年里，由 CRISPR/Cas 启发的技术已经优化成真核生物的基因组工程工具，这些技术迅速成为细胞系、动物和植物中促进靶向 DNA 修饰的首选方法（Jinek et al., 2012; Cong et al., 2013）。此外，这些研究工具还具有直接治疗的用途，包括用于治疗多种人类疾病的基因和细胞疗法（Barrangou and Doudna, 2016）。在 CRISPR 或其他基于位点特异性核酸内切酶活性的技术出现之前，例如，锌指核酸酶（zinc finger nuclease, ZFN）（Urnov et al., 2010）和转录激活因子样效应核酸酶（transcription activator-like effector nuclease, TALEN）（Boch, 2011），在模式生物中进行靶向诱变的唯一可能策略是基于胚胎干细胞中的同源重组，这是一种仅适用于哺乳动物的转基因方法（Thomas and Capecchi, 1987）。与此不同的是，基于 CRISPR/Cas9 系统建立了一些精确和位点特异的 DNA 修饰技术，扩展了在任何物种（包括斑马鱼）中生成复杂遗传模型的可能性。

基于 CRISPR/Cas9 的工具建立在两个主要组件的协调活动之上：一是单一向导 RNA（single guide RNA, sgRNA），它包含与目标基因组 DNA 互补的 20 个核苷酸（nt）组成的序列；另一个是 Cas9 核酸内切酶，它负责产生双链断裂（double-strand breaks, DSB）。当 Cas9 与 sgRNA 形成核糖核蛋白复合物后，Cas9 可以定位至目标基因组位点。因此，为了有效地切割目标 DNA，它需要存在一个由单一向导 RNA 识别的 20 个核苷酸的前间区序

[1] ZeClinics SL, IGTP (Germans Trias i Pujol Research Institute), Barcelona, Spain.

列邻近基序（the protospacer adjacent motif, PAM）。在通过Cas9诱导双链断裂后，目标位点容易发生突变，包括由非同源末端连接机制介导的易错修复机制诱导的小规模插入和缺失的产生（Jinek et al., 2012）。

在接下来的章节中，主要介绍多功能的斑马鱼CRISPR/Cas工具，它在过去几年实现了不同的目标，包括诱导目的基因的靶向破坏或沉默、精确整合报告基因或点突变、调控基因转录，以及为蛋白质动力学和谱系追踪研究引入标签。

18.2 敲除策略

基于CRISPR/Cas9的高效的基因编辑方法的建立和进一步完善，极大地促进了斑马鱼功能丧失模型的生成。迄今为止，任何基因组位点都可以被成功诱变，这要归功于能确保在早期胚胎发育过程中同时注入Cas9酶和一种（或多种）单一向导RNA（sgRNA）的多项实验程序（Simone et al., 2018; Cornet et al., 2018）。

在最初的报道中，通过将sgRNA和Cas9 mRNA共注入单细胞期胚胎实现了基因的破坏（Hwang et al., 2013a; Chang et al., 2013; Jao et al., 2013）。通过这种策略，Cas9核酸酶在体内翻译、折叠并与单一向导sgRNA结合后，被驱动到目标位点，并在那产生双链断裂。随后，携带突变型和野生型等位基因（在体细胞和生殖细胞中）的被注射的仔鱼，生长至性成熟。再鉴定出在生殖系中携带功能丧失突变的鱼（F0代）。然后将F0初代进行杂交，产生F1同基因杂合子个体，随后杂交产生具有孟德尔比率的纯合子F2代。虽然这代表了一种有价值的方法来产生用于深入研究基因功能的同基因稳定敲除系，但其需要很长的世代时间。

有趣的是，已经证明注射预先形成的sgRNA/Cas9核糖核蛋白复合物可以最大限度地提高系统的切割效率，既促进种系转基因，又可以轻松评估注射的F0仔鱼的功能丧失表型（Kotani et al., 2015; Burger et al., 2016; Shah et al., 2016）。事实上，在注射后，体外组装的核糖核蛋白复合物早在单细胞阶段就能够诱导断裂双链的形成。早期切割可能会促进目的基因位点的两个等位基因同时失活，从而导致产生一组携带高比例双等位基因体细胞突变的仔鱼（称为CRISPR介导的突变体或Crispant）（Burger et al., 2016）。随后，大多数Crispant动物将表现为纯合子突变体，从而可以直接识别和分析功能缺失的表型，因此，可以减少进一步将鱼培育为纯合子所需的时间和成本。

最近，Wu及其同事采用了一种基于Crispant的方法来快速可靠地筛选具假定功能的目标基因（Wu et al., 2016）。作者证明，用四种特定的sgRNA同时靶向基因组位点可以提高CRISPR/Cas9系统的诱变效率。具体来说，F0代动物可重复地显示出高于90%的突变率，最重要的是，重现了在同一基因的稳定敲除中观察到的表型。

尽管如此，有必要指出的是，一群Crispant仔鱼代表了一个基因异质性的群体，它们携带不同的突变等位基因。这一特点是Cas9酶诱导双链断裂的发育阶段中无法控制的必然结果。尽管注射是在单细胞胚胎阶段进行的，但Cas9内切酶活性可能在较晚的发育点开始，甚至在几轮细胞分裂之后。在这种情况下，修复机制会促使不同细胞形成不同的突变，从而产生遗传嵌合体，其中来自不同祖代的细胞会携带不同的突变等位基因。因此，使用同基因敲除系仍是更可靠的选择。

为了减少可能与注射有关的实验不稳定性，有人建议使用稳定的转基因斑马鱼系在早期发育阶段表达 *cas9*。为此，最近的研究描述了一种 *cas9* 表达受泛在启动子控制的转基因品系，包括泛素（ubiquitin, ubb）启动子（Shiraki and Kawakami, 2018; Yin et al., 2015）和延伸因子 1a（the elongation factor, ef1a）的调控区域（Yang et al., 2018）。

此外，另一类 CRISPR/Cas 系统被称为 CRISPR/cpf1（Cas12a）（Zetsche et al., 2015），该系统已成功用于诱导斑马鱼的突变（Moreno-Mateos et al., 2017）。Cas12a 核酸酶与 Cas9 在 sgRNA 结构、PAM 识别序列和 DNA 切割动力学方面都有所不同（Strohkendl et al., 2018），Cas12a 核酸酶尤其适合靶向富含 AT 的基因编码区和非翻译区。

总体而言，上述策略代表了大量基于 CRISPR/Cas 来成功诱导体细胞和生殖细胞等位基因的方法，它们为斑马鱼的反向遗传研究提供相对简单的方案，见图 18-1。

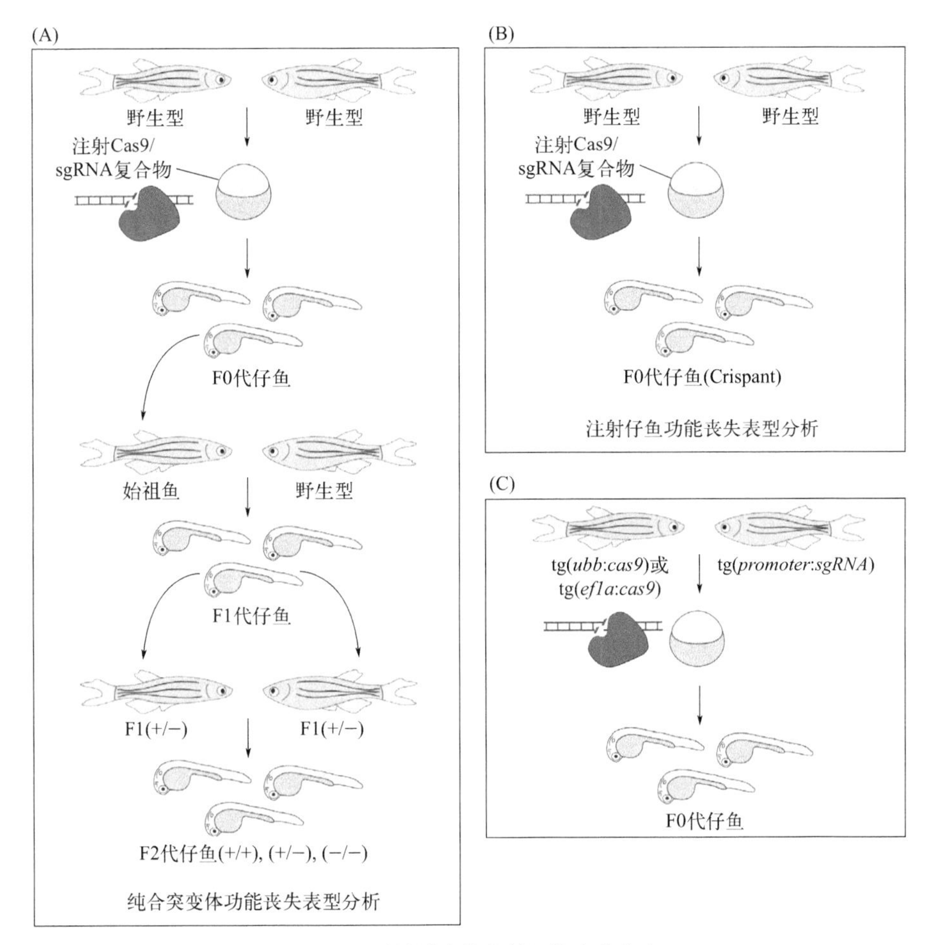

图 18-1 功能缺失等位基因的产生方法

（A）建立稳定敲除斑马鱼品系的流程。Cas9/sgRNA 复合物注射到单细胞期胚胎中。一旦鱼长到成年，就要对其进行筛选，确定一种将功能缺失突变传递给 F1 的始祖鱼。F1 兄弟姐妹近交产生 F2 代，可以分析基因失活引起的特定表型。（B）CRISPANT 方法。选择 sgRNA 靶向的目的基因组位点及之后的 Cas9/sgRNA 复合物。在 F0 代中可以实现高达 100% 的诱变效率，从而可以分析功能缺失表型。（C）通过 *cas9* 表达介导的基因失活。在 F0 代中，培育一个普遍表达 Cas9 的斑马鱼转基因系和一个在普遍或组织特异性启动子下表达 sgRNA 的转基因系可以使目的基因功能丧失

18.3 组织特异性基因破坏

尽管组成型基因失活有其优点，但组成型基因失活可能有一些局限性，如胚胎致死性、补偿机制和多效性表型等，这些可能会妨碍对功能丧失表型的评估。在这些情况下，以严格时空调节的方式诱导基因失活的可能性或许是一种优势。

在CRISPR/Cas9出现之前，在斑马鱼中产生条件敲除等位基因是一项难以实现的任务。现在，许多基于CRISPR/Cas9的技术为研究人员提供了在明确的细胞类型或组织的精确发育阶段诱导基因失活的可能性。

在第一份报告中，研究者开发了基于Cre/lox重组的方法（啮齿动物条件敲除金标准）用于斑马鱼（Bouabe and Okkenhaug, 2013）。该方法的核心是基于Cre酶在loxP序列两侧的基因组区域催化位点特异性重组的能力。转基因动物的产生，需要在目的序列的5′和3′端具有两个loxP位点（floxed目标等位基因），以允许基于Cre的条件基因敲除。事实上，时空调节的Cre表达促进了时空调控的目标等位基因的切除，进而精确控制基因敲除。基于这一原理，CRISPR/Cas9技术已被用于通过在所需基因组位置敲入loxP位点，从而在斑马鱼中生成floxed等位基因。下一节将进一步讨论通过CRISPR/Cas系统产生敲入等位基因的策略。

关于组织特异性基因敲除，Cre/lox策略的另一种可能性是CRISPR/Cas9基本组分Cas9和/或sgRNA的时空调控。实现这一目标的第一个策略是使用模块化的载体，其包含细胞类型特异性启动子来驱动*cas9*表达和*U6*启动子（RNA聚合酶Ⅲ依赖的组成性启动子）在体内转录一个sgRNA（Ablain et al., 2015）。同样，通过CRISPR/Cas9和Gal4/UAS系统（Di Donato et al., 2016; De Santis et al., 2016）的结合，成功诱导了条件突变，后者确保了位于上游激活序列（upstream activating sequence, UAS）下游基因的Gal4依赖性转录调控（Asakawa and Kawakami, 2008）。该方法基于2C-Cas9 (Cre介导的重组，用于克隆分析Cas9突变细胞)的转基因载体，调节UAS驱动的*cas9*表达和*U6*介导的靶向目标基因的两种sgRNA的产生。另一种策略是基于两个转基因品系的杂交，每个品系控制CRISPR/Cas9系统的一个组成部分的表达。第一个品系含有一个floxed等位基因（*hsp70:loxP-mCherry-STOP-loxP-cas9*），可以在Cre介导的重组后通过热休克诱导*cas9*表达；第二个品系包含多个*U6*启动子，转录不同的sgRNA（Yin et al., 2015）。最近，通过在Cas9核酸酶组成型激活的遗传背景下采用促进sgRNA组织特异性表达的策略，成功实现了CRISPR/Cas9的条件激活（Shiraki and Kawakami, 2018; Lee et al., 2016）。为此，单个转录本，包括多个sgRNA序列，由细胞类型特异性RNA聚合酶Ⅱ依赖性启动子驱动。转录成熟后，功能性sgRNA可用于与Cas9形成核糖核蛋白复合物。已有研究表明，核酶介导的（Lee et al., 2016）和基于tRNA的（Shiraki and Kawakami, 2018）多重sgRNA表达框都可以在体内进行加工，从而产生多种sgRNA。

综上所述，这些方法扩大了在目标组织内的精确时间窗口中对基因功能丧失的影响进行分析可能性（图18-2）。

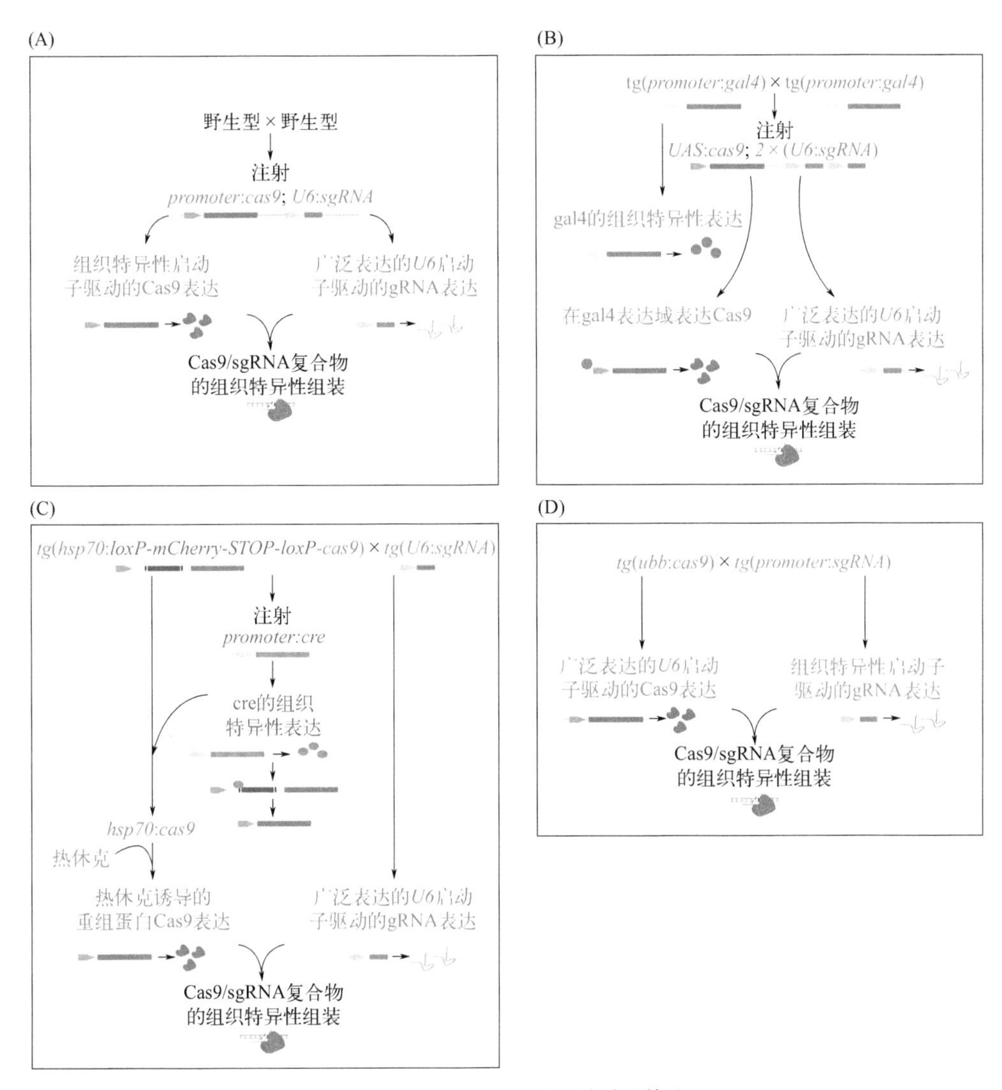

图 18-2 **斑马鱼组织特异性敲除策略**

（A）用于基因敲除的细胞特异性 Cas9 表达。载体系统允许同时表达由组织特异性启动子驱动的 Cas9 和 *U6* 启动子驱动普遍表达的靶向目的基因组位点的 sgRNAs。（B）通过 Gal4:UAS 系统进行组织特异性基因敲除。Gal4 的组织特异性表达激活了位于 UAS 序列下游的 Cas9 和 sgRNA 的转录，使 Gal4 表达细胞中的基因特异性失活。（C）通过 Cre/loxP 系统诱导组织特异性基因敲除。Cas9 的表达可以通过热休克诱导的转录和 Cre 重组酶的活性来进行时空调控。该方法是基于使用第一个携带 floxed 等位基因（*hsp70:loxP-mCherry-STOP-loxP-cas9*）的转基因系和第二个携带用于 sgRNA 普遍表达的转基因品系。Cre 介导重组后，通过基于热休克诱导 *cas9* 表达来实现基因失活。（D）细胞特异性 sgRNAs 表达实现基因失活。培育普遍表达 Cas9 的斑马鱼转基因品系和在组织特异性启动子下表达 sgRNA 的转基因品系会导致目的基因功能丧失

18.4 敲入策略

DSB 可以通过内源性 DNA 修复机制进行修复，包括非同源末端连接、微同源介导的末端连接和同源定向修复。如上所述，非同源末端连接确保在 DNA 损伤时，切割 DNA

的末端直接重新连接，通常会由于不完全修复而产生小的插入和缺失（Lieber, 2010）。如果微同源序列（3 ～ 30bp）存在于 DSB 附近，则会发生微同源介导的末端连接，并且导致可预测的缺失（McVey and Lee, 2008）。最后，同源定向修复是最精确的机制，自然地发生在姐妹染色单体之间，并且需要长同源序列位于修复位点的两侧（大小范围从约 30bp ～ >1000 bp）（Moynahan and Jasin, 2010）。

许多基于 CRISPR/Cas9 的敲入策略已经应用了这些机制，旨在将外源 DNA 插入明确的基因组位点。迄今为止，成功产生整合长和短的 DNA 序列的敲入系是有可能的。长敲入适合产生报告基因系或标记内源性基因位点（产生 N 端或 C 端融合蛋白）（Auer et al., 2014; Li et al., 2015; Luo et al., 2018; Hisano et al., 2015）；短敲入允许插入 loxP 位点（Barrangou and Doudna, 2016）、HA 标签（Hruscha et al., 2013）或点突变（Hwang et al., 2013b; Irion et al., 2014）。

一项开创性的研究利用非同源末端连接的优势来促进外源 DNA 在选定基因的 ATG 位点整合。为此，CRISPR/Cas9 被应用于体内切割靶向内源性位点和含有 *gal4*CDS 的供体质粒。通过 UAS 驱动报告基因的激活可很容易地检测到成功的框内插入介导的 Gal4 反式激活子的表达（Auer et al., 2014）。在随后的研究中，在 *gal4* 序列上游加入 hsp70 增强子大大提高了该系统的效率，因为插入报告基因的表达被独立于整合阅读框之外（Kimura et al., 2014）。使用这种方法，已经证明通过在基因的 ATG 处加入报告基因，可以同时产生功能缺失的等位基因和再现靶基因表达模式的转基因系（Ota et al., 2016）。此外，该策略还可以成功地用于产生 C 末端融合蛋白。为此，Li 和同事瞄准了目的基因的最后一个内含子，并设计了一个供体质粒，该质粒含有同一基因最后一个外显子的序列，并与荧光报告基因的开放阅读框融合（Li et al., 2015）。然而，基于非同源末端连接的方法不适合外源 DNA 的精确整合，因为在基因组和插入的 DNA 之间的连接处可能会产生许多突变。为了克服这一限制，基于微同源介导的末端连接和同源定向修复的各种技术已经建立。

据报道，已经有研究者使用基于微同源介导的末端连接的敲入策略精确整合供体 DNA。为了成功地应用这项技术，获得纯净的插入，有必要设计一个含有短同源臂（10 ～ 40bp）的 DNA 序列，与基因组目标位点的侧翼序列互补。CRISPR/Cas9 系统用于在目标位点诱导 DSB，并促进质粒线性化，允许在框架内插入相对较长的外源 DNA 序列，其中包含荧光标记的编码序列（Luo et al., 2018; Hisano et al., 2015）。

有趣的是，Hoshijima 等人采用了基于同源重组的敲入策略来实现长 DNA 序列的整合（Hoshijima et al., 2016）。最初的方案利用 TALEN 切割内源性 DNA。尽管使用了不同的机制来诱导 DNA 切割，我们可以预期这种策略与 CRISPR/Cas9 技术可以联合使用。作者提供了荧光报告基因编码序列在基因 ATG 处正整合的证据（从而产生 N 端融合蛋白），并报告了包含 loxP 位点的 DNA 盒与转基因标记的成功整合。与上述策略不同的是，含有 1kb 长同源臂的供体 DNA 在体外通过 ISce-I 介导的消化进行线性化。

基于同源性的敲入方法已经实现了短外源序列和单碱基替换的精确整合。大多数研究使用与目标 DNA 共享小同源臂（约 50bp）的单链寡核苷酸（ssODN）的短供体模板（Chang et al., 2013; Hwang et al., 2013b; Irion et al., 2014; Zhang et al., 2016）。此外，其他成功的策略是基于使用含有更长同源臂的供体质粒（Irion et al., 2014; Hoshijima et al., 2016）。

最近的一项研究系统地测试了不同的方法来优化基于寡核苷酸的点突变敲入效率

（Prykhozhij et al., 2018）。当使用反义不对称寡核苷酸时，成功率最高，这可能是 Cas9 和切割的 DNA 之间解离动力学的结果。同一项研究证实，通过在寡核苷酸末端加入硫代磷酸修饰以及将所需的碱基替换靠近 CRISPR 靶位点（越接近越好），敲入效率可能会大大提高。有趣的是，使用 CRISPR/Cpf1 系统（而不是 CRISPR/Cas9）可使基于 ssODN 的敲入效率提高四倍（Moreno-Mateos et al., 2017）。

在另一项研究中，Zhang 等人比较了多种通过同源性介导的敲入来诱导体细胞和生殖细胞点突变的策略的效率（Zhang et al., 2018）。分析的方法在四个参数上有所不同：①供体模板的类型（未消化质粒、线性化双链 DNA 和 ssODN）；②同源臂的长度（40bp 或 400bp）；③ Cas9（mRNA 或蛋白质）的递送方法；④使用抑制非同源末端连接或刺激同源定向修复的药物。有趣的是，在经过验证的组合中，当使用含有长同源臂且含有 CRISPR 靶点的供体质粒进行体内线性化时，获得了最佳敲入得分。作为 sgRNA 的核糖核蛋白复合物，Cas9 的注入进一步提高了系统的效率。通过使用促进同源定向修复或抑制非同源末端连接的化合物，也可提高效率。见图 18-3。

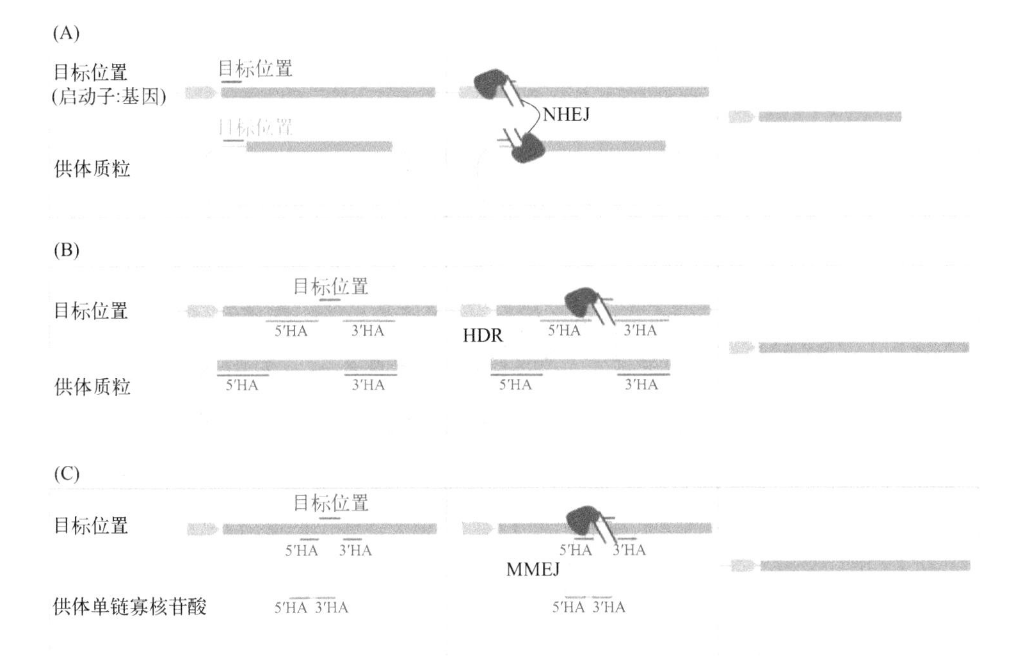

图 18-3 外源 DNA 敲入斑马鱼基因组的方法

（A）非同源末端连接（NHEJ）介导的敲入。基因组 DNA 双链断裂后，将含有目的序列的供体质粒插入斑马鱼基因组。插入序列的框架和方向将是随机的。（B）同源定向重组（HDR）介导的敲入。在基因组 DNA 双链断裂时，通过同源定向重组插入一个带有同源臂（HA，基因组中含有相同序列）的质粒。外源序列将插入与基因组序列相同的阅读框和方向。（C）微同源性介导的末端连接（MMEJ）敲入。短同源臂用于在双链断裂位点插入 DNA 片段（如 ssODN 单链寡核苷酸）

18.5 碱基编辑

如上一节所述，可以通过应用基于同源定向修复的敲入方法来生成单碱基修饰。尽管这些方法的效率在不断提高，但它们在斑马鱼和其他动物中的应用仍然具有挑战性。Zhang 及其同事最近提出了一种可能的替代方案（Zhang et al., 2017），他们使用可编程“碱基编辑器”（BEs）技术，在斑马鱼基因组的精确位置成功整合了点突变。该技术最初是在哺乳动物细胞中开发的（Komor et al., 2016），可实现从一个目标碱基到另一个目标碱基的直接和永久的转换，特别是胞嘧啶（C, cytidine）到胸腺嘧啶（T, thymine）或鸟嘌呤（G, guanine）到腺嘌呤（A, adenine），而这个过程中无须诱导 DSB 形成。该系统利用包含三种功能成分的融合蛋白：Cas9 切口酶（nCas9）（Shen et al., 2014）、胞苷脱氨酶（APOBEC）（Navaratnam and Sarwar, 2006）和尿嘧啶 DNA 糖基化酶抑制剂（UGI）（Mol et al., 1995）。nCas9 保留其以 sgRNA 依赖方式结合特定基因组区域的能力，但该酶只能催化单链 DNA 的切割（从而产生缺口）。在目标识别和缺口诱导后，目标碱基（始终为胞嘧啶）暴露于碱基编辑复合物的其他组分的催化活性下。更准确地说，APOBEC 蛋白促进胞嘧啶脱氨基，转化为尿嘧啶（U）。在自然界中，尿嘧啶不存在于 DNA 分子中。然后，这种碱基修饰激活了一种 DNA 修复机制，该机制基于尿嘧啶 DNA 糖苷酶促进尿嘧啶的切除（Kunz et al., 2009）。BE 复合物中 UGI 的存在保证了这种修复机制的失活，从而促进有针对性的 C-T 碱基转化。事实上，U-G 错配会转换成 U-A 对，进而再转换成 T-A 对（Komor et al., 2016）。在斑马鱼中应用该技术的基础是将 BE mRNA（包含密码子优化版本的 nCas9）以及靶向包含 C 的目标基因组区域的 sgRNA 注射到单细胞期胚胎中（Zhang et al., 2017）。这种碱基编辑方法的效率已经在不同的基因位点上进行了测试，报告的预期点突变的种系传播率为 7% ～ 37%（Zhang et al., 2017）。

最近，腺嘌呤碱基编辑器（ABEs）的实施进一步扩大了诱导靶向点突变的可能性，使 A-T 转化为 G-C 对成为可能（Gaudelli et al., 2017）。该系统利用大肠杆菌衍生的腺嘌呤脱氨酶（adenine deaminase, TadA），能够在 tRNA 分子中诱导腺嘌呤转化为肌苷（I, Inosine）（Wolf et al., 2002）。在体外进化的酶，意味着它可以识别一段作为底物的 DNA 序列，一个已经融合到 nCas9 的改良版本 TadA（TadA7.1），从而生成一个复合物，并在哺乳动物细胞中进行精确的碱基修饰（Gaudelli et al., 2017）。就像 BE 系统，ABE 系统也已经被改造后用于斑马鱼，并证明该系统能够促进不同基因组位置的碱基编辑（Qin et al., 2018）。

总的来说，对于对分析疾病相关点突变感兴趣的科学家来说，碱基编辑系统是强有力的工具，能够有效替代基于同源定向修复的基因敲入方法。

18.6 转录调控

严格的基因表达时空调控对于合适的细胞功能和组织稳态至关重要。CRISPR/Cas9 系统已成功应用于原核和真核系统中调控目的基因的转录（Qi et al., 2013; Bikard et al., 2013）。CRISPR 干扰（CRISPRi）和 CRISPR 增强（CRISPR-on）方法可分别用于抑制或激活靶基因的转录。这两种工具都是基于使用由靶向特异性 sgRNA 和催化非活性形式的 Cas9 内切酶［也称为 dead-Cas9（d-Cas9）］形成的复合物，该复合物结合到相关的调控基因组区域而不诱导 DSB。在 CRISPR 干扰的情况下，Kox1 的转录阻遏物结构域（Krüppel-associated box, KRAB）与 d-Cas9 融合以实现对靶基因的转录抑制。在 CRISPR 增强的情况下，转录激活剂 VP16 与 d-Cas9 的融合增加了靶基因的表达。在斑马鱼中，最近的一份研究证明了 CRISPR 干扰和增强方法的原理（Long et al., 2015）。作者通过用五个识别启动子内多个基因组序列的 sgRNA 和 dCAS9/VP16 或 dCAS9/KRAB 靶向调控 *fgf8a* 和 *foxi1* 基因的启动子区域，实现了这两个基因的显著上调或下调。正如预期的那样，这些基因的转录调控导致耳泡发育受损。

另一种影响基因调控的方法是调控增强子的突变或缺失。这在最近的一份研究报告中得到了证实，作者分析了菱脑原节边界的肌动蛋白聚合相关基因的潜在协同调节。为了理解基因协同调控，他们利用 CRISPR/Cas9 删除了一个含有潜在增强子的 2.5kb 基因组区域。正如预期的那样，增强子缺失的纯合子动物表现出肌动蛋白减少（Letelier et al., 2018）。上述工具的进一步改进有利于对体内转录调控进行更精确的时空控制，以深入描述生理过程中的基因调控。

18.7 谱系追踪

理解多细胞生物中细胞类型之间的谱系关系对于阐明发育过程以及阐明癌症等重要疾病的发病机制具有重要意义。尽管在过去几十年中已经发展了几种细胞谱系追踪策略（染料、报告基因表达、自然发生的突变），但细胞谱系的大规模重建仍然是一个挑战。最近，利用 CRISPR/Cas9 基因编辑克服了以往方法的局限性。McKenna 等人开发了一种用于谱系追踪的合成靶阵列基因编辑（GESTALT）方法，可标记祖细胞及其后代，以识别整个生物体（即斑马鱼）内的所有谱系和亚系，（McKenna et al., 2016）。该技术基于 Cas9/sgRNA 复合物诱导 DSB 导致在目标位点引入短插入和缺失这一事实（Jao et al., 2013; Varshney et al., 2015）。因此，包含特定插入和缺失模式的祖细胞谱系继承了相同的突变谱，变得可追溯。作者构建了一种斑马鱼品系，该品系普遍表达转基因序列（由 10 个 CRISPR/Cas9 靶点组成的紧凑序列）的转基因。然后，通过在单细胞阶段注射 Cas9/sgRNA 诱导该位点发生突变。在多轮细胞分裂过程中突变的逐渐积累导致条形码上的组合编辑，从而通过细胞之间共享的突变模式阐明谱系关系。为了证明通过靶向诱变谱系追踪是可行的，该研究重建了胚胎细胞对成年斑马鱼组织的谱系贡献，表明大多数器官来自少数祖细胞（McKenna et al., 2016）。

尽管信息非常丰富，但 GESTALT 方法仅检测早期谱系关系，因为 Cas9/sgRNA 靶向基因组条形码是在单细胞阶段诱导的，无法区分细胞类型。为了克服这些限制，最近有研究 GESTALT 改进为单细胞 GESTALT（single-cell GESTALT, scGESTALT）（Spanjaard et al., 2018）。scGESTALT 允许：①同时进行谱系追踪和单细胞转录组学，以检测细胞身份；②在发育后期诱导 Cas9 激活，以分析细胞在确定的发育阶段的命运选择。作者利用 scGESTALT 对斑马鱼大脑进行了最先进的高分辨率谱系追踪研究。重要的是，scGESTALT 是分析发育和疾病过程中细胞关系的宝贵工具。

18.8 结论和展望

本章描述了如何将斑马鱼模型的固有特征（快速胚胎发育、实时成像的可能性、转基因的难易程度）与 CRISPR/Cas9 系统相结合，为基础研究和转化研究开辟了多种途径。在过去几年中，基于 CRISPR/Cas9 技术的几种新方法在斑马鱼群体中得到了发展和优化。在许多情况下，这种共同的努力使得用于基因功能操控的成熟技术得以被替代。一个例子是基因敲除的产生，这在过去只能通过靶向诱导的基因组局部损伤（targeting induced local lesions in genomes, TILLING）来实现，不但耗时长而且价格昂贵。另一个例子是标记目标细胞群的转基因斑马鱼系或融合蛋白的产生，这在传统上是通过细菌人工染色体（bacterial artificial chromosome, BAC）转基因实现的。如今，实现这些目标的首选方法是 CRISPR/Cas9 介导的内源性位点报告基因敲入。外源 DNA 的靶向整合克服了旧技术的缺点，如基因的随机整合和非预期过度表达，从而提供了更准确的生物过程再现。除了提高实验观察的准确性外，在斑马鱼中使用 CRISPR/Cas9 显著减少了基因功能分析所需的时间。

在斑马鱼体内进行快速基因敲除和高通量表型分析在药物发现研究中非常有价值，特别是通过对携带某种突变体的患者进行转录组分析或全基因组分析来筛选确定潜在性治疗靶点方面。

为此，最近的报道表明，目标基因组位点的多重靶向能够很容易地在 F0 动物中诱导无效表型（Wu et al., 2018）。因此，当检测到目的基因位点的高突变率时，注射 Cas9/sgRNA 复合物的鱼可以作为事实上的突变体进行分析。这种方法适用于在短时间内测试几个基因在参与发育和疾病的特定过程中的潜在功能。在这种情况下，一旦获得更高的通量，斑马鱼很快就会取代基于细胞的系统进行目标验证和药物发现，补充来自小鼠模型的信息。一旦验证成功，这些相同的基因靶标可以通过将在患者中发现的点突变敲入斑马鱼同源基因位点来进行修改，以便分析从细胞水平到生物体水平的预期表型。这种方法代表了个性化医疗和药物开发的进一步发展。

尽管有上述优势，但需要指出的是，CRISPR/Cas9 系统的高切割效率可能导致非特异性和非预期的 DNA 修饰（非靶点突变），这可能会改变目的基因外的其他基因组位点的功能。事实上，脱靶位点的突变使得功能缺失表型的分析更加复杂（Zhang et al., 2015），并极大地限制了 CRISPR/Cas9 系统在人类基因治疗中的适用性。因此，基因组编辑领域的当务之急是对脱靶突变的检测和量化（Gabriel et al., 2015）。在过去的几年中，已经开发了几种预测不同 sgRNA 特异性评分的算法，来帮助研究人员设计基于 CRISPR/Cas9 的策略，以减少脱靶 DSB 的概率（Concordet and Haeussler, 2018; Abadi et al., 2017; Bae et al., 2014）。然而，这些算法的预测能力较低，因为它们主要基于序列相似性进行计算机分析，导致假阳性率较高（Lin and Wong, 2018）。最近，Wienert 等人建立了一种体外和体内无偏差检测脱靶基因修饰的方法（Wienert et al., 2019）。这种方法被命名为 DISCOVER-Seq（原位 Cas 脱靶和测序验证）。其基础是分析 Cas9 诱导 DSB 后，募集 DNA 修复蛋白到靶点和非靶点基因组位点，然后进行测序。重要的是，这项技术允许检测 Cas9 动力学和检测真实的切割事件。已经证明 DISCOVER-Seq 在人类和小鼠细胞以及体内病毒介导的基因编辑中是有效的，几乎适用于任何模式生物，并可能在体内或体外用于患者细胞编辑的脱靶鉴定。

尽管还有改进的空间，但靶向基因编辑的持续实施正在以前所未有的细节对基因功能进行表征，克服了以往方法所面临的效率、时间和精度问题。得益于 CRISPR/Cas9 新型方法的快速发展，可用于基因研究的资源将满足基础科学和转化科学领域研究人员迄今尚未满足的大部分需求。

致谢

这项工作得到了 H2020-SME 第 2 阶段基金的支持，基金协议号为 755988。VDD 由 Torres Quevedo-MINECO 项目（PTQ-16-08819）和 Marie Curie 协会和企业博士后奖学金（授权协议编号 845713）共同创立。FDS 得到了 Torres Quevedo-MINECO 项目博士后奖学金（PTQ-17-09249）的支持。

（石晶晶翻译　李丽琴审校）

参考文献

Abadi, S., Yan, W.X., Amar, D., Mayrose, I., 2017. A machine learning approach for predicting CRISPR-Cas9 cleavage efficiencies and patterns underlying its mechanism of action. PLoS Computational Biology. https://doi.org/10.1371/journal.pcbi.1005807.

Ablain, J., Durand, E.M., Yang, S., Zhou, Y., Zon, L.I., 2015. A CRISPR/Cas9 vector system for tissue-specific gene disruption in zebrafish. Developmental Cell. https://doi.org/10.1016/j.devcel.2015.01.032.

Asakawa, K., Kawakami, K., 2008. Targeted gene expression by the Gal4-UAS system in zebrafish. Development Growth and Differentiation. https://doi.org/10.1111/j.1440-169X.2008.01044.x.

Auer, T.O., Duroure, K., De Cian, A., Concordet, J.P., Del Bene, F., 2014. Highly efficient CRISPR/Cas9-mediated knock-in in zebrafish by homology-independent DNA repair. Genome Research. https://doi.org/10.1101/gr.161638.113.

Bae, S., Park, J., Kim, J.S., 2014. Cas-OFFinder: a fast and versatile algorithm that searches for potential off-target sites of Cas9 RNA-guided endonucleases. Bioinformatics. https://doi.org/10.1093/bioinformatics/btu048.

Barrangou, R., Doudna, J.A., 2016. Applications of CRISPR technologies in research and beyond. Nature Biotechnology. https://doi.org/10.1038/nbt.3659.

Bikard, D., et al., 2013. Programmable repression and activation of bacterial gene expression using an engineered CRISPR-Cas system. Nucleic Acids Research. https://doi.org/10.1093/nar/gkt520.

Boch, J., 2011. TALEs of genome targeting. Nature Biotechnology. https://doi.org/10.1038/nbt.1767.

Bouabe, H., Okkenhaug, K., 2013. Gene targeting in mice: a review. Methods in Molecular Biology. https://doi.org/10.1007/978-1-62703-601-6_23.

Burger, A., et al., 2016. Maximizing mutagenesis with solubilized CRISPR-Cas9 ribonucleoprotein complexes. Development. https://doi.org/10.1242/dev.134809.

Chang, N., et al., 2013. Genome editing with RNA-guided Cas9 nuclease in Zebrafish embryos. Cell Research. https://doi.org/10.1038/cr.2013.45.

Concordet, J.P., Haeussler, M., 2018. CRISPOR: Intuitive guide selection for CRISPR/Cas9 genome editing experiments and screens. Nucleic Acids Research. https://doi.org/10.1093/nar/gky354.

Cong, L., et al., 2013. Multiplex genome engineering using CRISPR/Cas systems. Science. https://doi.org/10.1126/science.1231143.

Cornet, C., Di Donato, V., Terriente, J., 2018. Combining zebrafish and CRISPR/Cas9: toward a more efficient drug discovery pipeline. Frontiers in Pharmacology 9, 1-11.

De Santis, F., Di Donato, V., Del Bene, F., 2016. Clonal analysis of gene loss of function and tissue-specific gene deletion in zebrafish via CRISPR/Cas9 technology. Methods in Cell Biology. https://doi.org/10.1016/bs.mcb.2016.03.006.

Di Donato, V., et al., 2016. 2C-Cas9: a versatile tool for clonal analysis of gene function. Genome Research. https://doi.org/10.1101/gr.196170.115.

Gabriel, R., von Kalle, C., Schmidt, M., 2015. Mapping the precision of genome editing. Nature Biotechnology. https://doi.org/10.1038/nbt.3142.

Gaudelli, N.M., et al., 2017. Programmable base editing of T to G C in genomic DNA without DNA cleavage. Nature 551, 464-471.

Hisano, Y., et al., 2015. Precise in-frame integration of exogenous DNA mediated by CRISPR/Cas9 system in zebrafish. Scientific Reports 5, 1-7.

Hoshijima, K., Jurynec, M.J., Grunwald, D.J., 2016. Precise editing of the zebrafish genome made simple and efficient. Developmental Cell. https://doi.org/10.1016/j.devcel.2016.02.015.

Hruscha, A., et al., 2013. Efficient CRISPR/Cas9 genome editing with low off-target effects in zebrafish. Development 140, 4982-4987.

Hwang, W.Y., et al., 2013. Efficient genome editing in zebrafish using a CRISPR-Cas system. Nature Biotechnology. https://doi.org/10.1038/nbt.2501.

Hwang, W.Y., et al., 2013. Heritable and precise zebrafish genome editing using a CRISPR-cas system. PLoS One 8,1-9.

Irion, U., Krauss, J., Nusslein-Volhard, C., 2014. Precise and efficient genome editing in zebrafish using the CRISPR/Cas9 system. Development 141, 4827-4830.

Jao, L.-E., Wente, S.R., Chen, W., 2013. Efficient multiplex biallelic zebrafish genome editing using a CRISPR nuclease system. Proceedings of the National Academy of Sciences of the United States of America. https://doi.org/10.1073/pnas.1308335110.

Jinek, M., et al., 2012. A programmable dual-RNA-guided DNA endonuclease in adaptive bacterial immunity. Science. https://doi.org/10.1126/science.1225829.

Kimura, Y., Hisano, Y., Kawahara, A., Higashijima, S.I., 2014. Efficient generation of knock-in transgenic zebrafish carrying reporter/driver genes by CRISPR/Cas9-mediated genome engineering. Scientific Reports. https:// doi.org/10.1038/srep06545.

Komor, A.C., Kim, Y.B., Packer, M.S., Zuris, J.A., Liu, D.R., 2016. Programmable editing of a target base in genomic DNA without double-stranded DNA cleavage. Nature. https://doi.org/10.1038/nature17946.

Kotani, H., Taimatsu, K., Ohga, R., Ota, S., Kawahara, A., 2015. Efficient multiple genome modifications induced by the crRNAs, tracrRNA and Cas9 protein complex in zebrafish. PLoS One. https://doi.org/10.1371/journal.pone.0128319.

Kunz, C., Saito, Y., Schär, P., 2009. DNA repair in mammalian cells: mismatched repair: variations on a theme. Cellular and Molecular Life Sciences. https://doi.org/10.1007/s00018-009-8739-9.

Lee, R.T.H., Ng, A.S.M., Ingham, P.W., 2016. Ribozyme mediated gRNA Generation for in vitro and in vivo CRISPR/Cas9 mutagenesis. PLoS One. https://doi.org/10.1371/journal.pone.0166020.

Letelier, J., et al., 2018. Evolutionary emergence of the rac3b / rfng / sgca regulatory cluster refined mechanisms for hindbrain boundaries formation. Proceedings of the National Academy of Sciences of the United States of America. https://doi.org/10.1073/pnas.1719885115.

Li, J., Zhang, B., Bu, J., Du, J., 2015. Intron-based genomic editing: a highly efficient method for generating knock in zebrafish. Oncotarget. https://doi.org/10.18632/oncotarget.4547.

Lieber, M.R., 2010. The mechanism of double-strand DNA break repair by the nonhomologous DNA end-joining pathway. Annual Review of Biochemistry. https://doi.org/10.1146/annurev.biochem.052308.093131.

Lin, J., Wong, K.C., 2018. Off-target predictions in CRISPR-Cas9 gene editing using deep learning. Bioinformatics 34(17). https://doi.org/10.1093/bioinformatics/bty554.

Long, L., et al., 2015. Regulation of transcriptionally active genes via the catalytically inactive Cas9 in *C. elegans* and *D. rerio*. Cell Research 25, 638.

Luo, J.J., et al., 2018. CRISPR/Cas9-based genome engineering of zebrafish using a seamless integration strategy. The FASEB Journal 32, 5132-5142.

Marraffini, L.A., Sontheimer, E.J., 2008. CRISPR interference limits horizontal gene transfer in staphylococci by targeting DNA. Science. https://doi.org/10.1126/science.1165771.

McKenna, A., et al., 2016. Whole-organism lineage tracing by combinatorial and cumulative genome editing. Science. https://doi.org/10.1126/science.aaf7907.

McVey, M., Lee, S.E., 2008. MMEJ repair of double-strand breaks (director's cut): deleted sequences and alternative endings. Trends in Genetics 24, 529-538.

Mol, C.D., et al., 1995. Crystal structure of human uracil-DNA glycosylase in complex with a protein inhibitor: protein mimicry of DNA. Cell. https://doi.org/10.1016/0092-8674(95)90467-0.

Moreno-Mateos, M.A., et al., 2017. CRISPR-Cpf1 mediates efficient homology-directed repair and temperature-controlled genome editing. Nature Communications. https://doi.org/10.1038/s41467-017-01836-2.

Moynahan, M.E., Jasin, M., 2010. Mitotic homologous recombination maintains genomic stability and suppresses tumorigenesis. Nature Reviews Molecular Cell Biology. https://doi.org/10.1038/nrm2851.

Navaratnam, N., Sarwar, R., 2006. An overview of cytidine deaminases. International Journal of Hematology. https://doi.org/10.1532/IJH97.06032.

Ota, S., et al., 2016. Functional visualization and disruption of targeted genes using CRISPR/Cas9-mediated eGFP reporter integration in zebrafish. Scientific Reports 6, 1-10.

Prykhozhij, S.V., et al., 2018. Optimized knock-in of point mutations in zebrafish using CRISPR/Cas9. Nucleic Acids Research 46, 1-15.

Qi, L.S., et al., 2013. Repurposing CRISPR as an RNA-guided platform for sequence-specific control of gene expression. Cell. https://doi.org/10.1016/j.cell.2013.02.022.

Qin, W., et al., 2018. Precise A-T to G-C base editing in the zebrafish genome. BMC Biology. https://doi.org/10.1186/s12915-018-0609-1.

Shah, A.N., Moens, C.B., Miller, A.C., 2016. Targeted candidate gene screens using CRISPR/Cas9 technology. Methods in Cell Biology 135. Elsevier Ltd.

Shen, B., et al., 2014. Efficient genome modification by CRISPR-Cas9 nickase with minimal off-target effects. Nature Methods. https://doi.org/10.1038/nmeth.2857.

Shiraki, T., Kawakami, K., 2018. A tRNA-based multiplex sgRNA expression system in zebrafish and its application to generation of transgenic albino fish. Scientific Reports. Rep. https://doi.org/10.1038/s41598-018-31476-5.

Simone, B.W., Martínez-Gálvez, G., WareJoncas, Z., Ekker, S.C., 2018. Fishing for understanding: unlocking the zebrafish gene editor's toolbox. Methods. https://doi.org/10.1016/j.ymeth.2018.07.012.

Spanjaard, B., et al., 2018. Simultaneous lineage tracing and cell-type identification using CRISPR-Cas9-induced genetic scars. Nature Biotechnology. https://doi.org/10.1038/nbt.4124.

Strohkendl, I., Saifuddin, F.A., Rybarski, J.R., Finkelstein, I.J., Russell, R., 2018. Kinetic basis for DNA target specificity of CRISPR-Cas12a. Molecular Cell. https://doi.org/10.1016/j.molcel.2018.06.043.

Thomas, K.R., Capecchi, M.R., 1987. Site-directed mutagenesis by gene targeting in mouse embryo-derived stem cells. Cell. https://doi.org/10.1016/0092-8674(87)90646-5.

Urnov, F.D., Rebar, E.J., Holmes, M.C., Zhang, H.S., Gregory, P.D., 2010. Genome editing with engineered zinc finger nucleases. Nature Reviews Genetics. https://doi.org/10.1038/nrg2842.

Varshney, G.K., et al., 2015. High-throughput gene targeting and phenotyping in zebrafish using CRISPR/Cas9. Genome Research. https://doi.org/10.1101/gr.186379.114.

Wienert, B., et al., 2019. Unbiased detection of CRISPR off-targets in vivo using DISCOVER-Seq. Science. https:// doi.org/10.1126/science.aav9023.

Wolf, J., Gerber, A.P., Keller, W., 2002. TadA, an essential tRNA-specific adenosine deaminase from *Escherichia coli*. The EMBO Journal. https://doi.org/10.1093/emboj/cdf362.

Wu, R.S., et al., 2018. A rapid method for directed gene knockout for screening in G0 zebrafish. Developmental Cell. https://doi.org/10.1016/j.devcel.2018.06.003.

Yang, Z., et al., 2018. Generation of Cas9 transgenic zebrafish and their application in establishing an ERV-deficient animal model. Biotechnology Letters. https://doi.org/10.1007/s10529-018-2605-5.

Yin, L., et al., 2015. Multiplex conditional mutagenesis using transgenic expression of Cas9 and sgRNAs. Genetics. https://doi.org/10.1534/genetics.115.176917.

Zetsche, B., et al., 2015. Cpf1 is a single RNA-guided endonuclease of a class 2 CRISPR-cas system. Cell. https:// doi.org/10.1016/j.cell.2015.09.038.

Zhang, X.H., Tee, L.Y., Wang, X.G., Huang, Q.S., Yang, S.H., 2015. Off-target effects in CRISPR/Cas9-mediated genome engineering. Molecular Therapy Nucleic Acids 4, e264.

Zhang, Y., Huang, H., Zhang, B., Lin, S., 2016. TALEN- and CRISPR-enhanced DNA homologous recombination for gene editing in zebrafish. Methods in Cell Biology. https://doi.org/10.1016/bs.mcb.2016.03.005.

Zhang, Y., et al., 2017. Programmable base editing of zebrafish genome using a modified CRISPR-Cas9 system. Nature Communications. https://doi.org/10.1038/s41467-017-00175-6.

Zhang, Y., Zhang, Z., Ge, W., 2018. An efficient platform for generating somatic point mutations with germline transmission in the zebrafish by CRISPR/Cas9-mediated gene editing. Journal of Biological Chemistry 293, 6611-6622.

第十九章

PhOTO 斑马鱼和启动转换：促进神经发育与疾病机制研究

Konstantinos Kalyviotis[1], Hanyu Qin[2], Periklis Pantazis[1]

19.1 引言：谱系追踪是精确理解机制不可或缺的方法

100 多年来，胚胎学家和发育生物学家一直在使用细胞追踪和谱系追踪技术来揭示和了解细胞之间的关系。单细胞子代的识别鉴定为健康组织发育和疾病进展过程中的细胞动力学提供了有价值的信息。一些最初的谱系追踪研究结果是采用光学显微镜对细胞的生命周期进行连续观察而获得的。研究者绘制了令人信服的各种模式生物的胚胎发育过程图。例如，Charles O. Whitman 和他的同事（Whitman, 1878）追踪了水蛭胚胎中单个细胞从未受精卵细胞到原肠胚形成的全过程。由于水蛭胚胎的每个细胞都会产生特定的细胞群，他们由此得出结论：每个细胞的命运在发育早期就已经决定了。谱系追踪方法也有助于理解疾病和功能紊乱（如癌症）发生时的细胞动力学。最近的研究（Lamprecht et al., 2017; Shimokawa et al., 2017）通过重组技术对基因进行多种颜色标记，揭示了结直肠癌中肿瘤干细胞的克隆构型和可塑性。目前，细胞追踪技术越来越多地与最先进的成像方法相结合，例如细胞追踪和光片显微成像技术（light-sheet microscopy）（Huisken et al., 2004）结合，从而全面捕获整个胚胎的谱系动态（Keller et al., 2008）。这种先进的成像实验需要更大的数据存储容量和用于细胞分割和跟踪的复杂谱系跟踪算法（Mikut et al., 2013）。

[1] Department of Bioengineering, Imperial College London, South Kensington Campus, London, United Kingdom.
[2] Department of Biosystems Science and Engineering, ETH Zürich, Basel, Switzerland.

19.2 谱系追踪的历史

19.2.1 细胞标记

发育生物学和疾病生物学的主要目标是鉴定单个细胞所有后代和重建细胞轨迹。过去，胚胎学家和发育生物学家采用了多种追踪方法，这些方法独特的光学特征决定了实验中细胞标记的质量和谱系追踪的有效性。第一种实验设计利用了组织的内源性对比（endogenous contrast）进行研究（见下文），这些工作后来也通过采用外源性标志物以随机或精确的方式标记细胞群或单细胞的方法进行了补充研究。无论采用什么类型的标志物，都应该在一段时间内对目标细胞进行标记，而非标记到无关的细胞。为了实现准确跟踪，理想的标记不应改变目标细胞、其子代和邻近细胞的特性。理想的标记应该能显示细胞分裂、细胞迁移和细胞形状的改变，从而揭示复杂的细胞和分子过程。总之，理想的标记应该能将细胞的状态与其各自的命运联系起来。

19.2.2 直接观察和内源性对比增强方法

最早的谱系追踪和原基分布图主要使用光学显微镜观察无脊椎动物（如水蛭和秀丽隐杆线虫）的细胞动力学。虽然光学显微镜能放大发育的胚胎，但组织对比度却很小。1952年，Georges Nomarski 发明了微分干涉对比显微镜（Nomarski, 1952），可以增强未染色和透明样品的对比度。Sulston 等人利用这种光学技术追踪了秀丽隐杆线虫（*C. elegans*）从合子到孵化幼虫的胚胎细胞谱系（Sulston et al., 1983），对比度的增加使研究者能够明确秀丽隐杆线虫完整的原基分布图，发现它的细胞谱系变异性较小。

另一种增强组织中细胞对比度的实验方法是异种移植。来自不同但相关物种的细胞或组织具有不同的光学特征，如细胞色素沉着不同。将这些细胞或组织用于靶向移植，从而增强对比度，提高细胞追踪的准确性。通过这种方法，Spemann 和 Mangold 利用三种含不同色素的蝾螈胚胎制造了种间嵌合体（Spemann and Mangold, 1924）。他们发现的 Spemanne-Mangold 组织将诱导概念引入脊椎动物胚胎发育中。同样，Conrad Hal Waddington 在体外用鸡和鸭胚胎进行种间移植，以研究原肠胚形成期间的器官形成区域（Waddington, 1932），Le Douarin 用鸟类种间移植物研究神经嵴细胞的迁移和分化（Le Douarin and Teillet, 1974）。尽管种间移植已经成功地用于增强细胞 / 组织对比，但手术本身和创伤的产生可能会影响移植物在其环境中的行为，这引起了对解释观察到的细胞动力学的重点关注。

19.2.3 活体染料

为了提高标记灵敏度并增强与内源性组织相比的对比度，活体染料是直接标记胚胎中细胞的有力手段。具有特定光学特性的有机、无细胞毒性染料通常借助直接注射或移植染料浸渍材料的方式转入目标细胞的细胞质中。1925 年，Walter Vogt 将一小块浸有活体染料尼罗蓝的琼脂植入两栖动物胚胎中，并使用光学显微镜成功跟踪了原肠胚形成过程中标记细胞的后代（Vogt, 1925）。为了标记活胚胎的磷脂双层膜，研究者引入了具有明显的荧光特性、可覆盖较大范围波长的亲脂性示踪剂，例如 DiI、DiO、DiD、DiA 和 DiR（Honig and Hume, 1986, 1989）。Baier 等人利用 DiI 和 DiO 标记了斑马鱼发育中的视网膜轴突，并发明了一种系统的筛选方法来鉴定影响斑马鱼视网膜顶盖投射的关键突变（Baier et al., 1996）。虽然活体染料可标记特定的细胞，但由于它在多轮细胞分裂中的稀释或扩散，对比度会随着时间的推移而持续降低。

19.2.4 基因编码标签和组合方法

从维多利亚水母中发现的绿色荧光蛋白（green fluorescent protein, GFP）（Shimomura et al., 1962）改变了生物成像，因为它易于用作永久基因表达或蛋白质标记，而且现有的生物成像平台可以采用传统的检测器直接观察到它。因此，GFP 及其衍生物荧光蛋白（fluorescent proteins, FP）成为生物化学和分子细胞生物学中广泛使用的基因编码成像探针（Chalfie et al., 1994）。通过直接注射 mRNA、转染、病毒感染、脂质体转染或电穿孔，可以很容易地标记基因。与活体染料相比，它们不会扩散到邻近细胞，如果稳定表达，就会被标记细胞的所有后代遗传。组织或细胞特异表达 FP 的转基因品系被广泛用于研究斑马鱼的发育过程。例如，Carney 等人构建了一种利用斑马鱼 *sox10* 启动子驱动 GFP 表达的稳定转基因系，研究了 *sox10* 基因在神经嵴衍生背根神经节感觉神经元中的作用（Carney et al., 2006）。这种标记方法使他们证明了 Sox10 的功能是通过调节原神经基因 *neurogenin1* 来促进感觉神经元前体的分化。

研究者通过引入 FP 表达的遗传重组的方法来短期控制细胞群体动态的标记。基因谱系示踪最常见的方法是采用组织或细胞特异性表达的位点特异性重组酶（例如 Cre）。当 Cre 重组酶促进 DNA 链交换时，它可以激活特定细胞或一组细胞及其后代中遗传标记的表达。此外，通过使用配体诱导型 $CreER^{T2}$ 和三苯氧胺（tamoxifen, TAM）或其活性代谢物 4- 羟基三苯氧胺（4-OHT），可以实现重组的时间控制。Pinzon-Olejua 等人为了选择性地标记髓鞘胶质细胞，构建了配体诱导的位点特异性重组斑马鱼转基因品系（Pinzon-Olejua et al., 2017）。通过随机 Cre 重组在不同细胞中差异表达在光谱上不同的 FP 组合，实现了多色单细胞标记和谱系追踪的并行化，从而创造出可以在较长时间内跟踪的“彩虹”色（Livet et al., 2007）。2013 年，Pan 等人（2013）培育出了能够多色标记的转基因斑马鱼品系 Zebrabow，并证明在斑马鱼胚胎眼睛发育期间，外周角膜细胞扩增形成的克隆替代了角膜上皮克隆。

尽管上述采用基因编码标记和组合事件的细胞标记方法是谱系追踪的有力工具，但仍然存在一些局限性。产生和建立具有组织或细胞特异性标记的斑马鱼转基因系是一项艰巨的工作，并且需要了解标记特异性的全部表征。此外，还需要了解特定组织特异性启动子的已有结构。“彩虹”式方法的随机标记排除了对单个目标细胞的靶向标记。在以细胞快速分裂和迁移为特征的高度动态生物系统中，达到稳态的荧光颜色组合所需的时间往往太长，导致无法识别克隆标记的单细胞起源。

19.3 PhOTO 系统和应用

为了克服这些局限性，建立快速、精确的细胞谱系追踪，可以使用可变光 FP。这些 FP 是一类蛋白质，通常在 405 nm 左右的紫外光照射下，可以可逆地（例如光切换 FP）或不可逆地（例如光激活或光转换 FP）改变其吸收和发射光谱。例如，FP Dropa（Ando et al., 2004）从非荧光状态转换到荧光状态的过程是可逆的，而光激活 GFP（paGFP）（Patterson and Lippincott-Schwartz, 2002）从非荧光或弱荧光状态转换到强绿色荧光状态的过程是不可逆的。光转换荧光蛋白如 Dendra2（Gurskaya et al., 2006）可将其发射光谱从绿色变为红色。与光激活 FP 相比，光转换 FP 可以在光转换前后观察到。实验者可通过这一特性观察到分散标记的红色荧光分子、结构、细胞或感兴趣的多细胞结构，并保留生物系统中非光转换绿色荧光群体的全程视图。Dempsey 等人（2012, 2014）利用了这一特点构建了可实现活体细胞核和细胞膜动态的光转换光学跟踪（Photoconvertible Optical Tracking Of nuclear and membrane dynamics in vivo）的转基因斑马鱼品系，命名为 PhOTO 斑马鱼，该品系结合了精确的单细胞标记和长期谱系追踪的特点。

19.3.1 PhOTO 斑马鱼的特征

PhOTO 斑马鱼是一种可广泛表达蓝色 FP Cerulean 和光转换 FP Dendra2 的靶向融合蛋白的转基因品系。具体而言，PhOTO 载体［图 19-1（A）］可通过组成型 β-actin2 驱动 Cerulean 和 Dendra2 的表达，通过蛋白质 N 末端棕榈酰化和肉豆蔻酰化脂肪酸底物序列靶向到细胞膜（memb），或者通过组蛋白 2B（H2B 融合）靶向到细胞核。位于两个编码区之间的明脉扁刺蛾病毒（*Thosea asigna* virus, TaV）2A 自剪切序列在翻译过程中将两种蛋白质以 1∶1 的化学计量比分配到各自的目标位置。研究者利用 Tol2 转位酶介导的基因组整合系统，建立了三个转基因斑马鱼品系：PhOTO-N［*Tg* (*bactin2*: *memb-Cerulean-2A-H2B-Dendra2*)］、PhOTO-M［*Tg* (*bactin2: memb-Dendra2-2A-H2B-Cerulean*)］和 PhOTO-C［*Tg*(*bactin2:cyto-Dendra2-2A-H2B-Cerulean*)］。PhOTO-N 斑马鱼品系［网址：https://www.addgene.org/92400/，欧洲斑马鱼资源中心（EZRC）（https://www.ezrc.kit.edu/）基因组特征：zh21Tg］表达 H2B-Dendra2 和 memb-Cerulean，分别选择性标记细胞核和细胞膜。互补的 PhOTO-M 斑马鱼品系［网址：https://www.addgene.org/92401/，EZRC 基因组特征：zh22Tg］可以表达 H2B-Cerulean 和 memb-Dendra2。PhOTO-C［网址：https://www.addgene.org/92402/，EZRC 基因组特征：zh23Tg］可以表达胞质 Dendra2 和核 Cerulean。在所有的 PhOTO 转基因品系中，Dendra2 的瞬时光转换可以精准定位、精确分割和重建高时空分辨率和信噪比的细胞轨迹，而细胞形态和行为的全程视图取决于 Cerulean 的区间特异性表达。

19.3.2 基于 PhOTO 品系的谱系追踪和细胞动力学

PhOTO 斑马鱼的整个生命周期中都持续表达 Cerulean 和 Dendra2，因此在转基因动物的任何阶段均可以成功地进行谱系追踪实验（Dempsey et al., 2014）。PhOTO-N 转基因斑马鱼成功地揭示了在原肠胚形成的高动态期间单个细胞发育过程的复杂性（Dempsey et al., 2012）。通过精确的光转换细胞核 Dendra2，可标记斑马鱼胚胎中的目的细胞或结构，如体节细胞或尾尖［见图 19-1（B）］，从而可以正确地进行细胞追踪和谱系追踪。与此同时，周围细胞的 memb-Cerulean 和未转换的 H2B-Dendra2 荧光数据为光转化细胞核的行为提供了背景，从而揭示了细胞的相互作用和细胞动力学。PhOTO-M 转基因斑马鱼可用于突

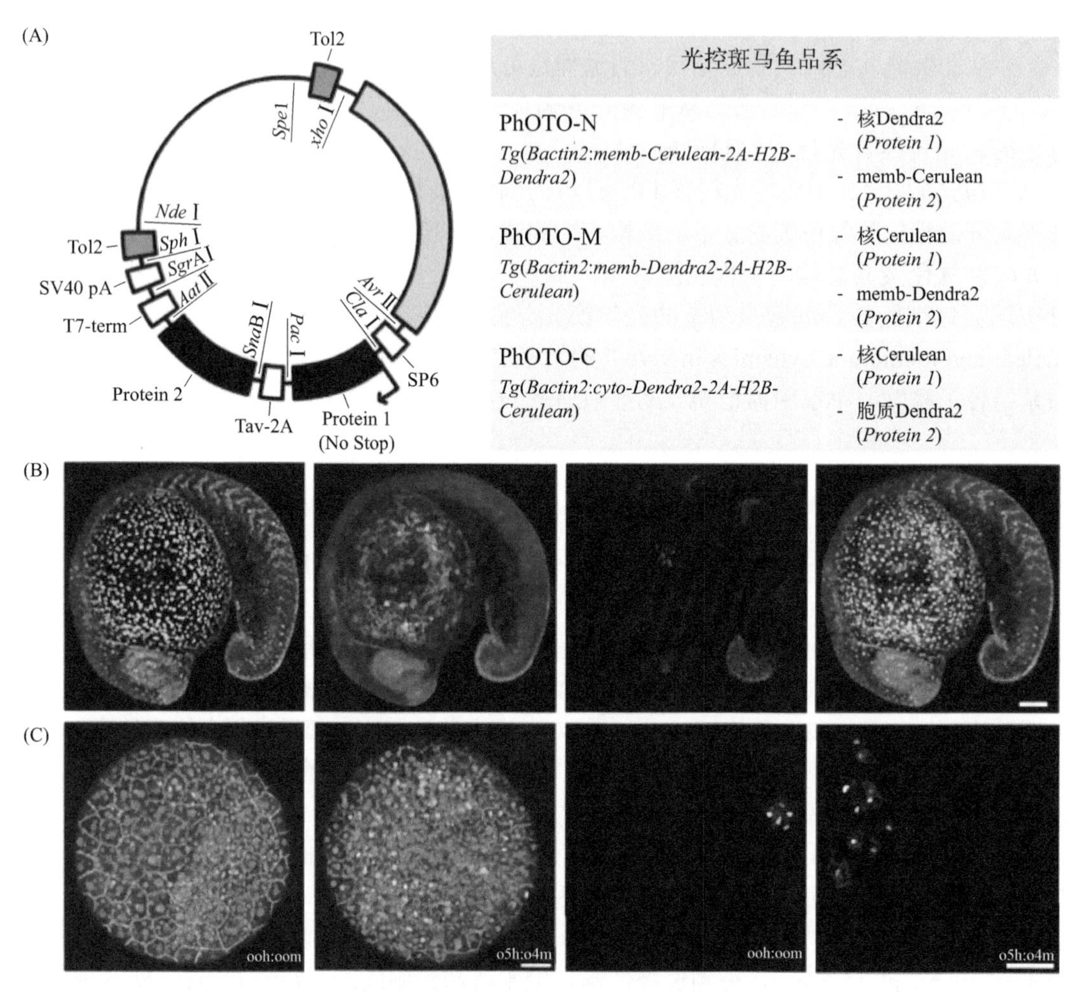

光控斑马鱼品系	
PhOTO-N *Tg(Bactin2:memb-Cerulean-2A-H2B-Dendra2)*	- 核Dendra2 (*Protein 1*) - memb-Cerulean (*Protein 2*)
PhOTO-M *Tg(Bactin2:memb-Dendra2-2A-H2B-Cerulean)*	- 核Cerulean (*Protein 1*) - memb-Dendra2 (*Protein 2*)
PhOTO-C *Tg(Bactin2:cyto-Dendra2-2A-H2B-Cerulean)*	- 核Cerulean (*Protein 1*) - 胞质Dendra2 (*Protein 2*)

图 19-1 PhOTO基因载体和光控转基因斑马鱼品系的说明

（A）PhOTO 基因载体示意图，表格中列出了三条 PhOTO 斑马鱼品系（PhOTO-N、PhOTO-M 和 PhOTO-C）的主要特征。载体的主要成分，包括：（i）启动子，（ii）TaV 2A 序列，（iii）Tol2 转座因子和（iv）FP 位置，分别标记于质粒环的外部区域。每个 PhOTO 基因载体的设计应确保每个组分（如启动子、FP 等）都可以轻易地使用列出的限制性内切酶位点（位于质粒环内）亚克隆，以进行交替的蛋白质融合。注意，表示编码序列的阻断区域的大小不可改变。（B）受精后 18 ～ 19h 的 F1 代转基因斑马鱼胚胎中 PhOTO-N 杂合子表达的典型图像。从左到右依次为：未转换的 H2B-Dendra2（绿色），memb-Cerulean（蓝色），四个体节中（包括尾巴尖端，眼睛中的一部分细胞，卵黄顶部的一部分细胞）光转换的 H2B-Dendra2（品红），以及所有三种颜色的合并图像。比例尺为 100μm。（C）杂合子 F1 代 PhOTO-M 斑马鱼从原肠胚晚期（>80% 表观代谢）到早期分裂 1 ～ 5h 首个 21mm（深度）的动物极视图最大强度投影图像。前两个图描绘了拍照时间的第一帧和最后一帧合并的最大强度投影，显示了未转换的 memb-Dendra2（绿色）、分区的光转换 memb-Dendra2（洋红色）、H2B-Cerulean（蓝色）标记的核，以及分区的 H2B-Cerulean 核（光转换细胞中的多色表面）。另外两个图显示了相同时间点的分区、光转换和迁移 memb-Dendra2 强度数据（洋红色）和分区 H2B-Cerulean 核（多色表面）的放大最大强度投影。比例尺为 50μm

显早期胚胎发育过程中细胞膜和细胞核的动态变化（Dempsey et al., 2012）[图 19-1（C）]。采集 memb-Dendra2 的未转换绿色荧光捕获了细胞膜水平细胞动力学的全程视图。memb-Dendra2 光转换分散的红色荧光展现了较详细的膜动力学，如细胞迁移过程中胚胎外周的短膜延伸。

Dendra2 和 Cerulean 在成年 PhOTO 斑马鱼细胞中高表达且分布范围较广，这有利于开展发育后谱系追踪实验。PhOTO-N 转基因品系尾鳍被切除后再生过程中，通过光转换被切除区域附近的细胞核，可以监测到斑马鱼尾鳍再生过程中细胞相应的变化（即细胞迁移和分裂）（Dempsey et al., 2012）。光转换红核的定位提供了关于细胞迁移范围的信息，区域红核的平均强度可反映再生过程中细胞分裂程度。再生 7 天后，无红色信号的细胞和比背景略有增加的细胞可组成新的再生区域，而一部分与潜水无关的亮红细胞则位于尾鳍的远端边缘。这些数据表明，随着鳍状结构的重建，新区域中大多数细胞是广泛增殖的结果。有趣的是，PhOTO-N 品系鱼鳍光电转换具有强烈的红色信号，据此可筛选出第二种非分裂细胞群。这一细胞群似乎沿着再生组织的生长轴排列并均匀分布，可能通过潜在的信号转导来引导细胞迁移和沉积，从而协调鳍的再生。

19.3.3 优点和局限性

通过用紫光对 Dendra2- 融合体进行光转换，PhOTO 斑马鱼品系能够在生命周期的任何时间对任何细胞亚群的细胞核或细胞膜进行即时和精确（即非随机）的标记。PhOTO-N、PhOTO-M 和 PhOTO-C 系结合了光转换荧光蛋白的优势，在发育和再生过程中提供散在的和全面的细胞追踪，以进行选择性谱系追踪。此外，在数据分析和解释过程中，区间特异性标记的存在致使不再需要外源性复染。然而，直到最近，光转换还受到强烈紫光照射的限制，这种照射可能具有潜在的光毒性效应（Waldchen et al., 2015）。此外，这种光照过程的轴向不受限制的特性阻止了活体复杂组织的单细胞和 / 或亚细胞区域标记。

19.3.4 有限的和无毒的光转换：启动转换

在紫光照射下，所有绿色到红色的光转换荧光蛋白可共享一个相同的光转换机制。该机制依赖于 N_α 和 C_α 原子之间的键断裂，以及发色团中组氨酸 C_α 和 C_β 原子之间的双键形成，从而导致发射和吸收的不可逆转变（Nienhaus et al., 2005; Mizuno et al., 2003）。最近，Dempsey 等人（2015）报道了一种不同的光转换机制，称为启动转换（primed conversion）。他们发现，同时使用 488nm 和 730nm 的光照射，Dendra2 可以从绿色光转换为红色光。绿色光首先吸收 488nm 的“启动”光，从而进入几毫秒的持久中间状态，这种状态可以吸收远红外“转换”光子，从而导致发射光谱从绿色到红色的移动。有趣的是，紫光光转换和启动转换都会产生光谱相同的红色光。最近，Mohr 等人（2017）进行了一系列同位素实验，以便更好地理解启动转换机制。他们发现该机制经历了三重中间态（Mohr et al., 2017）。研究表明，在序列 69 位置处引入单个苏氨酸或丝氨酸均可以促进这种转换。利用这一点，他们使大多数绿色到红色的光转化蛋白发生启动转换（网址：https://www.addgene.org/Periklis_Pantazis/）（见表 19-1）。创建的“pr”-（用于启动转换）

表 19-1 启动转换的荧光蛋白

种类	光转换蛋白
可转换荧光蛋白单聚体	Dendra2
	pr-mEosFP
	pr-mEos2
	pr-mEos3.1
	pr-mEos3.2
	pr-mEos4a
	pr-mEos4b
	pr-mKikGR
	mMaple
	mClavGR
可转换荧光蛋白寡聚体	pr-Kaede
光遗传转换工具	PhOCI
启动转换的 Ca^{2+} 控制器	pr-CaMPARI
启动转换的 Ca^{2+} 指示剂	GR-GECO1.1
	GR-GECO1.2

变体不仅可以在前两个阶段切换，也更加稳定。最近，Mohr 等人（2017）和 Klementieva 等人（2016）发现，每台商用荧光显微镜（630 ～ 650nm）的传统红色激光也可以实现这种效果。因此，无需大量技术改革，可使用任何荧光显微镜实现有限的启动转换，方便大量用户群使用。

与紫光照明相反，启动转换的轴向尺寸受限。只需添加一个市售的激发转换滤光片，同时使用双激光照明，共聚焦激光扫描显微镜（CLSM）即可实现启动转换（Mohr et al., 2016; Mohr and Pantazis, 2016）。通过两束光在一个共同焦点的选择性交叉，在密集 3D 环境中实现了单细胞和（/ 或）亚细胞区室和 / 或结构的标记（Mohr et al., 2016, 2017; Klementieva et al., 2016）[图 19-2（A）]。重要的是，与之前提到的精确标记发育中胚胎的双光子激发相比，限制性启动转换能够以更小的峰值和平均光照强度实现空间选择性光转换（Plachta et al., 2011）。

19.3.5 PhOTO 斑马鱼和启动转换

通过使用轴向限制的启动转换和高分辨率容积成像，可以长时间标记和跟踪发育中斑马鱼胚胎的单个细胞 [图 19-2（B）]。PhOTO-N 斑马鱼胚胎的单细胞核可以随时在普通 CLSM 中进行启动转换，然后转移到光片成像平台，完成高分辨率活体动物成像。限制性启动转换的独特优势在于它能够在斑马鱼胚胎发育的密集环境中，以高时空分辨率揭示其复杂的发育过程，甚至可达单个细胞水平。通过高信噪比可以很容易监测到标记细胞迁移和细胞重排的程度，从而揭示在发育胚胎的不同区域中单细胞或细胞群的不同动态。将详细的空间动态信息与成像过程中检索的时间数据相结合，可以辨别细胞分裂的速率和数量以及它们的同步性和方向，以便深入地分析和理解。

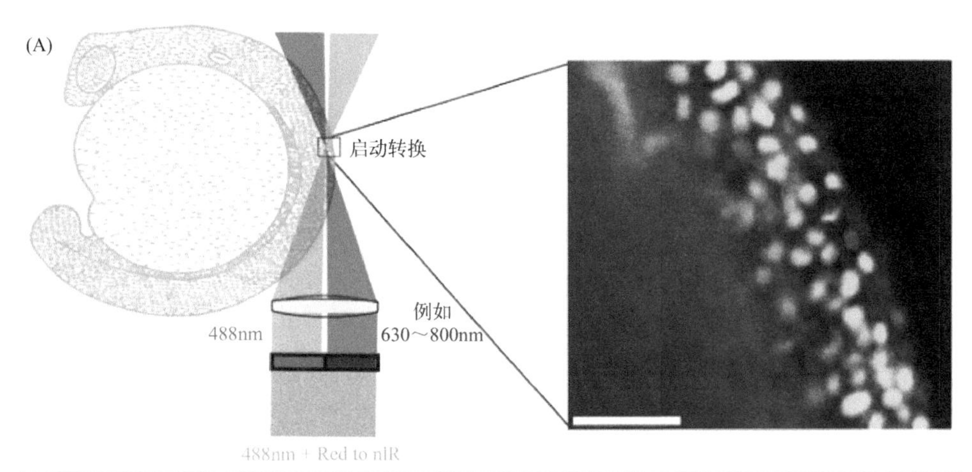

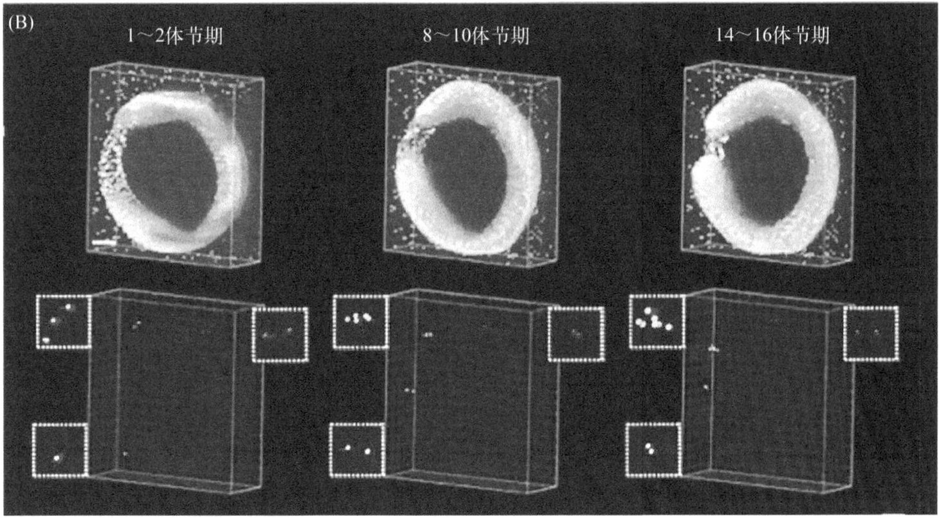

图 19-2 **轴向限制启动转换的扫描原理，用于标记活斑马鱼胚胎中的单个细胞，并跟踪它们在不同发育期间的动态**

(A) 启动转换的转换滤板（the primed conversion filter plate）由两个可用于启动（488 nm）或转换光束（从红色到近红外；630 ～ 800nm）的半圆形光学滤光片以及共聚焦激光扫描显微镜光路内的不透明分隔条所组成。激光束仅在焦点处重叠，从而在样品的轴向特定位置中产生启动转换成像，其上面和下面的细胞没有被启动转换。在 PHOTO-N 转基因斑马鱼胚胎的单细胞核中的 H2B-Dendra2 轴向限制启动转换。比例尺为 50μm。(B) 在光片显微镜下，对 PhOTO-N 斑马鱼从 1 ～ 2 体节期到 14 ～ 16 体节期的发育过程中，对启动的转化核进行体积成像跟踪（volumetric tracking of primed converted nuclei）。PhOTO-N 斑马鱼胚胎的单细胞细胞核可以很容易地在普通共聚焦激光扫描显微镜中进行启动转换，然后转移到光片成像平台上进行高分辨率活体体积成像。可以长时间监测单个细胞核，显示细胞分裂（光转换 Dendra2 信号的强度随着多轮细胞分裂而降低）和细胞迁移（红色信号的定位）。比例尺分别为 100μm 和 10μm

19.3.6 启动转换和活体神经元标记

了解神经元的组织结构可以揭示大脑功能的基本机制。神经系统中细胞紧密且高度聚集使得单个神经元的可视化具有挑战性。绘制神经元回路图的传统方法是采用连续切片进行电子显微镜观察（Hildebrand et al., 2017），它可以提供单个神经突起和突触的成像。连续切片电子显微镜视野小，需要计算机重建，要求组织固定和树脂包埋，这些都限制了该方法的应用。较新的电子显微镜，例如聚焦离子束和气体团簇离子束扫描电子显微镜

[FIB-SEM（Knott et al., 2008）和 GCIB-SEM（Hayworth et al., 2020）]，光束扫描有利于增强采集图像的质量和采集速度，但是仍然无法实现体内应用。

与电子显微镜相比，基于荧光标记的光学方法可以提供大视野，并且可以对长程连接进行成像。扩展显微成像技术（expansion microscopy）（Chen et al., 2015）利用凝胶聚合物在化学固定细胞和组织间吸水膨胀，可产生更高分辨率的大脑组织图像。然而，该方法需要固定生物样品，不能用于活体成像。与此相比，启动转换已成功地应用于精准标记活体单细胞。此前，Dempsey 等人（2015）利用启动转换标记并揭示了斑马鱼三叉神经感觉神经节、视顶盖和脊髓单个神经元的分支形态（图 19-3）。他们利用神经特异性 HuC 启动子在神经元胞质中表达 Dendra2，通过限制性启动转换在光转换单个神经元中表达 Dendra2。光转换红色形式的 Dendra2 通过胞质扩散到神经突，使细胞分支形态可视化。因此，限制性启动转换是研究活体结构连接的一个有效工具。

PhOTO-C 斑马鱼胞质表达 Dendra2，细胞核表达 Cerulean。因此，可以利用启动转换来绘制 PhOTO-C 斑马鱼活体胚胎和成鱼神经连接的全面图谱。此外，膜标记的 PhOTO-M 可用于揭示突触的膜可塑性和动力学。这种结构分析可以与兼容的有机荧光 Ca^{2+} 指示剂的功能读数相结合，反映中间神经元的活性。这种实验设计将有助于研究以神经元网络同步

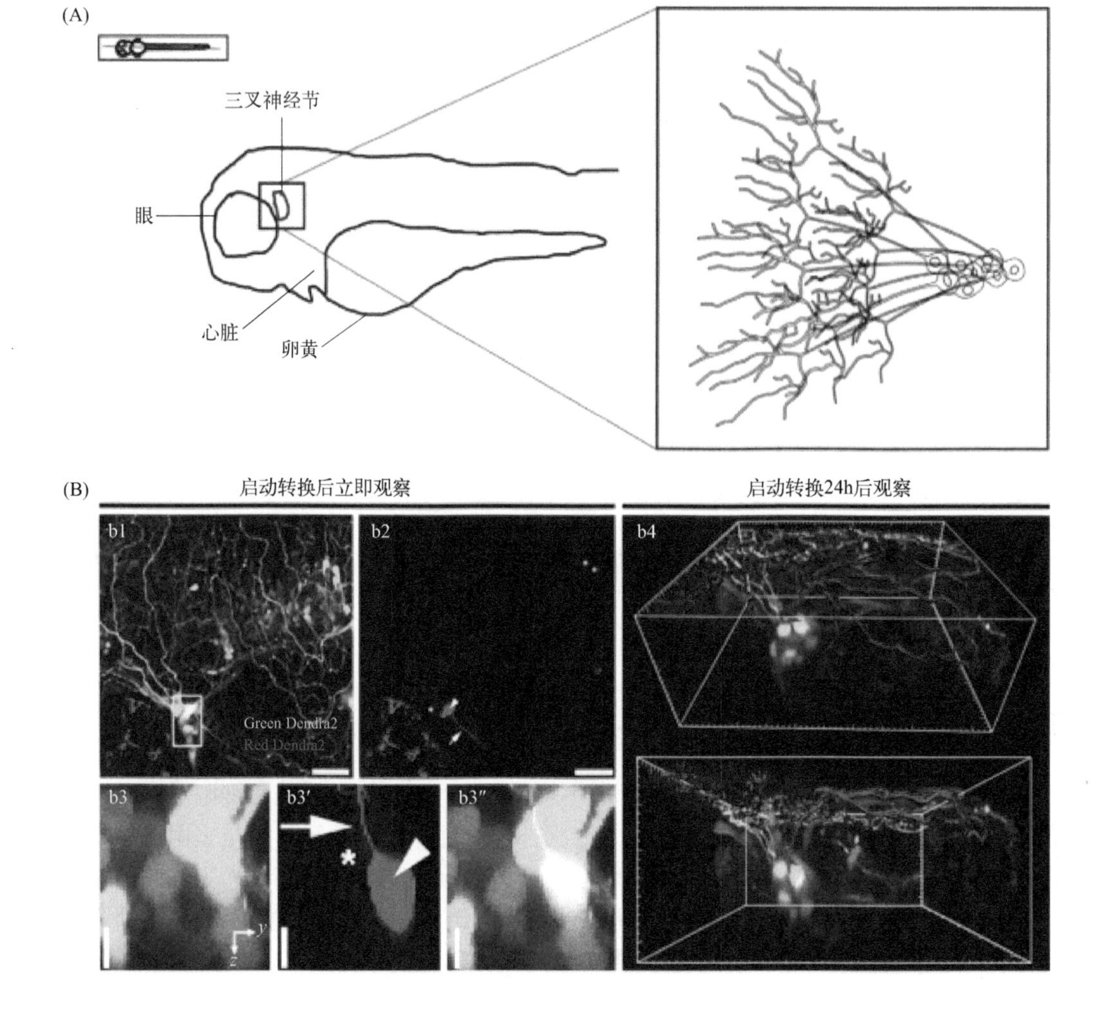

(C)

图 19-3 **限制性启动转换能够在活体斑马鱼仔鱼密集交织神经簇中进行单个神经元标记**

（A）三叉神经节或第五神经节是三叉神经或第五对脑神经触觉神经元的集合，位于眼睛和耳朵之间的大脑区域（Kimmel et al., 1995）。（B）b1, b2 图为受精后 3 天，三叉神经节单个神经元中 Dendra2 的启动转换（最大强度投影，深度约为 82μm），比例尺为 50μm，该图像是在光转换后立即获得的；b3、b3′ 和 b3″ 是 b1 中方框区域的放大和轴向图像，轴向比例尺为 10μm。有尾箭头标记的是从胞体延伸的神经突起，无尾箭头标记的是光转换细胞；b2 和 b3′ 中的星号表示早在启动转换前就有明显红色信号的细胞；b4 为胞体中 Dendra2 启动转换后 24h 神经突追踪的不同视角。（C）发育中的视顶盖（c1）和脊髓（c2）单个神经元的启动转换，比例尺为 20μm。c1 和 c2 中方框区域中的插图为在相同放大倍数下显示的对比度增强的 Dendra2 红色信号

化为特征的癫痫发作。了解神经元是如何连接并能够有效地相互交流，对于控制反复发作活动（即癫痫）至关重要。斑马鱼适用于高通量筛选，且与人类的神经系统和行为具有高度的相似性，这使其成为筛选神经特异性化合物的理想模型。在药物筛选中使用 PhOTO-C 和 PhOTO-M 斑马鱼，可以监测应用化合物后神经元网络的动态变化。产生疾病模型的已知化学物质（如导致帕金森病的 1-甲基-4-苯基-1,2,3,6-四氢吡啶）在光控斑马鱼品系中的应用研究，将为神经退行性疾病提供有价值的见解。此外，细胞核表达 Dendra2 和细胞膜标记 Cerulean 的 PhOTO-N 可用于了解成鱼神经发生和脑再生的过程。

19.4 展望

利用 PhOTO 斑马鱼转基因品系广泛表达蓝色荧光蛋白和光转换荧光蛋白 Dendra2 的靶向融合蛋白这一特征，可以在斑马鱼生命周期的任何阶段实现即时、准确和无创的谱系追踪。结合限制性启动转换的 PhOTO 品系，可以在活体胚胎和成年斑马鱼中更好地揭示复杂的发育、疾病和再生动力学。单细胞分辨率下映射神经元回路和神经元活动是理解大脑可塑性的关键步骤，即在学习和记忆相关的重要脑区中新的神经元连接何时以及如何形成。PhOTO 系统提供的时空数据可以弥补最先进的单细胞基因组学方法的缺憾，因为后者目前无法记录动态细胞过程。高保真谱系追踪和细胞成分分析的结合，将促进对生理过程机制的理解。为了明确单细胞基因组学方法鉴定出的这些因子的作用，可以使用启动转换和光遗传学工具如 PhOCl（Zhang et al., 2017），从而可以更加深入地在特定时间点分析这些因子在单个细胞中的功能。启动转换到光片成像平台的实现，也可以最大限度地减少实验内误差，从而促进了上述实验工作的进行。最后，“智能”成像平台的开发可以观察复杂的动力学，例如细胞运动性的变化和复杂的细胞重排，从而促进 PhOTO 系统在研究生理过程中的应用。最近的工作流程分析，如启动追踪（Welling et al., 2019），可以重建

精确的细胞轨迹，有望进一步促进对发育和疾病机制的认识。

（郭煊君翻译　赵丽审校）

参考文献

Ando, R., Mizuno, H., Miyawaki, A., 2004. Regulated fast nucleocytoplasmic shuttling observed by reversible protein highlighting. Science (New York, N.Y.) 306, 1370-1373. https://doi.org/10.1126/science.1102506.

Baier, H., Klostermann, S., Trowe, T., Karlstrom, R.O., Nusslein-Volhard, C., Bonhoeffer, F., 1996. Genetic dissection of the retinotectal projection. Development 123, 415-425.

Carney, T.J., Dutton, K.A., Greenhill, E., et al., 2006. A direct role for Sox10 in specification of neural crest-derived sensory neurons. Development 133 (23), 4619-4630.

Chalfie, M., Tu, Y., Euskirchen, G., Ward, W.W., Prasher, D.C., 1994. Green fluorescent protein as a marker for gene expression. Science 263 (5148), 802-805.

Chen, F., Tillberg, P.W., Boyden, E.S., 2015. Expansion microscopy. Science 347 (6221), 543-548.

Dempsey, W.P., Fraser, S.E., Pantazis, P., 2012. PhOTO zebrafish: a transgenic resource for in vivo lineage tracing during development and regeneration. PLoS One 7 (3), e32888.

Dempsey, W.P., Qin, H., Pantazis, P., 2014. In vivo cell tracking using PhOTO zebrafish. Methods in Molecular Biology 1148, 217-228.

Dempsey, W.P., Georgieva, L., Helbling, P.M., et al., 2015. In vivo single-cell labeling by confined primed conversion. Nature Methods 12 (7), 645-648.

Gurskaya, N.G., et al., 2006. Engineering of a monomeric green-to-red photoactivatable fluorescent protein induced by blue light. Nature biotechnology 24, 461-465. https://doi.org/10.1038/nbt1191.

Hayworth, K.J., Peale, D., Januszewski, M., et al., 2020. Gas cluster ion beam SEM for imaging of large tissue samples with 10 nm isotropic resolution. Nat Methods 17, 68-71. https://doi.org/10.1038/s41592-019-0641-2.

Hildebrand, D.G.C., Cicconet, M., Torres, R.M., et al., 2017. Whole-brain serial-section electron microscopy in larval zebrafish. Nature 545, 345.

Honig, M.G., Hume, R.I., 1986. Fluorescent carbocyanine dyes allow living neurons of identified origin to be studied in long-term cultures. The Journal of Cell Biology 103 (1), 171-187.

Honig, M.G., Hume, R.I., 1989. Dil and diO: versatile fluorescent dyes for neuronal labelling and pathway tracing. Trends in Neurosciences 12 (9), 333-335, 340-331.

Huisken, J., Swoger, J., Del Bene, F., Wittbrodt, J., Stelzer, E.H.K., 2004. Optical Sectioning Deep Inside Live Embryos by Selective Plane Illumination Microscopy. Science (765 New York, N.Y.) 305, 1007. https://doi.org/10.1126/ science.1100035.

Keller, P.J., Schmidt, A.D., Wittbrodt, J., Stelzer, E.H., 2008. Reconstruction of zebrafishearly embryonic development by scanned light sheet microscopy. Science (New York, N.Y.) 322, 1065-1069. https://doi.org/10.1126/ science.1162493.

Kimmel, C.B., Ballard, W.W., Kimmel, S.R., Ullmann, B., Schilling, T.F., 1995. Stages of embryonic development of the zebrafish. Developmental Dynamics: An Official Publication of the American Association of Anatomists 203 (3), 253-310.

Klementieva, N.V., Lukyanov, K.A., Markina, N.M., Lukyanov, S.A., Zagaynova, E.V., Mishin, A.S., 2016. Green-to-red primed conversion of Dendra2 using blue and red lasers. Chemical Communications 52 (89), 13144-13146.

Knott, G., Marchman, H., Wall, D., Lich, B., 2008. Serial section scanning electron microscopy of adult brain tissue using focused ion beam milling. Journal of Neuroscience: The Official Journal of the Society for Neuroscience 28 (12), 2959-2964.

Lam, C.S., Korzh, V., Strahle, U., 2005. Zebrafish embryos are susceptible to the dopaminergic neurotoxin MPTP. European Journal of Neuroscience 21 (6), 1758-1762.

Lamprecht, S., Schmidt, E.M., Blaj, C., et al., 2017. Multicolor lineage tracing reveals clonal architecture and dynamics in colon cancer. Nature Communications 8 (1), 1406.

Le Douarin, N.M., Teillet, M.A., 1974. Experimental analysis of the migration and differentiation of neuroblasts of the autonomic nervous system and of neurectodermal mesenchymal derivatives, using a biological cell marking technique. Developmental Biology 41 (1), 162-184.

Livet, J., Weissman, T.A., Kang, H., et al., 2007. Transgenic strategies for combinatorial expression of fluorescent proteins in the nervous system. Nature 450 (7166), 56-62.

Mikut, R., et al., 2013. Automated processing of zebrafish imaging data: a survey. Zebrafish 10, 401-421. https://doi.

org/10.1089/zeb.2013.0886.

Mizuno, H., Mal, T.K., Tong, K.I., et al., 2003. Photo-induced peptide cleavage in the green-to-red conversion of a fluorescent protein. Molecular Cell 12 (4), 1051-1058.

Mohr, M.A., Pantazis, P., 2016. Single neuron morphology in vivo with confined primed conversion. Methods in Cell Biology 133, 125-138.

Mohr, M.A., Argast, P., Pantazis, P., 2016. Labeling cellular structures in vivo using confined primed conversion of photoconvertible fluorescent proteins. Nature Protocols 11 (12), 2419-2431.

Mohr, M.A., Kobitski, A.Y., Sabater, L.R., et al., 2017. Rational engineering of photoconvertible fluorescent proteins for dual-color fluorescence nanoscopy enabled by a triplet-state mechanism of primed conversion. Angewandte Chemie 56 (38), 11628-11633.

Nienhaus, K., Nienhaus, G.U., Wiedenmann, J., Nar, H., 2005. Structural basis for photo-induced protein cleavage and green-to-red conversion of fluorescent protein EosFP. Proceedings of the National Academy of Sciences of the United States of America 102 (26), 9156-9159.

Nomarski, G., 1952. Lnterferornetre a Potartsenon. Brevet Francais No 1059123.

Pan, Y.A., Freundlich, T., Weissman, T.A., et al., 2013. Zebrabow: multispectral cell labeling for cell tracing and line-age analysis in zebrafish. Development 140 (13), 2835-2846.

Patterson, G.H., Lippincott-Schwartz, J., 2002. A photoactivatable GFP for selective photolabeling of proteins and cells. Science (New York, N.Y.) 297, 1873-1877. https://doi.org/10.1126/science.1074952.

Pinzon-Olejua, A., Welte, C., Chekuru, A., et al., 2017. Cre-inducible site-specific recombination in zebrafish oligodendrocytes. Developmental Dynamics: An Official Publication of the American Association of Anatomists. 246 (1), 41-49.

Plachta, N., Bollenbach, T., Pease, S., Fraser, S.E., Pantazis, P., 2011. Oct4 kinetics predict cell lineage patterning in the early mammalian embryo. Nature Cell Biology 13 (2), 117-123.

Shimokawa, M., Ohta, Y., Nishikori, S., et al., 2017. Visualization and targeting of LGR5(+) human colon cancer stem cells. Nature 545 (7653), 187-192.

Shimomura, O., Johnson, F.H., Saiga, Y., 1962. Extraction, purification and properties of aequorin, a bioluminescent protein from the luminous hydromedusan, Aequorea. Journal of Cellular and Comparative Physiology 59, 223-239.

Spemann, H., Mangold, H., 1924. Induction of embryonic primordia by implantation of organizers from a different species. Archiv für Mikroskopische Anatomie und Entwicklungsmechanik 100, 599-563.

Sulston, J.E., Schierenberg, E., White, J.G., Thomson, J.N., 1983. The embryonic cell lineage of the nematode *Caeno rhabditis elegans*. Developmental Biology 100 (1), 64-119.

Vogt, W., 1925. Gestaltungsanalyse am Amphibienkeim mit ortlicher Vitalfarbung. Vorwarts uber Wege und Ziele. I. Methodik und Wirkungsweise der Vitalfarbung mit Agar als Farbtrager 106.

Waddington, C.H., 1932. Experiments on the development of chick and duck embryos, cultivated in vitro. Philosophical Transactions of the Royal Society of London e Series B: Containing Papers of a Biological Character 221, 179-230.

Waldchen, S., Lehmann, J., Klein, T., van de Linde, S., Sauer, M., 2015. Light-induced cell damage in live-cell super-resolution microscopy. Scientific Reports 5, 15348.

Welling, M., Mohr, M.A., Ponti, A., et al., 2019. Primed Track, high-fidelity lineage tracing in mouse pre-implantation embryos using primed conversion of photoconvertible proteins. eLife 8.

Whitman, C.O., 1878. The Embryology of Clepsine. J.E. Adlard, London.

Zhang, W., Lohman, A.W., Zhuravlova, Y., et al., 2017. Optogenetic control with a photocleavable protein, PhoCl. Nature Methods 14 (4), 391-394.

第五部分

疾病模型和行为遗传学应用

第二十章

乙醇对斑马鱼的急性、慢性效应

Steven Tran[1]

20.1 乙醇成瘾的代价

目前，美国的乙醇消费量有所增加，超过12%的美国人存在乙醇使用障碍（Grant et al., 2017）。在过去的几十年里，随着乙醇滥用在年轻人中越来越普遍，偶发性超量饮酒率也有所增加（Linden-Carmichael et al., 2017）。在美国，仅就治疗相关花费和收入损失而言，乙醇滥用的社会成本每年就超过2500亿美元（Sacks et al., 2015）。乙醇滥用的疾病模型可用于开发药物疗法来治疗乙醇使用障碍。然而，尽管几十年来一直在研究用于治疗乙醇使用障碍的药物，但这些药物的疗效仍然是有限的（Franck and Jayaram-Lindstrom, 2013）。这很大程度上是由于人们对于乙醇对大脑的影响（急性乙醇效应）以及长期饮酒后乙醇使用障碍（慢性乙醇效应）的发生机制仍知之甚少。乙醇被认为是一种“肮脏”的药物，因为它可以通过多种机制发挥作用而引起大量的级联效应，其中许多效应尚未被完全了解（Costardi et al., 2015）。利用啮齿动物来研究乙醇对大脑的影响以及乙醇滥用的发展为乙醇使用障碍的研究提供了重要见解（Bell et al., 2016）。然而，灵长类动物模型的高昂成本及低顺从性阻碍了其在高通量行为筛选中的应用。果蝇（黑腹果蝇）和线虫（秀丽隐杆线虫）等无脊椎动物模式生物具有体积小、繁殖力高等高通量行为筛查的关键特征，因而克服了这些缺点（Park et al., 2017; Zhu et al., 2014）。与哺乳动物相比，使用无脊椎动物的缺点是后者中枢神经系统简单、行为反应的种类少以及遗传背景差距更大。

[1] Department of Biology and Biological Engineering, California Institute of Technology, Pasadena, CA, United States.

20.2 斑马鱼：研究乙醇效应的动物模型

斑马鱼（*Danio rerio*）体现了系统复杂性和实用简便性之间的折中平衡（Gerlai, 2014）。由于其体积小、繁殖力强、进化保守且中枢神经系统简单，最重要的是遗传易于控制，这种小鱼成为高通量遗传筛选的绝佳选择。一只雌性斑马鱼每隔一天可以产下200多个卵，受精后3天就能够孵化出可以自由游动的斑马鱼仔鱼。受精后7天，斑马鱼仔鱼表现出大量复杂的行为反应，包括习惯化、敏感化、经典条件反射、运动适应和社会学习。由于斑马鱼仔鱼体型小，受精后4天即可用于96孔板中进行的平行测试（每个孔中只放一条仔鱼），并且可以在其不进食的情况下监测其数天的运动和睡眠行为（Naumann et al., 2010）。利用这种方法，使用斑马鱼仔鱼对数千种药物化合物进行了高通量筛选，得到的药物效应与在人体上观察到的相似，证明了斑马鱼和人类之间的功能保守性（Rihel et al., 2010）。斑马鱼在行为药理学上的另一个经常被忽视的优势是药物递送的方法。利用斑马鱼通过鱼鳃吸收药物这一特点，可将成鱼和仔鱼简单地浸入药液中作为一种非侵入性的药物递送方法。斑马鱼仔鱼缺乏血脑屏障，这使得药物可以很容易进入大脑并对中枢神经系统产生影响（Rihel et al., 2010）。许多药物也可以穿过成年斑马鱼的血脑屏障，现已证明，外部水溶液和大脑中的乙醇浓度在暴露40min内可达到平衡，大脑乙醇浓度为外部浓度的10%～30%（Pan et al., 2011; Steven Tran, Chatterjee and Gerlai, 2015）。由于这种小硬骨鱼的遗传易控性，斑马鱼很适用于研究乙醇对中枢神经系统的急性效应（Lockwood et al., 2004），以及慢性乙醇暴露后的长期神经适应性（Tran et al., 2016）。

20.3 急性乙醇暴露

乙醇被归为中枢神经系统抑制剂，血液乙醇浓度（BAC）过高会导致镇静、昏迷甚至死亡。然而，乙醇具有双相效应，低至中等剂量的乙醇表现出兴奋和抗焦虑作用。一般来说，乙醇的兴奋性剂量与精力、躁动增加和抑制降低的主观感觉有关（Hendler et al., 2013）。在啮齿类动物中也有类似的发现，低到中等剂量的乙醇会增加自发活动（Fish et al., 2002）。在成年斑马鱼中，外部水溶液中乙醇浓度低于1%（体积分数）时通常会增加自发活动［以移动速度（cm/s）或总移动距离（cm）来衡量］，这证明了乙醇的兴奋性作用（Tran and Gerlai, 2013）。乙醇浓度超过1%（体积分数）时通常会降低成年斑马鱼的自发活动，表现出镇静作用（Mathur and Guo, 2011）。总之，这些研究结果表明，高浓度乙醇暴露与自发活动成倒U形剂量-效应关系，最终可导致死亡。重要的是，乙醇的镇静作用不仅具有剂量依赖性，而且具有时间依赖性。例如，将成年斑马鱼暴露于1%的乙醇中，其自发活动在暴露后的30min内会增加，然后逐渐降低到基线水平（Tran and Gerlai, 2013）。这种时间依赖性的倒U形曲线是由于随着大脑中乙醇浓度的增加，更高浓度的乙醇会抑制自发活动（Rosemberg et al., 2012; Tran et al., 2015a）。能够降低脑乙醇浓度的化合物，如牛磺酸，已被证明可以抑制乙醇的镇静作用，这表明乙醇的镇静作用取决于脑中乙醇浓度的增加（Rosemberg et al., 2012）。

人们用乙醇对动物模型的兴奋作用来代表其对人体的欣快和强化效应，这些效应可能是乙醇使用障碍形成的重要因素（Hendler et al., 2013）。大量的研究集中于阐明乙醇兴奋作用的分子机制（see Phillips, 2002; Phillips and Shen, 1996）。如前所述，由于乙醇可对大脑产生大量的影响，因此通常被认为是一种“肮脏”的药物。在斑马鱼中，现已证明，乙醇暴露能够引起大量神经化学及分子靶点的变化，包括但不限于多巴胺、血清素及其代谢物以及谷氨酸盐、阿斯巴甜、甘氨酸、牛磺酸和γ-氨基丁酸（Chatterjee et al., 2014）。本章将重点介绍急性乙醇暴露所改变的两个主要神经递质系统：多巴胺能和血清素能系统。

20.3.1　多巴胺与自发活动

在研究乙醇奖赏特性时通常会检测多巴胺。在啮齿类动物的研究中，乙醇暴露已被证明会增加伏隔核（哺乳动物大脑的主要奖励中枢）中多巴胺的释放，而阻断这种多巴胺的释放会减弱乙醇的奖励作用（Young et al., 2014）。尽管在斑马鱼脑中尚未发现与哺乳动物伏隔核同源的解剖结构（Rink and Wullimann, 2001），但多巴胺能细胞群及其上行 / 下行投射已得到很好的表征（Reinig et al., 2017; Tay et al., 2011）。例如，从斑马鱼间脑后结节到端脑皮层下的上行多巴胺能投射被认为与哺乳动物的中脑多巴胺系统功能相似（Rink and Wullimann, 2001）。与乙醇的奖赏效应相比较，多巴胺还可用于研究乙醇的运动兴奋效应（Phillips and Shen, 1996）。乙醇可以增加成年斑马鱼的自发活动和全脑多巴胺及其代谢物二羟基苯乙酸的水平（Tran et al., 2015）。多巴胺生物合成中的限速酶是酪氨酸羟化酶。乙醇暴露会增加斑马鱼酪氨酸羟化酶的 mRNA 表达（Puttonen et al., 2013）、蛋白表达（Tran et al., 2017）以及酶活性（Chatterjee et al., 2014）。值得注意的是，急性乙醇暴露并没有改变单胺氧化酶（一种将多巴胺分解为二羟基苯乙酸的酶）的酶活性速率，这表明乙醇会增加多巴胺的合成而不是代谢（Chatterjee et al., 2014）。此外，药物阻断磷酸化酪氨酸羟化酶（酪氨酸羟化酶的活化形式）可以抑制乙醇诱导的多巴胺增加并减弱乙醇引起的自发活动（Nowicki et al., 2015）。进一步研究表明，阻断多巴胺 D1 受体而非 D2 受体可以抑制乙醇引起的自发活动（Tran et al., 2015c）。总之，这些发现表明，乙醇可上调酪氨酸羟化酶的基因表达，从而增加多巴胺合成和释放，进而通过激活多巴胺 D1 受体来增加自发活动。

胚胎乙醇暴露实验进一步证明了乙醇能改变斑马鱼多巴胺能系统。大量研究表明，发育过程中特定时间点的胚胎乙醇暴露不仅会改变多巴胺能系统（Buske and Gerlai, 2011），还会改变与多巴胺能系统相关的行为，包括运动（Sterling et al., 2016）、焦虑反应（Baggio et al., 2018）和群聚行为（Fernandes and Gerlai, 2009）。

乙醇最初的兴奋作用可视为乙醇在成瘾发展过程中强化特性的体现。某些饮用少量或适量乙醇的人会有一种欣快感，进而增加这些人将来饮酒的可能性。人们认为，人体对乙醇兴奋作用的敏感性是日后患乙醇使用障碍风险的部分原因（Hendler et al., 2013）。例如，一个对乙醇的欣快和兴奋作用更敏感的人比那些没有这种欣快感的人更有可能再次饮酒。乙醇使用障碍的形成也有很大的遗传因素。例如，与继父有乙醇使用障碍的孩子相比，亲生父亲有乙醇使用障碍的孩子患乙醇使用障碍的风险更大（Kendler et al., 2017）。根据双胞胎和收养研究的聚类分析估计，乙醇使用障碍有50%是可遗传的，适度的环境因素也会有影响（Verhulst et al., 2015）。这些发现表明，潜在的遗传因素大大增加了乙醇使用障碍的风险。

20.3.2 血清素和焦虑

人们利用斑马鱼对行为和乙醇敏感性的个体差异进行了研究。多种动物模型被用于个体差异或个性研究（Pawlak et al., 2008）。人工选择具有不同行为特征（例如，焦虑反应或探索行为的差异）的斑马鱼，能够建立多个表现这些特定特征的品系，证明这些特征是可遗传的（Ariyomo et al., 2013）。对人类而言，抑郁症和焦虑症也有很大的遗传成分，并且与乙醇使用障碍共存。例如，对焦虑性药物使用和情绪障碍的共病聚类分析表明，乙醇使用障碍与焦虑障碍之间以及乙醇使用障碍和重度抑郁症之间存在很强的相关性（Lai et al., 2015）。在斑马鱼中，表现出焦虑样反应升高和自发活动减少（例如，在新环境中探索减少）的动物对乙醇的兴奋和抗焦虑（减轻焦虑）作用更为敏感（Tran et al., 2016c）。消除焦虑等负面刺激会产生负强化，可能是乙醇使用障碍形成的一个重要因素（Wise and Koob, 2014）。血清素水平的差异与抑郁症和焦虑症的病因有关。抑郁症和焦虑症的一种常见药物治疗方法是使用选择性血清素再摄取抑制剂（SSRIs），它可以选择性地阻断突触前膜上血清素的再摄取，从而增加突触内可用血清素的含量（Andrews et al., 2015）。与对人体的作用相似，SSRIs可减少斑马鱼焦虑反应（Pittman and Hylton, 2015）。研究表明，不同的血清素受体拮抗剂可以改变斑马鱼的焦虑样行为反应，从而证明血清素在调节斑马鱼焦虑样行为中的重要作用（Nowicki et al., 2014）。焦虑样行为反应中的个体和品种差异与血清素及其代谢物5-羟基吲哚-3-乙酸（5-HIAA）的全脑水平相关（Maximino et al., 2013; Tran et al., 2016c）。此外，实验诱导的应激（增加了斑马鱼焦虑样行为反应）还可以改变血清素能系统（Tran et al., 2016b），表明斑马鱼焦虑样行为的个体差异可能是由于血清素能神经传递的变化，这与人类相似。有趣的是，表现出高焦虑样行为反应（自然或实验诱导）的斑马鱼对乙醇的运动兴奋、抗焦虑和血清素能改变等效应更敏感（Tran et al., 2016b, 2016c）。此外，表现出较低焦虑样行为（通常被称为“大胆”）的斑马鱼也表现出表达较高水平的δ-阿片受体和多巴胺D2受体，这两种受体都与药物奖赏和强化的调节有关（Thornqvist et al., 2019）。这些发现表明，血清素能神经传递的差异可以通过改变斑马鱼对环境的基线行为反应（例如，高度焦虑样反应），使其对乙醇的正向强化作用更加敏感。这在斑马鱼焦虑样行为的神经化学相关物质和乙醇敏感性倾向之间建立了重要联系，与人类非常相似。

20.3.3 多巴胺和群聚行为

尽管乙醇的运动兴奋和焦虑改变效应通常是斑马鱼行为药理学研究的主要焦点，但也有研究表明乙醇会破坏斑马鱼的群聚行为。群聚被定义为斑马鱼没有特定方向性的聚集，这可能提供许多优势，包括觅食和免受捕食者的侵害（Miller and Gerlai, 2011）。斑马鱼的群聚行为可以通过活的或模拟的同种个体来人工诱导（Qin et al., 2014）。与同种个体距离的减少是可以测定的，从而为这些研究中的群聚行为提供了可操作性定义。事实证明，急性暴露于低剂量和中等剂量的乙醇会抑制这种群聚反应（Gerlai et al., 2009）。尽管斑马鱼群聚行为以及乙醇如何消除这种反应的分子机制仍然未知，但有研究表明这与多巴胺能系统相关。例如，乙醇和多巴胺 D1 受体拮抗剂 SCH-23390 都能够改变多巴胺能信号并抑制群聚行为（Scerbina et al., 2012）。尽管这两种药物都抑制群聚行为并改变多巴胺能系统，但它们不太可能通过相同的途径起作用，因为乙醇会增加全脑多巴胺水平（Chatterjee and Gerlai, 2009），而 SCH-23390 则会减少（Tran et al., 2015）全脑多巴胺水平。乙醇通过多巴胺能系统抑制群聚行为的进一步证据来自斑马鱼胚胎乙醇暴露模拟胎儿乙醇谱系障碍的研究。众所周知，鱼类的群聚行为早在受精后 2 周就随着年龄的增长而形成，同时伴随着全脑多巴胺水平的增加（Buske and Gerlai, 2012）。然而，研究表明，在受精 24h 后单次乙醇暴露 2h 会损害群聚行为，并以剂量依赖的方式降低全脑多巴胺水平并减少其代谢物 DOPAC（Fernandes et al., 2015）。这些发现表明，急性乙醇暴露通过改变多巴胺能系统来损害斑马鱼的群聚行为。然而，这个假设仅仅是基于可能并非因果关系的相关结果，因此需要更多的研究来充分了解其潜在的机制。

20.4 慢性乙醇效应

与急性乙醇暴露的即时效应（如过度活跃和焦虑减少）相反，长期饮酒也会导致耐受性的形成。乙醇耐受可以定义为，与最初达到或维持某种特定效应所需的量或剂量相比，需要更大量或更大剂量的乙醇来最初达到或维持这一效应。根据美国《精神障碍诊断与统计手册（第五版）》（DSM-V），乙醇耐受被列为诊断乙醇使用障碍的标准之一（American Psychiatry Association, 2013）。已有多种慢性 / 长期乙醇暴露动物模型被用于研究乙醇耐受性发展的潜在机制（McBride 和 Li, 1998）。就斑马鱼而言，常用的慢性乙醇暴露方法有两种。第一种方法是在低浓度乙醇中长时间饲养斑马鱼。这种方法首选使用较低浓度的乙醇，以降低因长期接触乙醇而导致的毒性相关的死亡率（Dlugos and Rabin, 2003）。斑马鱼也可以通过在多天内缓慢增加乙醇浓度来逐渐达到所需的乙醇浓度，以适应药物并进一步降低与乙醇毒性相关的死亡率（Gerlai et al., 2009）。第二种方法是将斑马鱼反复暴露于较高浓度的乙醇中，通常每天一次，持续数天至数周，简单来讲，这种暴露方法被称为反复乙醇暴露（Mathur and Guo, 2011）。有趣的是，反复接触乙醇会诱发两种与乙醇使用障碍相关的不同现象。第一种是耐受的形成，可以表现为行为耐受和生理耐受。第二种与直觉相反，反复接触乙醇会致敏，这是一种与耐受相反的反应，动物对乙醇的作用变得更加敏感，最常见的是兴奋作用。

尽管乙醇引起的耐受和致敏对乙醇使用障碍的形成都很重要，并且在啮齿动物中得到了证明，但这并不是斑马鱼研究的主要关注点。只有一项针对斑马鱼的研究表明，乙醇诱导的致敏作用是环境依赖的，需要特定的乙醇暴露方案（Blaser et al., 2010）。相比之下，人们进行了大量研究来解析慢性持续乙醇暴露导致乙醇耐受性的机制。然而，值得注意的是，持续乙醇暴露的方法不能模拟临床人群的实际饮酒情况。人类饮酒是不定期的，之后会有一段时间不饮酒，甚至那些被诊断为酗酒的人也不会不间断地饮酒。然而，尽管存在这些差异，持续乙醇暴露方法仍有一个主要优势。持续的乙醇暴露消除了反复戒酒带来的混乱，反复戒酒涉及一套不同的机制，会导致一种厌恶性反应，驱使个人摄入更多的乙醇以回到基线，是一种通过惩罚得到加强的行为（Hendler et al., 2013）。连续乙醇暴露方法的主要优点是该法可以直接分析乙醇诱导耐受的机制，而不会产生潜在的混淆。

代谢耐受的形成是乙醇耐受的一种有趣形式，即乙醇的代谢方式发生了改变。这就提出了一个重要的问题，行为的耐受和神经化学反应的迟钝是否仅是肝脏更有效地清除身体内的乙醇，从而减少血液和大脑中乙醇含量的结果。乙醇在斑马鱼肝脏中的代谢过程分为两步。首先，乙醇被乙醇脱氢酶（ADH）代谢成乙醛，然后被再被乙醛脱氢酶（ALDH）进一步分解成乙酸。ADH和ALDH都在肝脏中表达，这与人类代谢乙醇的方式相似（Lassen et al., 2005; Reimers et al., 2004）。在反复和持续的乙醇暴露下，斑马鱼的ADH和ALDH酶活性在更高剂量的乙醇刺激下均表现出代谢耐受（Tran et al., 2015c）。然而，与人类及其他哺乳动物的乙醇耐受模型不同，斑马鱼在这些实验中一直浸泡在乙醇中，使得乙醇持续扩散到血液和大脑中。例如，连续乙醇暴露3周后的大脑乙醇浓度与首次接触乙醇的斑马鱼仔鱼的大脑乙醇浓度相似（Tran et al., 2015b）。这些发现表明，行为耐受并非只是肝脏乙醇代谢增加的结果，而是由于长期接触乙醇所导致的中枢神经系统的适应。

20.4.1 行为和神经化学耐受

如前所述，连续三周暴露于0.5%乙醇可诱导斑马鱼的行为和神经化学耐受（Dlugos and Rabin, 2003; Egan et al., 2009; Gerlai et al., 2009）。例如，1%乙醇急性暴露会增加自发活动和全脑多巴胺水平。然而，连续暴露于0.5%乙醇3周后，当斑马鱼再暴露于1%乙醇时，乙醇诱导的自发活动和全脑多巴胺的增加显著减弱，表明行为和神经化学耐受的形成（Tran et al., 2015b）。同样地，行为和神经化学耐受的形成也适用于乙醇的焦虑和血清素能改变效应。乙醇暴露还会改变表现出焦虑样行为的行为反应，如与底部的距离（底部停留）和绝对转角（不稳定运动）以及全脑血清素水平，所有这些在持续乙醇暴露后表现出明显减弱（Tran et al., 2015b）。虽然这里特别关注了乙醇引起的多巴胺能和血清素能系统的变化，但需要注意的是，持续的乙醇暴露还会改变氨基酸神经递质的水平（Chatterjee et al., 2014）、酶活性（Agostini et al., 2018）、基因表达（Pan et al., 2011; Rico et al., 2018）、蛋白质表达（Damodaran et al., 2006）和嘌呤核苷酸水解（Rico et al., 2011）。尽管乙醇诱导斑马鱼行为和神经化学耐受形成的确切机制尚未阐明，但对基因表达差异的研究为这一现象提供了重要的见解。

20.5 斑马鱼品系和种群对乙醇暴露的差异

比较不同斑马鱼品系的基因差异是确定调节急性和慢性乙醇暴露效应的基因的起点。

迄今为止，讨论的许多研究都使用了 AB 品系斑马鱼，该品系是斑马鱼的高度近交系之一，纯合性可达到 70% ～ 80%。一般来说，AB 品系斑马鱼对乙醇的兴奋、抗焦虑和镇静作用都很敏感（de Esch et al., 2012; Gerlai et al., 2009; Guo et al., 2015）。然而，在斑马鱼研究中，比较其他准近交系和常用品系，发现乙醇产生的效应存在显著差异。例如，WIK 和 TU 品系是两个准近交品系，也经常用于遗传学研究，但与 TU 品系相比，WIK 品系对乙醇的兴奋和抗焦虑作用明显更敏感（Pannia et al., 2014），下文将对杂合性和乙醇效应敏感性之间的关系进行讨论。对野生型、长鳍条纹（LFS）和蓝色长鳍（BLF）这三种不同品系的斑马鱼进行比较，发现 LFS 品系对急性乙醇暴露最敏感，而 BLF 品系最不敏感（Dlugos and Rabin, 2003）。有趣的是，与 LFS 品系斑马鱼相比，野生型的雌性斑马鱼似乎对慢性乙醇暴露的效应更敏感（Dlugos et al., 2011）。胚胎乙醇暴露中也能够观察到类似的品系差异。例如，与 AB 品系斑马鱼相比，TU 品系似乎更能抵抗胚胎乙醇暴露的有害效应（Mahabir et al., 2014）。基于这些发现，特定斑马鱼品系（如 TU 品系）的遗传背景差异为对抗乙醇的影响提供了防护优势。遗憾的是，鉴定乙醇“抗性”品系（TU）和其他乙醇“敏感”品系（如 AB 和 WIK）之间差异表达基因的遗传分析尚未进行。不过，人们已经对不同斑马鱼种群（同样对乙醇有不同反应）的差异表达基因进行了研究。如前所述，上述许多研究中使用的准近交系——AB 斑马鱼品系已经经历了许多代的近交繁殖，与其他品系（如 TU）相比，对乙醇的影响更敏感。相比之下，从宠物店获得并在实验室中维持多代的远交短鳍野生型（SF）斑马鱼在遗传上是异质的，与 AB 品系相比，其对乙醇的影响不太敏感（Gerlai et al., 2009）。急性暴露于 1% 的乙醇可以消除 AB 品系斑马鱼的群聚行为，但对 SF 斑马鱼群体无效。有趣的是，AB 和 SF 品系斑马鱼在连续暴露后，对乙醇的群聚破坏效应均产生了耐受性。然而，当对两个群体都进行乙醇戒断时，戒断效应彻底消除了 AB 品系的群聚行为，但在 SF 品系群体中没有此现象（Gerlai et al., 2009）。此外，急性和慢性乙醇暴露对单胺类神经递质（多巴胺、血清素及其代谢产物）、氨基酸类神经递质（谷氨酸、天冬氨酸、甘氨酸、牛磺酸和 GABA）、酪氨酸羟化酶和单胺氧化酶的影响在 AB 和 SF 品系斑马鱼之间存在显著差异（Chatterjee et al., 2014）。这些发现表明，这两个不同的斑马鱼种群对乙醇的急性和慢性效应表现出不同的行为和神经化学反应。一种可能的解释是，SF 品系等遗传异质型斑马鱼种群对外部影响，如环境变化或者本文所述的乙醇暴露更具抵抗力，这种效应被称为杂交优势。然而，目前尚不清楚哪些基因导致了这些行为和神经化学上的差异。此后，人们利用 qPCR 分析比较了 10 个基因的表达，这些基因是根据已知的乙醇暴露对斑马鱼神经化学水平的影响而选择的。基因表达数据显示，与 AB 品系相比，在 SF 种群中，编码多巴胺受体的基因显著过表达，而编码 GABA-1B 受体和 SLC6（一种膜蛋白，负责运输包括神经递质在内的分子）的基因表达显著下调（Pan et al., 2012）。使用这种假设导向的方法，发现在两个表现出不同的乙醇应答反应的种群中，仅有 3/10 的候选基因存在差异表达。在这两个遗传背景不同的群体之间可能会有更多的基因表达差异，这解释了对乙醇反应的行为和神经化学差异。为了识别额外的基因，可以采用一种无偏见的、以发现为导向的方法来量化基因表达的变化，一种可能的方法是使用全基因组 DNA 微阵列分析。

20.6 用于慢性乙醇诱导基因表达研究的微阵列芯片

研究表明，长期饮酒会导致小鼠、大鼠、斑马鱼等动物以及包括人类在内的大脑基因表达持久性变化（Hoffman et al., 2003; Marballi et al., 2016; Pan et al., 2011）。这些持久的变化被认为是乙醇耐受性的基础，并导致成瘾性药物寻求行为（Mulholland et al., 2016）。然而，许多研究经常使用假设或数据导向的方法来量化候选基因的表达。使用全基因组DNA 微阵列系统（Bumgarner, 2013），可以以无偏倚的方式同时收集数千个基因的基因表达数据，以确定与慢性乙醇暴露相关的新候选基因（Kily et al., 2008; Pan et al., 2011）。DNA 微阵列由附着在表面（如玻璃芯片）上的大量 DNA 序列组成。芯片上的每个 DNA 序列（探针）代表一个基因的一小部分，用来杂交 cDNA 或 cRNA，芯片上收集的探针代表整个基因组。该技术背后的原理是基因过表达产生更多的 mRNA，而基因低表达产生更少的 mRNA。首先从未经处理以及乙醇处理的动物（通常是脑组织）中收集 RNA，逆转录合成标记的 cDNA，然后杂交到微阵列上。或者，可以将 cDNA 进一步转录为 cRNA 并进行标记，然后再与微阵列上的寡核苷酸杂交。然后分析乙醇处理组动物每个基因的表达水平，并与未处理的对照组进行比较。

如前所述，经常使用的慢性乙醇暴露方法有两种：重复间歇性暴露和连续暴露。在重复乙醇暴露 4 周后（每天 20min），使用微阵列芯片分析斑马鱼脑组织样本，该微阵列芯片可探测 16000 条斑马鱼表达的序列标签（ESTsd 短 cDNA 序列）。微阵列分析发现，与未经处理的动物相比，在长期反复乙醇暴露后 647 个基因的表达发生了显著改变（两个方向的变化 >1.5 倍）（Kily et al., 2008）。有趣的是，153 个过表达的基因在尼古丁长期反复间歇性暴露后也出现了过表达，这表明这些基因可能与长期成瘾药物暴露后的神经适应有关。其中包括神经适应通路的组成部分，例如谷氨酸受体（GRIA2、NMDAR1、GRIA2）和与突触可塑性相关的蛋白质（NCAM2 和钙调磷酸酶 B）。有趣的是，反复乙醇暴露也降低了外被体蛋白 beta2 亚基编码基因的表达，这是一种调节多巴胺 D1 受体转运的外被体复合物。该基因的下调暗示了多巴胺能系统与以往的研究相似（Puttonen et al., 2013; Tran et al., 2015c），可能是由于乙醇暴露期间多巴胺 D1 受体的反复活化。

与反复乙醇暴露相比，长期持续乙醇暴露同样改变了 DNA 微阵列检测到的许多基因的表达，同时消除了反复戒断的混杂因素。连续 3 周的乙醇暴露后，使用 DNA 微阵列（探测至少 24000 个基因和额外的 ESTs）分析了对照组和乙醇暴露组斑马鱼脑组织样本的基因表达变化。在长期持续乙醇暴露后，微阵列分析发现，与未处理的对照组相比，乙醇暴露组的 1914 个基因的表达发生了显著改变（两个方向的变化 >2 倍）（Pan et al., 2011）。值得注意的是，上调的基因比下调的多 30%，且大约三分之二的基因没有已知的功能。根据分子功能对差异表达基因进行分类和聚类。例如，大量编码电压门控离子通道（如 L 型电压依赖性钙通道 alpha 1F 和 1S 亚单位）的差异表达基因在持续乙醇暴露后下调（Pan et al., 2011）。与之相反，蛋白质组分析表明，电压依赖性阴离子通道 1 和 2（voltage-dependent anion channel 1 and 2, VDAC1/VDAC2）的表达在连续乙醇暴露 4 周后上调（Damodaran et al., 2006）。得注意的是，大量编码溶质载体（solute carrier, SLC）家族蛋白的基因在持续乙醇暴露后也会上调，包括许多 SLC6 家族中的基因，它们负责转运多巴胺、血清素、甘氨酸和 GABA 等神经递质（Pan et al., 2011）。这可以解释持续乙醇暴露后观察到的神经化学差异。与 Kily 等的报道（Kily et al., 2008）相反，谷氨酸受体（如 *α*-氨基 -3- 羟基 -5- 甲基 -4- 异噁唑受体和 *N*- 甲基 -D- 天冬氨酸受体）表达的变化在乙醇暴露组和未处理的对照组之间没有显著差异，这可能是由两个研究之间的乙醇暴露方法（即重复暴露和连续暴露）不同造成的。正如预期的那样，大量参与乙醇代谢的基因在连续暴露后上调，包括编码乙醇脱氢酶、乙醛脱氢酶和细胞色素 P450 酶超家族蛋白的基因（Pan et al., 2011）。更重要的是，研究者发现了许多非典型的功能性乙醇应答基因，这些基因可能作为调节急性和慢性乙醇效应的新候选基因。

20.7　反向遗传策略

无论使用假设导向还是发现导向的方法来确定候选基因，反向遗传技术的进步都允许对特定基因进行操作以确定其功能和作用。CRISPR/Cas9 等基因组编辑技术已可用于斑马鱼，以实现特定基因的高效定向突变，从而解决许多重要问题（Jao et al., 2013）。例如，药理学扰动表明，乙醇引起的自发活动是通过诱导酪氨酸羟化酶表达和多巴胺 D1 受体活化（Tran et al., 2015c），进而激活多巴胺能系统（Puttonen et al., 2013）来介导的。然而，行为药理学的一个主要缺点是大多数药物都具有脱靶效应，并且许多用于斑马鱼研究的药物最初是基于哺乳动物数据开发的，并没有在斑马鱼上得到验证。使用更具选择性的方法，将利用 CRISPR/Cas9 技术编辑产生的酪氨酸羟化酶和多巴胺受体 1 突变体斑马鱼暴露于乙醇中，以确定这些基因是否是乙醇的运动兴奋效应所必需的。然而，很少有研究对急性和慢性乙醇暴露对不同斑马鱼突变体的影响进行实验分析。

20.8　正向遗传学和诱变

最后，与反向遗传方法鉴定乙醇效应调节基因相比，正向遗传方法还可以用来筛选对乙醇作用敏感（或多或少）的突变体。使用致畸剂 *N*- 乙基 -*N*- 亚硝基脲（*N*-ethyl-*N*-nitrosourea, ENU）可诱导全基因组的随机点突变，在斑马鱼体内有效地实现随机诱变（de Bruijn et al., 2009）。采用正向遗传方法，利用 ENU 诱变的斑马鱼进行高通量筛选，已经鉴定出对发育至关重要的基因（Haffter et al., 1996）以及对乙醇影响胚胎发育更敏感的基因（Swartz et al., 2014）。例如，与杂合子和野生型对照相比，*vangl2* 突变的纯合子斑马鱼突变体会产生严重的乙醇诱导的胚胎缺陷（Swartz et al., 2014）。然而，很少有研究筛选对乙醇的急性和慢性效应敏感的突变体。一项研究使用 ENU 诱导的突变来筛选对乙醇调节的伪装反应敏感的仔鱼（Peng et al., 2009）。在光照条件下，由于黑素细胞聚集到细胞核中，受精 5 ～ 7 天的斑马鱼仔鱼体色变浅。在黑暗或乙醇暴露条件下，斑马鱼由于黑素细胞在细胞质中扩散而变黑，这种反应被称为乙醇调节的伪装反应。利用这种反应作为斑马鱼仔鱼的行为筛选方法，分离出一种形态与野生型相似的突变体，但没有表现出乙醇诱导的伪装反应（即乙醇暴露后不变黑）。该突变体被命名为*fantastma*（*fan*），原位克隆显示，一个点突变导致腺苷酸环化酶 5 编码基因的第一个外显子的终止密码子提前，而腺苷酸环化酶 5 是细胞外信号调节激酶（extracellular signal regulated kinase, ERK）磷酸化所必需的。值得注意的是，与野生型相比，纯合的*fan*突变斑马鱼对乙醇（1.5%）的刺激作用不太敏感，但对乙醇（3%）镇静作用的敏感性与野生型相同。与这一发现相一致的是，轻微的 ERK 磷酸化药理学抑制降低了乙醇对野生型仔鱼的兴奋效应，而强效抑制则使兴奋剂量的乙醇产生镇静作用（Peng et al., 2009）。使用这种正向遗传方法可以鉴定出调节急性乙醇暴露时运动反应敏感性的新基因。

20.9 结论

几十年来，人们一直在研究乙醇对中枢神经系统的急性效应及其对神经适应的长期慢性诱导作用，主要困难之一是乙醇所涉及的分子途径众多（Costardi et al., 2015）。来自斑马鱼的相关数据强调了这一难点，研究表明急性乙醇暴露能够改变激素（Oliveira et al., 2013）、神经肽（Coffey et al., 2013）、神经递质（Chatterjee et al., 2014）、信号转导途径（Peng et al., 2009）和基因表达（Puttonen et al., 2013）的水平。长期慢性乙醇暴露的影响可能更为复杂，会导致蛋白表达的显著变化（Damodaran et al., 2006）和近 2000 个基因的差异表达（Kily et al., 2008; Pan et al., 2011）。由于斑马鱼的遗传易控性，正向和反向遗传方法将有助于识别和确定调节急性和慢性乙醇暴露效应的新基因。这些研究将有助于阐明乙醇是如何影响中枢神经系统的，以及乙醇使用障碍的形成机制是什么。该研究领域的进展将有助于发现用于乙醇使用障碍治疗的新靶点和药物疗法（Gorini et al., 2014）。

（郭煊君翻译　石童审校）

参考文献

Agostini, J.F., Toe, H.C.Z.D., Vieira, K.M., Baldin, S.L., Costa, N.L.F., Cruz, C.U., et al., 2018. Cholinergic system and oxidative stress changes in the brain of a zebrafish model chronically exposed to ethanol. Neurotoxicity Research 33 (4), 749-758. https://doi.org/10.1007/s12640-017-9816-8.

American Psychiatric Association, 2013. Diagnostic and statistical manual of mental disorders, fifth ed. Washington, DC.

Andrews, P.W., Bharwani, A., Lee, K.R., Fox, M., Thomson, J.A., 2015. Is serotonin an upper or a downer? The evolution of the serotonergic system and its role in depression and the antidepressant response. Neuroscience & Biobehavioral Reviews 51, 164-188. https://doi.org/10.1016/j.neubiorev.2015.01.018.

Ariyomo, T.O., Carter, M., Watt, P.J., 2013. Heritability of boldness and aggressiveness in the zebrafish. Behavior Genetics 43 (2), 161-167. https://doi.org/10.1007/s10519-013-9585-y.

Baggio, S., Mussulini, B.H., de Oliveira, D.L., Gerlai, R., Rico, E.P., 2018. Embryonic alcohol exposure leading to social avoidance and altered anxiety responses in adult zebrafish. Behavioural Brain Research 352, 62-69. https://doi.org/10.1016/j.bbr.2017.08.039.

Bell, R.L., Hauser, S., Rodd, Z.A., Liang, T., Sari, Y., McClintick, J., et al., 2016. A genetic animal model of alcoholism for screening medications to treat addiction. International Review of Neurobiology 126, 179-261. https://doi.org/10.1016/

bs.irn.2016.02.017.

Blaser, R.E., Koid, A., Poliner, R.M., 2010. Context-dependent sensitization to ethanol in zebrafish (*Danio rerio*). Pharmacology Biochemistry and Behavior 95 (3), 278-284. https://doi.org/10.1016/j.pbb.2010.02.002.

Bumgarner, R., 2013. Overview of DNA microarrays: types, applications, and their future. Current Protocols in Molecular Biology. https://doi.org/10.1002/0471142727.mb2201s101 (Chapter 22, Unit 22.1).

Buske, C., Gerlai, R., 2011. Early embryonic ethanol exposure impairs shoaling and the dopaminergic and serotoninergic systems in adult zebrafish. Neurotoxicology and Teratology 33 (6), 698-707. https://doi.org/10.1016/j.ntt.2011.05.009.

Buske, C., Gerlai, R., 2012. Maturation of shoaling behavior is accompanied by changes in the dopaminergic and serotoninergic systems in zebrafish. Developmental Psychobiology 54 (1), 28-35. https://doi.org/10.1002/dev.20571.

Chatterjee, D., Gerlai, R., 2009. High precision liquid chromatography analysis of dopaminergic and serotoninergic responses to acute alcohol exposure in zebrafish. Behavioural Brain Research 200 (1), 208-213.

Chatterjee, D., Shams, S., Gerlai, R., 2014. Chronic and acute alcohol administration induced neurochemical changes in the brain: comparison of distinct zebrafish populations. Amino Acids 46 (4), 921-930. https://doi.org/10.1007/s00726-013-1658-y.

Coffey, C.M., Solleveld, P.A., Fang, J., Roberts, A.K., Hong, S.-K., Dawid, I.B., et al., 2013. Novel oxytocin gene expression in the hindbrain is induced by alcohol exposure: transgenic zebrafish enable visualization of sensitive neurons. PLoS One 8 (1), e53991. https://doi.org/10.1371/journal.pone.0053991.

Costardi, J.V.V., Nampo, R.A.T., Silva, G.L., Ribeiro, M.A.F., Stella, H.J., Stella, M.B., Malheiros, S.V.P., 2015. A review on alcohol: from the central action mechanism to chemical dependency. Revista da Associação Médica Brasileira 61 (4), 381-387. https://doi.org/10.1590/1806-9282.61.04.381.

Damodaran, S., Dlugos, C.A., Wood, T.D., Rabin, R.A., 2006. Effects of chronic ethanol administration on brain protein levels: a proteomic investigation using 2-D DIGE system. European Journal of Pharmacology 547 (1e3), 75-82. https://doi.org/10.1016/j.ejphar.2006.08.005.

de Bruijn, E., Cuppen, E., Feitsma, H., 2009. Highly efficient ENU mutagenesis in zebrafish. Methods in Molecular Biology 546, 3-12. https://doi.org/10.1007/978-1-60327-977-2_1.

de Esch, C., van der Linde, H., Slieker, R., Willemsen, R., Wolterbeek, A., Woutersen, R., De Groot, D., 2012. Locomotor activity assay in zebrafish larvae: influence of age, strain and ethanol. Neurotoxicology and Teratology 34 (4), 425-433. https://doi.org/10.1016/j.ntt.2012.03.002.

Dlugos, C.A., Brown, S.J., Rabin, R.A., 2011. Gender differences in ethanol-induced behavioral sensitivity in zebrafish. Alcohol 45 (1), 11e18. https://doi.org/10.1016/j.alcohol.2010.08.018.

Dlugos, C.A., Rabin, R.A., 2003. Ethanol effects on three strains of zebrafish: model system for genetic investigations. Pharmacology Biochemistry and Behavior 74 (2), 471-480.

Egan, R.J., Bergner, C.L., Hart, P.C., Cachat, J.M., Canavello, P.R., Elegante, M.F., et al., 2009. Understanding behaveioral and physiological phenotypes of stress and anxiety in zebrafish. Behavioural Brain Research 205 (1), 38e44. https://doi.org/10.1016/j.bbr.2009.06.022.

Fernandes, Y., Gerlai, R., 2009. Long-term behavioral changes in response to early developmental exposure to ethanol in zebrafish. Alcoholism: Clinical and Experimental Research 33 (4), 601-609. https://doi.org/10.1111/j.1530-0277.2008.00874.x.

Fernandes, Y., Rampersad, M., Gerlai, R., 2015. Embryonic alcohol exposure impairs the dopaminergic system and social behavioral responses in adult zebrafish. The International Journal of Neuropsychopharmacology 18 (6). https://doi.org/10.1093/ijnp/pyu089.

Fish, E.W., DeBold, J.F., Miczek, K.A., 2002. Repeated alcohol: behavioral sensitization and alcohol-heightened aggression in mice. Psychopharmacology 160 (1), 39-48. https://doi.org/10.1007/s00213-001-0934-9.

Franck, J., Jayaram-Lindstrom, N., 2013. Pharmacotherapy for alcohol dependence: status of current treatments. Current Opinion in Neurobiology 23 (4), 692-699. https://doi.org/10.1016/j.conb.2013.05.005.

Gerlai, R., Chatterjee, D., Pereira, T., Sawashima, T., Krishnannair, R., 2009. Acute and chronic alcohol dose: population differences in behavior and neurochemistry of zebrafish. Genes, Brain and Behavior 8 (6), 586-599. https://doi.org/10.1111/j.1601-183X.2009.00488.x.

Gerlai, R., 2014. Fish in behavior research: unique tools with a great promise! Journal of Neuroscience Methods 234, 54-58. https://doi.org/10.1016/j.jneumeth.2014.04.015.

Gorini, G., Harris, R.A., Mayfield, R.D., 2014. Proteomic approaches and identification of novel therapeutic targets for alcoholism. Neuropsychopharmacology: Official Publication of the American College of Neuropsychopharmacology 39 (1), 104-130. https://doi.org/10.1038/npp.2013.182.

Grant, B.F., Chou, S.P., Saha, T.D., Pickering, R.P., Kerridge, B.T., Ruan, W.J., 2017. Prevalence of 12-month alcohol use, high-risk drinking, and DSM-IV alcohol use disorder in the United States, 2001e2002 to 2012e2013: results from the national epidemiologic survey on alcohol and related conditions. JAMA Psychiatry 74 (9), 911-923. https://doi.org/10.1001/jamapsychiatry.2017.2161.

Guo, N., Lin, J., Peng, X., Chen, H., Zhang, Y., Liu, X., Li, Q., 2015. Influences of acute ethanol exposure on locomotor activities of zebrafish larvae under different illumination. Alcohol 49 (7), 727-737. https://doi.org/10.1016/j.alcohol.2015.08.003.

Haffter, P., Granato, M., Brand, M., Mullins, M.C., Hammerschmidt, M., Kane, D.A., et al., 1996. The identification of genes with unique and essential functions in the development of the zebrafish, *Danio rerio*. Development (Cam-bridge, England) 123, 1-36.

Hendler, R.A., Ramchandani, V.A., Gilman, J., Hommer, D.W., 2013. Stimulant and sedative effects of alcohol. Current Topics in Behavioral Neurosciences 13, 489-509. https://doi.org/10.1007/7854_2011_135.

Hoffman, P.L., Miles, M., Edenberg, H.J., Sommer, W., Tabakoff, B., Wehner, J.M., Lewohl, J., 2003. Gene expression in brain: a window on ethanol dependence, neuroadaptation, and preference. Alcoholism: Clinical and Experimental Research 27 (2), 155-168. https://doi.org/10.1097/01.ALC.0000060101.89334.11.

Jao, L.-E., Wente, S.R., Chen, W., 2013. Efficient multiplex biallelic zebrafish genome editing using a CRISPR nuclease system. Proceedings of the National Academy of Sciences of the United States of America 110 (34), 13904-13909. https://doi.org/10.1073/pnas.1308335110.

Kendler, K.S., Ohlsson, H., Edwards, A., Sundquist, J., Sundquist, K., 2017. The clinical features of alcohol use disorders in biological and step-fathers that predict risk for alcohol use disorders in offspring. American Journal of Medical Genetics Part B, Neuropsychiatric Genetics : The Official Publication of the International Society of Psychiatric Genetics 174 (8), 779-785. https://doi.org/10.1002/ajmg.b.32583.

Kily, L.J., Cowe, Y.C., Hussain, O., Patel, S., McElwaine, S., Cotter, F.E., Brennan, C.H., 2008. Gene expression changes in a zebrafish model of drug dependency suggest conservation of neuro-adaptation pathways. The Journal of Experimental Biology 211 (Pt 10), 1623-1634. https://doi.org/10.1242/jeb.014399.

Lai, H.M., Cleary, M., Sitharthan, T., Hunt, G.E., 2015. Prevalence of comorbid substance use, anxiety and mood disorders in epidemiological surveys, 1990e2014; a systematic review and meta-analysis. Drug and Alcohol Dependence 154, 1-13. https://doi.org/10.1016/j.drugalcdep.2015.05.031.

Lassen, N., Estey, T., Tanguay, R.L., Pappa, A., Reimers, M.J., Vasiliou, V., 2005. Molecular cloning, baculovirus expression, and tissue distribution of the zebrafish aldehyde dehydrogenase 2. Drug Metabolism and Disposition: The Biological Fate of Chemicals 33 (5), 649-656. https://doi.org/10.1124/dmd.104.002964.

Linden-Carmichael, A.N., Vasilenko, S.A., Lanza, S.T., Maggs, J.L., 2017. High-intensity drinking versus heavy episodic drinking: prevalence rates and relative odds of alcohol use disorder across adulthood. Alcoholism: Clinical and Experimental Research 41 (10), 1754-1759. https://doi.org/10.1111/acer.13475.

Lockwood, B., Bjerke, S., Kobayashi, K., Guo, S., 2004. Acute effects of alcohol on larval zebrafish: a genetic system for large-scale screening. Pharmacology Biochemistry and Behavior 77 (3), 647-654. https://doi.org/10.1016/j.pbb.2004.01.003.

Mahabir, S., Chatterjee, D., Gerlai, R., 2014. Strain dependent neurochemical changes induced by embryonic alcohol exposure in zebrafish. Neurotoxicology and Teratology 41, 1-7. https://doi.org/10.1016/j.ntt.2013.11.001.

Marballi, K., Genabai, N.K., Blednov, Y.A., Harris, R.A., Ponomarev, I., 2016. Alcohol consumption induces global gene expression changes in VTA dopaminergic neurons. Genes, Brain and Behavior 15 (3), 318-326. https://doi.org/10.1111/gbb.12266.

Mathur, P., Guo, S., 2011. Differences of acute versus chronic ethanol exposure on anxiety-like behavioral responses in zebrafish. Behavioural Brain Research 219 (2), 234-239. https://doi.org/10.1016/j.bbr.2011.01.019.

Maximino, C., Puty, B., Matos Oliveira, K.R., Herculano, A.M., 2013. Behavioral and neurochemical changes in the zebrafish leopard strain. Genes, Brain and Behavior 12 (5), 576-582. https://doi.org/10.1111/gbb.12047.

McBride, W.J., Li, T.K., 1998. Animal models of alcoholism: neurobiology of high alcohol-drinking behavior in rodents. Critical Reviews in Neurobiology 12 (4), 339-369.

Miller, N.Y., Gerlai, R., 2011. Shoaling in zebrafish: what we don't know. Reviews in the Neurosciences 22 (1), 17-25. https://doi.org/10.1515/RNS.2011.004.

Mulholland, P.J., Chandler, L.J., Kalivas, P.W., 2016. Signals from the fourth dimension regulate drug relapse. Trends in Neurosciences 39 (7), 472-485. https://doi.org/10.1016/j.tins.2016.04.007.

Naumann, E.A., Kampff, A.R., Prober, D.A., Schier, A.F., Engert, F., 2010. Monitoring neural activity with biolumi-nescence during natural behavior. Nature Neuroscience 13 (4), 513-520. https://doi.org/10.1038/nn.2518.

Nowicki, M., Tran, S., Chatterjee, D., Gerlai, R., 2015. Inhibition of phosphorylated tyrosine hydroxylase attenuates ethanol-induced hyperactivity in adult zebrafish (*Danio rerio*). Pharmacology Biochemistry and Behavior 138, 32-39. https://doi.org/10.1016/j.pbb.2015.09.008.

Nowicki, M., Tran, S., Muraleetharan, A., Markovic, S., Gerlai, R., 2014. Serotonin antagonists induce anxiolytic and anxiogenic-like behavior in zebrafish in a receptor-subtype dependent manner. Pharmacology Biochemistry and Behavior 126, 170-180. https://doi.org/10.1016/j.pbb.2014.09.022.

Oliveira, T.A., Koakoski, G., Kreutz, L.C., Ferreira, D., da Rosa, J.G., de Abreu, M.S., et al., 2013. Alcohol impairs predation risk response and communication in zebrafish. PLoS One 8 (10), e75780. https://doi.org/10.1371/journal.pone.0075780.

Pan, Y., Chatterjee, D., Gerlai, R., 2012. Strain dependent gene expression and neurochemical levels in the brain of zebrafish: focus on a few alcohol related targets. Physiology and Behavior 107 (5), 773-780. https://doi.org/10.1016/j.physbeh.2012.01.017.

Pan, Y., Kaiguo, M., Razak, Z., Westwood, J.T., Gerlai, R., 2011. Chronic alcohol exposure induced gene expression changes in the zebrafish brain. Behavioural Brain Research 216 (1), 66-76. https://doi.org/10.1016/j.bbr.2010.07.017.

Pannia, E., Tran, S., Rampersad, M., Gerlai, R., 2014. Acute ethanol exposure induces behavioural differences in two zebrafish (*Danio rerio*) strains: a time course analysis. Behavioural Brain Research 259, 174-185. https://doi.org/10.1016/j.bbr.2013.11.006.

Park, A., Ghezzi, A., Wijesekera, T.P., Atkinson, N.S., 2017. Genetics and genomics of alcohol responses in *Drosophila*. Neuropharmacology 122, 22-35. https://doi.org/10.1016/j.neuropharm.2017.01.032.

Pawlak, C.R., Ho, Y.-J., Schwarting, R.K., 2008. Animal models of human psychopathology based on individual differences in novelty-seeking and anxiety. Neuroscience and Biobehavioral Reviews 32 (8), 1544-1568. https://doi.org/10.1016/j.neubiorev.2008.06.007.

Peng, J., Wagle, M., Mueller, T., Mathur, P., Lockwood, B.L., Bretaud, S., Guo, S., 2009. Ethanol-modulated camouflage response screen in zebrafish uncovers a novel role for cAMP and extracellular signal-regulated kinase signaling in behavioral sensitivity to ethanol. Journal of Neuroscience: The Official Journal of the Society for Neuroscience 29 (26), 8408-8418. https://doi.org/10.1523/JNEUROSCI.0714-09.2009.

Phillips, T., 2002. Animal models for the genetic study of human alcohol phenotypes. Alcohol Research and Health: The Journal of the National Institute on Alcohol Abuse and Alcoholism 26 (3), 202-207.

Phillips, T.J., Shen, E.H., 1996. Neurochemical bases of locomotion and ethanol stimulant effects. International Review of Neurobiology 39, 243-282.

Pittman, J., Hylton, A., 2015. Behavioral, endocrine, and neuronal alterations in zebrafish (*Danio rerio*) following sub-chronic coadministration of fluoxetine and ketamine. Pharmacology Biochemistry and Behavior 139 (Pt B), 158-162. https://doi.org/10.1016/j.pbb.2015.08.014.

Puttonen, H.A., Sundvik, M., Rozov, S., Chen, Y.-C., Panula, P., 2013. Acute ethanol treatment upregulates Th1, Th2, and Hdc in larval zebrafish in stable networks. Frontiers in Neural Circuits 7, 102. https://doi.org/10.3389/fncir.2013.00102.

Qin, M., Wong, A., Seguin, D., Gerlai, R., 2014. Induction of social behavior in zebrafish: live versus computer animated fish as stimuli. Zebrafish 11 (3), 185-197. https://doi.org/10.1089/zeb.2013.0969.

Reimers, M.J., Hahn, M.E., Tanguay, R.L., 2004. Two zebrafish alcohol dehydrogenases share common ancestry with mammalian class Ⅰ, Ⅱ, Ⅳ, and Ⅴ alcohol dehydrogenase genes but have distinct functional characteristics. The Journal of Biological Chemistry 279 (37), 38303-38312. https://doi.org/10.1074/jbc.M401165200.

Reinig, S., Driever, W., Arrenberg, A.B., 2017. The descending diencephalic dopamine system is tuned to sensory stimuli. Current Biology: CB 27 (3), 318-333. https://doi.org/10.1016/j.cub.2016.11.059.

Rico, E.P., Rosemberg, D.B., Berteli, J.F.A., da Silveira Langoni, A., Souto, A.A., Bogo, M.R., et al., 2018. Adenosine deaminase activity and gene expression patterns are altered after chronic ethanol exposure in zebrafish brain. Neurotoxicology and Teratology 65, 14-18. https://doi.org/10.1016/j.ntt.2017.11.001.

Rico, E.P., Rosemberg, D.B., da Langoni, A.S., Souto, A.A., Dias, R.D., Bogo, M.R., et al., 2011. Chronic ethanol treatment alters purine nucleotide hydrolysis and nucleotidase gene expression pattern in zebrafish brain. Neurotoxicology 32 (6), 871-878. https://doi.org/10.1016/j.neuro.2011.05.010.

Rihel, J., Prober, D.A., Arvanites, A., Lam, K., Zimmerman, S., Jang, S., et al., 2010. Zebrafish behavioral profiling links drugs to biological targets and rest/wake regulation. Science 327 (5963), 348-351. https://doi.org/10.1126/science.1183090.

Rink, E., Wullimann, M.F., 2001. The teleostean (zebrafish) dopaminergic system ascending to the subpallium (striatum) is located in the basal diencephalon (posterior tuberculum). Brain Research 889 (1e2), 316-330. https://doi.org/10.1016/s0006-8993(00)03174-7.

Roberts, A.C., Bill, B.R., Glanzman, D.L., 2013. Learning and memory in zebrafish larvae. Frontiers in Neural Circuits 7, 126. https://doi.org/10.3389/fncir.2013.00126.

Rosemberg, D.B., Braga, M.M., Rico, E.P., Loss, C.M., Cordova, S.D., Mussulini, B.H.M., et al., 2012. Behavioral effects of taurine pretreatment in zebrafish acutely exposed to ethanol. Neuropharmacology 63 (4), 613-623. https://doi.org/10.1016/j.neuropharm.2012.05.009.

Sacks, J.J., Gonzales, K.R., Bouchery, E.E., Tomedi, L.E., Brewer, R.D., 2015. 2010 national and state costs of excessive alcohol consumption. American Journal of Preventive Medicine 49 (5), e73-e79. https://doi.org/10.1016/j.amepre.2015.05.031.

Scerbina, T., Chatterjee, D., Gerlai, R., 2012. Dopamine receptor antagonism disrupts social preference in zebrafish: a strain comparison study. Amino Acids 43 (5), 2059-2072. https://doi.org/10.1007/s00726-012-1284-0.

Sterling, M.E., Chang, G.-Q., Karatayev, O., Chang, S.Y., Leibowitz, S.F., 2016. Effects of embryonic ethanol exposure at low doses on neuronal development, voluntary ethanol consumption and related behaviors in larval and adult zebrafish: role of hypothalamic orexigenic peptides. Behavioural Brain Research 304, 125-138. https://doi.org/10.1016/j.bbr.2016.01.013.

Swartz, M.E., Wells, M.B., Griffin, M., McCarthy, N., Lovely, C.B., McGurk, P., et al., 2014. A screen of zebrafish mutants identifies ethanol-sensitive genetic loci. Alcoholism: Clinical and Experimental Research 38 (3), 694-703. https://doi.org/10.1111/acer.12286.

Tay, T.L., Ronneberger, O., Ryu, S., Nitschke, R., Driever, W., 2011. Comprehensive catecholaminergic projectome analysis reveals single-neuron integration of zebrafish ascending and descending dopaminergic systems. Nature Communications 2, 171. https://doi.org/10.1038/ncomms1171.

Thornqvist, P.-O., McCarrick, S., Ericsson, M., Roman, E., Winberg, S., 2019. Bold zebrafish (*Danio rerio*) express higher levels of delta opioid and dopamine D2 receptors in the brain compared to shy fish. Behavioural Brain Research 359, 927-934. https://doi.org/10.1016/j.bbr.2018.06.017.

Tran, S., Facciol, A., Gerlai, R., 2016a. The zebrafish, a novel model organism for screening compounds affecting acute and chronic ethanol-induced effects. International Review of Neurobiology 126, 467-484. https://doi.org/10.1016/bs.irn.2016.02.016.

Tran, S., Chatterjee, D., Gerlai, R., 2015a. An integrative analysis of ethanol tolerance and withdrawal in zebrafish (*Danio rerio*). Behavioural Brain Research 276, 161-170. https://doi.org/10.1016/j.bbr.2014.02.034.

Tran, S., Facciol, A., Nowicki, M., Chatterjee, D., Gerlai, R., 2017. Acute alcohol exposure increases tyrosine hydroxylase protein expression and dopamine synthesis in zebrafish. Behavioural Brain Research 317, 237-241. https://doi.org/10.1016/j.bbr.2016.09.048.

Tran, S., Gerlai, R., 2013. Time-course of behavioural changes induced by ethanol in zebrafish (*Danio rerio*). Behavioural Brain Research 252, 204e213. https://doi.org/10.1016/j.bbr.2013.05.065.

Tran, S., Nowicki, M., Chatterjee, D., Gerlai, R., 2015b. Acute and chronic ethanol exposure differentially alters alcohol dehydrogenase and aldehyde dehydrogenase activity in the zebrafish liver. Progress in Neuro-Psychopharmacology and Biological Psychiatry 56, 221-226. https://doi.org/10.1016/j.pnpbp.2014.09.011.

Tran, S., Nowicki, M., Fulcher, N., Chatterjee, D., Gerlai, R., 2016b. Interaction between handling induced stress and anxiolytic effects of ethanol in zebrafish: a behavioral and neurochemical analysis. Behavioural Brain Research 298 (Pt B), 278-285. https://doi.org/10.1016/j.bbr.2015.10.061.

Tran, S., Nowicki, M., Muraleetharan, A., Chatterjee, D., Gerlai, R., 2015. Differential effects of acute administration of SCH-23390, a D(1) receptor antagonist, and of ethanol on swimming activity, anxiety-related responses, and neurochemistry of zebrafish. Psychopharmacology 232 (20), 3709-3718. https://doi.org/10.1007/s00213-015-4030-y.

Tran, S., Nowicki, M., Muraleetharan, A., Chatterjee, D., Gerlai, R., 2016c. Neurochemical factors underlying individual differences in locomotor activity and anxiety-like behavioral responses in zebrafish. Progress in Neuro-Psychopharmacology and Biological Psychiatry 65, 25-33. https://doi.org/10.1016/j.pnpbp.2015.08.009.

Verhulst, B., Neale, M.C., Kendler, K.S., 2015. The heritability of alcohol use disorders: a meta-analysis of twin and adoption studies. Psychological Medicine 45 (5), 1061-1072. https://doi.org/10.1017/S0033291714002165.

Wise, R.A., Koob, G.F., 2014. The development and maintenance of drug addiction. Neuropsychopharmacology: Official Publication of the American College of Neuropsychopharmacology 39 (2), 254-262. https://doi.org/10.1038/npp.2013.261.

Young, E.A., Dreumont, S.E., Cunningham, C.L., 2014. Role of nucleus accumbens dopamine receptor subtypes in the learning and expression of alcohol-seeking behavior. Neurobiology of Learning and Memory 108, 28-37. https://doi.org/10.1016/j.nlm.2013.05.004.

Zhu, G., Zhang, F., Li, W., 2014. Nematodes feel a cravingeusing Caenorhabditis elegans as a model to study alcohol addiction. Neuroscience Bulletin 30 (4), 595-600. https://doi.org/10.1007/s12264-014-1451-7.

第二十一章

斑马鱼胚胎乙醇暴露的行为遗传学：FASD 模型

Steven Tran[1], Amanda Facciol[2], Robert T. Gerlai[3]

21.1 引言

众所周知，产前经历乙醇暴露可导致胎儿一系列认知、行为和形态缺陷（Easey et al., 2019; Jones et al., 1973; Sokol et al., 2003; Streissguth, 1997; Streissguth et al., 1978）。尽管如此，根据 McHugh 等人（2014）的描述，仍有多达 30% 的女性在妊娠期饮酒。胎儿乙醇谱系障碍（fetal alcohol spectrum disorder, FASD）是一个概括性术语，常用于描述母亲在妊娠期饮酒的胎儿个体可能发生的一系列情况。FASD 主要包括与乙醇相关的出生缺陷（alcohol-related birth defects, ARBD）、神经发育障碍（alcohol-related neuro-development disorders, ARND），以及最为严重的胎儿乙醇综合征（fetal alcohol syndrome, FAS）。FAS 的特点是严重的面部畸形和行为表达受损，包括社交行为、校园问题和后期的乙醇成瘾（Jones, 1975; Sokol et al., 2003）。ARBD 和 ARND 属于 FASD 较轻的症状，但可能包括颅面畸形和行为缺陷。然而，由于其性质温和，ARND 和 ARBD 经常不能被及时确诊或被误诊（Burd and Martsolf, 1989; Clark and Gibbard, 2013）。FASD 仍然是西方国家最普遍的可预防的精神残疾形式（McHugh et al., 2014）。

FASD 在同卵双胞胎中有 100% 的一致性，在异卵双胞胎中有 64% 的一致性，这表明遗传因素在 FASD 相关缺陷的严重程度和表达中起着重要作用（Streissguth and Dehaene, 1993）。这一显著的遗传结果，使人们对识别和理解影响产前乙醇暴露的遗传机制及其效应产生

[1] Department of Biology and Biological Engineering, California Institute of Technology, Pasadena, CA, United States.
[2] Department of Cell and Systems Biology, University of Toronto, Mississauga, ON, Canada.
[3] Department of Psychology, University of Toronto Mississauga, Mississauga, ON, Canada.

了极大的兴趣。然而乙醇是一种高度复杂的药物，具有多种分子作用，会影响许多生物过程，包括细胞分化、增殖、迁移和凋亡（Blader and Strahle, 1998; Goodlett et al., 2005; Michaelis, 1990; Shibley et al., 1999）。由于乙醇对发育中的中枢神经系统具有复杂的影响，人们对产前乙醇引起缺陷的遗传机制知之甚少。虽然人类研究提供了对这种疾病的深入了解，但动物研究对于理解产前乙醇暴露对中枢神经系统和行为影响的分子和遗传机制至关重要（Patten et al., 2014）。迄今为止，啮齿类动物模型是产前或胚胎乙醇暴露相关研究的金标准，然而，斑马鱼也开始在该领域被广泛应用（Fernandes and Gerlai, 2009; Baggio et al., 2018; Lovley et al., 2016）。

21.2 斑马鱼作为 FASD 的动物模型

斑马鱼传统上被用于发育生物学研究，最近成为行为遗传学和药理学研究的一种流行的动物模型。由于繁殖力强、体形小、易于饲养、大脑结构简单且进化上保守以及与人类基因组高度同源，斑马鱼被认为是合理兼顾了系统复杂性和实用简单性的动物模型（Ali et al., 2011a,b; Gerlai et al., 2000）。一对繁殖期的斑马鱼每周可以产生 200 ～ 300 个受精卵，后代将在 3 个月内性成熟。斑马鱼是一种遗传易处理的动物模型，其全基因组已完成测序，大约 70% 的人类基因在斑马鱼中至少有一个同源基因，意味着其与人类有很大的遗传相关性（Howe et al., 2013）。得益于在遗传学领域的广泛应用，正向和反向遗传技术也在斑马鱼上得到了很好的发展，例如，基于光遗传学的特定神经元回路的快速激活或失活，基于 CRISPR/Cas9 或 TALEN 的基因组编辑工具，以及可以在 mRNA 水平上敲低基因表达的 morpholinos。这些技术可产生斑马鱼突变体，进而重现人类中枢神经系统疾病的某些过程，也可以产生许多转基因品系，用来可视化和 / 或干扰斑马鱼大脑中的特定神经回路。

斑马鱼作为 FASD 相关研究的理想选择还有如下原因。首先，斑马鱼卵和发育中的胚胎 / 仔鱼是透明的，可以从受精那一刻起分析和观察形态发育（Ali et al., 2011a,b; Facciol et al., 2019）。其次，斑马鱼卵是体外受精的，因此可以无创地给予乙醇或其他药物（Ali et al., 2011a, b; Gerlai et al., 2000; Patten et al., 2014）。可以简单地将斑马鱼胚胎浸入乙醇溶液中，然后在预定的精确时间点从乙醇溶液中取出；同样，成年斑马鱼也可以被浸入药物或乙醇溶液中。斑马鱼研究中的药物浸泡法可对药物剂量、暴露持续时间和暴露发育时间点进行精确控制，同时避免了应激和侵入式给药，例如注射。再次，对斑马鱼采用的浸泡法还避免了与母体乙醇代谢以及与母体生理相关的其他复杂因素相混淆，这在啮齿动物研究中是不能实现的。最后，与发育分为产前和产后阶段的啮齿动物不同，斑马鱼的发育完全是外部的（Kimmel et al., 1995; Patten et al., 2014; West 1987）。斑马鱼受精后 120h（hpf）就可从受精卵发育成完全成型的鱼（Kimmel et al., 1995），斑马鱼胚胎的外部发育意味着在这个快速发育时期可以使用精确而一致的乙醇浸泡法进行乙醇暴露。

21.2.1 斑马鱼胚胎轻度乙醇暴露

虽然对斑马鱼的FASD相关研究较多，但这些研究大多侧重于模拟最严重的乙醇相关缺陷，即FAS。例如，在斑马鱼胚胎发育过程中长时间接触高浓度乙醇已被证明会导致显著的形态缺陷并损害行为反应（Ali et al., 2011a, b；Bilotta et al., 2004; Dlugos and Rabin, 2007; Parker et al., 2014）。但长时间接触高剂量乙醇并不是孕妇饮酒的典型特征。而且，FASD虽然以颅面畸形为典型特征，但也常伴有因轻度胚胎乙醇暴露所产生的无形态缺陷的行为异常。因此，我们认为研究更温和、更普遍的FASD非常重要，其通常表现为行为障碍，而无任何明显的形态变化。尽管斑马鱼FASD的行为遗传学研究仍处于起步阶段，但一些研究已经开始探讨斑马鱼胚胎轻度乙醇暴露的潜在机制（Baggio et al., 2018; Fernandes and Gerlai, 2009）。在本章中，我们讨论了许多不同的技术，这些技术可被用于研究轻度胚胎乙醇暴露引起斑马鱼行为缺陷的分子和遗传机制。

简而言之，在胚胎发育的特定时期，短时间接触低剂量的乙醇已被证明会改变成年斑马鱼的许多行为，包括运动（Bailey et al., 2015）、学习（Carvan et al., 2004；Fernandes et al., 2014）、社交行为（Buske and Gerlai, 2011; Fernandes and Gerlai, 2009; Fernandes et al., 2015a; Fernandes et al., 2018 a,b）、成瘾行为（Sterling et al., 2016）和焦虑样行为（Bailey et al., 2015; Baggio et al., 2018; Carvan et al., 2004），但不会引起明显的形态变化。例如，图21-1显示，在受精后16h内接触0.25%～1.0%（体积分数）乙醇2h会显著损害联想学习任务的认知表现（Fernandes et al., 2014）。许多研究已经开始分析这些改变的神经化学基础。例如，已证明斑马鱼胚胎低剂量乙醇暴露会降低多巴胺、5-羟色胺（Buske and Gerlai, 2011; Fernandes et al., 2015b）、腺苷（Haab Lutte et al., 2018）、谷氨酸的转运水平（Baggio et al., 2017），并增加成年斑马鱼大脑中牛磺酸的水平（Mahabir et al., 2018b）。但是关于斑马鱼胚胎低剂量乙醇暴露引起的成鱼行为障碍的基因和遗传机制仍不清楚。

21.2.2 品系差异

确定调控胚胎乙醇暴露影响的候选基因的一个目的是对遗传上不同的近交斑马鱼品系进行系统比较。斑马鱼基因组已经完成测序，某些准近交系的基因高达80%的位点处于纯合状态。这些遗传定义明确的斑马鱼品系由于遗传奠基者效应（品系起源的个体取样）以及随机遗传漂变［由于封闭（或内部）繁殖导致等位基因变异的随机丢失］，预计将显示出显著的基因组和表型差异。一些研究已经开始表征斑马鱼品系在行为和神经化学方面的差异。例如，图21-2和图21-3显示了两个相对高度近交系AB和TU之间的群聚行为和神经化学的显著差异（Mahabir et al., 2013）。AB品系斑马鱼的群聚行为以及全脑多巴胺及其代谢物3,4-二羟基苯乙酸（DOPAC）表现出线性的年龄依赖性增加（Buske and Gerlai, 2012），这些神经化学物质与斑马鱼群聚行为有关（Saif et al., 2013; Scerbina et al., 2012）。相比之下，TU品系斑马鱼的群聚行为以及全脑多巴胺和DOPAC水平呈现出年龄依赖性的逐步增加，在受精后40～70d之间出现明显的快速增加（Mahabir et al., 2013）。这些行为和大脑神经化学的差异可能是由AB和TU品系的不同遗传背景造成的。

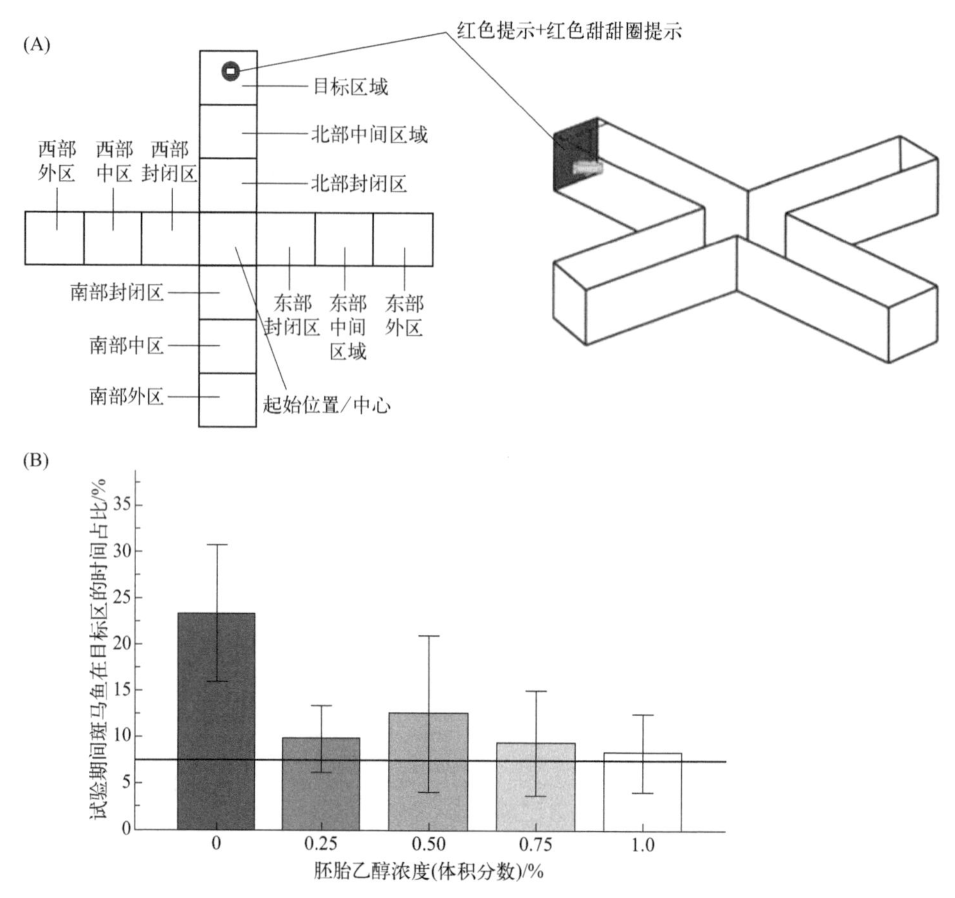

图21-1 **（A）十字迷宫示意图；（B）胚胎期暴露乙醇后显著损害了成年斑马鱼的联想学习能力**

（A）图中，需要注意的是，条件刺激（CS）是一个红色的塑料提示卡，放置在其中一个手臂（被任意指定为北手臂）的末端。还要注意的是，除了这个提示，一个红色的浮动环（“甜甜圈”）被放置在旁边的提示卡水面上。这个圆环既是食物输送装置（漂浮的片状食物放置在圆环内），又是一个CS。迷宫被划分成假想的区域，测量鱼在目标区域停留的时间，以及它们进入目标区域和所有其他区域的频率。（B）图显示了试验期间在目标区域花费的时间百分比（平均值 ±SEM）。实心水平线表示随机进入区域的性能。注意，对照组（不接触乙醇）鱼的表现明显好于接触乙醇的鱼，也明显高于随机进入区域时间，而接触乙醇的鱼进入目标区域与随机进入区域时间在统计学上无明显差异。同时注意到不同乙醇浓度组的鱼彼此之间没有显著差异。有关更多细节、结果的完整描述和统计分析，请参阅文献 Fernandes 等（2014）。参考以下文献并作了修改：Fernandes, Y., Tran, S., Abraham, E., Gerlai, R., 2014. Embryonic alcohol exposure impairs associative learning performance in adult zebrafish. Behavioural Brain Research 265, 181-187

胚胎在短时间内暴露于低浓度乙醇已被证明会损害 AB 品系（Fernandes and Gerlai, 2009）和 *casper* 突变体（Fernandes et al., 2019）的社交行为。图 21-4 显示胚胎乙醇暴露损害了由出现同种动物的动画图像引起的成鱼群聚行为。然而，胚胎乙醇暴露对脑神经化学的影响已被证明是品系依赖的，表明某些等位基因可能对胚胎乙醇暴露的有害影响发挥保护作用（Mahabir et al. submitted; Mahabir et al., 2018b）。例如，图 21-5 显示胚胎乙醇暴露剂量依赖性地降低了 AB 型斑马鱼的全脑多巴胺水平，而 TU 型没有。

在比较胚胎乙醇暴露对 AB 品系和 TU 品系脑神经化学的影响时，胚胎乙醇暴露仅在 AB 品系中剂量依赖性地降低了成鱼全脑多巴胺和 5- 羟色胺以及代谢物含量（Mahabir et al., 2014）。值得注意的是，图 21-6 显示，1% 乙醇暴露 2h 后，AB 和 TU 品系的胚胎中测定

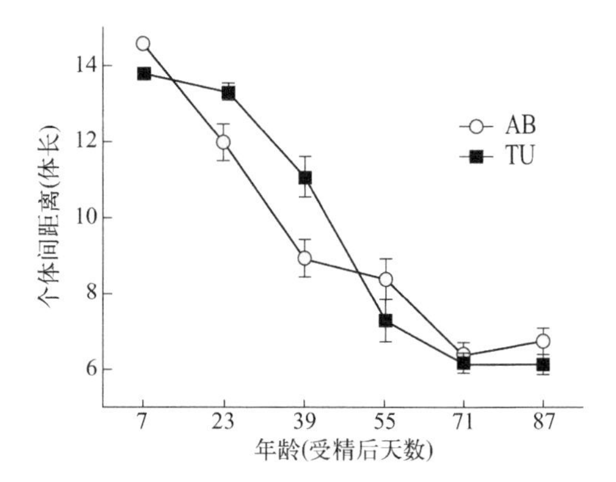

图21-2 **随着斑马鱼的成熟，鱼群成员之间个体间距离的应变依赖性减小**

Mean±SEM（均值 ± 标准误）如上图所示，黑色方块代表 TU 品系，灰色圆圈代表 AB 品系。请注意：AB 品系斑马鱼在受精后 7 ～ 71d 之间的个体间距离（鱼群凝聚力的增强）呈线性降低；在受精后 39 ～ 55d 期间 TU 品系斑马鱼中的倒 S 形轨迹变化最快；个体间距离是相对于实验对象的体长表示的。更多详细信息、结果的完整描述和统计分析，请参阅文献 Mahabir 等（2013）。参考以下文献并作了修改：Mahabir, S., Chatterjee, D., Buske, C., Gerlai, R., 2013. Maturation of shoaling in two zebrafish strains: a behavioural and neurochemical analysis. Behavioural Brain Research 247, 1-8

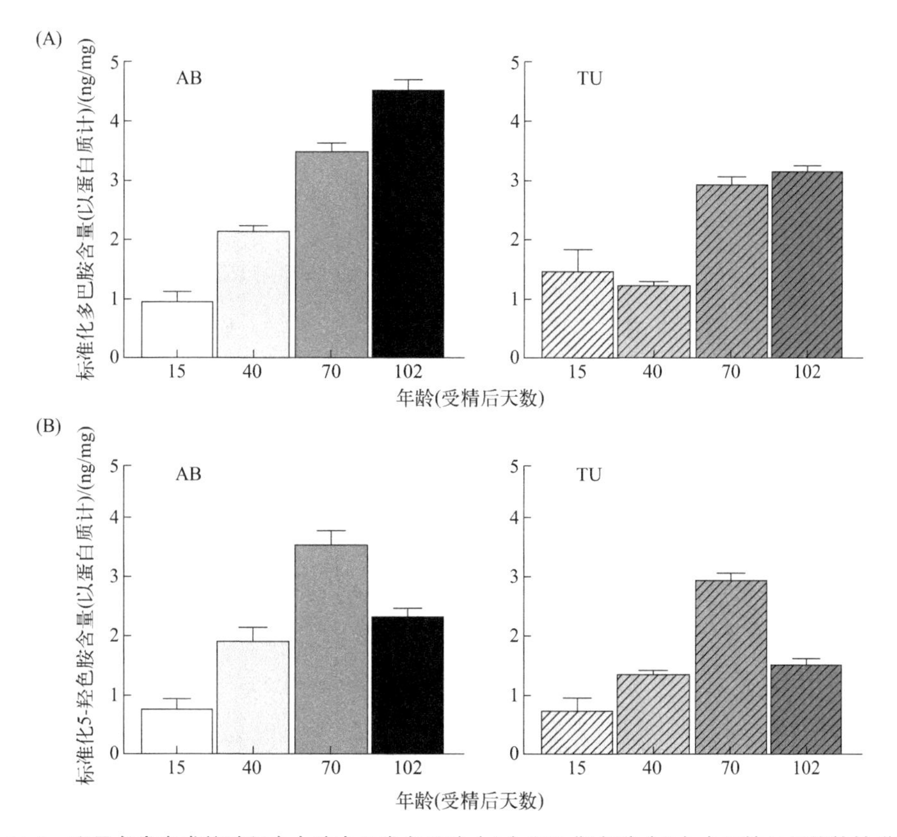

图21-3 **斑马鱼发育成熟过程中大脑中正常多巴胺（A）和5-羟色胺（B）水平的品系依赖性增加**

数据以 Mean±SEM 表示。左边的单色系代表 AB 品系获得的结果，右边的条纹代表 TU 品系的结果。请注意 AB 品系斑马鱼中相对多巴胺水平线性增加，TU 品系斑马鱼多巴胺水平在受精后 40 ～ 70d 呈阶梯状增加。还要注意两种品系的斑马鱼 5- 羟色胺水平呈倒 U 形发展轨迹。多巴胺和 5- 羟色胺水平表示为相对于总脑蛋白的质量。有关更多详细信息、结果的完整描述和统计分析，请参阅文献 Mahabir 等（2013）。参考以下文献并作了修改：Mahabir, S., Chatterjee, D., Buske, C., Gerlai, R., 2013. Maturation of shoaling in two zebrafish strains: a behavioural and neurochemical analysis. Behavioural Brain Research 247, 1-8

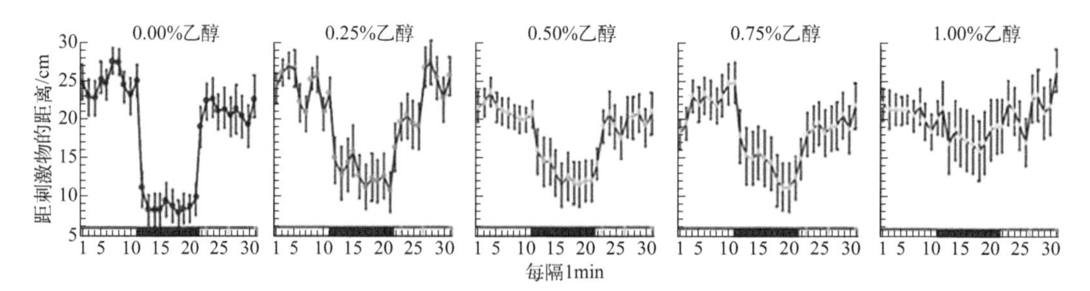

图21-4　**胚胎乙醇暴露损害了由出现同种动物的动画图像引起的成鱼聚集行为**

在30min的行为记录过程中，成年实验斑马鱼与呈现在计算机屏幕上的刺激之间的平均距离以1min间隔记录，数据以Mean±SEM表示。刺激呈现的时间轴由x轴上方的横线表示（白色表示无刺激；黑色表示由五个同族个体构成的图像刺激；灰色表示捕食者图像刺激）。乙醇处理浓度（从受精后24h开始，持续2h）如图所示。当屏幕上同种个体出现时，受试鱼与屏幕的距离显著减少，而屏幕上同种个体消失后，受试鱼与屏幕的距离显著增加。这种现象随着乙醇浓度的增加而减弱。更多细节、结果的完整描述和统计分析，见文献Fernandes和Gerlai（2009）。参考以下文献并作了修改：Fernandes, Y., Gerlai, R., 2009. Long-term behavioural changes in response to early developmental exposure to ethanol in zebrafish. Alcoholism: Clinical and Experimental Research 33, 601-609

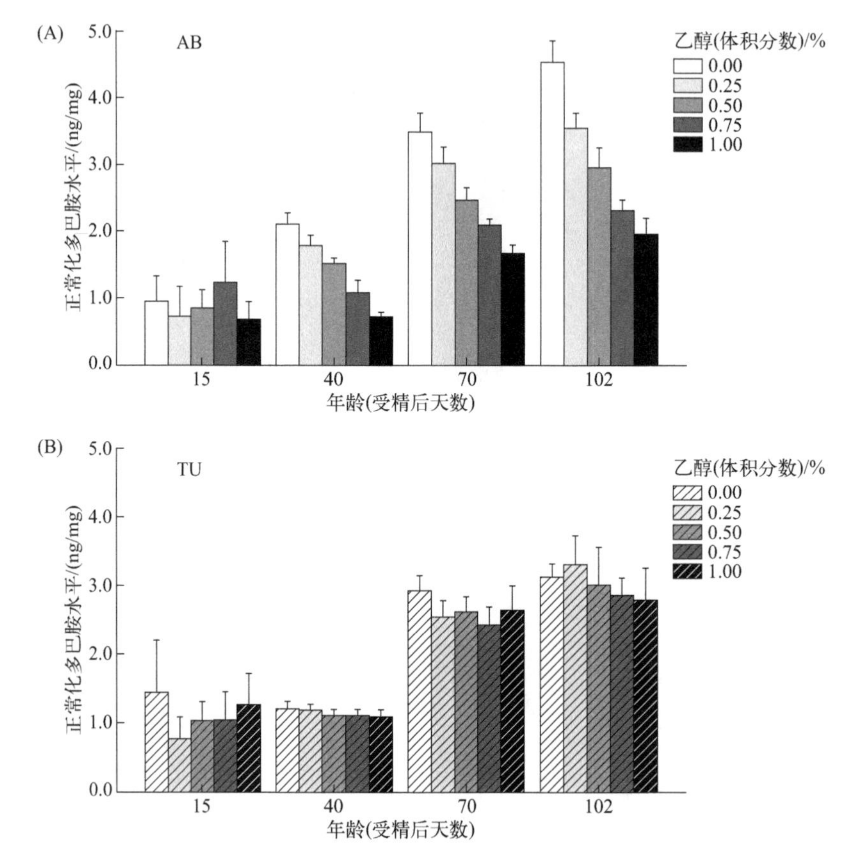

图21-5　**胚胎乙醇暴露导致AB品系斑马鱼（实心条形）脑内多巴胺水平显著的剂量依赖性降低（A），而TU品系（条纹条形）脑内多巴胺水平无显著降低（B）**

数据显示为Mean±SEM。多巴胺水平单位以总脑蛋白（计）为ng/mg。值得注意的是，随着AB品系鱼的成熟，乙醇暴露导致的组间差异变得更加显著。更多细节、结果的完整描述和统计分析，请参见文献Mahabir等人（2014）。参考以下文献并作了修改：Mahabir, S., Chatterjee, D., Gerlai, R., 2014. Strain dependent neurochemical changes induced by embryonic alcohol exposure in zebrafish. Neurotoxicology and Teratology 41, 1-7

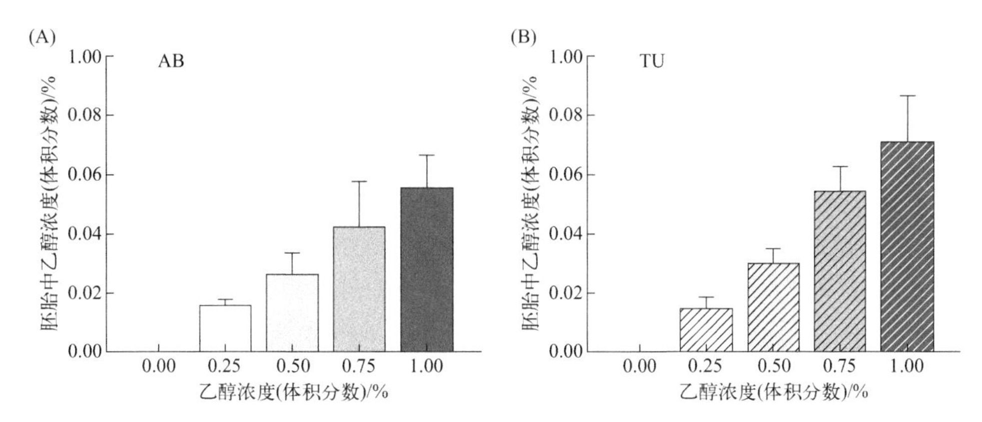

图21-6 **（A）AB（实心条形）和（B）TU（条纹条形）斑马鱼品系胚胎内的乙醇浓度未见差异，大约是胚胎浸泡2h时的乙醇溶液浓度的1/12**

数据以 Mean±SEM 表示。更多细节、结果的完整描述和统计分析，请参见文献 Mahabir 等人（2014）。参考以下文献并作了修改：Mahabir, S., Chatterjee, D., Gerlai, R., 2014. Strain dependent neurochemical changes induced by embryonic alcohol exposure in zebrafish. Neurotoxicology and Teratology 41, 1-7

的乙醇量没有显著差异（Mahabir et al., 2014）。这表明 TU 品系中缺乏乙醇效应不是因为乙醇通过 TU 品系鱼绒毛膜扩散的减少。值得注意的是，将胚胎浸泡在大量乙醇溶液中，使用推测的乙醇代谢或分泌相关机制不太可能解释品系间的差异。因此，作者认为观察到的差异可能是由 TU 和 AB 品系之间遗传差异导致的神经生物学机制（Mahabir et al., 2014）。

同样值得注意的是，胚胎乙醇暴露的影响不是全身的，而是对特定神经回路的选择性影响。例如，尽管胚胎暴露于乙醇会损害 AB 品系的多巴胺能和 5 - 羟色胺能系统，但它不会显著改变 AB 或 TU 品系鱼的几种氨基酸神经递质的水平，如谷氨酸、天冬氨酸、甘氨酸和 GABA（Mahabir et al., 2018b）。这些发现表明 TU 品系的遗传背景可能对胚胎乙醇暴露发挥保护作用。使用 DNA 微阵列对肝脏 mRNA 进行分析，发现 AB 和 TU 品系斑马鱼之间存在大量的差异表达基因。例如，AB 品系有 141 个基因在性别之间表达有差异，但在 TU 品系中没有观察到显著的性别依赖性差异。在雄性斑马鱼中，AB 品系和 TU 品系之间有 102 个基因不同，而在雌性斑马鱼中，这两个品系之间有 125 个基因不同（Holden and Brown, 2018）。然而，目前尚不清楚这些基因中的哪些（如果有）可能介导乙醇对斑马鱼中枢神经系统及其行为的致畸作用。鉴定哪些基因因胚胎乙醇暴露而改变的一种方法是确定哪些基因在胚胎乙醇暴露后发生了上调或下调。

21.2.3 基因表达变化

显色原位杂交（chromogenic in situ hybridization, CISH）和荧光原位杂交（fluorescence in situ hybridization, FISH）是检测斑马鱼大脑基因表达的一种半定量技术。CISH/FISH 可用于鉴定特定 mRNA 转录本的表达模式，并在较小程度上确定表达水平（Hauptmann et al., 2016; Thisse and Thisse, 2008）。由于斑马鱼仔鱼发育的透明性，CISH/FISH 可以优化用于整个斑马鱼仔鱼的检测（Hauptmann, et al., 2016; Thisse and Thisse, 2008）。为了对大脑结构深部成像和更好地实现原位探针穿透，可以在 CISH/FISH 之前解剖斑马鱼仔鱼的大脑。对斑马鱼仔鱼的 CISH 分析显示，当胚胎暴露于受精后 6 ～ 24h 2% 乙醇中，斑马鱼仔鱼的眼睛、前脑和后脑中的 *pax6a* 表达降低，而在菱脑原节（rhombomere）R1 边缘的 *epha4a* 表达降低（Zhang et al., 2017）。*pax6a* 属于 PAX 家族基因，其编码含有 DNA 结合配对结构域的转录因子，这些结构域在胚胎发育过程中起重要作用（Strachan and Read, 1994）。*pax6a* 是人类 *PAX6* 基因的斑马鱼同源基因，该基因在眼睛发育中具有重要作用，已被用作斑马鱼眼睛和前脑发育的标志物（Feiner et al., 2014）。应用全定量 CISH 方法已证明胚胎乙醇暴露可减少 *pax6a* 表达，并使斑马鱼仔鱼的眼睛变小（Zhang et al., 2011）。

相比之下，成年斑马鱼的大脑通常需要切片，CISH/FISH 可以在单个组织切片上进行（Kuhn and Koster, 2010）。对成年斑马鱼大脑切片中 *pax6a* 表达的 FISH 分析显示，*pax6a* 在嗅球、前端脑和视网膜中表达（Adolf et al., 2006; Feiner et al., 2014）。然而，目前尚不清楚胚胎乙醇暴露后成年斑马鱼大脑中 *pax6a* 的表达是如何改变的。尽管 FISH/CISH 对于通过检测不同的发育时间点来确定基因的表达部位和表达时间很重要，但该技术也只是半定量的。如果荧光强度已经饱和，量化 mRNA 表达水平的荧光强度变化可能会有问题。例如，如果 mRNA 转录本已经高度表达并进一步增加，荧光强度将难以检测。类似地，如果转录本以低水平表达，进一步降低的荧光也很难检测到。

反转录酶定量 PCR（RT-qPCR）有时也被称为 qPCR，是检测基因表达倍数变化的一种更为敏感的定量技术。利用这项技术，mRNA 首先被逆转录为 cDNA，然后作为 PCR 的模板，与 DNA 结合荧光染料一起使用。在每个 PCR 循环后测定荧光，荧光强度与扩增的起始 DNA 量成正比，可用于指示原始 mRNA 的表达水平。qPCR 的局限性在于，它是一种昂贵且耗时的技术，其可非特异性地与双链 DNA 结合，因此一次只能使用一种荧光染料，每个反应只能检测一个目的基因。另外，需要采用一种预计不会因实验操作而改变的管家基因（同条件下的乙醇暴露）作为阴性对照进行检验。然后将目的基因的表达计算为相对于管家基因的百分比变化。CISH 的定量结果表明，胚胎乙醇诱导的小眼症（眼睛尺寸减小）与 *pax6a* 表达的减少相关（Zhang et al., 2011）。后续 qPCR 分析显示，胚胎乙醇暴露使 *pax6a* 的表达量降低 1/2，并呈浓度依赖性和发育窗依赖性（Zhang et al., 2014）。

如前所述，胚胎乙醇暴露已被证明会损害斑马鱼的社交行为（Fernandes and Gerlai, 2009; Parker et al., 2014）。qPCR 分析表明，胚胎乙醇暴露还以年龄依赖性方式改变了几种社交行为相关基因的表达水平，包括 *oxtr*（催产素受体）、*slc6a4*（5- 羟色胺转运蛋白）、*htr1aa*（5- 羟色胺受体 1aa），但不会影响 *avpr*（精氨酸加压素受体）基因的表达（Parker et al., 2014）。

分析基因表达的变化是确定胚胎乙醇暴露如何影响大脑的重要方法。然而，考虑蛋白质的翻译后修饰也很重要，因为它最终会影响细胞信号通路。利用抗体和抗原结合亲和力对蛋白质进行免疫组化染色，对于揭示胚胎乙醇诱导缺陷的机制非常重要。如前所述，胚胎乙醇暴露可降低全脑多巴胺、5- 羟色胺及其代谢产物的水平（Buske and Gerlai, 2011），但不会降低谷氨酸、天冬氨酸、甘氨酸和 GABA 神经递质的水平（Mahabir et al., 2018b）。一种可能的解释是，胚胎乙醇暴露可诱导特定神经细胞类型的凋亡，例如多巴胺能和 5- 羟色胺能细胞，但不诱导谷氨酸能、甘氨酸能或 GABA 能细胞的凋亡。免疫组织学染色显示，胚胎乙醇暴露可减少成年斑马鱼大量脑区包括背侧端脑区的内侧和外侧区、腹侧端脑区的背侧和腹侧核、视前核小细胞部、腹侧缰核、小脑体和下网状结构中的脑源性神经营养因子（brain-derived neurotrophic factor, BDNF）和神经元细胞黏附分子（neuronal cell adhesion molecule, NCAM）阳性神经元的数量（Mahabir et al., 2018a）。然而，除视前区和小脑外，所有上述脑区均观察到突触素阳性细胞数量的减少，这在一定程度上支持了胚胎乙醇的细胞类型特异性效应（Mahabir et al., 2018a）。值得注意的是，在视前区发现了兴奋性谷氨酸能和抑制性 GABA 能神经元（Yu et al., 2016），腹侧视前核约 80% 的神经元是 GABA 能神经元（Lu et al., 2000）。相反，小脑中的大部分神经元是抑制性甘氨酸能和 GABA 能神经元（Simat et al., 2007）。胚胎乙醇暴露缺乏对视前区和小脑的突触素阳性细胞数量的影响与其缺乏对谷氨酸、甘氨酸和 GABA 神经递质水平的影响密切相关。

CISH/FISH 和 qPCR 等方法可用于确定基因在胚胎乙醇暴露后是上调还是下调，但这些方法不会就该基因的功能做出强有力的因果表述。例如，胚胎乙醇暴露已被证明会损害眼睛发育以及音猬因子（sonic hedgehog, *shh*）基因的表达（Brennan and Giles, 2013; Burton et al., 2017）。然而，这一发现本身并不意味着胚胎乙醇暴露的不利影响是由改变 *ssh* 基因信号引起的。通过使用反向遗传技术敲减 / 敲除和恢复 *shh* 基因表达，研究人员才能够得出相关的因果结论。

21.3 反向遗传学

21.3.1 基因敲减

吗啉反义寡核苷酸（morpholino）已被用于瞬时敲减基因表达。它们的工作原理类似于反义寡核苷酸，但与天然寡核苷酸相比，它们的化学性质有所改变，使其更稳定，在实验中更为有用。通常，它们的长度为 25 个核苷酸，其序列被设计成能与特定的 mRNA 结合，从而干扰（减少）目的基因的翻译。有研究在单细胞阶段将 morpholino 显微注射到胚胎中，成功地敲减了斑马鱼的基因表达（Pauli et al., 2015）。例如，morpholino 诱导的 *shh* 抑制已被证明会损害斑马鱼的发育，类似于胚胎乙醇暴露（Nasevicius and Ekker, 2000）。*shh* 基因是胚胎发育过程中参与器官发生、神经系统和骨骼发育的重要基因之一。*shh* 是在人类、啮齿动物、斑马鱼甚至昆虫中发现的高度保守的基因家族（Hammerschmidt et al., 1997），与包括FASD在内的人类疾病相关的形态和颅面异常有关（Zhang et al., 2014, 2015, 2017; li et al., 2007）。斑马鱼表达了人类 *SHH* 基因的两个相似基因：*shha* 和 *shhb*。morpholino 诱导的 *shha* 敲减已被证明会损害发育，而 morpholino 诱导的 *shhb* 敲减没有产生明显的缺陷。然而，morpholino 诱导的 *shha* 和 *shhb* 的同时敲减却会产生更严重的发育缺陷，表明二者的同时敲减有协同效应（Nasevicius and Ekker, 2000）。

除了 *shh* 信号通路的改变，视黄酸信号通路的中断也被认为是胚胎乙醇诱导的发育缺陷和 FASD 的潜在机制。视黄酸合成的限速步骤是由乙醇脱氢酶（alcohol dehydrogenase, ADH）催化的视黄醇氧化。然而，乙醇也是抗利尿激素（ADH）的底物，高水平的乙醇已被证明可以抑制视黄酸浓度（Deltour et al., 1996）。在斑马鱼中，morpholino 诱导的 *aldh3a*（一种视黄酸生物合成酶）的敲减产生了类似于胚胎乙醇暴露的表型（Yahyavi et al., 2013）。在受精后 2d（dpf）进行的定量研究表明，原肠形成过程中，视黄酸与乙醇共给药可以防止胚胎乙醇暴露诱导的Mauthner细胞突触缺陷形成（Ferdous et al., 2017）。同样，*aldh3a* mRNA 和乙醇的共同给药被证明可以挽救胚胎乙醇暴露所诱导的畸形发生（Burton et al., 2017）。尽管有上述视黄酸信号通路的研究，但是近几年才在斑马鱼中开展胚胎乙醇诱导的行为缺陷的背景下 *shh* 和视黄酸信号转导作用的研究。

Burton 等人（2017）研究了 *shh* 和视黄酸信号在斑马鱼早期发育期间（即 8 ～ 10hpf 或 24 ～ 27hpf）暴露于低剂量乙醇引起的焦虑行为和冒险行为中的作用。底栖是斑马鱼的一种正常的焦虑样行为反应，通常在暴露于一个新的鱼缸后的最初几分钟内观察到（Bencan et al., 2009; Nowicki et al., 2014）。在胚胎发育过程中暴露于 1% 或 3% 的乙醇显著减少了成年斑马鱼在新鱼缸中第一分钟内的底栖，表明焦虑样行为减少。虽然 morpholino 诱导的 *shh* 或 *aldh3a* 基因敲减均未加重乙醇诱导的行为改变，但 *shh* 而非 *aldh3a* mRNA 的共同给药挽救了行为损伤。这些结果表明，胚胎乙醇诱导的焦虑样行为的变化受到 *shh* 减少的调控，而不是视黄酸信号转导的调节（Burton et al., 2017）。

然而，使用 morpholino 有几个限制。首先，与基因组敲除基因的情况不同，morpholino 只是暂时性敲减基因表达，基因表达减少也是不完全的。因此，可能无法观察到需要完全敲除基因或长期抑制基因表达的表型。然而，基因敲减而非敲除，也可能被认为是一种优势，因为采用基因敲减方法可研究那些如果完全敲除就会致命的基因。其次，更重要的是，morpholino 诱导的表型可能被脱靶效应所混淆。例如，据估计，注射 morpholino 具有 10% ～ 20% 的概率诱发 p53 依赖性神经毒性，这可能掩盖或加剧预期的缺陷（Bedell et al., 2011）。同一基因的 morpholino 诱导敲减与基因敲除之间的系统比较发现，即使包括适当的对照，morpholino 诱导的表型也可能是由脱靶剪接效应所形成的（Gentsch et al., 2018）。

21.3.2 基因敲除

基因组编辑技术的进步，如锌指核酸酶（ZFN）、转录激活因子样效应物核酸酶（TALEN），以及最近的 CRISPR/Cas9，使斑马鱼基因组的高效靶向和操纵成为可能，与 morpholino 诱导的基因敲减相比，其脱靶效应显著减少（Schulte-Merker and Stainer, 2014）。CRISPR/Cas9 系统是一种古老的细菌防御系统，由两部分组成，即一种切割双链 DNA 的 Cas9 酶和一种支架 RNA 分子。后者由结合 Cas9 酶的 trace RNA 序列和可将 Cas9 酶引导至基因组上位于原间隔区相邻基序序列上游位置的 crRNA 序列组成。在 Cas9 酶切割诱导的双链断裂后，将启动 DNA 修复机制；但是，修复过程容易出错，导致引入插入和 / 或删除。重要的是，非三个核苷酸倍数的基因插入或删除或将导致移码突变，导致翻译氨基酸序列的完全改变，甚至可能导致提前终止。因此，这种移码突变导致蛋白质结构的显著改变和 / 或截短，使翻译的蛋白质高度功能失调，即功能缺失。CRISPR/Cas9 技术的进展使研究人员能够设计出一种向导 RNA（sgRNA），该 RNA 具有用于基因组靶向的可变的 20 个碱基对序列，以及作为 Cas9 酶支架的保守序列（Jinek et al., 2012）。CRISPR/Cas9 系统除了相对简单和高效之外，值得注意的是，这种基因打靶方法有望并已被证明可以诱导目标位点的两个等位基因发生预期的突变，从而替代昂贵和耗时的纯合子育种。

对 *shha* 突变体斑马鱼的分析显示，其发育缺陷类似于 morpholino 诱导的 *shha* 敲减。例如，纯合子 *shha* 突变体在体节、侧板细胞、胸鳍以及视网膜神经节细胞上表现出缺陷（Schauerte et al., 1998）。同样，纯合子 *shha* 突变体也表现出因视网膜中细胞分化和繁殖缺陷导致的小眼症（Stenkamp et al., 2002）。值得注意的是，使用反向遗传技术产生的斑马鱼突变体也为鉴定特定的候选基因是否对胚胎乙醇暴露的影响具有敏感性提供了便利。例如，人类基因 MED12 的突变会导致与 FAS 相似的精神发育迟滞和行为困难（Graham et al., 2010）。斑马鱼同源基因 *med12* 的突变也被证明会损害斑马鱼大脑的发育，其表型与人类相似（Hong et al., 2005）。值得注意的是，不改变野生型斑马鱼发育的低剂量乙醇暴露可显著加剧斑马鱼胚胎纯合子 $med12^{y82}$ 突变的发育缺陷，表明两者具有协同效应（Coffey et al., 2013）。与在胚胎乙醇暴露过程中检测 *med12* 和 *shh* 等基因特定作用的反向遗传技术相比，正向遗传学方法可利用斑马鱼的独特优势进行大规模遗传筛选以及鉴定在发育过程中对胚胎乙醇暴露敏感的重要基因。

21.3.3 正向遗传学和遗传筛选

N- 乙基 -*N*- 亚硝基脲（*N*-ethyl-*N*-nitrosourea, ENU）是一种用于在基因组中引入随机单点突变的诱变剂，是大规模遗传筛选中常用的随机诱变化学物质。与在小鼠中注射 ENU 不同，斑马鱼只需简单地浸泡在 ENU 溶液中，以一种非侵入性的方式诱导突变。对成年雄性斑马鱼开展连续 6 周的 1h 的 ENU 处理，可实现有效的 ENU 诱变，平均每 1×10^5 ～ 1.5×10^5 个碱基对发生一个种系突变（de Brujin et al., 2009）。在大规模 ENU 诱变筛选中，鉴定出约 4000 多个斑马鱼突变体，并对其进行了发育异常筛选。对这些突变体的后代在发育的第 2d、3d 和 6d 进行筛选，发现 372 个基因对早期斑马鱼的发育具有重要意义（Haffter et al., 1996）。由于斑马鱼的遗传易处理性和基因组编辑技术的进步，目前通过斑马鱼国际资源中心（Zebrafish International Resource Center, ZIRC）可以获得大量的突变斑马鱼品种。斑马鱼具有大量的可用突变系，且体形小、繁殖力高，使其也适合于高通量药物筛选。例如，使用药物浸泡法，在斑马鱼仔鱼身上测试了超过 5000 种化合物，其中许多药物引发的行为反应与报道的人类行为反应相似（Rihel et al., 2010）。

尽管规模要小得多，但已经有研究者针对胚胎乙醇暴露敏感基因进行了筛选。McCarthy 等人（2013）筛选出了五种不同的表现出颅面缺陷的突变品系，发现血小板衍生生长因子受体 a（platelet-derived growth factor receptor a，*pdgfra*）基因突变加剧了胚胎乙醇暴露的影响。*pdgfra* 是一种酪氨酸激酶受体，由其配体血小板衍生生长因子（platelet-derived growth factor, *pdgf*）激活，目前已知可以调节包括细胞迁移和增殖在内的发育过程（Tallquist and Kazlauskas, 2004），以及神经嵴迁移（Soriano, 1997）。研究发现，胚胎乙醇暴露可诱导纯合子 *pdgfra* 突变体的神经嵴细胞凋亡，表明 *pdgfra* 在正常情况下发挥保护作用（McCarthy et al., 2013）。Swartz 等人（2014）在更大规模上筛选了 20 个同样表现出发育结构异常的突变系。这 20 个突变系中有 5 个在阈下剂量的胚胎乙醇暴露后表现出更严重的发育缺陷，这表明这 5 个额外的基因在参与胚胎乙醇暴露中起到保护作用。这两项研究共同支持以下观点：斑马鱼可作为识别胚胎乙醇敏感基因的行为筛选的有用遗传模型。与携带突变基因的斑马鱼不同，建立转基因品系可对在胚胎乙醇暴露期间和应对乙醇暴露过程中不同的神经元群体进行检测和干扰。

21.3.4 报告基因表达的转基因系

斑马鱼仔鱼的主要优势之一是它的发育透明性，允许对发育中的大脑和不同的神经元群体进行活体成像。研究人员利用这一点，在特定神经元启动子（或启动子片段）的控制下，生成表达荧光标记的转基因斑马鱼，以再现或报告内源性基因表达用于活体成像。由于不同工程荧光蛋白的使用，一旦确定了特定的启动子或启动子片段，就可以产生在同一动物体内用荧光标记多个神经元群体的转基因斑马鱼系。通常，克隆靶基因上游的几千个碱基对（kb）（启动子片段）足以在克隆荧光蛋白［如增强型绿色荧光蛋白（eGFP）］上游时驱动内源性基因的表达（St John and Key, 2012）。然而，有时小的启动子片段不足以驱动内源性基因的表达，并且由于技术限制，不能克隆足够大的序列。最近开发的适用于斑马鱼的方法是使用细菌人工染色体，它可以携带高达350kb的基因组DNA，并已被成功地用于产生用荧光标记斑马鱼的不同神经元群体的稳定品系（Bussmann and Schulte-Merker, 2011）。

胚胎期乙醇暴露后，斑马鱼大脑中神经元的体内成像是确定乙醇如何影响特定神经系统的有力方法。例如，转基因斑马鱼系 *hcrt*：*eGFP* 在内源性表达下丘脑泌素（hypocretin gene, *hcrt*）基因的细胞中表达 eGFP（Prober et al., 2006）。下丘脑泌素 / 食欲素（hypocretin/orexin）是一种下丘脑神经肽，参与多种稳态系统，包括但不限于睡眠、觉醒、进食和疼痛调节（Li et al., 2017）。胚胎乙醇暴露已被证明可诱导斑马鱼仔鱼下丘脑前部的 *hcrt* 表达（Sterling et al., 2016）。胚胎乙醇暴露后，监测 *hcrt*：*eGFP* 斑马鱼的细胞荧光可以实时显示食欲素基因表达的变化，这是一项尚未进行的研究。使用同样的方法，Coffey 等人（2013）对不同神经元群体中表达 GFP 的转基因仔鱼进行了体内成像，发现胚胎乙醇暴露改变了 GFP 标记神经元特定亚群的表达。例如，胚胎时期的乙醇暴露可诱导后脑特异性表达催产素，但不会诱导催产素能系统的其他部分的催产素表达（Coffey et al., 2013）。

最后，已知乙醇通过多种机制发挥作用，因此可能应该首先确定胚胎乙醇暴露对大脑中哪些特定结构的激活和 / 或抑制作用最大。确定神经元激活 / 抑制的一种方法是检测 *c-fos*，这是一种对刺激作出反应而上调的即早基因，通常被用作神经元激活的标记（Kovacs, 2008）。*c-fos* mRNA 可通过原位杂交检测，而 cFos 蛋白可通过免疫组织化学检测，这两种方法均适用于斑马鱼（Baraban et al., 2005; Chatterjee et al., 2015）。使用即早基因来衡量神经活动的缺点之一是现有方法的时间分辨率较差。例如，*c-fos* mRNA 在 10 ～ 15min 后被诱导（Baraban et al., 2005），而蛋白质表达可能需要 1 ～ 2h（Chatterjee et al., 2015）。此外，斑马鱼大脑中的本底 *c-fos* 通常较低，这使得通过神经元抑制进一步降低 *c-fos* 而几乎不可能检测到。

最近，有人开发了一种更灵敏的新方法来鉴定斑马鱼仔鱼中的活跃神经元细胞群（Randlett et al., 2015）。在细胞去极化之后，钙的内流激活 Ras-Erk 通路，该通路涉及下游磷酸化的级联反应，例如 Erk 的磷酸化（pERK）。免疫组织化学方法检测 pERK 与总 ERK（tERK）阳性细胞的比值，为观察神经元的激活提供了一种更具时间敏感性的方法。一旦获得全脑神经元激活的共聚焦叠加图像，即可获得斑马鱼参考脑图谱（6dpf），以确定神经元活跃和抑制的细胞群（Randlett et al., 2015）。虽然这两种方法（即早基因和pERK/tERK）都能识别活跃/非活跃的神经元，但它们都有相同的局限性，即必须固定/处死动物，因此，在一个特定的时间点上，只能获取单一的神经活动图像。这一限制可以通过量化多个时间点的IEGs和pERK/tERK信号来部分规避，但更直接的解决方案是在活体对整体神经元活动进行实时活体可视化。

利用基因编码的钙指示剂GCaMP，可以实现神经元活动的活体可视化。GCAMPs是由循环排列的GFP、钙结合蛋白钙调素（calcium binding protein calmodulin, CaM）和肌球蛋白轻链上的Ca^{2+}/CaM结合基序M13组成的重组蛋白。Ca^{2+}、CaM及其结合基序之间的相互作用导致蛋白质构象变化，使得GFP在激光激发后可以荧光成像（Pologruto et al., 2004）。钙离子（Ca^{2+}）是所有神经元中普遍存在的第二信使，对信号转导至关重要。利用双光子激发并在斑马鱼泛神经元启动子*elavl3*的控制下表达GCaMP，研究人员已经能够以单细胞分辨率在时间和空间上可视化整个斑马鱼大脑的神经元活动（Ahrens et al., 2012）。这种强大的方法可以确定在发育中的斑马鱼胚胎中，哪些细胞群在胚胎乙醇暴露后被激活。GCaMP成像不仅是一种用来发现对胚胎乙醇暴露有反应的神经元群的工具，还可以用来证实现有的假说。例如，24～26hpf的胚胎乙醇暴露已被证明会减少斑马鱼大脑中的全脑多巴胺（Buske and Gerlai, 2011）。一种假设可能是，胚胎期乙醇暴露后，多巴胺能神经元的活动被改变或可能被抑制，从而导致以后多巴胺的合成减少。利用GCaMP已经可以在24hpf检测到间脑（DC2和DC4）中的多巴胺能神经元（Du et al., 2016）。制备*dat*：*GCaMP6s*转基因斑马鱼将使得GCaMP在多巴胺能神经元中表达，这些多巴胺能神经元的神经元活动可以在24hpf开始的胚胎期乙醇暴露中通过活体监测到。转基因和突变斑马鱼的应用在行为遗传学领域至关重要，并可为胚胎乙醇暴露的潜在机制提供重要的研究思路。

21.4 结论

总之，胚胎乙醇暴露对发育中的中枢神经系统的影响是非常复杂的。乙醇已被证明可以改变斑马鱼中许多基因的表达和大量不同的生化途径。重要的是，胚胎乙醇暴露的影响可能能检测到，也可能检测不到，这取决于胚胎暴露的阶段和/或分析进行的阶段（Parker et al., 2014; Zhang et al., 2014）。虽然许多研究分析了斑马鱼胚胎轻度乙醇暴露导致的基因表达变化和行为缺陷，但很少有研究专注于识别胚胎乙醇诱导行为障碍的遗传机制（Burton et al., 2017）。近年来斑马鱼正向和反向遗传技术的研究进展有助于研究胚胎乙醇诱导的行为损伤的遗传机制。尽管乙醇致畸作用的性质很复杂，但可确定的是无论是行为上的还是遗传上的，与胚胎乙醇暴露相关的缺陷从胚胎被放入乙醇溶液中的那一刻就已经产生了。因此，正常中枢神经系统发育的中断很可能为成年斑马鱼以及成年人的疾病行为和分子损伤机制研究奠定了基础。

（曹念念翻译　石童审校）

参考文献

Adolf, B., Chapouton, P., Lam, C., Topp, S., Tannhauser, B., Strahle, U., Gotz, M., Bally-cuif, L., 2006. Conserved and acquired features of adult neurogenesis in the zebrafish telencephalon. Developmental Biology 295, 278-293.

Ahrens, M., Li, J., Orger, M., Robson, D., Schier, A., Engert, F., Portugues, R., 2012. Brain-wide neuronal dynamics during motor adaptation in zebrafish. Nature 485, 471-477.

Ali, S., Champagne, D., Alia, A., Richardson, M., 2011a. Large-scale analysis of acute ethanol exposure in zebrafish development: a critical time window and resilience. PLoS One 6, -20037.

Ali, S., Champagne, D., Spaink, H., Richardson, M., 2011b. Zebrafish embryos and larvae: a new generation of disease models and drug screen. Birth Defects Research Part C: Embryo Today 92, 115-133.

Baggio, S., Mussulini, B., de Oliveira, D., Gerlai, R., Rico, E., 2018. Embryonic ethanol exposure leading to social avoidance and altered anxiety responses in adult zebrafish. Behavioural Brain Research 352, 62-69.

Baggio, S., Mussulini, B., de Oliveira, D., Zenki, K., Santos da Silva, E., Rico, E., 2017. Embryonic alcohol exposure promotes long-term effects on cerebral glutamate transport of adult zebrafish. Neuroscience Letters 636, 265-269.

Bailey, J., Oliveri, A., Zhang, C., Frazier, J., Mackinnon, S., Cole, G., Levin, E., 2015. Long-term behavioural impairment following acute embryonic ethanol exposure in zebrafish. Neurotoxicology and Teratology 48, 1-8.

Baraban, S., Taylor, M., Catro, P., Baier, H., 2005. Pentylenetetrazole induced changes in zebrafish behaviour, neural activity and c-fos expression. Neuroscience 131, 759-768.

Bedell, V., Westcot, S., Ekker, S., 2011. Lessons from morpholinos-based screening in zebrafish. Briefings in Functional Genomics 10, 181-188.

Bencan, Z., Sledge, D., Levin, E., 2009. Buspirone, chlordiazepoxide and diazepam effects in a zebrafish model of anxiety. Pharmacology Biochemistry and Behavior 94, 75-80.

Bilotta, J., Banett, J., Hancock, L., Saszik, S., 2004. Ethanol exposure alters zebrafish development: a novel model of fetal alcohol syndrome. Neurotoxicology and Teratology 26, 737-743.

Blader, P., Strahle, U., 1998. Ethanol impairs migration of the prechordal plate in zebrafish embryo. Developmental Biology 201, 185-201.

Brennan, D., Giles, S., 2013. Sonic hedgehog expression is disrupted following in ovo ethanol exposure during early chick eye development. Reproductive Toxicology 41, 49-56.

Burd, L., Martsolf, J., 1989. Fetal alcohol syndrome: diagnosis and syndromal variability. Physiology and Behavior 46,39-43.

Burton, D., Zhang, C., Boa-Amponsem, O., Mackinnon, S., Cole, G., 2017. Long-term behavioural change as a result of acute ethanol exposure in zebrafish: evidence for a role for sonic hedgehog but not retinoic acid signaling. Neurotoxicology and Teratology 61, 66-73.

Buske, C., Gerlai, R., 2011. Early embryonic ethanol exposure impairs shoaling and the dopaminergic and serotoninergic systems in adult zebrafish. Neurotoxicology and Teratology 33, 698-707.

Buske, C., Gerlai, R., 2012. Maturation of shoaling behaviour is accompanied by changes in the dopaminergic and serotoninergic systems in adult zebrafish. Developmental Psychobiology 54, 28-35.

Bussmann, J., Schulte-Merker, S., 2011. Rapid BAC selection for tol2-mediated transgenesis in zebrafish. Development 138, 4327-4332.

Carvan, M., Loucks, E., Weber, D., Williams, F., 2004. Ethanol effects on the developing zebrafish: neurobehaviour and skeletal morphogenesis. Neurotoxicology and Teratology 26, 757-768.

Chatterjee, D., Tran, S., Shams, S., Gerlai, R., 2015. A simple method for immunohistochemical staining of zebrafish brain sections for c-fos protein expression. Zebrafish 12, 414-420.

Clark, M.E., Gibbard, W.B., 2013. Overview of fetal alcohol spectrum disorders for mental health professionals. Canadian Child and Adolescent Psychiatry Review 12, 57-63.

Coffey, C., Solleveld, P., Fang, J., Roberts, A., Hong, S., Dawid, I., Laverriere, C., Glasgow, E., 2013. Novel oxytocin gene expression in the hindbrain is induced by alcohol exposure: transgenic zebrafish enable visualization of sensitive neurons. PLoS One 8, e53991.

De Bruijn, E., Cuppen, E., Feitsma, H., 2009. Highly efficient ENU mutagenesis in zebrafish. Methods in Molecular Biology 546, 3-12.

Deltour, L., Ang, H., Duester, G., 1996. Ethanol inhibition of retinoic acid synthesis as a potential mechanism for fetal alcohol syndrome. The FASEB Journal 10, 1050-1057.

Dlugos, C., Rabin, R., 2007. Ocular deficits associated with alcohol exposure during zebrafish development. Journal of Comparative Neurology 503, 497-506.

Du, Y., Guo, Q., Shan, M., Wu, Y., Huang, S., Zhao, H., Hong, H., et al., 2016. Spatial and temporal distribution of dopaminergic neurons during development in zebrafish. Frontiers in Neuroanatomy 10, 115.

Easey, K., Dyer, M., Timpson, N., Munafo, M., 2019. Prenatal alcohol exposure and offspring mental health: a systematic review. Drug and Alcohol Dependence 197, 344-353.

Facciol, A., Tsang, B., Gerlai, R., 2019. Alcohol exposure during embryonic development: an opportunity to conduct systematic developmental time course analyses in zebrafish. Neuroscience & Biobehavioral Reviews 98m, 185-193.

Feiner, N., Meyer, A., Kuraku, S., 2014. Evolution of the vertebrate Pax4/6 class of genes with focus on its novel member, the Pax10gene. Genome Biology and Evolution 6, 1635-1651.

Ferdous, J., Mukherjee, R., Ahmed, K., Ali, D., 2017. Retinoic acid prevents synaptic deficiencies induced by alcohol exposure during gastrulation in zebrafish embryos. Neurotoxicology 62, 100-110.

Fernandes, Y., Gerlai, R., 2009. Long-term behavioural changes in response to early developmental exposure to ethanol in zebrafish. Alcoholism: Clinical and Experimental Research 33, 601-609.

Fernandes, Y., Rampersad, M., Gerlai, R., 2015a. Impairment of social behaviour persists two years after embryonic alcohol exposure in zebrafish: a model of fetal alcohol spectrum disorders. Behavioural Brain Research 292,102-108.

Fernandes, Y., Rampersad, M., Gerlai, R., 2015b. Embryonic alcohol exposure impairs the dopaminergic system and social behavioural responses in adult zebrafish. International Journal of Neuropsychopharmacology 18, pyu089.

Fernandes, Y., Buckley, D., Eberhard, J., 2018a. Diving into the world of alcohol teratogenesis: a review of zebrafish models of fetal alcohol spectrum disorder. Biochemistry and Cell Biology 96, 88-97.

Fernandes, Y., Rampersad, M., Jones, E., Eberhard, J., 2018b. Social deficits following embryonic ethanol exposure arise in post-larval zebrafish. Addiction Biology 24, 898-907.

Fernandes, Y., Rampersad, Y., Eberhart, J., 2019. Social behavioural phenotyping of the zebrafish casper mutant following embryonic alcohol exposure. Behavioural Brain Research 356, 46-50.

Fernandes, Y., Tran, S., Abraham, E., Gerlai, R., 2014. Embryonic alcohol exposure impairs associative learning performance in adult zebrafish. Behavioural Brain Research 265, 181-187.

Gentsch, G., Spruce, T., Monteiro, R., Owens, N., Matrin, S., Smith, J., 2018. Innate immune response and off-target mis-splicing are common morpholinos-induced side effects of Xenopus. Developmental Cell 44, 597-610.

Gerlai, R., Lahav, M., Guo, S., Rosenthal, A., 2000. Drinks like a fish: zebra fish (*Danio rerio*) as a behaviour genetic model to study alcohol effects. Pharmacology Biochemistry and Behavior 67, 773-782.

Goodlett, C., Horn, K., Zhou, F., 2005. Alcohol teratogenesis: mechanisms of damage and strategies for intervention. Experimental Biology and Medicine 230, 394-406.

Graham, J., Clark, R., Moeschler, J., Rogers, R., 2010. Behavioural features in young adults with FG syndrome (opitzkaveggia syndrome). American Journal of Medical Genetics Part C Seminars in Medical Genetics 154, 477-485.

Haab Lutte, A., HuppesMajolo, J., Reali Nazario, L., Da Silva, R., 2018. Early exposure to ethanol is able to affect the memory of adult zebrafish: possible role of adenosine. Neurotoxicology 69, 17-22.

Haffter, P., Granato, M., Brand,M., Mullins, M., Hammerschmidt, M., Kane, D., Odenthal, J., et al., 1996. The identification of genes with unique essential functions in the development of the zebrafish, *Danio rerio*. Development 123, 1-36.

Hammerschmidt, M., Brook, A., McMahon, A., 1997. The world according to hedgehog. Trends in Genetics 13,14-21.

Hauptmann, G., Lauter, G., Soll, I., 2016. Detection and signal amplification in zebrafish RNA FISH. Methods 98,50-59.

Holden, L., Brown, K., 2018. Baseline mRNA expression differs widely between common laboratory strains of zebrafish, 8, 4780.

Hong, S., Haidin, C., Lawson, N., Weinstein, B., Dawid, I., Hukriede, N., 2005. The zebrafish kohtalo/trap230 gene is required for the development of the brain, neural crest, and pronephric kidney. Proceedings of the National Academy of Sciences of the United States of America 102, 18473-18478.

Howe, K., Clark, M., Torroja, C., Torrance, J., Berthelot, C., et al., 2013. The zebrafish reference genome sequence and its relationship to the human genome. Nature 496, 498-503.

Jinek, M., Chylinski, K., Fonfara, I., Hauer, M., Dondna, J., Charpentier, E., 2012. A programmable dual-RNA-guided DNA endonuclease in adaptive bacterial immunity. Science 337, 816-821.

Jones, K., 1975. The fetal alcohol syndrome. Addictive Diseases 2, 79-88.

Jones, K.L., Smith, D.W., Ulleland, C.N., Streissguth, P., 1973. Pattern of malformation in offspring of chronic alcoholic mothers. Lancet 1, 1267-1271.

Kimmel, C., Ballard, W., Kimmel, S., Ullmann, B., Schilling, T., 1995. Stages of embryonic development of the zebrafish. Developmental Dynamics 203, 253-310.

Kovacs, K., 2008. Measurement of immediate-early gene activation-c-fos and beyond. Journal of Neuroendocrinology 20, 665-672.

Kuhn, E., Koster, R., 2010. Analysis of gene expression by in situ hybridization on adult zebrafish brain sections. Cold Spring Harbour Protocols pdb.prot5382.

Li, S., Giardino, W., de Lecea, L., 2017. Hypocretins and arousal. Current Topics in Behavioral Neurosciences 33, 93-104.

Li, X., Yang, H., Zdanowicz, M., Sicklick, J., Qi, Y., Camp, T., Diehl, A., 2007. Fetal alcohol exposure impairs Hedgehog cholesterol modification and signaling. Laboratory Investigation 87, 231-240.

Lovley, C., Fernandes, Y., Eberhart, J., 2016. Fishing for fetal alcohol spectrum disorder: zebrafish as a model for ethanol teratogenesis. Zebrafish 13, 391-398.

Lu, J., Greco, M., Shiromani, P., Saper, C., 2000. Effect of lesions of the ventrolateral preoptic nucleus in NREM and REM sleep. Journal of Neuroscience 20, 3830-3842.

Mahabir, S., Chatterjee, D., Gerlai, R., 2018b. Short exposure to low concentrations of alcohol during embryonic development has only subtle and strain-dependent effect on the levels of five amino acid neurotransmitters in zebrafish. Neurotoxicology and Teratology 68, 91-96.

Mahabir, S., Chatterjee, D., Buske, C., Gerlai, R., 2013. Maturation of shoaling in two zebrafish strains: a behavioural and neurochemical analysis. Behavioural Brain Research 247, 1-8.

Mahabir, S., Chatterjee, D., Gerlai, R., 2014. Strain dependent neurochemical changes induced by embryonic alcohol exposure in zebrafish. Neurotoxicology and Teratology 41, 1-7.

Mahabir, S., Chatterjee, D., Misquitta, K., Chatterjee, D., Gerlai, R., 2018a. Lasting changes induced by mild alcohol exposure during embryonic development in brain derived neurotrophic factor, neuronal cell adhesion molecule and synaptophysin positive neurons quantified in adult zebrafish. European Journal of Neuroscience 47, 1457-1473.

McCarthy, N., Wetherill, L., Lovely, C., Swartz, M., Foroud, T., Eberhart, J., 2013. Pdgfra protects against ethanol-induced craniofacial defects in a zebrafish model of FASD. Development 140, 3254-3265.

McHugh, R.K., Wigderson, S., Greenfield, S.F., 2014. Epidemiology of substance use in reproductive-age women. Obstetrics and Gynecology Clinics of North America 41, 177-189.

Michaelis, E., 1990. Fetal alcohol exposure: cellular toxicity and molecular events involved in toxicity. Alcoholism:Clinical and Experimental Research 14, 819-826.

Nasevicius, A., Ekker, S., 2000. Effective targeted gene “knockdown” in zebrafish. Nature Genetics 26, 216-220.

Nowicki, M., Tran, S., Muraleetharan, A., Markovic, S., Gerlai, R., 2014. Serotonin antagonist induce anxiolytic and anxiogenic-like behaviour in zebrafish in a receptor-subtype dependent manner. Pharmacology Biochemistry and Behavior 126, 170-180.

Parker, M., Annan, L., Kanellopoulos, A., Brock, A., Combe, F., Baiamonte, M., The, M., Brennan, C., 2014. The utility of zebrafish to study the mechanisms by which ethanol affects social behaviour and anxiety during early brain development. Progress in Neuro-Psychopharmacology and Biological Psychiatry 55, 94-100.

Patten, A., Fontaine, C., Christie, B., 2014. A comparison of the different animal models of fetal alcohol spectrum disorders and their use in studying complex behaviours. Frontiers in Pediatrics 2, 93.

Pauli, A., Montague, T., Lennox, K., Behlke, M., Schier, A., 2015. Antisense oligonucleotide-mediated transcript knockdown in zebrafish. PLoS One 10, e139504.

Pologruto, T., Yasuda, R., Svoboda, K., 2004. Monitoring neural activity and [Ca^{2+}] with genetically encoded Ca^{2+} indicators. Journal of Neuroscience 24, 9572-9579.

Prober, D., Rihel, J., Onah, A., Sung, R., Schier, A., 2006. Hypocretin/orexin overexpression induces an insomnia-like phenotype in zebrafish. Journal of Neuroscience 26, 13400-13410.

Randlett, O., Wee, C., Naumann, E., Nnaemeka, O., Schoppik, D., Fitzgeral, J., Portugues, R., Lacoste, A., Riegler, C.,Engert, F., Schier, A., 2015. Whole-brain activity mapping onto a zebrafish brain atlas. Nature Methods 12,1039-1046.

Rihel, J., Prober, D., Arvanites, A., Lam, K., Zimmerman, S., Jang, S., Haggarty, S., et al., 2010. Zebrafish behavioural profiling links drugs to biological targets and rest/wake regulation. Science 327, 348-351.

Saif, M., Chatterjee, D., Buske, C., Gerlai, R., 2013. Sight of conspecific images induces changes in neurochemistry in zebrafish. Behavioural Brain Research 243, 294-299.

Scerbina, T., Chatterjee, D., Gerlai, R., 2012. Dopamine receptor antagonism disrupts social preference in zebrafish: a strain comparison study. Amino Acids 43, 2059-2072.

Schauerte, H., van Eeden, F., Fricke, C., Odenthal, J., Strahle, U., Haffter, P., 1998. Sonic hedgehog is not required for the induction of medial floor plate cells in zebrafish. Development 125, 2983-2993.

Schulte-Merker, S., Stainier, D., 2014. Out with the old, in with the new: reassessing morpholinos knockdowns in light of genome editing technology. Development 414, 3103-3104.

Shibley, I., McIntyre, T., Pennington, S., 1999. Experimental models used to measure direct and indirect ethanol teratogenicity. Alcohol and Alcoholism 34, 125-140.

Simat, M., Parpan, F., Fritschy, J., 2007. Heterogeneity of glycinergic and gabaergic interneurons in the granule cell later of mouse cerebellum. Journal of Comparative Neurology 500, 71-83.

Sokol, R., Delaney-Black, V., Nordstrom, B., 2003. Fetal alcohol spectrum disorder. Journal of the American Medical Association 290, 2996-2999.

Soriano, P., 1997. The PDGF alpha receptor is required for neural crest cell development and for normal patterning of the somites. Development 124, 2691-2700.

Stenkamp, D., Frey, R., Mallory, D., Shupe, E., 2002. Embryonic retinal gene expression in sonic-you mutant zebrafish. Developmental Dynamics 225, 334-350.

St John, J., Key, B., 2012. HuC-eGFP mosaic labeling of neurons in zebrafish enables in vivo live cell imaging of growth cones. Journal of Molecular Histology 43, 615-623.

Sterling, M., Chang, G., Karatayev, S., Chang, S., Leibowitz, S., 2016. Effects of embryonic ethanol exposure at low doses on neuronal development, voluntary ethanol consumption and related behaviours in larval and adult zebrafish: role of hypothalamic orexigenic peptides. Behavioural Brain Research 304, 125-138.

Strachan, T., Read, A., 1994. PAX genes. Current Opinion in Genetics and Development 4, 427-438.

Streissguth, A., 1997. Fetal Alcohol Syndrome: A Guide for Families and Communities. Brookes Publishing, Baltimore. Streissguth, A., Dehaene, P., 1993. Fetal alcohol syndrome of alcoholic mothers: concordance of diagnosis and IQ. American Journal of Medical Genetics 47, 857-861.

Streissguth, A.P., Herman, C.S., Smith, D.W., 1978. Intelligence, behaviour, and dysmorphogenesis in the fetal alcohol syndrome: a report on 20 patients. The Journal of Pediatrics 92, 363-367.

Swartz, M., Wells, M., Griffin, M., McCarthy, N., Lovely, C., McGurk, P., Rozacky, J., Eberhart, J., 2014. A screen of zebrafish mutants identifies ethanol-sensitive genetic loci. Alcoholism: Clinical and Experimental Research 38,694-703.

Tallquist, M., Kazlauskas, A., 2004. PDGF signaling in cells and mice. Cytokine and Growth Factor Reviews 15,205-213.

Thisse, C., Thisse, B., 2008. High-resolution in situ hybridization to whole-mount zebrafish embryos. Nature Protocols 3, 59-69.

West, J., 1987. Fetal alcohol-induced brain damage and the problem of determining temporal vulnerability: a review. Alcohol and Drug Research 7, 423-441.

Yahyavi, M., Abouzeid, H., Gawdat, G., de Preux, A., Xiao, T., Bardakjian, T., Schneider, A., et al., 2013. ALDH1A3 loss of function causes bilateral anophthalmia/microphthalmia and hypoplasia of the optic nerve and optic chiasm. Human Molecular Genetics 22, 2350-3258.

Yu, S., Qualls-Creekmore, E., Rezai-Zadeh, K., Jiang, Y., Berthoud, H., Morrison, C., Derbenev, A., Zsombok, A., Munzberg, H., 2016. Glutamatergic preoptic area neurons that express leptin receptors drive temperature-dependent body weight homeostasis. Journal of Neuroscience 36, 5034-5046.

Zhang, C., Anderson, A., Cole, G., 2015. Analysis of crosstalk between retinoic acid and sonic hedgehog pathways following ethanol exposure in embryonic zebrafish. Birth Defects Research Part A: Clinical and Molecular Teratology 103, 1046-1057.

Zhang, C., Boa-Amponsem, O., Cole, G., 2017. Comparison of molecular marker expression in early zebrafish brain development following chronic ethanol or morpholinos treatment. Experimental Brain Research 235, 2413-2423.

Zhang, C., Frazier, J., Chen, H., Liu, Y., Lee, J., Cole, G., 2014. Molecular and morphological changes in zebrafish following transient ethanol exposure during defined developmental stages. Neurotoxicology and Teratology 44, 70-80.

Zhang, C., Turton, Q., Mackinnon, S., Sulik, K., Cole, G., 2011. Agrin function associated with ocular development is a target of ethanol exposure in embryonic zebrafish. Birth Defects Research Part A: Clinical and Molecular Teratology 91, 129-141.

第二十二章

斑马鱼多巴胺能神经元的存活、死亡和再生

Khang Hua[1], Marc Ekker[1]

22.1 引言

脊椎动物的基因和基因通路是具有保守性的，但硬骨鱼受伤后细胞、组织和器官再生的能力明显超过哺乳动物，这是非常令人惊讶的。因此人们对鱼类，特别是斑马鱼（*Danio rerio*）的再生研究可能会有助于识别鱼类对损伤后细胞损失作出反应的细胞和分子机制的关键差异，对治疗神经退行性疾病（如帕金森病）有所帮助。

由于再生过程的复杂性，要取代那些损失的细胞需要进行适当的发育和分化程序，这对中枢神经系统的再生提出了特殊的挑战。细胞须准确定位以建立适当的连接，以重建损失的神经回路。在进行比较研究时，不同脊椎动物大脑的解剖差异可能会带来额外的困难。目前，对鱼脑中特定的多巴胺能（DA）神经元群与海马黑质 / 腹侧被盖区（帕金森病患者DA神经元损失的主要部位）之间的细胞和功能的直接相关性缺乏理论论证。此外，与哺乳动物相比，成年鱼大脑中持续的神经发生可能对杀死DA神经元的实验策略造成妨碍，这是研究神经元再生的必要步骤。

在本章中，我们总结了斑马鱼DA神经元发育的解剖学和分子机制研究进展、目前用于触发DA神经元丢失的一些实验方法，以及DA神经元再生潜能的可用证据。

22.2 斑马鱼DA神经元的发育

斑马鱼和其他脊椎动物一样，拥有DA神经元，这些神经元的功能之一是控制运动。

[1] Department of Biology, University of Ottawa, Ottawa, ON, Canada.

研究斑马鱼 DA 神经元，主要是采用酪氨酸羟化酶（tyrosine hydroxylase, TH）免疫组织化学方法对 DA 神经元的定位进行研究。TH 是多巴胺合成的第一步骤的催化酶和限速酶。由于 TH 参与多巴胺和去甲肾上腺素的合成，DA 神经元的特点是缺乏多巴胺 β- 羟化酶，该酶可将多巴胺转化为去甲肾上腺素，使其成为去甲肾上腺素能神经元的标记。尽管多巴胺转运体 *dat*（*slc6a3*）并不是在所有 DA 神经元中都存在，但其免疫组化却是多巴胺神经元的另一个标记（Holzschuh et al., 2001; Xi et al., 2011）。

在斑马鱼胚胎和仔鱼的发育过程中，DA 神经元分布在视网膜、嗅球、大脑皮质、视前区、前丘脑、腹侧间脑、背侧前顶盖及后脑（图 22-1）（Rink and Wullimann, 2002; McLean and Fetcho, 2004; Kastenhuber et al., 2010; Schweitzer et al., 2012）。腹侧间脑的 DA 神经元因其与哺乳动物黑质的同源性而备受关注（Rink and Wullimann, 2001；综述见 Smeets and González, 2000）。这些间脑神经元从后结节（posterior tuberculum）延伸到下丘脑，从 DC1 到 DC7 排列成一组。有丝分裂后的第一批 DA 神经元是 DC2 神经元，发生在受精后 16 ～ 19h（hpf）左右（Holzschuh et al., 2001; Filippi et al., 2012; Schweitzer et al., 2012; Du et al., 2016）。DC1 和 DC4/5 在这些早期阶段也被检测到，但显示出不太均匀的外观。DC3 与 DC6 在受精后第 3 天可以首次检测到。DC7 出现在受精后第 4 天（Du et al., 2016）。因此，到受精后 3 天时，腹侧间脑内大部分 DA 神经元细胞群已经形成。

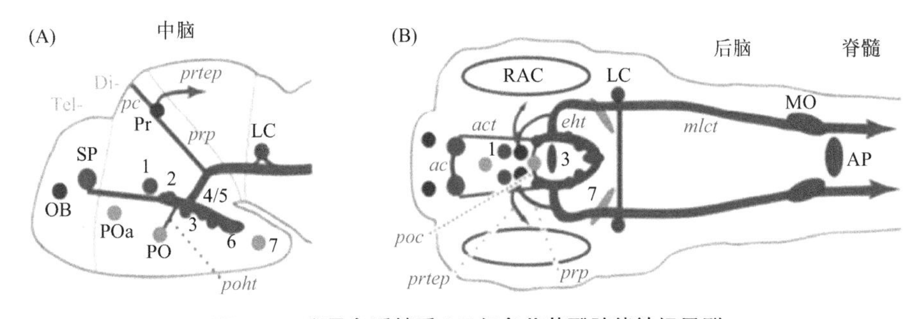

图 22-1 **斑马鱼受精后 4 天仔鱼儿茶酚胺能神经元群**

（A）侧视图；（B）背视图。间脑多巴胺能神经元群编号为 1 ～ 7，文中称为 DC 1 ～ 7。AP（area postrema）：第四脑室腹侧面极后区；LC（locus coeruleus）：蓝斑；MO（medulla oblongata）：延髓；OB（olfactory bulb）：嗅球；PO（preoptic region）：视前区；POa（anterior preoptic region）：前视前区；Pr（dorsal pretectum）：背侧前顶盖；RAC（retinal amacrine cells）：视网膜无长突细胞；SP（subpallium）：大脑皮质。主要 CA 轴突束的缩写如下。ac（anterior commissure）：前连合；act（anterior CA tract）：前 CA 束；eht（endohypothalamic tract）：下丘脑内束；mlct（medial longitudinal CA tract）：内侧纵 CA 束；pc（posterior commissure）：后连合［（B）图未显示］；poc（postoptic commissure）：视后连合；poht（preopticohypothalamic tract）：前庭丘脑束；prp（pretectal projections）：顶盖前投射；prtep（pretectotectal projection）：前顶盖投射。参考以下文献并做了修改，Schweitzer, J. et al., 2012. Dopaminergic and noradrenergic circuit development in zebrafish. Developmental Neurobiology 72(3), 256-268

22.3 DA 神经元发育中的信号通路

目前已经利用突变体研究了斑马鱼 DA 神经元发育中经典信号通路的参与。在缺乏转化生长因子 / 淋巴结途径某些成分的突变胚胎中某些特定群中 DA 神经元的数量有所减少（Holzschuh et al., 2003）。典型 Wnt 信号拮抗剂 *dkk1* 的过表达可导致腹侧间脑神经元数量增加，因此 Wnt 信号通路也可能参与 DA 神经元的发育（Russek-Blum et al., 2008）。*dkk1* 过表达效应的时间依赖性表明，典型的 Wnt 信号可能限制 DA 神经元前体的大小。DA 前体表达 Olig2 和 Ngn1 的维持依赖于通过 DeltaA 和 DeltaD 配体的 Notch 信号（Mahler et al., 2010）。到目前为止，现有的证据仍然不足以证明 Shh 和 Fgf8 在腹侧间脑 DA 神经元形成中的作用（Holzschuh et al., 2003）。

22.4 转录因子与 DA 神经元的发育

许多转录因子在哺乳动物 DA 神经元的发育中起着重要作用（表 22-1）。其中包括 *Nurr1*、*Pitx3* 和 *Lmx1b*（Smidt and Burbach, 2007）。它们在斑马鱼 DA 发育中的作用主要是通过使用吗啉反义寡核苷酸（morpholino）的敲低实验来进行研究的（Filippi et al., 2007; Blin et al., 2008; Luo et al., 2008）。斑马鱼的 *Nurr1*、*nr4a2a* 在视前区（preoptic）、顶盖前区（pretectal）和视网膜无长突 DA 神经元中表达，而第二种类似物 *nr4a2b* 只在视前区和视网膜 DA 神经元中表达。吗啉反义寡核苷酸介导的 *nr4a2a* 基因敲低表明，*nr4a2a* 是通常存在 *nr4a2a* 转录本的神经元中 *th* 和 *dat* 表达所必需的。相比之下，敲低 *nr4a2b* 只产生了一个弱表型，这表明 *nr4a2b* 可能不是 DA 神经元分化所必需的。因此，*nr4a2/Nurr1* 在调节晚期分化和神经递质表型的表达方面具有保守作用（Filippi et al., 2007）。在斑马鱼中，*lmx1b.1* 和 *lmx1b.2* 在成熟 DA 神经元中不表达，但在可能包括 DA 神经元前体的间脑区域有表达。敲低 *lmx1b.1* 和 *lmx1b.2* 可导致具有上行投射的间脑多巴胺能簇神经元数量的特异性减少（Filippi et al., 2007）。尽管 Pitx3 转录因子控制着哺乳动物中脑 DA 神经元晚期的分化（Nunes et al., 2003; Smidt, 2004），斑马鱼 *pitx3* 同源基因似乎在任何 DA 组中都没有表达，但可能像 *lmx1b* 基因一样在 DA 神经元前体中表达。敲低 *pitx3* 基因并没有导致斑马鱼 DA 神经元发育的任何损伤（Filippi et al., 2007）。在一项全面的表达研究中，

表 22-1 **多巴胺能神经元发育过程中一些转录因子的表达和功能**

转录因子	鼠	斑马鱼
Nurr1/Nr4a2	E, P	*nr4a2a* E, P
		nr4a2b E
Pitx3	E, P①	E
Lmx1b	E, P	*lmx1b.1* E, P
		Lmx1b.2 E,P
Arx1	E, P②	E, P
Isl1	E, P②	E, P
Otp	E, P	*otpa* E, P
		Otpb E

① Maxwell 等（2005）。

② Andrews 等（2003）。请参阅斑马鱼参考文献。

注：E 表示表达于 DA 神经元 / 前体；P 表示表型 = 功能丧失影响 DA 神经元发育。

我们发现了与每群斑马鱼 DA 神经元相关的一系列转录因子，该研究还揭示了在丘脑前 DA 神经元发育中对 *Arx1* 和 *Isl1* 的严格需求（Filippi et al., 2012）。对 *Orthopedia*（*Otp*）同源结构域编码基因家族的成员也进行了广泛的研究。在 DC2 和 DC4 ～ 6 组斑马鱼 DA 神经元的发育中，*Otp* 是选择性必需的（Ryu et al., 2007）。DA 神经元的发育需要 Arnt2 和 Sim1a 蛋白的作用（Löhr et al., 2009），尽管它们的作用途径似乎与 Otp 蛋白相似。斑马鱼基因组编码了两个 *otp* 类似物 *otpa* 和 *otpb*。缺乏 *otpa* 功能的突变体大大减少了腹侧间脑区的 DA 神经元数量。相比之下，缺乏 *otpb* 的突变体并没有显示 DA 神经元数量的显著减少（Fernandes et al., 2013）。突变体缺乏两种 *otp* 类似物的功能，会出现特定的 DA 神经元群和大量神经内分泌细胞的完全丧失。这些观察结果表明，这两个基因存在部分功能冗余，其中 *otpa* 在腹侧间脑 DA 神经元的发育中起着极其重要的作用。

22.5 DA 神经回路

儿茶酚胺能神经回路的形成发生在受精后 1 ～ 3 d 之间（McLean and Fetcho, 2004; Kastenhuber et al., 2010; Tay et al., 2011）。随着第一个 TH 阳性神经元在受精后 18 ～ 24h 产生，轴突很快开始投射。到受精后 24h，这些纵向轴突从腹侧中脑的神经元投射到中脑、后脑，并进入脊髓（图 22-1）（Kastenhuber et al., 2010）。在受精后 2d 时，DA 神经元数量继续增加，第一个横切投影出现在中脑边界和蓝斑（LC）之间的区域。此外，还可以检测到视前区 DA 细胞与腹侧间脑之间的投射（Kastenhuber et al., 2010）。在受精后 3d 时，细胞核和束的整体模式已经完整了，与成鱼大脑的模式一致。

已经有研究者在细胞分辨率水平下对完整的 DA 神经元投射（“投射组”）进行了描述（Tay et al., 2011）。嗅觉和视网膜 DA 神经元分别在嗅球和视网膜内进行局部投射。下丘脑前部视前区、视前区、下丘脑 DC3 和 DC7 以及丘脑腹侧 DC1 也投射到局部或邻近的脑区。来自 DC2、DC4、DC5 和 DC6 的 *Otp* 依赖性 DA 神经元因其复杂性和投射范围而受到广泛关注。这些群体是下丘脑内束以及后脑和脊髓束的主要贡献者。DC2 和 DC4 DA 神经元随着向下丘脑、顶盖、后脑和脊髓的投射，向下丘脑下皮层发送上行投射。DC5 神经元也像 DC2 和 DC4 一样投射到下丘脑、顶盖、后脑和脊髓。此外，DC5 神经元也向 LC 投射。然而，DC6 DA 神经元不仅优先向下丘脑局部投射，而且也优先向后脑投射。皮层下 DA 神经元在皮层下局部投射，并延伸至对侧皮层下、丘脑和下丘脑（Tay et al., 2011）。

22.6 斑马鱼 DA 神经元缺失模型的研究

斑马鱼在建立与神经退行性疾病相关的 DA 神经元丢失模型中已成为广泛研究的实验对象。在这里，我们将讨论利用神经毒素和化学遗传方法构建 DA 神经元损失的仔鱼和成年斑马鱼模型。

22.6.1 1-甲基-4-苯基-1,2,3,6-四氢吡啶（MPTP）

1-甲基-4-苯基-1,2,3,6-四氢吡啶（1-methyl-4-phenyl-1,2,3,6-tetrahydropyridine, MPTP）是一种广泛用于模拟帕金森病（PD）和DA神经元损失的强效毒素（Meredith and Rademacher, 2011; Tieu, 2011; Hare et al., 2013）。由于MPTP的亲脂性，它可以很容易地穿过血脑屏障（BBB）并在胶质细胞中积累，然后通过单胺氧化酶（MAO）活性代谢为1-甲基-4-苯基吡啶离子（MPP^+）。而后MPP^+通过多巴胺转运体进入DA神经元，并在线粒体中积累，在线粒体中它作为一种有效的复合物Ⅰ抑制剂发挥作用，导致活性氧（ROS）合成增加，随后引起细胞死亡（Przedborski et al., 2004; Hare et al., 2013）。虽然已证明MPTP是诱导非人类灵长类动物和小鼠DA神经元损失最有效的方法（Tieu, 2011），但MPTP对斑马鱼的作用产生了不同的结果。

用MPTP处理斑马鱼仔鱼通常包括在鱼缸水中溶解药物，浓度从0.1～1000μmol/L不等进行暴露实验。通过使用*th*和*dat* RNA探针进行原位杂交评估，显示MPTP处理后间脑DA神经元显著减少（Bretaud et al., 2004; Lam et al., 2005; McKinley et al., 2005; Sallinen et al., 2009）。此外，笔者实验室的研究通过使用转基因品系Tg（*dat: eGFP*）证实了腹侧间脑DA神经元的缺失，这使得DA神经元的数量得以量化（Xi et al., 2011）。在Souza等人（2011）的一项研究中，0.1～1μmol/L的低浓度MPTP可影响间脑的TH^+ DA神经元。在不同的研究中，其DA簇内的神经元损失程度差异很大。位于前脑、视前区、丘脑、下丘脑和嗅球区域的DA神经元簇也对MPTP表现出敏感性（Wen et al., 2008; Sallinen et al., 2009; Souza et al., 2011）。相比之下，仔鱼端脑和LC中都没有DA神经元的损失（Bretaud et al., 2004），或者在LC中只有少量损失（McKinley et al., 2005）。

据报道，在对仔鱼施用MPTP后，伴随着DA神经元的损失，运动功能也出现了障碍。总的来说，经MPTP处理的仔鱼表现出以下特点：游泳速度下降（Bretaud et al., 2004），总行程减少（Sallinen et al., 2009），触摸刺激后躯干运动较弱且持续时间较短（Lam et al., 2005），不动且对触摸缺乏反应（McKinley et al., 2005），以及运动开始减少（Souza et al., 2011）。这些研究中显示的运动能力低下的表型与PD相关的运动症状相似，如运动迟缓（运动缓慢）和无法开始运动（Jankovic 2008; Rosner et al., 2008）。

在成年斑马鱼中，腹腔注射MPTP导致多巴胺水平下降，运动能力明显下降，出现类似帕金森病的症状，如游泳速度和距离下降，以及僵直次数增加（Anichtchik et al., 2004; Bretaud et al., 2004; Sarath Babu et al., 2016）。然而，通过免疫组织化学或Western blotting确定腹腔注射MPTP并没有导致DA神经元或TH蛋白的损失（Anichtchik et al., 2004; Bretaud et al., 2004; Sarath Babu et al., 2016）。尽管没有DA神经元损失，可能是MPTP影响了TH的活性，从而导致多巴胺水平和自发活动水平的下降（Serra et al., 2002; Vaz et al., 2018）。成年斑马鱼对MPTP的敏感性较低，缺乏DA神经元的死亡，这可能是由于相对于仔鱼而言，药物对成年鱼大脑的接触和渗透性有所降低。

22.6.2 1-甲基-4-苯基吡啶离子（MPP^+）

MPP^+是MPTP的代谢物，但与MPTP不同，MPP^+不能轻易穿过哺乳动物的血脑屏障

（BBB）（Perry et al., 1985）。由于 MPP^+ 的极性，其被 DA 神经元吸收需要多巴胺转运体的功能（Dauer and Przedborski, 2003）。在斑马鱼中，血脑屏障的发育发生在受精后 3d 左右（Fleming et al., 2013）。自受精后 1 ～ 4d 用 MPP^+ 处理斑马鱼胚胎 / 仔鱼，导致 DA 神经元减少，但仅表现在间脑内（Lam et al., 2005; Sallinen et al., 2009）。MPP^+ 治疗还降低了总游动距离（salinen et al., 2009）、平均速度和移动时间百分比（Farrell et al., 2011）。与仔鱼的研究相反，成年斑马鱼腹腔注射 MPP^+ 不会导致 DA 神经元损失或运动缺陷（Bretaud et al., 2004）。因此，MPP^+ 有效性的差异可能是由于药物无法通过成年斑马鱼的血脑屏障。

22.6.3 6-羟多巴胺

6- 羟多巴胺（6-hydroxydopamine, 6-OHDA）是一种羟基化多巴胺化合物，可通过多巴胺或去甲肾上腺素转运体进入儿茶酚胺能神经元（Dauer and Przedborski, 2003）。6-OHDA 可致 PD 鼠模型 DA 神经元损失，参见 Bové 和 Perier（2012）的综述。由于 6-OHDA 不能轻易穿过血脑屏障，通常以立体定位方式将其注射进入大脑（Blesa et al., 2012; Vijayanathan et al., 2017）。6-OHDA 还与去甲肾上腺素转运蛋白抑制剂共同使用，以增加对 DA 神经元神经毒性的特异性（Bové and Perier, 2012）。虽然在小鼠、猫、狗、猴子和大鼠等多种哺乳动物中证明 6-OHDA 可以有效地诱导 DA 神经元损失（Blesa et al., 2012），但很少有研究显示斑马鱼 DA 神经元对 6-OHDA 的敏感性。

在对仔鱼进行的一项研究中（Feng et al., 2014），斑马鱼从受精后 2 ～ 4d 暴露于 6-OHDA 浓度为 10 ～ 250μmol/L 的水中 48h。暴露于 6-OHDA 可导致 TH 免疫反应性降低、游泳距离缩短和焦虑样行为增加。通过联合服用维生素 E（一种抗氧化剂）、二甲胺四环素（抗神经炎症）和复方卡比多巴（Sinemet，治疗帕金森病使用最广泛的药物之一），可以预防运动损伤和 DA 神经元的损失。

在成年斑马鱼中，肌内注射 6-OHDA 可导致大脑内多巴胺和去甲肾上腺素浓度的显著降低。然而，TH 免疫组化未发现 DA 神经元数量的任何变化，利用免疫印迹也未发现蛋白质水平的下降。尽管缺乏神经元的损失或 TH 蛋白水平的降低，但经注射的鱼却显示出游动总距离和速度的显著下降和转弯角度的增加（Anichtchik et al., 2004）。与肌内注射相比，将 6-OHDA 靶向脑室内注射到成年斑马鱼的中脑腹侧，可以显著消融嗅球、端脑、视前区、后结节和下丘脑内的 DA 神经元。随着 DA 神经元的减少，总游动距离和平均速度也会减少（Vijayanathan et al., 2017）。因此，在将 6-OHDA 注射进肌肉（Anichtchik et al., 2004）或直接注射进脑室（Vijayanathan et al., 2017）的成年斑马鱼中，6-OHDA 诱导的神经元死亡之间的表型差异可能归因于 6-OHDA 无法穿过血脑屏障。这表明神经毒素直接暴露于成年斑马鱼的大脑可以提高药物诱导神经元细胞死亡的效果。

22.6.4 鱼藤酮

鱼藤酮是一种杀虫剂，目前已证明其与患 PD 的风险升高有关（Tanner et al., 2011）。由于鱼藤酮的亲脂性，它可以很容易地穿过 BBB 并进入细胞，而不需要特定的运输工具（Tieu, 2011）。鱼藤酮的毒性机制是通过抑制线粒体呼吸链复合物Ⅰ。鱼藤酮对复合物Ⅰ

的抑制会诱发线粒体ROS的产生和细胞程序性死亡（Li et al., 2003）。尽管已经证明鱼藤酮暴露在小鼠和大鼠中会诱发路易体包涵体、神经变性和帕金森类似症状（Betarbet et al., 2000; Dauer and Przedborski, 2003; Liu et al., 2015），但在斑马鱼中的研究产生了相矛盾的结果（Vaz et al., 2018）。

一项对斑马鱼仔鱼的研究表明，鱼藤酮暴露不会导致DA神经元缺失或帕金森病类似症状运动功能障碍（Bretaud et al., 2004）。然而，我们发现斑马鱼仔鱼暴露于鱼藤酮6d后可导致间脑DA神经元发生剂量依赖性的损失。处理后的仔鱼在腹侧间脑和嗅球内的细胞数量显著减少，并且出现严重的认知偏差（Martel et al., 2015）。

最初在Bretaud的研究（Bretaud et al., 2004）中，经鱼藤酮处理后，成年斑马鱼未观察到明显的运动功能障碍或DA神经元损失。相反，Wang等人（2017）的一项研究报告称，斑马鱼体内多巴胺和TH水平下降，伴有运动障碍、焦虑和抑郁样行为以及嗅觉功能障碍。

22.6.5 百草枯

百草枯是最主要的商用联吡啶除草剂（Bromilow, 2003）。流行病学研究表明，百草枯的化学结构与MPP^{+}相似，长时间接触百草枯与帕金森病风险增加之间存在联系（Dhillon et al., 2008）。目前证明百草枯可以穿过大鼠血脑屏障并诱导多巴胺能毒性（Shimizu et al., 2001）。

在一项初步研究中，将百草枯溶解于暴露斑马鱼仔鱼的水中，并没有观察到DA神经元和自发活动的变化（Bretaud et al., 2004）。然而，在最近的一项研究中，持续暴露于50%致死浓度（LC_{50}）的百草枯对斑马鱼仔鱼造成了DA神经元的累积损伤，并降低了多巴胺和5-羟色胺神经递质的水平（Nellore and Nandita, 2015）。

成年斑马鱼的百草枯暴露实验则出现了不同的结果。在鱼缸水中添加百草枯并不影响DA神经元的数量，也未能诱导成年斑马鱼运动受损表型（Bretaud et al., 2004）。在另一项研究中，在成年斑马鱼中重复腹腔注射百草枯导致中线穿越次数、移动总距离和平均速度以剂量依赖的方式减少（Bortolotto et al., 2014）。虽然在运动曲线上有显著的变化，但通过Western印迹实验观察到TH蛋白水平没有变化。此外，多巴胺水平升高，多巴胺代谢物3,4-二羟基苯乙酸（DOPAC）下降。*dat*基因的表达也受到了负面影响（Bortolotto et al., 2014）。多巴胺和DOPAC水平的变化表明，百草枯暴露可促进斑马鱼大脑中的多巴胺代谢。这些结果与之前在大鼠研究中观察到的结果相似（Ossowska et al., 2005）。最后，与Bortolotto等人（2014）观察到的运动表型相反，最近的一项研究表明，反复腹腔注射百草枯并不能诱导运动表型（Nunes et al., 2017）变化。相反，将成年斑马鱼反复全身性暴露在百草枯中，通过增加攻击行为和减少防御行为，导致非运动模式的改变。百草枯暴露导致表型的巨大差异可能是由使用的浓度、给药方法和使用的鱼的品系不同所导致的。

22.6.6 硝基还原酶系统对DA神经元的化学消融

由于神经毒素对动物具有毒性，且效果不稳定，作为用神经毒素消融DA神经元的替代方法，一种条件性的和特异性的遗传学细胞消融方法被开发用于消融 DA 神经元。以前的化学和遗传学方法对特定细胞类型的消融有许多局限性：特异性差、效果有限、不可逆、对启动子泄漏的敏感性高、对增殖的细胞的限制和诱导缓慢。基于硝基还原酶（nitroreductase, NTR）表达的化学消融方法允许对特定细胞类型进行条件性和细胞特异性消融，限制较少（Curado et al., 2008）。该技术依赖于在以组织特异性方式表达的基因调控元件的控制下表达的 NTR 酶。也可以采用 Gal4-UAS 双转基因系统。最初在大肠杆菌中发现的 NTR 酶被 NADH 或 NADPH 还原。NTR（以还原形式）催化无害的前药甲硝唑（metronidazole, MTZ）还原为强效的 DNA 交联剂，导致细胞死亡（Curado et al., 2007）。由于 NTR 的表达是细胞特异性的，MTZ 的毒性仅限于表达 NTR 的细胞，没有旁侧效应（Curado et al., 2008）。

NTR 在 DA 神经元中的表达是通过特定表达基因的调控元件实现的。该方法导致了斑马鱼 DA 神经元的有效消融（Fernandes et al., 2012; Lambert et al., 2012; Duncan et al., 2015; Godoy et al., 2015; McPherson et al., 2015）。Fernandes 及其同事使用基于 Gal4-UAS 系统的三个转基因的组合，观察到用 7.5mmol/L MTZ 处理后，DC2 和 DC4 的 DA 神经元被有效消融（Fernandes et al., 2012）。在另一项研究中，*otpb* 调控元件被用来驱动 NTR 在尾部腹侧间脑（caudal ventral diencephalon）的 *otpb* 阳性 DA 神经元中的表达，在那里 *th* 和 *dat* 共同表达（Lambert et al., 2012）。将 1 ～ 4d 龄的斑马鱼暴露于 10mmol/L MTZ，引起腹侧脑内 DA 神经元细胞的明显死亡，表现为 *dat* mRNA 信号和 TH/GFP 免疫活性细胞数量的减少（Lambert et al., 2012）。*th2* 基因的调控元件也被用来驱动 NTR 在 Tg（th2: Gal4-VP16; UAS: NTR-mCherry）双转基因鱼 DA 神经元中的表达。这些鱼在 5 ～ 7 d 龄时暴露于 5mmol/L MTZ，可导致下丘脑表达 *th2* 的 DA 神经元明显减少（Duncan et al., 2015; McPherson et al., 2015），每秒游泳次数减少，游泳时间百分比减少（McPherson et al., 2015）。然而，作者并没有调查其他 $th2^{+}$ DA 神经元簇中 DA 神经元的损失。我们使用 *dat* 基因的调控区域来驱动 NTR 在 DA 神经元中的表达（Godoy et al., 2015）。用 5mmol/L MTZ 处理 1 ～ 2d 龄的仔鱼，用 7.5mmol/L MTZ 处理 2 ～ 4d 龄的仔鱼，导致几个区域的 DA 神经元数量减少，包括嗅球（OB）、背侧前顶盖（Pr）、大脑皮质（SP）、间脑簇 2/3（DC 2/3）和尾部下丘脑（HC）。DA 神经元的减少与多巴胺水平的下降和运动障碍相吻合（Godoy et al., 2015）。因此，一些研究表明，硝基还原酶介导的化学遗传系统在幼年斑马鱼中消减 DA 神经元方面具有有效性，在成年动物中进行的类似研究还有待报道。

22.7 斑马鱼 DA 神经元的再生

成体神经发生的过程，或从成体神经前体产生新的神经元的过程，可以在哺乳动物的一生中发生，但仅限于大脑中的特定区域（Ming and Song, 2011）。尽管在哺乳动物大脑中存在神经干细胞簇，但它们无法再生中枢神经系统细胞来从神经元损失中恢复（Kyritsis et al., 2014; Jessberger, 2016）。与成年哺乳动物有限的神经发生和再生潜能相比，硬骨鱼具有巨大的神经发生和神经再生能力（Zupanc, 2009; Kizil et al., 2012; Kyritsis et al., 2014）。

特别是斑马鱼，由于其广泛的干细胞活性和再生能力，已成为研究成体神经发生和大脑再生的热门实验生物（Kizil et al., 2012）。然而，只有少数研究涉及斑马鱼再生 DA 神经元的能力。此外，控制多巴胺能神经元再生的分子机制尚不清楚。

目前已有研究证明斑马鱼能够在仔鱼（Duncan et al., 2015; Godoy et al., 2015; McPherson et al., 2015）或成鱼（Vijayanathan et al., 2017）的细胞特异性消融后再生 DA 神经元。在成年斑马鱼中通过脑室内注射（ICV）6-OHDA 诱导 DA 神经元损失后，在病变后 30 d 观察到损失的神经元的恢复。从损伤后 14 d 开始，下丘脑（Hyp）、视前区（preoptic area, POA）和后结节（posterior tuberculum, PT）损失的神经元数量减少。损伤后 30 d, POA、PT 和 Hyp 的 DA 神经元数量完全恢复到化学消融前的水平。此外，DA 神经元在斑马鱼的 POA、PT 和 Hyp 中的分布与对照组相当（Vijayanathan et al., 2017）。

在幼年斑马鱼中，在 NTR 诱导的 DA 神经元损失 2 周后，下丘脑内的 DA 神经元出现了惊人的恢复（McPherson et al., 2015）。此外，与对照组相比，观察到新生 DA 神经元的比例增加，这与再生反应是一致的（McPherson et al., 2015）。观察到的 $th2^{+}$DA 神经元的替换也与游泳障碍的恢复有关。

22.8 展望与挑战

斑马鱼 DA 神经元发育的特征表明，硬骨鱼和哺乳动物在信号转导机制和相关因素方面存在很大的保守性。后者包括但不限于一些转录调控因子。在成年斑马鱼的大脑中，是否同样的成分也参与了新 DA 神经元的正常产生和再生的过程，还有待确定。

在斑马鱼仔鱼和成鱼体内，已经通过多种方法实现了 DA 神经元的消融。仍然需要回答的问题包括：在暴露于各种神经毒素和化学消融后，导致 DA 神经元细胞死亡的事件是否相同？导致 DA 神经元死亡的途径是否会对再生机制的触发产生影响还有待确定。

在研究 DA 神经元的再生时，还面临一些挑战。首先，成年斑马鱼大脑中一个或几个神经元区域是否有特定的刺激？参与 DA 神经元再生的基因和信号通路是否与负责 DA 神经元再生的基因和信号通路相同？那些负责胚胎 / 幼体发育的基因和信号通路是否相同？前体迁移到其最终位置的机制是什么？新细胞是否能够成功地整合到神经元回路中？表型分析，包括正常参数的恢复，如与运动有关的参数，表明情况可能是这样，然而，更详细的表型分析也可能揭示出成年斑马鱼与幼年斑马鱼相比在再生能力方面的局限性或不同的再生能力。

致谢

作者实验室的研究得到了加拿大自然科学和工程研究委员会的支持。我们感谢 Gary Hatch 对手稿的评论。

（曹念念翻译　石童审校）

参考文献

Andrews, G.L., Yun, K., Rubenstein, J.L., Mastick, G.S., 2003. Dlx transcription factors regulate differentiation of dopaminergic neurons of the ventral thalamus. Molecular and Cellular Neuroscience 23, 107-120.

Anichtchik, O.V., et al., 2004. Neurochemical and behavioural changes in zebrafish *Danio rerio* after systemic administration of 6-hydroxydopamine and 1-methyl-4-phenyl-1,2,3,6-tetrahydropyridine. Journal of Neurochemistry 88,443-453. https://doi.org/10.1046/j.1471-4159.2003.02190.x.

Betarbet, R., et al., 2000. Chronic systemic pesticide exposure reproduces features of Parkinson's disease. Nature Neuroscience. https://doi.org/10.1038/81834.

Blesa, J., et al., 2012. Classic and new animal models of Parkinson's disease. Journal of Biomedicine and Biotechnology 2012, 1-10. https://doi.org/10.1155/2012/845618.

Blin, M., et al., 2008. NR4A2 controls the differentiation of selective dopaminergic nuclei in the zebrafish brain. Molecular and Cellular Neuroscience 39 (4), 592-604. https://doi.org/10.1016/j.mcn.2008.08.006. Elsevier Inc.

Bortolotto, J.W., et al., 2014. Long-term exposure to paraquat alters behavioral parameters and dopamine levels in adult zebrafish (*Danio rerio*). Zebrafish 11 (2), 142-153. https://doi.org/10.1089/zeb.2013.0923.

Bové, J., Perier, C., 2012. Neurotoxin-based models of Parkinson's disease. Neuroscience 211, 51-76. https://doi.org/10.1016/j.neuroscience.2011.10.057. Elsevier Inc.

Bretaud, S., Lee, S., Guo, S., 2004. Sensitivity of zebrafish to environmental toxins implicated in Parkinson's disease. Neurotoxicology and Teratology 26 (6), 857-864. https://doi.org/10.1016/j.ntt.2004.06.014.

Bromilow, R.H., 2003. Paraquat and sustainable agriculture. Pest Management Science 60 (4), 340-349. https://doi.org/10.1002/ps.823.

Curado, S., et al., 2007. Conditional targeted cell ablation in zebrafish: a new tool for regeneration studies. Developmental Dynamics 236 (4), 1025-1035. https://doi.org/10.1002/dvdy.21100.

Curado, S., Stainier, D.Y., Anderson, R.M., 2008. Nitroreductase-mediated cell/tissue ablation in zebrafish: a spatially and temporally controlled ablation method with applications in developmental and regeneration studies. Nature Protocols 3 (6), 948-954. https://doi.org/10.1038/nprot.2008.58.

Dauer, W., Przedborski, S., 2003. Parkinson's disease: mechanisms and models. Neuron 39, 889-909.

Dhillon, A.S., et al., 2008. Pesticide/environmental exposures and Parkinson's disease in East Texas. Journal of Agromedicine 13 (1), 37-48. https://doi.org/10.1080/10599240801986215.

Du, Y., et al., 2016. Spatial and temporal distribution of dopaminergic neurons during development in zebrafish. Frontiers in Neuroanatomy 10, 1-7. https://doi.org/10.3389/fnana.2016.00115.

Duncan, R.N., et al., 2015. Hypothalamic radial glia function as self-renewing neural progenitors in the absence of Wnt/b-catenin signaling. Development 143, 45-53. https://doi.org/10.1242/dev.126813.

Farrell, T.C., et al., 2011. Evaluation of spontaneous propulsive movement as a screening tool to detect rescue of Parkinsonism phenotypes in zebrafish models. Neurobiology of Disease 44 (1), 9-18. https://doi.org/10.1016/j.nbd.2011.05.016. Elsevier B.V.

Feng, C.-W., et al., 2014. Effects of 6-hydroxydopamine exposure on motor activity and biochemical expression in zebrafish (*Danio rerio*) larvae. Zebrafish 11 (3), 227-239. https://doi.org/10.1089/zeb.2013.0950.

Fernandes, A.M., et al., 2012. Deep brain photoreceptors control light-seeking behavior in zebrafish larvae. Current Biology 22 (21), 2042-2047. https://doi.org/10.1016/j.cub.2012.08.016.

Fernandes, A.M., et al., 2013. Orthopedia transcription factor otpa and otpb paralogous genes function during dopaminergic and neuroendocrine cell specification in larval zebrafish. PLoS One. https://doi.org/10.1371/journal.pone.0075002.

Filippi, A., et al., 2007. Expression and function of nr4a2, lmx1b, and pitx3 in zebrafish dopaminergic and noradrenergic neuronal development. BMC Developmental Biology 7, 1-21. https://doi.org/10.1186/1471-213X-7-135.

Filippi, A., Jainok, C., Driever, W., 2012. Analysis of transcriptional codes for zebrafish dopaminergic neurons reveals essential functions of Arx and Isl1 in prethalamic dopaminergic neuron development. Developmental Biology. https://doi.org/10.1016/j.ydbio.2012.06.010.

Fleming, A., Diekmann, H., Goldsmith, P., 2013. Functional characterisation of the maturation of the blood-brain barrier in larval zebrafish. PLoS One 8 (10), 1-12. https://doi.org/10.1371/journal.pone.0077548.

Godoy, R., et al., 2015. Chemogenetic ablation of dopaminergic neurons leads to transient locomotor impairments in zebrafish larvae. Journal of Neurochemistry 135 (2), 249-260. https://doi.org/10.1111/jnc.13214.

Hare, D.J., et al., 2013. Metallobiology of 1-methyl-4-phenyl-1,2,3,6-tetrahydropyridine neurotoxicity. Metallomics 5 (2), 91-109.

https://doi.org/10.1039/c2mt20164j.

Holzschuh, J., et al., 2001. Dopamine transporter expression distinguishes dopaminergic neurons from other catecholaminergic neurons in the developing zebrafish embryo. Mechanisms of Development 101 (1e2), 237-243. https://doi.org/10.1016/S0925-4773(01)00287-8.

Holzschuh, J., Hauptmann, G., Driever, W., 2003. Genetic analysis of the roles of Hh, FGF8, and nodal signaling during catecholaminergic system development in the zebrafish brain. Journal of Neuroscience 23 (13), 5507-5519. https://doi.org/10.1523/jneurosci.23-13-05507.2003.

Jankovic, J., 2008. Parkinson's disease: clinical features and diagnosis. Journal of Neurology, Neurosurgery and Psychiatry 79 (4), 368-376. https://doi.org/10.1136/jnnp.2007.131045.

Jessberger, S., 2016. Stem cell-mediated regeneration of the adult brain. Transfusion Medicine and Hemotherapy 43(5), 321-326. https://doi.org/10.1159/000447646.

Kastenhuber, E., et al., 2010. Genetic dissection of dopaminergic and noradrenergic contributions to catecholaminergic tracts in early larval zebrafish. The Journal of Comparative Neurology 518 (4), 439-458. https://doi.org/10.1002/cne.22214.

Kizil, C., et al., 2012. Adult neurogenesis and brain regeneration in zebrafish. Developmental Neurobiology 72 (3),429-461. https://doi.org/10.1002/dneu.20918.

Kyritsis, N., Kizil, C., Brand, M., 2014. Neuroinflammation and central nervous system regeneration in vertebrates. Trends in Cell Biology 24 (2), 128-135. https://doi.org/10.1016/j.tcb.2013.08.004. Elsevier Ltd.

Lam, C.S., Korzh, V., Strahle, U., 2005. Zebrafish embryos are susceptible to the dopaminergic neurotoxin MPTP. European Journal of Neuroscience 21 (6), 1758-1762. https://doi.org/10.1111/j.1460-9568.2005.03988.x.

Lambert, A.M., Bonkowsky, J.L., Masino, M.A., 2012. The conserved dopaminergic diencephalospinal tract mediates vertebrate locomotor development in zebrafish larvae. Journal of Neuroscience 32 (39), 13488e13500. https://doi.org/10.1523/jneurosci.1638-12.2012.

Li, N., et al., 2003. Mitochondrial complex I inhibitor rotenone induces apoptosis through enhancing mitochondrial reactive oxygen species production. Journal of Biological Chemistry 278 (10), 8516-8525. https://doi.org/10.1074/jbc.M210432200.

Liu, Y., et al., 2015. Environment-contact administration of rotenone: a new rodent model of Parkinson's disease. Behavioural Brain Research. https://doi.org/10.1016/j.bbr.2015.07.058.

Löhr, H., Ryu, S., Driever, W., 2009. Zebrafish diencephalic A11-related dopaminergic neurons share a conserved transcriptional network with neuroendocrine cell lineages. Development 136 (6), 1007-1017.https:// doi.org/10.1242/dev.033878.

Luo, G.R., et al., 2008. Nr4a2 is essential for the differentiation of dopaminergic neurons during zebrafish embryogenesis. Molecular and Cellular Neuroscience 39 (2), 202-210. https://doi.org/10.1016/j.mcn.2008.06.010.

Mahler, J., Filippi, A., Driever, W., 2010. DeltaA/DeltaD regulate multiple and temporally distinct phases of notch signaling during dopaminergic neurogenesis in zebrafish. Journal of Neuroscience 30 (49), 16621-16635. https://doi.org/10.1523/jneurosci.4769-10.2010.

Martel, S., Keow, J.Y., Ekker, M., 2015. Rotenone neurotoxicity causes dopamine neuron loss in zebrafish. University of Ottawa Journal of Medicine 5 (2), 16-21.

Maxwell, S.L., Ho, H.Y., Kuehner, E., Zhao, S., Li, M., 2005. Pitx3 regulates tyrosine hydroxylase expression in the substantia nigra and identifies a subgroup of mesencephalic dopaminergic progenitor neurons during mouse development. Developmental Biology 282, 467-479.

McKinley, E.T., et al., 2005. Neuroprotection of MPTP-induced toxicity in zebrafish dopaminergic neurons. Molecular Brain Research 141 (2), 128-137. https://doi.org/10.1016/j.molbrainres.2005.08.014.

McLean, D.L., Fetcho, J.R., 2004. Ontogeny and innervation patterns of dopaminergic, noradrenergic, and serotonergic neurons in larval zebrafish. Journal of Comparative Neurology 480 (1), 38-56. https://doi.org/10.1002/cne.20280.

McPherson, A.D., et al., 2015. Motor behavior mediated by continuously generated dopaminergic neurons in the zebrafish hypothalamus recovers after cell ablation. Current Biology 26 (2), 263-269. https://doi.org/10.1016/j.cub.2015.11.064. Elsevier.

Meredith, G.E., Rademacher, D.J., 2011. MPTP mouse models of Parkinson's disease: an update. Journal of Parkinson's Disease 1 (1), 19-33. https://doi.org/10.3233/JPD-2011-11023.

Ming, G. li, Song, H., 2011. Adult neurogenesis in the mammalian brain: significant answers and significant questions. Neuron 70 (4), 687-702. https://doi.org/10.1016/j.neuron.2011.05.001. Elsevier Inc.

Nellore, J., Nandita, P., 2015. Paraquat exposure induces behavioral deficits in larval zebrafish during the window of dopamine neurogenesis. Toxicology Reports 2, 950-956. https://doi.org/10.1016/j.toxrep.2015.06.007. Elsevier Ireland Ltd.

Nunes, I., et al., 2003. Pitx3 is required for development of substantia nigra dopaminergic neurons. Proceedings of the National Academy of Sciences of the United States of America 100 (7), 4245-4250. https://doi.org/10.1073/pnas.0230529100.

Nunes, M.E., et al., 2017. Chronic treatment with paraquat induces brain injury, changes in antioxidant defenses system, and modulates behavioral functions in zebrafish. Molecular Neurobiology 54 (6), 3925-3934. https://doi.org/10.1007/s12035-016-9919-x.

Ossowska, K., Wardas, J., Kuter, K., Nowak, P., Dabrowska, J., Bortel, A., Labus, Ł., Kwiecinski, A., Krygowska-Wajs, A., Wolfarth, S., 2005. Influence of paraquat on dopaminergic transporter in the rat brain. Pharmacological Reports 57 (3), 330-335.

Perry, T.L., et al., 1985. Effects of N-methyl-4-phenyl-1, 2, 3, 6-tetrahydropyridine and its metabolite, N-methyl-4-phenylpyridinium ion, on dopaminergic nigrostriatal neurons in the mouse. Neuroscience Letters 58, 321-326.

Przedborski, S., et al., 2004. MPTP as a mitochondrial neurotoxic model of Parkinson's disease. Journal of Bioenergetics and Biomembranes 36 (4), 375-379. https://doi.org/10.1023/B:JOBB.0000041771.66775.d5.

Rink, E., Wullimann, M.F., 2001. The teleostean (zebrafish) dopaminergic system ascending to the subpallium (striatum) is located in the basal diencephalon (posterior tuberculum). Brain Research 889 (1-2), 316-330. https://doi.org/10.1016/S0006-8993(00)03174-7.

Rink, E., Wullimann, M.F., 2002. Development of the catecholaminergic system in the early zebrafish brain: an immunohistochemical study. Developmental Brain Research 137 (1), 89-100. https://doi.org/10.1016/S0165-3806(02)00354-1.

Rosner, S., Giladi, N., Orr-Urtreger, A., 2008. Advances in the genetics of Parkinson's disease. Acta Pharmacologica Sinica 29 (1), 21-34. https://doi.org/10.1111/j.1745-7254.2008.00731.x.

Russek-Blum, N., et al., 2008. Dopaminergic neuronal cluster size is determined during early forebrain patterning. Development 135, 3401-3413. https://doi.org/10.1242/dev.024232.

Ryu, S., et al., 2007. Orthopedia homeodomain protein is essential for diencephalic dopaminergic neuron development. Current Biology 17 (10), 873-880. https://doi.org/10.1016/j.cub.2007.04.003. Cell Press.

Sallinen, V., et al., 2009. MPTP and MPPþ target specific aminergic cell populations in larval zebrafish. Journal of Neurochemistry 108 (3), 719-731. https://doi.org/10.1111/j.1471-4159.2008.05793.x.

Sarath Babu, N., et al., 2016. 1-Methyl-4-phenyl-1,2,3,6-tetrahydropyridine induced Parkinson's disease in zebrafish. Proteomics 16 (9), 1407-1420. https://doi.org/10.1002/pmic.201500291.

Schweitzer, J., et al., 2012. Dopaminergic and noradrenergic circuit development in zebrafish. Developmental Neurobiology 72 (3), 256-268. https://doi.org/10.1002/dneu.20911.

Serra, P.A., et al., 2002. The neurotoxin 1-methyl-4-phenyl-1,2,3,6-tetrahydropyridine induces apoptosis in mouse nigrostriatal glia. Relevance to nigral neuronal death and striatal neurochemical changes. Journal of Biological Chemistry 277 (37), 34451-34461. https://doi.org/10.1074/jbc.M202099200.

Shimizu, K., et al., 2001. Carrier-mediated processes in blood-brain barrier penetration and neural uptake of paraquat. Brain Research 906 (1-2), 135-142. https://doi.org/10.1016/S0006-8993(01)02577-X.

Smeets, W.J.A.J., González, A., 2000. Catecholamine systems in the brain of vertebrates: new perspectives through a comparative approach. Brain Research Reviews 33 (2-3), 308-379. https://doi.org/10.1016/S0165-0173(00)00034-5.

Smidt, M.P., 2004. Early developmental failure of substantia nigra dopamine neurons in mice lacking the homeodomain gene Pitx3. Development 131 (5), 1145-1155. https://doi.org/10.1242/dev.01022.

Smidt, M.P., Burbach, J.P.H., 2007. How to make a mesodiencephalic dopaminergic neuron. Nature Reviews Neuroscience 8 (1), 21-32. https://doi.org/10.1038/nrn2039.

Souza, B.R., Romano-Silva, M.A., Tropepe, V., 2011. Dopamine D2 receptor activity modulates Akt signaling and alters GABAergic neuron development and motor behavior in zebrafish larvae. Journal of Neuroscience 31 (14), 5512-5525. https://doi.org/10.1523/jneurosci.5548-10.2011.

Tanner, C.M., et al., 2011. Rotenone, paraquat, and Parkinson's disease. Environmental Health Perspectives 119 (6), 866-872. https://doi.org/10.1289/ehp.1002839.

Tay, T.L., et al., 2011. Comprehensive catecholaminergic projectile analysis reveals single-neuron integration of zebrafish ascending and descending dopaminergic systems. Nature Communications 2 (1), 171. https://doi.org/10.1038/ncomms1171. Nature Publishing Group.

Tieu, K., 2011. A guide to neurotoxic animal models of Parkinson's disease. Cold Spring Harbor Perspectives in Medicine 1 (1), 1-20.

Vaz, R.L., Outeiro, T.F., Ferreira, J.J., 2018. Zebrafish as an animal model for drug discovery in Parkinson's disease and other movement disorders: a systematic review. Frontiers in Neurology 9 (June), 347. https://doi.org/10.3389/fneur.2018.00347.

Vijayanathan, Y., et al., 2017. 6-OHDA-Lesioned adult zebrafish as a useful Parkinson's disease model for dopaminergic neuroregeneration. Neurotoxicity Research 32 (3), 496-508. https://doi.org/10.1007/s12640-017-9778-x.

Wang, Y., et al., 2017. Parkinson's disease-like motor and non-motor symptoms in rotenone-treated zebrafish. Neuro Toxicology 58, 103-109. Elsevier B.V.

Wen, L., et al., 2008. Visualization of monoaminergic neurons and neurotoxicity of MPTP in live transgenic zebrafish. Developmental Biology 314 (1), 84-92. https://doi.org/10.1016/j.ydbio.2007.11.012.

Xi, Y., et al., 2011. Transgenic zebrafish expressing green fluorescent protein in dopaminergic neurons of the ventral diencephalon. Developmental Dynamics 240 (11), 2539-2547. https://doi.org/10.1002/dvdy.22742.

Zupanc, G.K.H., 2009. Towards brain repair: insights from teleost fish. Seminars in Cell and Developmental Biology 20 (6), 683-690. https://doi.org/10.1016/j.semcdb.2008.12.001.

第二十三章

帕金森病

W. Philip Bartel[1], Victor S. Van Laar[1], Edward A. Burton[1]

23.1 引言

帕金森病（Parkinson's disease, PD）是最常见的神经退行性运动障碍疾病，也是一种主要的老年病，40 岁至 49 岁的人群中每 100000 人中就有 41 人患有 PD，而 80 岁以上的人群中每 100000 人中患有 PD 的人数可达 1930 人（Cacabelos, 2017; Pringsheim et al., 2014）。黑质致密部（substantia nigra pars compacta, SNc）的多巴胺能神经元退化引起的主要运动体征为运动迟缓、强直和震颤，这些症状可作为 PD 的临床诊断标准。但是，在中枢、外周和肠神经系统的其他部分也会产生相关的病理变化，这也是 PD 患者经常出现许多其他症状的原因，这些症状包括痴呆、抑郁、焦虑、精神病、跌倒、语言和吞咽问题、睡眠障碍、嗅觉缺陷、自主功能衰竭和便秘。这种疾病是渐进性的，具有显著的发病率，可导致慢性残疾。

James Parkinson 于 1817 年在他的专著《震颤性麻痹随笔》（*An Essay on the Shaking Palsy*）中首次描述了 PD。在随后的 200 年间，PD 的对症治疗取得了重大突破。最值得注意的是多巴胺代谢前体 L- 二羟基苯乙酸（L-dihydroxyphenylacetic acid, L-DOPA）的应用，L-DOPA 经口服后可进入大脑，并经脱羧作用形成多巴胺，进而部分改善由黑质变性引起的神经化学缺陷。在 L-DOPA 进入临床应用半个多世纪后，其仍然是最有效的 PD 对症治疗药物。然而，目前仍没有阐明 PD 神经退行性病变的潜在机制并缺乏阻止疾病进展的治疗方法。这是 PD 治疗中的一个关键空白，迫切需要开发有效的神经保护治疗方法。

过去人们已经采用了许多不同的实验系统来阐明 PD 神经退行性病变的细胞和分子作用机制，这些实验系统包括人类遗传学和模式生物，以及培养细胞和对分离蛋白的生化分析。本章对帕金森病的发病机制进行了简要介绍，并对使用斑马鱼模型研究帕金森发病机制的最新进展进行综述，同时还对斑马鱼模型在实验方面的便利性及其驱动科学发现的适

[1] Department of Neurology, University of Pittsburgh, Pittsburgh, PA, United States.

用性方面的潜在优势，以及使用水生生物模拟人类神经系统疾病方面存在的一些挑战进行了讨论。

23.2 关于帕金森病神经退行性病变的原因我们知道些什么？

PD 患者脑组织的组织病理学检查结果显示，包括 SNc 在内的易损神经元群发生退化，并在存活的神经元细胞质内发现嗜酸性蛋白包涵体路易小体（Lewy 小体）的形成。星形胶质细胞增生和小胶质细胞激活也常见于受影响的大脑区域。一些因素对定义 PD 终末期变化的机制提出了挑战：PD 是偶发性的，通常发生在晚年；其病理变化可能在出现临床症状时才得到确认；组织病变出现与临床症状出现之间可能间隔很多年。关于 PD 的病因和发病机制的线索主要来自以下几个不同的研究方向。

- 通过比较 PD 样本与对照组的研究发现，在 PD 样本的中枢神经系统和外周组织中存在多种异常，如线粒体功能障碍、氧化应激和内源性抗氧化剂的消耗，以及细胞大分子的氧化损伤，神经炎症反应，蛋白质的错误折叠、聚集和堆积，泛素 - 蛋白酶体蛋白降解系统的功能障碍（Xie et al., 2017; Irwin et al., 2017; Lill, 2016; Rokad et al., 2017; Toledo et al., 2013）。目前尚不清楚哪些异常是发病的原因，哪些是发病后的结果。
- 目前已克隆出多个罕见的单基因 PD 表型相关基因。尽管遗传性 PD 与偶发性 PD 之间存在一些差异（如发病年龄往往较低，有些类型的病情进展迅速或出现早期痴呆），但在临床和病理上都具有很大的相似性，因此人们预言，阐明这些单基因疾病能够为理解 PD 提供大量的参考信息。这在某些情况下已得到证实。例如，在 PD 患者的 Lewy 小体病理中发现 α- 突触核蛋白（见下文），随后证明 α- 突触核蛋白编码基因 SNCA 基因的突变是导致大家族中常染色体显性遗传性 PD 的主要原因（Polymeropoulos et al., 1997）。有趣的是，许多基因编码的蛋白质的主要功能似乎集中于少数途径，包括维持线粒体功能和清除错误折叠的蛋白质。
- 全基因组关联研究已经确定了许多非编码遗传变异，这些变异似乎可以调节偶发性 PD 的发病风险（Satake et al., 2009; Simon-Sanchez et al., 2009）。一些风险相关的多态性与单基因型 PD 的表型相关基因存在联系，例如 SNCA 可能会调节基因表达。但是，对于许多已经明确的风险变异，其机制仍然不清楚。
- 一些急性 PD 病例是由 1- 甲基 -4- 苯基 -1,2,3,6- 四氢吡啶（1-methyl-4-phenyl-1,2,3,6-tetrahydropyridine, MPTP）暴露引起的（Langston et al., 1983），MPTP 是一种非法麻醉品合成的副产品（见下文）。MPTP 很容易进入中枢神经系统，并被氧化成 1- 甲基 -4- 苯基吡啶离子（MPP^+），这是一种线粒体复合物 I 抑制剂，通过与多巴胺共享相同的摄取机制而聚集在多巴胺能神经元中。尽管仅有很少的 MPTP 病例发生，且并非所有的 MPTP 病例都存在路易小体病变，但这一发现非常重要，因为它证明了线粒体抑制剂可以引发神经退行性病变，这与临床上 PD 相类似，并促使进一步研究发现在偶发性 PD 患者的中枢神经系统和外周组织中都存在线粒体功能障碍。随后的研究表明，大鼠全身暴露于另一种复合物 I 抑制剂——杀虫剂鱼藤酮，会导致黑质纹状体变性和产生类似路易小体的 α- 突触核蛋白包涵体（Betarbet et al., 2000）。尽管鱼藤酮会引起系统性复合体 I 抑制，但是黑质纹状体系统的病理结果仍具有相对的选择性，这表明黑质纹状体系统特别容易受到线粒体功能紊乱的影响。

• 暴露于某些环境毒物，包括农药和工业化学品，在流行病学上与患 PD 风险增加有关（Pezzoli and Cereda, 2013; Pouchieu et al., 2018; Tanner et al., 2011）。例如职业性鱼藤酮暴露可使患 PD 风险增加 2.5 倍。其他一些 PD 相关暴露似乎具有线粒体抑制和氧化应激的相同潜在作用机制。其中一些化学物质还能够在实验动物身上引发 PD 样病理变化（Gash et al., 2008）。

• 在路易小体（以及 PD 中的其他蛋白质聚集体，包括路易轴突和苍白小体）中检测 α-突触核蛋白聚集的高度灵敏的免疫组化方法的开发使人们观察到突触核蛋白病在许多病例（但不是全部病例）中具有特征性发病进程（Braak et al., 2003, 2004）。在疾病早期，α-突触核蛋白聚集于嗅球和脑干核中，包括迷走神经背侧运动核（dorsal motor nucleus of the vagus, DMNV）、蓝斑（locus coeruleus, LC）和中缝背核（dorsal raphe nucleus, DRN）。随后累及黑质致密部（substantia nigra compacta, SNc），并伴有明显的神经退行性病变，最终导致特征性的运动障碍。随着病情的进一步发展，α- 突触核蛋白聚集和神经性退化也可累及大脑皮层和其他区域。这种解剖学进展衍生出了下面一种假说，即错误折叠的突触核蛋白可以引导其他突触核蛋白分子的错误折叠，从而导致扩张性病理变化，类似于 Creutzfeldte Jakob 疾病中 PrP^{c} 到 PrP^{sc} 的转化。有一些实验支持这一假说，但没有证据表明 PD 可以传播，而且也不确定突触核蛋白病变和神经性退化之间的关系。目前，“传播病理学”的观点仍存在争议。

因此，人们对于 PD 的病因因素和发病机制还只是部分理解，仍有几个主要问题尚未解决。目前的主流观点认为，多种遗传易感性因素与环境暴露的相互作用诱发了该疾病，并通过线粒体功能障碍、蛋白质错误折叠等一系列复杂的致病机制最终导致临床表型的出现。

23.3 帕金森病动物模型的作用和斑马鱼的潜在用途

人们利用实验模型来确定各种遗传和环境因素是如何单独或联合诱发介导 PD 的生化异常，并且开发了多种模型来复制疾病发病机制或病理生理学的各个方面，以便能够发现和优化新的治疗方法。在这两种情况下，体内模型可能是必要的，原因如下。

• 特定群体的神经元参与 PD 病理过程。了解细胞内在的因素（如基因表达或细胞形态）对这种选择性易损性的影响，可能需要专门研究易损性神经元群体，以确保潜在的发病机制的存在。

• 越来越多的证据表明，PD 的发生与其他细胞外部因素有关，如神经胶质细胞相互作用、神经炎症、血脑屏障或其他因素，如神经连接等，这需要研究完整的中枢神经系统来描述相关的疾病机制。

• 神经表型是由中枢神经系统环路活动紊乱引起的，目前尚不清楚神经元损失前的生理紊乱机制是否可以作为治疗靶点。模式生物也许能够提供神经行为或神经生理的表型输出，从而允许对关注的可能性进行测试，此外还更为广泛地用于开发新的对症治疗方法。

- 药物和毒物渗透到中枢神经系统的作用部位取决于其吸收、分布、代谢和排泄。这些过程对于确定假定的环境毒物或实验药物是否能够到达中枢神经系统至关重要，因此最好在体内进行研究。

一系列的模型动物，从无脊椎动物（如果蝇和秀丽线虫）到哺乳动物（如啮齿动物和灵长类动物），已被用于研究帕金森病的遗传学、分子机制、药理学和其他方面。每种模型动物对特定类型的研究都有其独特的优势。下文将着重阐述斑马鱼模型在研究 PD 发病机制方面的优势。

- 斑马鱼胚胎在母体外生长，通身透明。因此利用荧光报告分子可在活体显微镜下对细胞群、细胞器和感兴趣的生化过程进行观察。这为 PD 致病机制（特别是 PD 中易受损的神经元群体以及细胞或亚细胞水平）提供了非常有力的评估方法。
- 由于可对斑马鱼进行广泛的基因操作，如瞬时或稳定的转基因表达、瞬时基因敲减、种系突变体构建，使其成为研究脊椎动物发育的模式生物。这些方法为在体内评估孟德尔帕金森病表型相关基因的生理功能和确定其变异导致的致病机制提供了有力的手段，已被用于了解与 PD GWAS 风险位点相邻的基因如何影响相关中枢神经系统神经元群体的生物学原理。
- 成年斑马鱼可以定期繁殖，每次都会产生相当数量的后代。由于斑马鱼可以进行大量饲养，因此可实现罕见生物事件的高通量筛选，包括应用无偏倚的基因筛选，以分离与发病机制相关的突变；通过化合物库筛选发现缓解疾病的分子修饰剂，并可能为药物治疗学的发展提供基础。
- 作为水生生物，斑马鱼可以很容易地暴露于溶解于水的化学物质中。对于斑马鱼仔鱼，还可以在多孔板中进行研究，从而允许在统计可靠的样本中测试大量的化学物质。由于利用斑马鱼模型可以单独或组合研究发病机制相关的环境化学暴露条件，因此在帕金森病研究中非常重要。此外，斑马鱼可用于化学文库的筛选以找到能够减轻 PD 病理的分子。在早期胚胎发生后，化学物质进入中枢神经系统依赖于药物的胃肠道吸收率和血脑屏障透过率，这与药物开发以及确定疾病发病过程中环境暴露因素研究都紧密相关。
- 斑马鱼模型的高通量体现在可在多个动物中并行进行自动化表型分析。PD 在临床上主要表现为运动机能减退的运动障碍，因此能够测量斑马鱼的运动功能，这是斑马鱼作为 PD 模型开发和利用的重要优势。多孔板运动功能的自动测量也为化学和遗传修饰研究提供了一个直接、公正且与疾病相关的表型终点。除了提高以发现为目标的技术方法的通量外，多孔板运动功能的自动测量还有助于实验重复和获得可靠的统计样本，因此斑马鱼模型对提高神经行为实验结果的再现性具有非常大的潜力。
- 斑马鱼大量饲养的成本低且容易实现，相对于哺乳动物模型而言可证实更多的假设。这对于 PD 研究非常重要，特别是在需要评估大量候选基因的情况下（例如，目前已确定的风险位点附近的基因超过 40 GWAS），或者组合的因素可能是关键的情况下，例如基因环境相互作用被认为是 PD 发病的核心。

虽然斑马鱼模型在 PD 研究这一领域的应用才刚刚开始，但人们已经利用其对 PD 开展了许多探索性研究，下节内容将着重进行阐述。

23.4 斑马鱼作为帕金森病研究模型的适用性

开发人类疾病模型是为了阐明疾病发病机制，以便为发现新的治疗方法提供指导。这是可行的，特别是当模型复制了对发病机制至关重要的因素时，例如介导细胞病理的特定基因和生化途径、与疾病相关的神经和胶质细胞群的功能障碍，以及疾病中导致相关表型的神经化学系统和回路异常。毫无疑问，斑马鱼和人类的中枢神经系统和多巴胺能系统具有很多的相似之处，但与 PD 相关的一些差异非常值得深入讨论。

23.4.1 遗传学

尽管人类和斑马鱼之间存在显著的进化距离（最后一个共同祖先是在 4.5 亿年前），但它们基因组之间有着惊人的高度保守性（Howe et al., 2013; Kumar and Hedges, 1998）。大约 70% 的人类基因在斑马鱼中至少有一个明显的同源基因，由于硬骨鱼特有的全基因组复制周期，许多基因有两个同源基因（Howe et al., 2013; Meyer and Schartl, 1999）。人类孟德尔遗传在线（Online Mendelian Inheritance in Man, OMIM）列出的与人类疾病相关的基因中有 82% 至少有一个斑马鱼同源基因，这使得斑马鱼成为一个极具吸引力的人类遗传疾病研究模型（Howe et al., 2013）。

斑马鱼有许多孟德尔帕金森病表型相关基因的保守同源基因，包括 *PINK1*、*PARKIN*、*PARLA/B*、*DJ-1*、*LRRK2*、*FBXO7* 和 *ATP13A2*（表 23-1），以及包括 *GBA1* 在内的关键 PD 风险基因。斑马鱼可表达突触核蛋白家族的三个成员，包括 β- 突触核蛋白和 γ- 突触核蛋白的两个旁系同源物（Milanese et al., 2012; Sun and Gitler, 2008）。它们对多巴胺能系统的发育和正常的运动功能非常重要（Milanese et al., 2012）。但是，一个古老的 SNCA 基因从斑马鱼的基因组中消失了，这导致斑马鱼不表达 α- 突触核蛋白同系物（Milanese et al., 2012; Sun and Gitler, 2008）（图 23-1）。目前尚不清楚 α- 突触核蛋白的缺失是否会成为斑马鱼 PD 模型的一个问题。含有聚集的 α- 突触核蛋白的路易小体是帕金森病诊断的神经病理标志，因此人们对斑马鱼能否复制帕金森病关键病理生理机制产生了疑虑。但是在斑马鱼多巴胺能神经元中表达的 γ1- 突触核蛋白，与人类 α- 突触核蛋白具有显著的序列相

表 23-1 与遗传性帕金森病有关的基因以及在斑马鱼上的同源性

蛋白质	遗传性 PD 的可突变基因类型	人与斑马鱼间的氨基酸一致性百分率（同源性）
α- 突触核蛋白	常染色体显性遗传基因	47（56）①
LRRK2（ROCO 家族蛋白）	常染色体显性遗传基因	48（50）
PINK1（一种蛋白激酶）	常染色体隐性遗传基因	54（62）
Parkin	常染色体隐性遗传基因	62（75）
DJ-1（肽酶 C56 蛋白质家族成员）	常染色体隐性遗传基因	83（89）
GBA1（溶酶体酶葡糖脑苷脂酶）	风险等位基因	57（68）

① 与斑马鱼 γ1- 突触核蛋白相比较。

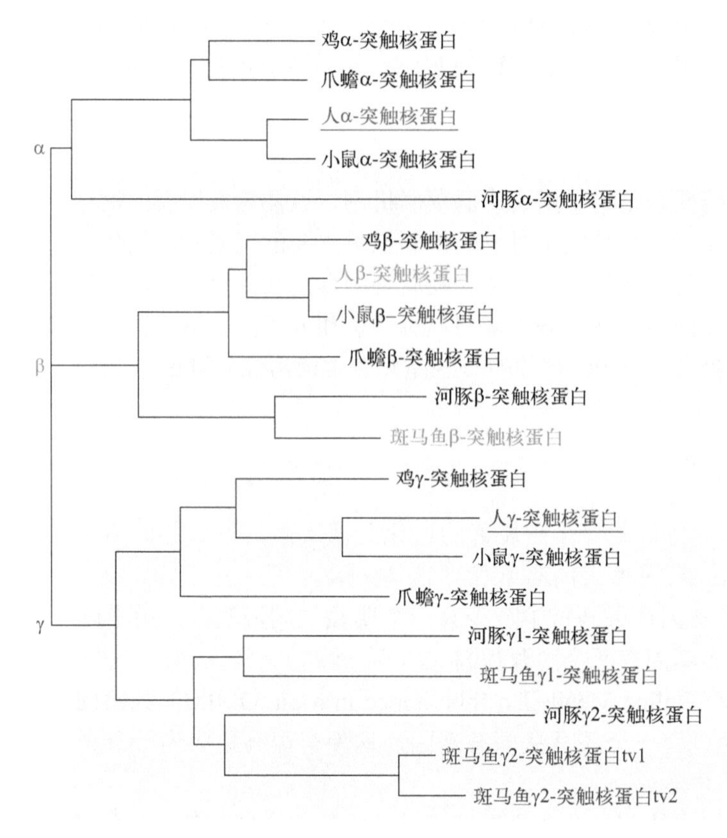

图23-1　**显示不同脊椎动物物种间突触核蛋白进化关系的树状图**

利用ClustalW软件对来自人类、小鼠、鸡、爪蟾、河豚和斑马鱼的突触核蛋白进行氨基酸序列比对，并用邻接法构建树突图。注意：斑马鱼缺失α-突触核蛋白；斑马鱼存在两种γ-突触核蛋白旁系同源物，其中一个同源物有两种亚型，来自*sncg2*转录本的不同剪接体

似性，而且人类α-突触核蛋白可以恢复由于斑马鱼β-突触核蛋白和γ1-突触核蛋白缺失而产生的表型，这些结果表明人类和斑马鱼的蛋白质功能具有相似性。此外，由于α-突触核蛋白在PD中的作用可能代表一种毒性功能的获得，而不是生理功能的丧失，因此在斑马鱼神经元表达人类α-突触核蛋白可能会解决这一问题（可能在$sncb^{-/-}$、$sncg1^{-/-}$的遗传背景基础上产生以药物发现为目的的“人源化”模型）。

23.4.2　中枢神经系统结构和多巴胺能系统

作为脊椎动物，人类和斑马鱼都有基本的中枢神经系统组织，包括前脑、中脑、后脑和脊髓（图23-2）（Rupp et al., 1996）。这种结构的保守性，以及特殊神经元群、神经胶质亚群和回路结构的相似性，使斑马鱼模型能够帮助人们更深刻地理解脊椎动物神经系统发育和生理学，并可能阐明类似的疾病机制（Sager et al., 2010）。

SNc在PD中的显著参与（在保留相邻神经元群的背景下）强烈表明SNc多巴胺能神经元的特定特性是PD发病机制的关键。因此，人类和斑马鱼多巴胺能系统之间的关系对于PD病理学的关键机制能否在斑马鱼中复制这一问题是非常有意义的。与哺乳动

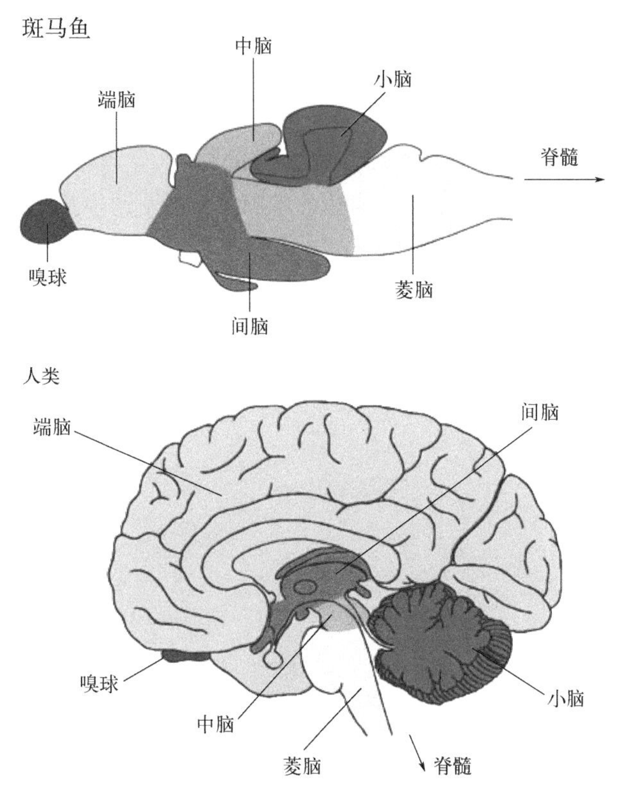

图 23-2　**人类和斑马鱼中枢神经系统组织之间的保守性**

人类和斑马鱼大脑图。在每张图中，颜色相似的区域基因同源。尽管人类和斑马鱼的大脑在数量和复杂程度上存在很多差异，但它们却具有相似的基本组织

物不同，斑马鱼没有中脑多巴胺能系统（Du et al., 2016）。与包括人类在内的大多数脊椎动物的外凸相比，斑马鱼前脑中发生的发育外凸使直接的神经解剖学比较变得复杂（Wullimann and Rink, 2002）。对比分析表明斑马鱼腹侧端脑的背层核在解剖学上与人类纹状体同源。斑马鱼腹侧端脑的多巴胺能神经支配来自端脑细胞，也来自位于间脑后结节中的多巴胺能神经元的上行投射（图 23-3）。从解剖学上讲，从间脑群 DC2 和 DC4 到腹侧端脑的多巴胺能投射可能是人类黑质纹状体系统的同源物（Rink and Wullimann, 2001, 2002; Wullimann and Rink, 2002）。但是，需要注意的是，投射到斑马鱼前脑的 DC2 和 DC4 多巴胺能神经元也投射到脊髓，而哺乳动物的多巴胺能间脑脊髓投射与黑质纹状体系统是分开的。这些差异表明，斑马鱼 DC2 和 DC4 神经元可能与人类的 SNc 神经元不同。然而，在各种模型中的研究表明（下文讨论），导致 PD 患者 SNc 多巴胺神经元易受损的关键因素也存在于 DC2 和 DC4 神经元中。此外，DC2 和 DC4 的长脊髓轴突投射的活体成像使人们可以直接在体内观察多巴胺能神经元的轴突线粒体动力学（Dukes et al., 2016）。

23.4.3　多巴胺能神经化学

多巴胺能信号的分子生理学在人和斑马鱼之间是保守的（图 23-4）。例如，多巴胺转运蛋白（DAT）是多巴胺从突触间隙再摄取到突触前末端所必需的，在人和斑马鱼之间高

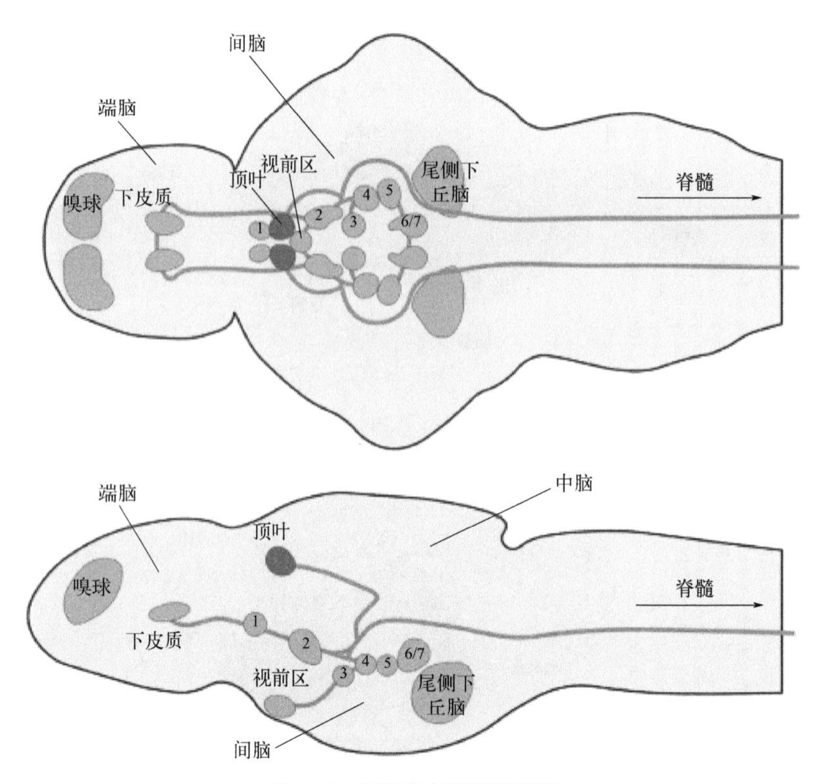

图23-3 **斑马鱼多巴胺能系统**

该图显示了斑马鱼受精 5d 后斑马鱼仔鱼脑中各主要多巴胺能神经元组的位置，从上方（上图）和侧面（下图）观察。使用 Rink 和 Wullimann（2002）提出的系统将间脑群编号为 DC1 ～ DC7。从解剖学角度讲，认为向端脑发送轴突投射的 DC2/4 神经元是哺乳动物黑质纹状体系统的同源物

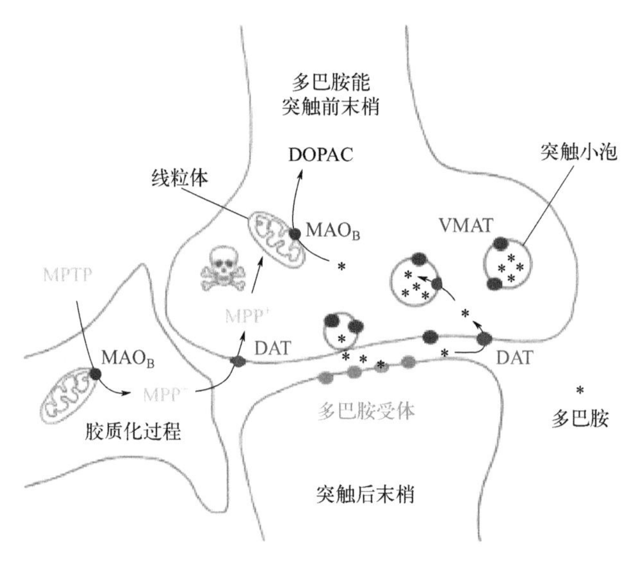

图23-4 **中枢神经系统多巴胺能突触**

图中示意了多巴胺能突触前末梢、突触后神经元和相邻神经胶质细胞，说明了文中讨论的关键特征。多巴胺从突触前末端释放，与突触后膜上的受体结合，然后通过多巴胺转运蛋白 DAT 转运回突触前末端，并通过囊泡单胺转运蛋白 VMAT 包装成突触小泡。多巴胺降解部分由 MAO_B 介导。许多相同的机制解释了 MPTP 的多巴胺能神经元特异性毒性，其被神经胶质 MAO_B 转化为线粒体抑制剂 MPP^+。然后通过 DAT 将 MPP^+ 富集于多巴胺能神经末梢

度保守（Bai and Burton, 2009; Holzschuh et al., 2001）。在表达酪氨酸羟化酶（TH）的多巴胺能神经元中，这种相似性延伸到基因组结构和编码 DAT 的 *slc6a3* mRNA 的表达。除了提供多巴胺能神经元的组织化学标志物和用于在多巴胺能神经元中表达转基因的顺式调节元件（源自 *slc6a3* 基因）之外，DAT 保守性的重要意义在于其在介导 MPP^+（一种与人类帕金森病有关的线粒体抑制剂，常用于实验动物模拟 PD 模型）的细胞摄取中发挥重要作用。尽管有些多巴胺受体在斑马鱼中有多个类似物，但是其在人类和斑马鱼之间还是高度保守的（Boehmler et al., 2004, 2007）。

相比之下，单胺氧化酶（monoamine oxidase, MAO）在斑马鱼中只有一种亚型，而在人类中有两种亚型（Anichtchik et al., 2006; Setini et al., 2005）。在人类中，MAO_B 有助于多巴胺的降解，其抑制作用在治疗上用于增加 PD 中的纹状体多巴胺水平。尽管斑马鱼 MAO 在氨基酸水平上与人类 MAO_A 和 MAO_B 具有 68% 的一致性，但它的大脑活动仅被 selegiline（一种抑制食欲的药物）部分抑制（Anichtchik et al., 2006; Setini et al., 2005）。斑马鱼仔鱼暴露于 selegiline 后，仔鱼的 5-HT 水平增加，而多巴胺水平没有变化，这表明斑马鱼仔鱼的多巴胺可能通过其他途径代谢，如 COMT，COMT 有助于人体多巴胺代谢（Sallinen et al., 2009a）。

人类和斑马鱼多巴胺能神经化学的另一个区别与 TH 有关（Candy and Collet, 2005; Panula et al., 2010）。TH 是儿茶酚胺合成中的限速酶，在哺乳动物研究中常被用作多巴胺能神经元和其他单胺能神经元的免疫组化标志物。斑马鱼基因组包含两个 TH 旁系同源物。斑马鱼 *th1* 和 *th2* 的表达与多巴胺能神经元中编码单胺氧化酶、5- 羟色胺转运蛋白和 DAT 的 mRNA 转录本的表达重叠（Filippi et al., 2010）。重组 TH2 体外实验表明，它有色氨酸羟化酶活性，但没有 TH 活性（Ren et al., 2013），而体内 TH2 基因的下调降低了多巴胺水平，但对 5- 羟色胺没有影响（Chen et al., 2016）。因此，TH2 的功能尚不清楚，TH1 通常用作标记多巴胺能神经元的组织化学标志物，尽管其可能不会标记所有合成和释放多巴胺的神经元。

23.4.4 运动功能的多巴胺能调节

帕金森病最突出的临床特征通常是运动障碍，所以运动功能恢复是帕金森病模型中衡量预后的重要指标，在斑马鱼中尤其如此，自动化的高通量神经行为筛选分析可能是其用来阐明发病机制和开发治疗方法的核心。斑马鱼 PD 模型的研究主要集中在几种类型的运动反应。斑马鱼仔鱼的自发活动是不连续的、随机的和刻板的；也就是说，不活动的状态会被相对较短的刻板推进和不可预测的转向运动打断。此外，有人还报道了在机械或光刺激下发生的其他几种类型的运动（Kalueff and Cachat, 2011）。斑马鱼仔鱼基本运动轨迹的运动学及其生理学基础都是有据可查的（Budick and O'malley, 2000; Burgess and Granato, 2007）。每一种运动都是脊髓运动神经元活动的独特模式产物，这种神经元活动由仔鱼脑干所调节（Liu and Fetcho, 1999; Orger et al., 2008）。大多数类型的斑马鱼仔鱼的运动在 500ms 内完成，通常使用高速视频显微成像技术对斑马鱼仔鱼的运动进行分析，但该技术尚未针对高通量应用进行优化。相比较而言，简单的分析方法对于 PD 模型更为实用。

使用专有设备或带有开源软件的简单摄像机，通过动物在多孔板中游动的相对低帧率视频记录来测量斑马鱼仔鱼自发运动的频率和幅度（Cario et al., 2011; Zhou et al., 2014）。

利用目标检测的算法确定斑马鱼在每个视频帧中的位置，计算出基本参数。例如在一段时间内移动的距离、每次运动的幅度和平均速度、运动起始事件的频率以及发生位移时的帧转换比例。这些试验对多巴胺能功能敏感：总体而言，公开的数据表明，多巴胺调节了斑马鱼仔鱼自发运动的起始，而不是执行。例如，斑马鱼暴露于D2受体拮抗剂，如抗精神病药氯丙嗪和氟哌啶醇，会降低运动起始事件的频率（但不会改变其运动速度），从而导致自发性位移的总体减少（Farrell et al., 2011; Giacomini et al., 2006; Irons et al., 2013）。相比之下，D2受体激动剂，如罗匹尼罗（ropinirole）和阿扑吗啡（apomorphine），会导致运动次数减少但运动机能亢进（Farrell et al., 2011; Irons et al., 2013）。使用线粒体抑制剂MPP$^+$会破坏多巴胺能神经末梢和神经元（图23-5）并导致运动起始的频率降低，但却不会影响运动速度（Farrell et al., 2011）。在用化学遗传学消融多巴胺系统后也能看到类似的结果。将表达细菌硝基还原酶的斑马鱼暴露于甲硝唑下，可导致多巴胺能神经元的损失，该硝基还原酶的表达受一个长度约为35kb的调控元件调控，该调控元件源自编码DAT的*sllc6a3*基因。由于移动时间的减少，这些动物的整体位移减少，但对游动速度影响不大（Godoy et al., 2015）。最后，采用激光消融的方法破坏Et（*vmat*: *egfp*）标记的DC2和DC4/5神经元，从而消除脊髓DA信号的转导（Jay et al., 2015）。这导致游动的时间比例显著下降，尽管个体游动事件的持续时间和峰值速度是正常的。上述研究结果表明，多巴胺只调节斑马鱼仔鱼自发游泳运动的起始。倘若可较为容易地在多孔板中实现斑马鱼仔鱼自发运动的自动化测量，那么自发运动功能就可用于斑马鱼PD模型的疾病相关表型分析，这是非常令人

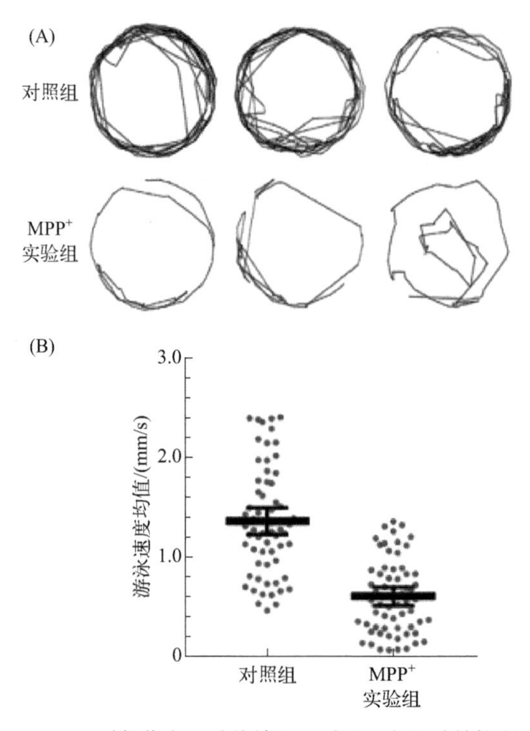

图23-5 **MPP$^+$损伤多巴胺能神经元对斑马鱼运动的抑制作用**

（A）受精后5d斑马鱼仔鱼游动路径的矢量轨迹，在96孔板中记录2min。对照组包含三条对照仔鱼，每条仔鱼在记录期间完成了多次孔缘环行。MPP$^+$实验组显示了在实验前暴露于MPP$^+$ 48h的三条仔鱼，活动明显减少。（B）1h内的总位移显示为每组48条仔鱼的平均游泳速度。数据点代表斑马鱼，标尺代表95%可置信区间

振奋的。事实上也确实如此，如果表型足够严重，仔鱼自发运动的自动测量可作为筛选工具对帕金森病表型的部分恢复进行分析，这具有重要的应用价值（Farrell et al., 2011）。

利用改进的方法评估斑马鱼仔鱼对环境光变化的反应，这很容易通过红外摄像记录来实现。视觉运动反应（visual motor response, VMR）包括突然明 - 暗转换时的刻板“O 形”反应，随后在黑暗中运动起动事件的频率增加，并随着时间的推移而下降，在突然的暗 - 明转换时运动受到抑制，然后回到稳定的运动频率（Burgess and Granato, 2007）。运动起动事件概率的这些变化可以通过多个明暗周期内平均游泳速度的变化来测量，并使用分析自发运动的视频跟踪软件进行分析（Cario et al., 2011）。多巴胺拮抗剂显著抑制了这种对光照变化的特征性反应（Irons et al., 2013）。

触摸逃逸反应提供了一种测量 PD 模型中运动功能的手动方法。在该方法中，细丝与仔鱼头部的物理接触引发了一个固定的大角度转向，随后是远离刺激的推进运动（Eaton et al., 1977）。这种反射由位于后脑的 Mauthner 细胞及其同源物（MiD2cm 和 MiD3cm）所介导。间脑多巴胺能神经元的轴突靠近 Mauthner 神经元及其同源物的腹侧树突（McLean and Fetcho, 2004），并且 Mauthner 细胞接受调节性多巴胺能输入（Korn and Faber, 2005）。因此，减少的多巴胺能信号可能会减弱这一反应。与使用自动化方法测量自发运动和光诱发运动相比，触摸逃逸反应需投入更多的劳动力，而且易受观察者之间和观察者主动性的影响。然而，这一反应可在受精后 48h 内被诱发（自发运动的可靠测量则需在受精后 5 ～ 6d 才可进行），并且已成功地应用于斑马鱼 PD 模型。

23.4.5 超越多巴胺：与帕金森病相关的非运动表型

斑马鱼仔鱼中与 PD 非运动症状相关的其他行为表型的测定。

- 早期嗅觉缺陷是 PD 的特征（Doty, 2012），并且嗅球病理在尸检中非常普遍。硬骨鱼的嗅球要大于大脑的其他部分。在斑马鱼仔鱼中评估了其对某些氨基酸的喜好和厌恶反应，进而测量嗅觉的辨别能力（Lindsay and Vogt, 2004）。
- 尽管睡眠障碍潜在的病理学尚不清楚，但它几乎普遍存在于 PD 中。斑马鱼活动中表现出长期的特征性变化，这在很大程度上受到环境光线的影响。已有文献描述了斑马鱼仔鱼睡眠和失眠的定量分析方法（Rihel et al., 2010; Sveinbjornsdottir, 2016; Yokogawa et al., 2007）。
- 胃肠动力受损导致便秘在 PD 中几乎普遍存在。有人报道了 PD 肌间神经丛的路易体病理学，并且认为动力丧失是由肠神经系统受累引起的。斑马鱼仔鱼的肠道神经系统已得到部分表征，并且最近的研究描述了肠道转运时间的高通量测量方法（Cassar et al., 2018）。
- 焦虑和抑郁在 PD 中出现的频率高于其他可导致类似程度身体残疾的神经系统疾病，是 PD 患者常见的发病症状。斑马鱼仔鱼不太可能经历焦虑或情绪变化。然而，有人报道了斑马鱼仔鱼中可由已知的人类焦虑和抑郁治疗药物调节的典型运动反应。如果这些运动反应与人类焦虑和抑郁的潜在机制相通，那么使用斑马鱼模型则是解决 PD 中焦虑和抑郁等问题的有用工具。

迄今为止，这些非运动检测在斑马鱼 PD 模型中的应用仍然受限。但是随着越来越多的人认识到 PD 的非运动症状对总体发病率有重要影响，这可能会是一个发展潜力巨大的领域。

23.5 利用斑马鱼模型阐明孟德尔帕金森病表型和偶发性帕金森病相关基因的功能

据报道，多个不同基因的突变会导致罕见的单基因帕金森病表型。尽管该表型估计仅占PD病例总数的5%到15%（Deng et al., 2018），但是这些明确的神经退行性疾病的病变原因可为阐明常见的散发性疾病的潜在细胞机制提供基础。对斑马鱼PD连锁基因操控的大量研究可帮助人们破译这些基因在PD发病机制中的作用，并确定PD的潜在治疗途径（表23-1）。

斑马鱼作为一种优良的脊椎动物遗传模型，非常适用于旨在了解基因表达改变后果的研究。在斑马鱼仔鱼中，通常采用多种方法来控制基因表达，这些方法都可以用于研究PD神经退行性变的机制：

- 与mRNA前体结合并阻止剪接或与成熟mRNA结合并阻止翻译的吗啉反义寡核苷酸已被用于早期开发过程中的瞬时基因敲减研究。
- TALEN和Cas9/CRISPR已被用于在目的基因中产生可遗传的无效突变。
- mRNA显微注射已被用于在斑马鱼胚胎中瞬时表达基因。
- 通过质粒注射、巨核酶介导的转基因和转座子介导的转基因方法，已经产生了包含特定基因表达盒的稳定转基因品系。

基因敲除和基因敲减技术非常适合于检测基因产物的功能，此外，基因敲除和敲低技术还适用于构建由隐性突变和显性负突变或单倍不足突变导致表型的疾病模型，这些疾病的潜在机制涉及基因功能丧失。转基因技术已被应用于标记神经元群和特定靶细胞成分以进行活体显微镜检查，以及模拟潜在机制涉及获得毒性功能的显性突变诱导的疾病。由于斑马鱼可大量繁殖，因此产生复杂的复合基因型相对简单，例如，将报告基因与致病突变基因相结合，通过活体显微镜确定致病机制。这是斑马鱼模型的一个主要优势，可能会越来越多地用于帕金森病研究。

迄今为止，大多数研究都是检测单基因突变，或更为普遍的是采用吗啉反义寡核苷酸进行瞬时基因敲减来研究基因功能。虽然这是一种有应用前景的方法（快速且廉价），但使用吗啉反义寡核苷酸有以下几点需要引起注意。首先，吗啉反义寡核苷酸诱导的表型与遗传无效突变导致的表型相关性较差（Kok et al., 2015）。其次，众所周知，某些吗啉反义寡核苷酸以非特异性的方式引起神经退行性病变（Robu et al., 2007），这显然是神经退行性疾病研究中的一大难题。因此，在某些情况下，对同一基因的吗啉反义寡核苷酸的研究结果可能会相反。随着TALEN和CRISPR/Cas9的广泛应用，生成和分析PD相关基因的稳定无效突变是该领域的优先事项。

23.5.1 常染色体显性帕金森病

23.5.1.1 α-突触核蛋白

编码α-突触核蛋白的*SNCA*基因中的错义突变会导致常染色体显性进行性PD

(Polymeropoulos et al., 1997)。有意思的是，*SNCA* 基因的扩增增加了野生型 α- 突触核蛋白的表达，同样引起了家族性 PD，这表明野生型 α- 突触核蛋白可能也参与了偶发性 PD（Singleton et al., 2003）。该结果在偶发性 PD 患者的路易小体内存在的 α- 突触核蛋白（Spillantini et al., 1997），以及全基因组关联研究中也得到证实，全基因组关联研究结果显示 *SNCA* 基因附近的多态性标记中存在非编码风险变异，这可能会调节 α- 突触核蛋白的表达水平。此外，内源性 α- 突触核蛋白可促进大鼠 PD 样神经病理状态（暴露于与 PD 相关的环境毒物——鱼藤酮引起的）的发生（Zharikov et al., 2015）。但是，目前尚不清楚 α- 突触核蛋白在 PD 发病机制中的作用。

如上所述，斑马鱼基因组丢失了一个编码 α- 突触核蛋白的起源基因（Milanese et al., 2012; Sun and Gitler, 2008）。相反，斑马鱼表达 β- 突触核蛋白和两个 γ- 突触核蛋白旁系同源物（图 23-6），这是多巴胺能系统和运动功能正常发育所必需的（Milanese et al., 2012）。人 α- 突触核蛋白能够替代斑马鱼突触核蛋白，表明二者功能相似（Milanese et al., 2012）。主要体现在斑马鱼 γ1- 突触核蛋白的疏水结构域与人 α- 突触核蛋白具有同源性（Milanese et al., 2012）。且与人 α- 突触核蛋白相似，斑马鱼 γ1- 突触核蛋白可在体外发生聚集（Lulla et al., 2016），斑马鱼 γ1- 突触核蛋白的瞬时过表达可引起细胞内聚集和发育毒性（Lulla et al., 2016）。这些数据表明斑马鱼 PD 模型中的 γ1- 突触核蛋白可能与人类 α- 突触核蛋白具有相似的功能。

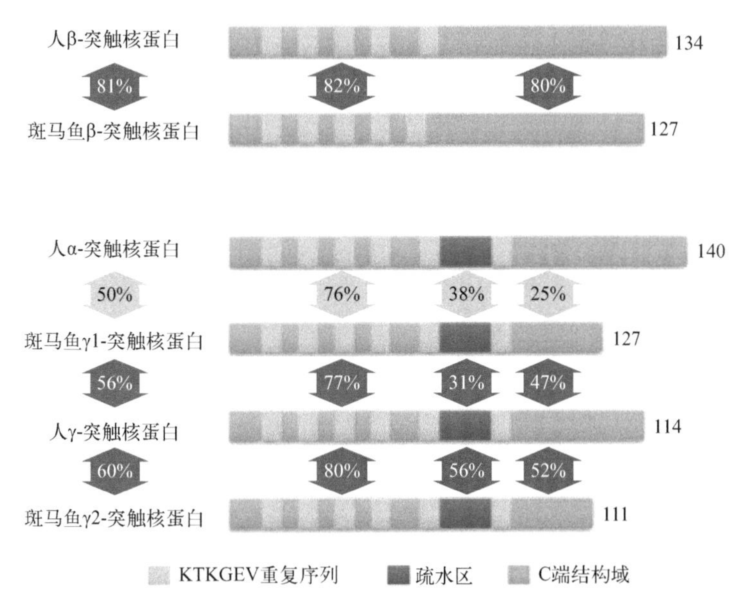

图23-6　**人类和斑马鱼突触核蛋白的结构域和系统发育高度保守**

人类和斑马鱼突触核蛋白的比对示意图显示突触核蛋白的结构域相当保守。人和斑马鱼蛋白质之间的氨基酸相似性以百分比表示（左下方百分数代表整个蛋白质的相似性，右下方百分数代表各区段的每个结构域的相似性）。每种蛋白质中氨基酸数量在右侧显示

目前普遍认为遗传性突触核蛋白病可能与毒性功能获得有关，因此在斑马鱼模型中过表达人源突触核蛋白是阐明潜在致病机制的可行方法。人 α- 突触核蛋白在斑马鱼 Rohon-Beard 感觉细胞中的短暂过表达导致轴突病变和细胞死亡增加（O'Donnell et al., 2014）。通过活体延时显微镜可以观察到轴突变化的进展，并且异常部分可通过 Wallerian 变性抑制

剂 WldS 恢复。在该模型中，即使是正常的轴突，其线粒体运输也受到抑制，且细胞死亡和轴突异常都可以通过过表达 PGC1a（线粒体发生的主要调控因子）来恢复。这些数据表明，α- 突触核蛋白可能通过对线粒体运输产生不利影响进而导致神经元轴突功能障碍（O'Donnell et al., 2014）。人源 α- 突触核蛋白在泛神经元模式中的瞬时过表达会导致严重的形态缺陷和死亡，并被一种可防止蛋白质聚集的化学物质恢复（Prabhudesai et al., 2012）。虽然这些观察结果很有趣，但需要在一个具有稳定转基因表达特性和更能反映 PD 表型的模型中进行验证。

虽然 α- 突触核蛋白在 PD 中发挥重要作用，但斑马鱼模型中枢神经系统多巴胺能神经元可稳定表达 α- 突触核蛋白相关研究尚未见报道。许多 α- 突触核蛋白转基因小鼠仅表现出轻微的表型，这就说明了 α- 突触核蛋白在接近生理水平表达但并不会引起瞬时过表达产生严重反应。由于 α- 内源性突触核蛋白在 PD 中的作用可能是调节细胞对环境毒物的反应，因此使用稳定表达 α- 突触核蛋白的细胞系可用于阐明基因与环境相互作用。

23.5.1.2 富亮氨酸重复激酶 2

富亮氨酸重复激酶 2（leucine-rich repeat kinase 2，LRRK2）是一种大蛋白，具有多个结构域，包括激酶结构域、GTPase 结构域、RAS 结构域以及与其他蛋白质相互作用相关的 WD40 结构域。在某些病例中，LRRK2 的错义突变导致常染色体显性 PD 表型，在临床和病理上与偶发性 PD 难以区分（Zimprich et al., 2004）。LRRK2 突变是 PD 最常见的遗传形式，在某些人群中，LRRK2 突变在明显偶发性 PD 病例中占很大比例（Lesage et al., 2006）。LRRK2 的正常细胞功能尚不清楚。最常见的突变 G2019S 似乎能够组成性激活激酶活性，可导致 LRRK2 的自磷酸化和下游靶点的磷酸化。有趣的是，偶发性 PD 的 LRRK2 也会发生组成性磷酸化，这表明即使在没有突变的情况下，激酶活性的增强也具有致病效应（Di Maio et al., 2018）。目前正在开发和评估 LRRK2 抑制剂分子，该分子可作为 PD 的治疗药物。因此，研究学者更多关注 LRRK2 的正常的细胞功能（可能被药理学抑制破坏）和 LRRK2 突变导致发病的下游机制研究。

斑马鱼 *lrrk2* 基因包含 51 个外显子，其编码 2533 个氨基酸组成的预测蛋白（图 23-7）（Sheng et al., 2010）。该蛋白质含有与人类 LRRK2 同源的功能域，虽然 LRRK2 的激酶结构域高度保守，同源率高达 71%，但人类和斑马鱼 LRRK2 的氨基酸同源率总体为 47%。RNA 原位杂交实验结果表明，LRRK2 mRNA 转录本在整个胚胎中表达程度很弱。通过使用吗啉反义寡核苷酸靶向调控 *lrrk2* 的表达来阐明其功能，但却产生了相互矛盾的结果。

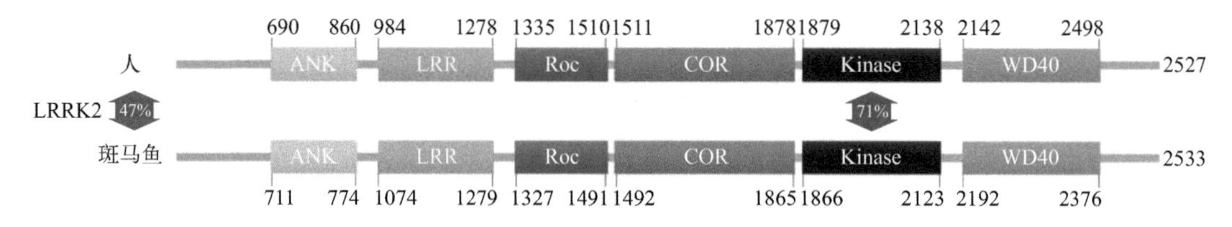

图 23-7 **人类和斑马鱼 LRRK2 的结构域和系统发育保守性**

图中给出了 LRRK2 的关键结构域：ANK, LRR, Roc, COR, Kinase（激酶），WD40；图中数字代表每个结构域的起始和结束位置；百分数表示人类和斑马鱼蛋白质之间的氨基酸相似性（左侧 47% 代表整个蛋白质的相似性，右侧 71% 代表激酶结构域的相似性）

两项不同的研究报告了靶向 ATG 翻译起始位点或上游外显子剪接位点的吗啉反义寡核苷酸，这导致了严重的胚胎形态异常，这几乎肯定是由脱靶基因表达变化或非特异性吗啉反义寡核苷酸毒性引起的（Prabhudesai et al., 2016; Sheng et al., 2010）。相反，据报道，一个剪接位点吗啉反义寡核苷酸引起 WD40 结构域上游的开放阅读框（ORF）被截断，从而减少间脑多巴胺能神经元的数量（Sheng et al., 2010）。斑马鱼或人类 LRKK2 的过表达则可使这种情况得到部分改善。但是，另一项研究结果却截然不同，使用完全相同的吗啉反义寡核苷酸，以及其他几种不同的吗啉反义寡核苷酸，可使 *lrrk2* mRNA 的转录下调，但并没有发现对多巴胺能系统有任何影响（Ren et al., 2011）。这些令人困惑的结果可能要通过开发生殖细胞 *lrrk2* 无效突变种系（germline null *lrrk2* mutants）来解决。另外 *LRRK2*$^{-/-}$ 敲除小鼠并没有表现出明显的形态学改变或多巴胺能缺失，这一结果提示在一些吗啉反义寡核苷酸实验中报道的表型改变可能不是由斑马鱼的 LRRK2 缺失而引起的。

值得注意的是，LRRK2 突变通过功能获得机制引起常染色体显性帕金森病，但目前尚无转基因斑马鱼过表达野生型或突变型 LRRK2 的报道。这可能是由 LRRK2 cDNA 太大所带来的技术障碍以及没有可用的大型转基因质粒造成的。随着 CRISPR/Cas9 敲入技术的发展，人们可将显性突变基因引入斑马鱼内源性转座子，从而构建转基因品系。应用该技术构建的转基因斑马鱼具有在内源性调控元件下的生理表达优势，并可用于阐明特定神经元群的选择性损伤机制。

23.5.2 常染色体隐性帕金森病

23.5.2.1 PTEN 诱导激酶 1

PTEN 诱导激酶 1（PTEN-induced putative kinase 1, PINK1）是一种线粒体丝氨酸 / 苏氨酸蛋白激酶（McWilliams and Muqit, 2017），可磷酸化线粒体自噬关键途径中的 Parkin（一种泛素连接酶，也与 PD 相关）。PINK1 功能的丧失会导致常染色体隐性 PD，通常在年轻时发病，但发病机制不明确，可能是有缺陷的线粒体降解导致受损线粒体积累，从而产生活性氧（Barodia et al., 2017）。

斑马鱼和人类 PINK1 和 PINK1 激酶结构域同源率分别为 53% 和 73%（图 23-8）。这种高度保守序列表明 PINK1 功能也是保守的，因此可以用斑马鱼研究 PINK1 的生理作用及其缺失的致病机制（Zhang et al., 2017）。与 LRRK2 一样，斑马鱼 PINK1 的吗啉反义寡核苷酸研究也产生了相互矛盾的结果，从几乎可以肯定是由非特异性毒性（而非 PINK1

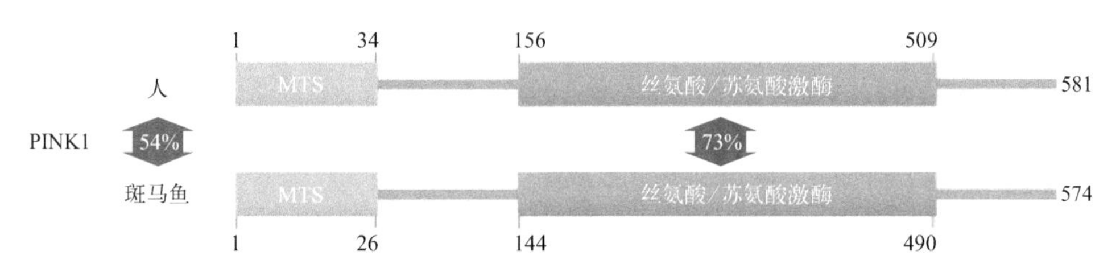

图 23-8 **人类和斑马鱼 PINK1 的结构域和系统发育保守性**

PINK1 关键结构域示意图，数字代表每个结构域的开始和结束位置。百分数表示人类和斑马鱼蛋白质之间的氨基酸相似性（左侧 54% 代表整个蛋白质的相似性，右侧 73% 代表激酶结构域的相似性）。MTS ：线粒体靶向序列损失）引起的严重形态缺陷（Anichtchik et al., 2008）到更微妙的表型，包括间脑多巴胺神经元的模式改变、触觉受损诱发逃逸反应，以及 MPTP 敏感性增加等（Sallinen et al., 2010; Xi et al., 2010）。

从定向诱导基因组局部突变技术（targeting induced local lesions in genomes, TILLING）筛选中发现了导致激酶结构域内提前产生终止密码子的 *pink1* 基因突变（Flinn et al., 2013）。通过标记探针 *toth1* 和 *slc6a3* 的 RNA 原位杂交实验发现，纯合的 *pink1* 缺失（$pink1^{-/-}$）斑马鱼的 DC1、DC2、DC4 和 DC5 多巴胺能神经元数量减少。虽然线粒体的数量和密度保持正常，但相对于对照组，单个线粒体平均增大 40%，每个切片的线粒体面积更大。生化分析显示呼吸链复合物Ⅰ和Ⅲ的活性降低。虽然多巴胺能神经元的数量减少了，但 $pink1^{-/-}$ 斑马鱼模型的运动表型并没有发生异常，提示（与 PINK1 相关的帕金森病不同）该表型似乎是非进行性的。$pink1^{-/-}$ 斑马鱼的基因表达结果显示 TigarB 上调，TigarB 是一种线粒体自噬的负调控因子，同时在细胞凋亡和糖酵解中发挥作用。使用吗啉反义寡核苷酸敲减 TigarB 可使 $pink1^{-/-}$ 斑马鱼受损的多巴胺能神经元和线粒体恢复功能。这项出色的研究利用斑马鱼遗传学技术，在体内 PD 相关神经元群体中定义了一条参与 PINK1 缺失反应的新通路（Flinn et al., 2013）。

另一项精心设计的研究利用 TALEN 靶向斑马鱼 *pink1* 基因的外显子 6，引起终止密码子的提前出现，破坏了 PINK1 激酶结构域（Zhang et al., 2017）。突变体显示 *pink1* mRNA 表达减少，表明存在无义介导的降解（nonsense-mediated decay）。与 *pink1* TILLING 突变体类似，这些斑马鱼的 TH 免疫反应性神经元数量略有减少，但基线运动正常。然而，低浓度的鱼藤酮不足以引起野生型斑马鱼仔鱼的表型改变，但却可以导致 TH^{+} 神经元的损失，并引起 $pink1^{-/-}$ 斑马鱼触摸诱发逃避反应受损。这是鱼藤酮（抑制线粒体复合物Ⅰ）特有的作用，与复合物Ⅲ抑制剂抗霉素 A 的敏感性没有差异。与 *pink1* 和鱼藤酮单独干预相比，*pink1* 的缺失与鱼藤酮相结合导致动物氧消耗量显著降低。通过对化学物质的筛选，发现了三种吩噻嗪，它们可恢复触摸诱发逃避反应，同时也部分挽救了线粒体功能和损失的 TH^{+} 神经元。这些化合物可能增强了自噬，但不能恢复受损线粒体表面的 Parkin 易位。这项研究表明，增强的自噬可能弥补 PINK1 的缺失，而不一定恢复线粒体功能或 PINK1/parkinson 依赖性的线粒体质量控制。另外，在 PINK1 缺失的斑马鱼模型中发现吩噻嗪可作为神经保护剂，这一结果说明在 PD 研究中将化学生物学应用于斑马鱼模型具有非常重要的作用。确定这些吩噻嗪的保护特性是否可以与其作为多巴胺受体拮抗剂的药理学特性分离开来很有意义，这是上述初步发现最终可用于 PD 治疗的前提。

Tg（*pink1*：*egfp*）品系可在 2kb 的 *pink1* 5′ 调节序列下表达 EGFP，已被用于研究氧化应激对 *pink1* 表达的调控（Priyadarshini et al., 2013a）。将转基因仔鱼暴露于 H_2O_2 可导致大脑中 EGFP 的表达增加，而用抗氧化剂谷胱甘肽处理则将 EGFP 的表达降低到正常水平。这些数据表明，*pink1* 2kb 的调控区包含一个氧化还原反应元件（RedOx response element），该元件可能在氧化应激条件下增加神经元 *pink1* 的表达。将对照组与 *pink1* 吗啉反义寡核苷酸敲低的斑马鱼进行基因表达分析，发现 HIF、TGFb、RAR 以及与线粒体功能和生物发生相关的基因表达均发生了改变（Priyadarshini et al., 2013b），表明 *pink1* 在细胞对氧化应激的反应中发挥关键作用。

在生理条件下，PINK1 进入线粒体，在线粒体内被一种位于线粒体膜内的蛋白酶——早老素相关菱形样蛋白（presenilin-associated rhomboid-like protein, PARL）切割（Shi et al., 2011）。斑马鱼有两个 PARL 类似物，都在中枢神经系统中表达（Noble et al., 2012）。其中一个基因的吗啉反义寡核苷酸敲低可导致间脑中 *th1* 和 Tg（*slc6a3*：*egfp*）标记的多巴胺能神经元数量减少，同时敲低这两个基因则导致更严重的异常和更高的死亡率，如基因敲除小鼠缺失 PARL 可引起小鼠死亡，人源 PARL 或斑马鱼 *pink1* mRNA 可修复上述异常。目前已经鉴定出了几种 PARL 的底物，包括线粒体裂变蛋白 OPA1。斑马鱼模型的研究结果发现，PARL 缺失所引起的不良细胞后果可以通过 PINK1 的不相关功能来补偿，表明 PARL 和 PINK1 可能在相同的通路中发挥作用。

23.5.2.2 Parkin

Parkin 是一种 E3 泛素连接酶，在 PINK1 下游发挥作用，调节受损线粒体的降解（Barodia et al., 2017）。编码 Parkin 基因的多种功能缺失突变导致常染色体隐性 PD，可能发生在青少年时期。Parkin 相关性 PD 的病理学不同于偶发性 PD。在大多数报告的案例中，并没有观察到路易小体。然而，与 PD 相似，黑质和蓝斑退化及由此引起的运动障碍可对左旋多巴（L-DOPA）产生反应。

斑马鱼和人类 Parkin 基因有 62% 的同源率，关键功能域的同源率则增加到 80%（图 23-9）（Fett et al., 2010; Flinn et al., 2009）。斑马鱼 *prkn* 基因组成性表达于整个胚胎发生过程中（Flinn et al., 2009）。细胞实验表明，与人类同源基因相似，斑马鱼 Parkin 也具有 E3 泛素连接酶活性（Fett et al., 2010）。鱼藤酮或海藻碱可诱导斑马鱼的 *prkn* mRNA 上调，并且该蛋白可能具有神经保护功能（Fett et al., 2010）。斑马鱼 Parkin 的上述许多特性与人类 Parkin 一样，这表明斑马鱼模型可能有助于确定 Parkin 功能丧失导致神经变性的机制。

在斑马鱼中，使用吗啉反义寡核苷酸瞬时敲减 *prkn* 产生了相互矛盾的结果。吗啉反义寡核苷酸导致初级 *prkn* 转录本外显子 9 被切除，最终导致无义突变和 *prkn* mRNA 表达缺失。在注射胚胎中发现了较少的表达 TH1 的间脑多巴胺能神经元，这也表明了线粒体复合体 Ⅰ 活性的全身性缺陷（Flinn et al., 2009）。线粒体复合物 Ⅰ 抑制剂 MPP^{+} 可进一步减少多巴胺能神经元的数量，但是并不能观察到运动功能的异常。相反，阻止外显子 3 被剪切的肽核酸寡核苷酸可导致 *prkn* mRNA 总水平减少 70%，但不会引起多巴胺神经元损失或线粒体异常的发生（Fett et al., 2010）。同样的研究表明，过表达 Parkin 对热休克具有一定的保护作用。稳定的 *prkn* 缺失突变体的产生有望解决 Parkin 的缺失是否可以引起斑马鱼 PD 相关表型的难题。

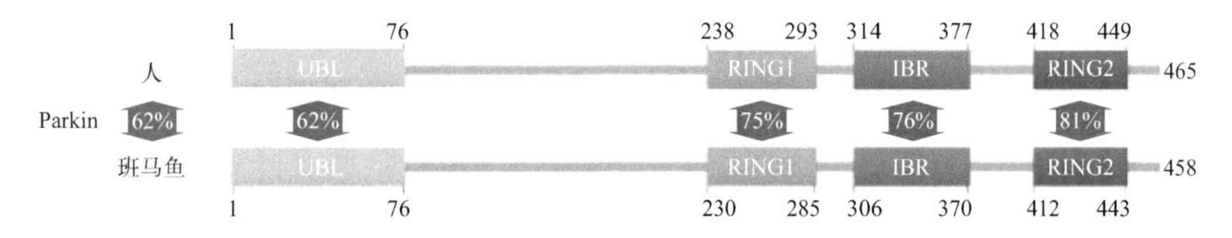

图 23-9 **人类和斑马鱼 Parkin 的域结构和系统发育保守性**

Parkin 的关键结构域示意性图；数字代表每个结构域的起始和结束位置。百分数表示人类和斑马鱼蛋白质之间的氨基酸相似性（左侧 62% 代表整个蛋白质的相似性，其余 % 代表单个结构域的相似性）

23.5.2.3 DJ-1

DJ-1是一种高度保守的蛋白质，在细菌、酵母、无脊椎动物和脊椎动物中都有同源蛋白质。其功能尚不完全清楚，已知其具有抗氧化特性、转录调节功能和分子伴侣活性等（Hijioka et al., 2017）。DJ-1编码基因功能缺失突变可引起常染色体隐性早发型PD（Bonifati et al., 2003）。主要病理特征与PD类似，包括神经元大量丢失和路易小体形成（Antipova and Bandopadhyay, 2017）。

人类与斑马鱼DJ-1之间存在显著的序列保守性，氨基酸同源率为83%，相似性为89%。人类蛋白中受致病性突变影响的每一个残基在斑马鱼中都是保守的。斑马鱼DJ-1在前脑和间脑的多巴胺能神经元中表达（Bai et al., 2006）。吗啉反义寡核苷酸敲除斑马鱼DJ-1可上调*p53*的表达，但本身并不引起神经元死亡（Bretaud et al., 2007）。然而，同时使用蛋白酶体抑制剂可引起广泛的神经元死亡，表明DJ-1在蛋白毒性应激反应中发挥作用。*dj1*缺失突变的斑马鱼种系尚未有报道。

在胶质细胞中DJ-1可能发挥关键作用，其对神经元的保护作用可能是非细胞自主性的（De Miranda et al., 2018）。在神经胶质*gfap*调控元件下表达DJ-1的转基因斑马鱼并不能产生自发表型。但表达DJ-1的转基因斑马鱼对MPP^+有抗性，已知MPP^+是一种线粒体复合物Ⅰ抑制剂，特定集中在多巴胺能神经元中。与对照组相比，MPP^+暴露后，Tg（*gfap:dj1*）仔鱼表现出iNOS诱导减少、蛋白S-亚硝基化以及TH表达缺失（Froyset et al., 2018）。对照组仔鱼在MPP^+暴露后出现的运动减退可通过神经胶质*dj1*表达来恢复。对流式细胞分选得到的对照组和Tg（*gfap:dj1*）组仔鱼神经胶质细胞进行蛋白质组学分析，发现参与氧化还原调节、炎症和线粒体呼吸的蛋白质均被上调。这些发现加深了我们对星形胶质细胞介导的体内神经保护分子基础以及DJ-1缺失引发神经变性机制的了解。

23.5.3 调节帕金森病风险的基因

如前所述，偶发性PD的病因可能包括环境和遗传因素，遗传易感性很复杂，涉及多个基因。全基因组关联研究已经确定了许多与PD相关的非编码遗传变异（Satake et al., 2009; Simon-Sanchez et al., 2009）。利用CRISPR技术引入生殖系突变，是斑马鱼反向遗传学的一个令人兴奋的应用，它将分析这些基因在体内的功能及其与PD中其他基因和环境毒素的相互作用。目前该方法仍在开发中，除了面对多突变体的产生和组合分析固有的复杂性的挑战外，还可能面临单个基因变异对PD风险的影响相对较小的挑战。在这方面，分析对PD风险有实质性影响的稀有基因突变可能是改进该方法的最有效手段。GBA1基因就是一个很好的例子。

葡糖脑苷脂酶（glucocerebrosidase, GCase）由GBA1基因编码，是一种主要参与糖脂分解代谢的溶酶体酶。人类功能性缺失GBA1纯合突变可导致戈谢病（Gaucher’s disease，又称高雪病），一种常染色体隐性溶酶体储存障碍症（Grabowski, 2012）。然而有趣的是，杂合GBA1突变携带者患PD的风险约为正常人的20倍（McNeill et al., 2012），GBA1突变在PD患者中的发生率为正常人的5倍（Sidransky et al., 2009）。GBA1突变可能的致病机制涉及GCase活性的丧失导致溶酶体蛋白质降解受损引起的α-突触核蛋白累积，以及GCase底物葡萄糖神经酰胺促进α-突触核蛋白的聚集（Mazzulli et al., 2011）。利用这种功能丧失机制可开发靶向治疗策略。

斑马鱼 *gba1* 编码的蛋白质与人类 GCase 的同源率为 57%（Keatinge et al., 2015）。利用 TALEN 核酸酶产生了 *gba1* 突变体，该突变体含有一个因提前终止翻译产生的外显子 7，并导致 *gba1* mRNA 和 GCase 活性降低 50% 以上（Keatinge et al., 2015）。纯合的 $gba1^{-/-}$ 斑马鱼重现了戈谢病的许多表型，包括鞘脂代谢物的积累、进行性神经缺陷和生存能力受损，组织学结果显示全脑有严重的神经炎症，并在脑室附近积聚了充盈着脂质的巨噬细胞。受精后 12 周，$gba1^{-/-}$ 斑马鱼后结节和下丘脑尾侧的多巴胺能神经元明显减少，线粒体复合物Ⅲ和Ⅳ的脑内活性降低 50%，复合物Ⅰ、Ⅳ的编码组件 *ndufa9*、*cox4i1* 和 *atp5a* 的表达减少。同时还观察到 LC3-Ⅱ的增加，表明自噬增加或溶酶体对自噬体的清除减少。在杂合的 $gba1^{+/-}$ 斑马鱼中并未出现相似表型，该模型模拟了携带者可增加患 PD 风险的状态。目前尚不清楚是斑马鱼大脑中 α- 突触核蛋白的缺失，还是斑马鱼实验时间尺度的压缩，又或是其他因素导致了 $gba1^{+/-}$ 表型的缺失，因此将 $gba1^{+/-}$ 斑马鱼与表达 α- 突触核蛋白的转基因动物杂交具有非常重要的意义。

23.6 利用斑马鱼模型阐明环境毒物与帕金森病相关的作用机制

虽然单基因型 PD 表型的复制为阐明神经退行性变的分子机制提供了一个极具潜力的研究途径，但大多数 PD 患者在上述基因中并没有致病突变。普遍认为，在大多数病例中，环境暴露是 PD 发病机制的关键部分，但是相关毒物的作用很多仍然不清楚。线粒体功能障碍和氧化应激在 PD 组织中反复出现。线粒体抑制剂和氧化毒物的职业暴露（和无意中的自我摄入）与帕金森病和急性帕金森病有关。在动物模型中，已知某些化学物质与毒性诱发的帕金森病有关（图 23-10），可导致多巴胺能神经元的缺失。考虑到采用无偏见的发现导向型方法（unbiased discovery-driven approaches）和活体成像方法描述病理生理学所展现出的潜力，人们对使用斑马鱼仔鱼开发毒物 PD 模型产生了浓厚的兴趣（表 23-2）。

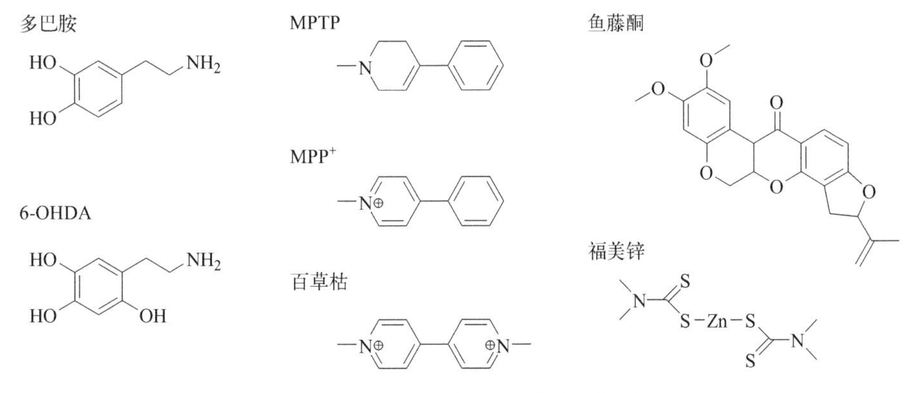

图 23-10 **与帕金森病相关的化学毒物**

1- 甲基 -4- 苯基 1,2,3,6- 四氢吡啶（MPTP）/1- 甲基 -4- 苯基吡啶（MPP^+）、百草枯、鱼藤酮和福美锌的化学结构均与人类 PD 有关，这些化合物的结构与神经递质多巴胺及其衍生物 6-OHDA（6- 羟多巴胺）的结构对比展示，后者常被用作氧化毒物来制作帕金森病模型

表 23-2 已发表的将斑马鱼暴露于与 PD 相关化学毒物的研究摘要

化合物	年龄	染毒途径	主要表型	参考文献
6-羟多巴胺	成鱼	肌肉	运动和多巴胺（DA）水平减少	Anichtchik et al., (2004)
	成鱼	脑室	运动和多巴胺神经元减少	Vijayanathan et al., (2017)
	仔鱼	水体	运动减少，基因表达改变	Feng et al., (2014)
1-甲基-4-苯基-1,2,3,6-四氢吡啶	仔鱼	水体	运动和多巴胺神经元减少	Lam et al., (2005)
	仔鱼	水体	运动和多巴胺神经元减少，可通过 DAT 敲减后恢复	McKinley et al., (2005)
	成鱼	腹腔	运动减少	Bretaud et al., (2004)
	仔鱼	水体	发育缺陷	
	成鱼	肌肉	运动和多巴胺水平减少	Anichtchik et al., (2004)
	成鱼	腹腔	僵直期，基因表达改变	Sarath babu et al., (2016)
	仔鱼	水体	表达 Et(*vmat:egfp*) 的多巴胺神经元减少	Wen et al., (2008)
	仔鱼	水体	运动减少，多巴胺神经元和多巴胺水平降低	Sal.,inen et al., (2009b)
1-甲基-4-苯基吡啶	仔鱼	水体	运动减少，多巴胺神经元和多巴胺水平降低	Sal.,inen et al., (2009b)
	仔鱼	水体	线粒体逆行和顺行转运改变	Dukes et al., (2016)
	仔鱼	水体	自发运动的频率减少	Farrell et al., (2011)
百草枯	成鱼	腹腔	运动减少、DAT 水平降低，空间记忆受损	Bortolotto et al., (2014)
	成鱼	腹腔	行为和基因表达改变	Nunes et al., (2017)
	仔鱼	水体	无行为和 DA 神经元改变	Bretaud et al., (2004)
	仔鱼	水体	发育异常、运动能力和 DA 水平降低，细胞死亡	Nellore and Nandita., (2015)
	仔鱼	水体	氧化应激、凋亡和巨噬细胞侵袭	Wang et al., (2016)
鱼藤酮	成鱼	腹腔	无表型	Bretaud et al., (2004)
	仔鱼	水体		
	成鱼	水体	运动、嗅觉和 DA 水平降低	Wang et al., (2017)

23.6.1 6-羟多巴胺

6-羟多巴胺（6-hydroxydopamine, 6-OHDA）是第一个被证明能引起多巴胺能神经元特异性中毒的化学物质（Jonsson and Sachs, 1975; Ungerstedt, 1968）。在此之后，人们使用 6-OHDA 进行体内、外多巴胺能神经元变性的机制研究。儿茶酚胺能神经元的特异性可归因于 6-OHDA 是 DAT 和去甲肾上腺素转运蛋白（norepinephrine transporter, NET）的底物（Luthman et al., 1989）。已知 6-OHDA 通过产生氧化应激而发挥作用（Cohen, 1984）。但由于 6-OHDA 在哺乳动物中不能跨过血脑屏障，因此它必须以特定方式给药才能对大脑多巴胺能结构产生影响（Javoy et al., 1976）。

在受精后 2 ～ 4 天（血脑屏障完全发育之前），在含有 10 ～ 250μmol/L 6-OHDA 的水中孵育斑马鱼仔鱼，仔鱼呈现出浓度和时间依赖性活动减少（Feng et al., 2014）。联合应用左旋多巴、小胶质细胞抑制剂米诺环素和抗氧化剂维生素 E 可以恢复运动缺陷，但是对于多巴胺能神经元的完整性尚无相关报道。有其他研究在成年斑马鱼中使用了 6-OHDA，成年斑马鱼肌内注射 6-OHDA 后，脑内儿茶酚胺浓度降低，游动减少（Anichtchik et al., 2004）。虽然没有量化多巴胺能神经元的数量，但 TH 表达没有变化，没有 TUNEL 染色或 caspase-3 激活的证据（提示没有发生细胞凋亡）。此外，由于成年斑马鱼具有血脑屏障，尚不清楚 6-OHDA 在中枢神经多巴胺能系统中能否起到显著作用。有意思的是，6-OHDA 注射后 7 ～ 9 天斑马鱼出现了行为恢复的趋势；这与其他 6-OHDA 模型中报道的运动功能和神经化学恢复类似（Vijayanathan et al., 2017）。将 6-OHDA 直接注射到成年斑马鱼的腹侧间脑，可在 3 天内导致视前区、后结节和下丘脑约 85% 的多巴胺能神经元缺失，但不引起死亡（Vijayanathan et al., 2017）。与多巴胺功能的丧失相对应，暴露于 6-OHDA 的斑马鱼表现出游动速度和活性的降低。值得注意的是，在成年斑马鱼大脑受损 30 天后，多巴胺能系统完全再生，运动功能完全恢复（Vijayanathan et al., 2017）。多位学者的研究成果证实成年斑马鱼中枢神经系统的再生能力（Becker and Becker, 2014; Dias et al., 2012; Goldshmit et al., 2012; Reimer et al., 2008, 2009），斑马鱼模型中成熟多巴胺能系统再生的分子机制对利用中枢神经系统修复途径治疗 PD 的研究具有重大意义。

23.6.2 线粒体抑制剂：MPTP、MPP^+ 和鱼藤酮

1- 甲基 -4- 苯基 -1,2,3,6- 四氢吡啶（1-methyl 4-phenyl 1,2,3,6-tetrahydropyridine, MPTP）是偶然发现的非法麻醉品合成的副产品。使用受 MPTP 污染的毒品可导致急性 PD，并伴随 SNc 多巴胺能神经元退化（Davis et al., 1979; Langston et al., 1983, 1999）。MPTP 是亲脂性的，可透过血脑屏障，进入中枢神经系统后，被神经胶质细胞中的单胺氧化酶 B（MAO_B）氧化，形成 1- 甲基 -4- 苯基吡啶（MPP^+），这是一种非亲脂性化合物，作为 DAT 的底物，被多巴胺能神经元选择性地摄取（Javitch et al., 1985; Javitch and Snyder, 1984; Mayer et al., 1986）。MPP^+ 是一种线粒体复合物 I 抑制剂，会导致电子传输受损和活性氧的产生（Mizuno et al., 1987; Nicklas et al., 1985; Ramsay et al., 1986）。自发现以来，MPTP 已被广泛应用于选择性诱导多巴胺能神经元氧化应激以阐明下游机制，并消融多巴胺能系统以生成用于测试对症治疗的动物模型。

多项研究评估了 MPTP 和 MPP^+ 对斑马鱼仔鱼的影响（Bretaud et al., 2004; Farrell et al., 2011; Lam et al., 2005; Sallinen et al., 2009b；Zhou et al., 2014）。这些研究发现斑马鱼仔鱼暴露于含有 0.8 ～ 1.0mmol/L MPTP 或 MPP^+ 的水中 2 ～ 4 天的过程中，反复出现自发运动和触摸诱发逃避反应的缺陷（Bretaud et al., 2004; Farrell et al., 2011; Lamet al., 2005; Sallinen et al., 2009b；Zhou et al., 2014）。与其他降低多巴胺功能的干预措施一致，MPP^+ 暴露后自发运动的异常表现是总游泳距离减少，这归因于运动时间减少，而运动的速度并没有改变（Farrell et al., 2011）。也有报道称，暴露于 MPP^+ 会导致多巴胺能神经元的缺失，特别是在腹侧间脑和前脑区（Bretaud et al., 2004; Lam et al., 2005; Sallinen et al., 2009b）。与 MPP^+ 相比，MPTP 会导致更多多巴胺能神经元的缺失，仔鱼总多巴胺水平下降 60% 以上（Sallinen et al., 2009b）。有趣的是，在停止 MPTP 暴露后，运动功能和表达 TH 的神经元数量都恢复了，这表明可能是神经元在暴露期间抑制了它们的神经化学表型，又或者是在暴露后新的细胞再生了。MPTP 在斑马鱼仔鱼中的毒性机制似乎与先前在哺乳动物模型中证实的机制一致。MAO_B 抑制剂司来吉兰（selegiline）部分阻止了细胞缺失、神经化学和神经行为缺陷，这表明 MPTP 氧化为 MPP^+ 是斑马鱼仔鱼产生毒性所必需的（Lam et al., 2005; McKinley et al., 2005; Sallinen et al., 2009b）。此外，使用药理学抑制剂诺米芬辛

（nomifensine）或使用吗啉反义寡核苷酸通过靶向基因敲低 DAT，也可以改善暴露于 MPTP 后的细胞缺失（McKinley et al., 2005）。这表明 MPP^+ 可被 DAT 转运至斑马鱼的多巴胺能神经末梢，类似于哺乳动物。通过给成年斑马鱼注射 MPTP，也发现了大致相似的结果，包括对儿茶酚胺水平和运动活性的影响（Anichtchik et al., 2004; Bretaud et al., 2004）。

已在斑马鱼仔鱼中将 MPTP 和 MPP^+ 与荧光素报告基因及活体成像相结合，为检测体内细胞缺失和了解病理生理学提供了强大的模型。利用编码 GFP 的报告基因与最小启动子连接构建重组体，增强子诱捕筛选重新获得了插入到编码囊泡单胺转运蛋白的 *vmat2* 基因位点的斑马鱼转基因品系（Wen et al., 2008）。由此产生的 ET（*vmat2:egfp*）斑马鱼在整个中枢神经系统中显示出单胺能神经元的 GFP 荧光标记，包括多巴胺能系统。MPTP 暴露 4 天后，GFP 标记的腹侧间脑神经元数量显著减少，而其他神经元基本不受影响。该模型提供了对多巴胺能神经元缺失的快速体内测定方法，可能实现对推定的导致神经变性的化学和遗传修饰物进行简单评估。使用转基因斑马鱼品系的研究也得出了类似的结果，在该品系斑马鱼中，GFP 在编码 DAT 的 *sl6a3* 基因的调控元件下表达，进而可检测 MPTP 暴露后体内细胞的缺失（Xi et al., 2011）。最近的研究中，标记了转基因斑马鱼多巴胺能神经元的线粒体（Dukes et al., 2016; Noble et al., 2015），使用延时共聚焦活体显微镜，首次实现了对体内多巴胺能神经元中的轴突线粒体转运的直接成像（Dukes et al., 2016）。该转基因品系与短暂暴露于低浓度 MPP^+ 相结合，以检测氧化应激反应后多巴胺能神经元中发生的早期事件。在受精后 5 天的仔鱼中，用 0.5mmol/L MPP^+ 孵育 16h 不会导致运动功能或多巴胺水平变化、多巴胺能细胞损失或死亡率发生变化。但是这种浓度的 MPP^+ 暴露引起了轴突线粒体转运的显著变化，包括逆行移动的线粒体比例增加了两倍以上。这些发现表明，神经元线粒体动力学的改变可能是发病机制的早期事件，并且与之前在体外进行的研究结果大致相符（Arnold et al., 2011; Cartelli et al., 2010; Kim-Han et al., 2011）。目前尚不清楚线粒体逆行运输是否是处理受损线粒体的一种代偿机制。

与 MPP^+ 一样，鱼藤酮是一种线粒体抑制剂，被用作农用杀虫剂，职业性暴露于鱼藤酮发生偶发性 PD 的风险增加两倍以上（Pouchieu et al., 2018; Tanner et al., 2011）。鱼藤酮高度亲脂，易于透过细胞膜和血脑屏障。在大鼠中，亚急性全身暴露于鱼藤酮会导致运动障碍、黑质纹状体多巴胺能投射的相对选择性退化和类似路易小体的 α- 突触核蛋白免疫反应包涵体的形成（Betarbet et al., 2000; Cannon et al., 2009）。人们对利用斑马鱼 - 鱼藤酮模型了解其潜在的机制产生浓厚的兴趣。鱼藤酮也是一种高效的杀鱼剂，在极低浓度时对许多成年鱼都会产生毒性（Melo et al., 2015）。最早的研究报道将受精后 1 ～ 5 天的斑马鱼仔鱼暴露于 5 ～ 10μg/L（约 12 ～ 25nmol/L）的鱼藤酮不会导致运动异常或多巴胺能神经元缺失（Bretaud et al., 2004）。随后，有报道称，从早期发育到受精后 96h 的斑马鱼胚胎暴露于鱼藤酮的 LD_{50} 为 12.2μg/L（约 30nmol/L）。幸存的仔鱼表现出一系列全身性异常，包括平衡丧失、色素沉着缺乏和严重的体轴畸形，这些都不可能反映多巴胺能系统的异常（Melo et al., 2015）。相比之下，成年斑马鱼暴露于 2μg/L（约 5nmol/L）鱼藤酮 4 周后，出现 TH 表达和多巴胺水平轻度降低，伴有与高速游泳相关的细微运动缺陷，以及气味辨别和避黑检测异常（Wang et al., 2017）。这些被认为可以反映 PD 的运动和非运动症状，但没有组织学分析或尝试用多巴胺能药物来恢复运动异常。研究还发现 PD 相关基因 mRNA 表达发生变化，包括 *dj1* 表达下调和 *lrrk2* 表达上调（Wang et al., 2017）。需要注意的是，在大鼠中，鱼藤酮引起的 PD 样神经病理学依赖于 α- 突触核蛋白的表达（Zharikov et al., 2015）。而野生型斑马鱼并不表达突触核蛋白，目前尚不清楚斑马鱼 γ1- 突触核蛋白和人 α- 突触核蛋白之间的功能相似性是否可以扩展说明鱼藤酮诱导的多巴胺能神经元死亡过程中 γ1- 突触核蛋白发挥类似的作用。因此，鱼藤酮可能只会在表达其他物种（如人或大鼠）的 α- 突触核蛋白的转基因斑马鱼中引发 PD 样病理。

23.6.3 百草枯

百草枯（1,10-dimethyl-4,40-bipyridine，1,10- 二甲基 -4,40- 联吡啶）是世界上应用最广泛的农业除草剂。在流行病学上，接触百草枯及相关除草剂与 PD 风险增加有关（Pezzoli and Cereda, 2013; Pouchieu et al., 2018; Tanner et al., 2011）。百草枯与 MPP^+ 在结构上有一些相似之处，在哺乳动物模型中，已证明能诱发与 PD 相关的神经元损伤，包括氧化应激增加、α- 突触核蛋白聚集和选择性黑质损伤（Day et al., 1999; Kuter et al., 2010; McCormack et al., 2002）。

初步研究表明，在受精后 24h 加入百草枯 5 ～ 10mg/L（约 19 ～ 38μmol/L），在受精后 5 天不会导致斑马鱼仔鱼多巴胺能神经元的缺失，在受精后 7 天也不会导致可检测到的运动功能异常（Bretaud et al., 2004）。最近，人们对浓度较低的百草枯进行了更多的研究。受精后 18h 的斑马鱼暴露于 0.02×10^{-6} ～ 0.08×10^{-6}（约 0.08 ～ 0.31μmol/L）百草枯时，尾卷和自发位移减少，多巴胺水平在受精后 48 ～ 72h 略有降低。随后证实了氧化应激反应和可能严重影响运动功能变化的形态学严重异常的发生（Nellore and Nandita, 2015）。受精后 3 天的仔鱼暴露于 0.1 ～ 1000μg/mL（约 0.38 ～ 3.8μmol/L）百草枯，72h 的 LD_{50} 为 0.38μmol/L，用于后续研究。对接触百草枯的仔鱼进行一系列分析，结果显示氧化应激反应的发生、细胞死亡率增加、血管发育缺陷、巨噬细胞迁移以及相关基因表达发生变化（Wang et al., 2016）。长期或反复低浓度暴露于百草枯是否会在不引起严重全身毒性的情况下引发斑马鱼的 PD 样变化，这一问题有待解决。

23.6.4 其他化合物

在工作和生活场所长期暴露于二硫代氨基甲酸酯（dithiocarbamate, DTC）类化合物——灭菌剂福美锌（ziram）的人群，患 PD 的风险增加 3 倍（Wang et al., 2011）。斑马鱼仔鱼 ET（*vmat2:egfp*）暴露于 50nmol/L 福美锌，从受精后 24h 到受精后 6 天进行观察，发现端脑和间脑中表达 GFP 的神经元数量减少，但中腹核神经元和感觉神经元则没有变化（Lulla et al., 2016）。在受精后 7 天时，暴露于福美锌的仔鱼出现运动功能异常，主要表现为在黑暗中位移减少，这一运动功能异常可通过使用多巴胺激动剂阿扑吗啡（apomorphine）恢复。吗啉反义寡核苷酸敲低 γ1- 突触核蛋白也可使 *nvmat2:egfp* 表达神经元的减少得到恢复，这表明福美锌对儿茶酚胺能神经元的毒性可能是突触核蛋白依赖性的。这项重要的研究首次证明福美锌可以在体内激发与 PD 相关的表型，并阐述了可获得的模型在检测可能对环境产生毒性的毒物过程中的重要性。

23.7 结论

斑马鱼模型在阐明PD机制方面具有很大的潜力。脊椎动物神经系统与基因操作、活体成像和化学暴露的强大结合，已经在基因功能、新型环境毒物的作用、细胞和生化病理生理学以及PD的神经保护方面产生了许多建设性的研究成果。在大多数病例中，遗传和环境因素都可能是PD的病因。因此，建立一个能够同时改变基因表达和化学暴露的模型来阐明相关机制非常重要。而斑马鱼模型可实现这些基因操作，并可进行多因素组合的高通量筛选，以及统计样本的相关控制。人类受试者的遗传和流行病学研究揭示出一系列令人困惑的可能的基因环境组合，这都需要进行大量的生物学验证和组织解剖，这使得斑马鱼的这种能力将变得越来越重要。此外，最近发表的首个利用PD模型进行的无偏化学修饰剂筛选，为斑马鱼PD模型在药物发现中的可能应用提供了研究基础（Zhang et al., 2017）。开发可模拟潜在病理生理学的模型，并对化学修饰剂筛选的相关表型进行高通量分析仍然是当前发展的一个重要目标。

PD模型的建立有几点需要优先考虑。首先是可重复性问题，在人类神经系统疾病建模的许多领域，包括鱼PD模型，可重复性一直是一个问题。例如，与PD相关的化学品暴露的标准化方案可能有助于提高毒物模型的可靠性并消除不同研究之间的明显差异。对吗啉反义寡核苷酸实验的正确控制达成共识，以确定诸如神经退行性表型变化是由靶基因的表达缺失引起的，而不是由不相关的人为因素造成的，这有助于避免产生由脱靶效应引起的错误结论。采用标准化的神经行为测试方法也有助于可重复性的提高，并且对于基于行为的无偏性筛选绝对重要。用于视频跟踪和运动分析的开源软件有助于可重复性的实现（Zhou et al., 2014）。其次，应当优先考虑的是在斑马鱼与PD相关基因的同源物中，转基因模型和种系突变体的系统开发和适当验证。CRISPR/Cas9的最新应用为斑马鱼的反向遗传学提供了直接的方法；除了可模拟导致单基因PD表型的基因突变外，还可使下游通路成为可靠的靶点，为PD病理生理学的理解提供可靠信息。活体显微镜和转基因品系表达的基因产物（涉及细胞生物学的多个方面，如溶酶体或线粒体功能）的最新进展，有助于在单细胞甚至亚细胞分辨率下研究易感神经元群退化的分子进程。

从首次使用斑马鱼模型对PD进行研究到目前取得的优异成绩，仅仅用了大约15年。随着新工具和资源的不断涌现，以及由此产生的关于PD生物学基础的发现，也许我们最终可能会实现James Parkinson在1817年时的预言："有充分的理由期待，在不久的将来，人们就能发现某种治疗 / 补救方法，至少可能阻止疾病的发展。"

（刘艳丽翻译 石童审校）

参考文献

Anichtchik, O., Diekmann, H., Fleming, A., Roach, A., Goldsmith, P., Rubinsztein, D.C., 2008. Loss of PINK1 function affects development and results in neurodegeneration in zebrafish. Journal of Neuroscience 28, 8199-8207.

Anichtchik, O., Sallinen, V., Peitsaro, N., Panula, P., 2006. Distinct structure and activity of monoamine oxidase in the brain of zebrafish (*Danio rerio*). Journal of Comparative Neurology 498, 593-610.

Anichtchik, O.V., Kaslin, J., Peitsaro, N., Scheinin, M., Panula, P., 2004. Neurochemical and behavioural changes in zebrafish

Danio rerio after systemic administration of 6-hydroxydopamine and 1-methyl-4-phenyl-1,2,3,6 tetrahydropyridine. Journal of Neurochemistry 88, 443-453.

Antipova, D., Bandopadhyay, R., 2017. Expression of DJ-1 in neurodegenerative disorders. Advances in Experimental Medicine and Biology 1037, 25-43.

Arnold, B., Cassady, S.J., VanLaar, V.S., Berman, S.B., 2011. Integrating multiple aspects of mitochondrial dynamics in neurons: age-related differences and dynamic changes in a chronic rotenone model. Neurobiology of Disease 41, 189-200.

Bai, Q., Burton, E.A., 2009. Cis-acting elements responsible for dopaminergic neuron-specific expression of zebrafish slc6a3 (dopamine transporter) in vivo are located remote from the transcriptional start site. Neuroscience 164, 1138-1151.

Bai, Q., Mullett, S.J., Garver, J.A., Hinkle, D.A., Burton, E.A., 2006. Zebrafish DJ-1 is evolutionarily conserved and expressed in dopaminergic neurons. Brain Research 1113, 33-44.

Barodia, S.K., Creed, R.B., Goldberg, M.S., 2017. Parkin and PINK1 functions in oxidative stress and neurodegeneration. Brain Research Bulletin 133, 51-59.

Becker, T., Becker, C.G., 2014. Axonal regeneration in zebrafish. Current Opinion in Neurobiology 27, 186-191.

Betarbet, R., Sherer, T.B., MacKenzie, G., Garcia-Osuna, M., Panov, A.V., Greenamyre, J.T., 2000. Chronic systemic pesticide exposure reproduces features of Parkinson's disease. Nature Neuroscience 3, 1301-1306.

Boehmler, W., Carr, T., Thisse, C., Thisse, B., Canfield, V.A., Levenson, R., 2007. D4 Dopamine receptor genes of zebrafish and effects of the antipsychotic clozapine on larval swimming behaviour. Genes, Brain and Behavior 6, 155-166.

Boehmler, W., Obrecht-Pflumio, S., Canfield, V., Thisse, C., Thisse, B., Levenson, R., 2004. Evolution and expression of D2 and D3 dopamine receptor genes in zebrafish. Developmental Dynamics 230, 481-493.

Bonifati, V., Rizzu, P., van Baren, M.J., Schaap, O., Breedveld, G.J., Krieger, E., Dekker, M.C., Squitieri, F., Ibanez, P., Joosse, M., et al., 2003. Mutations in the DJ-1 gene associated with autosomal recessive early-onset parkinsonism. Science 299, 256-259.

Bortolotto, J.W., Cognato, G.P., Christoff, R.R., Roesler, L.N., Leite, C.E., Kist, L.W., Bogo, M.R., Vianna, M.R., Bonan, C.D., 2014. Long-term exposure to paraquat alters behavioral parameters and dopamine levels in adult zebrafish (*Danio rerio*). Zebrafish 11, 142-153.

Braak, H., Del Tredici, K., Rub, U., de Vos, R.A., Jansen Steur, E.N., Braak, E., 2003. Staging of brain pathology related to sporadic Parkinson's disease. Neurobiology of Aging 24, 197-211.

Braak, H., Ghebremedhin, E., Rub, U., Bratzke, H., Del Tredici, K., 2004. Stages in the development of Parkinson's disease-related pathology. Cell and Tissue Research 318, 121-134.

Bretaud, S., Allen, C., Ingham, P.W., Bandmann, O., 2007. p53-dependent neuronal cell death in a DJ-1-deficient zebrafish model of Parkinson's disease. Journal of Neurochemistry 100, 1626-1635.

Bretaud, S., Lee, S., Guo, S., 2004. Sensitivity of zebrafish to environmental toxins implicated in Parkinson's disease. Neurotoxicology and Teratology 26, 857-864.

Budick, S.A., O'Malley, D.M., 2000. Locomotor repertoire of the larval zebrafish: swimming, turning and prey capture. Journal of Experimental Biology 203, 2565-2579.

Burgess, H.A., Granato, M., 2007. Modulation of locomotor activity in larval zebrafish during light adaptation. Journal of Experimental Biology 210, 2526-2539.

Cacabelos, R., 2017. Parkinson's disease: from pathogenesis to pharmacogenomics. International Journal of Molecular Sciences 18.

Candy, J., Collet, C., 2005. Two tyrosine hydroxylase genes in teleosts. Biochimica et Biophysica Acta 1727, 35-44.

Cannon, J.R., Tapias, V., Na, H.M., Honick, A.S., Drolet, R.E., Greenamyre, J.T., 2009. A highly reproducible rotenone model of Parkinson's disease. Neurobiology of Disease 34, 279-290.

Cario, C.L., Farrell, T.C., Milanese, C., Burton, E.A., 2011. Automated measurement of zebrafish larval movement. Journal of Physiology 589, 3703-3708.

Cartelli, D., Ronchi, C., Maggioni, M.G., Rodighiero, S., Giavini, E., Cappelletti, G., 2010. Microtubule dysfunction precedes transport impairment and mitochondria damage in MPP^+-induced neurodegeneration. Journal of Neurochemistry 115, 247-258.

Chen, Y.C., Semenova, S., Rozov, S., Sundvik, M., Bonkowsky, J.L., Panula, P., 2016. A Novel Developmental Role for Dopaminergic Signaling to Specify Hypothalamic Neurotransmitter Identity. J Biol Chem 291, 21880-21892.

Cassar, S., Huang, X., Cole, T., 2018. High-throughput measurement of gut transit time using larval zebrafish. Journal of Visualized Experiments 140, 58497.

Cohen, G., 1984. Oxy-radical toxicity in catecholamine neurons. Neurotoxicology 5, 77-82.

Chronic Parkinsonism secondary to intravenous injection of meperidine analogues. Psychiatry Research 1, 249-254.

Davis, G.C., Williams, A.C., Markey, S.P., Ebert, M.H., Caine, E.D., Reichert, C.M., Kopin, I.J., 1979. Chronic Parkinsonism

secondary to intravenous injection of meperidine analogues. Psychiatry Research 1, 249-254.
Day, B.J., Patel, M., Calavetta, L., Chang, L.Y., Stamler, J.S., 1999. A mechanism of paraquat toxicity involving nitric oxide synthase. Proc Natl Acad Sci U S A 96, 12760-12765.
De Miranda, B.R., Rocha, E.M., Bai, Q., El Ayadi, A., Hinkle, D., Burton, E.A., Greenamyre, J.T., 2018. Astrocytespecific DJ-1 overexpression protects against rotenone-induced neurotoxicity in a rat model of Parkinson's disease. Neurobiology of Disease 115, 101-114.
Deng, H., Wang, P., Jankovic, J., 2018. The genetics of Parkinson disease. Ageing Research Reviews 42, 72-85.
Di Maio, R., Hoffman, E.K., Rocha, E.M., Keeney, M.T., Sanders, L.H., De Miranda, B.R., Zharikov, A., Van Laar, A., Stepan, A.F., Lanz, T.A., et al., 2018. LRRK2 activation in idiopathic Parkinson's disease. Science Translational Medicine 10.
Dias, T.B., Yang, Y.J., Ogai, K., Becker, T., Becker, C.G., 2012. Notch signaling controls generation of motor neurons in the lesioned spinal cord of adult zebrafish. Journal of Neuroscience 32, 3245-3252.
Doty, R.L., 2012. Olfactory dysfunction in Parkinson disease. Nature Reviews Neurology 8, 329-339.
Du, Y., Guo, Q., Shan, M., Wu, Y., Huang, S., Zhao, H., Hong, H., Yang, M., Yang, X., Ren, L., et al., 2016. Spatial and Temporal Distribution of Dopaminergic Neurons during Development in Zebrafish. Front Neuroanat 10, 115.
Dukes, A.A., Bai, Q., Van Laar, V.S., Zhou, Y., Ilin, V., David, C.N., Agim, Z.S., Bonkowsky, J.L., Cannon, J.R., Watkins, S.C., et al., 2016. Live imaging of mitochondrial dynamics in CNS dopaminergic neurons in vivo demonstrates early reversal of mitochondrial transport following MPP(+) exposure. Neurobiology of Disease 95, 238-249.
Eaton, R.C., Bombardieri, R.A., Meyer, D.L., 1977. The Mauthner-initiated startle response in teleost fish. Journal of Experimental Biology 66, 65-81.
Farrell, T.C., Cario, C.L., Milanese, C., Vogt, A., Jeong, J.H., Burton, E.A., 2011. Evaluation of spontaneous propulsive movement as a screening tool to detect rescue of Parkinsonism phenotypes in zebrafish models. Neurobiology of Disease 44, 9-18.
Feng, C.W., Wen, Z.H., Huang, S.Y., Hung, H.C., Chen, C.H., Yang, S.N., Chen, N.F., Wang, H.M., Hsiao, C.D., Chen, W.F., 2014. Effects of 6-hydroxydopamine exposure on motor activity and biochemical expression in zebra-fish (*Danio rerio*) larvae. Zebrafish 11, 227-239.
Fett, M.E., Pilsl, A., Paquet, D., van Bebber, F., Haass, C., Tatzelt, J., Schmid, B., Winklhofer, K.F., 2010. Parkin is protective against proteotoxic stress in a transgenic zebrafish model. PLoS One 5, e11783.
Filippi, A., Mahler, J., Schweitzer, J., Driever, W., 2010. Expression of the paralogous tyrosine hydroxylase encoding genes th1 and th2 reveals the full complement of dopaminergic and noradrenergic neurons in zebrafish larval and juvenile brain. Journal of Comparative Neurology 518, 423-438.
Flinn, L., Mortiboys, H., Volkmann, K., Koster, R.W., Ingham, P.W., Bandmann, O., 2009. Complex I deficiency and dopaminergic neuronal cell loss in parkin-deficient zebrafish (*Danio rerio*). Brain 132, 1613-1623.
Flinn, L.J., Keatinge, M., Bretaud, S., Mortiboys, H., Matsui, H., De Felice, E., Woodroof, H.I., Brown, L., McTighe, A., Soellner, R., et al., 2013. TigarB causes mitochondrial dysfunction and neuronal loss in PINK1 deficiency. Annals of Neurology 74, 837-847.
Froyset, A.K., Edson, A.J., Gharbi, N., Khan, E.A., Dondorp, D., Bai, Q., Tiraboschi, E., Suster, M.L., Connolly, J.B., Burton, E.A., et al., 2018. Astroglial DJ-1 over-expression up-regulates proteins involved in redox regulation and is neuroprotective in vivo. Redox Biology 16, 237-247.
Gash, D.M., Rutland, K., Hudson, N.L., Sullivan, P.G., Bing, G., Cass, W.A., Pandya, J.D., Liu, M., Choi, D.Y., Hunter, R.L., et al., 2008. Trichloroethylene: parkinsonism and complex 1 mitochondrial neurotoxicity. Annals of Neurology 63, 184-192.
Giacomini, N.J., Rose, B., Kobayashi, K., Guo, S., 2006. Antipsychotics produce locomotor impairment in larval zebra-fish. Neurotoxicology and Teratology 28, 245-250.
Godoy, R., Noble, S., Yoon, K., Anisman, H., Ekker, M., 2015. Chemogenetic ablation of dopaminergic neurons leads to transient locomotor impairments in zebrafish larvae. Journal of Neurochemistry 135, 249-260.
Goldshmit, Y., Sztal, T.E., Jusuf, P.R., Hall, T.E., Nguyen-Chi, M., Currie, P.D., 2012. Fgf-dependent glial cell bridges facilitate spinal cord regeneration in zebrafish. Journal of Neuroscience 32, 7477-7492.
Grabowski, G.A., 2012. Gaucher disease and other storage disorders. Hematology-American Society of Hematology Education Program 2012, 13-18.
Hijioka, M., Inden, M., Yanagisawa, D., Kitamura, Y., 2017. DJ-1/PARK7: a new therapeutic target for neurodegenerative disorders. Biological and Pharmaceutical Bulletin 40, 548-552.
Holzschuh, J., Ryu, S., Aberger, F., Driever, W., 2001. Dopamine transporter expression distinguishes dopaminergic neurons from other catecholaminergic neurons in the developing zebrafish embryo. Mechanisms of Development 101, 237-243.

Howe, K., Clark, M.D., Torroja, C.F., Torrance, J., Berthelot, C., Muffato, M., Collins, J.E., Humphray, S., McLaren, K., Matthews, L., et al., 2013. The zebrafish reference genome sequence and its relationship to the human genome. Nature 496, 498-503.

Irons, T.D., Kelly, P.E., Hunter, D.L., Macphail, R.C., Padilla, S., 2013. Acute administration of dopaminergic drugs has differential effects on locomotion in larval zebrafish. Pharmacology Biochemistry and Behavior 103, 792-813.

Irwin, D.J., Grossman, M., Weintraub, D., Hurtig, H.I., Duda, J.E., Xie, S.X., Lee, E.B., Van Deerlin, V.M., Lopez, O.L., Kofler, J.K., et al., 2017. Neuropathological and genetic correlates of survival and dementia onset in synucleinopathies: a retrospective analysis. Lancet Neurol 16, 55-65.

Javitch, J.A., D'Amato, R.J., Strittmatter, S.M., Snyder, S.H., 1985. Parkinsonism-inducing neurotoxin, N-methyl-4-phenyl-1,2,3,6-tetrahydropyridine: uptake of the metabolite N-methyl-4-phenylpyridine by dopamine neurons explains selective toxicity. Proceedings of the National Academy of Sciences of the United States of America 82, 2173-2177.

Javitch, J.A., Snyder, S.H., 1984. Uptake of MPP(+) by dopamine neurons explains selectivity of parkinsonism-inducing neurotoxin, MPTP. European Journal of Pharmacology 106, 455-456.

Javoy, F., Sotelo, C., Herbet, A., Agid, Y., 1976. Specificity of dopaminergic neuronal degeneration induced by intracerebral injection of 6-hydroxydopamine in the nigrostriatal dopamine system. Brain Research 102, 201-215.

Jay, M., De Faveri, F., McDearmid, J.R., 2015. Firing dynamics and modulatory actions of supraspinal dopaminergic neurons during zebrafish locomotor behavior. Current Biology 25, 435-444.

Jonsson, G., Sachs, C., 1975. Actions of 6-hydroxydopamine quinones on catecholamine neurons. Journal of Neurochemistry 25, 509-516.

Kalueff, A.V., Cachat, J.M., 2011. Zebrafish Models in Neurobehavioral Research. Humana Press, New York.

Kastenhuber, E., Kratochwil, C.F., Ryu, S., Schweitzer, J., Driever, W., 2010. Genetic dissection of dopaminergic and noradrenergic contributions to catecholaminergic tracts in early larval zebrafish. The Journal of Comparative Neurology 518, 439-458.

Keatinge, M., Bui, H., Menke, A., Chen, Y.C., Sokol, A.M., Bai, Q., Ellett, F., Da Costa, M., Burke, D., Gegg, M., et al., 2015. Glucocerebrosidase 1 deficient *Danio rerio* mirror key pathological aspects of human Gaucher disease and provide evidence of early microglial activation preceding alpha-synuclein-independent neuronal cell death. Human Molecular Genetics 24, 6640-6652.

Kim-Han, J.S., Antenor-Dorsey, J.A., O'Malley, K.L., 2011. The parkinsonian mimetic, MPP+, specifically impairs mitochondrial transport in dopamine axons. Journal of Neuroscience 31, 7212-7221.

Kok, F.O., Shin, M., Ni, C.W., Gupta, A., Grosse, A.S., van Impel, A., Kirchmaier, B.C., Peterson-Maduro, J., Kourkoulis, G., Male, I., et al., 2015. Reverse genetic screening reveals poor correlation between morpholinoinduced and mutant phenotypes in zebrafish. Developmental Cell 32, 97-108.

Korn, H., Faber, D.S., 2005. The Mauthner cell half a century later: a neurobiological model for decision-making? Neuron 47, 13e28.

Kumar, S., Hedges, S.B., 1998. A molecular timescale for vertebrate evolution. Nature 392, 917-920.

Kuter, K., Nowak, P., Golembiowska, K., Ossowska, K., 2010. Increased reactive oxygen species production in the brain after repeated low-dose pesticide paraquat exposure in rats. A comparison with peripheral tissues. Neurochemical Research 35, 1121-1130.

Lam, C.S., Korzh, V., Strahle, U., 2005. Zebrafish embryos are susceptible to the dopaminergic neurotoxin MPTP. European Journal of Neuroscience 21, 1758-1762.

Langston, J.W., Ballard, P., Tetrud, J.W., Irwin, I., 1983. Chronic Parkinsonism in humans due to a product of meperidine-analog synthesis. Science 219, 979-980.

Langston, J.W., Forno, L.S., Tetrud, J., Reeves, A.G., Kaplan, J.A., Karluk, D., 1999. Evidence of active nerve cell degeneration in the substantia nigra of humans years after 1-methyl-4-phenyl-1,2,3,6-tetrahydropyridine exposure. Annals of Neurology 46, 598-605.

Lesage, S., Durr, A., Tazir, M., Lohmann, E., Leutenegger, A.L., Janin, S., Pollak, P., Brice, A., French Parkinson's Disease Genetics Study Group, 2006. LRRK2 G2019S as a cause of Parkinson's disease in north African Arabs. New England Journal of Medicine 354, 422-423.

Lill, C.M., 2016. Genetics of Parkinson's disease. Molecular and Cellular Probes 30, 386e396.

Lindsay, S.M., Vogt, R.G., 2004. Behavioral responses of newly hatched zebrafish (*Danio rerio*) to amino acid chemostimulants. Chemical Senses 29, 93-100.

Liu, K.S., Fetcho, J.R., 1999. Laser ablations reveal functional relationships of segmental hindbrain neurons in zebra-fish. Neuron 23, 325-335.

Lulla, A., Barnhill, L., Bitan, G., Ivanova, M.I., Nguyen, B., O'Donnell, K., Stahl, M.C., Yamashiro, C., Klarner, F.G., Schrader, T., et al., 2016. Neurotoxicity of the Parkinson disease-associated pesticide ziram is synuclein-dependent in zebrafish embryos. Environmental Health Perspectives 124, 1766-1775.

Luthman, J., Fredriksson, A., Sundstrom, E., Jonsson, G., Archer, T., 1989. Selective lesion of central dopamine or noradrenaline neuron systems in the neonatal rat: motor behavior and monoamine alterations at adult stage. Behavioural Brain Research 33, 267-277.

Mayer, R.A., Kindt, M.V., Heikkila, R.E., 1986. Prevention of the nigrostriatal toxicity of 1-methyl-4-phenyl-1,2,3,6-tetrahydropyridine by inhibitors of 3,4-dihydroxyphenylethylamine transport. Journal of Neurochemistry 47, 1073-1079.

Mazzulli, J.R., Xu, Y.H., Sun, Y., Knight, A.L., McLean, P.J., Caldwell, G.A., Sidransky, E., Grabowski, G.A., Krainc, D., 2011. Gaucher disease glucocerebrosidase and alpha-synuclein form a bidirectional pathogenic loop in synucleinopathies. Cell 146, 37-52.

McCormack, A.L., Thiruchelvam, M., Manning-Bog, A.B., Thiffault, C., Langston, J.W., Cory-Slechta, D.A., Di Monte, D.A., 2002. Environmental risk factors and Parkinson's disease: selective degeneration of nigral dopaminergic neurons caused by the herbicide paraquat. Neurobiology of Disease 10, 119-127.

McKinley, E.T., Baranowski, T.C., Blavo, D.O., Cato, C., Doan, T.N., Rubinstein, A.L., 2005. Neuroprotection of MPTP-induced toxicity in zebrafish dopaminergic neurons. Brain Research Molecular Brain Research 141, 128-137.

McLean, D.L., Fetcho, J.R., 2004. Relationship of tyrosine hydroxylase and serotonin immunoreactivity to sensorimotor circuitry in larval zebrafish. Journal of Comparative Neurology 480, 57-71.

McNeill, A., Duran, R., Hughes, D.A., Mehta, A., Schapira, A.H., 2012. A clinical and family history study of Parkinson's disease in heterozygous glucocerebrosidase mutation carriers. Journal of Neurology Neurosurgery and Psychiatry 83, 853-854.

McWilliams, T.G., Muqit, M.M., 2017. PINK1 and Parkin: emerging themes in mitochondrial homeostasis. Current Opinion in Cell Biology 45, 83-91.

Melo, K.M., Oliveira, R., Grisolia, C.K., Domingues, I., Pieczarka, J.C., de Souza Filho, J., Nagamachi, C.Y., 2015. Short-term exposure to low doses of rotenone induces developmental, biochemical, behavioral, and histological changes in fish. Environmental Science and Pollution Research International 22, 13926-13938.

Meyer, A., Schartl, M., 1999. Gene and genome duplications in vertebrates: the one-to-four (-to-eight in fish) rule and the evolution of novel gene functions. Current Opinion in Cell Biology 11, 699-704.

Milanese, C., Sager, J.J., Bai, Q., Farrell, T.C., Cannon, J.R., Greenamyre, J.T., Burton, E.A., 2012. Hypokinesia and reduced dopamine levels in zebrafish lacking beta-and gamma1-synucleins. Journal of Biological Chemistry 287, 2971-2983.

Mizuno, Y., Sone, N., Saitoh, T., 1987. Effects of 1-methyl-4-phenyl-1,2,3,6-tetrahydropyridine and 1-methyl-4phenylpyridinium ion on activities of the enzymes in the electron transport system in mouse brain. Journal of Neurochemistry 48, 1787-1793.

Nellore, J., Nandita, P., 2015. Paraquat exposure induces behavioral deficits in larval zebrafish during the window of dopamine neurogenesis. Toxicology Reports 2, 950-956.

Nicklas, W.J., Vyas, I., Heikkila, R.E., 1985. Inhibition of NADH-linked oxidation in brain mitochondria by 1-methyl4-phenyl-pyridine, a metabolite of the neurotoxin, 1-methyl-4-phenyl-1,2,5,6-tetrahydropyridine. Life Sciences 36, 2503-2508.

Noble, S., Godoy, R., Affaticati, P., Ekker, M., 2015. Transgenic zebrafish expressing mCherry in the mitochondria of dopaminergic neurons. Zebrafish 12, 349-356.

Noble, S., Ismail, A., Godoy, R., Xi, Y., Ekker, M., 2012. Zebrafish Parla-and Parlb-deficiency affects dopaminergic neuron patterning and embryonic survival. Journal of Neurochemistry 122, 196-207.

Nunes, M.E., Muller, T.E., Braga, M.M., Fontana, B.D., Quadros, V.A., Marins, A., Rodrigues, C., Menezes, C., Rosemberg, D.B., Loro, V.L., 2017. Chronic Treatment with Paraquat Induces Brain Injury, Changes in Antioxidant Defenses System, and Modulates Behavioral Functions in Zebrafish. Mol Neurobiol 54, 3925-3934.

O'Donnell, K.C., Lulla, A., Stahl, M.C., Wheat, N.D., Bronstein, J.M., Sagasti, A., 2014. Axon degeneration and PGC1alpha-mediated protection in a zebrafish model of alpha-synuclein toxicity. Disease Models and Mechanisms 7, 571-582.

Orger, M.B., Kampff, A.R., Severi, K.E., Bollmann, J.H., Engert, F., 2008. Control of visually guided behavior by distinct populations of spinal projection neurons. Nature Neuroscience 11, 327-333.

Panula, P., Chen, Y.C., Priyadarshini, M., Kudo, H., Semenova, S., Sundvik, M., Sallinen, V., 2010. The comparative neuroanatomy and neurochemistry of zebrafish CNS systems of relevance to human neuropsychiatric diseases. Neurobiology of Disease 40, 46-57. Parkinson, J., 1817. An Essay on the Shaking Palsy.

Pezzoli, G., Cereda, E., 2013. Exposure to pesticides or solvents and risk of Parkinson disease. Neurology 80, 2035-2041.

Polymeropoulos, M.H., Lavedan, C., Leroy, E., Ide, S.E., Dehejia, A., Dutra, A., Pike, B., Root, H., Rubenstein, J., Boyer, R., et al., 1997. Mutation in the alpha-synuclein gene identified in families with Parkinson's disease. Science 276, 2045-2047.

Pouchieu, C., Piel, C., Carles, C., Gruber, A., Helmer, C., Tual, S., Marcotullio, E., Lebailly, P., Baldi, I., 2018. Pesticide use in agriculture and Parkinson's disease in the AGRICAN cohort study. International Journal of Epidemiology 47, 299-310.

Prabhudesai, S., Bensabeur, F.Z., Abdullah, R., Basak, I., Baez, S., Alves, G., Holtzman, N.G., Larsen, J.P., Moller, S.G.,

2016. LRRK2 knockdown in zebrafish causes developmental defects, neuronal loss, and synuclein aggregation. Journal of Neuroscience Research 94, 717-735.

Prabhudesai, S., Sinha, S., Attar, A., Kotagiri, A., Fitzmaurice, A.G., Lakshmanan, R., Ivanova, M.I., Loo, J.A., Klarner, F.G., Schrader, T., et al., 2012. A novel "molecular tweezer" inhibitor of alpha-synuclein neurotoxicity in vitro and in vivo. Neurotherapeutics 9, 464-476.

Pringsheim, T., Jette, N., Frolkis, A., Steeves, T.D., 2014. The prevalence of Parkinson's disease: a systematic review and meta-analysis. Movement Disorders: Official Journal of the Movement Disorder Society 29, 1583-1590.

Priyadarshini, M., Orosco, L.A., Panula, P.J., 2013a. Oxidative stress and regulation of Pink1 in zebrafish (*Danio rerio*). PLoS One 8, e81851.

Priyadarshini, M., Tuimala, J., Chen, Y.C., Panula, P., 2013b. A zebrafish model of PINK1 deficiency reveals key pathway dysfunction including HIF signaling. Neurobiology of Disease 54, 127-138.

Ramsay, R.R., Dadgar, J., Trevor, A., Singer, T.P., 1986. Energy-driven uptake of N-methyl-4-phenylpyridine by brain mitochondria mediates the neurotoxicity of MPTP. Life Sciences 39, 581-588.

Reimer, M.M., Kuscha, V., Wyatt, C., Sorensen, I., Frank, R.E., Knuwer, M., Becker, T., Becker, C.G., 2009. Sonic hedgehog is a polarized signal for motor neuron regeneration in adult zebrafish. Journal of Neuroscience 29, 15073-15082.

Reimer, M.M., Sorensen, I., Kuscha, V., Frank, R.E., Liu, C., Becker, C.G., Becker, T., 2008. Motor neuron regeneration in adult zebrafish. Journal of Neuroscience 28, 8510-8516.

Ren, G., Xin, S., Li, S., Zhong, H., Lin, S., 2011. Disruption of LRRK2 does not cause specific loss of dopaminergic neurons in zebrafish. PLoS One 6, e20630.

Ren, G., Li, S., Zhong, H., Lin, S., 2013. Zebrafish tyrosine hydroxylase 2 gene encodes tryptophan hydroxylase. J Biol Chem 288, 22451-22459.

Rihel, J., Prober, D.A., Arvanites, A., Lam, K., Zimmerman, S., Jang, S., Haggarty, S.J., Kokel, D., Rubin, L.L., Peterson, R.T., et al., 2010. Zebrafish behavioral profiling links drugs to biological targets and rest/wake regulation. Science 327, 348-351.

Rink, E., Wullimann, M.F., 2001. The teleostean (zebrafish) dopaminergic system ascending to the subpallium (striatum) is located in the basal diencephalon (posterior tuberculum). Brain Research 889, 316-330.

Rink, E., Wullimann, M.F., 2002. Development of the catecholaminergic system in the early zebrafish brain: an immunohistochemical study. Brain Research Developmental Brain Research 137, 89-100.

Robu, M.E., Larson, J.D., Nasevicius, A., Beiraghi, S., Brenner, C., Farber, S.A., Ekker, S.C., 2007. p53 activation by knockdown technologies. PLoS Genetics 3, e78.

Rokad, D., Ghaisas, S., Harischandra, D.S., Jin, H., Anantharam, V., Kanthasamy, A., Kanthasamy, A.G., 2017. Role of neurotoxicants and traumatic brain injury in alpha-synuclein protein misfolding and aggregation. Brain Research Bulletin 133, 60-70.

Rupp, B., Wullimann, M.F., Reichert, H., 1996. The zebrafish brain: a neuroanatomical comparison with the goldfish. Anatomy and Embryology 194, 187-203.

Sager, J.J., Bai, Q., Burton, E.A., 2010. Transgenic zebrafish models of neurodegenerative diseases. Brain Structure and Function 214, 285-302.

Sallinen, V., Kolehmainen, J., Priyadarshini, M., Toleikyte, G., Chen, Y.C., Panula, P., 2010. Dopaminergic cell damage and vulnerability to MPTP in Pink1 knockdown zebrafish. Neurobiology of Disease 40, 93e101.

Sallinen, V., Sundvik, M., Reenila, I., Peitsaro, N., Khrustalyov, D., Anichtchik, O., Toleikyte, G., Kaslin, J., Panula, P., 2009a. Hyperserotonergic phenotype after monoamine oxidase inhibition in larval zebrafish. Journal of Neurochemistry 109, 403-415.

Sallinen, V., Torkko, V., Sundvik, M., Reenila, I., Khrustalyov, D., Kaslin, J., Panula, P., 2009b. MPTP and MPP^+ target specific aminergic cell populations in larval zebrafish. Journal of Neurochemistry 108, 719-731.

Sarath Babu, N., Murthy Ch, L., Kakara, S., Sharma, R., Brahmendra Swamy, C.V., Idris, M.M., 2016. 1-Methyl-4phenyl-1,2,3,6-tetrahydropyridine induced Parkinson's disease in zebrafish. Proteomics 16, 1407-1420.

Satake, W., Nakabayashi, Y., Mizuta, I., Hirota, Y., Ito, C., Kubo, M., Kawaguchi, T., Tsunoda, T., Watanabe, M., Takeda, A., et al., 2009. Genome-wide association study identifies common variants at four loci as genetic risk factors for Parkinson's disease. Nature Genetics 41, 1303-1307.

Setini, A., Pierucci, F., Senatori, O., Nicotra, A., 2005. Molecular characterization of monoamine oxidase in zebrafish (*Danio rerio*). Comparative Biochemistry and Physiology Part B: Biochemistry and Molecular Biology 140, 153-161.

Shi, G., Lee, J.R., Grimes, D.A., Racacho, L., Ye, D., Yang, H., Ross, O.A., Farrer, M., McQuibban, G.A., Bulman, D.E., 2011. Functional alteration of PARL contributes to mitochondrial dysregulation in Parkinson's disease. Hum Mol Genet 20, 1966-1974.

Sheng, D., Qu, D., Kwok, K.H., Ng, S.S., Lim, A.Y., Aw, S.S., Lee, C.W., Sung, W.K., Tan, E.K., Lufkin, T., et al., 2010.

Deletion of the WD40 domain of LRRK2 in Zebrafish causes Parkinsonism-like loss of neurons and locomotive defect. PLoS Genetics 6, e1000914.

Sidransky, E., Nalls, M.A., Aasly, J.O., Aharon-Peretz, J., Annesi, G., Barbosa, E.R., Bar-Shira, A., Berg, D., Bras, J., Brice, A., et al., 2009. Multicenter analysis of glucocerebrosidase mutations in Parkinson's disease. New England Journal of Medicine 361, 1651-1661.

Simon-Sanchez, J., Schulte, C., Bras, J.M., Sharma, M., Gibbs, J.R., Berg, D., Paisan-Ruiz, C., Lichtner, P., Scholz, S.W., Hernandez, D.G., et al., 2009. Genome-wide association study reveals genetic risk underlying Parkinson's disease. Nature Genetics 41, 1308-1312.

Singleton, A.B., Farrer, M., Johnson, J., Singleton, A., Hague, S., Kachergus, J., Hulihan, M., Peuralinna, T., Dutra, A., Nussbaum, R., et al., 2003. Alpha-synuclein locus triplication causes Parkinson's disease. Science 302, 841.

Spillantini, M.G., Schmidt, M.L., Lee, V.M., Trojanowski, J.Q., Jakes, R., Goedert, M., 1997. Alpha-synuclein in Lewy bodies. Nature 388, 839840.

Sun, Z., Gitler, A.D., 2008. Discovery and characterization of three novel synuclein genes in zebrafish. Developmental Dynamics 237, 2490-2495.

Sveinbjornsdottir, S., 2016. The clinical symptoms of Parkinson's disease. Journal of Neurochemistry 139 (Suppl. 1), 318-324.

Tanner, C.M., Kamel, F., Ross, G.W., Hoppin, J.A., Goldman, S.M., Korell, M., Marras, C., Bhudhikanok, G.S., Kasten, M., Chade, A.R., et al., 2011. Rotenone, paraquat, and Parkinson's disease. Environmental Health Perspectives 119, 866-872.

Toledo, J.B., Arnold, S.E., Raible, K., Brettschneider, J., Xie, S.X., Grossman, M., Monsell, S.E., Kukull, W.A., Trojanowski, J.Q., 2013. Contribution of cerebrovascular disease in autopsy confirmed neurodegenerative disease cases in the National Alzheimer's Coordinating Centre. Brain 136, 2697-2706.

Ungerstedt, U., 1968. 6-Hydroxy-dopamine induced degeneration of central monoamine neurons. European Journal of Pharmacology 5, 107-110.

Vijayanathan, Y., Lim, F.T., Lim, S.M., Long, C.M., Tan, M.P., Majeed, A.B.A., Ramasamy, K., 2017. 6-OHDA-Lesioned adult zebrafish as a useful Parkinson's disease model for dopaminergic neuroregeneration. Neurotoxicity Research 32, 496-508.

Wang, A., Costello, S., Cockburn, M., Zhang, X., Bronstein, J., Ritz, B., 2011. Parkinson's disease risk from ambient exposure to pesticides. European Journal of Epidemiology 26, 547-555.

Wang, Q., Liu, S., Hu, D., Wang, Z., Wang, L., Wu, T., Wu, Z., Mohan, C., Peng, A., 2016. Identification of apoptosis and macrophage migration events in paraquat-induced oxidative stress using a zebrafish model. Life Sci 157, 116-124.

Wang, Y., Liu, W., Yang, J., Wang, F., Sima, Y., Zhong, Z.M., Wang, H., Hu, L.F., Liu, C.F., 2017. Parkinson's disease-like motor and non-motor symptoms in rotenone-treated zebrafish. Neurotoxicology 58, 103-109.

Wen, L., Wei, W., Gu, W., Huang, P., Ren, X., Zhang, Z., Zhu, Z., Lin, S., Zhang, B., 2008. Visualization of monoaminergic neurons and neurotoxicity of MPTP in live transgenic zebrafish. Developmental Biology 314, 84-92.

Wullimann, M.F., Rink, E., 2002. The teleostean forebrain: a comparative and developmental view based on early proliferation, Pax6 activity and catecholaminergic organization. Brain Research Bulletin 57, 363-370.

Xie, Y., Feng, H., Peng, S., Xiao, J., Zhang, J., 2017. Association of plasma homocysteine, vitamin B12 and folate levels with cognitive function in Parkinson's disease: A meta-analysis. Neurosci Lett 636, 190-195.

Xi, Y., Ryan, J., Noble, S., Yu, M., Yilbas, A.E., Ekker, M., 2010. Impaired dopaminergic neuron development and locomotor function in zebrafish with loss of pink1 function. Eur J Neurosci 31, 623-633.

Xi, Y., Yu, M., Godoy, R., Hatch, G., Poitras, L., Ekker, M., 2011. Transgenic zebrafish expressing green fluorescent protein in dopaminergic neurons of the ventral diencephalon. Developmental Dynamics 240, 2539-2547.

Yokogawa, T., Marin, W., Faraco, J., Pezeron, G., Appelbaum, L., Zhang, J., Rosa, F., Mourrain, P., Mignot, E., 2007. Characterization of sleep in zebrafish and insomnia in hypocretin receptor mutants. PLoS Biology 5, e277.

Zhang, Y., Nguyen, D.T., Olzomer, E.M., Poon, G.P., Cole, N.J., Puvanendran, A., Phillips, B.R., Hesselson, D., 2017. Rescue of Pink1 deficiency by stress-dependent activation of autophagy. Cell Chemical Biology 24, 471-480 e474.

Zharikov, A.D., Cannon, J.R., Tapias, V., Bai, Q., Horowitz, M.P., Shah, V., El Ayadi, A., Hastings, T.G., Greenamyre, J.T., Burton, E.A., 2015. shRNA targeting alpha-synuclein prevents neurodegeneration in a Parkinson's disease model. Journal of Clinical Investigation 125, 2721-2735.

Zhou, Y., Cattley, R.T., Cario, C.L., Bai, Q., Burton, E.A., 2014. Quantification of larval zebrafish motor function in multiwell plates using open-source MATLAB applications. Nature Protocols 9, 1533-1548.

Zimprich, A., Biskup, S., Leitner, P., Lichtner, P., Farrer, M., Lincoln, S., Kachergus, J., Hulihan, M., Uitti, R.J., Calne, D.B., et al., 2004. Mutations in LRRK2 cause autosomal-dominant parkinsonism with pleomorphic pathology. Neuron 44, 601-607.

第二十四章

痫性发作和癫痫

Rosane Souza Da Silva[1], Monica Ryff Moreira Roca Vianna[2], Carla Denise Bonan[1]

24.1 引言

癫痫是神经系统疾病第三大常见病，有明显的精神和全身性症状，其特征是反复、无故的癫痫发作。癫痫是大脑出现突发性异常放电，可影响患者运动、行为和意识水平（世界卫生组织）。癫痫发作是一组脑细胞过度放电的结果，可表现为短暂的意识丧失或肌肉痉挛，甚至剧烈抽搐。癫痫发作可分为全身性发作（同时影响两侧大脑半球或者说大脑两侧的细胞群）、局灶性发作（大脑一侧一个区域或一组细胞先被激活），或未分类发作（由于资料不充足或不完整而不能分类）（Scheffer et al., 2017）。

抗癫痫新药的获批和对该疾病机制的破译是过去几年的重要进展（Myers et al., 2019）。然而经过了几十年的研究和新型抗癫痫药物的应用，对癫痫患者的治疗依然很复杂，其中30% 的患者在试用多种抗癫痫药物后仍然出现癫痫发作，并被诊断为难治性癫痫（Younus and Reddy, 2018）。此外，药物不良反应造成患者依从性差和治疗中止，因此仍需要探寻更多新的抗癫痫药物治疗难治性癫痫（Franco et al., 2016; Griffin et al., 2018）。癫痫实验动物模型的建立是研究该疾病发病机制和治疗方法的重要步骤。抗癫痫药物的临床前筛选研究包括：药理学方法和电诱发癫痫模型、自发反复发作相关的策略改善以及对治疗无反应动物的监测（Löscher, 2017）。这些模型虽然用于探索癫痫耐药性相关因素，但仍不常用于药物测试（Grone and Baraban, 2015）。此外，诱发癫痫模型无法说明遗传性癫痫的发病机制。虽然近年来使用转基因实验动物研究神经系统疾病，但是传统抗癫痫药物筛选仍不包括遗传性癫痫模型（Griffin et al., 2018）。测序技术的最新发展导致了癫痫相关新基因的发现，这些基因可作为体内遗传性癫痫模型研究的靶点，进而开发疾病特异性和个

[1] Laboratório de Neuroquímica e Psicofarmacologia, Escola de Ciências da Saúde e da Vida, Pontifícia Universidade Católica do Rio Grande do Sul, Porto Alegre, Rio Grande do Sul, Brazil.

[2] Laboratório de Biologia e Desenvolvimento do Sistema Nervoso, Escola de Ciências da Saúde e da Vida, Pontifícia Universidade Católica do Rio Grande do Sul, Porto Alegre, Rio Grande do Sul, Brazil.

性化治疗方案。

斑马鱼为药物筛选提供了另一种模型，而且即使在早期发育阶段，斑马鱼也能快速评价药物毒性和新型抗癫痫药物对中枢神经系统的影响（Sakai et al., 2018）。此外，由于人类癫痫相关基因与斑马鱼同源基因之间的高度保守性，斑马鱼的用途也越来越多（Howe et al., 2013）。斑马鱼胚胎外表透明且发育迅速，便于人们清楚地观察其神经系统完整的神经发育过程，这对于研究产前损伤引发的神经系统疾病非常重要（Sakai et al., 2018）。此外，与电生理分析相关的对惊厥药敏感的全部行为均可使用斑马鱼模型研究其机制和治疗方法（Da Silva et al., 2016; Cunliffe, 2016; Griffin et al., 2016）。本章旨在讨论与斑马鱼药理性癫痫、遗传性癫痫以及癫痫模型相关的内容，并重点介绍斑马鱼在阐明癫痫机制和鉴定潜在候选药物中的作用。

24.2 化学诱导癫痫发作

没有任何模型能够完全复制人类癫痫的发病情况。然而，斑马鱼急性化学诱导癫痫模型在明确癫痫的发病机制、表型特征和治疗反应等方面有较大意义。Baraban 的研究发现，斑马鱼可以反映出化学药品的作用，在模拟癫痫中具有一定的优势（Baraban et al., 2005）。啮齿类动物癫痫模型常用的经典化学类惊厥药 γ- 氨基丁酸（GABA）受体拮抗剂［戊四唑、荷包牡丹碱（bicuculline）和印防己毒素（picrotoxin）］、谷氨酸受体激动剂（红藻氨酸和喹啉酸）、胆碱能受体激动剂（毛果芸香碱）以及胆碱酯酶抑制剂（毒扁豆碱）均可造成斑马鱼惊厥反应（Baraban et al., 2005; Alfaro et al., 2011; Berghmans et al., 2007; Cassar et al., 2017; Wong et al., 2010）。此外，K^+ 通道阻滞剂 4- 氨基吡啶（4-aminopyridine, 4AP）、二氢叶酸还原酶抑制剂氨甲蝶呤，以及谷氨酸能受体激动剂软骨藻酸，也可引起斑马鱼化学性惊厥（Cassar et al., 2017; Lefebvre et al., 2009; Winter et al., 2017; Tiedeken et al., 2005）。

任何发育阶段的斑马鱼癫痫发作的表型特征都与啮齿类动物惊厥特征相似，很容易识别。虽然戊四唑可促进斑马鱼仔鱼和成鱼的癫痫样行为，但红藻氨酸在仔鱼和成鱼阶段具有不同的作用，可能是谷氨酸能系统发育不成熟所致（Baraban et al., 2005; Cassar et al., 2017; Wong et al., 2010; Menezes et al., 2014, 2015）。斑马鱼在几种促惊厥药物（如戊四唑和红藻氨酸）作用下出现癫痫发作，表现为游泳活动增加，进而发展为阵挛样惊厥，并最终可能导致死亡（表 24-1）（Baraban et al., 2005; Alfaro et al., 2011）。斑马鱼性别和体重会影响戊四唑引起的行为反应，而水温会增加斑马鱼对戊四唑的敏感性（Menezes and Da Silva., 2017）。

虽然斑马鱼行为可以表现出惊厥反应（Wong et al., 2010），但这还不足以证明癫痫发作时大脑的过度活动。即使在斑马鱼的仔鱼阶段也可以记录到异常的大脑活动（Kim et al., 2010; Afrikanova et al., 2013）。可以通过使用侵入性和非侵入性微（多）电极的方法检测戊四唑、红藻氨酸和毒扁豆碱诱导的癫痫样放电（Baraban et al., 2005; Kim et al., 2010; Cho et al., 2017; Meyer et al., 2016）。计算机模型数据能够记录戊四唑诱发斑马鱼仔鱼急性

表 24-1 **几种化学惊厥剂对斑马鱼的作用比较**

药品	行为活动	电生理结果	分子和细胞结果
4-氨基吡啶（Cassar et al., 2017; Winter et al., 2017）			
K^+ 通道阻滞剂	鱼类游泳活动急剧增加	强烈爆发神经活动	n.d.a
荷包牡丹碱（Cassar et al., 2017; Menezes and Da Silva, 2017）			
$GABA_A$ 受体竞争性拮抗剂	鱼类游泳活动急剧增加	爆发性电诱发的神经反应	n.d.a
软骨藻酸（Lefebvre et al., 2009; Tiedeken et al., 2005）			
离子型谷氨酸受体激动剂	不受控制的胸鳍运动 强直 - 阵挛型惊厥	n.d.a	几种转录因子和信号转导分子上调
红藻氨酸（Alfaro et al., 2011; Menezes and Da Silva, 2017; Kim et al., 2010）			
离子型谷氨酸受体激动剂	**第一阶段**：不动和过度换气；**第二阶段**：快速的“漩涡状”盘旋行为；**第三阶段**：从右到左的快速运动；**第四阶段**：异常和痉挛性肌肉收缩；**第五阶段**：快速的全身阵挛样抽搐；**第六阶段**：痉挛下沉到底部；**第七阶段**：死亡	癫痫样放电	GFAP 细胞减少 S100ß 水平降低 谷氨酸摄取减少
戊四唑（Baraban et al., 2005; Menezes and Da Silva, 2017; Mussulini et al., 2018; Pineda et al., 2011; Gupta et al., 2014）			
（PTX） γ- 氨基丁酸 A 型受体位点	**第一阶段**：鱼类游泳活动急剧增加； **第二阶段**：快速的“漩涡状”盘旋行为； **第三阶段**：一连串简单姿势直至行为丧失	癫痫样放电	*c-fos* 表达升高 神经发生减少 *bdnf* 表达减少
毒扁豆碱（Menezes and Da Silva, 2017; Lopes et al., 2016）			
可逆胆碱酯酶抑制剂	快速，突然运动	癫痫样放电	神经发生减少
毛果芸香碱（Winter et al., 2017; Kundap et al., 2017; Vermoesen et al., 2011; Pitkänen et al., 2005）			
胆碱能受体激动剂	蹒跚 / 头部撞击、头部与孔壁接触、头尾起伏、增加嘴巴动作、震颤、身体扭曲、硬挺、姿态丧失	癫痫样放电	*c-fos* 表达升高 *bdnf* 和 *slc1a2b* 表达长期下降

注：n.d.a 表示“没有相关资料”。

癫痫发作期间的大脑活动和神经元动力学（Hunyadi et al., 2017; Rosch et al., 2018）。快速、灵敏的基因编码钙离子指示剂（如 GCaMP）的光学测量显示，戊四唑等化学物质诱发斑马鱼惊厥发作时，大脑活动与尾部运动之间存在很好的相关性（Turrini et al., 2017）。

化学惊厥药物影响斑马鱼大脑活动的同时，伴随着神经发生的减少和即早基因 *c-fos* 表达的增加（Baraban et al., 2005; Kim et al., 2010）。戊四唑、红藻氨酸和毒扁豆碱对斑马鱼大脑发育的影响似乎非常广泛，这些化学惊厥剂会导致大脑皮层、皮质下区、丘脑、顶盖和小脑增殖区、视前区和中脑增殖区的细胞增殖减少（Kim et al., 2010）。

化学诱发斑马鱼急性癫痫样发作是筛选并评估各种化合物抗惊厥特性的最简单方法之一。通过戊四唑诱导 96 孔板中斑马鱼仔鱼癫痫模型，发现抗癫痫药物在哺乳动物和斑马鱼中的抗惊厥特性具有相似性（Berghmans et al., 2007）。经典的抗癫痫药物如拉莫三嗪、苯妥英和丙戊酸可以减少戊四唑引起的癫痫样行为（Baraban et al., 2005; Martinez et al., 2018）。高浓度丙戊酸具有有效的抗惊厥作用，减少癫痫样放电（Baraban et al., 2005）。卡马西平对斑马鱼仔鱼的癫痫样行为和神经元活动几乎没有影响（Baraban et al., 2005; Martinez et al.,

2018)。通过斑马鱼筛选的具有抗惊厥特性的新药，包括合成、天然和已获批准的化合物(药物再利用)。抗炎药吲哚美辛降低了戊四唑暴露后的炎症标志物，使幼年期斑马鱼 *c-fos* 表达降低、癫痫相关行为减少 (Barbalho et al., 2016)。天然化合物，如牛磺酸、甘丙肽、肌苷、靛玉红、伪神经素 A (pseurotin A)、氮杂螺呋喃 (azaspirofuran)、微粉化姜黄素、天麻素、柚皮素 DM 和青阳参苷 N (otophylloside N) 均被发现具有抗癫痫特性 (Aourz et al., 2018; Bertoncello et al., 2018; Brillatz et al., 2018; Copmans et al., 2018 a, b; Fontana et al., 2019; Jin et al., 2018; Podlasz et al., 2018; Sheng et al., 2016)。

在早期发育阶段和因误诊而使用抗癫痫药物治疗，造成了药物对人类致畸的问题。斑马鱼已被用于评估发育期间抗癫痫药物作用后的行为和形态学改变 (Kundap et al., 2017)。卡马西平、拉莫三嗪、左乙拉西坦和苯巴比妥对斑马鱼胚胎的发育没有产生任何改变 (Martinez et al., 2018)。然而，实验证明苯妥英会影响斑马鱼的孵化，而丙戊酸强烈影响斑马鱼孵化、存活和形态 (Martinez et al., 2018)。

癫痫伴有多种并发症，包括：一般生理影响，如慢性疼痛和胃肠道疾病；精神影响，如焦虑和抑郁；认知影响，如注意缺陷多动障碍和智力低下 (Seidenberg et al., 2009)。化学药物诱发斑马鱼癫痫模型已被用于判断和了解癫痫的并发症。单次应用戊四唑会诱发攻击性行为和记忆障碍 (Kundap et al., 2017; Canzian et al., 2018)。戊四唑诱发癫痫发作，表现出相关认知障碍的同时，往往伴随着神经化学方面的改变，比如 GABA 和乙酰胆碱等神经递质水平下降、谷氨酸水平升高，以及 *bdnf*、*creb1* 和 *npy* 表达的变化 (Kundap et al., 2017)。

24.3 诱发癫痫发作的非化学或遗传方法

创伤、颅内感染、缺氧、低血糖、急性电解质紊乱和高热也可诱发癫痫发作 (Dadas and Janigro, 2019; Laino et al., 2018)。事实上，高热惊厥是儿童时期最常见的癫痫形式 (Leung et al., 2018)。斑马鱼是研究高热惊厥发病机制的最佳模型，仅需提高仔鱼浴温就会导致其脑电图痫样放电 (Hunt et al., 2012)。斑马鱼脑电图 (EEG) 活动增加的机制与 TRPV4 通道有关。TRPV4 是一种具有高 Ca^{2+} 通透性的非选择性阳离子通道 (Hunt et al., 2012; Kiyatkin, 2007)。由于这些通道的激活通常会促进谷氨酸的释放，因此即使有戊四唑等致惊厥药物作用，MK-801 和苄哌酚醇阻断谷氨酸受体后仍会减弱癫痫发作活动 (Menezes and Da Silva, 2017; Hunt et al., 2012)。

24.4 癫痫的遗传学特征

几十年来，人们对癫痫的遗传学基础进行了假设，20 世纪 80 年代第一次有流行病学研究报告提出，癫痫患者的亲属患病风险增加 (Annegers et al., 1982; Ottman, 1989)。接下来的十余年中，研究人员对编码离子通道亚基基因的突变进行了鉴定并提出了一种假设，即功能失调的离子通道和由此产生的大脑活动电生理模式的改变是癫痫的发病机制，这导致了使用“通道病”一词来描述疾病形式，如 Dravet 综合征和新生儿良性癫痫发作，通常分别与电压门控钠通道和钾通道亚基突变有关。

与基因突变相关的癫痫可分为单基因孟德尔遗传疾病和多基因综合征。可以说，没有一种疾病是真正的单基因疾病，因为特定基因引起的病理变化受环境风险因素和个体遗传背景的影响（Demarest and Brooks-Kayal, 2018）。然而，目前也发现一些特定基因足以导致疾病表型出现。有趣的是，对变异不耐受的基因在进化上高度保守，这一事实支持利用动物模型研究该基因在疾病中的特定作用。

大规模全基因组关联研究，例如由国际抗癫痫联盟和 Feenstra 团队（Feenstra et al., 2014）进行的研究，连同之前几项较小规模的研究，发现了几种可提高不同类型癫痫发病风险的常见遗传变异。这些研究还证实了可引起这种疾病的基因之间的联系（Myers and Mefford, 2015）。

新一代测序揭示了几种可引起癫痫的单基因，这需要在临床前模型中进行研究，以促进我们对基因型 - 表型相关性的理解。利用基因组和外显子组测序技术，最新发现了近 1000 个与癫痫相关的基因，这些基因很多与离子通道编码基因无关（Demarest and Brooks-Kayal, 2018; Myers and meford, 2015）。这项技术可用于研究表型和评估潜在的新靶向疗法，其发现率超过了传统癫痫实验模型研究，如小鼠模型和异源表达细胞系统。在这种情况下，高通量模型对于利用潜在的治疗药物平行筛选特定基因变化的基因型和表型至关重要。与癫痫相关的新基因类别包括与染色质重塑和转录调控、突触发生、突触小泡运输和信号通路等相关的基因（Myers and mefford, 2015）。

斑马鱼是一个理想的系统，可用于研究越来越多与这种疾病相关的基因，并促进人们对已有基因的了解。癫痫的遗传模型有助于更好地理解发病机制，包括识别与发病风险增加相关的潜在过程（Grone et al., 2016）。转基因斑马鱼可以用来测试促惊厥药物的癫痫易感性，筛选抗癫痫候选药物，评估癫痫样放电和大脑活动，从而将基因型与表型联系起来（Griffin et al., 2016）。这些方法可以与本章前面描述的诱发癫痫表型的药理学和行为方式相结合。此外，正如最近对 Dravet 综合征进行的研究一样，与人类表型具有良好相关性的斑马鱼遗传模型有助于探寻潜在的新靶向疗法。

斑马鱼基因组与人类基因组有 70% 的同源率（Howe et al., 2013），即与大多数人类基因有同源性。重要的是，斑马鱼基因组包含 85% 与人类癫痫相关的基因（Hortopan et al., 2010）。与其他硬骨鱼类一样，斑马鱼也有一个重复的基因组，这通常导致它们有至少两个旁系同源基因与人类相同的基因同源。

24.4.1　斑马鱼的基因靶向和基因敲减研究

在过去的几十年里，采用了不同的方法诱导基因修饰，以研究斑马鱼癫痫的机制和癫痫易感性（表 24-2）。正向遗传策略是在斑马鱼研究中最早使用的方法，以此发现了数百个基因。1996 年，来自美国波士顿和德国图宾根的科学家们在第一批大量收集的脊椎动物突变体中使用了诱导单核苷酸突变的化学物质 *N*- 乙基 -*N*- 亚硝基脲（*N*-ethyl-*N*-nitrosourea,

表 24-2　**斑马鱼癫痫遗传模型**

癫痫综合征	方式	斑马鱼基因	人类同源蛋白	编码蛋白	参考文献
Dravet 综合征	ENU 诱变	*scn1lab*	SCN1A	电压门控钠通道亚基 α Nav1.1	Schoonheim et al. (2010); Baraban et al. (2013), Grone et al. (2017), Griffin et al. (2017), Sourbron et al. (2017)
	吗啉反义寡核苷酸	*scn1Lab*	SCN1A		Ceulemans et al. (2012)
非典型 Dravet 综合征	吗啉反义寡核苷酸	*cdh2*	CHD2	染色质解旋酶 /DNA 结合蛋白 2	Suls et al. (2013)
婴儿癫痫伴游走性局灶性发作	TALEN	*kcc2a and kcc2b*	SLC12A5	氯化钾协同转运蛋白 KCC2	Stödberg et al. (2015)
良性家族性癫痫发作和抽搐	吗啉反义寡核苷酸	*kcnq3*	KCNQ2 KCNQ3	电压门控钾通道 K(v)7.3	Chege et al. (2012)
具有听觉特征的常染色体显性遗传性癫痫	吗啉反义寡核苷酸	*lgi1a and lgi1b*	LGI1	富含亮氨酸的胶质瘤失活蛋白 1	Teng et al. (2010, 2011)
PRICKLE1 相关进行性肌阵挛性癫痫伴共济失调	合成 mRNA 以诱导过度表达	*pk1a*	PK1	刺痛感蛋白同系物 1	Teng et al. (2011)
	吗啉反义寡核苷酸	*pk1a*	PK1	刺痛感蛋白同系物 1	Mei et al. (2013)
EAS 或 SeSAME 综合征	吗啉反义寡核苷酸	*kcnj10*	KCNJ10	钾电压门控通道亚家族 J 成员 10 Kir4.1	Mahmood et al. (2013)
遗传性癫痫伴热性惊厥	吗啉反义寡核苷酸	*stx1b*	STX1B	突触融合蛋白 1B	Schubert et al. (2014)
STXBP1 相关性癫痫	CRISPR/Cas9	*stxbp1a* *stxbp1b*	STXBP1	突触结合蛋白 1	Grone et al. (2016)
吡哆醇依赖性癫痫（PDE）	CRISPR/Cas9	*aldh7a1*	ALDH7A1	*α*- 氨基己二酸半醛脱氢酶	Zabinyakov et al. (2017) Pena et al. (2017)
Lowe 综合征	插入逆转录病毒	*ocrl1*	OCRL1	眼脑肾综合征蛋白（OCRL）	Oltrabella et al. (2015)
Batten 病 *CLN3* 脂褐质沉积症	吗啉反义寡核苷酸	*cln3*	CNL3	巴泰宁（Battenin）	Wager et al. (2016)
CLN2 晚期婴儿神经元蜡样脂褐质沉积症	ENU 诱变	*tpp1*	TPP1	三肽基肽酶 1	Mahmood et al. (2013)

ENU）（Haffter et al., 1996; Driever et al., 1996）。ENU 诱导的化学突变是一种有效的方法，但在大规模筛选中，只有一小部分已知基因表型改变。在功能研究和表型分析中，敲除突变是最常用的选择，该法可以使基因失去功能。

基因组编辑技术，如锌指核酸酶（ZFN）和转录激活因子样效应物核酸酶（TALEN），代表了靶向遗传方法的新时代，可以在培养细胞和整体动物（包括斑马鱼）中产生不同的基因组修饰。尽管 ZFN 和 TALEN 对定向诱变做出了贡献，但它们也有一些限制，阻碍了其广泛应用，包括效率预测、外显子内的靶点识别、时间和成本。最近发现了 CRISPR/Cas9 技术，并于 2013 年首次用于斑马鱼，从根本上减少了靶向基因组工程的困难（Jao et al., 2013）。CRISPR/Cas9 基因组编辑技术在过去几年发展迅速，现在可以相对快速且低成本地产生功能丧失或模仿患者突变的突变体。

此外，对癫痫患者受影响的直系同源基因在斑马鱼中进行 mRNA 敲减研究，有助于研究癫痫常见的细胞和分子机制，并验证斑马鱼可用于筛查基因功能丧失方面的研究。尽管通常被描述为“遗传学方法”，胚胎注射选择性吗啉反义寡核苷酸（morpholino oligonucleotide）不会干扰基因组序列（Bill et al., 2009），而是通过阻断转录或mRNA 剪接，暂时敲除斑马鱼胚胎和仔鱼中的基因，因此被称为“形态突变”（morphants）。

24.4.2　斑马鱼癫痫疾病模型

24.4.2.1　Dravet 综合征

Dravet 综合征是一种致命、难治性儿童癫痫，从出生第一年开始，患者就面临危及生命的持续性癫痫发作、严重的智力和运动障碍、睡眠障碍、焦虑以及社会交往能力受损。Claes 的研究发现（Claes et al., 2001），大约 75% 的 Dravet 综合征患者 *SCN1A* 基因中携带了一个新的功能缺失突变（Mastrangelo and Leuzzi, 2012）。*SCN1A* 基因编码电压门控钠通道亚基 αNav1.1，其功能缺失突变导致通道功能不足。

大规模突变筛选中发现动眼神经缺陷的仔鱼在眼动反应中无法维持扫视运动，从中发现了 ENU 诱导的纯合 *scn1lab*^{s552}（Schoonheim et al., 2010），这是 Dravet 综合征斑马鱼突变体中最具特征的一种突变。该品系在斑马鱼 *SCN1A* 直系同源 *scn1lab* 中有一个单点突变。纯合子 *scn1lab*^{s552} 突变体与 Dravet 综合征患者表现相似，包括从 4 dpf 开始自发性癫痫发作以及通过前脑细胞外场记录的行为和 EEG 模式（Baraban et al., 2013）。该突变体还表现出睡眠、焦虑和运动障碍等特征（Grone et al., 2017）。与 Dravet 综合征患者一样，该突变体癫痫发作对已有抗癫痫药物有抵抗力，并逐渐恶化，直到动物在 14dpf 时死亡（Schoonheim et al., 2010; Baraban et al., 2013）。

由于硬骨鱼额外的基因组重复，斑马鱼有两个旁系同源 *scn1a* 基因，它们与人类 *SCN1A* 基因同源。*scn1laa* 和 *scn1lab* 分别与 *SCN1A* 序列具有 67% 和 76% 的同源率，二者也在中枢神经系统中高度表达（Baraban et al., 2013）。由于重复的旁系同源物 *scn1laa* 的表达，导致 *scn1labs*^{552} 纯合子系 Nav1.1 钠通道单倍缺陷，也再现了大多数 Dravet 综合征患者的遗传病因。

Baraban 团队在基于表型筛选 320 种化合物时使用了 *scn1lab*^{s552}，发现美国食品和药品管理局（FDA）批准的克立咪唑（clemizole）可以抑制惊厥行为和电诱导癫痫发作（Baraban et al., 2013; Griffin et al., 2017）。克立咪唑是一种组胺受体（H1）拮抗剂，它也与 5- 羟色胺能 $5HT_{2A}$ 和 $5HT_{2B}$ 受体结合，后一种机制与预期治疗作用相关。Baraban 团队使用临床批准的 5- 羟色胺能受体激动剂盐酸氯卡色林（Lorcaserin）（Belviq）治疗五名医学上难治的 Dravet 综合征患者后，发现其癫痫发作频率和 / 或严重程度有所降低（Baraban et al., 2013）。研究结果表明，斑马鱼可以成为临床前发现和鉴定靶点的有效模型，并指出 5- 羟色胺调节剂是一类有待探索的新型抗癫痫药物。氟苯丙胺是一种 5- 羟色胺能释放的诱导剂，Sourbron 及其同事使用另一个 *scn1a* 突变体证明了氟苯丙胺具有抗癫痫机制，可以改善 Dravet 综合征患者症状（Sourbron et al., 2017; Ceulemans et al., 2012）。

利用 *scn1Lab* 吗啉反义寡聚体建立了 Dravet 综合征的斑马鱼模型，发现斑马鱼出现癫痫自发发作，并伴有类似 Dravet 综合征患者的癫痫样放电（Zhang et al., 2015）。有趣的是，丙戊酸钠和芬氟拉明减少了癫痫样放电。*scn1Lab* 基因敲减的仔鱼对高温也很敏感，这与人类 Dravet 综合征患者相似。

据估计，患 Dravet 综合征的个体中约有 25% 可能并没有 *SCN1A* 基因突变而是其他基因受到影响。已知原钙黏蛋白 19（*PCDH19*）基因突变会导致女性出现类似 Dravet 综合征的表型。然而，对于一些确诊患者，*SCN1A* 和 *PCDH19* 突变与他们的症状无关。

通过对非典型 Dravet 综合征患者和其家庭成员进行外显子组测序，发现有三名患者伴有染色体解旋酶 DNA 结合蛋白 2（*CHD2*）基因突变，同时出现智力障碍、对发热敏感的全身性和肌阵挛性癫痫发作（Suls et al., 2013）。Suls 及其同事在不同发病年龄的癫痫患者中发现了 *CHD2* 新发突变和缺失（Suls et al., 2013）。他们使用吗啉反义寡核苷酸造成 *cdh2* 功能丧失，结果发现 *cdh2* 基因敲减导致异常的行为模式。基因沉默的仔鱼也有癫痫样放电，类似于行为改变相关场电位测量的癫痫发作。

24.4.2.2 婴儿癫痫伴游走性局灶性发作

婴儿癫痫伴游走性局灶性发作（EIMFS），也称为婴儿游走性部分性癫痫，是一种以多灶性发作、发育停滞或倒退以及不同的异常脑电图（EEG）为特征的儿童期癫痫性脑病。EIMFS 遗传因素各种各样，难以确定。有研究在来自两个无关家庭的四名 EIMFS 儿童中发现了一种新的遗传基因：隐性 *SLC12A5* 突变（Stödberg et al., 2015）。*SLC12A5* 编码氯化钾协同转运蛋白 KCC2，它对神经元抑制和皮质树突棘成熟至关重要。斑马鱼有两个 KCC2 直系同源体，*kcc2a* 和 *kcc2b*。在同一项研究中，他们使用 TALEN 介导的基因组编辑技术诱导 *kcc2a* 和 *kcc2b* 的外显子 4 缺失，从而破坏这两个基因。单敲除突变体没有症状，但 *kcc2a-kcc2b* 双基因敲除突变体出现早期运动障碍（Claes et al., 2004）。

24.4.2.3 良性家族性癫痫发作和惊厥

良性家族性新生儿惊厥或癫痫发作的特征是新生儿或幼儿反复出现癫痫发作。对于新生儿来说，这种癫痫大约在出生后的第三天开始，并在 1 ～ 4 个月内消退。父母或一级亲属的类似癫痫病史是诊断的关键。这些婴儿通常智力发育正常，并有新生儿或婴儿癫痫发作家族史，预后良好。然而，部分婴儿受影响后会出现智力障碍，大约 15% 的患儿在以后的生活中会出现癫痫。电压门控钾通道亚基编码基因 K（v）7.2（*KCNQ2*）和 K（v）7.3（*KCNQ3*）

是调节神经元兴奋性的M电流的基础，大多数良性家族性新生儿癫痫伴有这两个基因突变（Claes et al., 2004）。*KCNQ2* 基因突变在患者和家庭中很常见，并会导致一定程度的功能丧失。这些突变与电压感应、孔形成和细胞内结合域相关。较少 *KCNQ2* 突变可能会通过稳定通道激活状态（Rikee 项目）恢复原来功能。*kcnq2* 和 *kcnq3* 早在 2dpf 时就在斑马鱼体内表达（Chege et al., 2012）。Chege 等的研究证明，这些通道的抑制剂会诱导电生理上爆发式放电和癫痫样行为，而通道开放剂可以阻止上述变化。与通道抑制剂作用相似，反义敲减 *kcnq3* 基因也可引起多脉冲爆发式放电（Chege et al., 2012）。

24.4.2.4 具有听觉特征的常染色体显性遗传癫痫

具有听觉特征的常染色体显性遗传癫痫（ADEAF），也称为常染色体显性遗传颞叶外侧癫痫（ADLTE），临床特征为局灶性癫痫发作，伴有明显的听觉或失语先兆，且无脑结构异常。疾病发病不一，临床病程良性，大多数患者可以通过药物治疗控制癫痫发作。大约一半患者家族中发现有基因突变，最常受影响的基因是富亮氨酸胶质瘤失活基因 1（LGI1）（Pippucci et al., 2015）。基因敲减斑马鱼旁系同源基因 *lgi1a* 和 *lgi1b*，能够重现一些类似 ADEAF 患者的症状，但未应用电生理技术确认癫痫发作（Teng et al., 2010, 2011）。对于 *lgi1a* 基因沉默较少的突变体，戊四唑更容易诱发癫痫发作；而较多 *lgi1a* 基因沉默后，有形态缺陷和多动行为。*lgi1b* 基因沉默突变体没有表现出类似癫痫发作的行为，但对戊四唑诱导的多动敏感（Teng et al., 2011）。

24.4.2.5 *PRICKLE1* 基因相关的进行性肌阵挛性癫痫伴共济失调癫痫发作

PRICKLE1（*PK1*）基因是平面细胞极性网络的一个元素，当该基因纯合子突变时会导致常染色体隐性进行性肌阵挛癫痫 - 共济失调综合征（Bassuk et al., 2008）。然而，*PK1* 致病机制尚不清楚。斑马鱼直系同源基因 *pk1a* 在胚胎发育过程中协调细胞运动、神经元迁移和轴突生长。Bassuk 团队使用合成 mRNA 诱导野生型 *prickle1* 和与人相似突变的 *prickle1* 过度表达，发现原肠胚形成变化和诱导产生的解剖畸形，这些变化在注射 mRNA 野生型动物中更为明显（Bassuk et al., 2008）。Mei 团队利用 MO 敲减 *pk1a* 活性后发现，斑马鱼仔鱼对戊四唑敏感（Mei et al., 2013）。*pk1a* 基因沉默可导致神经突生长缺陷。*pk1a* 基因敲减引起的有害作用可以被野生型 mRNA 逆转，但不能被患者突变 mRNA 逆转。

24.4.2.6 EAST 或 SeSAME 综合征

EAST（癫痫、共济失调、感音神经性耳聋、小管病变）或 SeSAME（癫痫、感音神经性耳聋、共济失调、精神发育迟缓和电解质失衡）综合征是一种罕见的常染色体隐性综合征，于 2009 年首次被描述。该神经系统症状包括全身性癫痫，通常在幼年发作（Celmina et al., 2019）。它是由编码内向整流型钾通道家族成员 Kir 4.1 的钾电压门控通道亚家族 J 成员 10（*KCNJ10*）突变引起的。*KCNJ10* 有 16 种不同突变与 EAST/SeSAME 综合征有关。吗啉反义寡核苷酸剪接和翻译阻断 *kcnj10* 后，出现了该疾病的所有主要症状（Mahmood et al., 2013a, b）。这些症状被共同注射的人源野生型而非突变型的 *KCNJ10* mRNA 逆转，为 EAST 研究建立了一个新系统。

24.4.2.7 遗传性癫痫伴热性惊厥

遗传性癫痫伴热性惊厥附加症（GEFS+）是一种范围广泛、严重程度各异的癫痫疾病。GEFS+ 是在由高热引发的热性惊厥和其他类型癫痫反复发作相结合的家系中诊断出来的。GEFS+ 中最常见疾病是单纯性热性惊厥，通常从婴儿期开始，在儿童期结束的。当 5 岁后持续热性惊厥时，这种情况称为热性惊厥附加症（FS+），通常会在青春期早期消失。持续和严重的 GEFS+ 谱系包括 Dravet 综合征和其他癫痫。由于 GEFS+ 谱系的多样性，前面提到的基因突变与它们有因果关系，包括电压门控钠通道亚基 *SCN1A* 和 *SCN1B* 以及 $GABA_A$ 受体亚基 *GABRG2*。对发热相关癫痫综合征新基因的遗传研究发现，突触融合蛋白 1B（STX1B）与热性惊厥和癫痫都有关（Schubert et al., 2014）。全外显子组测序确定了 *STX1B* 的突变，包括截短或插入以及缺失。在同一项研究中，他们使用吗啉反义寡核苷酸敲减 *stx1b*，发现出现癫痫样行为和对温度升高敏感的癫痫样放电。

24.4.2.8 *STXBP1* 基因相关性癫痫

STXBP1 基因，也称为 *MUNC18-1*，编码突触融合蛋白结合蛋白 1，它是神经传递和与突触融合蛋白的相互作用所必需的，而突触融合蛋白是突触囊泡融合机制的一个组成部分。常染色体显性 *STXBP1* 基因突变与临床癫痫相关，包括 *STXBP1* 脑病伴癫痫、早发癫痫性脑病（EIEE 或 Ohtahara 综合征）以及 Lennox-Gastaut 和 Dravet 综合征。突变在不同程度上影响功能性蛋白的数量，损害神经递质释放和神经元兴奋性，从而导致癫痫发作。斑马鱼有两个旁系同源体 Stxbp1a 和 Stxbp1b，它们与人类 STXBP1 的氨基酸序列同源性约为 80%。为了研究 STXBP1 在斑马鱼中的功能，Grone 团队使用 CRISPR/Cas9 基因编辑生成 *stxbp1a* 和 *stxbp1b* 稳定突变系（Grone et al., 2016）。*stxbp1a* 纯合子突变导致与表型无关的严重癫痫，而 *stxbp1a* 杂合突变的影响程度从轻微到相对正常。有趣的是，*stxbp1b* 纯合突变引起的癫痫发作与正常移动、心率、代谢和总体形态以及轻度色素沉着相关。

24.4.2.9 吡哆醇依赖性癫痫

吡哆醇依赖性癫痫（PDE）是一种常染色体隐性遗传的神经代谢疾病，表现为新生儿或婴儿早期癫痫发作，且目前可用的抗癫痫药物难以治疗。大剂量吡哆醇（维生素 B6）或 5′- 磷酸吡哆醛可缓解 PDE 的癫痫发作，但不能预防出现神经发育症状。如果未治疗，PDE 会导致患者死亡。PDE 是由 *ALDH7A1* 基因编码的 *α*- 氨基己二酸半醛脱氢酶（赖氨酸降解酶）的突变引起的。ALDH7A1 是赖氨酸分解代谢途径的一部分，酶缺乏会导致赖氨酸在包括中枢神经系统在内的组织中积累。2017 年，利用 CRISPR/Cas9 生成斑马鱼 *aldh7a1* 突变体，出现吡哆醇和 5′- 磷酸吡哆醛可缓解的癫痫样电活动，类似于 PDE 患者（Zabinyakov et al., 2017; Pena et al., 2017）。Pena 等的研究还证明，补充赖氨酸会加重其症状并诱发早发性癫痫和死亡（Pena et al., 2017）。

24.4.2.10 影响癫痫易感性的其他人类疾病的斑马鱼模型

以癫痫发作风险增加相关基因产物为目标的基因编辑斑马鱼品系或变异体是增加了癫痫发作易感性的其他神经系统疾病的模型。

（1）Lowe 综合征

Lowe 综合征（Lowe syndrome），也称为眼脑肾综合征（OCRL），是一种罕见的 X 染色体隐性遗传病，其特征是肾脏、眼睛和神经系统的异常。大约 50% 的患者在 18 岁以上出现癫痫发作，高达 9% 的患者出现热性惊厥，其寿命很少超过 40 年。导致该疾病的原因是编码磷脂酰肌醇 (4,5)- 二磷酸 -5- 磷酸酶［PtdIns(4,5)P2］的 *OCRL1* 基因的突变。许多患者体内均检测到 *OCRL1* 基因的突变（Oltrabella et al., 2015）。该疾病的潜在机制尚不清楚，但 *OCRL1* 突变导致酶活性显著降低。2012 年，Ramirez 及其合作者报告了第一个 Lowe 综合征动物模型，这是一种纯合的 *ocrl* 基因敲除斑马鱼（Oltrabella et al., 2015; Ramirez et al., 2012）。与斑马鱼 *ocrl1* 基因相关的结构域组织、组织特异性剪接和表达具有保守性。Ramirez 报告的模型出现了 Lowe 综合征的几个神经学特征，包括对热致癫痫和白质损伤的易感性增加（Oltrabella et al., 2015; Ramirez et al., 2012）。

（2）Batten 病（Batten disease）

Batten 病又称为神经元蜡样脂褐质沉积症（NCL），是一类溶酶体贮存障碍疾病，具有遗传多样性，也是儿童最常见、具有致命性的神经退行性疾病。目前该病分型从 CNL1 到 CNL10，发病年龄和严重程度各不相同。每个分型都有一共同机制，即溶酶体对储存物质降解功能障碍，但这些储存囊泡的容量和形态各异，与不同表型相关的突变基因也不尽相同。大多数 NCL 患者儿童时期开始发病，临床症状通常包括视力丧失、癫痫、痴呆和运动异常。总体而言，症状和癫痫发作会随着疾病发展而恶化，早期死亡的风险取决于发病类型和发病年龄。

Batten 病或 CLN3 是常染色体隐性突变（大部分是 CLN3 基因的小缺失）引起的。CLN3 蛋白的主要功能尚不明确，可能与细胞运输、自噬、细胞存活、增殖、迁移和形态学有关。在 CLN3 患者中，线粒体 ATP 合酶的 c 亚基是溶酶体储存物质的主要成分。斑马鱼中 *cln3* 基因敲减再现了患者的几种病理特征，包括生存率降低、视网膜病变、神经退行性病变、癫痫样活动以及随着 c 亚基积累而增大的溶酶体（Wager et al., 2016）。

虽然 Batten 病特指青少年发病的 NCL，但现在该术语更多地用于描述所有形式的 NCL。CLN2 晚期婴儿型疾病（或经典晚期婴儿神经元蜡样脂褐质沉积症）是由三肽基肽酶 1（*TPP1*）突变引起的，患者表现出广泛脑萎缩、失明、共济失调、认知缺陷和癫痫，由于缺乏有效的治疗，最终在青春期过早死亡。Mahmood 及其同事研究了 ENU 造成斑马鱼 $tpp1^{sa0011}$ 纯合突变的 CLN2 疾病（Mahmood et al., 2013 a, b）。该突变使斑马鱼出现严重的、进行性的、早发神经退行性病变，视网膜和头部体积减小，存活率降低，在 5dpf 时死亡（Mahmood et al., 2013 a, b）。与人类患者一样，突变的斑马鱼体内溶酶体肥大。变异体异常的凋亡和神经性活动为 CLN2 疾病的发病机制提供新的见解。

24.5 癫痫发作和癫痫高级研究的模型——斑马鱼模型

斑马鱼的特征有助于更好地了解癫痫和癫痫发作相关的致病机制。人类癫痫致病基因和斑马鱼同源基因之间的高度保守性、高度药物摄取的相关性以及斑马鱼表现出癫痫样活动的能力，使之成为模拟人类癫痫的理想模型（MacRae and Peterson, 2015; Phillips and Westerfield, 2014; Khan et al., 2017）。斑马鱼模型的使用须加强以下方面的研究：①癫痫和癫痫新生标志物的发现；②癫痫相关行为参数分析的改进；③斑马鱼遗传模型的使用。

斑马鱼遗传性癫痫模型可能揭示一些未知的癫痫发病机制，并为临床前高通量药物筛选提供新方法。遗传性癫痫会影响生化和炎症通路，这些可作为该疾病新的生物标志物或治疗靶点（Pearson-Smith et al., 2017; Pearson-Smith and Patel, 2017; Kumar et al., 2016）。

研究还观察了抗癫痫药物对斑马鱼行为的影响，以及增加斑马鱼癫痫发作潜伏期或改善异常行为的能力。尽管胚胎和幼体斑马鱼的使用比成年斑马鱼更为普遍，但仍应考虑到它们的局限性及其对研究结果的影响。斑马鱼幼体未成熟脑区可能会影响对行为表型和神经系统疾病其他方面的评估。另一点需要考虑的是，斑马鱼受精后 10 天以前血脑屏障尚未完全发育。因此，在这个早期发育阶段，增加药物渗透性，即增加药物进入大脑的能力，降低了斑马鱼模型对药物效应预测的相关性（Fleming et al., 2013）。此外，某些化合物的溶解度有限，会影响其在培养器皿中的给药，因此必须使用其他给药途径，例如微量注射（Truong et al., 2011）。综上，应考虑多方面因素确定使用何种发育阶段的斑马鱼。

DNA 测序技术的快速发展加深了我们对疾病的理解，并发现需要鉴别的新基因。与癫痫相关的新基因类别包括染色质重塑和转录调控、突触小泡运输，以及促使对所有癫痫类型核心通道编码基因进行更正的信号通路。在研究靶点在癫痫发病机制中的作用时，必须考虑到不同基因型和表型相关性在患者中存在显著差异。在这种情况下，来自临床前模型的可信数据必须与来自临床试验的患者信息相结合。近年来开发的斑马鱼遗传模型（Mahmood et al., 2013a, b; Pena et al., 2017; Wager et al., 2016; Cao et al., 2017）对于理解疾病基本发病机制和探寻潜在的靶向疗法是很重要的（Griffin et al., 2017）。斑马鱼为高通量筛选提供了一种重要的替代手段，可以与异源表达系统互补。虽然在斑马鱼模型中对几种化合物进行高通量筛选很有研究前景，但对众多不同突变的筛选仍具有挑战性。开发新的斑马鱼突变体需要大量的时间和成本。此外，斑马鱼也无法重现人类患者的遗传背景、相似的寿命时间和复杂的神经结构（Demarest and Brooks-Kayal, 2018）。

虽然没有一个模型可以完整地模拟疾病的各个方面，但斑马鱼癫痫模型是对传统模型的补充，在其他模型受到限制的情况下提供了新的研究途径。在同一生物体中同时结合不同的方法，例如在新候选药物的作用下，利用转基因品系进行化学惊厥剂的药理学试验，可以加速确定该疾病的新靶点。新的药理学和基因发现策略可能针对耐药性癫痫患者亚群开展，而不是在所有类型癫痫中寻找有效的替代方案。因此，与任何一种单独模型相比，斑马鱼模型可以帮助人们更完整地了解癫痫的遗传发病机制，并有望在未来十年将治疗方法转化为临床实践。

（刘东鑫翻译　赵丽审校）

参考文献

Afrikanova, T. Serruys, A.S. Buenafe, O.E. et al., 2013. Validation of the zebrafish pentylenetetrazol seizure model:locomotor versus electrographic responses to antiepileptic drugs. PLoS One 8 (1), e54166. https://doi.org/10.1371/journal.pone.0054166.

Alfaro, J.M. Ripoll-Gómez, J. Burgos, J.S. 2011. Kainate administered to adult zebrafish causes seizures similar to those in rodent models. European Journal of Neuroscience 33 (7), 1252-1255. https://doi.org/10.1111/ j.1460-9568.2011.07622.x.

Annegers, J.F. Hauser, W.A. Anderson, V.E. Kurland, L.T. 1982. The risks of seizure disorders among relatives of patients with childhood onset epilepsy. Neurology 32 (2), 174-179.

Aourz, N. Serruys, A.K. Chabwine, J.N. et al., 2018. Identification of GSK-3 as a potential therapeutic entry point for epilepsy. ACS Chemical Neuroscience. https://doi.org/10.1021/acschemneuro.8b00281.

Baraban, S.C. Taylor, M.R. Castro, P.A. Baier, H. 2005. Pentylenetetrazole induced changes in zebrafish behavior, neural activity and *c-fos* expression. Neuroscience 131 (3), 759-768.

Baraban, S.C. Dinday, M.T. Hortopan, G.A. 2013. Drug screening in Scn1a zebrafish mutant identifies clemizole as a potential Dravet syndrome treatment. Nature Communications 4, 2410. https://doi.org/10.1038/ncomms3410.

Barbalho, P.G. Lopes-Cendes, I. Maurer-Morelli, C.V. 2016. Indomethacin treatment prior to pentylenetetrazole-induced seizures downregulates the expression of *il1b* and *cox2* and decreases seizure-like behavior in zebrafish larvae. BMC Neuroscience 17, 12. https://doi.org/10.1186/s12868-016-0246-y.

Bassuk, A.G. Wallace, R.H. Buhr, A. et al., 2008. A homozygous mutation in human PRICKLE1 causes an autosomal-recessive progressive myoclonus epilepsy-ataxia syndrome. The American Journal of Human Genetics 83 (5), 572-581. https://doi.org/10.1016/j.ajhg.2008.10.003.

Berghmans, S. Hunt, J. Roach, A. Goldsmith, P. 2007. Zebrafish offer the potential for a primary screen to identify a wide variety of potential anticonvulsants. Epilepsy Research 75 (1), 18-28.

Bertoncello, K.T. Aguiar, G.P.S. Oliveira, J.V. Siebel, A.M. 2018. Micronization potentiates curcumin's anti-seizure effect and brings an important advance in epilepsy treatment. Scientific Reports 8 (1), 2645. https://doi.org/ 10.1038/s41598-018-20897-x.

Bill, B.R. Petzold, A.M. Clark, K.J. Schimmenti, L.A. Ekker, S.C. 2009. A primer for morpholino use in zebrafish. Zebrafish 6 (1), 69-77. https://doi.org/10.1089/zeb.2008.0555.

Brillatz, T. Lauritano, C. Jacmin, M. et al., 2018. Zebrafish-based identification of the antiseizure nucleoside inosine from the marine diatom Skeletonema marinoi. PLoS One 13 (4), e0196195. https://doi.org/10.1371/journal.pone.0196195.

Canzian, J. Fontana, B.D. Quadros, V.A. Müller, T.E. Duarte, T. Rosemberg, D.B. 2018. Single pentylenetetrazole exposure increases aggression in adult zebrafish at different time intervals. Neuroscience Letters 692, 27-32. https://doi.org/10.1016/j.neulet.2018.10.045.

Cao, S. Smith, L.L. Padilla-Lopez, S.R. et al., 2017. Homozygous EEF1A2 mutation causes dilated cardiomyopathy, failure to thrive, global developmental delay, epilepsy and early death. Human Molecular Genetics 26 (18), 3545-3552. https://doi.org/10.1093/hmg/ddx239.

Cassar, S. Breidenbach, L. Olson, A. et al., 2017. Measuring drug absorption improves interpretation of behavioral responses in a larval zebrafish locomotor assay for predicting seizure liability. Journal of Pharmacological and Toxicological Methods 88 (Pt 1), 56-63. https://doi.org/10.1016/j.vascn.2017.07.002.

Celmina, M. Micule, I. Inashkina, I. et al., 2019. EAST/SeSAME syndrome: review of the literature and introduction of four new Latvian patients. Clinical Genetics 95 (1), 63-78. https://doi.org/10.1111/cge.13374.

Ceulemans, B. Boel, M. Leyssens, K. et al., 2012. Successful use of fenfluramine as an add-on treatment for Dravet syndrome. Epilepsia 53, 1131-1139. https://doi.org/10.1111/j.1528-1167.2012.03495.x.

Chege, S.W. Hortopan, G.A. T Dinday, M. Baraban, S.C. 2012. Expression and function of KCNQ channels in larval zebrafish. Developmental Neurobiology 72 (2), 186-198. https://doi.org/10.1002/dneu.20937.

Cho, S.J. Byun, D. Nam, T.S. et al., 2017. Zebrafish as an animal model in epilepsy studies with multichannel EEG recordings. Scientific Reports 7 (1), 3099. https://doi.org/10.1038/s41598-017-03482-6.

Claes, L. Del-Favero, J. Ceulemans, B. Lagae, L. Van Broeckhoven, C. De Jonghe, P. 2001. De novo mutations in the sodium-channel gene SCN1A cause severe myoclonic epilepsy of infancy. The American Journal of Human Genetics 68 (6), 1327-1332.

Claes, L.R. Ceulemans, B. Audenaert, D. Deprez, L. Jansen, A. Hasaerts, D. 2004. De novo KCNQ2 mutations in patients with benign neonatal seizures. Neurology 63, 2155-2158. https://doi.org/10.1212/ 01.WNL.0000145629.94338.89.

Copmans, D. Orellana-Paucar, A.M. Steurs, G. et al., 2018a. Methylated flavonoids as anti-seizure agents: naringenin 4',7-dimethyl ether attenuates epileptic seizures in zebrafish and mouse models. Neurochemistry International 112, 124-133. https://doi.org/10.1016/j.neuint.2017.11.011.

Copmans, D. Rateb, M. Tabudravu, J.N. et al., 2018b. Zebrafish-based discovery of antiseizure compounds from the red sea: pseurotin A2 and azaspirofuran A. ACS Chemical Neuroscience 9 (7), 1652-1662. https://doi.org/ 10.1021/acschemneuro.8b00060.

Cunliffe, V.T. 2016. Building a zebrafish toolkit for investigating the pathobiology of epilepsy and identifying new treatments

for epileptic seizures. Journal of Neuroscience Methods 260, 91-95. https://doi.org/10.1016/ j.jneumeth.2015.07.015.
Da Silva, R.S. Bonan, C.D. Vianna, M.R.M. 2016. Zebrafish as a platform for studies on seizures and epilepsy. Current Psychopharmacology 5, 211-220. https://doi.org/10.2174/2211556005666160526112030.
Dadas, A. Janigro, D. 2019. Breakdown of blood brain barrier as a mechanism of post-traumatic epilepsy. Neurobiology of Disease 123, 20-26. https://doi.org/10.1016/j.nbd.2018.06.022.
Demarest, S.T. Brooks-Kayal, A. 2018. From molecules to medicines: the dawn of targeted therapies for genetic epilepsies. Nature Reviews Neurology 14 (12), 735-745. https://doi.org/10.1038/s41582-018-0099-3.
Driever, W. Solnica-Krezel, L. Schier, A.F. et al., 1996. A genetic screen for mutations affecting embryogenesis in zebrafish. Development 123, 37-46.
Feenstra, B. Pasternak, B. Geller, F. et al., 2014. Common variants associated with general and MMR vaccine-related febrile seizures. Nature Genetics 46, 1274-1282. https://doi.org/10.1038/ng.3129.
Fleming, A. Diekmann, H. Goldsmith, P. 2013. Functional characterisation of the maturation of the blood-brain barrier in larval zebrafish. PLoS One 8 (10), e77548. https://doi.org/10.1371/journal.pone.0077548.
Fontana, B.D. Ziani, P.R. Canzian, J. et al., 2019. Taurine protects from pentylenetetrazole-induced behavioral and neurochemical changes in zebrafish. Molecular Neurobiology 56 (1), 583-594. https://doi.org/10.1007/s12035- 018-1107-8.
Franco, V. French, J.A. Perucca, E. 2016. Challenges in the clinical development of new antiepileptic drugs. Pharmacological Research 103, 95-104. https://doi.org/10.1016/j.phrs.2015.11.007.
Griffin, A. Krasniak, C. Baraban, S.C. 2016. Advancing epilepsy treatment through personalized genetic zebrafish models. Progress in Brain Research 226, 195-207. https://doi.org/10.1016/bs.pbr.2016.03.012.
Griffin, A. Hamling, K.R. Knupp, K. Hong, S. Lee, L.P. Baraban, S.C. 2017. Clemizole and modulators of serotonin signalling suppress seizures in Dravet syndrome. Brain 140 (3), 669-683. https://doi.org/10.1093/ brain/aww342.
Griffin, A. Hamling, K.R. Hong, S. Anvar, M. Lee, L.P. Baraban, S.C. 2018. Preclinical animal models for Dravet syndrome: seizure phenotypes, comorbidities and drug screening. Frontiers in Pharmacology 9, 573. https:// doi.org/10.3389/ fphar.2018.00573.
Grone, B.P. Baraban, S.C. 2015. Animal models in epilepsy research: legacies and new directions. Nature Neuroscience 18 (3), 339-343. https://doi.org/10.1038/nn.3934.
Grone, B.P. Marchese, M. Hamling, K.R. et al., 2016. Epilepsy, behavioral abnormalities, and physiological comorbidities in syntaxin-binding protein 1 (STXBP1) mutant zebrafish. PLoS One 11 (3), e0151148. https://doi.org/ 10.1371/journal. pone.0151148.
Grone, B.P. Qu, T. Baraban, S.C. 2017. Behavioral comorbidities and drug treatments in a zebrafish scn1lab model of Dravet syndrome. eNeuro 4 (4). https://doi.org/10.1523/ENEURO.0066-17.2017.
Gupta, P. Khobragade, S.B. Shingatgeri, V.M. 2014. Effect of various antiepileptic drugs in zebrafish PTZ-seizure model. Indian Journal of Pharmaceutical Sciences 76 (2), 157-163.
Haffter, P. Granato, M. Brand, M. et al., 1996. The identification of genes with unique and essential functions in the development of the zebrafish, Danio rerio. Development 123, 1-36.
Hortopan, G.A. Dinday, M.T. Baraban, S.C. 2010. Zebrafish as a model for studying genetic aspects of epilepsy. Disease Models and Mechanisms 3, 144-148. https://doi.org/10.1242/dmm.002139.
Howe, K. Clark, M.D. Torroja, C.F. et al., 2013. The zebrafish reference genome sequence and its relationship to the human genome. Nature 496 (7446), 498-503. https://doi.org/10.1038/nature12111.
Hunt, R.F. Hortopan, G.A. Gillespie, A. Baraban, S.C. 2012. A novel zebrafish model of hyperthermia-induced seizures reveals a role for TRPV4 channels and NMDA-type glutamate receptors. Experimental Neurology 237 (1), 199-206. https://doi. org/10.1016/j.expneurol.2012.06.013.
Hunyadi, B. Siekierska, A. Sourbron, J. Copmans, D. de Witte, P.A.M. 2017. Automated analysis of brain activity for seizure detection in zebrafish models of epilepsy. Journal of Neuroscience Methods 287, 13-24. https:// doi.org/10.1016/ j.jneumeth.2017.05.024.
International League Against Epilepsy Consortium on Complex Epilepsies, 2014. Genetic determinants of common epilepsies: a meta-analysis of genome-wide association studies. The Lancet Neurology 13 (9), 893-903. https://doi.org/10.1016/S1474-4422(14)70171-1.
Jao, L.-E. Wente, S.R. Chen, W. 2013. Efficient multiplex biallelic zebrafish genome editing using a CRISPR nuclease system. Proceedings of the National Academy of Sciences 110, 13904-13909. https://doi.org/10.1073/ pnas.1308335110.

Jin, M. He, Q. Zhang, S. Cui, Y. Han, L. Liu, K. 2018. Gastrodin suppresses pentylenetetrazole-induced seizures progression by modulating oxidative stress in zebrafish. Neurochemical Research 43 (4), 904-917. https:// doi.org/10.1007/s11064-018-2496-9.

Khan, K.M. Collier, A.D. Meshalkina, D.A. et al., 2017. Zebrafish models in neuropsychopharmacology and CNS drug discovery. British Journal of Pharmacology 174 (13), 1925-1944. https://doi.org/10.1111/ bph.13754.

Kim, Y.H. Lee, Y. Lee, K. Lee, T. Kim, Y.J. Lee, C.J. 2010. Reduced neuronal proliferation by proconvulsant drugs in the developing zebrafish brain. Neurotoxicology and Teratology 32 (5), 551-557. https://doi.org/10.1016/ j.ntt.2010.04.054.

Kiyatkin, E.A. 2007. Physiological and pathological brain hyperthermia. Progress in Brain Research 162, 219-243. https://doi.org/10.1016/S0079-6123(06)62012-8.

Kumar, M.G. Rowley, S. Fulton, R. Dinday, M.T. Baraban, S.C. Patel, M. 2016. Altered glycolysis and mitochondrial respiration in a zebrafish model of Dravet syndrome. eNeuro 3 (2). https://doi.org/10.1523/ENEURO.0008- 16.2016.

Kundap, U.P. Kumari, Y. Othman, I. Shaikh, M.F. 2017. Zebrafish as a model for epilepsy-induced cognitive dysfunction: a pharmacological, biochemical and behavioral approach. Frontiers in Pharmacology 8, 515. https://doi.org/10.3389/fphar.2017.00515.

Laino, D. Mencaroni, E. Esposito, S. 2018. Management of pediatric febrile seizures. International Journal of Environmental Research and Public Health 15 (10), 2232. https://doi.org/10.3390/ijerph15102232.

Lee, Y. Lee, B. Jeong, S. Park, J.W. Han, I.O. Lee, C.J. 2016. Increased cell proliferation and neural activity by physostigmine in the telencephalon of adult zebrafish. Neuroscience Letters 629, 189-195. https://doi.org/ 10.1016/j.neulet.2016.07.00134.

Lefebvre, K.A. Tilton, S.C. Bammler, T.K. et al., 2009. Gene expression profiles in zebrafish brain after acute exposure to domoic acid at symptomatic and asymptomatic doses. Toxicological Sciences 107 (1), 65-77. https:// doi.org/10.1093/toxsci/kfn207.

Leung, A.K. Hon, K.L. Leung, T.N. 2018. Febrile seizures: an overview. Drugs in Context 7, 212536. https:// doi.org/10.7573/dic.212536.

Lopes, M.W. Sapio, M.R. Leal, R.B. Fricker, L.D. 2016. Knockdown of carboxypeptidase A6 in zebrafish larvae reduces response to seizure-inducing drugs and causes changes in the level of mRNAs encoding signaling molecules. PLoS One 11 (4), e0152905. https://doi.org/10.1371/journal.pone.0152905.

Löscher, W. 2017. Animal models of seizures and epilepsy: past, present, and future role for the discovery of antiseizure drugs. Neurochemistry Research 42 (7), 1873-1888. https://doi.org/10.1007/s11064-017- 2222-z.

MacRae, C.A. Peterson, R.T. 2015. Zebrafish as tools for drug discovery. Nature Reviews Drug Discovery 14 (10), 721-731. https://doi.org/10.1038/nrd4627.

Mahmood, F. Mozere, M. Zdebik, A.A. et al., 2013a. Generation and validation of a zebrafish model of EAST (epilepsy, ataxia, sensorineural deafness and tubulopathy) syndrome. Disease Models and Mechanisms 6 (3), 652-660. https://doi.org/10.1242/dmm.009480.

Mahmood, F. Fu, S. Cooke, J. Wilson, S.W. Cooper, J.D. Russell, C. 2013b. A zebrafish model of CLN2 disease is deficient in tripeptidyl peptidase 1 and displays progressive neurodegeneration accompanied by a reduction in proliferation. Brain 136 (5), 1488-1507. https://doi.org/10.1093/brain/awt043.

Martinez, C.S. Feas, D.A. Siri, M. et al., 2018. In vivo study of teratogenic and anticonvulsant effects of antiepileptics drugs in zebrafish embryo and larvae. Neurotoxicology and Teratology 66, 17-24. https://doi.org/10.1016/ j.ntt.2018.01.008.

Mastrangelo, M. Leuzzi, V. 2012. Genes of early-onset epileptic encephalopathies: from genotype to phenotype. Pediatric Neurology 46, 24-31. https://doi.org/10.1016/j.pediatrneurol.2011.11.003.

Mei, X. Wu, S. Bassuk, A.G. Slusarski, D.C. 2013. Disease Mechanisms of prickle1a function in zebrafish epilepsy and retinal neurogenesis. Disease Models and Mechanisms 6 (3), 679-688.

Menezes, F.P. Da Silva, R.S. 2017. The influence of temperature on adult zebrafish sensitivity to pentylenetetrazole. Epilepsy Research 135, 14-18. https://doi.org/10.1016/j.eplepsyres.2017.05.009.

Menezes, F.P. Rico, E.P. Da Silva, R.S. 2014. Tolerance to seizure induced by kainic acid is produced in a specific period of zebrafish development. Progress in Neuro-Psychopharmacology and Biological Psychiatry 55, 109-112. https://doi.org/10.1016/j.pnpbp.2014.04.004.

Menezes, F.P. Kist, L.W. Bogo, M.R. Bonan, C.D. Da Silva, R.S. 2015. Evaluation of age-dependent response to NMDA receptor antagonism in zebrafish. Zebrafish 12 (2), 137-143. https://doi.org/10.1089/zeb.2014.1018.

Meyer, M. Dhamne, S.C. LaCoursiere, C.M. Tambunan, D. Poduri, A. Rotenberg, A. 2016. Microarray noninvasive neuronal seizure recordings from intact larval zebrafish. PLoS One 11 (6), e0156498. https://doi.org/10.1371/ journal.pone.0156498.

Published correction appears in PloS One 11(7): e0159472. doi.org/10.1371/ journal.pone.0159472.

Mussulini, B.H.M. Vizuete, A.F.K. Braga, M. et al., 2018. Forebrain glutamate uptake and behavioral parameters are altered in adult zebrafish after the induction of status epilepticus by kainic acid. Neurotoxicology 67, 305-312. https://doi.org/10.1016/ j.neuro.2018.04.007.

Myers, C.T. Mefford, H.C. 2015. Advancing epilepsy genetics in the genomic era. Genome Medicine 7, 91. https:// doi. org/10.1186/s13073-015-0214-7.

Myers, K.A. Johnstone, D.L. Dyment, D.A. 2019. Epilepsy genetics: current knowledge, applications, and future directions. Clinical Genetics 95 (1), 95-111. https://doi.org/10.1111/cge.13414.

Oltrabella, F. Pietka, G. Ramirez, I.B. et al., 2015. The Lowe syndrome protein OCRL1 is required for endocytosis in the zebrafish pronephric tubule. PLoS Genetics 11 (4), e1005058. https://doi.org/10.1371/journal.pgen.1005058.

Ottman, R. 1989. Genetics of the partial epilepsies: a review. Epilepsia 30 (1), 107-111.

Pearson-Smith, J.N. Patel, M. 2017. Metabolic dysfunction and oxidative stress in epilepsy. International Journal of Molecular Science 18, E2365. https://doi.org/10.3390/ijms18112365.

Pearson-Smith, J.N. Liang, L.P. Rowley, S.D. Day, B.J. Patel, M. 2017. Oxidative stress contributes to status epilepticus associated mortality. Neurochemistry Research 42, 2024-2032. https://doi.org/10.1007/s11064-017-2273-1.

Pena, I.A. Roussel, Y. Daniel, K. et al., 2017. Pyridoxine-dependent epilepsy in zebrafish caused by Aldh7a1 deficiency. Genetics 207 (4), 1501-1518. https://doi.org/10.1534/genetics.117.300137.

Phillips, J.B. Westerfield, M. 2014. Zebrafish models in translational research: tipping the scales toward advancements in human health. Disease Models and Mechanisms 7 (7), 739-743. https://doi.org/10.1242/dmm.015545.

Pineda, R. Beattie, C.E. Hall, C.W. 2011. Recording the adult zebrafish cerebral field potential during pentylenetetrazole seizures. Journal of Neuroscience Methods 200 (1), 20-28. https://doi.org/10.1016/j.jneumeth.2011.06.001.

Pippucci, T. Licchetta, L. Baldassari, S. et al., 2015. Epilepsy with auditory features: a heterogeneous clinicomolecular disease. Neurology Genetics 1 (1), e5. https://doi.org/10.1212/NXG.0000000000000005.

Pitkänen, A. Schwartzkroin, P. Moshé, S. 2005. Models of Seizures and Epilepsy, first ed. Academic Press, California, USA. Podlasz, P. Jakimiuk, A. Kasica-Jarosz, N. Czaja, K. Wasowicz, K. 2018. Neuroanatomical localization of galanin in zebrafish telencephalon and anticonvulsant effect of galanin overexpression. ACS Chemical Neuroscience. https://doi. org/10.1021/acschemneuro.8b00239.

Ramirez, I.B. Pietka, G. Jones, D.R. et al., 2012. Impaired neural development in a zebrafish model for Lowe syndrome. Human Molecular Genetics 21 (8), 1744-1759. https://doi.org/10.1093/hmg/ddr608.

Rosch, R.E. Hunter, P.R. Baldeweg, T. Friston, K.J. Meyer, M.P. 2018. Calcium imaging and dynamic causal modelling reveal brain-wide changes in effective connectivity and synaptic dynamics during epileptic seizures. PLoS Computational Biology 14 (8), e1006375. https://doi.org/10.1371/journal.pcbi.1006375.

Sakai, C. Ijaz, S. Hoffman, E.J. 2018. Zebrafish models of neurodevelopmental disorders: past, present, and future. Frontiers in Molecular Neuroscience 11, 294. https://doi.org/10.3389/fnmol.2018.00294.

Scheffer, I.E. Berkovic, S. Capovilla, G. et al., 2017. ILAE classification of the epilepsies: position paper of the ILAE commission for classification and terminology. Epilepsia 58, 512-521. https://doi.org/10.1111/epi.13709.

Schoonheim, P.J. Arrenberg, A.B. Del Bene, F. Baier, H. 2010. Optogenetic localization and genetic perturbation of saccade-generating neurons in zebrafish. Journal of Neuroscience 30, 7111-7120. https://doi.org/10.1523/ JNEUROSCI.5193-09.2010.

Schubert, J. Siekierska, A. Langlois, M. et al., 2014. Mutations in STX1B, encoding a presynaptic protein, cause feverassociated epilepsy syndromes. Nature Genetics 46 (12), 1327-1332.

Seidenberg, M. Pulsipher, D.T. Hermann, B. 2009. Association of epilepsy and comorbid conditions. Future Neurology 4 (5), 663-668. https://doi.org/10.2217/fnl.09.32.

Sheng, F. Chen, M. Tan, Y. et al., 2016. Protective effects of otophylloside N on pentylenetetrazol-induced neuronal injury in vitro and in vivo. Frontiers in Pharmacology 7, 224. https://doi.org/10.3389/fphar.2016.00224.

Sourbron, J. Smolders, I. de Witte, P. Lagae, L. 2017. Pharmacological analysis of the anti-epileptic mechanisms of fenfluramine in scn1a mutant zebrafish. Frontiers in Pharmacology 8, 191. https://doi.org/10.3389/fphar.2017.00191.

Stödberg, T. McTague, A. Ruiz, J. et al., 2015. Mutations in SLC12A5 in epilepsy of infancy with migrating focal seizures. Nature Communications 6, 8038. https://doi.org/10.1038/ncomms9038.

Suls, A. Jaehn, J.A. Kecskés, A. et al., 2013. De novo loss-of-function mutations in CHD2 cause a fever-sensitive myoclonic epileptic encephalopathy sharing features with Dravet syndrome. The American Journal of Human Genetics 93 (5), 967-975.

https://doi.org/10.1016/j.ajhg.2013.09.017.

Teng, Y. Xie, X. Walker, S. et al., 2010. Knockdown of zebrafish lgi1a results in abnormal development, brain defects and a seizure-like behavioral phenotype. Human Molecular Genetics 19 (22), 4409-4420. https://doi.org/ 10.1093/hmg/ddq364.

Teng, Y. Xie, X. Walker, S. et al., 2011. Loss of zebrafish lgi1b leads to hydrocephalus and sensitization to pentylenetetrazol induced seizure-like behavior. PLoS One 6 (9), e24596. https://doi.org/10.1371/journal.pone.0024596.

The Rikee project: Rational Intervention for KCNQ2/3 Epileptic Encephalopathy. http://www.rikee.org. Accessed 2 January 2019.

Tiedeken, J.A. Ramsdell, J.S. Ramsdell, A.F. 2005. Developmental toxicity of domoic acid in zebrafish (*Danio rerio*). Neurotoxicology and Teratology 27, 711-717. https://doi.org/10.1016/j.neuro.2010.05.003.

Truong, L. Harper, S.L. Tanguay, R.L. 2011. Evaluation of embryotoxicity using the zebrafish model. Methods in Molecular Biology 691, 271-279. https://doi.org/10.1007/978-1-60761-849-2_16.

Turrini, L. Fornetto, C. Marchetto, G. et al., 2017. Optical mapping of neuronal activity during seizures in zebrafish. Scientific Reports 7 (1), 3025. https://doi.org/10.1038/s41598-017-03087-z.

Vermoesen, K. Serruys, A.S. Loyens, E. et al., 2011. Assessment of the convulsant liability of antidepressants using zebrafish and mouse seizure models. Epilepsy and Behavior 22 (3), 450-460. https://doi.org/10.1016/j.yebeh.2011.08.016.

Wager, K. Zdebik, A.A. Fu, S. Cooper, J.D. Harvey, R.J. Russell, C. 2016. Neurodegeneration and epilepsy in a zebrafish model of CLN3 disease (batten disease). PLoS One 11 (6), e0157365. https://doi.org/10.1371/ journal.pone.0157365.

World Health Organization (WHO). Epilepsy. https://www.who.int/news-room/fact-sheets/detail/epilepsy. Accessed 2 January 2019.

Winter, M.J. Windell, D. Metz, J. et al., 2017. 4-dimensional functional profiling in the convulsant-treated larval zebrafish brain. Scientific Reports 7 (1), 6581. https://doi.org/10.1038/s41598-017-06646-6. Published correction appears in Scientific Reports. 2018:15903. doi:10.1038/s41598-017-06646-6.

Wong, K. Stewart, A. Gilder, T. et al., 2010. Modeling seizure-related behavioral and endocrine phenotypes in adult zebrafish. Brain Research 1348, 209-215. https://doi.org/10.1016/j.brainres.2010.06.012.

Younus, I. Reddy, D.S. 2018. A resurging boom in new drugs for epilepsy and brain disorders. Expert Review of Clinical Pharmacology 11 (1), 27-45. https://doi.org/10.1080/17512433.2018.1386553.

Zabinyakov, N. Bullivant, G. Cao, F. et al., 2017. Characterization of the first knock-out aldh7a1 zebrafish model for pyridoxine-dependent epilepsy using CRISPR-Cas9 technology. PLoS One 12 (10), e0186645. https://doi.org/ 10.1371/ journal.pone.0186645.

Zhang, Y. Kecskés, A. Copmans, D. et al., 2015. Pharmacological characterization of an antisense knockdown zebrafish model of Dravet syndrome: inhibition of epileptic seizures by the serotonin agonist fenfluramine. PLoS One 10 (5), e0125898. https://doi.org/10.1371/journal.pone.0125898.

第二十五章

衰老、生物钟和神经发生研究：斑马鱼法

Irina V. Zhdanova[1], Alexander Stankiewicz[1]

25.1 引言

20 世纪的技术发展和科学进步显著延长了人类平均寿命。下一个挑战就是延长人类健康期，即在漫长的一生中保持正常认知和身体功能。除了流行病学和临床研究外，解决这个问题还需要有效的整体动物转化模型。这是因为在一个复杂的多细胞有机体中，单个细胞、组织或器官的衰老是一个相互依赖的过程，其中整合系统（integrative systems）至关重要。确定哪些年龄相关因素会损害整合系统、它们之间的相互作用以及周期性变化的环境，对于开发预防和治疗疾病的方法以增加健康期非常重要。斑马鱼是一种小型且有研究优势的脊椎动物，本章介绍了它在研究衰老关键因素中所起的作用。

在过去 30 年的研究中，斑马鱼已经成为生物医学研究中的一种流行且高效的实验动物。最初，由于斑马鱼胚胎和仔鱼具有全身透明、体积小、繁殖力高和相对容易饲养等优点，成为发育生物学家最喜欢的脊椎动物（Ingham, 1997）。然而，斑马鱼仔鱼甚至成鱼不仅体形小，而且存在复杂的个体和社会行为等，因此也便于人们在行为和睡眠研究中采用高通量方法进行研究（Granato et al., 1996; Gerlai et al., 2000; Zhdanova et al., 2001）。目前，斑马鱼基因组的详细表征、多种突变和转基因表型的发现以及分子遗传学技术的出现，使得斑马鱼成为研究与人类正常和病理状况下相关问题的良好模型（Santoriello and Zon, 2012; Saleem and Kannan, 2018），其中包括衰老机制方面的研究（Kishi et al., 2009）。在转化研究方面，斑马鱼衰老模型较其他常见脊椎动物的衰老模型具有一些优点：斑马鱼是一种白天活动、夜间睡眠的脊椎动物，寿命相对较长且逐渐衰老，这些特征与人类相符。

[1] BioChron LLC, Worcester, MA, United States.

25.2 昼行性脊椎动物模型

过去转化生物医学研究中，忽视了昼行性动物模型的重要性，但在近年来的研究中，学者们越来越关注这个问题。物种对不同生态位的适应取决于动物在一天中的活动和觅食时间，而不是休息和睡眠时间。夜间、白天或傍晚的生活方式决定了不同分子过程、代谢、认知、免疫、生殖和其他基本身体功能、相互作用以及对周期性变化环境适应的特定时间序列。这些时间序列是通过一种复杂的整合机制——生物钟系统（Reppert and Weaver, 2002）进行编排的。

生物钟的重要性主要体现在其进化保守性，在微生物、植物和动物中存在相似的生物钟。生物钟的分子成分通过作为多个细胞内过程的调节蛋白、不同基因的转录因子（Kondratov et al., 2006; McDearmon et al., 2006）以及表观遗传过程（包括染色质重塑）的直接调节剂，完成其关键的生理和行为效应（Doi et al., 2006）。

越来越多的证据表明细胞分裂周期性与生物钟之间存在分子联系（Johnson, 2010; Shostak, 2017）。生物钟可直接调节细胞增殖，其机制可能是核心生物钟基因的产物可作为调节细胞周期的转录因子。或者，生物钟可以通过周期性地调节细胞环境来调节细胞分裂周期。体外单细胞或组织模型为了解细胞分裂周期和生物钟之间相互作用的基本机制提供了独特的见解（Bjarnason and Jordan, 2002; Dekens et al., 2003; Feillet et al., 2015; Malik et al., 2015; Matsuo et al., 2003）。此外，干预生物钟分子机制的遗传模型显示细胞增殖异常或中断（Bouchard-Cannon et al., 2013; Rakai et al., 2014），并可促进癌症的发展（Fu and Kettner, 2013）。这进一步证明了生物钟在细胞增殖中的重要作用。

鉴于生物钟的这些普遍功能，夜间和日间物种的主要昼夜节律过程与睡眠或代谢功能之间的时间关系正好相反，其机制仍有待充分阐明（Challet, 2007; Smale et al., 2003）。事实上，无论一个物种的昼夜适应性如何，主要的生物钟——下丘脑视交叉上核（SCN）神经元活动在白天达到峰值。松果体是生物钟的主要组成部分，它在夜间产生褪黑激素，或表达核心生物钟蛋白（如 BMAL1 和 CLOCK），这与人类的睡眠有关，但与小鼠的高自发活动、社会互动和觅食有关。这就是为什么昼行性动物模型可以更好地研究衰老和年龄依赖性疾病人体功能的时间节律，以及生物钟调节（Tranah et al., 2011）。

睡眠 - 觉醒是一种重要的昼夜节律，它对整个有机体具有重要的整合作用。睡眠通过稳态和昼夜节律机制影响发育、成熟和衰老。2001 年我们通过记录斑马鱼夜间睡眠的所有经典行为特征，并证明斑马鱼对褪黑激素和常用催眠药物的促眠作用高度敏感，从而建立了斑马鱼作为睡眠研究的模型（Zhdanova et al., 2001）。此后，发现斑马鱼是研究其他睡眠现象的绝佳模型，包括研究食欲肽的作用（Appelbaum et al., 2009; Oikonomou and Prober, 2017; Zhdanova, 2011）。最近的研究利用斑马鱼模型来记录睡眠在 DNA 修复中的作用，这是睡眠生物学中一项新颖而令人兴奋的进展（Zada et al., 2019）。

我们还讨论了斑马鱼衰老过程中睡眠改变的程度（Zhdanova, 2005; Zhdanova et al., 2008）。与其他身体功能类似，睡眠和昼夜节律的年龄依赖性变化从 2 岁左右开始，然后逐渐增加。这些变化与睡眠碎片化逐渐增加、睡眠节律日振幅下降、褪黑激素产生减少以及核心生物钟基因表达有关（Zhdanova et al., 2008; Tsai et al., 2007）。核心生物钟基因Bmal1的表达在衰老斑马鱼中下降极其显著。生物钟在斑马鱼衰老过程中的作用与在其他物种中获得的数据是一致的。从昆虫到哺乳动物的各种动物模型中，昼夜节律幅度和日常模式的年龄依赖性变化或生物钟基因的突变与过早衰老有关（Kondratov, 2007）。

总之，斑马鱼是生物医学研究中使用的为数不多的昼行性脊椎动物，它们在其生命的不同阶段都有很好的特征，并且易于进行高通量实验。因此，当研究身体功能时间节律的主要作用、生物钟和睡眠的整合作用以及它们随着衰老如何变化时，可以优先考虑使用斑马鱼模型。

25.3 逐渐衰老的脊椎动物

研究人类衰老的困难之一在于其渐进性。随着时间推移，衰老变化逐渐积累，并增加了患帕金森病或阿尔茨海默病等疾病的风险，而这些疾病可能在数年甚至数十年后才表现出来（Lane et al., 2018; Reeve et al., 2014）。这些需要历经数年的病理过程，不会在寿命短的物种中出现，这些物种通常表现出身体功能相对快速的年龄依赖性下降。相比之下，斑马鱼表现出明显的逐渐衰老模式（Kishi et al., 2009）。斑马鱼早期发育很快，受精6天后仔鱼成为活跃的猎手，在3个月时进入繁殖活跃阶段，这与小鼠相似。斑马鱼的第一个明显衰老迹象可以在2岁时检测到［图25-1（A）］。然而，在良好的实验室条件和喂养计划下，斑马鱼可以存活长达7年（在鱼类设施中记录在册的年龄）。它们的死亡率在4岁左右达到峰值［图25-1（B）］，而小鼠或大鼠的平均寿命为2年。在此期间，斑马鱼睡眠、认知功能或昼夜节律逐渐恶化。这表明，斑马鱼相对较长的寿命中至少有一半时间处于逐渐衰老的状态。

我们发现，与人类相似，斑马鱼的逐渐衰老与生殖能力降低、皮肤中β-半乳糖苷酶活性增加、肌肉中氧化蛋白积累、肝脏中脂褐素（“老化色素”）积累、睡眠改变、昼夜节律异常、焦虑增加和认知能力下降有关（Kishi et al., 2008; 2009; Tsai et al., 2007; Yu et al., 2006）。随着年龄增长，斑马鱼经常表现出脊柱弯曲（脊柱侧弯）和肌肉萎缩（Gerhard et al., 2002），这也是人类衰老的典型特征（Boswell and Ciruna, 2017）［图25-1（A）］。老年斑马鱼视网膜色素上皮中也经常出现脂褐素积累和玻璃膜疣样病变，类似于人类与年龄相关的黄斑变性。而且，几乎所有的老年斑马鱼在4岁时都会患上白内障，而且大部分老年斑马鱼会出现视网膜萎缩（Kishi et al., 2008）。

斑马鱼和许多其他低等脊椎动物一样，具有非常活跃的组织修复和再生能力。我们和其他研究者均发现，哺乳动物的再生能力随着年龄的增长而逐渐下降（Tsai et al., 2007; Kishi et al., 2003, 2008）。例如，成年斑马鱼鳍再生是一个高度协调的过程（Poleo et al., 2001）。这种再生能力随着年龄的增长而下降，导致斑马鱼形态受损和鱼鳍形状扭曲（Tsai et al., 2007）。而1.5岁或更年轻的鱼，尾鳍再生通常会恢复整个结构，没有可见的缺陷。尾鳍缺陷数量在2岁时显著增加，此后，鱼鳍再生能力逐渐下降，导致鳍变短或显著变形，或者一些动物年老时完全停止再生。

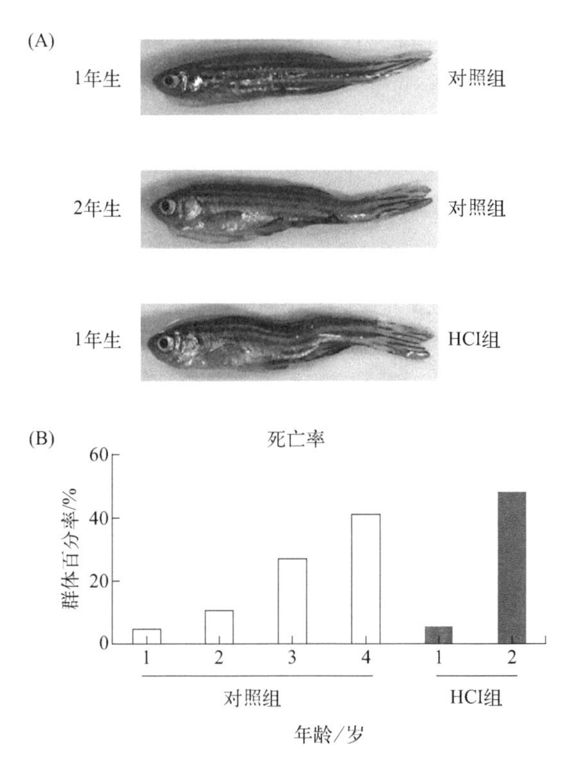

图25-1 **高热量摄入（high caloric intake, HCI）处理对斑马鱼老化表型和死亡率的影响**
图（A）：与对照组1岁斑马鱼相比，对照组2岁斑马鱼出现脊柱侧凸体征，而HCI组1岁斑马鱼出现脊柱侧凸和肌肉萎缩。图（B）：对照组1～4岁斑马鱼和HCI组1～2岁斑马鱼的死亡率。获准摘自以下文献并作了修改：Stankiewicz, A.J., McGowan, E.M., Yu, L., Zhdanova, I.V, 2017. Impaired sleep, circadian rhythms and neurogenesis in diet-induced premature aging. International Journal of Molecular Sciences 18(11), 3

25.4 代谢挑战引起的斑马鱼加速衰老

许多病理因素可以加速正常衰老过程，并且一个最佳的转化模型应该对已知会加速人类衰老的那些因素表现出类似的反应。根据流行病学研究，代谢异常与人类早衰之间存在密切联系（Bonomini et al., 2015; Sargent, 2015）。在某种程度上，这是由饮食引起的疾病，例如代谢综合征、2-型糖尿病、睡眠和心血管疾病（Akinnusi et al., 2012; Drager et al., 2013; Guh et al., 2009; Jensen et al., 2014; Leitner et al., 2017; Poirier et al., 2006）。同样，研究发现高热量摄入（HCI）会加速其他物种的衰老过程，而热量限制则会产生相反的效果（Arslan-Ergul et al., 2013; Balasubramanian et al., 2017; Liang et al., 2018; Weindruch et al., 1986）。

为了确定斑马鱼是否可以成为研究代谢异常对寿命和健康影响的有效模型，我们对低浓度HCI处理的动物进行了一系列研究（Stankiewicz et al., 2017, 2019）。从早期仔鱼阶段开始，我们向鱼类提供过多的食物，并在斑马鱼1岁及以上时记录其作用效果。与正常饮食的对照组相比，在相同的环境条件下HCI处理的斑马鱼生长速度更快，体形也更大。然而，HCI处理的斑马鱼寿命缩短，死亡率峰值从4岁降低至2岁［图25-1（A）］。

此外，在1岁时，几种类型的评估均显示HCI喂养加速了斑马鱼老化表型的出现（Stankiewicz et al., 2017）。这体现在与年龄相关的早期异常中，包括骨骼肌系统畸形［图25-2（A）］、焦虑样行为、底栖增加［图25-2（B）］、睡眠改变［图25-3（A）］以及核心生物钟基因表达的昼夜节律幅度降低［图25-3（B）～（D）］。

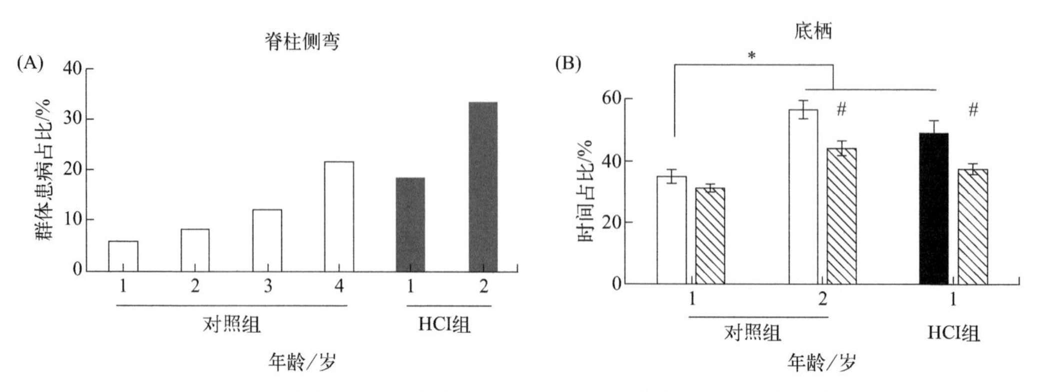

图25-2 **高热量摄入（HCI）加速斑马鱼衰老的脊柱侧弯和焦虑样表型出现**

（A）各种情况下斑马鱼患脊柱侧弯占比。白色：1～4岁对照组；黑色：1～2岁HCI处理组。（B）底栖反映斑马鱼焦虑行为，统计了斑马鱼在池底三分之一处游泳的时间占比。实心柱：没有用地西泮处理；条纹柱：地西泮处理。每组6～8尾鱼。结果用平均数 ± 标准误表示。*P<0.05，与1岁对照组相比；#P<0.05，地西泮处理组与未处理组相比。获准摘自以下文献并作了修改：Stankiewicz, A.J., McGowan, E.M., Yu, L., Zhdanova, I.V, 2017. Impaired sleep, circadian rhythms and neurogenesis in diet-induced premature aging. International Journal of Molecular Sciences 18(11), 3

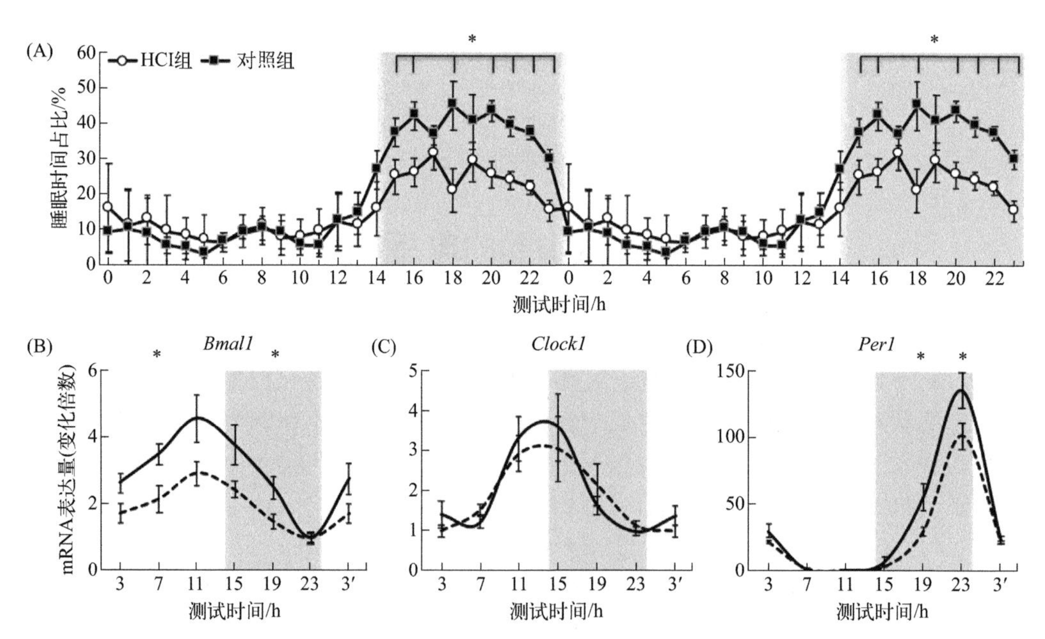

图25-3 **饮食诱导的早衰斑马鱼夜间睡眠减少和核心生物钟基因表达降低**

（A）24h内睡眠时间占比的双倒数图。黑框为对照组，白圈为HCI处理组。平均每组8尾鱼。（B）～（D）相对于组内最小表达，1岁斑马鱼HCI处理组和对照组中核心生物钟基因（*Bmal1*, *Clock1*, *Per1*）24h mRNA丰度差异倍数。每组5～6尾鱼，每个时间点，实线为对照组，虚线为HCI处理组。结果用均值 ± 标准误表示。1岁幼年斑马鱼的HCI组和对照组在各个时间点的对比，*P<0.05。灰色背景表示黑夜；14∶10的光-暗（LD）循环。获准摘自以下文献并作了修改：Stankiewicz, A.J., McGowan, E.M., Yu, L., Zhdanova, I.V, 2017. Impaired sleep, circadian rhythms and neurogenesis in diet-induced premature aging. International Journal of Molecular Sciences 18(11), 4

代谢途径和生理系统复杂性导致的功能年龄依赖性下降，以及大量活跃的代偿过程，使得很难明确早衰的主要原因。例如，研究发现生物钟系统的改变既可以加速衰老，又可以导致代谢异常（Hood and Amir, 2017; Maury et al., 2014; Sato et al., 2017）。这与生物钟进化为高度整合的计时系统一致，该系统将多种身体功能和动物行为联系起来。具体到新陈代谢，生物钟控制食欲的时间模式、消化器官的活动、葡萄糖敏感性、线粒体功能和许多新陈代谢其他方面（Reinke and Asher, 2019）。研究发现，HCI 可以降低斑马鱼核心生物钟基因每日表达幅度［图 25-3（B）～（D）］，表明慢性代谢负荷可导致生物钟故障（Stankiewicz et al., 2017）。尽管生物钟基因的表达与 HCI 斑马鱼的明暗周期保持正常的时间对应。但后者可能反映了斑马鱼生物钟对主要环境时间线索——环境光的高度敏感性（Tamai et al., 2005）。总之，我们的研究选用斑马鱼作为正常衰老和代谢诱导早衰的模型。这种昼行性脊椎动物模型为衰老转化研究和药物开发提供辅助平台。因此，斑马鱼有助于寻找新的延长人类健康期的预防和治疗方式。

25.5　成年斑马鱼神经发生受昼夜节律控制

具有开创性的研究发现成年啮齿动物存在神经元增殖（Altman and Das, 1965），并证实了新生细胞的神经元命运，这为进一步研究成体神经发生奠定了基础（Kaplan and Hinds, 1977）。现在人们认为无脊椎动物和脊椎动物普遍存在细胞增殖、成熟、迁移和新细胞与现有神经回路的整合，虽然发生程度因物种而异，例如在蟋蟀（Cayre et al., 1994）、甲壳类动物（Harzsch and Dawirs, 1996）、斑马鱼（Zupanc et al., 2005）、鸟类（Paton and Nottebohm, 1984）、啮齿动物（Gould et al., 1992）、猴子（Gould et al., 1999）和人类（Eriksson et al., 1998）中均发现此现象。在哺乳动物中，有两个主要的神经发生区：海马齿状回的颗粒下区（subgranular zone, SGZ）和侧脑室的脑室下区（subventricular zone, SVZ），其中神经母细胞可通过吻侧迁移流（rostral migratory stream, RMS）迁移到嗅球（Eriksson et al., 1998; Gould, 2007）。一些研究还表明成年哺乳动物下丘脑（Cheng, 2013; Evans et al., 2002; Xu et al., 2005）甚至新皮质中（Bhardwaj et al., 2006; Dayer et al., 2005）也可能存在神经发生。

因为斑马鱼中枢神经系统有很强的增殖能力，它成为研究成体神经发生最有力的脊椎动物模型之一。斑马鱼大脑中有 16 个分散的神经性生态位（neurogenic niches），其中神经干细胞（neural stem cells, NSC）可以自我复制并产生瞬时神经祖细胞（neural progenitor cells, NPC）或有丝分裂后细胞（postmitotic cells, PMC）（Grandel et al., 2006）。生成的有丝分裂后细胞、神经元和神经胶质细胞对个体神经性小室具有特异性，这与哺乳动物相似（Zupanc et al., 2005）。这些新细胞迁移到小室周围或更远的结构，例如沿着 RMS 向嗅觉基板（Kishimoto et al., 2011）迁移。这些新细胞整合到现有的大脑回路中，补充因细胞凋亡而损失的细胞，在斑马鱼大脑中逐渐适应并终生生长（Grandel et al., 2006）。

斑马鱼对于 NSC 实时成像具有明显优势。在其他脊椎动物中，端脑外侧壁向内弯曲（内翻）发育成端脑。在斑马鱼中，可沿着脑室壁观察到大脑内部的神经发生。与此不同的是，鳐鱼（辐鳍鱼）在发育过程中前脑的侧壁向外折叠（外翻）。由于这种不寻常的结构，对应于心室壁的增殖区域覆盖在斑马鱼端脑的外表面。通过使用转基因鱼，Barbosa 等人（2015）成功地对活斑马鱼的端脑 NSC、NPC 和新生神经元进行了成像（Barbosa et al., 2015）。

为了了解大脑如何控制成体神经发生，并揭示在不同大脑区域增加或抑制细胞增殖的小室特异性因素的类型，我们同时研究了多个大脑小室（Akle et al., 2017）。我们首先在 24h 或更长时间内，每隔 4h 记录 S 期细胞数量，探讨在年轻斑马鱼大脑细胞分裂周期进展的时间模式。我们还监测了关键细胞分裂周期基因的时间表达模式，例如细胞周期蛋白 A、B、D、E、G1/S 抑制剂（p20）和斑马鱼大脑中的核心生物钟基因。这两种方法都揭示了斑马鱼成体神经存在显著的时序调控（Akle et al., 2017）。

在大脑中，细胞周期蛋白和 p20 的 mRNA 丰度呈现稳定的日常模式，每个基因表达峰值都有其特定的时间（图 25-4）。总之，这些模式表明细胞分裂周期的 S 期发生在白天，G1/S 转换在晚上被抑制，而 G2 和 M 期在傍晚和夜间被促进。这些假设通过直接观察经历 S 期的细胞得到了证实（图 25-5）。

收集脑样本前 2h，在复制细胞中导入胸苷类似物（BrdU）。因此，每个时间点都反映了当时处于 S 期的细胞数量。在记录有 BrdU 阳性细胞的所有五个神经性小室中，S 期细胞数量白天逐渐增多，晚上达到峰值，然后夜间逐渐减少。尽管从端脑到小脑，这些不同神经性小室的整体细胞分裂周期模式相似，但它们之间仍存在一些明显差异。通过计算分析显示不同小室 S 期平均长度有所不同，从松果体缰核的 10.6h 到小脑的 16.1h 不等（$P<0.01$）。

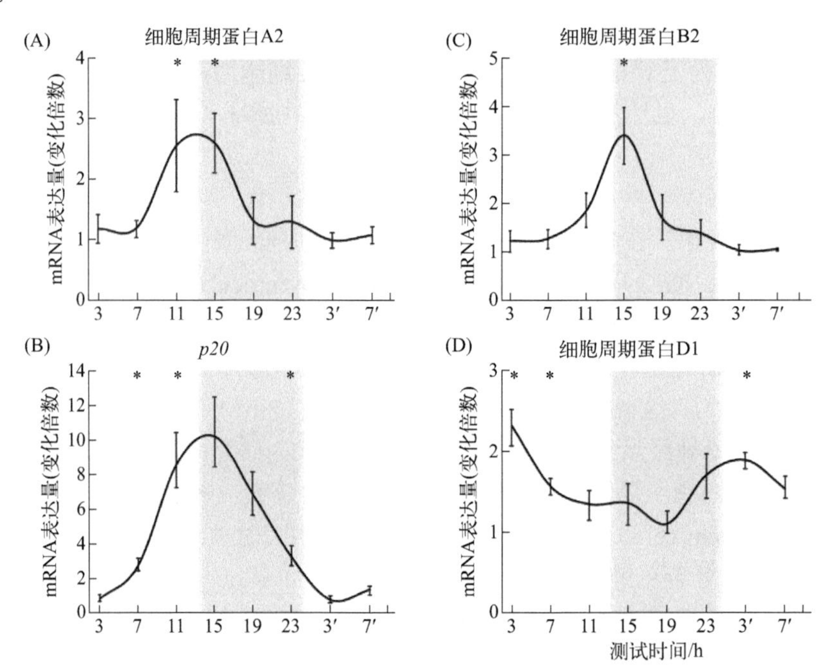

图 25-4　**成年斑马鱼脑细胞周期调节因子 mRNA 的日表达周期**

细胞周期蛋白 A2（A）、p20（B）、细胞周期蛋白 B2（C）和细胞周期蛋白 D1（D）的转录表达量。* 表示各时间点相对于每个基因 mRNA 表达低谷的表达倍数（$P < 0.05$）。以蛋白质低谷的表达量 =1（细胞周期蛋白 A2、p20、细胞周期蛋白 B2 在 ZT3 时间点的表达量为 1，而细胞周期蛋白 D1 在 ZT19 时间点的表达量为 1）计算。每个时间点 5 ～ 6 尾鱼，结果表示为平均数 ± 标准误。灰色背景表示夜晚，14∶10 的光 - 暗（LD）循环。获准摘自以下文献并作了修改：Akle, V., Stankiewicz, A.J., Kharchenko, V., Yu, L., Kharchenko, P.V., Zhdanova, I.V., 2017. Circadian kinetics of cell cycle progression in adult neurogenic niches of a diurnal vertebrate. Journal of Neuroscience: The Official Journal of the Society for Neuroscience 37(7), 1900-1909

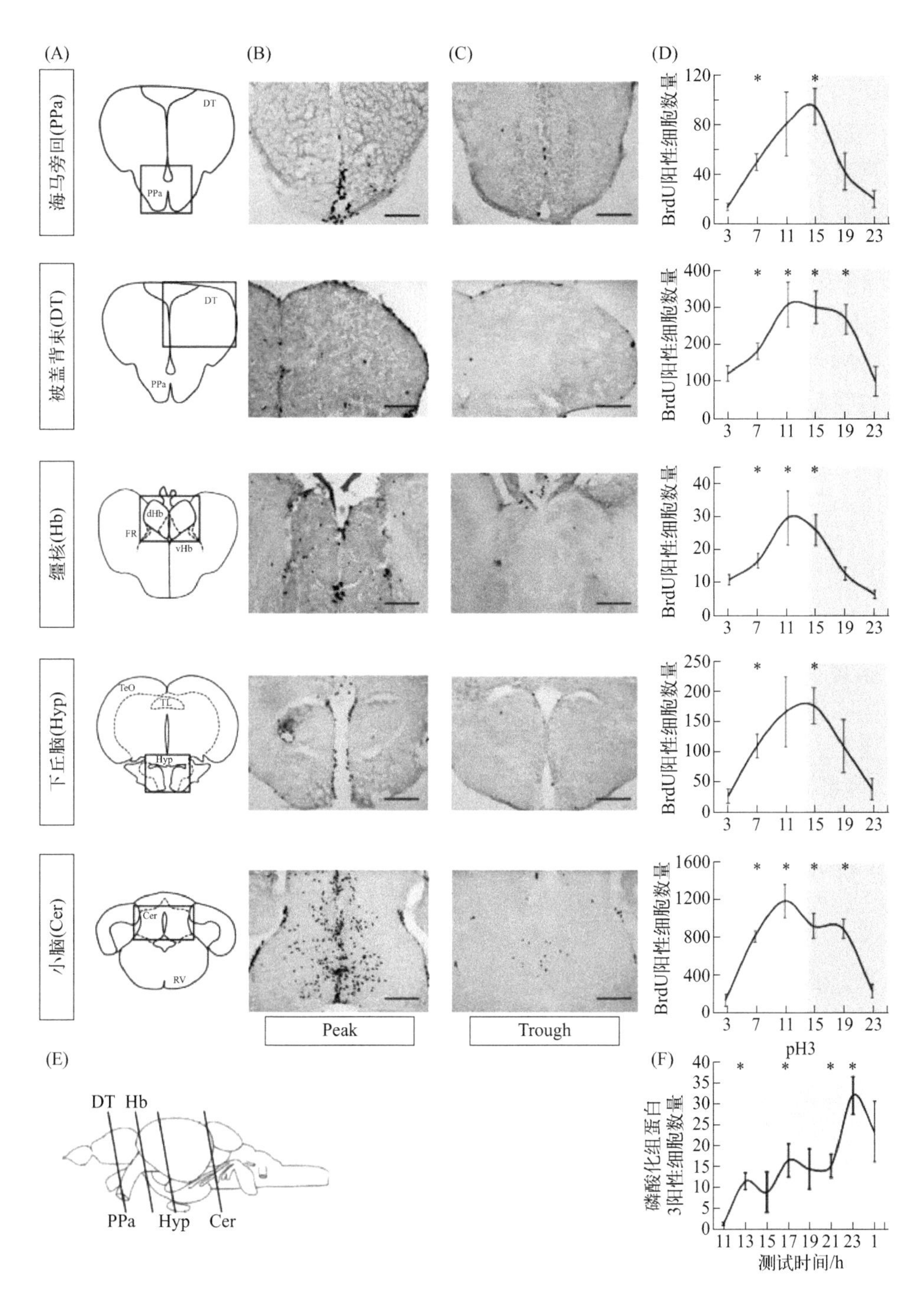

图 25-5 **成年斑马鱼大脑五个神经性小室细胞分裂周期 S 期的日变化规律**

(A) 成年斑马鱼大脑五个神经性小室示意图，方框标出（B）和（C）中的观察区域。(B, C) 图 B 为波谷，图 C 为峰值，为每个生态位中 BrdU 细胞数量日变化的典型图像，标尺为 50μm，数据呈现在（D）图中。(D) 24h 内每隔 4h 记录的每个神经性小室中处于 S 期的 BrdU 细胞数量。(E) 斑马鱼大脑通过脑组织切片评估（B）～（D）图中 BrdU 细胞的冠状平面示意图。(F) 小脑中表达 pH3（G2 晚期和 M 早期的标志）细胞数量和 BrdU（S 期的标志）细胞数量。每个时间点 4 ～ 6 尾鱼，结果表示为平均数 ± 标准误。*$P<0.05$，相对于波谷特定小室 BrdU 和 pH3 细胞数量。灰色背景表示夜晚，在 14∶10 的光 - 暗（LD）循环，体内应用 BrdU 4h 后，进行测试。获准摘自以下文献并作了修改：Akle, V., Stankiewicz, A.J., Kharchenko, V., Yu, L., Kharchenko, P.V., Zhdanova, I.V., 2017. Circadian kinetics of cell cycle progression in adult neurogenic niches of a diurnal vertebrate. Journal of Neuroscience: The Official Journal of the Society for Neuroscience 37(7), 1900e1909

随后研究中发现了成年斑马鱼神经性小室中磷酸化组蛋白 3（pH3）的存在。该组蛋白磷酸化标志着 G2/M 转变，即发生在接近染色体分离的时间。这种转变发生快，因此抗体一次只能在几个处于细胞分裂周期阶段的细胞中捕获 pH3。这种对单个细胞细胞周期的研究非常有价值。我们发现 pH3 阳性细胞数量在夜间上升，早上下降。

总之，这些研究表明，昼行性脊椎动物的神经发生遵循一定的细胞分裂昼夜节律模式。尽管不同神经性小室的一般模式相似，但也存在内部差异，这些差异可能是由每个小室的特定环境、周围细胞产生的代谢、神经递质或旁分泌因子，以及它们对斑马鱼大脑控制或受其影响的内部或外部过程的反应引起的。此外，神经性小室的老化和干细胞的耗竭可能对脊椎动物衰老的整个过程至关重要。考虑到斑马鱼是研究这些问题的代表模型，我们进一步的研究将集中在斑马鱼的正常衰老和加速衰老是否是改变神经发生的因素之一。

25.6 正常衰老和加速衰老中的神经发生

大脑是负责认知、感觉和运动功能的主要调节器官，这些功能会随着年龄的增长而衰退。大脑代谢改变、神经元连接受损以及神经元和神经胶质细胞的损失会危及整个生物体。事实上，在正常衰老过程中，大多数物种的神经发生都会下降（Katsimpardi and Lledo, 2018），包括鱼（Edelmann et al., 2013）、啮齿动物（Kempermann et al., 1998; Kuhn et al., 1996; Seki and Arai, 1995）、非人灵长类动物（Gould et al., 1999）和人类（Eriksson et al., 1998; Bergmann et al., 2015; Spalding et al., 2013）。在斑马鱼中，我们还观察到所有神经性小室中进行 DNA 复制的细胞数量逐渐减少（Stankiewicz et al., 2019; Akle et al., 2017），然而，在一些保持高性能且没有明显衰老表型的动物中（例如，没有脊柱侧弯或肌肉萎缩、活动量大、睡眠充足），神经形成的速度可以保持相对较高水平。

神经退行性疾病是延缓衰老的主要障碍。记忆力、情绪状态、肌肉协调性和其他身体功能的变化会给患有这些疾病的人及其护理人员带来重大负担（Mattson et al., 2002）。现有药物的功效有限，并且基本上缺乏能够显著抑制神经退行性疾病发展的预防措施。阐明渐进性年龄依赖性神经变性所涉及的机制并找到其预防方法，具有至关重要的社会意义。

目前研究已经发现，代谢和神经退行性疾病之间存在潜在联系（Mazon et al., 2017; Rojas-Gutierrez et al., 2017）。流行病学研究表明，热量摄入增加是阿尔茨海默病和帕金森病发展的一个重要危险因素（Rojas-Gutierrez et al., 2017; Ashrafian et al., 2013; Beydoun et al., 2008; Whitmer et al., 2008）。此外，研究表明，大量摄入高脂肪食物会对夜行性啮齿动物海马神经发生产生不利影响（Lindqvist et al., 2006），而热量限制可能会增加成年海马中细胞增殖、细胞存活和神经元活性（Kempermann et al., 1998; Kuhn et al., 1996; Seki and Arai, 1995; Bondolfi et al., 2004; Kumar et al., 2009; Lee et al., 2000, 2002a,b）。因此，我们的研究基于以下假设：随着年龄增长，成体神经发生的病理性变化至少在一定程度上是神经退行性疾病和代谢疾病之间关联的基础。

饮食因素或其代谢结果能够影响成体神经发生的高度动态和复杂过程。神经性小室中细胞动力学的改变包括 NSC 损失、其静止或分裂模式增加、瞬时祖细胞及其子代存活率降低、迁移能力改变，或无法融入现有的神经网络。寻找改变神经发生的预防或治疗解决方案需要清楚地了解疾病中哪些细胞和哪个阶段是问题的核心，从而可以作为有效的药物靶点。反过来，这需要使用高通量定量方法和数学建模来模拟神经干细胞和祖细胞的几种分裂模式以及细胞最终结局。

在正常饲养和慢性 HCI 加速衰老斑马鱼中，我们使用了数天 BrdU-EdU 脉冲追踪，该方案考虑了 DNA 复制的昼夜节律模式（Akle et al., 2017）。我们还采用了一个描述细胞密度随时间变化的微分方程模型。这种综合方法揭示了与 HCI 诱导老化加速相关的多种变化（Stankiewicz et al., 2019）。对照组斑马鱼成体神经发生的细胞动力学模式是有序的，与此相比，HCI 处理组的鱼在细胞增殖方面表现出高度日间差异［图 25-6（A）和（B）］。这种尚待解释的神经性过程不稳定性与 NSC 和瞬态 NPC 数量极低有关，与干细胞不同的是，NSC 和瞬态 NPC 应该进行多次连续分裂。当涉及瞬时祖细胞，以及其有丝分裂后子代的缓慢迁移时，低数量的分裂细胞往往寿命短暂。我们最有趣的发现也许是 HCI 可以干扰神经元祖细胞的主要分裂模式。在对照动物中，这些瞬时祖细胞更倾向于分裂成有丝分裂后的子细胞，数学模型表明，在 HCI 处理组鱼类中，NPC 更倾向于自我更新，即产生更多的瞬时祖细胞［图 25-6（C）］。虽然这可能是对增殖细胞池整体下降的一种代偿反应，但最终结果是在加速衰老的个体中，新分化神经元的缺陷增加。总体而言，我们的研究表明，慢性 HCI 可能通过多种分子机制在多个层面上干扰成体神经发生。

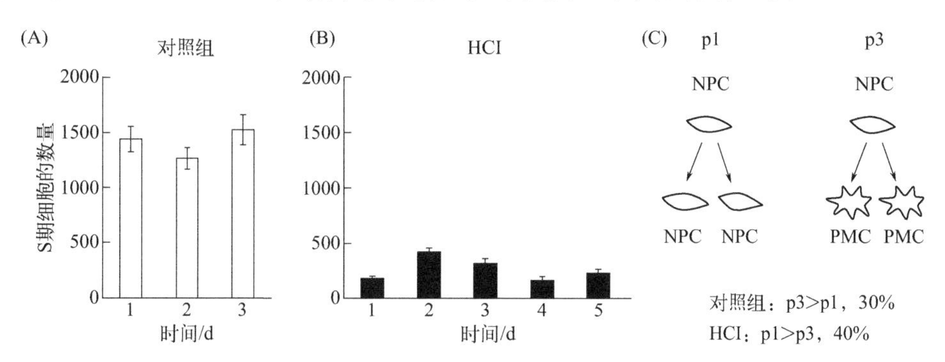

图 25-6 **年龄为 1 岁的斑马鱼长期摄入高热量后脑中的增殖活性下降**

连续 3 ～ 5 天定量评估小脑 S 期细胞的数量。图（A）为对照组，图（B）为 HCI 处理组，斑马鱼在每日高峰期 ZT 9 ～ 11（在 ZT 9 ～ 11，进行 EdU 暴露 2h 后）的 S 期细胞数量，对照组为白色柱，HCI 组为黑色柱。每个时间点 4 ～ 6 尾鱼，结果表示为平均数 ± 标准误。（C）图为在对照组和 NIH 鱼中 NPC 分裂不同模式的相对概率，p1 为对称分裂导致两个 NPC 细胞的概率；p3 为一个对称分裂导致两个有丝分裂后细胞（postmitotic cells, PMC）的概率。获准摘自以下文献并作了修改：Stankiewicz, A.J., Mortazavi, F., Kharchenko, P.V., McGowan, E.M., Kharchenko, V., Zhdanova, I.V.,2019. Cell kinetics in the adult neurogenic niche and impact of dietinduced accelerated aging. Journal of Neuroscience: The Official Journal of the Society for Neuroscience 39(15), 2810-2822

25.7 结论

斑马鱼是一个可用于研究整合系统及其在整个生命周期中调节不同分子、生理和行为过程机制的全能生物模型。斑马鱼是一种小型且特征鲜明的脊椎动物，易于采用高通量方法研究，且它还有其他一些特性有助于老龄化研究。斑马鱼日常生活方式与人类时间关系相似，其核心生物钟可调节分子、生理现象和行为功能。与人类相似，斑马鱼逐渐衰老，它较长的寿命也方便我们研究与年龄依赖性相关人类疾病逐渐发展的病理过程。另一方面，斑马鱼的高增殖能力，包括成体神经发生的高增殖能力，超过了在人类中观察到的相关能力，可能为基础研究、寻找神经退行性疾病和其他神经疾病的有效治疗方法开辟新的可能性。新研究方法的使用，如全脑透明辅助免疫染色技术，以及各种细胞标志物3D可视化和自动量化，可以进一步丰富斑马鱼实验，为利用复杂的数学建模进行大数据分析提供支持（Stankiewicz et al., 2019）。总之，斑马鱼模型为相关研究带来了希望。或许有一天，一条小鱼可以帮助揭开衰老和与之相关的神经退行性疾病的奥秘。

（刘东鑫翻译　赵丽审校）

参考文献

Akinnusi, M.E. Saliba, R. Porhomayon, J. El-Solh, A.A. 2012. Sleep disorders in morbid obesity. European Journal of Internal Medicine 23(3), 219-226.

Akle, V. Stankiewicz, A.J. Kharchenko, V. Yu, L. Kharchenko, P.V. Zhdanova, I.V. 2017. Circadian kinetics of cell cycle progression in adult neurogenic niches of a diurnal vertebrate. Journal of Neuroscience: The Official Journal of the Society for Neuroscience 37(7), 1900-1909.

Altman, J. Das, G.D. 1965. Autoradiographic and histological evidence of postnatal hippocampal neurogenesis in rats. The Journal of Comparative Neurology 124(3), 319-335.

Appelbaum, L. Wang, G.X. Maro, G.S. et al., 2009. Sleep-wake regulation and hypocretin-melatonin interaction in zebrafish. Proceedings of the National Academy of Sciences of the United States of America 106(51), 21942-21947.

Arslan-Ergul, A. Ozdemir, A.T. Adams, M.M. 2013. Aging, neurogenesis, and caloric restriction in different model organisms. Aging and disease 4(4), 221-232.

Ashrafian, H. Harling, L. Darzi, A. Athanasiou, T. 2013. Neurodegenerative disease and obesity: what is the role of weight loss and bariatric interventions? Metabolic Brain Disease 28(3), 341-353.

Balasubramanian, P. Howell, P.R. Anderson, R.M. 2017. Aging and caloric restriction research: a biological perspective with translational potential. EBioMedicine 21, 37-44.

Barbosa, J.S. Sanchez-Gonzalez, R. Di Giaimo, R. et al., 2015. Neurodevelopment. Live imaging of adult neural stem cell behavior in the intact and injured zebrafish brain. Science 348(6236), 789-793.

Bergmann, O. Spalding, K.L. Frisen, J. 2015. Adult neurogenesis in humans. Cold Spring Harbor perspectives in biology 7(7), a018994.

Beydoun, M.A. Gary, T.L. Caballero, B.H. Lawrence, R.S. Cheskin, L.J. Wang, Y. 2008. Ethnic differences in dairy and related nutrient consumption among US adults and their association with obesity, central obesity, and the metabolic syndrome. American Journal of Clinical Nutrition 87(6), 1914-1925.

Bhardwaj, R.D. Curtis, M.A. Spalding, K.L. et al., 2006. Neocortical neurogenesis in humans is restricted to development. Proceedings of the National Academy of Sciences of the United States of America 103(33), 12564-12568.

Bjarnason, G.A. Jordan, R. 2002. Rhythms in human gastrointestinal mucosa and skin. Chronobiology International 19(1), 129-140.

Bondolfi, L. Ermini, F. Long, J.M. Ingram, D.K. Jucker, M. 2004. Impact of age and caloric restriction on neurogenesis in the dentate gyrus of C57BL/6 mice. Neurobiology of Aging 25(3), 333-340.

Bonomini, F. Rodella, L.F. Rezzani, R. 2015. Metabolic syndrome, aging and involvement of oxidative stress. Aging and disease 6(2), 109-120.

Boswell, C.W. Ciruna, B. 2017. Understanding idiopathic scoliosis: a new zebrafish school of thought. Trends in Genetics 33(3), 183-196.

Bouchard-Cannon, P. Mendoza-Viveros, L. Yuen, A. Kaern, M. Cheng, H.Y. 2013. The circadian molecular clock regulates adult hippocampal neurogenesis by controlling the timing of cell-cycle entry and exit. Cell Reports 5(4), 961-973.

Cayre, M. Strambi, C. Strambi, A. 1994. Neurogenesis in an adult insect brain and its hormonal control. Nature 368(6466), 57-59.

Challet, E. 2007. Minireview: entrainment of the suprachiasmatic clockwork in diurnal and nocturnal mammals. Endocrinology

148(12), 5648-5655.
Cheng, M.F. 2013. Hypothalamic neurogenesis in the adult brain. Frontiers in Neuroendocrinology 34(3), 167-178.
Dayer, A.G. Cleaver, K.M. Abouantoun, T. Cameron, H.A. 2005. New GABAergic interneurons in the adult neocortex and striatum are generated from different precursors. The Journal of Cell Biology 168(3), 415-427.
Dekens, M.P. Santoriello, C. Vallone, D. Grassi, G. Whitmore, D. Foulkes, N.S. 2003. Light regulates the cell cycle in zebrafish. Current Biology: CB 13(23), 2051-2057.
Doi, M. Hirayama, J. Sassone-Corsi, P. 2006. Circadian regulator CLOCK is a histone acetyltransferase. Cell 125(3), 497-508.
Drager, L.F. Togeiro, S.M. Polotsky, V.Y. Lorenzi-Filho, G. 2013. Obstructive sleep apnea: a cardiometabolic risk in obesity and the metabolic syndrome. Journal of the American College of Cardiology 62(7), 569-576.
Edelmann, K. Glashauser, L. Sprungala, S. et al., 2013. Increased radial glia quiescence, decreased reactivation upon injury and unaltered neuroblast behavior underlie decreased neurogenesis in the aging zebrafish telencephalon. The Journal of Comparative Neurology 521(13), 3099-3115.
Eriksson, P.S. Perfilieva, E. Bjork-Eriksson, T. et al., 1998. Neurogenesis in the adult human hippocampus. Nature Medicine 4(11), 1313-1317.
Evans, J. Sumners, C. Moore, J. et al., 2002. Characterization of mitotic neurons derived from adult rat hypothalamus and brain stem. Journal of Neurophysiology 87(2), 1076-1085.
Feillet, C. van der Horst, G.T. Levi, F. Rand, D.A. Delaunay, F. 2015. Coupling between the circadian clock and cell cycle oscillators: implication for healthy cells and malignant growth. Frontiers in Neurology 6, 96.
Fu, L. Kettner, N.M. 2013. The circadian clock in cancer development and therapy. Progress in Molecular Biology and Translational Science 119, 221-282.
Gerhard, G.S. Kauffman, E.J. Wang, X. et al., 2002. Life spans and senescent phenotypes in two strains of Zebrafish (*Danio rerio*). Experimental Gerontology 37(8-9), 1055-1068.
Gerlai, R. Lahav, M. Guo, S. Rosenthal, A. 2000. Drinks like a fish: zebra fish (*Danio rerio*) as a behavior genetic model to study alcohol effects. Pharmacology Biochemistry and Behavior 67(4), 773-782.
Gould, E. Cameron, H.A. Daniels, D.C. Woolley, C.S. McEwen, B.S. 1992. Adrenal hormones suppress cell division in the adult rat dentate gyrus. Journal of Neuroscience: The Official Journal of the Society for Neuroscience 12(9), 3642-3650.
Gould, E. Reeves, A.J. Fallah, M. Tanapat, P. Gross, C.G. Fuchs, E. 1999. Hippocampal neurogenesis in adult old world primates. Proceedings of the National Academy of Sciences of the United States of America 96(9), 5263-5267.
Gould, E. 2007. How widespread is adult neurogenesis in mammals? Nature Reviews Neuroscience 8(6), 481-488.
Granato, M. van Eeden, F.J. Schach, U. et al., 1996. Genes controlling and mediating locomotion behavior of the zebrafish embryo and larva. Development 123, 399-413.
Grandel, H. Kaslin, J. Ganz, J. Wenzel, I. Brand, M. 2006. Neural stem cells and neurogenesis in the adult zebrafish brain: origin, proliferation dynamics, migration and cell fate. Developmental Biology 295(1), 263-277.
Guh, D.P. Zhang, W. Bansback, N. Amarsi, Z. Birmingham, C.L. Anis, A.H. 2009. The incidence of co-morbidities related to obesity and overweight: a systematic review and meta-analysis. BMC Public Health 9, 88.
Harzsch, S. Dawirs, R.R. 1996. Neurogenesis in the developing crab brain: postembryonic generation of neurons persists beyond metamorphosis. Journal of Neurobiology 29(3), 384-398.
Hood, S. Amir, S. 2017. Neurodegeneration and the circadian clock. Frontiers in Aging Neuroscience 9, 170.
Ingham, P.W. 1997. Zebrafish genetics and its implications for understanding vertebrate development. Human Molecular Genetics 6(10), 1755-1760.
Jensen, M.D. Ryan, D.H. Apovian, C.M. et al., 2014. 2013 AHA/ACC/TOS guideline for the management of overweight and obesity in adults: a report of the American college of cardiology/American heart association task force on practice guidelines and the obesity society. Circulation 129(25 Suppl. 2), S102-S138.
Johnson, C.H. 2010. Circadian clocks and cell division: what's the pacemaker? Cell Cycle 9(19), 3864-3873.
Kaplan, M.S. Hinds, J.W. 1977. Neurogenesis in the adult rat: electron microscopic analysis of light radioautographs. Science 197(4308), 1092-1094.
Katsimpardi, L. Lledo, P.M. 2018. Regulation of neurogenesis in the adult and aging brain. Current Opinion in Neurobiology 53, 131-138.
Kempermann, G. Kuhn, H.G. Gage, F.H. 1998. Experience-induced neurogenesis in the senescent dentate gyrus. Journal of Neuroscience: The Official Journal of the Society for Neuroscience 18(9), 3206-3212.
Kishi, S. Uchiyama, J. Baughman, A.M. Goto, T. Lin, M.C. Tsai, S.B. 2003. The zebrafish as a vertebrate model of functional

aging and very gradual senescence. Experimental Gerontology 38(7), 777-786.

Kishi, S. Bayliss, P.E. Uchiyama, J. et al., 2008. The identification of zebrafish mutants showing alterations in senescence-associated biomarkers. PLoS Genetics 4(8), e1000152.

Kishi, S. Slack, B.E. Uchiyama, J. Zhdanova, I.V. 2009. Zebrafish as a genetic model in biological and behavioral gerontology: where development meets aging in vertebratesea mini-review. Gerontology 55(4), 430-441.

Kishimoto, N. Alfaro-Cervello, C. Shimizu, K. et al., 2011. Migration of neuronal precursors from the telencephalic ventricular zone into the olfactory bulb in adult zebrafish. The Journal of Comparative Neurology 519(17), 3549-3565.

Kondratov, R.V. Kondratova, A.A. Gorbacheva, V.Y. Vykhovanets, O.V. Antoch, M.P. 2006. Early aging and agerelated pathologies in mice deficient in BMAL1, the core component of the circadian clock. Genes & Development 20(14), 1868-1873.

Kondratov, R.V. 2007. A role of the circadian system and circadian proteins in aging. Ageing Research Reviews 6(1), 12-27.

Kuhn, H.G. Dickinson-Anson, H. Gage, F.H. 1996. Neurogenesis in the dentate gyrus of the adult rat: age-related decrease of neuronal progenitor proliferation. Journal of Neuroscience: The Official Journal of the Society for Neuroscience 16(6), 2027-2033.

Kumar, S. Parkash, J. Kataria, H. Kaur, G. 2009. Interactive effect of excitotoxic injury and dietary restriction on neurogenesis and neurotrophic factors in adult male rat brain. Neuroscience Research 65(4), 367-374.

Lane, C.A. Hardy, J. Schott, J.M. 2018. Alzheimer's disease. European Journal of Neurology 25(1), 59-70.

Lee, J. Duan, W. Long, J.M. Ingram, D.K. Mattson, M.P. 2000. Dietary restriction increases the number of newly generated neural cells, and induces BDNF expression, in the dentate gyrus of rats. Journal of Molecular Neuroscience 15(2), 99-108.

Lee, J. Duan, W. Mattson, M.P. 2002a. Evidence that brain-derived neurotrophic factor is required for basal neurogenesis and mediates, in part, the enhancement of neurogenesis by dietary restriction in the hippocampus of adult mice. Journal of Neurochemistry 82(6), 1367-1375.

Lee, J. Seroogy, K.B. Mattson, M.P. 2002b. Dietary restriction enhances neurotrophin expression and neurogenesis in the hippocampus of adult mice. Journal of Neurochemistry 80(3), 539-547.

Leitner, D.R. Fruhbeck, G. Yumuk, V. et al., 2017. Obesity and type 2 diabetes: two diseases with a need for combined treatment strategies e EASO can lead the way. Obesity Facts 10(5), 483-492.

Liang, Y. Liu, C. Lu, M. et al., 2018. Calorie restriction is the most reasonable anti-ageing intervention: a meta-analysis of survival curves. Scientific Reports 8(1), 5779.

Lindqvist, A. Mohapel, P. Bouter, B. et al., 2006. High-fat diet impairs hippocampal neurogenesis in male rats. European Journal of Neurology 13(12), 1385-1388.

Malik, A. Kondratov, R.V. Jamasbi, R.J. Geusz, M.E. 2015. Circadian clock genes are essential for normal adult neurogenesis, differentiation, and fate determination. PLoS One 10(10), e0139655.

Matsuo, T. Yamaguchi, S. Mitsui, S. Emi, A. Shimoda, F. Okamura, H. 2003. Control mechanism of the circadian clock for timing of cell division in vivo. Science 302(5643), 255-259.

Mattson, M.P. Chan, S.L. Duan, W. 2002. Modification of brain aging and neurodegenerative disorders by genes, diet, and behavior. Physiological Reviews 82(3), 637-672.

Maury, E. Hong, H.K. Bass, J. 2014. Circadian disruption in the pathogenesis of metabolic syndrome. Diabetes and Metabolism 40(5), 338-346.

Mazon, J.N. de Mello, A.H. Ferreira, G.K. Rezin, G.T. 2017. The impact of obesity on neurodegenerative diseases. Life Sciences 182, 22-28.

McDearmon, E.L. Patel, K.N. Ko, C.H. et al., 2006. Dissecting the functions of the mammalian clock protein BMAL1 by tissue-specific rescue in mice. Science 314(5803), 1304-1308.

Oikonomou, G. Prober, D.A. 2017. Attacking sleep from a new angle: contributions from zebrafish. Current Opinion in Neurobiology 44, 80-88.

Paton, J.A. Nottebohm, F.N. 1984. Neurons generated in the adult brain are recruited into functional circuits. Science 225(4666), 1046-1048.

Poirier, P. Giles, T.D. Bray, G.A. et al., 2006. Obesity and cardiovascular disease: pathophysiology, evaluation, and effect of weight loss: an update of the 1997 American heart association scientific statement on obesity and heart disease from the obesity committee of the council on nutrition, physical activity, and metabolism. Circulation 113(6), 898-918.

Poleo, G. Brown, C.W. Laforest, L. Akimenko, M.A. 2001. Cell proliferation and movement during early fin regeneration in zebrafish. Developmental Dynamics: An Official Publication of the American Association of Anatomists 221(4), 380-390.

Rakai, B.D. Chrusch, M.J. Spanswick, S.C. Dyck, R.H. Antle, M.C. 2014. Survival of adult generated hippocampal neurons is altered in circadian arrhythmic mice. PLoS One 9(6), e99527.

Reeve, A. Simcox, E. Turnbull, D. 2014. Ageing and Parkinson's disease: why is advancing age the biggest risk factor.Ageing Research Reviews 14, 19-30.

Reinke, H. Asher, G. 2019. Crosstalk between metabolism and circadian clocks. Nature Reviews Molecular Cell Biology 20(4), 227-241.

Reppert, S.M. Weaver, D.R. 2002. Coordination of circadian timing in mammals. Nature 418(6901), 935-941.

Rojas-Gutierrez, E. Munoz-Arenas, G. Trevino, S. et al., 2017. Alzheimer's disease and metabolic syndrome: a link from oxidative stress and inflammation to neurodegeneration. Synapse 71(10), e21990.

Saleem, S. Kannan, R.R. 2018. Zebrafish: an emerging real-time model system to study Alzheimer's disease and neurospecific drug discovery. Cell Death & Disease 4, 45.

Santoriello, C. Zon, L.I. 2012. Hooked! Modeling human disease in zebrafish. Journal of Clinical Investigation 122(7), 2337-2343.

Sargent, J. 2015. Neurodegenerative disease: balancing BMIerethinking the relationships between obesity, ageing and risk of dementia. Nature Reviews Endocrinology 11(6), 315.

Sato, S. Solanas, G. Peixoto, F.O. et al., 2017. Circadian reprogramming in the liver identifies metabolic pathways of aging. Cell 170(4), 664-677 e611.

Seki, T. Arai, Y. 1995. Age-related production of new granule cells in the adult dentate gyrus. NeuroReport 6(18), 2479-2482.

Shostak, A. 2017. Circadian clock, cell division, and cancer: from molecules to organism. International Journal of Molecular Sciences 18(4).

Smale, L. Lee, T. Nunez, A.A. 2003. Mammalian diurnality: some facts and gaps. Journal of Biological Rhythms 18(5), 356-366.

Spalding, K.L. Bergmann, O. Alkass, K. et al., 2013. Dynamics of hippocampal neurogenesis in adult humans. Cell 153(6), 1219-1227.

Stankiewicz, A.J. McGowan, E.M. Yu, L. Zhdanova, I.V. 2017. Impaired sleep, circadian rhythms and neurogenesis in diet-induced premature aging. International Journal of Molecular Sciences 18(11).

Stankiewicz, A.J. Mortazavi, F. Kharchenko, P.V. McGowan, E.M. Kharchenko, V. Zhdanova, I.V. 2019. Cell kinetics in the adult neurogenic niche and impact of diet-induced accelerated aging. Journal of Neuroscience: The Official Journal of the Society for Neuroscience 39(15), 2810-2822.

Tamai, T.K. Carr, A.J. Whitmore, D. 2005. Zebrafish circadian clocks: cells that see light. Biochemical Society Transactions 339Pt 5), 962-966.

Tranah, G.J. Blackwell, T. Stone, K.L. et al., 2011. Circadian activity rhythms and risk of incident dementia and mild cognitive impairment in older women. Annals of Neurology 70(5), 722-732.

Tsai, S.B. Tucci, V. Uchiyama, J. et al., 2007. Differential effects of genotoxic stress on both concurrent body growth and gradual senescence in the adult zebrafish. Aging Cell 6(2), 209-224.

Weindruch, R. Walford, R.L. Fligiel, S. Guthrie, D. 1986. The retardation of aging in mice by dietary restriction: longevity, cancer, immunity and lifetime energy intake. Journal of Nutrition 116(4), 641-654.

Whitmer, R.A. Gustafson, D.R. Barrett-Connor, E. Haan, M.N. Gunderson, E.P. Yaffe, K. 2008. Central obesity and increased risk of dementia more than three decades later. Neurology 71(14), 1057-1064.

Xu, Y. Tamamaki, N. Noda, T. et al., 2005. Neurogenesis in the ependymal layer of the adult rat 3rd ventricle. Experimental Neurology 192(2), 251-264.

Yu, L. Tucci, V. Kishi, S. Zhdanova, I.V. 2006. Cognitive aging in zebrafish. PLoS One 1, e14.

Zada, D. Bronshtein, I. Lerer-Goldshtein, T. Garini, Y. Appelbaum, L. 2019. Sleep increases chromosome dynamics to enable reduction of accumulating DNA damage in single neurons. Nature Communications 10(1), 895.

Zhdanova, I.V. Wang, S.Y. Leclair, O.U. Danilova, N.P. 2001. Melatonin promotes sleep-like state in zebrafish. Brain Research 903(1-2), 263-268.

Zhdanova, I.V. Yu, L. Lopez-Patino, M. Shang, E. Kishi, S. Guelin, E. 2008. Aging of the circadian system in zebrafish and the effects of melatonin on sleep and cognitive performance. Brain Research Bulletin 75(2-4), 433-441.

Zhdanova, I.V. 2005. Melatonin as a hypnotic: pro. Sleep Medicine Reviews 9(1), 51-65.

Zhdanova, I.V. 2011. Sleep and its regulation in zebrafish. Reviews in the Neurosciences 22(1), 27-36.

Zupanc, G.K. Hinsch, K. Gage, F.H. 2005. Proliferation, migration, neuronal differentiation, and long-term survival of new cells in the adult zebrafish brain. The Journal of Comparative Neurology 488(3), 290-319.

第二十六章

自闭症谱系障碍斑马鱼模型

Elena Dreosti[1], Ellen J. Hoffman[2], Jason Rihel[3]

26.1 自闭症谱系障碍：临床特征和研究挑战

自闭症谱系障碍（autism spectrum disorder, ASD）是一组神经发育障碍，其特征包括社交及沟通障碍、兴趣范围狭窄和动作行为刻板。根据美国《精神障碍诊断与统计手册（第五版）》（Diagnostic and Statistical Manual of Mental Disorder，美国精神病学会从1952年起制定，2013年更新到第五版，简称DSM-Ⅴ）标准，自闭症谱系障碍包括两个核心领域的缺陷：①社会沟通包括社交情感互惠、非语言交流以及维持和理解关系方面存在的困难；②兴趣受限和重复行为，包括刻板的动作，例行公事般地僵化专注于日常生活，以及对感官刺激反应过度或过低［美国精神病学会（American Psychiatric Association），2013］。虽然ASD有这些核心特征，但ASD具有高度异质性，与不同程度功能障碍相关，并且经常与其他合并症如智力障碍、注意力缺陷-多动障碍（attention-deficit hyperactivity disorder, ADHD）和焦虑症等同时存在（Lord et al., 2018; Muhle et al., 2018; Simonoff et al., 2008）。近年来，由于这些疾病的患病率不断增加，越来越多研究开始关注ASD。最近研究发现，约1%～2%人群患有ASD，男性发病可能性是女性的四倍（Baio et al., 2018; Elsabbagh et al., 2012; Lord et al., 2018; Lyall et al., 2017; Muhle et al., 2018）。尽管世界上越来越多人认识到ASD的危害和影响，但对ASD患者的治疗手段仍然有限。虽然早期行为干预如应用行为分析法（Applied Behavioral Analysis, ABA）对ASD有显著效果，但目前还没有针对其核心症状的药物疗法。相比之下，药物疗法通常用于治疗ASD患者的并发症或相关症状，例如ADHD注意力不集中和多动，或攻击性和易怒（Lord et al., 2018）。

[1] Wolfson Institute for Biomedical Research, University College London, London, United Kingdom.
[2] Child Study Center, Program on Neurogenetics, Yale School of Medicine, Yale University, New Haven, CT, United States.
[3] Department of Cell and Developmental Biology, University College London, London, United Kingdom.

目前，开发针对 ASD 核心症状的药物疗法的主要挑战之一是我们对导致这些症状的基本生物学机制的了解有限。因此，人们对开发独特的研究方法非常感兴趣，这将促进对 ASD 生物学产生新的认识和了解，并将该领域推向“精准医学”模式，从而可以针对特定个体定制药物治疗策略（Muhle et al., 2018）。

有几项研究旨在应对这些挑战，包括动物模型和人源诱导多能干细胞（induced pluripotent stem cells, iPSC），这些系统为研究 ASD 生物学机制提供了重要的方法和手段，但每个系统都有其缺点。例如，人源 iPSC 可以研究患者自身遗传背景中的细胞机制，并且非常适合大规模药物筛选，但作为体外系统，iPSC 缺乏完整的神经系统，不利于研究神经信号或行为反应。同样，许多用于研究 ASD 生物学的动物模型，从秀丽隐杆线虫、果蝇、斑马鱼到小鼠和猕猴，每一种动物都有不同的优势和局限性。虽然小鼠和人类在进化上普遍保守，但由于小鼠的体积、成本和复杂性，不太可能进行大规模的药理学筛选。较不复杂的系统，比如斑马鱼，具有更大的可操作性，方便进行大规模的药理学筛选（McCammon and Sive, 2015）。特别是，最新 ASD 遗传学进展（下文讨论）表明，越来越多 ASD 动物模型用于高通量研究，这导致鉴定的风险基因列表快速扩大，突显了多基因平行研究的重要性。因此，斑马鱼开始成为研究 ASD 风险基因功能和潜在的生物学机制的重要模型（Ijaz and Hoffman, 2016; Kozol et al., 2016; McCammon and Sive, 2015; Sakai et al., 2018; Shams et al., 2018）。具体而言，斑马鱼与人类进化上是保守的，有高度的易处理和易于遗传操作的特性（McCammon and Sive, 2015），这得益于 CRISPR/Cas9 技术的引入。因此，斑马鱼模型与其他动物和体外模型一起用于阐明 ASD 基本生物学机制，有助于开发更好的、有针对性的药物疗法。

26.2 自闭症谱系障碍遗传学的研究进展

近来，ASD 遗传学领域发生了一场革命，导致鉴定出的与风险密切相关的基因数量迅速增加，这些基因开始趋向于共同的机制途径。同时，这一进展对 ASD 模型中风险基因的功能分析提出了新的挑战，比如一次只能研究一个风险基因的传统方法无法跟上基因发现的速度。为了更好地应对这些挑战，我们概述了 ASD 遗传学的研究进展，即从早期基因发现开始，到最终鉴定出“高置信度”ASD 风险基因。

基于对双胞胎的研究，ASD 一直被认为具有高度遗传性。研究发现，单卵双胞胎的一致性比双卵双胞胎更高，约为 50% ～ 90%（Bailey et al., 1995; Ritvo et al., 1985; Rosenberg et al., 2009; Sandin et al., 2014; Steffenburg et al., 1989）。ASD 的大部分遗传风险可能来自常见变异，在一般人群中这些变异发生率大于 1%，但由于这些变异单独产生的效应量非常小，可能需要联合发挥作用（Anney et al., 2012; Gaugler et al., 2014; Klei et al., 2012），可能会为在模型系统中解释它们的功能带来挑战。然而，迄今为止，ASD 遗传学大部分进展都是在识别罕见变异（发生在 <1% 的人群中）方面取得的，这是因为基因组技术的进步，为可视化观察 DNA 序列和染色体结构的变化提供了越来越高的分辨率（Hoffman and State, 2010）。在这些技术可用之前，ASD 风险基因首先通过与 ASD 密切相关的罕见单基因综合征［如

脆性 X 染色体综合征（Fragile X syndrome）、Rett 综合征（Rett syndrome）和结节性硬化症（tuberous sclerosis）］的关联来识别鉴定。这些疾病中受损基因的鉴定在早期促进了人们对于 ASD 相关生物学机制的认识。例如，脆性 X 综合征是由基因 *FMR1* 的突变引起的，*FMR1* 编码 RNA 结合蛋白（FMRP），该蛋白被发现在代谢型谷氨酸受体介导的突触可塑性中起关键作用，提示该信号通路可作为一个潜在的药理学靶点（Huber et al., 2002）。

ASD 遗传学的下一个重大突破是微阵列技术的出现，它可以使染色体的亚微观结构可视化。在一项开创性研究中，Sebat 及其同事研究表明，包括染色体微缺失和微重复在内的新发拷贝数变异（copy number variants, CNV）在来自单一家族（该家族有一位患者，占 10%）的 ASD 患者中比在多发家族（3%）或未受影响个体（1%）中更常见（Sebat et al., 2007），后续的研究证实了这一发现，而且导致鉴定出与风险相关的特定染色体区域，例如 16p11.2，15q11.2-13.1，7q11.23 和 22q11.2，并在这些区域发现新的风险基因（Glessner et al., 2009; Marshall et al., 2008; Sanders et al., 2011, 2015）。有趣的是，这些区域中的 CNV 具有相当大的多效性，易导致一系列神经发育异常。例如，虽然 16p11.2 的微缺失和微重复都与 ASD 密切相关（Kumar et al., 2008; Marshall et al., 2008; Weiss et al., 2008），该区域微缺失与大头畸形、智力障碍和肥胖有关（D'Angelo et al., 2016; Hanson et al., 2015; Shinawi et al., 2010），该区域交互微重复与小头畸形、智力障碍和精神分裂症有关（D'Angelo et al., 2016; McCarthy et al., 2009; Shinawi et al., 2010）。然而，特定 CNV 易导致不同神经发育障碍的机制尚不清楚。

下一代测序开启了 ASD 遗传学研究的新时代，它为患者编码区或外显子组中每个碱基对测序提供了更高的分辨率。ASD 中的第一个全外显子组测序研究具有开创性意义，因为它们揭示了新生单核苷酸变异（single nucleotide variants, SNV）与新发 CNV 类似，并且在 ASD 中出现的频率更高（Neale et al., 2012; O'Roak et al., 2012; Sanders et al., 2012）。具体来说，与未受影响的兄弟姐妹相比，来自单患家庭的 ASD 患者中新发的、具有破坏性的 SNV 发生率更高（Sanders et al., 2012）。此外，通过识别反复被破坏的基因，即那些无血缘关系的 ASD 患者携带的新生破坏性突变的基因，可为基因发现开辟一条新的道路，导致识别出至少 65 个“高置信度”ASD 风险基因［误报率 <0.1（false discovery rate<0.1）］（Sanders et al., 2015）。有趣的是，这些基因很多被发现在共同的途径上聚合，包括转录调控、染色质修饰和突触形成（De Rubeis et al., 2014; Iossifov et al., 2014; Sanders et al., 2015）。这些研究还为 ASD 风险基因和与精神分裂症或智力障碍相关基因之间的多效性提供了证据，表明神经发育障碍可能涉及类似的途径（De Rubeis et al., 2014; Iossifov et al., 2014）。总之，全外显子组测序研究彻底改变了我们对 ASD 基因“结构”的理解，揭示了来自单显性家族个体新生变异的核心作用，并提供了数百个基因和区域可能导致患病风险的证据。具体而言，超过 200 个 CNV 风险位点和 800 个风险基因被评估为 ASD 中“易发生新生突变”（Sanders et al., 2015）。随着测序技术不断进步和成本下降，该领域现在正朝着全基因组测序方向发展，并已经开始阐明非编码区新发突变的潜在贡献（An et al., 2018）。未来几年，大规模全基因组测序的开展加上更强大的全基因组关联研究（genome wide association studies, GWAS）很可能会识别更多的高置信度风险基因以及 ASD 相关的常见变异，从而进一步阐明 ASD 的遗传基础（Grove et al., 2019; Sanders et al., 2018）。

既然可以准确识别出风险基因，那么下一步最重要的就是了解这些基因在发育中的大脑中是如何汇聚到共同通路的。共表达网络分析在这方面发挥了重要作用（Parikshak et al., 2013; Willsey et al., 2013）。具体而言，该方法分析了人脑转录组数据集中的ASD风险基因的表达模式，得出了ASD风险基因在大脑发育过程中的何时何地发挥作用的时空信息（Parikshak et al., 2013; Willsey et al., 2013）。这些研究将谷氨酸能投射神经元确定为一个汇聚点（Parikshak et al., 2013; Willsey et al., 2013），揭示了这种方法在识别与ASD生物学相关的特定细胞类型和发育时期方面的能力。下一步关键步骤是对发育中的大脑和行为环路中不断增加的ASD风险基因的功能进行建模，并确定潜在的药理学候选基因。鉴于风险基因数据很多，并且发现基因的速度越来越快，目前研究更需要低成本且适合高通量方法的体内模型，斑马鱼模型正是在这方面占有优势。

26.3 斑马鱼中保守的ASD相关神经通路

斑马鱼具有易处理、易于遗传操作以及易于高通量筛选的特性，在ASD风险基因的功能分析中发挥重要作用。此外，由于斑马鱼的胚胎和幼体是透明的，可以在体外将斑马鱼的发育事件和神经信号可视化。尽管斑马鱼和人类之间存在进化距离，但在遗传、结构、药理学和环路水平上存在明显的保守性。因此，正如McCammon和Sive的观察，斑马鱼可以被认为是神经发育障碍相关的基因建模的最佳系统，因为它们表现出高度易处理性和合理保守性之间的“平衡”（McCammon and Sive, 2015）。首先，强有力的证据表明斑马鱼遗传水平的保守性。斑马鱼80%以上基因是人类疾病相关基因直系同源体（McCammon and Sive, 2015）。然而，许多与神经发育障碍相关基因在斑马鱼中重现（Kozol et al., 2016），这可能会使特定人类风险基因的建模过程复杂化。其次，斑马鱼是脊椎动物，与哺乳动物具有相同的大脑分支，包括前脑、中脑、后脑和脊髓（Guo, 2009）。虽然斑马鱼缺乏新皮层，端脑的形成过程也与哺乳动物（外翻与内翻）不同，但斑马鱼的许多大脑区域显示出与哺乳动物的结构同源性，包括下丘脑、视顶盖和小脑（详见Kozol et al., 2016）。此外，斑马鱼具有与哺乳动物相同的神经递质系统，包括GABA、谷氨酸、多巴胺、去甲肾上腺素、5-羟色胺、组胺和乙酰胆碱，并显示出神经发育标志物保守的表达模式（Guo, 2009）。再次，许多研究发现了斑马鱼药理途径的保守性。例如，Rihel及其同事对500多种精神活性化合物进行了筛选，以确定它们对睡眠行为的影响，并发现针对特定神经递质系统的化合物可以对斑马鱼和哺乳动物的睡眠产生类似的影响（Rihel et al., 2010）。这项研究和其他基于行为的大规模筛选研究均发现，精神活性化合物完全可以根据它们在斑马鱼胚胎和幼体中行为效应的分子作用机制正确分类（Kokel et al., 2010; Rihel et al., 2010）。最后，越来越多的证据表明，斑马鱼和哺乳动物的基本行为（例如听觉惊吓、脉冲前抑制、睡眠和觉醒）的神经环路是保守的（Barlow and Rihel, 2017; Burgess and Granato, 2007a, 2007b; Leung et al., 2019; Lovett-Barronet et al., 2017; Prober et al., 2006; Schoonheim et al., 2010）。例如，Lovett-Barron等人报道，采用双光子成像技术记录了在警觉状态下的斑马鱼仔鱼全脑，以识别参与这种行为的神经调节细胞类型

（Lovett-Barron et al., 2017）。他们的研究表明，对相关细胞类型的直接光遗传学操作在小鼠中引发了类似的行为效应，为斑马鱼和哺乳动物唤醒行为背后的神经环路的保守性提供了显著证据（Lovett-Barron et al., 2017）。同时，需要强调的是，因为表面效度的限制，斑马鱼或任何动物系统都不可能重现人类ASD复杂的临床特征，研究斑马鱼的目的是要利用该系统的独特功能性进行ASD风险基因在神经系统发育基本过程和基本行为涉及的神经环路中的功能研究。总之，有多种证据分别在分子、细胞和环路水平上支持斑马鱼在研究ASD风险基因生物机制方面的平移相关性。

26.4 ASD相关的斑马鱼行为

ASD的一个核心症状是正常社交互动障碍。社交行为是人类活动的重要组成部分，但我们对人类的社交大脑环路知之甚少，更不了解这些环路如何发生故障。最复杂的社会行为是人类独有的，例如语言，但它们的基本构成要素在所有社会物种中都是保守的。斑马鱼是高度社会性的，表现出大量的简单的社交行为，这些行为可能与ASD研究相关。

以前人们认为社会行为只存在于成年斑马鱼中。然而，早在斑马鱼发育第一周，已经发现其存在一些基本的社交行为（Romero-Ferrero et al., 2019），并且在仔鱼时期（大约3周）社交行为已经非常突出（Dreosti et al., 2015)。发育早期斑马鱼社交行为的发现，使人们有可能利用透明仔鱼和幼鱼全套光学技术及鱼体限制性制剂，来研究正常和受损社交行为的神经基础。是否也可以在幼年斑马鱼中研究更复杂的社交行为，如社交学习、攻击性和群体决策，仍有待确定。然而，新的技术例如基于机器学习的跟踪系统和虚拟鱼显示器的使用，为研究幼年斑马鱼的复杂社交行为提供了广阔的前景。

斑马鱼中的大量社交行为可能与ASD相关。鉴于社交互动缺陷是ASD的核心特征，下面将更详细地探讨这些行为的基本原理、方法和特征，重点是这些行为研究中的最新技术进展。欲了解更多详细信息，我们推荐最近几篇涵盖斑马鱼社交行为众多方面的综述供读者阅读（Geng and Peterson, 2019; Maruska et al., 2019; Shams et al., 2018; Orger and de Polavieja, 2017）。

26.4.1 社交偏好

社交偏好定义为个体观察、模仿和接近同类的驱动力，社交偏好是所有其他社交行为的基础。社交偏好在社交型脊椎动物中高度保守（O'Connell and Hofmann, 2012）。人类的社交偏好是天生的，可以在新生儿中观察到，他们表现出对人脸的直接偏好。斑马鱼是在子宫外发育的，它们的社交偏好在出生1周左右首次出现，并在3周时完全建立（Dreosti et al., 2015），因此提供了一个很好的机会来研究社交环路最初是如何发展的。斑马鱼的社交偏好已经在成鱼（最近也在仔鱼）中进行了测试。

成鱼的社交偏好已经通过使用各种不同的活动场和社交刺激进行了研究。在有两个或三个室的水箱（Engeszer et al., 2004; Liu et al., 2018）中，位于一个室中的测试鱼与位于另一个室中的社交刺激物分开。两室分析法通过不同的方式对测试鱼社交偏好进行评分，例如通过测量与社交线索的平均距离或在社交刺激附近花费的时间［图 26-1（A）左］。为了获得可靠的测量结果，测试鱼通过透明屏障与同种鱼隔离，但这显然限制了在自然环境中可能发生的相互作用。三室分析法的优点在于通过将非社交刺激物（例如食物）放置在第三室中，可以比较不同的社交线索［图 26-1（A）中］或其他视觉偏好。Dreosti 等人最近开发了一种略微改进的三室分析法，即三室 U 形室［图 26-1（A）右］。在这个房间里，具有弱社交偏好或反社交偏好的鱼可以完全隐藏在远离一侧社交线索的地方，从而可以更完整地量化它们的行为表型。

在这些测定中使用了几种不同的社交刺激，例如静态图像、动画图像（Gerlai, 2017）、真鱼视频、机器控制的斑马鱼（Butail et al., 2013; Kopman et al., 2013; Polverino and Porfiri, 2013; Ruberto et al., 2016, 2017）、3D 打印的鱼模型（Bartolini et al., 2016），或相邻水箱中的活鱼（Zimmermann et al., 2015）。可以将一或两个刺激物放置在与测试鱼相邻的隔间中，以比较社交偏好。例如，野生型（wild-type, WT）鱼通常更喜欢同种野生型鱼（Braida et al., 2012）［图 26-1（A）中］。虽然真正的斑马鱼似乎最适合引发社交反应，但虚拟图像似乎也能发挥同样作用。Larsch 和 Baier（2018）研究表明，测试鱼面对活体斑马鱼和视频记录图像，对同种呈现的偏好是相似的。这也表明视觉刺激足以诱导社交偏好，而嗅觉、听觉和侧线刺激并非必不可少，这为功能成像实验的设计提供了巨大的实用性（Larsch and Baier, 2018）。

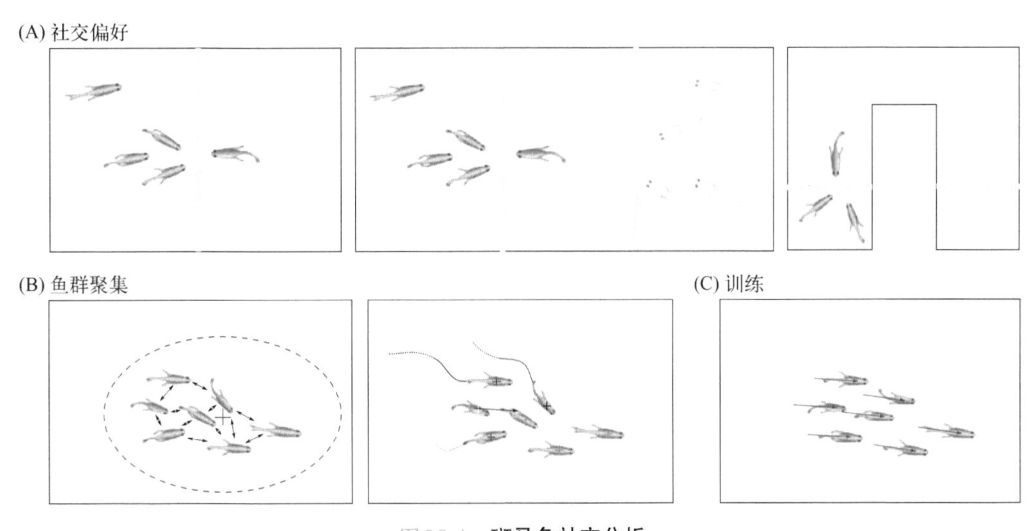

图26-1　**斑马鱼社交分析**

（A）社交偏好分析。左图：在一个两室社交偏好分析中，一个室中的测试鱼与另一个室中的社交线索之间的距离是确定的。中图：在一个三室试验中，测量了对同种鱼类与其他线索（例如不同颜色的鱼）的社交偏好。右图：在改进的 U 形三室试验中，可以发现不喜欢社交或反社交的鱼，因为测试鱼可以隐藏在测试室的空臂中。（B）鱼群聚集实验。左图：可以通过测量一段时间间隔内鱼与鱼之间平均距离来评估鱼群的凝聚力。右图：机器学习算法能够跟踪鱼群中的单条鱼，提供了关于鱼群中每条鱼参与情况的更详细测量数据。（C）训练实验。与鱼群聚集不同，鱼群是一群协调自身运动方向的鱼。鱼群的定向程度可以通过机器学习算法来确定，该算法可以跟踪鱼的相对位置以及它们随时间的变化方向

一项关于幼年斑马鱼社交偏好发生的研究发现，斑马鱼在 2 ～ 3 周龄左右开始表现出强烈的社交偏好（Dreosti et al., 2015）［图 26-1（A）右］。研究者发现，幼年斑马鱼对年

龄匹配的同种鱼表现出强烈的偏好。此外，一项使用两室的研究表明，受精后 5 ～ 6 天（dpf）的仔鱼已经表现出接近同种鱼群体的趋势（Patowary et al., 2019），3 ～ 4 周龄的斑马鱼也是如此。该测试在包含圆形和透明塑料孔的 12 孔塑料板中进行（Ruzzo et al., 2019）。这些研究结果令人鼓舞，因为仔鱼和幼年斑马鱼是透明的，能够使用为斑马鱼开发的各种光学方法，并且社交偏好发生的基础——神经环路的早期发育，正好在这些光学方法可操作的时间窗内发生。因此，可以在完整的、功能正常的斑马鱼大脑中，从单细胞分辨率水平极其详细地观察到基因和环境影响神经环路，尤其是与人类自闭症相关的神经环路的正常发育。

26.4.2 群聚

趋同驱动力与社交偏好导致社交群体的形成。在斑马鱼中，这种形成群体的倾向被称为群聚。群聚对单条鱼的好处是提高了捕食者探测 / 防御能力，并增加了定位 / 获取食物供应和配偶的机会（Eguiraun et al., 2018）。由于群聚可以通过简单地将一群鱼放入水箱中并观察它们的行为来测定，因此这已成为一种广泛使用的分析方法［图 26-1（B）］。

最近研究发现在受精后 6 ～ 7 天左右，仔鱼转向附近同种鱼的趋势增加，这说明此时斑马鱼之间已经出现吸引力（Hinz and de Polavieja, 2017），而其他群体在受精后 12 天才表现出同种偏好（Engeszer et al., 2007）。其他研究发现，虽然受精后 7 天仔鱼之间的平均距离（13 倍体长）仍然远高于在成鱼中常规观察到的平均距离（4 ～ 6 倍体长），但这种个体间距离已经比预期距离短（Buske and Gerlai, 2011b; Miller and Gerlai, 2008, 2012）。然而，群聚研究受益于新分析方法的灵敏性，这些分析法使用开源计算机视觉程序来检测斑马鱼群聚特征，能可靠地同时跟踪许多鱼类个体行为（Franco-Restrepo et al., 2019）。这些工具有望以前所未有的细节水平，在发育的最早期阶段，对斑马鱼行为中的细微变化进行重新审视。

群体跟踪软件不需要对个体进行识别就能测量鱼类行为参数［图 26-1（B）左］。软件提取整个群体的数据，例如个体间距离和最邻近距离，以量化鱼群的紧密程度（Miller and Gerlai, 2007），并通过跟踪鱼群中心点来量化鱼群的轨迹（Eguiraun et al., 2018）。为了对成鱼进行 3D 记录，研究者们使用了镜子（Audira et al., 2018; Maaswinkel et al., 2013）、多个摄像头（Bishop et al., 2016; Butail and Paley, 2012; Macri et al., 2017; Qian et al., 2016; Wang et al., 2017）以及具有深度感应能力的单摄像机（Kuroda, 2018）。

除了群体测量，为了获得个体之间更复杂的互动和相互依赖的准确信息（Katz et al., 2011），必须及时跟踪每条鱼［图 26-1（B）右］。机器学习方法最近使之成为可能，比如，Polavieja 小组开发的软件，该软件通过给每条鱼生成独特数字指纹来识别和跟踪个体（Perez-Escudero et al., 2014）。他们最近更新的方法 idTracker.ai，使用深度学习同时跟踪 100 个个体，具有超过 99.9% 的识别准确率（Romero-Ferrero et al., 2019）。此外，人们最近还开发出了其他相关软件，用于多个体识别和跟踪（Bai et al., 2018; Qian et al., 2016; Wang et al., 2016）。

社交凝聚力降低通常被解释为群体成员之间社会互动减少。因此，群聚分析代表了一种模拟社交互动变化的方法，这可能与 ASD 有关，而新技术的发展极大地提高了收集此类数据及其细节的便利性。

26.4.3　集群

ASD 患者表现出在模仿、无意识模仿他人行为方面存在缺陷（Neufeld et al., 2019）。已经证明，斑马鱼群可以产生随时间发展的极化队形和同步运动（Miller and Gerlai, 2012）（Dreosti et al., 2015）[图 26-1（C）]。使用不同的处理方法可以使鱼群由集群状态（鱼个体之间呈现定向紧密分布）变成较为松散的群聚状态（鱼个体之间呈现分散分布）。例如，急性乙醇暴露似乎会影响群聚，但不会影响群体极化（Shelton et al., 2015）。相反，尼古丁会影响极化，但不会影响群聚凝聚力（Miller et al., 2013）。最近开发了一种无监督学习方法来可靠地识别鱼类的同步模式（Tang et al., 2018）。这些结果都表明，斑马鱼社会协调分析可以更细致，从而发现假定 ASD 模型的社交行为中可能存在的细微缺陷。

26.4.4　攻击性

根据研究结果，15% ～ 56% 的 ASD 患者存在攻击行为，这可能是由于其无法处理和回应社交线索（Fitzpatrick et al., 2016）。成年斑马鱼也存在攻击性，并对其进行了深入研究（Teles and Oliveira, 2016; Way et al., 2015），诱导斑马鱼攻击行为的刺激物通常包括镜子中自己的影像（Zabegalov et al., 2019b）、仿制鱼或另一条鱼的视频记录（Way et al., 2015）。然而，迄今为止，只有平面镜测试主要用于成年斑马鱼的 ASD 建模（Zimmermann et al., 2015）。最近，开发了一种以亚秒级精度标注典型斗争行为的软件（Laan et al., 2018）。目前尚不清楚攻击行为在发育过程中何时出现，但按照理想的做法，使用无监督机器学习来识别未成年斑马鱼的社会行为和攻击性行为，这对明确其个体发育将是有用的。

26.4.5　社交可塑性

社交可塑性是根据他人的行为调整自身社交行为的能力，ASD 患者的这种能力也会受损（Boggio et al., 2015）。斑马鱼可以根据它们在群体中的地位调整它们的行为，因此表现出社交可塑性（Clements et al., 2018）。通常使用类似的攻击行为分析来研究这种行为（Guayasamin et al., 2017; Maruska et al., 2019; Sykes et al., 2018; Teles et al., 2016）。如果有新的方法来区分社交可塑性和纯粹的攻击性则会更好。

26.4.6　社交性推断

ASD 儿童通过观察社交相关信息和推测他人心理状态从而推断他人行为的能力有限（Zhou et al., 2019）。以类似的方式，斑马鱼可以观察其他个体之间的互动，并利用这些信息来调整自己未来的行为。Abril-de-Abreu 等最近开发了一种社交推断测试软件，两室由透明窗口分隔开，一条测试鱼被放置其中一室，并在不与其他鱼或视频鱼互动的情况下观看对面室的打斗（Abril-de-Abreu et al., 2015）。通过观察鱼在窗户附近的停留时间来衡量它的注意力，它的社会推断是通过观察实验动物随后的支配或顺从行为来衡量。该实验是否与在 ASD 中观察到的行为相关仍有待确定。

26.4.7 集体决策

机器学习方法现在可以识别单条鱼，并以高精度和分辨率描述这条鱼如何在群体中做出行为决策。一些实验室已经改进了对个体的连续跟踪，然后使用最优控制理论（Laan et al., 2017）、深度注意网络（Heras et al., 2019）、转移熵（Butail et al., 2016）和其他数据驱动方法（Zienkiewicz et al., 2018）对斑马鱼成对互动进行计算建模，以揭示成对的个体如何吸引、排斥和彼此结盟。其他研究应用计算和机器学习方法来模拟鱼的个体刻板运动（Marques et al., 2018; Mwaffo et al., 2017; Zienkiewicz et al., 2015）。研究发现斑马鱼个体的运动在“被动”行为模式（个体的行为不受其他群体成员的影响）和“主动”模式（个体的行为根据群体的社交输入进行调整）之间交替。该框架高精度地预测了群体中个体的行为（Harpaz et al., 2017）。

26.4.8 社交认可

ASD 儿童可能以不同的方式感知社交刺激。例如，ASD 患者的眼睛注视他人面部的不同特征，眼神接触较少，而更多地关注嘴巴（Klin and Jones, 2008）。斑马鱼 ASD 模型也可能以不同的方式处理社交刺激，并被视觉社交刺激的不同特征所吸引。修改社交输入的最佳方法是使用易于修改的虚拟刺激。有研究表明，鱼类既能跟随真实的同种鱼，也能跟随虚拟的图像（Larsch and Baier, 2018）。此外，人们开始使用闭环实验进行更详细的探索。在闭环实验中，测试鱼的行为会影响虚拟刺激，以确定斑马鱼能够识别同种鱼的最重要的视觉特征。例如，一个投射的虚拟物体以模仿斑马鱼游动的方式移动，足以触发幼鱼数小时的强烈群聚行为（Larsch and Baier, 2018）。该检测方法可与突变体鱼结合使用，以测试突变体鱼是否对鱼类特征有不同的偏好，例如色素沉着或特定模式。

26.4.9 限制性或重复性的行为

ASD 的另一个核心行为特征是兴趣受限和重复的刻板行为。对斑马鱼重复行为的研究比较有限，但当将动物圈养在小池中，一些药物处理和突变确实会导致趋触性和转圈行为增加（Zabegalov et al., 2019a）。最近，研究者将缺乏多巴胺转运体（dopamine transporter, dat）的成年突变体鱼的运动描述为在水箱底部呈现重复搜索运动（Wang et al., 2019）。然而，由于这些行为通常伴随着游泳速率或位置偏好的变化，因此很难将这些行为明确识别为真正的重复行为，而与其他行为特征无关。在最近的一项研究中，仔鱼连续多个昼夜的行为被划分为一连串的动作，称为行为基序（behavioral motifs）（Ghosh and Rihel, 2019）。通过分析行为基序的相对时间结构，可以计算出重复行为的总体指标，而无须假设哪些行为会在重复性任务中多次出现。这种重复性指标是否会在 ASD 模型中发生变化仍有待确定。

26.5 斑马鱼 ASD 风险基因模型——建立和表型分析

26.5.1 基因靶向方法

直到最近，利用斑马鱼进行风险基因建模的主要挑战之一是在目的基因中生成靶向突变的有效且经济的可用方法有限。多年来，斑马鱼主要用于正向遗传筛选，以确定参与发育基本过程的基因功能（Granato and Nusslein-Volhard, 1996），但靶向核酸酶［例如锌指核酸酶、转录激活因子样效应物核酸酶（transcription activator-like effector nucleases, TALENs)］的引入开创了斑马鱼反向遗传学的新时代（Dahlem et al., 2012; Doyon et al., 2008; Meng et al., 2008; Sander et al., 2011）。CRISPR/Cas9技术使该领域发生了革命性变化，使研究人员能够以高效率、高灵活性和合理成本靶向斑马鱼中几乎所有目的基因（Hwang et al., 2013; Jinek et al., 2012; Moreno-Mateos et al., 2015）。在这些方法出现之前，许多斑马鱼研究使用吗啉反义寡核苷酸（morpholinos），这是一种经过修饰的反义寡核苷酸，通过阻断 mRNA 剪接或翻译来降低靶基因表达，以研究目的基因的功能（Draper et al., 2001; Nasevicius and Ekker, 2000）。迄今为止，许多关于斑马鱼 ASD 风险基因的研究都使用这种方法来“敲减”基因表达。虽然吗啉反义寡核苷酸有助于研究早期发育过程中基因的作用，但有几个显著的缺点限制了它们在神经发育障碍基因研究中的应用。一个缺点是吗啉反义寡核苷酸诱发瞬时效应，这限制了它们在斑马鱼受精后 72h 研究基因对仔鱼行为的影响，而受精后 72h 是许多可量化的行为表型首次出现的时间。另一个重要的缺点是，吗啉反义寡核苷酸已被证明会诱导非靶向效应，例如激活肿瘤抑制因子 p53，并且可能与非特异性表型相关，包括非特异性神经表型，例如头部大小或大脑结构的变化（Eisen and Smith, 2008; Robu et al., 2007）。此外，越来越多的研究报告表明，吗啉反义寡核苷酸诱导的表型和同一基因的突变体观察到的表型之间缺乏一致性，这种情况很多是由脱靶效应造成的（Kok et al., 2015; Lawson, 2016）。虽然遗传补偿也可能解释吗啉反义寡核苷酸诱导表型和突变体表型之间的差异（El-Brolosy and Stainier, 2017; Rossi et al., 2015），但明确基因突变体中吗啉反义寡核苷酸诱导的表型对于区分脱靶效应和遗传补偿至关重要（Lawson, 2016）。

Eisen 和 Smith 早期制定了斑马鱼中吗啉反义寡核苷酸使用的指南，包括以下建议：①通过共注射缺乏吗啉反义寡核苷酸靶位点的 mRNA 进行表型拯救；②使用两个独立的吗啉反义寡核苷酸靶向同一基因的不同位点；③与基因突变的直接比较（Eisen and Smith, 2008）。鉴于吗啉反义寡核苷酸可能诱发脱靶效应，使用吗啉反义寡核苷酸对斑马鱼 ASD 风险基因的早期研究应按照该指南进行评估，并谨慎解释其结果（Lawson, 2016; Shams et al., 2018）。例如，研究发现吗啉反义寡核苷酸诱导的斑马鱼 *FMR1*（脆性 X 综合征中被破坏的基因）直系同源基因敲减会导致许多发育表型改变，包括神经突分支的改变和颅面异常（Tucker et al., 2006），这些表型在缺乏 Fmr 蛋白表达的基因突变中没有出现（den Broeder et al., 2009）。另一项研究发现，16p11.2 区间的 *kctd13* 基因敲减会导致大头畸形和大脑中细胞增殖增加（Golzio et al., 2012），而该基因的斑马鱼（和小鼠）突变体并未显示出头部大小或细胞增殖的变化（Escamilla et al., 2017）。这些研究突显了在与基因突变进行比较之前，谨慎解释吗啉反义寡核苷酸诱导的表型（特别是神经表型）的重要性。

最近，Stainier 及其同事（包括许多斑马鱼研究领域的领军人物），制定了新的指南，建议在基因突变中确定吗啉反义寡核苷酸诱导的表型（Stainier et al., 2017）。这些指南还建议通过将特定的吗啉反义寡核苷酸注射到同一基因的突变体中来验证其有效性，如果突变体基础上存在吗啉反义寡核苷酸诱导的表型表明存在脱靶效应，在突变体中抑制吗啉反义寡核苷酸诱导的表型可能是由于补偿机制（Stainier et al., 2017）。此外，随着有关突变体补偿机制的新证据出现，最近一项研究揭示了转录适应的作用（El-Brolosy et al., 2019），再加上 CRISPR/Cas9 技术的进步，在研究与人类疾病相关的基因功能丧失时，产生补偿较少的突变体将变得越来越可行。总之，可以预料，基因突变将成为斑马鱼风险基因功能分析的“金标准”，尤其是与神经发育障碍相关的基因（Shams et al., 2018）。

26.5.2 发育表型

越来越多的研究使用斑马鱼作为研究 ASD 相关基因功能的模型（详细综述见 Kozol et al., 2016; Sakai et al., 2018; Shams et al., 2018）。虽然这些研究中有许多使用了吗啉反义寡核苷酸，但在未来的几年里，由于 CRISPR 的广泛使用，我们可能会看到更多关于基因突变的研究。迄今为止，对斑马鱼 ASD 风险基因的研究突出了该系统在研究神经发育过程和简单行为以及进行药理学筛选方面的优势。一些研究调查了 ASD 风险基因破坏对大脑大小和结构以及运动功能的影响程度，这与 ASD 高度相关，因为 ASD 与特定发育时期的头部大小和大脑发育异常有关（Courchesne et al., 2011）。例如，一项研究使用斑马鱼来筛选 16p11.2 区间内 >20 个基因对发育的影响（Blaker-Lee et al., 2012），因为该区域的 CNV 与 ASD 密切相关，但该区域的单个基因在临床表现的作用尚不清楚。该研究发现，这些基因的大多数被敲减后会导致大脑异常，例如中脑 - 后脑边界和脑室大小的改变，以及某些基因敲减后运动和触觉反应降低（Blaker-Lee et al., 2012）。此外，使用该区域两个基因的单突变体和双突变体进行后续研究发现，头部大小稳定性、多动症和癫痫敏感性表型与基因拷贝数有关，这表明该区域的基因相互作用可能会调节形态和行为表型（McCammon et al., 2017）。在另一项研究中，*shank3a* 和 *syngap1b*（分别是高置信度 ASD 风险基因 *SHANK3* 和 *SYNGAP1* 的斑马鱼直系同源基因）的表达降低，导致胚胎出现小头畸形、发育迟缓、中脑 - 后脑边界和脑室大小改变，以及斑马鱼仔鱼异常逃逸反应和癫痫样行为（Kozol et al., 2015）。此外，*CHD8* 是一种与 ASD 密切相关的高置信度风险基因，其直系同源基因 *chd8* 的破坏导致斑马鱼仔鱼头部大小增加（包括吗啉突变体和 F0 CRISPR 嵌合体），这与该基因突变的 ASD 患者中发现的大头畸形率增加一致（Bernier et al., 2014）。虽然本研究中使用了 F0 CRISPR 嵌合体和多个吗啉反义寡核苷酸，但该基因的大尺寸使得 mRNA 拯救实验无法进行（Bernier et al., 2014）。虽然这些研究揭示了斑马鱼在研究基因功能丧失对神经系统发育影响中的潜力，但考虑到吗啉类药物的潜在缺陷（如上所述），确认这些研究中描述的种系突变的表型是下一步工作的关键。

26.5.3 活动亢进 / 癫痫

鉴于发育中大脑兴奋性和抑制性信号的不平衡被认为是 ASD 和癫痫的潜在机制，一些研究已经使用斑马鱼来研究 ASD 风险基因受损如何影响兴奋性和抑制性神经元的发育（Rubenstein and Merzenich, 2003）。例如，研究发现 *shank3a* 基因敲减导致受精后 48h 中脑和后脑抑制性 GABA 神经元减少，后脑兴奋性谷氨酸能神经元减少，而 *syngap1* 基因敲减也会导致中脑 GABA 能神经元和后脑谷氨酸能神经元减少（Kozol et al., 2015）。在另一项研究中，研究发现 *CNTNAP2* 基因（与 ASD 和癫痫综合征相关）的两个斑马鱼直系同源基因的双纯合突变体（Strauss et al., 2006），受精后 4 天在前脑 GABA 能神经元和前体细胞中表现出明显的缺陷，但在谷氨酸能神经元中未发现区域性缺失（Hoffman et al., 2016）。有趣的是，小鼠 *CNTNAP2* 敲除时也发现 GABA 能缺陷（Penagarikano et al., 2011）。

研究还发现，斑马鱼 *cntnap2* 突变体对药物诱发癫痫的易感性增加（Hoffman et al., 2016），这与基因敲除小鼠癫痫发作的表现一致（Penagarikano et al., 2011），为两个系统之间的保守性和斑马鱼翻译相关性提供了进一步的证据。

26.5.4 药理学筛选

斑马鱼系统的一个显著优势是能够进行大规模的、基于行为的药理学筛选（Rihel and Ghosh, 2015）。例如，一项研究旨在使用斑马鱼 *scn1lab* 突变体作为 Dravet 综合征（一种严重的难治性癫痫，主要由 *SCN1A* 基因突变引起）模型来识别具有抗癫痫活性的化合物（Baraban et al., 2013）。*scn1lab* 基因缺失纯合子突变斑马鱼在受精后 4 天开始表现出自发性癫痫发作（Baraban et al., 2013）。该研究通过筛选 320 种化合物抑制癫痫样行为的能力，发现一种化合物——克立咪唑（FDA 批准的抗组胺药物，也具有 5- 羟色胺活性）（Griffin et al., 2017），可以抑制癫痫样行为和脑电图痫性放电（Baraban et al., 2013）。其他后续研究利用这种斑马鱼模型筛选与 Dravet 综合征相关的抗癫痫化合物（Dinday and Baraban, 2015; Griffin et al., 2017; Sourbron et al., 2016）。在另一项研究中，Hoffman 及其同事进行了靶向药理学筛选，鉴定了斑马鱼 *CNTNAP2* 突变体行为表型的抑制因子（Hoffman et al., 2016）。具体而言，对斑马鱼受精后 4 ～ 7 天的休息 - 觉醒循环中的自发活动进行跟踪研究，发现 *cntnap2* 突变体表现出夜间活动亢进（Rihel et al., 2010）。通过比较 *cntnap2* 突变体在一系列休息 - 觉醒参数中的行为表型与 >500 种精神活性化合物诱发野生型鱼类的行为特征谱（Rihel et al., 2010），该研究预测并测试了潜在抑制因子逆转夜间活动亢进的能力（Hoffman et al., 2016）。有趣的是，研究发现雌激素化合物能够选择性抑制 *cntnap2* 突变行为表型，确定了一种过去未与该基因相关联的新的药理学通路，可供进一步研究（Hoffman et al., 2016）。总之，这些研究突显了斑马鱼高通量、基于行为的小分子化合物筛选的在发现与 ASD 和癫痫相关的新通路和药理学候选物方面的潜力。

26.5.5 胃肠功能障碍

因为 ASD 患者经常出现胃肠（gastrointestinal, GI）症状，包括腹痛、便秘和腹泻，其病因尚不清楚，斑马鱼ASD风险基因的模型也已用于研究GI功能障碍的潜在机制。斑马鱼仔鱼的透明度高使其成为实现这一目的的最佳系统，因为可以给仔鱼喂养荧光珠，然后通过胃肠道跟踪荧光珠的运动。使用这种方法，发现*SHANK3*的两个斑马鱼直系同源基因的双纯合和杂合突变体表现出胃肠动力降低，这可能与该基因突变人群经常出现的胃肠道症状有关（James et al., 2019）。有趣的是，该研究还发现突变体肠道中5-羟色胺阳性肠内分泌细胞减少，这可能导致胃肠道动力降低（James et al., 2019）。此外，在大多数CHARGE综合征（与ASD相关，以结肠瘤、心脏缺陷、后肛门闭锁、生长迟缓、生殖器和耳朵异常为特征）病例中，*CHD7*基因受损，研究发现该基因的斑马鱼突变体（以及吗啡啉突变体）表现出GI排空减少和前肠迷走神经控制减少，以及其他发育异常，如心包水肿和心脏肥大（Cloney et al., 2018）。斑马鱼*chd8*吗啉反义寡核苷酸突变体也发现了胃肠道蠕动降低和肠神经元数量减少，这与*CHD8*基因突变人群中高发的胃肠道症状一致（Bernier et al., 2014）。重要的是，这些研究显示了斑马鱼在阐明ASD患者经常同时出现的相关症状的潜在机制方面的作用。

26.5.6 社交互动

随着基因突变越来越多地用于ASD风险基因的建模，研究开始探讨这些基因在成鱼更复杂行为中的作用。例如，*dyrk1aa*（高置信风险基因*DYRK1A*的一个直系同源基因）突变斑马鱼在社交偏好和群聚试验方面表现出缺陷，并且在放入新水箱时焦虑行为减少（Kim et al., 2017b）。*shank3b*是*SHANK3*的第二个直系同源体，*shank3b*突变斑马鱼成年后也表现出行为异常，包括自发活动减少、趋触性降低、社交缺陷和重复行为（Liu et al., 2018）。有趣的是，这两项研究都发现突变体的形态表型取决于所检测的发育阶段。例如，*dyrk1a*突变体在胚胎和仔鱼阶段表现出正常发育，但在受精后3周表现出大脑细胞凋亡增加和成年后小头畸形（Kim et al., 2017b）。*shank3b*突变体在胚胎和仔鱼阶段表现出短暂的发育延迟，在发育过程中逐渐减少，而成年突变体表现出大脑体积有所增加（Liu et al., 2018）。因此，这些研究强调了在发育过程中观察突变体表型的必要性，以及研究基因受损与成年斑马鱼复杂行为表型发育之间联系机制的潜力。

除了对高置信度的ASD风险基因进行建模外，最近一项研究还利用斑马鱼研究一个新发现的ASD风险基因*NR3C2*，该基因是通过对多重（多个患者）家族的全基因组进行从头测序和遗传变异分析而发现的（Ruzzo et al., 2019）。有趣的是，我们发现斑马鱼*nr3c2*纯合突变体在仔鱼时期表现出活动亢进和夜间睡眠减少，成年后社交偏好降低，这表明*nr3c2*功能丧失会影响ASD相关行为的发展。除了社交缺陷外，个体还经常出现睡眠障碍（Ruzzo et al., 2019）。如表26-1所示，斑马鱼模型中与神经发育障碍相关的其他风险基因已在各种社交分析中进行了测试。

除了基因操纵外，我们还知道早期发育过程中经历的环境对人类和斑马鱼的社交行为有强烈影响。例如，胎儿乙醇谱系障碍较轻的患者可以在解剖学没有变化的情况下表现出社交缺陷（Seguin and Gerlai, 2018）。同样，短暂（2h）暴露于低水平（最高1%）乙醇的斑马鱼胚胎发育成成鱼时，也没有明显的解剖学变化，但对虚拟（Fernandes and Gerlai, 2009）或活体（Buske and Gerlai, 2011a）社交刺激的社交偏好表现出剂量依赖性

表 26-1 遗传和药理学操作对斑马鱼社交互动的影响

遗传模型				
斑马鱼基因	人类基因	操作	效果	参考文献
shank3b	*SHANK3*	突变	群聚和社交偏好缺陷	Durand 等 (2007)
shank3b	*SHANK3*	突变	亲缘认知缺陷	Liu 等 (2018)
sam2	*FAM19A2*	突变	群聚缺陷	Choi 等 (2018)
sam2	*FAM19A2*	突变	社交偏好缺陷	Ariyasiri 等 (2019)
cep41	*CEP41*	吗啉反义寡核苷酸突变体	社交偏好缺陷	Patowary 等 (2019)
immp2l	*IMMMP2L*	突变	群聚加强	Tang 等 (2018)
scn1lab	*SCN1A*	突变	社交互动抑制	Tang 等 (2018)
dyrk1aa	*DYRK1A*	突变	群聚和社交偏好缺陷	Kim 等 (2017b)
fmr1	*FMR1*	突变	社交行为早熟发展	Wu 等 (2017)
disc1	*DISC1*	突变	群聚受损	Eachus 等 (2017)
adra1aa/ adra1ab	*ADRA1A*	突变	不动时间增加	Tang 等 (2018)
nr3c2	*NR3C2*	突变	社交偏好减少	Ruzzo 等 (2019)

药物诱导模型				
药物	暴露时间	剂量	效应	参考文献
氯胺酮	2hpf 的鱼暴露 24h	50 ～ 90mg	无效果	Felix 等 (2017)
丙戊酸	0 ～ 5d	0.5 ～ 50μmol/L	社交偏好缺陷	Bailey 等 (2016), Baronio 等 (2018), Dwivedi 等 (2019), Zimmermann 等 (2015)
视黄酸	0 ～ 5d	0.02 ～ 2nmol/L	社交偏好减少	Bailey 等 (2016)
硒代甲硫氨酸	慢性	2.1 ～ 31.5μg/g	群聚抑制	Attaran 等 (2019)
催产素	急性	0.001 ～ 40ng/kg	群聚相互距离改变	Braida 等 (2012)
SCH23390（多巴胺 D1 受体拮抗剂）	急性	0.1mg/L 和 1.0mg/L	社交偏好减少	Scerbina 等 (2012)
氟西汀	急性	50μg/L	群聚抑制	Giacomini 等 (2016)
苯二氮卓类	急性	0.5mg/L, 1.5mg/L, 5.0mg/L	群聚抑制	Schaefer 等 (2015)
MK-801	急性	100mmol/L	社交偏好减少	Dreosti 等 (2015)
	急性	10μmol/L	群聚抑制	Maaswinkel 等 (2013)
乙醇	急性	1%	好奇性减少	Ariyasiri 等 (2019)
	1h	0.12 ～ 0.5	社交偏好减少	Dreosti 等 (2015)
	急性	0.1% ～ 0.5%	社交性的个体差异	Araujo-Silva 等 (2018)
	1h	0.25%	社交抑制	Fontana 等 (2018)
	慢性 8d	0.50%	社交凝聚力增加	Dewari 等 (2016)
	24hpf 的鱼暴露 2h	低剂量（最大 1%）	真实和虚拟刺激的社交偏好减少	Buske 和 Gerlai (2011a), Fernandes 和 Gerlai (2009)
LSD	急性	100μg/L	群聚抑制	Green 等 (2012), Grossman 等 (2010)

降低。各种鱼类对乙醇的反应也不同，这取决于它们的胆量。虽然“害羞”鱼类通常比“胆大”鱼类在鱼群附近花费更多时间，但乙醇会增加“胆大”鱼类的群聚，而抑制“害羞”鱼类的聚集（Araujo-Silva et al., 2018）。虽然急性乙醇暴露会轻度抑制群聚，但令人惊讶的是，长期（8 天）暴露于乙醇会增加群聚凝聚力（Muller et al., 2017）。丙戊酸是另一种在妊娠期间服用会导致人类行为改变的药物（胎儿丙戊酸盐综合征），并且与胎儿时期 ASD 风险增加有关（Christensen et al., 2013）。在斑马鱼中，胚胎暴露于丙戊酸或丙戊酸钠会导致社交偏好行为缺陷（Bailey et al., 2016; Baronio et al., 2018; Dwivedi et al., 2019; Zimmermann et al., 2015），但不会引起攻击性行为（Zimmermann et al., 2015）。其他在发育过程中会影响斑马鱼社交行为的药物包括视黄酸，其应用会减少斑马鱼对群聚视频的接近（Bailey et al., 2016），以及在活化的嗜水气单胞菌中加入福尔马林会减少它们的社交偏好行为（Kirsten et al., 2018）。

控制神经调节通路对哺乳动物和斑马鱼的社交行为产生一系列影响。例如，催产素、加压素和苯丙胺衍生物可以逆转社交偏好（Braida et al., 2012; Busnelli et al., 2016; Ponzoni et al., 2016）。用多巴胺 D1 受体拮抗剂 SCH23390 控制多巴胺 D1 受体显著降低了野生型（WT）AB 斑马鱼品系的社交偏好，而不会改变其视觉或运动功能。有趣的是，另一种 WT 斑马鱼品系没有表现出这种效果，这表明不同斑马鱼品系对神经活性化学物质的行为反应存在自然差异（Scerbina et al., 2012）。强效的谷氨酸能 *N*- 甲基 -D- 天冬氨酸（NMDA）受体拮抗剂 MK-801，常用于破坏记忆形成，也可引起斑马鱼的社交缺陷，包括减少所有测量得到的鱼群凝聚力（Maaswinkel et al., 2013）、社交偏好（Dreosti et al., 2015）、群聚（Maaswinkel et al., 2013）和攻击性的指标（Zimmermann et al., 2016），这种效应被催产素和催产素受体激动剂卡贝缩宫素逆转（Zimmermann et al., 2016 ）。NMDA 拮抗剂氯胺酮也会破坏鱼群凝聚力（Riehl et al., 2011）。表 26-1 中总结了其他神经调节药物的作用。

最后，发育过程中的社交隔离会降低社交偏好（Dreosti，数据未发表），但是否会减少群聚（Shams et al., 2018）仍然存在争议（Fulcher et al., 2017; Fulcher et al., 2018）。

26.6 斑马鱼 ASD 模型的现状和未来方向

到目前为止我们在本章中概述的大多数斑马鱼 ASD 研究都强调对单个 ASD 风险基因进行建模并检查相对有限的发育、环路功能和行为表型。展望未来，今后需要进行更全面的综合研究，包括系统地检查大脑形态学一系列变化，主要神经递质和神经调节系统的发育，ASD 相关行为任务期间神经元环路的功能成像和控制，以及斑马鱼仔鱼、幼鱼和成鱼阶段的多维行为表型。

同时将多个平行的 ASD 风险基因表型分析关联起来也很重要，这可能会在未来几年内正式开始。Thyme 及其同事成功地进行了这项研究，他们调查了全基因组关联分析研究（GWAS）发现了与精神分裂症相关基因组区域中的基因（Thyme et al., 2019）。他们通过使用 CRISPR 从 132 个候选基因的直系同源系列中产生突变体，并对大脑结构、活动和行为进行高通量分析，该研究确定了一些突变基因的趋同表型，有助于在含有多个基因的区域中优先对前 30 个候选基因排序（Thyme et al., 2019）。这项研究还强调了如何利用斑马鱼仔鱼简单行为（例如，再醒活动、惊吓、习惯化）和大脑活动图谱的定量分析来阐明风险基因在基本神经发育过程中的作用，以及基本行为背后的神经环路，这对推进我们对 ASD 生物学的理解是非常重要的。

斑马鱼 ASD 研究的另一个需要更多实验关注的方面是，更好地了解斑马鱼哪些类型的社交行为最能模拟人类 ASD 表型（例如，代表核心的和高度保守的行为，这些行为在 ASD 遗传模型中一直受到影响）。一系列不同社交行为分析方法的出现，以及基于机器学习的群体追踪技术的快速发展，大大提高了这些分析的通量和灵敏度，因此斑马鱼 ASD 模型最终将成为预筛选化合物的最佳候选模型，而筛选出的这些化合物可能会逆转 ASD 社交行为的改变。

尽管目前仍处于起步阶段，但同样重要的是阐明各种社交行为背后的神经环路，以及研究这些环路如何受到 ASD 风险基因突变的干扰。绘制斑马鱼幼鱼神经活动（特别是在自由游泳时）的技术的发展，将允许人们对产生社会行为的神经回路进行非常详细的剖析。这种特征对于了解 ASD 如何影响社交大脑以及新治疗方法如何影响这些环路至关重要，并且可能会提出新的治疗靶点。揭示社交相关神经环路的几个关键实验方向值得在这里详细讨论，因为它们印证了斑马鱼模型将 ASD 基因与环路功能和行为输出关联起来的潜力。

26.6.1 绘制社交脑图谱

相对而言，关于社交神经环路是如何形成的，以及它们是如何处理信息和如何产生行为输出的，我们知之甚少。使用（幼年透明）斑马鱼提供了许多优势，例如贯穿整个大脑的单细胞光学分辨率以及简单基因操作，可以揭示基本神经环路详细结构和功能。粗略地说，我们知道人类身上发现的许多社交区域在斑马鱼中是保守的（O'Connell and Hofmann, 2012），这表明任何斑马鱼实验发现的结果都与 ASD 和以许多其他社交行为受影响为特征的神经系统疾病相关。

26.6.2 社交大脑区域的消融

脑区损毁是一种评估不同大脑区域对特定社会行为必要性的方法。Watanabe 团队发现金鱼端脑腹侧而不是背侧损伤，会导致其社交行为减少（Shinozuka and Watanabe, 2004）。最近，Washbourne 团队已经证明腹侧端脑是参与斑马鱼社交互动和社交偏好的重要脑区之一。手术去除和化学消融腹侧端脑（一个类似于外侧隔的区域）都会抑制社交互动，量化为无法调整相对于社交刺激鱼的方向（Stednitz et al., 2018）。此外，他们还证明这些细胞大多数是胆碱能的。目前，尚不清楚这些细胞从何处接收信息以及它们在何处投射 / 交流，也不清楚它们是否参与处理视觉社交刺激或行为控制。

26.6.3 神经元活动的功能成像

功能研究需要监测一条完好无损的鱼的活动。最简单的方法是将斑马鱼固定在琼脂糖中，通过双光子（Renninger and Orger, 2013）或光片显微镜（Ahrens et al., 2013），使用钙或其他神经元活动指示剂来观察大脑活动。可以将鱼完全嵌入琼脂糖中，其眼睛和尾巴也可以自由活动，从而对它们的行为进行有限测量。眼睛或尾巴的运动会引起成像伪影，因此用电极记录瘫痪斑马鱼的有效游泳已经被开发出来，并成功地用于研究其他行为，例如捕获猎物。这些制剂主要用于仔鱼；然而，一些研究表明，将幼鱼限制在琼脂糖中并对功能反应（如气味和味道反应以及尾部运动）进行成像是有可能的（Jetti et al., 2014; Vendrell-Llopis and Yaksi, 2015）。此外，我们知道，虚拟刺激可以成功地刺激鱼类，从而引发社交反应（Larsch and Baier, 2018）。这表明，利用采取限制性调节的成像方法来研究鱼类社交反应中的功能反应是可能的，而且很可能即将实现。

26.6.4　解剖功能影像

虽然利用对鱼进行活动限制的技术可以在单细胞分辨率下进行全脑成像，并同时操纵神经元活动，但嵌入琼脂糖的鱼无法表现出所有自然社交行为的全部信息。因此，最好观察自由游动的鱼并记录它们的活动。一种方法是使用即早基因 *c-fos*（Baraban et al., 2005）、基因编码钙指示剂如 Campari（Fosque et al., 2015）或神经元活动读数技术 pERK（Randlett et al., 2015）。这些技术不是实时测量活动，而是测定长期累积的活动。然而，它们对于识别在鱼类执行特定任务或（社交）行为时表现出强烈活动的细胞和大脑区域非常有用。这些技术已成功用于识别与社交偏好有关的社交领域（Tunbak et al., 研究中），并且可能与上一节中描述的所有社交分析相兼容。

26.6.5　自由游动鱼的成像

最近的一些研究表明，可能很快就可以实时对自由游动仔鱼的大脑活动进行成像。Muto 和 Kawakami 的研究结果表明，在捕获猎物期间，可以使用简单的广角荧光显微镜对仔鱼大脑活动进行成像（Muto and Kawakami, 2016）。然而，神经元活动只能在两次运动之间记录，例如鱼静止很短一段时间。最近，Kim 等人开发了一种跟踪显微镜，能够在自由游动的斑马鱼仔鱼中以细胞分辨率进行全脑钙成像（Kim et al., 2017a）。该技术结合了使用显微镜和红外成像来跟踪行为领域中的目标动物以及一个自动工作台系统来对大脑运动进行三维补偿。此外，Cong 等人已经开发出一种类似的技术，通过高速摄像机跟踪鱼在三维空间的运动，通过计算机软件在显微镜下自动移动腔室，使动物的大脑始终处于聚焦状态（Cong et al., 2017）。这些技术从仔鱼扩展到幼鱼，将为理解社交行为环路创造巨大的潜力。

26.7　结论

尽管斑马鱼 ASD 研究仍处于早期阶段，但现有技术的集合是任何其他脊椎动物模型系统所无法比拟的。一系列使用新方法和新技术的研究表明，斑马鱼作为ASD社交障碍和基本神经通路及基本行为回路缺陷的模型将迅速崛起。总之，本章提到的研究和技术强调了斑马鱼模型在阐明与神经发育障碍相关的基本机制方面的应用前景。因此，在不久的将来，应用该模型来研究越来越多的ASD风险基因可能会对ASD生物学机制产生重要的见解。

（宋云扬翻译　赵丽审校）

参考文献

Abril-de-Abreu, R., Cruz, J., Oliveira, R.F., 2015. Social eavesdropping in zebrafish: tuning of attention to social interactions. Scientific Reports 5, 12678.

Ahrens, M.B., Orger, M.B., Robson, D.N., Li, J.M., Keller, P.J., 2013. Whole-brain functional imaging at cellular resolution using light-sheet microscopy. Nature Methods 10, 413-420.

American Psychiatric Association, 2013. Diagnostic and Statistical Manual of Mental Disorders, fifth ed. American Psychiatric Publishing, Arlington, VA.

An, J.Y., Lin, K., Zhu, L., Werling, D.M., Dong, S., Brand, H., Wang, H.Z., Zhao, X., Schwartz, G.B., Collins, R.L.,et al., 2018. Genome-wide de novo risk score implicates promoter variation in autism spectrum disorder. Science 362.

Anney, R., Klei, L., Pinto, D., Almeida, J., Bacchelli, E., Baird, G., Bolshakova, N., Bolte, S., Bolton, P.F., Bourgeron, T.,et al., 2012. Individual common variants exert weak effects on the risk for autism spectrum disorders. Human Molecular Genetics 21, 4781-4792.

Araujo-Silva, H., Pinheiro-da-Silva, J., Silva, P.F., Luchiari, A.C., 2018. Individual differences in response to alcohol exposure in zebrafish (*Danio rerio*). PLoS One 13, e0198856.

Ariyasiri, K., Choi, T.I., Kim, O.H., Hong, T.I., Gerlai, R., Kim, C.H., 2019. Pharmacological (ethanol) and mutation (sam2 KO) induced impairment of novelty preference in zebrafish quantified using a new three-chamber social choice task. Progress in Neuro-Psychopharmacology and Biological Psychiatry 88, 53-65.

Attaran, A., Salahinejad, A., Crane, A.L., Niyogi, S., Chivers, D.P., 2019. Chronic exposure to dietary selenomethionine dysregulates the genes involved in serotonergic neurotransmission and alters social and antipredator behaviours in zebrafish (*Danio rerio*). Environmental Pollution 246, 837-844.

Audira, G., Sarasamma, S., Chen, J.R., Juniardi, S., Sampurna, B.P., Liang, S.T., Lai, Y.H., Lin, G.M., Hsieh, M.C.,Hsiao, C.D., 2018. Zebrafish mutants carrying leptin a (lepa) gene deficiency display obesity, anxiety, less aggression and fear, and circadian rhythm and color preference dysregulation. International Journal of Molecular Sciences 19.

Bai, Y.X., Zhang, S.H., Fan, Z., Liu, X.Y., Zhao, X., Feng, X.Z., Sun, M.Z., 2018. Automatic multiple zebrafish tracking based on improved HOG features. Scientific Reports 8, 10884.

Bailey, A., Le Couteur, A., Gottesman, I., Bolton, P., Simonoff, E., Yuzda, E., Rutter, M., 1995. Autism as a strongly genetic disorder: evidence from a British twin study. Psychological Medicine 25, 63-77.

Bailey, J.M., Oliveri, A.N., Karbhari, N., Brooks, R.A., De La Rocha, A.J., Janardhan, S., Levin, E.D., 2016. Persistent behavioral effects following early life exposure to retinoic acid or valproic acid in zebrafish. Neurotoxicology 52, 23-33.

Baio, J., Wiggins, L., Christensen, D.L., Maenner, M.J., Daniels, J., Warren, Z., Kurzius-Spencer, M., Zahorodny, W., Robinson Rosenberg, C., White, T., et al., 2018. Prevalence of autism spectrum disorder among children aged 8 Years e autism and developmental disabilities monitoring network, 11 sites, United States, 2014. (MMWR) Surveillance Summaries 67, 1-23.

Baraban, S.C., Taylor, M.R., Castro, P.A., Baier, H., 2005. Pentylenetetrazole induced changes in zebrafish behavior,neural activity and c-fos expression. Neuroscience 131, 759-768.

Baraban, S.C., Dinday, M.T., Hortopan, G.A., 2013. Drug screening in Scn1a zebrafish mutant identifies clemizole as a potential Dravet syndrome treatment. Nature Communications 4, 2410.

Barlow, I.L., Rihel, J., 2017. Zebrafish sleep: from geneZZZ to neuronZZZ. Current Opinion in Neurobiology. https://doi.org/10.1016/j.conb.2017.02.009.

Baronio, D., Puttonen, H.A.J., Sundvik, M., Semenova, S., Lehtonen, E., Panula, P., 2018. Embryonic exposure to valproic acid affects the histaminergic system and the social behaviour of adult zebrafish (*Danio rerio*). British Journal of Pharmacology 175, 797-809.

Bartolini, T., Mwaffo, V., Showler, A., Macri, S., Butail, S., Porfiri, M., 2016. Zebrafish response to 3D printed shoals of conspecifics: the effect of body size. Bioinspiration and Biomimetics 11, 026003.

Bernier, R., Golzio, C., Xiong, B., Stessman, H.A., Coe, B.P., Penn, O., Witherspoon, K., Gerdts, J., Baker, C., Vulto-van Silfhout, A.T., et al., 2014. Disruptive CHD8 mutations define a subtype of autism earlyin development. Cell 158, 263-276.

Bishop, B.H., Spence-Chorman, N., Gahtan, E., 2016. Three-dimensional motion tracking reveals a diving component to visual and auditory escape swims in zebrafish larvae. Journal of Experimental Biology 219, 3981-3987.

Blaker-Lee, A., Gupta, S., McCammon, J.M., De Rienzo, G., Sive, H., 2012. Zebrafish homologs of genes within 16p11.2, a genomic region associated with brain disorders, are active during brain development, and include two deletion dosage sensor genes. Disease Models and Mechanisms 5, 834-851.

Boggio, P.S., Asthana, M.K., Costa, T.L., Valasek, C.A., Osorio, A.A., 2015. Promoting social plasticity in developmental disorders with non-invasive brain stimulation techniques. Frontiers in Neuroscience 9, 294.

Braida, D., Donzelli, A., Martucci, R., Capurro, V., Busnelli, M., Chini, B., Sala, M., 2012. Neurohypophyseal hormones manipulation modulate social and anxiety-related behavior in zebrafish. Psychopharmacology 220, 319-330.

Burgess, H.A., Granato, M., 2007a. Modulation of locomotor activity in larval zebrafish during light adaptation. Journal of Experimental Biology 210, 2526e2539.

Burgess, H.A., Granato, M., 2007b. Sensorimotor gating in larval zebrafish. Journal of Neuroscience 27, 4984-4994.

Buske, C., Gerlai, R., 2011a. Early embryonic ethanol exposure impairs shoaling and the dopaminergic and serotoninergic systems in adult zebrafish. Neurotoxicology and Teratology 33, 698-707.

Buske, C., Gerlai, R., 2011b. Shoaling develops with age in zebrafish (*Danio rerio*). Progress in Neuro-Psychopharmacology & Biological Psychiatry 35, 1409-1415.

Busnelli, M., Dagani, J., de Girolamo, G., Balestrieri, M., Pini, S., Saviotti, F.M., Scocco, P., Sisti, D., Rocchi, M., Chini, B., 2016. Unaltered Oxytocin and Vasopressin Plasma Levels in Patients with Schizophrenia After 4 Months of Daily Treatment with Intranasal Oxytocin. Journal of Neuroendocrinology 28.

Butail, S., Paley, D.A., 2012. Three-dimensional reconstruction of the fast-start swimming kinematics of densely schooling fish. Journal of the Royal Society Interface 9, 77-88.

Butail, S., Bartolini, T., Porfiri, M., 2013. Collective response of zebrafish shoals to a free-swimming robotic fish. PLoS One 8, e76123.

Butail, S., Mwaffo, V., Porfiri, M., 2016. Model-free information-theoretic approach to infer leadership in pairs of zebrafish. Physical Review E 93, 042411.

Choi, J.H., Jeong, Y.M., Kim, S., Lee, B., Ariyasiri, K., Kim, H.T., Jung, S.H., Hwang, K.S., Choi, T.I., Park, C.O., et al., 2018. Targeted knockout of a chemokine-like gene increases anxiety and fear responses. Proceedings of the National Academy of Sciences of the United States of America 115, E1041-E1050.

Christensen, J., Gronborg, T.K., Sorensen, M.J., Schendel, D., Parner, E.T., Pedersen, L.H., Vestergaard, M., 2013. Prenatal valproate exposure and risk of autism spectrum disorders and childhood autism. Journal of the American Medical Association 309 (16), 1696-1703.

Clements, K.N., Miller, T.H., Keever, J.M., Hall, A.M., Issa, F.A., 2018. Social status-related differences in motor activity between wild-type and mutant zebrafish. The Biological Bulletin 235, 71-82.

Cloney, K., Steele, S.L., Stoyek, M.R., Croll, R.P., Smith, F.M., Prykhozhij, S.V., Brown, M.M., Midgen, C., Blake, K.,Berman, J.N., 2018. Etiology and functional validation of gastrointestinal motility dysfunction in a zebrafish model of CHARGE syndrome. FEBS Journal 285 (11), 2125-2140.

Cong, L., Wang, Z., Chai, Y., Hang, W., Shang, C., Yang, W., Bai, L., Du, J., Wang, K., Wen, Q., 2017. Rapid whole brain imaging of neural activity in freely behaving larval zebrafish (*Danio rerio*). elife 6.

Courchesne, E., Campbell, K., Solso, S., 2011. Brain growth across the life span in autism: age-specific changes in anatomical pathology. Brain Research 1380, 138-145.

D'Angelo, D., Lebon, S., Chen, Q., Martin-Brevet, S., Snyder, L.G., Hippolyte, L., Hanson, E., Maillard, A.M., Faucett, W.A., Mace, A., et al., 2016. Defining the effect of the 16p11.2 duplication on cognition, behavior, andmedical comorbidities. JAMA Psychiatry 73, 20-30.

Dahlem, T.J., Hoshijima, K., Jurynec, M.J., Gunther, D., Starker, C.G., Locke, A.S., Weis, A.M., Voytas, D.F., Grunwald, D.J., 2012. Simple methods for generating and detecting locus-specific mutations induced with TALENs in the zebrafish genome. PLoS Genetics 8, e1002861.

De Rubeis, S., He, X., Goldberg, A.P., Poultney, C.S., Samocha, K., Cicek, A.E., Kou, Y., Liu, L., Fromer, M., Walker, S., et al., 2014. Synaptic, transcriptional and chromatin genes disrupted in autism. Nature 515, 209-215.

den Broeder, M.J., van der Linde, H., Brouwer, J.R., Oostra, B.A., Willemsen, R., Ketting, R.F., 2009. Generation and characterization of FMR1 knockout zebrafish. PLoS One 4, e7910.

Dewari, P.S., Ajani, F., Kushawah, G., Kumar, D.S., Mishra, R.K., 2016. Reversible loss of reproductive fitness in zebrafish on chronic alcohol exposure. Alcohol 50, 83-89.

Dinday, M.T., Baraban, S.C., 2015. Large-scale phenotype-based antiepileptic drug screening in a zebrafish model of Dravet syndrome(1,2,3). eNeuro 2.

Doyon, Y., McCammon, J.M., Miller, J.C., Faraji, F., Ngo, C., Katibah, G.E., Amora, R., Hocking, T.D., Zhang, L., Rebar, E.J., et al., 2008. Heritable targeted gene disruption in zebrafish using designed zinc-finger nucleases. Nature Biotechnology 26, 702-708.

Draper, B.W., Morcos, P.A., Kimmel, C.B., 2001. Inhibition of zebrafish fgf8 pre-mRNA splicing with morpholino oligos: a quantifiable method for gene knockdown. Genesis 30, 154-156.

Dreosti, E., Lopes, G., Kampff, A.R., Wilson, S.W., 2015. Development of social behavior in young zebrafish. Frontiers in Neural Circuits 9, 39.

Durand, C.M., Betancur, C., Boeckers, T.M., Bockmann, J., Chaste, P., Fauchereau, F., Nygren, G., Rastam, M., Gillberg, I.C., Anckarsater, H., et al., 2007. Mutations in the gene encoding the synaptic scaffolding protein SHANK3 are associated with autism spectrum disorders. Nature Genetics 39, 25-27.

Dwivedi, S., Medishetti, R., Rani, R., Sevilimedu, A., Kulkarni, P., Yogeeswari, P., 2019. Larval zebrafish model for studying the effects of valproic acid on neurodevelopment: an approach towards modeling autism. Journal of Pharmacological and Toxicological Methods 95, 56-65.

Eachus, H., Bright, C., Cunliffe, V.T., Placzek, M., Wood, J.D., Watt, P.J., 2017. Disrupted-in-Schizophrenia-1 is essential for normal hypothalamic-pituitary-interrenal (HPI) axis function. Human Molecular Genetics 26,1992-2005.

Eguiraun, H., Casquero, O., Sorensen, A.J., Martinez, I., 2018. Reducing the number of individuals to monitor shoaling fish systems e application of the shannon entropy to construct a biological warning system model. Frontiers in Physiology 9, 493.

Eisen, J.S., Smith, J.C., 2008. Controlling morpholino experiments: don't stop making antisense. Development 135,1735-1743.

El-Brolosy, M.A., Stainier, D.Y.R., 2017. Genetic compensation: a phenomenon in search of mechanisms. PLoS Genetics 13, e1006780.

El-Brolosy, M.A., Kontarakis, Z., Rossi, A., Kuenne, C., Gunther, S., Fukuda, N., Kikhi, K., Boezio, G.L.M., Takacs, C.M., Lai, S.L., et al., 2019. Genetic compensation triggered by mutant mRNA degradation. Nature 568, 193-197.

Elsabbagh, M., Divan, G., Koh, Y.J., Kim, Y.S., Kauchali, S., Marcin, C., Montiel-Nava, C., Patel, V., Paula, C.S., Wang, C., et al., 2012. Global prevalence of autism and other pervasive developmental disorders. Autism Research 5, 160-179.

Engeszer, R.E., Ryan, M.J., Parichy, D.M., 2004. Learned social preference in zebrafish. Current Biology 14, 881-884.

Engeszer, R.E., Barbiano, L.A., Ryan, M.J., Parichy, D.M., 2007. Timing and plasticity of shoaling behaviour in the zebrafish, *Danio rerio*. Animal Behaviour 74, 1269-1275.

Escamilla, C.O., Filonova, I., Walker, A.K., Xuan, Z.X., Holehonnur, R., Espinosa, F., Liu, S., Thyme, S.B., Lopez-Garcia, I.A., Mendoza, D.B., et al., 2017. Kctd13 deletion reduces synaptic transmission via increased RhoA. Nature 551, 227-231.

Felix, L.M., Antunes, L.M., Coimbra, A.M., Valentim, A.M., 2017. Behavioral alterations of zebrafish larvae after early embryonic exposure to ketamine. Psychopharmacology 234, 549-558.

Fernandes, Y., Gerlai, R., 2009. Long-term behavioral changes in response to early developmental exposure to ethanol in zebrafish. Alcoholism: Clinical and Experimental Research 33, 601-609.

Fitzpatrick, S.E., Srivorakiat, L., Wink, L.K., Pedapati, E.V., Erickson, C.A., 2016. Aggression in autism spectrum disorder: presentation and treatment options. Neuropsychiatric Disease and Treatment 12, 1525-1538.

Fontana, B.D., Mezzomo, N.J., Kalueff, A.V., Rosemberg, D.B., 2018. The developing utility of zebrafish models of neurological and neuropsychiatric disorders: a critical review. Experimental Neurology 299, 157-171.

Fosque, B.F., Schreiter, E.R., Sun, Y., Dana, H., Yang, C.-T., Ohyama, T., Tadross, M.R., Patel, R., Zlatic, M., Kim, D.S., Ahrens, M.B., Jayaraman, V., Looger, L.L., 2015. Labeling of active neural circuits in vivo with designed calcium integrators. Science 347, 755-760.

Franco-Restrepo, J.E., Forero, D.A., Vargas, R.A., 2019. A review of freely available, open-source software for the automated analysis of the behavior of adult zebrafish. Zebrafish 16, 223-232.

Fulcher, N., Tran, S., Shams, S., Chatterjee, D., Gerlai, R., 2017. Neurochemical and Behavioral Responses to Unpredictable Chronic Mild Stress Following Developmental Isolation: The Zebrafish as a Model for Major Depression.Zebrafish 14, 23-34.

Gaugler, T., Klei, L., Sanders, S.J., Bodea, C.A., Goldberg, A.P., Lee, A.B., Mahajan, M., Manaa, D., Pawitan, Y., Reichert, J., et al., 2014. Most genetic risk for autism resides with common variation. Nature Genetics 46, 881-885.

Geng, Y., Peterson, R.T., 2019. The zebrafish subcortical social brain as a model for studying social behavior disorders. Disease Models and Mechanisms 12.

Gerlai, R., 2017. Animated images in the analysis of zebrafish behavior. Current Zoology 63, 35-44.

Ghosh, M., Rihel, J., 2019. Hierarchical compression reveals sub-second to day-long structure in larval zebrafish behaviour. bioRxiv 694471.

Giacomini, A., Abreu, M.S., Giacomini, L.V., Siebel, A.M., Zimerman, F.F., Rambo, C.L., Mocelin, R., Bonan, C.D., Piato, A.L., Barcellos, L.J.G., 2016. Fluoxetine and diazepam acutely modulate stress induced-behavior. Behavioural Brain Research 296, 301-310.

Glessner, J.T.,Wang, K., Cai, G., Korvatska, O., Kim, C.E., Wood, S., Zhang, H., Estes, A., Brune, C.W., Bradfield, J.P., et al., 2009. Autism genome-wide copy number variation reveals ubiquitin and neuronal genes. Nature 459, 569-573.

Golzio, C., Willer, J., Talkowski, M.E., Oh, E.C., Taniguchi, Y., Jacquemont, S., Reymond, A., Sun, M., Sawa, A., Gusella, J.F., et al., 2012. KCTD13 is a major driver of mirrored neuroanatomical phenotypes of the 16p11.2 copy number variant. Nature 485, 363-367.

Granato, M., Nusslein-Volhard, C., 1996. Fishing for genes controlling development. Current Opinion in Genetics and Development 6, 461-468.

Green, J., Collins, C., Kyzar, E.J., Pham, M., Roth, A., Gaikwad, S., Cachat, J., Stewart, A.M., Landsman, S., Grieco, F., et al., 2012. Automated high-throughput neurophenotyping of zebrafish social behavior. Journal of Neuroscience Methods 210, 266-271.

Griffin, A., Hamling, K.R., Knupp, K., Hong, S., Lee, L.P., Baraban, S.C., 2017. Clemizole and modulators of serotonin signalling suppress seizures in Dravet syndrome. Brain 140, 669-683.

Grossman, L., Utterback, E., Stewart, A., Gaikwad, S., Chung, K.M., Suciu, C., Wong, K., Elegante, M., Elkhayat, S., Tan, J., et al., 2010. Characterization of behavioral and endocrine effects of LSD on zebrafish. Behavioural Brain Research 214, 277-284.

Grove, J., Ripke, S., Als, T.D., Mattheisen, M., Walters, R.K., Won, H., Pallesen, J., Agerbo, E., Andreassen, O.A., Anney, R., et al., 2019. Identification of common genetic risk variants for autism spectrum disorder. Nature Genetics 51, 431-444.

Guayasamin, O.L., Couzin, I.D., Miller, N.Y., 2017. Behavioural plasticity across social contexts is regulated by the directionality of inter-individual differences. Behavioural Processes 141, 196-204.

Guo, S., 2009. Using zebrafish to assess the impact of drugs on neural development and function. Expert Opinion on Drug Discovery 4, 715-726.

Hanson, E., Bernier, R., Porche, K., Jackson, F.I., Goin-Kochel, R.P., Snyder, L.G., Snow, A.V., Wallace, A.S., Campe, K.L., Zhang, Y., et al., 2015. The cognitive and behavioral phenotype of the 16p11.2 deletion in a clinically ascertained population. Biological Psychiatry 77, 785-793.

Harpaz, R., Tkacik, G., Schneidman, E., 2017. Discrete modes of social information processing predict individual behavior of fish in a group. Proceedings of the National Academy of Sciences of the United States of America 114, 10149-10154.

Heras, F.J.H., Romero-Ferrero, F., Hinz, R.C., de Polavieja, G.G., 2019. Deep attention networks reveal the rules of collective motion in zebrafish. PLoS Computational Biology 15, e1007354.

Hinz, R.C., de Polavieja, G.G., 2017. Ontogeny of collective behavior reveals a simple attraction rule. Proceedings of the National Academy of Sciences of the United States of America 114, 2295-2300.

Hoffman, E.J., State, M.W., 2010. Progress in cytogenetics: implications for child psychopathology. Journal of the American Academy of Child & Adolescent Psychiatry 49, 736-751 quiz 856-737.

Hoffman, E.J., Turner, K.J., Fernandez, J.M., Cifuentes, D., Ghosh, M., Ijaz, S., Jain, R.A., Kubo, F., Bill, B.R., Baier, H., et al., 2016. Estrogens suppress a behavioral phenotype in zebrafish mutants of the autism risk gene, CNTNAP2. Neuron 89, 725-733.

Huber, K.M., Gallagher, S.M., Warren, S.T., Bear, M.F., 2002. Altered synaptic plasticity in a mouse model of fragile X mental retardation. Proceedings of the National Academy of Sciences of the United States of America 99, 7746-7750.

Hwang, W.Y., Fu, Y., Reyon, D., Maeder, M.L., Kaini, P., Sander, J.D., Joung, J.K., Peterson, R.T., Yeh, J.R., 2013. Heritable and precise zebrafish genome editing using a CRISPR-Cas system. PLoS One 8, e68708.

Ijaz, S., Hoffman, E.J., 2016. Zebrafish: a translational model system for studying neuropsychiatric disorders. Journal of the American Academy of Child and Adolescent Psychiatry 55, 746-748.

Iossifov, I., O'Roak, B.J., Sanders, S.J., Ronemus, M., Krumm, N., Levy, D., Stessman, H.A., Witherspoon, K.T., Vives, L., Patterson, K.E., et al., 2014. The contribution of de novo coding mutations to autism spectrum disorder. Nature 515, 216-221.

James, D.M., Kozol, R.A., Kajiwara, Y., Wahl, A.L., Storrs, E.C., Buxbaum, J.D., Klein, M., Moshiree, B., Dallman, J.E.,2019. Intestinal dysmotility in a zebrafish (*Danio rerio*) shank3a;shank3b mutant model of autism. Molecular Autism 10, 3.

Jetti, S.K., Vendrell-Llopis, N., Yaksi, E., 2014. Spontaneous activity governs olfactory representations in spatially organized habenular microcircuits. Current Biology 24, 434-439.

Jinek, M., Chylinski, K., Fonfara, I., Hauer, M., Doudna, J.A., Charpentier, E., 2012. A programmable dual-RNA guided DNA endonuclease in adaptive bacterial immunity. Science 337, 816-821.

Katz, Y., Tunstrom, K., Ioannou, C.C., Huepe, C., Couzin, I.D., 2011. Inferring the structure and dynamics of interactions in schooling fish. Proceedings of the National Academy of Sciences of the United States of America 108,18720-18725.

Kim, D.H., Kim, J., Marques, J.C., Grama, A., Hildebrand, D.G.C., Gu, W., Li, J.M., Robson, D.N., 2017a. Pan-neuronal calcium imaging with cellular resolution in freely swimming zebrafish. Nature Methods 14, 1107-1114.

Kim, O.H., Cho, H.J., Han, E., Hong, T.I., Ariyasiri, K., Choi, J.H., Hwang, K.S., Jeong, Y.M., Yang, S.Y., Yu, K., et al.,2017b. Zebrafish knockout of Down syndrome gene, DYRK1A, shows social impairments relevant to autism. Molecular Autism 8, 50.

Kirsten, K., Soares, S.M., Koakoski, G., Carlos Kreutz, L., Barcellos, L.J.G., 2018. Characterization of sickness behavior in zebrafish. Brain, Behavior, and Immunity 73, 596-602.

Klei, L., Sanders, S.J., Murtha, M.T., Hus, V., Lowe, J.K., Willsey, A.J., Moreno-De-Luca, D., Yu, T.W., Fombonne, E., Geschwind, D., et al., 2012. Common genetic variants, acting additively, are a major source of risk for autism. Molecular Autism 3, 9.

Klin, A., Jones, W., 2008. Altered face scanning and impaired recognition of biological motion in a 15-month-old infant with autism. Developmental Science 11, 40-46.

Kok, F.O., Shin, M., Ni, C.W., Gupta, A., Grosse, A.S., van Impel, A., Kirchmaier, B.C., Peterson-Maduro, J., Kourkoulis, G., Male, I., et al., 2015. Reverse genetic screening reveals poor correlation between morpholino-induced and mutant phenotypes in zebrafish. Developmental Cell 32, 97-108.

Kokel, D., Bryan, J., Laggner, C., White, R., Cheung, C.Y., Mateus, R., Healey, D., Kim, S., Werdich, A.A., Haggarty, S.J., et al., 2010. Rapid behavior-based identification of neuroactive small molecules in the zebrafish. Nature Chemical Biology 6, 231-237.

Kopman, V., Laut, J., Polverino, G., Porfiri, M., 2013. Closed-loop control of zebrafish response using a bioinspired robotic-fish in a preference test. Journal of the Royal Society Interface 10, 20120540.

Kozol, R.A., Cukier, H.N., Zou, B., Mayo, V., De Rubeis, S., Cai, G., Griswold, A.J., Whitehead, P.L., Haines, J.L., Gilbert, J.R., et al., 2015. Two knockdown models of the autism genes SYNGAP1 and SHANK3 in zebrafish produce similar behavioral phenotypes associated with embryonic disruptions of brain morphogenesis. Human Molecular Genetics 24, 4006-4023.

Kozol, R.A., Abrams, A.J., James, D.M., Buglo, E., Yan, Q., Dallman, J.E., 2016. Function over form: modeling groups of inherited neurological conditions in zebrafish. Frontiers in Molecular Neuroscience 9, 55.

Kumar, R.A., KaraMohamed, S., Sudi, J., Conrad, D.F., Brune, C., Badner, J.A., Gilliam, T.C., Nowak, N.J., Cook Jr., E.H., Dobyns, W.B., et al., 2008. Recurrent 16p11.2 microdeletions in autism. Human Molecular Genetics 17, 628-638.

Kuroda, T., 2018. A system for the real-time tracking of operant behavior as an application of 3D camera. Journal of the Experimental Analysis of Behavior 110, 522-544.

Laan, A., Gil de Sagredo, R., de Polavieja, G.G., 2017. Signatures of optimal control in pairs of schooling zebrafish. Proceedings of the Biological Sciences 284.

Laan, A., Iglesias-Julios, M., de Polavieja, G.G., 2018. Zebrafish aggression on the sub-second time scale: evidence for mutual motor coordination and multi-functional attack manoeuvres. Royal Society Open Science 5, 180679.

Larsch, J., Baier, H., 2018. Biological motion as an innate perceptual mechanism driving social affiliation. Current Biology 28, 3523-3532 e3524.

Lawson, N.D., 2016. Reverse genetics in zebrafish: mutants, morphants, and moving forward. Trends in Cell Biology 26, 77-79.

Leung, L.C., Wang, G.X., Madelaine, R., Skariah, G., Kawakami, K., Deisseroth, K., Urban, A.E., Mourrain, P., 2019. Neural signatures of sleep in zebrafish. Nature 571, 198-204.

Liu, C.X., Li, C.Y., Hu, C.C., Wang, Y., Lin, J., Jiang, Y.H., Li, Q., Xu, X., 2018. CRISPR/Cas9-induced shank3b mutant zebrafish display autism-like behaviors. Molecular Autism 9, 23.

Lord, C., Elsabbagh, M., Baird, G., Veenstra-Vanderweele, J., 2018. Autism spectrum disorder. Lancet 392, 508-520.

Lovett-Barron, M., Andalman, A.S., Allen, W.E., Vesuna, S., Kauvar, I., Burns, V.M., Deisseroth, K., 2017. Ancestral circuits for the coordinated modulation of brain state. Cell 171 (6), 1411-1423.

Lyall, K., Croen, L., Daniels, J., Fallin, M.D., Ladd-Acosta, C., Lee, B.K., Park, B.Y., Snyder, N.W., Schendel, D.,Volk, H., et al., 2017. The changing epidemiology of autism spectrum disorders. Annual Review of Public Health 38, 81-102.

Maaswinkel, H., Zhu, L., Weng, W., 2013. Using an automated 3D-tracking system to record individual and shoals of adult zebrafish. Journal of Visualized Experiments (82), 50681.

Macri, S., Neri, D., Ruberto, T., Mwaffo, V., Butail, S., Porfiri, M., 2017. Three-dimensional scoring of zebrafish behavior unveils biological phenomena hidden by two-dimensional analyses. Scientific Reports 7, 1962.

Marques, J.C., Lackner, S., Felix, R., Orger, M.B., 2018. Structure of the zebrafish locomotor repertoire revealed with unsupervised behavioral clustering. Current Biology 28, 181-195 e185.

Marshall, C.R., Noor, A., Vincent, J.B., Lionel, A.C., Feuk, L., Skaug, J., Shago, M., Moessner, R., Pinto, D., Ren, Y., et al., 2008. Structural variation of chromosomes in autism spectrum disorder. The American Journal of Human Genetics 82, 477-488.

Maruska, K., Soares, M.C., Lima-Maximino, M., Henrique de Siqueira-Silva, D., Maximino, C., 2019. Social plasticity in the fish brain: neuroscientific and ethological aspects. Brain Research 1711, 156-172.

McCammon, J.M., Sive, H., 2015. Challenges in understanding psychiatric disorders and developing therapeutics: a role for zebrafish. Disease Models and Mechanisms 8, 647-656.

McCammon, J.M., Blaker-Lee, A., Chen, X., Sive, H., 2017. The 16p11.2 homologs fam57ba and doc2a generate certain brain and body phenotypes. Human Molecular Genetics 26, 3699-3712.

McCarthy, S.E., Makarov, V., Kirov, G., Addington, A.M., McClellan, J., Yoon, S., Perkins, D.O., Dickel, D.E., Kusenda, M., Krastoshevsky, O., et al., 2009. Microduplications of 16p11.2 are associated with schizophrenia. Nature Genetics 41, 1223-1227.

Meng, X., Noyes, M.B., Zhu, L.J., Lawson, N.D., Wolfe, S.A., 2008. Targeted gene inactivation in zebrafish using engineered zinc-finger nucleases. Nature Biotechnology 26, 695-701.

Miller, N., Gerlai, R., 2007. Quantification of shoaling behaviour in zebrafish (*Danio rerio*). Behavioural Brain Research 184, 157-166.

Miller, N.Y., Gerlai, R., 2008. Oscillations in shoal cohesion in zebrafish (*Danio rerio*). Behavioural Brain Research 193,148-151.

Miller, N., Gerlai, R., 2012. From schooling to shoaling: patterns of collective motion in zebrafish (*Danio rerio*). PLoS One 7, e48865.

Miller, N., Greene, K., Dydinski, A., Gerlai, R., 2013. Effects of nicotine and alcohol on zebrafish (*Danio rerio*) shoaling. Behavioural Brain Research 240, 192-196.

Moreno-Mateos, M.A., Vejnar, C.E., Beaudoin, J.D., Fernandez, J.P., Mis, E.K., Khokha, M.K., Giraldez, A.J., 2015. CRISPRscan: designing highly efficient sgRNAs for CRISPR-Cas9 targeting in vivo. Nature Methods 12, 982-988.

Muhle, R.A., Reed, H.E., Stratigos, K.A., Veenstra-VanderWeele, J., 2018. The emerging clinical neuroscience of autism spectrum disorder: a review. JAMA Psychiatry 75, 514-523.

Muller, T.E., Nunes, S.Z., Silveira, A., Loro, V.L., Rosemberg, D.B., 2017. Repeated ethanol exposure alters social behavior and oxidative stress parameters of zebrafish. Progress in Neuro-Psychopharmacology & Biological Psychiatry 79, 105-111.

Muto, A., Kawakami, K., 2016. Calcium imaging of neuronal activity in free-swimming larval zebrafish. Methods in Molecular Biology 1451, 333-341.

Mwaffo, V., Zhang, P., Romero Cruz, S., Porfiri, M., 2017. Zebrafish swimming in the flow: a particle image velocimetry study. PeerJ 5, e4041.

Nasevicius, A., Ekker, S.C., 2000. Effective targeted gene 'knockdown' in zebrafish. Nature Genetics 26, 216-220.

Neale, B.M., Kou, Y., Liu, L., Ma'ayan, A., Samocha, K.E., Sabo, A., Lin, C.F., Stevens, C., Wang, L.S., Makarov, V., et al., 2012. Patterns and rates of exonic de novo mutations in autism spectrum disorders. Nature 485, 242-245.

Neufeld, J., Hsu, C.T., Chakrabarti, B., 2019. Atypical reward-driven modulation of mimicry-related neural activity in autism. Frontiers in Psychiatry 10, 327.

O'Connell, L.A., Hofmann, H.A., 2012. Evolution of a vertebrate social decision-making network. Science 336,1154-1157.

O'Roak, B.J., Vives, L., Girirajan, S., Karakoc, E., Krumm, N., Coe, B.P., Levy, R., Ko, A., Lee, C., Smith, J.D., et al., 2012. Sporadic autism exomes reveal a highly interconnected protein network of de novo mutations. Nature 485, 246-250.

Orger, M.B., de Polavieja, G.G., 2017. Zebrafish behavior: opportunities and challenges. Annual Review of Neuroscience 40, 125-147.

Parikshak, N.N., Luo, R., Zhang, A., Won, H., Lowe, J.K., Chandran, V., Horvath, S., Geschwind, D.H., 2013. Integrative functional genomic analyses implicate specific molecular pathways and circuits in autism. Cell 155, 1008-1021.

Patowary, A., Won, S.Y., Oh, S.J., Nesbitt, R.R., Archer, M., Nickerson, D., Raskind, W.H., Bernier, R., Lee, J.E.,Brkanac, Z., 2019. Family-based exome sequencing and case-control analysis implicate CEP41 as an ASD gene.Translational Psychiatry 9, 4.

Penagarikano, O., Abrahams, B.S., Herman, E.I., Winden, K.D., Gdalyahu, A., Dong, H., Sonnenblick, L.I., Gruver, R.,Almajano,

J., Bragin, A., et al., 2011. Absence of CNTNAP2 leads to epilepsy, neuronal migration abnormalities,and core autism-related deficits. Cell 147, 235-246.

Perez-Escudero, A., Vicente-Page, J., Hinz, R.C., Arganda, S., de Polavieja, G.G., 2014. idTracker: tracking individuals in a group by automatic identification of unmarked animals. Nature Methods 11, 743-748.

Polverino, G., Porfiri, M., 2013. Zebrafish (*Danio rerio*) behavioural response to bioinspired robotic fish and mosquitofish (*Gambusia affinis*). Bioinspiration and Biomimetics 8, 044001.

Ponzoni, L., Sala, M., Braida, D., 2016. Ritanserin-sensitive receptors modulate the prosocial and the anxiolytic effect of MDMA derivatives, DOB and PMA, in zebrafish. Behavioural Brain Research 314, 181-189.

Prober, D.A., Rihel, J., Onah, A.A., Sung, R.J., Schier, A.F., 2006. Hypocretin/orexin overexpression induces an insomnia-like phenotype in zebrafish. Journal of Neuroscience 26, 13400-13410.

Qian, Z.M., Wang, S.H., Cheng, X.E., Chen, Y.Q., 2016. An effective and robust method for tracking multiple fish in video image based on fish head detection. BMC Bioinformatics 17, 251.

Randlett, O., Wee, C.L., Naumann, E.A., Nnaemeka, O., Schoppik, D., Fitzgerald, J.E., Portugues, R., Lacoste, A.M.,Riegler, C., Engert, F., et al., 2015. Whole-brain activity mapping onto a zebrafish brain atlas. Nature Methods 12,1039-1046.

Renninger, S.L., Orger, M.B., 2013. Two-photon imaging of neural population activity in zebrafish. Methods 62,255-267.

Rihel, J., Ghosh, M., 2015. Zebrafish. In: Hock, F.J. (Ed.), Drug Discovery and Evaluation: Pharmacological Assays. Springer, Berlin Heidelberg, pp. 1-102.

Rihel, J., Prober, D.A., Arvanites, A., Lam, K., Zimmerman, S., Jang, S., Haggarty, S.J., Kokel, D., Rubin, L.L.,Peterson, R.T., et al., 2010. Zebrafish behavioral profiling links drugs to biological targets and rest/wake regulation. Science 327, 348-351.

Riehl, R., Kyzar, E., Allain, A., Green, J., Hook, M., Monnig, L., Rhymes, K., Roth, A., Pham, M., Razavi, R., et al.,2011. Behavioral and physiological effects of acute ketamine exposure in adult zebrafish. Neurotoxicology and Teratology 33, 658-667.

Ritvo, E.R., Freeman, B.J., Mason-Brothers, A., Mo, A., Ritvo, A.M., 1985. Concordance for the syndrome of autism in 40 pairs of afflicted twins. American Journal of Psychiatry 142, 74-77.

Robu, M.E., Larson, J.D., Nasevicius, A., Beiraghi, S., Brenner, C., Farber, S.A., Ekker, S.C., 2007. p53 activation by knockdown technologies. PLoS Genetics 3, e78.

Romero-Ferrero, F., Bergomi, M.G., Hinz, R.C., Heras, F.J.H., de Polavieja, G.G., 2019. idtracker.ai: tracking all individuals in small or large collectives of unmarked animals. Nature Methods 16, 179-182.

Rosenberg, R.E., Law, J.K., Yenokyan, G., McGready, J., Kaufmann, W.E., Law, P.A., 2009. Characteristics and concordance of autism spectrum disorders among 277 twin pairs. Archives of Pediatrics and Adolescent Medicine 163, 907-914.

Rossi, A., Kontarakis, Z., Gerri, C., Nolte, H., Holper, S., Kruger, M., Stainier, D.Y., 2015. Genetic compensation induced by deleterious mutations but not gene knockdowns. Nature 524, 230-233.

Rubenstein, J.L., Merzenich, M.M., 2003. Model of autism: increased ratio of excitation/inhibition in key neural systems. Genes, Brain and Behavior 2, 255-267.

Ruberto, T., Mwaffo, V., Singh, S., Neri, D., Porfiri, M., 2016. Zebrafish response to a robotic replica in three dimensions. Royal Society Open Science 3, 160505.

Ruberto, T., Polverino, G., Porfiri, M., 2017. How different is a 3D-printed replica from a conspecific in the eyes of a zebrafish? Journal of the Experimental Analysis of Behavior 107, 279-293.

Ruzzo, E.K., Perez-Cano, L., Jung, J.Y., Wang, L.K., Kashef-Haghighi, D., Hartl, C., Singh, C., Xu, J., Hoekstra, J.N.,Leventhal, O., et al., 2019. Inherited and de novo genetic risk for autism impacts shared networks. Cell 178,850-866 e826.

Sakai, C., Ijaz, S., Hoffman, E.J., 2018. Zebrafish models of neurodevelopmental disorders: past, present, and future. Frontiers in Molecular Neuroscience 11, 294.

Sander, J.D., Cade, L., Khayter, C., Reyon, D., Peterson, R.T., Joung, J.K., Yeh, J.R., 2011. Targeted gene disruption in somatic zebrafish cells using engineered TALENs. Nature Biotechnology 29, 697-698.

Sanders, S.J., Ercan-Sencicek, A.G., Hus, V., Luo, R., Murtha, M.T., Moreno-De-Luca, D., Chu, S.H., Moreau, M.P., Gupta, A.R., Thomson, S.A., et al., 2011. Multiple recurrent de novo CNVs, including duplications of the 7q11.23 Williams syndrome region, are strongly associated with autism. Neuron 70, 863-885.

Sanders, S.J., Murtha, M.T., Gupta, A.R., Murdoch, J.D., Raubeson, M.J., Willsey, A.J., Ercan-Sencicek, A.G., DiLullo, N.M., Parikshak, N.N., Stein, J.L., et al., 2012. De novo mutations revealed by whole-exome sequencing are strongly associated with autism. Nature 485, 237-241.

Sanders, S.J., He, X., Willsey, A.J., Ercan-Sencicek, A.G., Samocha, K.E., Cicek, A.E., Murtha, M.T., Bal, V.H., Bishop, S.L., Dong, S., et al., 2015. Insights into autism spectrum disorder genomic architecture and biology from 71 risk loci. Neuron 87, 1215-1233.

Sanders, S.J., Neale, B.M., Huang, H., Werling, D.M., An, J.Y., Dong, S., Abecasis, G., Arguello, P.A., Blangero, J., Boehnke, M., et al., 2018. Publisher correction: whole genome sequencing in psychiatric disorders: the WGSPD consortium. Nature Neuroscience 21, 1017.

Sandin, S., Lichtenstein, P., Kuja-Halkola, R., Larsson, H., Hultman, C.M., Reichenberg, A., 2014. The familial risk of autism. Journal of the American Medical Association 311, 1770-1777.

Scerbina, T., Chatterjee, D., Gerlai, R., 2012. Dopamine receptor antagonism disrupts social preference in zebrafish: a strain comparison study. Amino Acids 43, 2059-2072.

Schaefer, I.C., Siebel, A.M., Piato, A.L., Bonan, C.D., Vianna, M.R., Lara, D.R., 2015. The side-by-side exploratory test: a simple automated protocol for the evaluation of adult zebrafish behavior simultaneously with social interaction.Behavioural Pharmacology 26, 691-696.

Schoonheim, P.J., Arrenberg, A.B., Del Bene, F., Baier, H., 2010. Optogenetic localization and genetic perturbation of saccade-generating neurons in zebrafish. Journal of Neuroscience 30, 7111-7120.

Sebat, J., Lakshmi, B., Malhotra, D., Troge, J., Lese-Martin, C., Walsh, T., Yamrom, B., Yoon, S., Krasnitz, A., Kendall, J., et al., 2007. Strong association of de novo copy number mutations with autism. Science 316, 445-449.

Seguin, D., Gerlai, R., 2018. Fetal alcohol spectrum disorders: Zebrafish in the analysis of the milder and more prevalent form of the disease. Behavioural Brain Research 352, 125-132.

Shams, S., Rihel, J., Ortiz, J.G., Gerlai, R., 2018. The zebrafish as a promising tool for modeling human brain disorders: a review based upon an IBNS Symposium. Neuroscience and Biobehavioral Reviews 85, 176-190.

Shelton, D.S., Price, B.C., Ocasio, K.M., Martins, E.P., 2015. Density and group size influence shoal cohesion, but not coordination in zebrafish (*Danio rerio*). Journal of Comparative Psychology 129, 72-77.

Shinawi, M., Liu, P., Kang, S.H., Shen, J., Belmont, J.W., Scott, D.A., Probst, F.J., Craigen, W.J., Graham, B.H., Pursley, A., et al., 2010. Recurrent reciprocal 16p11.2 rearrangements associated with global developmental delay, behavioural problems, dysmorphism, epilepsy, and abnormal head size. Journal of Medical Genetics 47, 332-341.

Shinozuka, K., Watanabe, S., 2004. Effects of telencephalic ablation on shoaling behavior in goldfish. Physiology and Behavior 81, 141-148.

Simonoff, E., Pickles, A., Charman, T., Chandler, S., Loucas, T., Baird, G., 2008. Psychiatric disorders in children with autism spectrum disorders: prevalence, comorbidity, and associated factors in a population-derived sample. Journal of the American Academy of Child and Adolescent Psychiatry 47, 921-929.

Sourbron, J., Schneider, H., Kecskes, A., Liu, Y., Buening, E.M., Lagae, L., Smolders, I., de Witte, P., 2016. Serotonergic modulation as effective treatment for Dravet syndrome in a zebrafish mutant model. ACS Chemical Neuroscience 7, 588-598.

Stainier, D.Y.R., Raz, E., Lawson, N.D., Ekker, S.C., Burdine, R.D., Eisen, J.S., Ingham, P.W., Schulte-Merker, S.,Yelon, D., Weinstein, B.M., et al., 2017. Guidelines for morpholino use in zebrafish. PLoS Genetics 13, e1007000.

Stednitz, S.J., McDermott, E.M., Ncube, D., Tallafuss, A., Eisen, J.S., Washbourne, P., 2018. Forebrain control of behaviorally driven social orienting in zebrafish. Current Biology 28, 2445-2451 e2443.

Steffenburg, S., Gillberg, C., Hellgren, L., Andersson, L., Gillberg, I.C., Jakobsson, G., Bohman, M., 1989. A twin study of autism in Denmark, Finland, Iceland, Norway and Sweden. Journal of Child Psychology and Psychiatry 30,405-416.

Strauss, K.A., Puffenberger, E.G., Huentelman, M.J., Gottlieb, S., Dobrin, S.E., Parod, J.M., Stephan, D.A.,Morton, D.H., 2006. Recessive symptomatic focal epilepsy and mutant contactin-associated protein-like 2. New England Journal of Medicine 354, 1370-1377.

Sykes, D.J., Suriyampola, P.S., Martins, E.P., 2018. Recent experience impacts social behavior in a novel context by adult zebrafish (*Danio rerio*). PLoS One 13, e0204994.

Tang, W., Zhang, G., Serluca, F., Li, J., Xiong, X., Coble, M., Tsai, T., Li, Z., Molind, G., Zhu, P., et al., 2018. Genetic architecture of collective behaviors in zebrafish. bioRxiv 350314.

Teles, M.C., Oliveira, R.F., 2016. Quantifying aggressive behavior in zebrafish. Methods in Molecular Biology 1451,293-305.

Teles, M.C., Cardoso, S.D., Oliveira, R.F., 2016. Social plasticity relies on different neuroplasticity mechanisms across the brain social decision-making network in zebrafish. Frontiers in Behavioral Neuroscience 10, 16.

Thyme, S.B., Pieper, L.M., Li, E.H., Pandey, S., Wang, Y., Morris, N.S., Sha, C., Choi, J.W., Herrera, K.J., Soucy, E.R., et al.,

2019. Phenotypic landscape of schizophrenia-associated genes defines candidates and their shared functions. Cell 177, 478-491.e420.

Tucker, B., Richards, R.I., Lardelli, M., 2006. Contribution of mGluR and Fmr1 functional pathways to neurite morphogenesis, craniofacial development and fragile X syndrome. Human Molecular Genetics 15, 3446-3458.

Vendrell-Llopis, N., Yaksi, E., 2015. Evolutionary conserved brainstem circuits encode category, concentration and mixtures of taste. Scientific Reports 5, 17825.

Wang, S.H., Cheng, X.E., Qian, Z.M., Liu, Y., Chen, Y.Q., 2016. Automated planar tracking the waving bodies of multiple zebrafish swimming in shallow water. PLoS One 11, e0154714.

Wang, X., Cheng, E., Burnett, I.S., Huang, Y., Wlodkowic, D., 2017. Automatic multiple zebrafish larvae tracking in unconstrained microscopic video conditions. Scientific Reports 7, 17596.

Wang, G., Zhang, G., Li, Z., Fawcett, C.H., Coble, M., Sosa, M.X., Tsai, T., Malesky, K., Thibodeaux, S.J., Zhu, P., et al., 2019. Abnormal behavior of zebrafish mutant in dopamine transporter is rescued by clozapine. iScience 17,325-333.

Way, G.P., Ruhl, N., Snekser, J.L., Kiesel, A.L., McRobert, S.P., 2015. A comparison of methodologies to test aggression in zebrafish. Zebrafish 12, 144-151.

Weiss, L.A., Shen, Y., Korn, J.M., Arking, D.E., Miller, D.T., Fossdal, R., Saemundsen, E., Stefansson, H.,Ferreira, M.A., Green, T., et al., 2008. Association between microdeletion and microduplication at 16p11.2 and autism. New England Journal of Medicine 358, 667-675.

Willsey, A.J., Sanders, S.J., Li, M., Dong, S., Tebbenkamp, A.T., Muhle, R.A., Reilly, S.K., Lin, L., Fertuzinhos, S.,Miller, J.A., et al., 2013. Coexpression networks implicate human midfetal deep cortical projection neurons in the pathogenesis of autism. Cell 155, 997-1007.

Wu, Y.J., Hsu, M.T., Ng, M.C., Amstislavskaya, T.G., Tikhonova, M.A., Yang, Y.L., Lu, K.T., 2017. Fragile X mental retardation-1 knockout zebrafish shows precocious development in social behavior. Zebrafish 14, 438-443.

Zabegalov, K.N., Khatsko, S.L., Lakstygal, A.M., Demin, K.A., Cleal, M., Fontana, B.D., McBride, S.D., Harvey, B.H.,de Abreu, M.S., Parker, M.O., et al., 2019a. Abnormal repetitive behaviors in zebrafish and their relevance to human brain disorders. Behavioural Brain Research 367, 101-110.

Zabegalov, K.N., Kolesnikova, T.O., Khatsko, S.L., Volgin, A.D., Yakovlev, O.A., Amstislavskaya, T.G., Friend, A.J.,Bao, W., Alekseeva, P.A., Lakstygal, A.M., et al., 2019b. Understanding zebrafish aggressive behavior. Behavioural Processes 158, 200-210.

Zhou, P., Zhan, L., Ma, H., 2019. Understanding others' minds: social inference in preschool children with autism spectrum disorder. Journal of Autism and Developmental Disorders 49 (11), 4523-4534.

Zienkiewicz, A., Barton, D.A., Porfiri, M., di Bernardo, M., 2015. Data-driven stochastic modelling of zebrafish locomotion. Journal of Mathematical Biology 71, 1081-1105.

Zienkiewicz, A.K., Ladu, F., Barton, D.A.W., Porfiri, M., Bernardo, M.D., 2018. Data-driven modelling of social forces and collective behaviour in zebrafish. Journal of Theoretical Biology 443, 39-51.

Zimmermann, F.F., Gaspary, K.V., Leite, C.E., De Paula Cognato, G., Bonan, C.D., 2015. Embryological exposure to valproic acid induces social interaction deficits in zebrafish (*Danio rerio*): a developmental behavior analysis. Neurotoxicology and Teratology 52, 36-41.

Zimmermann, F.F., Gaspary, K.V., Siebel, A.M., Bonan, C.D., 2016. Oxytocin reversed MK-801-induced social interaction and aggression deficits in zebrafish. Behavioural Brain Research 311, 368-374.

第二十七章

斑马鱼攻击行为研究

William HJ. Norton[1], Svante Winberg[2]

27.1 攻击行为和对抗行为

攻击行为可以定义为：对同种动物产生有害或潜在有害刺激的一系列复杂行为（Huntingford et al., 1987）。这种行为有众多生存目的，如确保获得生存资源、保护后代、建立社会等级以及防御保护（Koolhaas et al., 1999）。发起主动性攻击（进攻性攻击）与防御类的攻击（防御性攻击）不同。“对抗行为”一词用以形容防御性行为，如防御性威胁、屈服和逃跑（King, 1973）。然而，在现代用法中，术语“对抗性”（agonistic）和“攻击性”（aggressive）往往被模糊定义为同义词，包括与进攻和防守相关的行为。攻击性行为在代谢上消耗高，战斗有可能导致受伤或死亡（Koolhaas et al., 1999）。这意味着这种行为需要受到严格控制。

27.2 斑马鱼攻击性研究

利用鱼类研究攻击性有着悠久而丰富的历史。鱼类是地球上最丰富的脊椎动物物种，它们表现出各种不同的社交行为，包括攻击性特征，这使它们成为研究比较行为的理想选择（Maruska et al., 2013）。斑马鱼（*Danio rerio*）（Hamilton, 1822）是一种原产于恒河和雅鲁藏布江流域的鲤鱼，高度群居，通常成群结队地生活，它们有各种栖息地，如季风期间沿着溪流形成的浅而缓慢流淌的溪流和池塘等（Spence et al., 2008）。这些栖息地的温度、水位、浊度和捕食压力受季节变化影响较大。斑马鱼也生长在稻田、排水沟和其他由

[1] Department of Neuroscience, Psychology and Behaviour, University of Leicester, Leicester, United Kingdom.
[2] Department of Neuroscience, Uppsala University, Uppsala, Sweden.

人类活动形成的浅水区。因此，斑马鱼在攻击性等行为上表现出较大的种内差异也就不足为奇了。这种差异既与表型可塑性相关（Ricci et al., 2013; Bhat et al., 2015），又与遗传相关（如 Filby et al., 2010; Norton and Bally-Cuif, 2010）。在本章中，我们将讨论斑马鱼是如何产生对抗行为的，以及遗传因素和包括既往社交经验在内的环境因素对这种行为的影响。

27.3 斑马鱼在配对争斗中的对抗行为

与其他硬骨鱼一样，在斑马鱼中，配对争斗的互动通常会从最初的相互展示（盘旋）阶段升级到相互公开的攻击行为阶段，在这一阶段，对抗双方会评估彼此的战斗能力，包括攻击、冲撞和追逐（表 27-1）（Chou et al., 2016）。当失败者持续逃跑，并且不对获胜对手的攻击进行报复时，这场战斗就结束了。在决胜后阶段，胜利者将继续攻击和追逐其对手，失败者只能表现出防御行为，经常试图逃跑或躲藏（表 27-1）。失败者通常占据靠近水面或靠近水缸底部的位置，保持不动（僵直），并采取顺从的姿势，鱼鳍缩回，尾体向下（尾巴下垂）。身体颜色也可能出现变化，变得苍白，带有不太明显的条纹，这种效应可能是由交感神经激活引起的皮肤黑素细胞收缩引起的。占优势的胜利者仍然高度活跃地巡视水箱（鱼缸），偶尔挑战或攻击顺从的对手。

表 27-1 相互攻击过程中斑马鱼的对抗行为

行为	描述
展示	鱼缓慢移动，身体侧面朝向对手，背鳍和尾鳍竖起
环行	双方用竖立的鳍从相反的方向相互接近，并以反平行的姿势彼此环绕
突袭	鱼在没有身体接触的情况下迅速游向对手
撕咬	鱼会撕咬它的对手，通常是其对手身体的后部和腹部
追逐	打击对手，但对手在被侵略者积极追击时逃跑
撤退	对攻击或撕咬作出反应而迅速远离侵略者的动作
逃跑	继续快速逃离侵略者，以应对追捕
僵直	鱼保持不动，所有鳍都缩回靠近底部或靠近鱼缸表面，尾鳍通常向下

27.4 对抗行为的量化方法

攻击性的量化可能很复杂，因为个体表现出的攻击性水平取决于对手的行为反应。攻击性也高度依赖环境，例如，鱼在配对斗争中的对抗行为可能会受到观战鱼群的影响，这种现象被称为观众效应（Abril-de-Abreu et al., 2015）。用于量化斑马鱼对抗行为的方法有很多，每种方法都有各自的优缺点，因此，在具体研究中使用的最佳方法取决于需要解决的问题。

Way 等（2015）比较了两种不同的镜像测试（平面镜和倾斜镜）、活体同物种刺激、黏土模型刺激和视频刺激，以量化野生斑马鱼的攻击行为。结果表明，在所有测试中，个体鱼在总攻击性方面的排名相似，表明攻击性是一种持久稳定的个性特征，可以使用各种分析来监测总攻击性的差异。然而，不同测试中个体攻击行为的频率不同。平面镜像或活体刺激引发了相似的攻击行为，撕咬和展示的次数较多，但突袭的次数很少。在黏土模型刺激和视频刺激试验中也表现出类似的行为特征，该特征与在平面镜和活体同物种刺激试验中观察到的特征明显不同（Way et al., 2015）。

重要的是，在上面讨论的测试中，受试鱼没有受到对手的任何身体接触或顺从行为。因此，尽管上述测试比配对争斗更容易标准化，但使用镜像、视频刺激或人工鱼模型进行攻击性测试的生物学相关性可能会受到质疑。斑马鱼和其他群居动物一样，必须根据其社会地位改变其行为特征，这意味着同一基因型可以表达多种不同的行为表型（Burmeister et al., 2005; Cardoso et al., 2015）。在实际的对抗性互动中，鱼的内在状态和战斗力不断受到自身行为输出和对手反应的影响（Oliveira et al., 2016）。不同的行为表型之间转换以响应社会地位变化，这种变化的介导机制包括基因表达的影响和神经递质释放的快速变化。在斑马鱼的大脑中，对抗获胜者的基因表达模式和单胺类神经递质信号的变化明显不同于其对抗自身镜像时的变化，尽管两者表现出同等程度的攻击性（Oliveira et al., 2011; Teles et al., 2013; Teles and Oliveira, 2016）。这意味着需要谨慎地解释这两类实验的结果。

27.5 支配地位的发展：社会压力与对抗行为

以小群体或低密度饲养的斑马鱼可以形成基于支配的社会等级。支配地位在一定程度上与体形有关，体形较大的鱼在争夺社交支配地位方面具有优势。然而，竞争能力也与勇敢等性格特征有关。Dahlbom 等人（Dahlbom et al., 2011）的研究表明，勇敢（由配对斗争前的行为筛选判定）常决定阶段性配对争斗的结果，勇敢的鱼比不太勇敢（害羞）的对手更有可能成为社交统治者。尚不清楚为什么与勇敢相关的行为在配对斗争中很重要。一种可能性是，勇敢者具有决定社交地位的能力，可能与勇敢和攻击性之间潜在的积极关系有关，即行为综合征（Norton et al., 2011b; Kiesel et al., 2012; Norton and Bally-Cuif, 2012; Ariyomo et al., 2013）。

相互对抗也可能对个体的行为和生理产生巨大影响。例如，赢得一场战斗会极大地增加在未来战斗中获胜的机会，而失败则会降低未来战斗中获胜的机会。因此，赢家会继续赢，输家会继续输，这种效应被称为赢者 - 输者效应（Oliveira et al., 2011, 2016）。

在配对争斗中争夺支配地位的初始阶段，两条鱼都受到强烈的压力。然而，随着支配关系的牢固建立，胜利者，社交上占支配地位的鱼，不再表现出压力，而它的从属对手承受长期的社交压力。社交从属关系会导致大脑 5- 羟色胺能系统的慢性激活，伴随行为的普遍抑制（Theodoridi et al., 2017）。因此，大脑 5- 羟色胺能活性升高似乎是介导从属鱼行为抑制的一个因素（Winberg and Thornqvist, 2016; Backstrom and Winberg, 2017; Theodoridi et al., 2017）。

27.6 攻击性的神经基础

脑成像和处理工具的出现使得斑马鱼成为研究攻击性神经基础的良好模式生物。控制斑马鱼攻击行为的神经回路尚未被完全确定。然而，作为一种社交行为，攻击行为很可能被社会决策网络（相互连接的皮质下的脑区群）的组成部分所改变（Goodson and Kabelik, 2009）。社交决策网络的节点包括下丘脑前部视前区、外侧隔、杏仁核内侧、下丘脑前部和腹内侧、中脑导水管周围灰质、终纹床核和被盖（Newman, 1999; Goodson and Kabelik, 2009; O'Connell and Hofmann, 2011）。行为是整个网络中动态活动模式的一种涌现特性，而不是单个节点上的活动（Newman, 1999）。神经递质、神经肽和性类固醇激素可以改变节点之间的连通性，使其能够调节社交互动（Goodson and Kabelik, 2009）。

社交决策网络的几个节点已经被证明与斑马鱼的攻击行为有关。例如视前区（Larson et al., 2006; Carreno Gutierrez et al., 2019）。精氨酸催产素（精氨酸加压素的鱼类同系物）是一种在视前区表达并能改变脊椎动物社交行为的肽类神经递质。占支配地位的野生型斑马鱼视前区的背侧大细胞精氨酸催产素阳性神经元比其下属神经元多，后者则表达更多的腹侧小细胞精氨酸催产素神经元（Larson et al., 2006）。与精氨酸催产素的基础水平进行比较发现，配对斗争失败者在端脑和间脑、视顶盖和脑干的精氨酸催产素水平升高。配对斗争胜利者的端脑和间脑精氨酸催产素水平升高，而镜像实验的鱼端脑精氨酸催产素水平升高（Teles et al., 2016）。这些研究表明，表征社交决策网络的各个节点可以提供有关其控制攻击能力的信息。然而，进一步的研究应分析多个节点的同时激活，以证明该网络对攻击行为的重要性。

斑马鱼的前脑结构不同于其他脊椎动物，因此比较不同物种的大脑区域具有挑战性。人类控制攻击行为的神经回路包括内侧杏仁核、终纹体、内侧下丘脑和背侧导水管周围灰质。它由内侧、眶内和下额叶皮质调节，这些皮质在注意力和反应抑制中起着重要作用（Blair, 2010）。在斑马鱼中，杏仁核同源物可能是背侧端脑的内侧区（Rodriguez et al., 2002; Portavella et al., 2004; Northcutt, 2006; Mueller et al., 2011; von Trotha et al., 2014）。腹侧端脑的腹侧区（ventral zone of the ventral telencephalon, Vv）可能与其他脊椎动物的外侧隔同源（Teles et al., 2015），并且 Vv 区与蝴蝶鱼的攻击性相关（Dewan and Tricas, 2011）。斑马鱼终纹床核尚未确定，但可能是构成隆起丘脑的一部分（Wullimann and Mueller, 2004; Jesuthasan, 2012），然而，跨物种下丘脑内侧区的详细比对尚未完成。

27.7 攻击性的遗传基础

表征调节攻击的基因是确定调节这种行为的大脑区域和信号通路的重要的第一步。斑马鱼攻击性的中度遗传力指数为 0.36，母体效应占这一遗传变异的 9%（Ariyomo et al., 2013）。这意味着环境影响可能在决定个体攻击性水平方面发挥重要作用。尽管斑马鱼与包括人类在内的其他脊椎动物物种具有高度的遗传同源性（Howe et al., 2013），但约450万年前发生的一次额外的全基因组复制事件使该物种的遗传研究变得复杂化（Postlethwait et al., 1998）。斑马鱼通常含有一种以上在其他脊椎动物中发现的同源基因，这意味着在某些情况下，功能冗余可以掩盖行为表型。目前已构建大量的斑马鱼突变系和可用于操纵该物种的遗传工具（Del Bene and Wyart, 2012; Prykhozhij et al., 2017）使斑马鱼成为发现行为改变相关新基因的良好模型。

可以影响斑马鱼行为的基因最初是通过随机诱变和目的表型筛选来确定的（Driever et al., 1996; Haffter et al., 1996; Golling et al., 2002）。虽然这种方法允许以无偏倚的方式鉴别新基因，但通过位置克隆定位突变可能很耗时。基因组工程的最新进展，包括锌指核酸酶、TALEN和CRISPR/Cas9技术，已经可以实现已知基因的突变（Simone et al., 2018）。这种可控的基因组操作，包括敲入序列，使携带人源疾病基因模型的建立成为可能。这表明，在未来，斑马鱼可能被用作个体医疗的转化工具，从而对患者特异性突变进行工程改造和研究。

目前，已经确定了斑马鱼中与攻击相关的几个基因，包括室管膜蛋白（ependymin）（Sneddon et al., 2011）、成纤维细胞生长因子受体1a（*fgfr1a*）（Norton et al., 2011a）、一氧化氮合酶1（Carreno Gutierrez et al., 2017）、内皮素受体aa（*ednraa*）基因（Carreno Gutierrez et al., 2019）。*fgfr1a*$^{-/-}$和*ednraa*$^{-/-}$突变系尤为引人关注。Fgfr1a活性的降低导致镜像诱导的攻击水平的强烈增加，使得突变体的攻击性从起始阶段就不会降低（Norton et al., 2011a）。这种表型是由大脑中组胺信号的整体减少引起的，突出了这种神经递质对攻击性的重要性。此外，在德国慕尼黑饲养的*fgfr1a*$^{-/-}$斑马鱼比在法国Gif-sur-Yvette饲养的*fgfr1a*$^{-/-}$斑马鱼更具攻击性，表明环境会对这种行为产生长期影响。有趣的是，最近的一项研究表明，尽管*fgfr1a*$^{-/-}$表现出稳健的镜像表型，但在配对争斗中并不具有攻击性。这些突变体表现出的高度攻击性取决于它们遇到的刺激（Mustafa et al., 2019）。*ednraa*$^{-/-}$突变的斑马鱼，攻击性增加的同时导致该品系的社交互动减少。事实上，*ednraa*$^{-/-}$似乎是组成型显性的，这种表型与视前下丘脑中的小细胞精氨酸血管收缩素阳性神经元的减少相关（Carreno Gutierrez et al., 2019）。

尽管这些研究已经证明，识别和表征攻击性相关基因是可能的，但迄今为止，报道的新候选基因相对较少。这种方法的一个局限是，在大多数情况下，攻击性的变化是偶然发现的，而不是特异筛选的。在最近的一篇论文中，Carreno Gutierrez及其同事描述了一种自动量化幼年（1个月大）斑马鱼攻击性的方法，能实现高通量的攻击性研究（Carreno Gutierrez et al., 2018）。因此，现在应该可以通过对诱变的动物家族进行筛选，以更系统的方式发现新基因。发现攻击性行为相关新基因的第二种方法是，从野生型鱼类种群开始，有选择地培育高攻击性和低攻击性系，这种方法已成功用于啮齿动物（例如Malki et al., 2016）。使用下一代测序技术就可以用来揭示影响攻击性的整套基因。

27.8 攻击性的药理学基础

除了作为一种重要的适应性行为外，攻击性也是一些人类疾病的常见组成部分，包括间歇性狂暴症、注意力缺陷 / 多动症、自闭症谱系障碍、药物滥用、双相障碍、品行障碍和精神分裂症等（Coccaro et al., 1998; Monuteaux et al., 2009; Nevels et al., 2010; Malki et al., 2016）。通常会采用行为、认知和药物治疗相结合的方法控制人类攻击性行为（Vitiello and Stoff, 1997）。然而，用于治疗攻击性的药物疗效有限，并且可能具有非预期的副作用（Wernicke et al., 2003; Polzer et al., 2007; Nevels et al., 2010; Scotto Rosato et al., 2012）。近期的研究表明，斑马鱼可以用于筛选作用于行为模式的活性药物（Rihel and Schier, 2012; MacRae and Peterson, 2015），提示这可能是寻找攻击性治疗新方法的途径。

斑马鱼作为行为药理学模型的一项优势是：可以通过将药物溶解在饲养水箱（鱼缸）的水中进行测试。虽然这消除了对潜在的疼痛性注射给药的需要，但几乎没有研究比较这两种用药方法及其对行为的影响。此外，大多数用于斑马鱼的药物最初是在其他物种（如人类）中开发的，这意味着缺乏证据表明它们能够结合斑马鱼的受体。行为实验通过结合配体或再摄取研究来改进，提供关于其对靶蛋白选择性的信息。

尽管存在这些限制，但近期的几项研究表明，斑马鱼的攻击性可以通过药物治疗来改变。应用选择性 5- 羟色胺（5-HT）再摄取抑制剂氟西汀可以降低斑马鱼的攻击性（Norton et al., 2011a），而注射 5-HT_{1A} 受体拮抗剂 WAY-100635 则效果相反（Filby et al., 2010）。将斑马鱼浸泡在乙醇中会增加其攻击性（Gerlai et al., 2000; Fontana et al., 2016; Carreno Gutierrez et al., 2018），乙醇和氨基磺酸牛磺酸的合用使其攻击性更强。有趣的是，在这些实验中，斑马鱼的运动没有改变，这意味着攻击性的诱因可能已经改变（Fontana et al., 2016）。用非竞争性 *N*- 甲基 -D- 天冬氨酸受体拮抗剂 MK-801 治疗成年斑马鱼证实了谷氨酸信号对攻击性的重要性。急性摄入 MK-801 可降低攻击性和社交行为，这种效果可被催产素或卡贝缩宫素所逆转（Zimmermann et al., 2016）。由此推测谷氨酸和催产素信号在攻击性控制中似乎存在联系。

27.9 结论

最近的几项研究表明，斑马鱼是研究基因、神经回路和攻击性调节药物的良好模型。目前，已经建立了可靠的量化攻击水平的方法，包括可以追踪鱼群的新软件，可实现对动态社会互动的研究（Laan et al., 2018）。现在需要进一步的研究来更好地理解环境影响和压力对攻击性的影响，以及影响这种行为的神经回路。

（陈学军翻译　王陈审校）

参考文献

Abril-De-Abreu, R., Cruz, J., Oliveira, R.F., 2015. Social eavesdropping in zebrafish: tuning of attention to social interactions. Scientific Reports 5, 12678.

Ariyomo, T.O., Carter, M., Watt, P.J., 2013. Heritability of boldness and aggressiveness in the zebrafish. Behavior Genetics 43, 161-167.

Backstrom, T., Winberg, S., 2017. Serotonin coordinates responses to social stress-what we can learn fromfish. Frontiers in Neuroscience 11, 595.

Bhat, A., Greulich, M.M., Martins, E.P., 2015. Behavioral plasticity in response to environmental manipulation among zebrafish

(*Danio rerio*) populations. PLoS One 10.

Blair, R.J., 2010. Neuroimaging of psychopathy and antisocial behavior: a targeted review. Current Psychiatry Reports 12, 76-82.

Burmeister, S.S., Jarvis, E.D., Fernald, R.D., 2005. Rapid behavioral and genomic responses to social opportunity. PLoS Biology 3, e363.

Cardoso, S.D., Teles, M.C., Oliveira, R.F., 2015. Neurogenomic mechanisms of social plasticity. Journal of Experimental Biology 218, 140-149.

Carreno Gutierrez, H., Colanesi, S., Cooper, B., Reichmann, F., Young, A.M.J., Kelsh, R.N., Norton, W.H.J., 2019. Endothelin neurotransmitter signalling controls zebrafish social behaviour. Scientific Reports 9, 3040.

Carreno Gutierrez, H., O'leary, A., Freudenberg, F., Fedele, G., Wilkinson, R., Markham, E., Van Eeden, F., Reif, A., Norton, W.H.J., 2017. Nitric oxide interacts with monoamine oxidase to modulate aggression and anxiety-likebehaviour. European Neuropsychopharmacology. https://doi.org/10.1016/j.euroneuro.2017.09.004.

Carreno Gutierrez, H., Vacca, I., Pons, A.I., Norton, W.H.J., 2018. Automatic quantification of juvenile zebrafish aggression. Journal of Neuroscience Methods 296, 23-31.

Chou, M.Y., Amo, R., Kinoshita, M., Cherng, B.W., Shimazaki, H., Agetsuma, M., Shiraki, T., Aoki, T., Takahoko, M., Yamazaki, M., Higashijima, S., Okamoto, H., 2016. Social conflict resolution regulated by two dorsal habenular subregions in zebrafish. Science 352, 87-90.

Coccaro, E.F., Kavoussi, R.J., Berman, M.E., Lish, J.D., 1998. Intermittent explosive disorder-revised: development, reliability, and validity of research criteria. Comprehensive Psychiatry 39, 368e376.

Dahlbom, S.J., Lagman, D., Lundstedt-Enkel, K., Sundstrom, L.F., Winberg, S., 2011. Boldness predicts social status in zebrafish (*Danio rerio*). PLoS One 6, e23565.

Del Bene, F., Wyart, C., 2012. Optogenetics: a new enlightenment age for zebrafish neurobiology. Developmental Neurobiology 72, 404-414.

Dewan, A.K., Tricas, T.C., 2011. Arginine vasotocin neuronal phenotypes and their relationship to aggressive behavior in the territorial monogamous multiband butterflyfish, *Chaetodon multicinctus*. Brain Research 1401, 74-84.

Driever, W., Solnica-Krezel, L., Schier, A.F., Neuhauss, S.C., Malicki, J., Stemple, D.L., Stainier, D.Y., Zwartkruis, F., Abdelilah, S., Rangini, Z., Belak, J., Boggs, C., 1996. A genetic screen for mutations affecting embryogenesis in zebrafish. Development 123, 37-46.

Filby, A.L., Paull, G.C., Hickmore, T.F., Tyler, C.R., 2010. Unravelling the neurophysiological basis of aggression in a fish model. BMC Genomics 11, 498.

Fontana, B.D., Meinerz, D.L., Rosa, L.V., Mezzomo, N.J., Silveira, A., Giuliani, G.S., Quadros, V.A., Filho, G.L., Blaser, R.E., Rosemberg, D.B., 2016. Modulatory action of taurine on ethanol-induced aggressive behavior in zebrafish. Pharmacology Biochemistry and Behavior 141, 18-27.

Gerlai, R., Lahav, M., Guo, S., Rosenthal, A., 2000. Drinks like afish: zebrafish (*Danio rerio*) as a behavior genetic model to study alcohol effects. Pharmacology Biochemistry and Behavior 67, 773-782.

Golling, G., Amsterdam, A., Sun, Z., Antonelli, M., Maldonado, E., Chen, W., Burgess, S., Haldi, M., Artzt, K., Farrington, S., Lin, S.Y., Nissen, R.M., Hopkins, N., 2002. Insertional mutagenesis in zebrafish rapidly identifies genes essential for early vertebrate development. Nature Genetics 31, 135-140.

Goodson, J.L., Kabelik, D., 2009. Dynamic limbic networks and social diversity in vertebrates: from neural context to neuromodulatory patterning. Frontiers in Neuroendocrinology 30, 429-441.

Haffter, P., Granato, M., Brand, M., Mullins, M.C., Hammerschmidt, M., Kane, D.A., Odenthal, J., Van Eeden, F.J., Jiang, Y.J., Heisenberg, C.P., Kelsh, R.N., Furutani-Seiki, M., Vogelsang, E., Beuchle, D., Schach, U., Fabian, C., Nusslein-Volhard, C., 1996. The identification of genes with unique and essential functions in the development of the zebrafish,Danio rerio. Development 123, 1-36.

Howe, K., Clark, M.D., Torroja, C.F., Torrance, J., Berthelot, C., Muffato, M., Collins, J.E., Humphray, S., Mclaren, K., Matthews, L., Mclaren, S., Sealy, I., Caccamo, M., Churcher, C., Scott, C., Barrett, J.C., Koch, R., Rauch, G.J., White, S., Chow, W., Kilian, B., Quintais, L.T., Guerra-Assuncao, J.A., Zhou, Y., Gu, Y., Yen, J., Vogel, J.H., Eyre, T., Redmond, S., Banerjee, R., Chi, J., Fu, B., Langley, E., Maguire, S.F., Laird, G.K., Lloyd, D., Kenyon, E., Donaldson, S., Sehra, H., Almeida-King, J., Loveland, J., Trevanion, S., Jones, M., Quail, M., Willey, D., Hunt, A., Burton, J., Sims, S., Mclay, K., Plumb, B., Davis, J., Clee, C., Oliver, K., Clark, R., Riddle, C., Elliot, D., Threadgold, G., Harden, G., Ware, D., Begum, S., Mortimore, B., Kerry, G., Heath, P., Phillimore, B., Tracey, A., Corby, N., Dunn, M., Johnson, C., Wood, J., Clark, S., Pelan, S., Griffiths, G., Smith, M., Glithero, R., Howden, P., Barker, N., Lloyd, C., Stevens, C., Harley, J., Holt, K., Panagiotidis,

G., Lovell, J., Beasley, H., Henderson, C., Gordon, D., Auger, K., Wright, D., Collins, J., Raisen, C., Dyer, L., Leung, K., Robertson, L., Ambridge, K., Leongamornlert, D., Mcguire, S., Gilderthorp, R., Griffiths, C., Manthravadi, D., Nichol, S., Barker, G., et al., 2013. The zebrafish reference genome sequence and its relationship to the human genome. Nature 496, 498-503.

Huntingford, F.A., Turner, A.K., Downie, L.M., 1987. Animal Conflict. Chapman and Hall Animal Behaviour Series.

Jesuthasan, S., 2012. Fear, anxiety, and control in the zebrafish. Developmental Neurobiology 72, 395-403.

Kiesel, A.L., Snekser, J.L., Ruhl, N., Mcrobert, S.P., 2012. Behavioural syndromes and shoaling: connections between aggression, boldness and social behaviour in three different Danios. Behaviour 149, 1155-1175.

King, J.A., 1973. The ecology of aggressive behavior. Annual Review of Ecology and Systematics 4, 117-138.

Koolhaas, J.M., Korte, S.M., De Boer, S.F., Van Der Vegt, B.J., Van Reenen, C.G., Hopster, H., De Jong, I.C., Ruis, M.A., Blokhuis, H.J., 1999. Coping styles in animals: current status in behavior and stress-physiology. Neuroscience and Biobehavioral Reviews 23, 925-935.

Laan, A., Iglesias-Julios, M., De Polavieja, G.G., 2018. Zebrafish aggression on the sub-second time scale: evidence for mutual motor coordination and multi-functional attack manoeuvres. Royal Society Open Science 5, 180679.

Larson, E.T., O'malley, D.M., Melloni Jr., R.H., 2006. Aggression and vasotocin are associated with dominant-subordinate relationships in zebrafish. Behavioural Brain Research 167, 94e102.

Macrae, C.A., Peterson, R.T., 2015. Zebrafish as tools for drug discovery. Nature Reviews Drug Discovery 14, 721-731.

Malki, K., Du Rietz, E., Crusio, W.E., Pain, O., Paya-Cano, J., Karadaghi, R.L., Sluyter, F., De Boer, S.F., Sandnabba, K., Schalkwyk, L.C., Asherson, P., Tosto, M.G., 2016. Transcriptome analysis of genes and gene networks involved in aggressive behavior in mouse and zebrafish. American Journal of Medical Genetics B Neuropsychiatric Genetics 171, 827-838.

Maruska, K.P., Zhang, A., Neboori, A., Fernald, R.D., 2013. Social opportunity causes rapid transcriptional changes in the social behaviour network of the brain in an African cichlidfish. Journal of Neuroendocrinology 25, 145-157.

Monuteaux, M.C., Biederman, J., Doyle, A.E., Mick, E., Faraone, S.V., 2009. Genetic risk for conduct disorder symptom subtypes in an ADHD sample: specificity to aggressive symptoms. Journal of the American Academy of Child and Adolescent Psychiatry 48, 757-764.

Mueller, T., Dong, Z., Berberoglu, M.A., Guo, S., 2011. The dorsal pallium in zebrafish, *Danio rerio* (Cyprinidae, Teleostei). Brain Research 1381, 95-105.

Mustafa, A., Thornqvist, P.O., Roman, E., Winberg, S., 2019. The aggressive spiegeldanio, carrying a mutation in the fgfr1agene, has no advantage in dyadicfights with zebrafish of the AB strain. Behavioural Brain Research 370, 111942.

Nevels, R.M., Dehon, E.E., Alexander, K., Gontkovsky, S.T., 2010. Psychopharmacology of aggression in children and adolescents with primary neuropsychiatric disorders: a review of current and potentially promising treatment options. Experimental and Clinical Psychopharmacology 18, 184-201.

Newman, S.W., 1999. The medial extended amygdala in male reproductive behavior. A node in the mammalian social behavior network. Annals of the New York Academy of Sciences 877, 242-257.

Northcutt, R.G., 2006. Connections of the lateral and medial divisions of the goldfish telencephalic pallium. Journal of Comparative Neurology 494, 903-943.

Norton, N., Li, D., Rieder, M.J., Siegfried, J.D., Rampersaud, E., Zuchner, S., Mangos, S., Gonzalez-Quintana, J., Wang, L., Mcgee, S., Reiser, J., Martin, E., Nickerson, D.A., Hershberger, R.E., 2011a. Genome-wide studies of copy number variation and exome sequencing identify rare variants in BAG3 as a cause of dilated cardiomyopathy. The American Journal of Human Genetics 88, 273-282.

Norton, W., Bally-Cuif, L., 2010. Adult zebrafish as a model organism for behavioural genetics. BMC Neuroscience 11, 90.

Norton, W.H., Stumpenhorst, K., Faus-Kessler, T., Folchert, A., Rohner, N., Harris, M.P., Callebert, J., Bally-Cuif, L., 2011b. Modulation of *Fgfr1a* signaling in zebrafish reveals a genetic basis for the aggression-boldness syndrome. Journal of Neuroscience 31, 13796-13807.

Norton, W.H.J., Bally-Cuif, L., 2012. Unravelling the proximate causes of the aggression-boldness behavioural syndrome in zebrafish. Behaviour 149, 1063-1079.

O'connell, L.A., Hofmann, H.A., 2011. The vertebrate mesolimbic reward system and social behavior network: acomparative synthesis. Journal of Comparative Neurology 519, 3599-3639.

Oliveira, R.F., Silva, J.F., Simoes, J.M., 2011. Fighting zebrafish: characterization of aggressive behavior and winner-loser effects. Zebrafish 8, 73-81.

Oliveira, R.F., Simoes, J.M., Teles, M.C., Oliveira, C.R., Becker, J.D., Lopes, J.S., 2016. Assessment offight outcome isneeded to activate socially driven transcriptional changes in the zebrafish brain. Proceedings of the National Academy of Sciences of the U S A 113, E654-E661.

Polzer, J., Bangs, M.E., Zhang, S., Dellva, M.A., Tauscher-Wisniewski, S., Acharya, N., Watson, S.B., Allen, A.J.,Wilens, T.E., 2007. Meta-analysis of aggression or hostility events in randomized, controlled clinical trials of atomoxetine for ADHD. Biological Psychiatry 61, 713-719.

Portavella, M., Torres, B., Salas, C., Papini, M.R., 2004. Lesions of the medial pallium, but not of the lateral pallium,disrupt spaced-trial avoidance learning in goldfish (*Carassius auratus*). Neuroscience Letters 362, 75-78.

Postlethwait, J.H., Yan, Y.L., Gates, M.A., Horne, S., Amores, A., Brownlie, A., Donovan, A., Egan, E.S., Force, A.,Gong, Z., Goutel, C., Fritz, A., Kelsh, R., Knapik, E., Liao, E., Paw, B., Ransom, D., Singer, A., Thomson, M.,Abduljabbar, T.S., Yelick, P., Beier, D., Joly, J.S., Larhammar, D., Rosa, F., Westerfield, M., Zon, L.I.,Johnson, S.L., Talbot, W.S., 1998. Vertebrate genome evolution and the zebrafish gene map. Nature Genetics18, 345-349.

Prykhozhij, S.V., Caceres, L., Berman, J.N., 2017. New developments in CRISPR/Cas-based functional genomics andtheir implications for research using zebrafish. Current Gene Therapy 17, 286-300.

Ricci, L.A., Morrison, T.R., Melloni Jr., R.H., 2013. Adolescent anabolic/androgenic steroids: aggression and anxietyduring exposure predict behavioral responding during withdrawal in Syrian hamsters (*Mesocricetus auratus*). Hormones and Behavior 64, 770-780.

Rihel, J., Schier, A.F., 2012. Behavioral screening for neuroactive drugs in zebrafish. Developmental Neurobiology 72,373-385.

Rodriguez, F., Lopez, J.C., Vargas, J.P., Broglio, C., Gomez, Y., Salas, C., 2002. Spatial memory and hippocampal pallium through vertebrate evolution: insights from reptiles and teleostfish. Brain Research Bulletin 57, 499-503.

Scotto Rosato, N., Correll, C.U., Pappadopulos, E., Chait, A., Crystal, S., Jensen, P.S., 2012. Treatment of maladaptive aggression in youth: CERT guidelines Ⅱ. Treatments and ongoing management. Pediatrics 129, e1577-1586.

Simone, B.W., Martinez-Galvez, G., Warejoncas, Z., Ekker, S.C., 2018. Fishing for understanding: unlocking the zebrafish gene editor's toolbox. Methods 150, 3-10.

Sneddon, L.U., Schmidt, R., Fang, Y., Cossins, A.R., 2011. Molecular correlates of social dominance: a novel role for ependymin in aggression. PLoS One 6, e18181.

Spence, R., Gerlach, G., Lawrence, C., Smith, C., 2008. The behaviour and ecology of the zebrafish, *Danio rerio*. Biological Reviews of the Cambridge Philosophical Society 83, 13-34.

Teles, M.C., Almeida, O., Lopes, J.S., Oliveira, R.F., 2015. Social interactions elicit rapid shifts in functional connectivity in the social decision-making network of zebrafish. Proceedings of Biological Sciences 282, 20151099.

Teles, M.C., Dahlbom, S.J., Winberg, S., Oliveira, R.F., 2013. Social modulation of brain monoamine levels in zebrafish. Behavioural Brain Research 253, 17-24.

Teles, M.C., Gozdowska, M., Kalamarz-Kubiak, H., Kulczykowska, E., Oliveira, R.F., 2016. Agonistic interactionselicit rapid changes in brain nonapeptide levels in zebrafish. Hormones and Behavior 84, 57e63.

Teles, M.C., Oliveira, R.F., 2016. Quantifying aggressive behavior in zebrafish. Methods in Molecular Biology 1451,293-305.

Theodoridi, A., Tsalafouta, A., Pavlidis, M., 2017. Acute exposure tofluoxetine alters aggressive behavior of zebrafishand expression of genes involved in serotonergic system regulation. Frontiers in Neuroscience 11, 223.

Vitiello, B., Stoff, D.M., 1997. Subtypes of aggression and their relevance to child psychiatry. Journal of the AmericanAcademy of Child and Adolescent Psychiatry 36, 307-315.

Von Trotha, J.W., Vernier, P., Bally-Cuif, L., 2014. Emotions and motivated behavior converge on an amygdala-like structure in the zebrafish. European Journal of Neuroscience 40, 3302-3315.

Way, G.P., Ruhl, N., Snekser, J.L., Kiesel, A.L., Mcrobert, S.P., 2015. A comparison of methodologies to test aggressionin zebrafish. Zebrafish 12, 144-151.

Wernicke, J.F., Faries, D., Girod, D., Brown, J., Gao, H., Kelsey, D., Quintana, H., Lipetz, R., Michelson, D.,Heiligenstein, J., 2003. Cardiovascular effects of atomoxetine in children, adolescents, and adults. Drug Safety26, 729-740.

Winberg, S., Thornqvist, P.O., 2016. Role of brain serotonin in modulating fish behavior. Current Zoology 62, 317-323.

Wullimann, M.F., Mueller, T., 2004. Identification and morphogenesis of the eminentia thalami in the zebrafish. Journal of Comparative Neurology 471, 37-48.

Zimmermann, F.F., Gaspary, K.V., Siebel, A.M., Bonan, C.D., 2016. Oxytocin reversed MK-801-induced social interaction and aggression deficits in zebrafish. Behavioural Brain Research 311, 368-374.

第二十八章

甲基汞诱导影响斑马鱼行为的跨代传递的表观遗传改变

Michael J. Carvan Ⅲ[1]

28.1 引言

美国国家环境保护局（The United States Environmental Protection Agency, US EPA）将神经毒性定义为："在神经化学、行为学、神经生理学或解剖学水平上检测到的中枢和 / 或周围神经系统结构和 / 或功能的不良变化（Tilson et al., 1995）"。神经行为学是研究行为及其与神经系统功能之间关系的学科。神经行为作为一种非常有效的实验终点反映了由机体所产生的各种复杂的内部（如化学、分子、细胞、生理）刺激和外部（如环境）刺激的整合效果。生物体的神经行为反应可能会受到环境的危害，例如暴露于神经毒剂。从神经行为毒性研究中获得的知识有助于理解基因（或基因组）与环境、基因与行为、基因与神经通路、表观遗传学（或表观基因组学）和行为等多个层面的相互作用。

神经行为学筛选分析为确定遗传操作对斑马鱼和其他生物神经系统功能的影响提供了良好的平台。其中一些分析方法由于其可信度高，而且有可能以自动化和高通量（或更高）的方式用于筛选大量潜在的具有神经毒性的化学物质，从而使得这些分析方法在毒理学和药理学研究中非常有用（Reif et al., 2016）。神经行为分析方法通常作为最灵敏的终点检测法，可鉴定出采用解剖、生化或组织病理学筛查方法未检测到的效应（Detrich et al., 2009）。据估计，大约 30% 的商业化的化学品（目前＞80000 种化学品）具有神经毒性（Basu, 2015），然而，对这些化学品开展广泛研究并将其鉴定为神经毒性物质的却很少（Bal-Price et al., 2015）。

[1] University of Wisconsin-Milwaukee, School of Freshwater Sciences, Milwaukee, WI, United States.

本章将重点讨论甲基汞对斑马鱼神经行为的影响。这项工作起源于与美国威斯康星州（Wisconsin）拉博湖（Lac du Flambau）五大湖部落间委员会（Great Lakes Intertribal Council）的美洲原住民健康项目研究中心（Native American Research Centers for Health Program）的合作，并继续激励着我们现在所从事的工作。这项合作以斑马鱼作为实验模型，探索母鱼食用传统鱼类和其他食物资源（如富硒野生稻）对其后代鱼神经系统的潜在影响（Liu et al., 2016; Weber et al., 2008; Smith et al., 2010）。我们也参与了另一个合作项目，以了解甲基汞对野生鱼类种群的影响，这项研究以斑马鱼和其他鱼类作为实验模型，以模拟与重要种群参数有关的个体行为的影响，并将这些种群水平的影响与有害结局路径的分子水平的变化关联起来（Mora-Zamorano et al., 2016, 2018）。

28.2 甲基汞与行为学

大量文献描述了甲基汞直接暴露可引起人类（Yorifuji et al., 2015; Peplow and Augustine, 2014）和实验动物（Bisen-Hersh et al., 2014; Onishchenko et al., 2008; Weber et al., 2012）的神经毒性。汞是一种普遍存在于环境中的毒物，已经发现在所有鱼类和食鱼物种的组织中都有残留，主要以甲基汞的形式存在。人类主要因为食用这些受到污染的鱼类从而接触到甲基汞（Mahaffey, 2004）。众所周知，由于甲基汞能够透过血脑屏障，因此产前暴露高剂量的甲基汞可引起广泛的脑损伤和神经发育受损，并产生多种缺陷，例如，严重脑瘫和认知缺陷以及运动和感觉功能受损。而低剂量甲基汞暴露所带来的影响更加微妙，损害的范围包括：运动功能损伤、感觉缺失、轻微的学习和记忆损伤（Karagas et al., 2012; Grandjean et al., 2015）等。甲基汞对健康的影响众所周知，然而，除了对甲基汞可与细胞硫醇反应以及甲基汞具有促氧化应激作用具有一定的了解外，其发挥毒性作用的具体生化和分子机制尚未明确（Farina and Aschner, 2017; Unoki et al., 2018）。

甲基汞容易与体内大分子上的巯基反应，可能干扰所有细胞或亚细胞结构功能。人们早就知道，汞可以直接与DNA结合，并干扰DNA转录和蛋白质合成（Gruenwedel and Lu, 1970），包括内质网破坏和核糖体消失导致大脑发育过程中的蛋白质合成受到干扰（Herman et al., 1973）。强有力的证据表明，中枢神经系统和其他器官中的许多亚细胞成分会被其破坏，特别是线粒体（Aschner and Aschner, 2007；Huang et al., 2016）；它对血红素的合成（Fowler and Woods, 1977）、细胞膜的完整性（Nordberg et al., 2007）、自由基的产生（Crespo-López et al., 2009; Yee and Choi, 1996; Olivieri et al., 2000）均有影响，并干扰神经递质的传递、刺激神经性兴奋毒素的产生（Nordberg et al., 2007），进而引起大脑和外周神经系统多部位的结构和功能损伤（Nordberg et al., 2007）。

汞化合物已被证明通过干扰受体结合位点而破坏谷氨酸、GABA和甘氨酸的调节（Colón-Rodríguez et al., 2017; Fitsanakis and Aschner, 2005; Atchison, 2005）。发育过程中暴露于甲基汞还会诱导扣带皮层、丘脑和下丘脑的神经元变性（Kakita et al., 2000）。研究证明甲基汞影响多巴胺的释放水平，抑制多巴胺转运蛋白功能，并导致多巴胺能神经元的损失（Huang et al., 2016; Dreiem et al., 2009）。Newland 等人最近的一篇综述指出，多巴胺的传递对甲基汞特别敏感，这可能是造成大鼠选择性障碍、优势反应抑制能力差、辨别力倒退的原因。这些观察结果是相互关联的，但它们在甲基汞行为效应中的确切作用还有待阐明。

28.3　发育期甲基汞暴露对斑马鱼行为的影响

我们的实验室采用了持续的实验方法来研究半衰期较长的持久性环境化学品（如甲基汞、四氯二苯并对二噁英）的直接影响。在斑马鱼产卵后 2h 内收获胚胎，化合物持续处理到受精后 4 ～ 24h。处理后，冲洗胚胎 3 次，并在清水中培育。使用长半衰期的持久性化学品时，暴露不会随着水源性接触的结束而结束。这些化学物质最初分布到卵黄，然后随着储存在卵黄中的能量被消耗，而转移至胚胎。例如，根据激光取样等离子质谱（Laser Ablation Inductively Coupled Plasma Mass Spectrometry, LS-ICP-MS）（Carvan 实验室，未发表）测定，经甲基汞处理的 6 日龄斑马鱼胚胎在眼、脑、肌肉和肝中均含有高浓度的汞。处理浓度从 EC10（只有 10% 的胚胎受到明显影响）起跨越至少 3 个数量级，低剂量达到我们预期的与对照组无差异的剂量水平。在特定年龄评估斑马鱼，以检查相关行为（如自发运动、进食、学习和记忆、视觉惊吓反应）。

前期有研究表明，发育期汞暴露（如 $HgCl_2$）会导致对振动刺激的惊吓反应出现缺陷（Weber, 2006）。我们在现有的 Li 和 Dowling 方法的基础上，利用成年鱼开发了一个类似的惊吓反射反应评估方法，其中斑马鱼对旋转黑条刺激做出惊吓逃逸反应（Li and Dowling, 1997）。在发育过程中于受精后 4 ～ 24h 暴露于甲基汞的成年斑马鱼（4 ～ 6 月龄）在低至 10nmol/L（试验的最低浓度）的浓度下表现出显著的（$P < 0.05$）视觉惊吓缺陷（Weber et al., 2008）。对此结果有多种解释，包括失明、习惯化（联想学习）和运动缺陷。每条鱼对黑条都有反应，表明这些鱼没有失明，组织学分析显示视网膜没有结构问题（Weber et al., 2008）。另一项使用同一鱼类种群的研究表明，在低至 10nmol/L（测试的最低浓度）的处理下，发育中暴露于甲基汞会严重影响空间交替任务中的学习（Smith et al., 2010）。然后，我们在与视觉惊吓反应试验所用相同的 10cm 玻璃皿中评估了鱼的自发运动。与对照组相比，经暴露的鱼在所有浓度下（Carvan 实验室，未发表数据）均表现为活动亢进（$P < 0.05$）。

上述实验中的胚胎直接暴露于水中的甲基汞。我们还探讨了母体转移甲基汞的影响。给成年雌性斑马鱼喂食含甲基汞的食物，直到胚胎的含汞量与直接处理的胚胎的含汞量一致。行为分析发现，在自发运动、视觉运动反应（另一种普遍的斑马鱼行为分析）和觅食效率方面与对照组存在显著性差异，这表明，甲基汞对母体转移胚胎的影响与胚胎直接暴露在水中相似（Mora-Zamorano et al., 2016）。

28.4 功能障碍的跨代遗传

之后，我们探讨了直接暴露于甲基汞造成的发育缺陷可能会遗传给子代（代际）及其子代的后代（跨代）的可能性。跨代遗传是指在没有导致表型变异的直接环境影响的情况下，由种系介导的代间遗传（图 28-1 中给出了人类和斑马鱼的平行对比示例）。当斑马鱼卵直接（F0）暴露于具有长半衰期的持久性化学物质时，没有直接接触化学物质的第一代为 F2 代，而 F1 被认为是经过暴露的，因为在将来成为 F1 的生殖细胞的发育过程中存在持久性化学物质暴露。

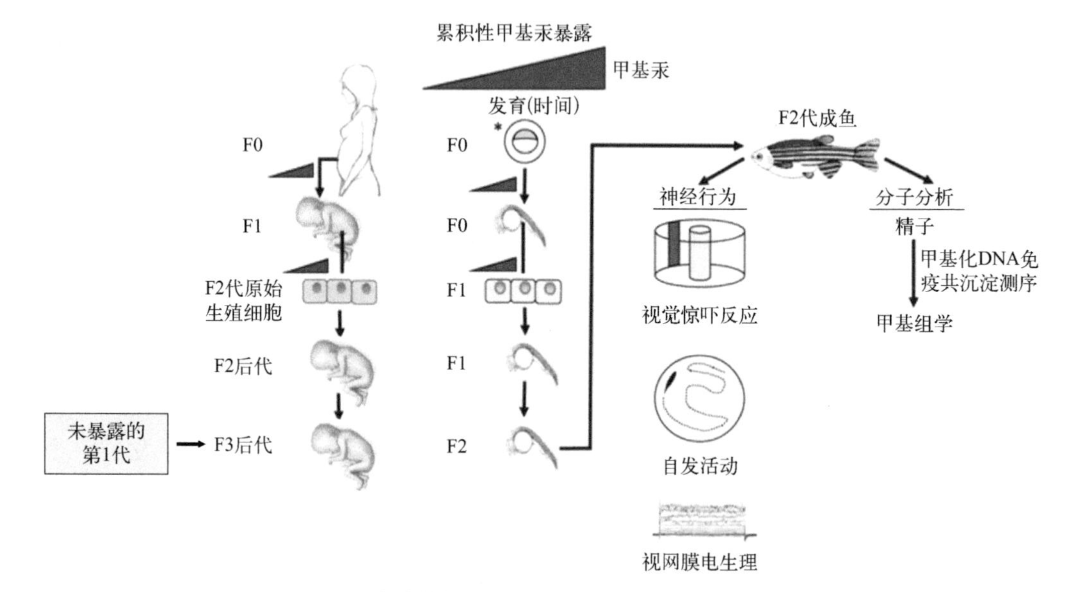

图 28-1 与人体暴露场景平行对比的实验设计示意图

繁殖 / 暴露方案和分析总结。* 在我们的范例中，F0 代斑马鱼胚胎体外发育暴露于甲基汞，用来模拟人类母体暴露。

我们在阅读文献时必须注意，许多研究声称具有“跨代”表型（F0 ～ F2），但他们的实验设计只揭示了“代际”遗传（F0 ～ F1）。已经发表的几项研究证明了环境诱导表型在人类和实验动物中的跨代遗传（表 28-1）。最常见的哺乳动物例子涉及影响衰老和寿命的代谢和心血管疾病，包括人类、啮齿动物、豚鼠和绵羊（Aiken et al., 2016），在长寿和新陈代谢研究方面，大量的工作集中于线虫（Greer et al., 2011; Rechavi et al., 2014; Rechavi and Lev, 2017）。斑马鱼是一种引人注目且稳定的系统，其高效的饲养实践可以实现种群的生物学研究和实验性复制，可用于探索环境诱发（或影响）疾病或功能障碍的跨代遗传。作为一种脊椎动物，斑马鱼与人类以及超过 80% 的致病基因有许多共同的调节和发育途径（Varga et al., 2018）。

在我们关于甲基汞的研究中，已经证明了视觉惊吓反应缺陷和活动亢进具有跨代遗传神经行为表型（Carvan et al., 2017）。图 28-2 中（A）和（B）的数据显示祖代在发育过程中接受甲基汞处理的鱼类谱系相对于未接受处理的谱系的反应性。反应性数据以 0nmol/L

表 28-1 环境诱导表型的跨代遗传示例（仅包括符合上述标准的“跨代”研究）

物种	跨代表型	引文
小鼠	F0 代母体肥胖引起的肥胖，通过补充甲基供体来预防	Waterland 等（2008）
小鼠	F0 代母体高脂高糖饮食致心肌线粒体功能障碍	Ferey 等（2019）
兔子	宫内染毒长春花碱和甲氧氯胺对雄性大鼠（F1）精子发生的多重影响	Anway 等（2005）
人	祖父辈饥荒导致的新生儿肥胖增加及晚年健康不佳	Painter 等（2008）
人	祖父辈在连续几年间食物摄入量的急剧变化引起心血管死亡率增加	Bygren 等（2001）
斑马鱼	F0 代发育时暴露于二噁英引起性别比、骨骼发育和男性生殖的改变	Baker 等（2014）
秀丽隐杆线虫	F0 病原暴露诱导的病原回避	Moore 等（2019）

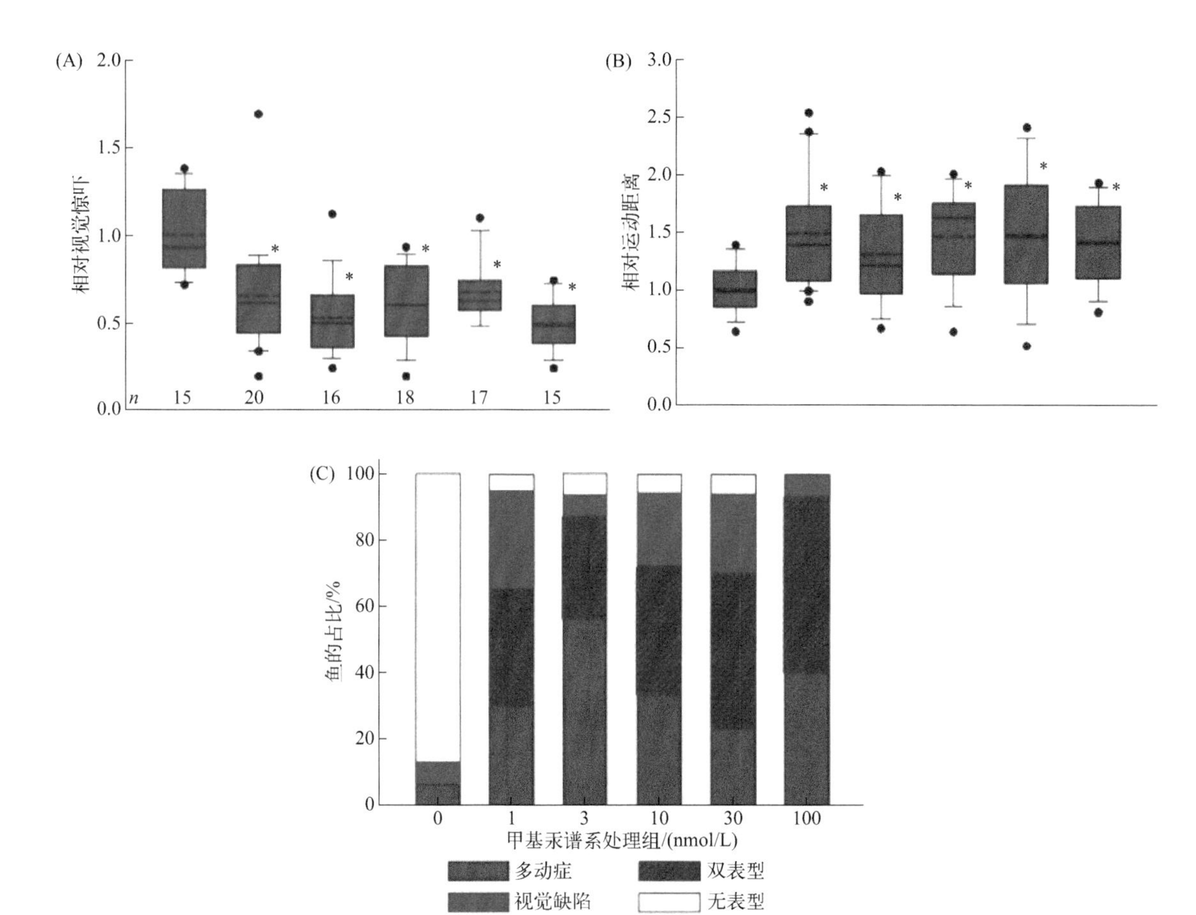

图 28-2 **跨代表型**

F2 代斑马鱼中观察到由于祖先暴露于甲基汞而导致的（A）视觉缺陷和（B）多动症（$^*P<0.001$）。（A）通过方差分析（ANOVA）和通过 Dunn 检验进行后处理分析显示，与对照组相比，甲基汞谱系的视觉惊吓反应具有统计学差异（df=5，H=34.596，P<0.001）。（B）通过 One-way ANOVA 和 Holm-Sidak 后处理分析显示，与对照组相比，所有甲基汞谱系中鱼的自发活动增加，并具有统计学意义（df=102，H=3.498，P=0.006）。n 为成年斑马鱼数。与 0nmol/L 组相比进行了相对反应标准化校正。甲基汞谱系（nmol/L）处理组指对 F0 胚胎进行的甲基汞处理。实水平线表示中位值，虚水平线表示平均值，方框表示第 25% 和第 75%，虚线表示第 5% 和第 95%，异常值用点表示。（C）表现出跨代遗传表型的 F2 代谱系的占比。CC-BY https://doi.org/10.1371/journal.pone.0176155.g002

种群（对照组）进行标准化校正。通过将所有个体行为响应值除以对照组种群的平均值来计算相对响应。在两项测定中，经甲基汞处理过的种群的响应与未处理的种群存在统计学差异。我们假设这两种表型都可能是单个基因位点发生变化的结果，因此我们重新分析了在两种测定中评估为“正常”或“异常”的每个种群个体的数据。未处理组中小于 5% 或大于 95% 的值判定为异常表型。基于孟德尔遗传学的估算，我们使用卡方分析来确定表型是否独立遗传。所观察到的表现出两种表型的鱼类比例与基于每个个体神经行为异常比例的预期值没有显著差异（Carvan et al., 2017），可见这些表型似乎是独立遗传的。我们预测，随着分析的表型增多，我们将揭示更多的独立基因位点。

我们得到的数据显示，在斑马鱼中观察到的视觉系统缺陷从发育暴露的 F0 代传递到从未暴露于甲基汞的 F2 代，这表明存在跨代遗传模式。一些研究表明，环境毒物可以诱导跨代表型（Nilsson et al., 2018; Anway and Skinner, 2006; Manikkam et al., 2012）。这些跨代表型极不可能仅仅是毒物诱导的突变从一代传递到下一代，因为暴露会导致极其一致的表型特征（Manikkam et al., 2012; Skinner, 2007）。此外，在群体研究中观察到每种特定表型出现的频率往往大大超过暴露于诱变剂时的预期频率，然而，其中任何一种表型的发生率往往非常低（Manikkam et al., 2012; Skinner et al., 2010）。基于这些观察结果和这些化学物质很少具有诱变作用的事实，有人认为跨代表型是由基因表达的改变产生的，而不改变 DNA 的基本核苷酸序列（即表观遗传机制）（Manikkam et al., 2012；Skinner et al., 2010; Berger et al., 2009）。

28.5 表观遗传学

已经证明，甲基汞暴露可影响个别基因和整个表观基因组的几种表观遗传途径（microRNA 表达，组蛋白修饰和 DNA 甲基化）（Culbreth and Aschner, 2019）。通过脐带血中的甲基汞水平和持续的认知表现可以确定，表观基因组的改变（即表观突变）与产前甲基汞暴露有关（Cardenas et al., 2015, 2017）。必须记住，这些表观遗传修饰的功能后果并不完全明显（Culbreth and Aschner, 2019）。然而，这些例子也表明甲基汞可能影响表观基因组并影响环境相关表型和疾病易感性。

表观遗传改变有几种机制，包括 DNA 甲基化和羟甲基化、组蛋白修饰、染色质重塑（例如转座子）和非编码 RNA（ncRNA）（图 28-3）。每个表观遗传标记都需要一组独特的蛋白质来协调读取、写入和删除，以保持对细胞功能的精确调控。这涉及脊椎动物基因组中的 100 多个基因。目前，至少已鉴定出 8 种不同类型的组蛋白修饰。大量研究描述了组蛋白乙酰化的作用，这通常会激活基因转录，而甲基化、磷酸化和泛素化通常是组蛋白和残基依赖性的（Heintzman et al., 2009）。组蛋白葡糖酰化、瓜氨酸化、巴豆酰化和异构化尚未有很好的解释。ncRNA 通常按其大小和功能分类：具有 mRNA 靶向的 miRNA（21 ～ 23bp）（O’Brien et al., 2018），通过序列互补靶向特定基因的 siRNA（20 ～ 25bp）（Dana et al., 2017），调节染色质和介导转座子沉默的 piRNA（27 ～ 30bp）（Czech et al., 2018），控制 X 染色体甲基化和失活的 XiRNA（24 ～ 42bp）（Kanduri et al., 2009），以及

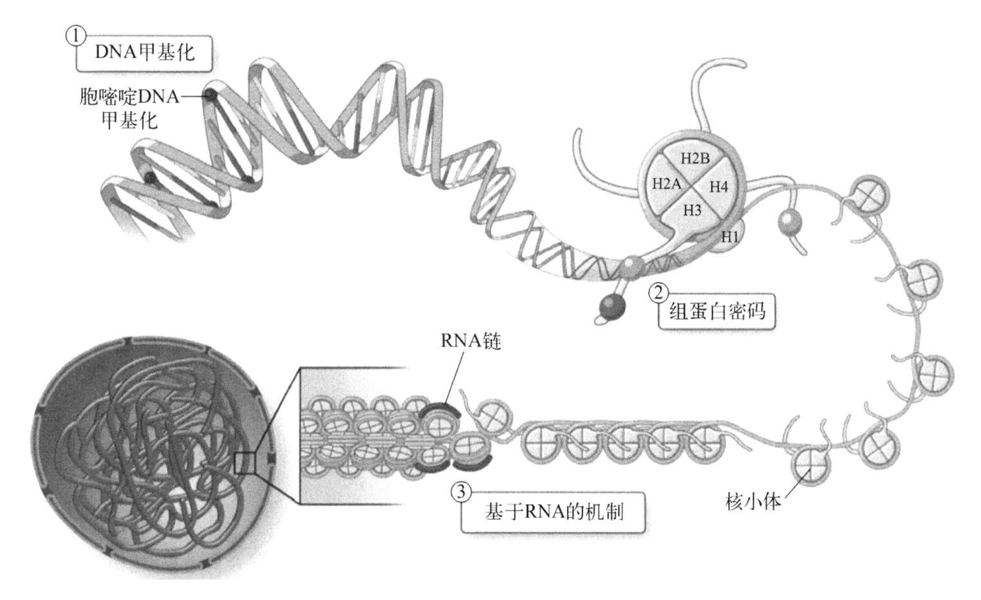

图 28-3 **表观遗传基因调控的三种基本机制**

基因表达的表观遗传机制由三种不同但高度相关的机制进行。① DNA 甲基化是指在 CpG 二核苷酸中胞嘧啶的 5 位增加一个甲基，称为“DNA 的第五个碱基”。②染色质的基本重复单位是核小体，核小体由核心组蛋白的八聚体组成，组蛋白氨基末端尾部（浅蓝色和深蓝色球）的翻译后修饰以及这些蛋白质在单位长度 DNA 上的密度可以对染色质结构产生重要影响，并构成假定的“组蛋白密码”。③最近证明基于 RNA 的机制也会影响染色质的高级结构。引自：Yan, M.S., Matouk, C.C., Marsden, P.A., 2010. Journal of Applied Physiology 109, 916-926（经允许后重新构图）

包括具有靶向特定基因等多种功能的 lncRNA（>200bp）（Sookoian et al., 2017）。

迄今为止，大多数探索环境毒物诱导的脊椎动物跨代表型的研究已经确定了遗传的分子机制是 DNA 甲基化的变化（Nilsson et al., 2012, 2018; Guerrero-Bosagna et al., 2012; Bernal and Jirtle, 2010; Ben Maamar et al., 2018）。DNA 甲基化是指从通用甲基供体 *S*- 腺苷甲硫氨酸（*S*-adenosylmethionine, SAM）中移除甲基，并将其添加到 DNA 链中的胞嘧啶残基上，生成 5- 甲基胞嘧啶的过程，该反应由 DNA 甲基转移酶催化（图 28-4）（Hales et al., 2011）。DNA 甲基化通常发生在整个基因组的特定 CpG 二核苷酸处，并与调控基因表达密切相关。DNA 甲基化模式主要建立在胚胎发生期间，这使发育特别容易受到干扰甲基化途径的环境的影响（Burdge et al., 2009）。当 DNA 甲基化在发育过程中发生变化时，发育中细胞和组织的基因表达谱发生永久改变，从而改变个体的基本生理或代谢（Hales et al., 2011）。

哺乳动物和斑马鱼在胚胎发生早期经历基因组去甲基化，随后使用保守机制并以类似模式进行从头甲基化（Feng et al., 2010; Mhanni and McGowan, 2004; MacKay et al., 2007; Fang et al., 2013），一个主要的区别是，在斑马鱼受精和基因组去甲基化之后，再甲基化是基于父系甲基化模式（Jiang et al., 2013; Potok et al., 2013）。基因组的连续去甲基化和再甲基化发生在早期胚胎发生和原始生殖细胞的产生过程中（Potok et al., 2013; Ortega-recalde and Hore, 2019）。DNA 甲基化模式重建的机制尚未揭示，这使我们对环境诱导的表型跨代遗传中 DNA 甲基化标记遗传的理解变得复杂。这也使人们怀疑 DNA 甲基化在跨代遗传中的作用，尤其是在哺乳动物中（Ortega-recalde and Hore, 2019; Xin et al., 2015; Horsthemke, 2018）。

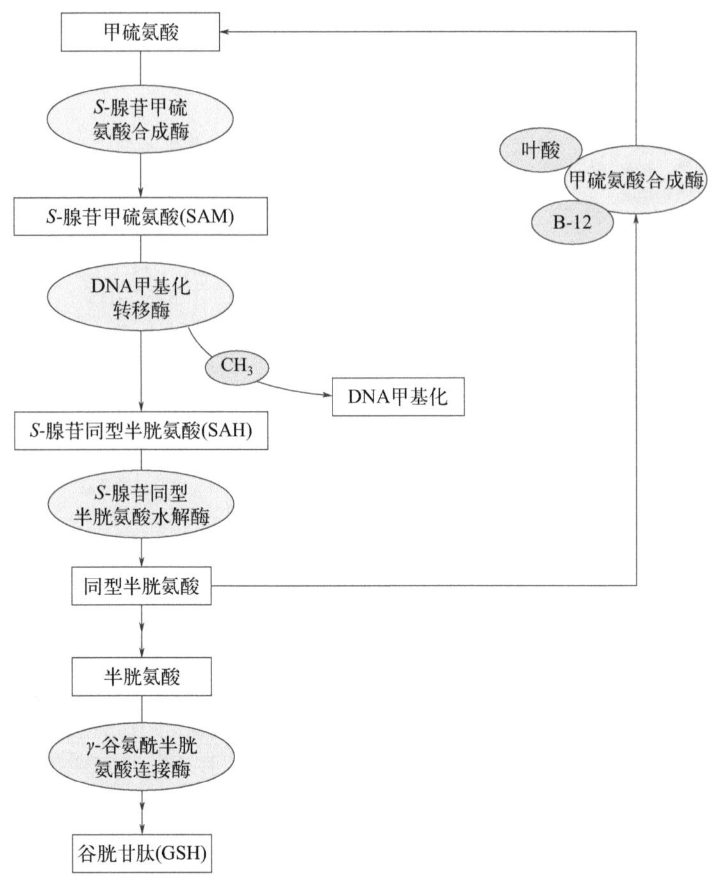

图 28-4 **DNA 甲基化通路与谷胱甘肽合成**

在正常情况下，同型半胱氨酸参与甲硫氨酸和 GSH 合成。在消耗 GSH 的反应性化学物质（如甲基汞）存在的情况下，由于需要合成 GSH 来对抗氧化应激，甲硫氨酸途径减弱。因此，DNA 甲基化模式可能会受到干扰，导致 DNA 甲基化水平低

来自人、动物和体外研究的多项证据表明，暴露于甲基汞能够在整体以及单个基因启动子水平上改变 DNA 甲基化（Carvan et al., 2017; Bose et al., 2012; Hanna et al., 2012; Onishchenko et al., 2007; Arai et al., 2011; Vandegehuchte and Janssen, 2014; Desaulniers et al., 2009; Basu et al., 2013; Ceccatelli et al., 2013）。例如，对北极熊脑组织的分析表明，随着汞浓度的增加，雄性熊的基因组 DNA 甲基化减少；而对发育中暴露于甲基汞的小鼠幼崽的研究表明，与对照动物相比，BDNF 基因启动子的甲基化模式发生了显著变化（Onishchenko et al., 2008; Richard Pilsner et al., 2010）。我们未发表的数据也表明，与对照组相比，直接暴露于 100nmol/L 甲基汞的斑马鱼胚胎（F0）在受精后 24h 整体 DNA 甲基化减少 30%。

28.6 *S*-腺苷甲硫氨酸、甲基汞和 DNA 甲基化

甲基汞以谷胱甘肽结合物的形式从体内排出。谷胱甘肽（GSH）是一种在体内发现的重要抗氧化剂，对甲基汞的代谢至关重要，因为甲基汞的汞原子可直接与谷胱甘肽的

巯基结合（Farina et al., 2011）。然后，这种复合物 GS-$HgCH_3$ 可被转运出细胞并排泄到胆汁中（Dutczak and Ballatori, 1994）。有人认为，暴露于包括甲基汞在内的毒物会耗尽细胞 GSH 水平，造成活性氧在细胞内积聚并导致氧化损伤（Farina et al., 2011; Lee et al., 2009; Thompson et al., 2000; Kaur et al., 2006， 2011. Deng et al., 2012）。暴露于甲基汞引起谷胱甘肽耗竭，导致招募同型半胱氨酸到谷胱甘肽合成途径中。由于同型半胱氨酸通常对甲硫氨酸和谷胱甘肽合成途径都有贡献，细胞对谷胱甘肽需求的增加会降低同型半胱氨酸对甲硫氨酸合成的利用率，反过来会导致细胞储存的 *S*- 腺苷甲硫氨酸耗竭，并可能减少 DNA 甲基化（Caudill et al., 2001; Tsuchiya et al., 2012）。这一假设性机制（图 28-4）常被引用来解释饮食与表观遗传变化之间的关系（Lee et al., 2009）。发育期甲基汞暴露导致成人视觉系统缺陷，这似乎是跨代遗传的，支持甲基化循环的观点。

28.7 甲基汞诱导的跨代表观遗传

为了确定基因组中对甲基汞所致行为障碍的跨代遗传有潜在影响的区域，我们在精子 DNA 中定位了对照组和 30nmol/L 甲基汞暴露的 F0 代和 F2 代谱系之间的差异 DNA 甲基化区域（DNA methylation regions, DMR）。DNA 片段化后，用甲基胞嘧啶抗体对甲基化的 DNA 进行免疫沉淀（MeDIP）（Carvan et al., 2017; Guerrero-Bosagna et al., 2010）。然后用第二代测序技术对免疫沉淀 DNA 进行分析，以确定斑马鱼基因组中的 DMR。将对照组和甲基汞组的谱系进行比较，发现了数千个不同的 DMR。为了降低数据的复杂性和保证可信度，我们选择将重点放在同一 DMR 中的多个位点（≥ 2100bp 窗口），并将 $p<10^{-7}$ 的显著性用于后续使用和数据展示。更严格鉴定的 DMR（多位点 $p < 10^{-7}$）提供了一个假定存在于所有动物中的高度可重复性的集合（即签名）（Carvan et al., 2017）。

斑马鱼基因组上精子多位点 DMR 的染色体位置如图 28-5 所示，图 28-5（A）为 F0 代 DMR，图 28-5（B）为 F2 代 DMR。F0 代 DMR 数量较少，大多数染色体有 DMR，但有 5 条染色体没有 DMR [图 28-5（A）]。在 4 号染色体的右臂上观察到一种奇怪的 DMR 过表达现象。这可能是由如前所述的斑马鱼染色体结构性质所致，与其基因组的其他部分相比，其染色体结构富含重复元件，GC 密度高，基因贫乏，延迟复制，且异染色质化（Wilson et al., 2014）。因此，由于对表观遗传程序的易感性，直接接触甲基汞改变了该区域的 DNA 甲基化。F2 代精子的所有染色体都出现 DRM，数量高于 F0 代精子的 DMR [图 28-5（B）]。前期研究已经证明，DMR 可以聚集在基因组上，以在统计上过度代表 DMR 群体（Skinner et al., 2012）。研究表明，DMR 通常具有低密度的 CpG 含量，存在于“CpG 沙漠”中（Skinner and Guerrero-Bosagna, 2014）。对本研究中鉴定的 DMR 的 CpG 密度分析表明，由甲基汞诱导的祖代 DMR 的 CpG 密度也较低。F0 和 F2 代上的 DMR 密度均为每 100bp 存在 2 ～ 4 个 CpG。在由几千个碱基组成的“CpG 沙漠”中，CpG 通常聚集在一起，可能充当调控位点的作用（Skinner and Guerrero-Bosagna, 2014）。因此，鉴定出的斑马鱼精子 DMR 的基因组特征与以前在其他物种中鉴定的 DMR 相似（Manikkam et al., 2012; Nozawa et al., 2007; Skinner, 2011）。

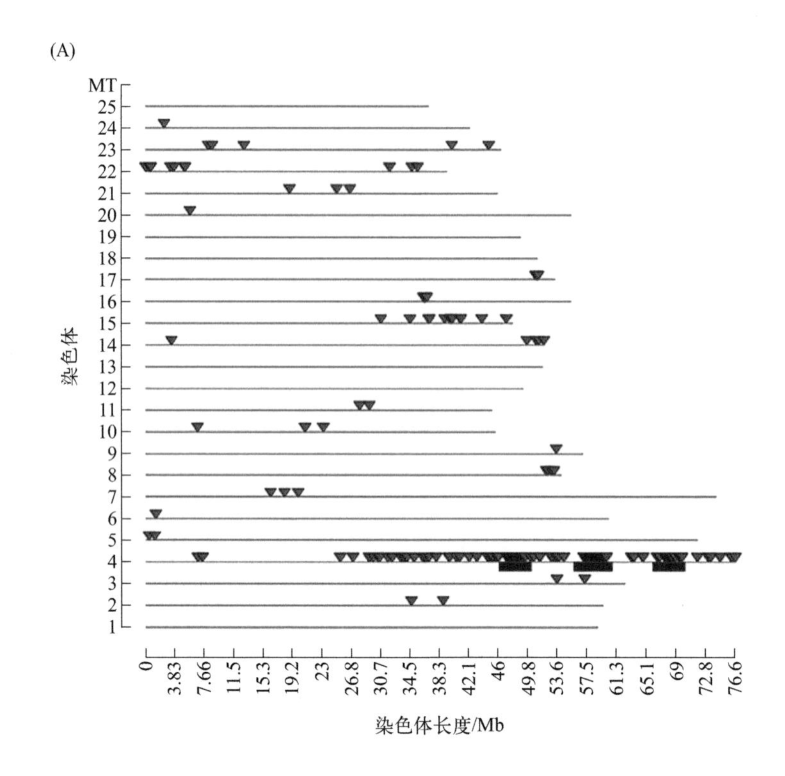

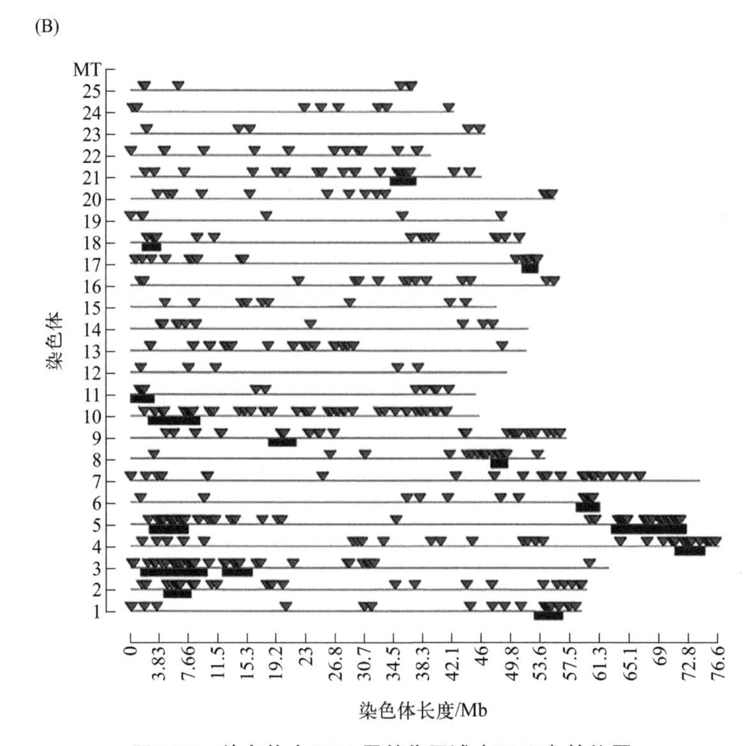

图28-5 染色体中DNA甲基化区域（DMR）的位置

（A）F0 代精子在个体染色体上的 DMR 位置。仅显示 P 值阈值为 10^{-7} 的多窗口 DMR。（B）F2 代精子在个体染色体上的 DMR 位置。此处仅显示 P 值阈值为 10^{-7} 的多窗口 DMR。箭头标识 DMR 的位置，方框标识 DMR 集群位点。

28.8 结论和下一步计划

本项研究工作证明了甲基汞能够促进斑马鱼表型改变和精子表观突变的表观遗传跨代遗传能力。获得的分子数据表明，表观遗传学比遗传学更有显著性和可再现性。这些观察结果与甲基汞暴露所致的表型改变相关。虽然发现了DMR的可重复特征，但这些DMR的功能影响还需要进一步研究。显然，目前的研究支持表观遗传学在甲基汞跨代遗传中发挥关键作用。然而，其结果是表观遗传学和遗传学的结合，将影响对于表型变异和健康的分子调控（Skinner et al., 2013; Haque et al., 2016）。目前的研究首次观察到重金属可以引起健康效应的表观遗传跨代遗传。汞等重金属应该被列入不断增加的可能促进跨代效应的环境暴露物清单中（Jirtle and Skinner, 2007）。斑马鱼是一个非常强大的实验系统，可用来开发将表观遗传变化与基因组中特定位点的基因表达联系起来的模型，这是推动这一领域向前发展的必要条件。表观遗传学分析必须在纯细胞型群体中进行，因为每类细胞都有独特的表观遗传模式。有许多转基因斑马鱼以特定细胞类型的方式表达荧光蛋白。这些细胞类型可以通过多轮荧光激活细胞分选技术来分离，以获得高丰度样本（>98%，Carvan未发表），用于同步开展全基因组重亚硫酸盐测序的DNA甲基化分析和全基因组测序的转录组分析。这必须在F0、F1和F2代的几种表型相关的细胞类型中进行，以全面了解位点特异性DNA甲基化在持续几代的环境诱导表型功能障碍中的作用。专注于阐明遗传环境对健康影响的科学研究的不断涌现，将强化代际环境公平问题，为了公共健康，这些问题需要在未来的公共政策和立法中加以考虑。

致谢

这项工作得到了日本国家医药品食品卫生研究所（the National Institute of Health Sciences）、美国五大湖美洲原住民健康研究中心（the Great Lakes Native American Research Centers for Health）、大密尔沃基基金会Shaw基金（the Shaw Fund of the Greater Milwaukee Foundation）和威斯康星大学密尔沃基分校儿童环境健康科学核心中心（University of Wisconsine Milwaukee Children's Environmental Health Sciences Core Center）（2P30ES004184）的资助。作者感谢John Dellinger对汞研究的贡献，感谢Dan Weber对鱼类行为观察的贡献，感谢Cheryl Murphy、Nil Basu、Jessica Head对环境应用部分的贡献。如果没有Carvan实验室众多研究生、博士后和本科生的奉献和辛勤工作，这项工作是不可能完成的，我对他们表示感谢。

（陈学军翻译　石晶晶审校）

参考文献

Aiken, C.E., Tarry-Adkins, J.L., Ozanne, S.E., 2016. Transgenerational effects of maternal diet on metabolic and reproductive ageing. Mammalian Genome. https://doi.org/10.1007/s00335-016-9631-1.

Anway, M.D., Skinner, M.K., 2006. Epigenetic transgenerational actions of endocrine disruptors. Endocrinology 147 (6 Suppl. 1), S43-S49. https://doi.org/10.1210/en.2005-1058.

Anway, M.D., Cupp, A.S., Uzumcu, M., et al., 2005. Epigenetic transgenerational actions of endocrine disruptors and male fertility. Science 308 (5727), 1466-1469. https://doi.org/10.1126/science.1108190.

Arai, Y., Ohgane, J., Yagi, S., et al., 2011. Epigenetic assessment of environmental chemicals detected in maternal peripheral and cord blood samples. Journal of Reproduction and Development. https://doi.org/10.1262/jrd.11-034A.

Aschner, J.L., Aschner, M., 2007. Methylmercury neurotoxicity: exploring potential novel targets. The Open Toxicology Journal 1(1), 1-10. https://doi.org/10.2174/1874340400701011040.

Atchison, W.D., 2005. Is chemical neurotransmission altered specifically during methylmercury-induced cerebellar dysfunction? Trends in Pharmacological Sciences. https://doi.org/10.1016/j.tips.2005.09.008.

Baker, T.R., King-Heiden, T.C., Peterson, R.E., Heideman, W., 2014. Dioxin induction of transgenerational inheritance of disease in zebrafish. Molecular and Cellular Endocrinology 398 (1-2), 36-41. https://doi.org/10.1016/j.mce.2014.08.011.

Bal-Price, A., Crofton, K.M., Leist, M., et al., 2015. International STakeholder NETwork (ISTNET): creating a developmental neurotoxicity (DNT) testing road map for regulatory purposes. Archives of Toxicology. https://doi.org/10.1007/s00204-015-1464-2.

Basu, N., 2015. Applications and implications of neurochemical biomarkers in environmental toxicology. Environmental Toxicology and Chemistry. https://doi.org/10.1002/etc.2783.

Basu, N., Head, J., Nam, D.-H., et al., 2013. Effects of methylmercury on epigenetic markers in three model species: mink, chicken and yellow perch. Comparative Biochemistry and Physiology e Part C: Toxicology and Pharmacology 157 (3), 322-327.

Ben Maamar, M., Sadler-Riggleman, I., Beck, D., Skinner, M.K., 2018. Epigenetic transgenerational inheritance of altered sperm histone retention sites. Scientific Reports 8 (1), 5308. https://doi.org/10.1038/s41598-018-23612-y.

Berger, S.L., Kouzarides, T., Shiekhattar, R., Shilatifard, A., 2009. An operational definition of epigenetics. Genes and Development 23 (7), 781-783. https://doi.org/10.1101/gad.1787609.

Bernal, A.J., Jirtle, R.L., 2010. Epigenomic disruption: the effects of early developmental exposures. Birth Defects Research Part A: Clinical and Molecular Teratology 88 (10), 938-944. https://doi.org/10.1002/bdra.20685.

Bisen-Hersh, E.B., Farina, M., Barbosa, F., Rocha, J.B.T., Aschner, M., 2014. Behavioral effects of developmental methylmercury drinking water exposure in rodents. Journal of Trace Elements in Medicine and Biology 28 (2), 117-124. https://doi.org/10.1016/j.jtemb.2013.09.008.

Bose, R., Onishchenko, N., Edoff, K., et al., 2012. Inherited effects of low-dose exposure to methylmercury in neural stem cells. Toxicological Sciences 130 (2), 383-390. https://doi.org/10.1093/toxsci/kfs257.

Burdge, G.C., Lillycrop, K.A., Jackson, A.A., 2009. Nutrition in early life, and risk of cancer and metabolic disease: alternative endings in an epigenetic tale? British Journal of Nutrition. https://doi.org/10.1017/S0007114508145883.

Bygren, L.O., Kaati, G., Edvinsson, S., 2001. Longevity determined by paternal ancestors' nutrition during their slow growth period. Acta Biotheoretica. https://doi.org/10.1023/A:1010241825519.

Cardenas, A., Koestler, D.C., Houseman, E.A., et al., 2015. Differential DNA methylation in umbilical cord blood of infants exposed to mercury and arsenic in utero. Epigenetics 10 (6), 508-515. https://doi.org/10.1080/15592294.2015.1046026.

Cardenas, A., Rifas-Shiman, S.L., Agha, G., et al., 2017. Persistent DNA methylation changes associated with prenatal mercury exposure and cognitive performance during childhood. Scientific Reports. https://doi.org/10.1038/s41598-017-00384-5.

Carvan, M.J., Kalluvila, T.A., Klingler, R.H., et al., 2017. Mercury-induced epigenetic transgenerational inheritance of abnormal neurobehavior is correlated with sperm epimutations in zebrafish. PLoS One 12 (5). https://doi.org/10.1371/journal.pone.0176155.

Caudill, M.A., Wang, J.C., Melnyk, S., et al., 2001. Intracellular S-adenosylhomocysteine concentrations predict global DNA hypomethylation in tissues of methyl-deficient cystathionine β-synthase heterozygous mice. Journal of Nutrition 131 (11), 2811-2818. https://doi.org/10.1093/jn/131.11.2811.

Ceccatelli, S., Bose, R., Edoff, K., Onishchenko, N., Spulber, S., 2013. Long-lasting neurotoxic effects of exposure to methylmercury during development. Journal of Internal Medicine. https://doi.org/10.1111/joim.12045.

Colón-Rodríguez, A., Hannon, H.E., Atchison, W.D., 2017. Effects of methylmercury on spinal cord afferents and efferentsda review. Neurotoxicology. https://doi.org/10.1016/j.neuro.2016.12.007.

Crespo-López, M.E., Macêdo, G.L., Pereira, S.I.D., et al., 2009. Mercury and human genotoxicity: critical considerations and possible molecular mechanisms. Pharmacological Research. https://doi.org/10.1016/j.phrs.2009.02.011.

Culbreth, M., Aschner, M., 2019. Methylmercury epigenetics. Toxics 7 (4), 56. https://doi.org/10.3390/toxics7040056.

Czech, B., Munafò, M., Ciabrelli, F., et al., 2018. piRNA-guided genome defense: from biogenesis to silencing. Annual Review of Genetics 52 (1), 131-157. https://doi.org/10.1146/annurev-genet-120417-031441.

Dana, H., Chalbatani, G.M., Mahmoodzadeh, H., et al., 2017. Molecular mechanisms and biological functions of siRNA. International Journal of Biomedical Sciences 13 (2), 48-57. http://www.ncbi.nlm.nih.gov/pubmed/28824341.

Deng, Y., Xu, Z.F., Liu, W., Xu, B., Yang, H.B., Wei, Y.G., 2012. Riluzole-triggered GSH synthesis via activation of glutamate transporters to antagonize methylmercury-induced oxidative stress in rat cerebral cortex. Oxidative Medicine and Cellular Longevity. https://doi.org/10.1155/2012/534705.

Desaulniers, D., Xiao, G.H., Lian, H., et al., 2009. Effects of mixtures of polychlorinated biphenyls, methylmercury, and organochlorine pesticides on hepatic DNA methylation in prepubertal female sprague-dawley rats. International Journal of Toxicology. https://doi.org/10.1177/1091581809337918.

Detrich, H.W., Zon, L.I., Westerfield, M. (Eds.), 2009. Essential Zebrafish Methods: Genetics and Genomics. Academic Press.

Dreiem, A., Shan, M., Okoniewski, R.J., Sanchez-Morrissey, S., Seegal, R.F., 2009. Methylmercury inhibits dopaminergic function in rat pup synaptosomes in an age-dependent manner. Neurotoxicology and Teratology. https://doi.org/10.1016/j.ntt.2009.05.001.

Dutczak, W.J., Ballatori, N., 1994. Transport of the glutathione-methylmercury complex across liver canalicular membranes on reduced glutathione carriers. Journal of Biological Chemistry 269 (13), 9746-9751.

Fang, X., Corrales, J., Thornton, C., Scheffler, B.E., Willett, K.L., 2013. Global and gene specific DNA methylation changes during zebrafish development. Comparative Biochemistry and Physiology Part B: Biochemistry and Molecular Biology. https://doi.org/10.1016/j.cbpb.2013.07.007.

Farina, M., Aschner, M., 2017. Methylmercury-induced neurotoxicity: focus on pro-oxidative events and related consequences. Advances in Neurobiology 18, 267-286. https://doi.org/10.1007/978-3-319-60189-2_13.

Farina, M., Rocha, J.B.T., Aschner, M., 2011. Mechanisms of methylmercury-induced neurotoxicity: evidence from experimental studies. In: Life Sciences, 89. Elsevier Inc., pp. 555-563. https://doi.org/10.1016/j.lfs.2011.05.019

Feng, S., Cokus, S.J., Zhang, X., et al., 2010. Conservation and divergence of methylation patterning in plants and animals. Proceedings of the National Academy of Sciences of the United States of America. https://doi.org/10.1073/pnas.1002720107.

Ferey, J.L.A., Boudoures, A.L., Reid, M., et al., 2019. A maternal high-fat, high-sucrose diet induces transgenerational cardiac mitochondrial dysfunction independently of maternal mitochondrial inheritance. American Journal of Physiology - Heart and Circulatory Physiology. https://doi.org/10.1152/ajpheart.00013.2019.

Fitsanakis, V.A., Aschner, M., 2005. The importance of glutamate, glycine, and g-aminobutyric acid transport and regulation in manganese, mercury and lead neurotoxicity. Toxicology and Applied Pharmacology. https://doi.org/10.1016/j.taap.2004.11.013.

Fowler, B.A., Woods, J.S., 1977. Ultrastructural and biochemical changes in renal mitochondria during chronic oral methyl mercury exposure. The relationship to renal function. Experimental and Molecular Pathology. https://doi.org/10.1016/0014-4800(77)90010-7.

Grandjean, P., Barouki, R., Bellinger, D.C., et al., 2015. Life-long implications of developmental exposure to environmental stressors: new perspectives. Endocrinology. https://doi.org/10.1210/EN.2015-1350.

Greer, E.L., Maures, T.J., Ucar, D., et al., 2011. Transgenerational epigenetic inheritance of longevity in Caenorhabditis elegans. Nature. https://doi.org/10.1038/nature10572.

Gruenwedel, D.W., Lu, D.S., 1970. Changes in the sedimentation characteristics of DNA due to methylmercuration. Biochemical and Biophysical Research Communications. https://doi.org/10.1016/0006-291X(70)90936-8.

Guerrero-Bosagna, C., Settles, M., Lucker, B., Skinner, M.K., 2010. Epigenetic transgenerational actions of vinclozolin on promoter regions of the sperm epigenome. PLoS One 5 (9), 1-17. https://doi.org/10.1371/journal.pone.0013100.

Guerrero-Bosagna, C., Covert, T.R., Haque, M.M., et al., 2012. Epigenetic transgenerational inheritance of vinclozolin induced mouse adult onset disease and associated sperm epigenome biomarkers. Reproductive Toxicology. https://doi.org/10.1016/j.reprotox.2012.09.005.

Hales, B.F., Grenier, L., Lalancette, C., Robaire, B., 2011. Epigenetic programming: from gametes to blastocyst. Birth Defects Research Part A: Clinical and Molecular Teratology. https://doi.org/10.1002/bdra.20781.

Hanna, C.W., Bloom, M.S., Robinson, W.P., et al., 2012. DNA methylation changes in whole blood is associated with exposure to the environmental contaminants, mercury, lead, cadmium and bisphenol A, in women undergoing ovarian stimulation for IVF. Human Reproduction 27 (5), 1401-1410. https://doi.org/10.1093/humrep/des038.

Haque, M.M., Nilsson, E.E., Holder, L.B., Skinner, M.K., 2016. Genomic Clustering of differential DNA methylated regions (epimutations) associated with the epigenetic transgenerational inheritance of disease and phenotypic variation. BMC

Genomics 17 (1). https://doi.org/10.1186/s12864-016-2748-5.

Heintzman, N.D., Hon, G.C., Hawkins, R.D., et al., 2009. Histone modifications at human enhancers reflect global cell-type-specific gene expression. Nature. https://doi.org/10.1038/nature07829.

Herman, S.P., Klein, R., Talley, F.A., Krigman, M.R., 1973. An ultrastructural study of methylmercury-induced primary sensory neuropathy in the rat. Laboratory Investigation 28 (1), 104-118.

Horsthemke, B., 2018. A critical view on transgenerational epigenetic inheritance in humans. Nature Communications. https://doi.org/10.1038/s41467-018-05445-5.

Huang, S.S.Y., Noble, S., Godoy, R., Ekker, M., Chan, H.M., 2016. Delayed effects of methylmercury on the mitochondria of dopaminergic neurons and developmental toxicity in zebrafish larvae (*Danio rerio*). Aquatic Toxicology 175, 73-80. https://doi.org/10.1016/j.aquatox.2016.03.004.

Jiang, L., Zhang, J., Wang, J.J., et al., 2013. Sperm, but not oocyte, DNA methylome is inherited by zebrafish early embryos. Cell. https://doi.org/10.1016/j.cell.2013.04.041.

Jirtle, R.L., Skinner, M.K., 2007. Environmental epigenomics and disease susceptibility. Nature Reviews Genetics 8 (4), 253-262. https://doi.org/10.1038/nrg2045.

Kakita, A., Wakabayashi, K., Su, M., Sakamoto, M., Ikuta, F., Takahashi, H., 2000. Distinct pattern of neuronal degeneration in the fetal rat brain induced by consecutive transplacental administration of methylmercury. Brain Research. https://doi.org/10.1016/S0006-8993(00)01964-8.

Kanduri, C., Whitehead, J., Mohammad, F., 2009. The long and the short of it: RNA-directed chromatin asymmetry in mammalian X-chromosome inactivation. FEBS Letters 583 (5), 857-864. https://doi.org/10.1016/j.febslet.2009.02.004.

Karagas, M.R., Choi, A.L., Oken, E., et al., 2012. Evidence on the human health effects of low-level methylmercury exposure. Environmental Health Perspectives 120 (January), 799-806. https://doi.org/10.1289/ehp.1104494.

Kaur, P., Aschner, M., Syversen, T., 2006. Glutathione modulation influences methyl mercury induced neurotoxicity in primary cell cultures of neurons and astrocytes. Neurotoxicology 27 (4), 492-500. https://doi.org/10.1016/j.neuro.2006.01.010.

Kaur, P., Aschner, M., Syversen, T., 2011. Biochemical factors modulating cellular neurotoxicity of methylmercury. Journal of Toxicology 2011. https://doi.org/10.1155/2011/721987.

Lee, D.-H., Jacobs, D.R., Porta, M., 2009. Hypothesis: a unifying mechanism for nutrition and chemicals as lifelong modulators of DNA hypomethylation. Environmental Health Perspectives 117 (12), 1799-1802. https://doi.org/10.1289/ehp.0900741.

Li, L., Dowling, J.E., 1997. A dominant form of inherited retinal degeneration caused by a non-photoreceptor cell-specific mutation. Proceedings of the National Academy of Sciences of the United States of America. https://doi.org/10.1073/pnas.94.21.11645.

Liu, Q., Klingler, R.H., Wimpee, B., et al., 2016. Maternal methylmercury from a wild-caught walleye diet induces developmental abnormalities in zebrafish. Reproductive Toxicology. https://doi.org/10.1016/j.reprotox.2016.08.010.

MacKay, A.B., Mhanni, A.A., McGowan, R.A., Krone, P.H., 2007. Immunological detection of changes in genomic DNA methylation during early zebrafish development. Genome. https://doi.org/10.1139/G07-055.

Mahaffey, K.R., 2004. Fish and shellfish as dietary sources of methylmercury and the u-3 fatty acids, eicosahexaenoic acid and docosahexaenoic acid: risks and benefits. Environmental Research. https://doi.org/10.1016/j.envres.2004.02.006.

Manikkam, M., Guerrero-bosagna, C., Tracey, R., Haque, M.M., Skinner, M.K., 2012. Transgenerational actions of environmental compounds on reproductive disease and identification of epigenetic biomarkers of ancestral exposures. PLoS One 7 (2), e31901. https://doi.org/10.1371/journal.pone.0031901.

Mhanni, A.A., McGowan, R.A., 2004. Global changes in genomic methylation levels during early development of the zebrafish embryo. Development Genes and Evolution. https://doi.org/10.1007/s00427-004-0418-0.

Moore, R.S., Kaletsky, R., Murphy, C.T., 2019. Piwi/PRG-1 argonaute and TGF-b mediate transgenerational learned pathogenic avoidance. Cell 177 (7), 1827-1841. https://doi.org/10.1016/j.cell.2019.05.024.

Mora-Zamorano, F., Klingler, R., Murphy, C., Basu, N., Head, J., Carvan III, M.J., 2016. Parental whole life cycle exposure to dietary methylmercury in zebrafish (*Danio rerio*) affects the behavior of offspring. Environmental Science and Technology 50 (9), 4808-4816. https://doi.org/10.1021/acs.est.6b00223.

Mora-Zamorano, F.X., Larson, J.K., Carvan, M.J., 2018. Neurobehavioral analysis methods for adverse outcome pathway (AOP) models and risk assessment. In: A Systems Biology Approach to Advancing Adverse Outcome Pathways for Risk Assessment. https://doi.org/10.1007/978-3-319-66084-4_8.

Nilsson, E., Larsen, G., Manikkam, M., Guerrero-Bosagna, C., Savenkova, M.I., Skinner, M.K., 2012. Environmentally induced epigenetic transgenerational inheritance of ovarian disease. PLoS One. https://doi.org/10.1371/journal.pone.0036129.

Nilsson, E.E., Sadler-Riggleman, I., Skinner, M.K., 2018. Environmentally induced epigenetic transgenerational inheritance of disease. Environmental Epigenetics 4 (2). https://doi.org/10.1093/eep/dvy016.

Nordberg, G.F., Fowler, B.A., Nordberg, M., Friberg, L.T., 2007. Handbook on the Toxicology of Metals. https://doi.org/10.1016/B978-0-12-369413-3.X5052-6.

Nozawa, M., Kawahara, Y., Nei, M., 2007. Genomic drift and copy number variation of sensory receptor genes in humans. Proceedings of the National Academy of Sciences of the United States of America 104 (51), 20421-20426. https://doi.org/10.1073/pnas.0709956104.

Olivieri, G., Brack, C., Müller-Spahn, F., et al., 2000. Mercury induces cell cytotoxicity and oxidative stress and increases β-amyloid secretion and tau phosphorylation in SHSY5Y neuroblastoma cells. Journal of Neurochemistry. https://doi.org/10.1046/j.1471-4159.2000.0740231.x.

Onishchenko, N., Tamm, C., Vahter, M., et al., 2007. Developmental exposure to methylmercury alters learning and induces depression-like behavior in male mice. Toxicological Sciences 97 (2), 428-437. https://doi.org/10.1093/toxsci/kfl199.

Onishchenko, N., Karpova, N., Sabri, F., Castrén, E., Ceccatelli, S., 2008. Long-lasting depression-like behavior and epigenetic changes of BDNF gene expression induced by perinatal exposure to methylmercury. Journal of Neurochemistry 106 (3), 1378-1387. https://doi.org/10.1111/j.1471-4159.2008.05484.x.

Ortega-recalde, O., Hore, T.A., 2019. DNA methylation in the vertebrate germline: balancing memory and erasure. Essays in Biochemistry 63 (6), 649-661. https://doi.org/10.1042/EBC20190038.

O'Brien, J., Hayder, H., Zayed, Y., Peng, C., 2018. Overview of microRNA biogenesis, mechanisms of actions, and circulation. Frontiers in Endocrinology 9 (402). https://doi.org/10.3389/fendo.2018.00402.

Painter, R.C., Osmond, C., Gluckman, P., Hanson, M., Phillips, D.I.W., Roseboom, T.J., 2008. Transgenerational effects of prenatal exposure to the Dutch famine on neonatal adiposity and health in later life. BJOG: An International Journal of Obstetrics and Gynaecology. https://doi.org/10.1111/j.1471-0528.2008.01822.x.

Peplow, D., Augustine, S., 2014. Neurological abnormalities in a mercury exposed population among indigenous Wayana in Southeast Suriname. Environmental Science: Processes and Impacts 16 (10), 2415-2422. https://doi.org/10.1039/c4em00268g.

Potok, M.E., Nix, D a, Parnell, T.J., Cairns, B.R., 2013. Reprogramming the maternal zebrafish genome after fertilization to match the paternal methylation pattern. Cell 153 (4), 759-772. https://doi.org/10.1016/j.cell.2013.04.030.

Rechavi, O., Lev, I., 2017. Principles of transgenerational small RNA inheritance in *Caenorhabditis elegans*. Current Biology. https://doi.org/10.1016/j.cub.2017.05.043.

Rechavi, O., Houri-Ze'Evi, L., Anava, S., et al., 2014. Starvation-induced transgenerational inheritance of small RNAs in C. elegans. Cell. https://doi.org/10.1016/j.cell.2014.06.020.

Reif, D.M., Truong, L., Mandrell, D., Marvel, S., Zhang, G., Tanguay, R.L., 2016. High-throughput characterization of chemical-associated embryonic behavioral changes predicts teratogenic outcomes. Archives of Toxicology. https://doi.org/10.1007/s00204-015-1554-1.

Richard Pilsner, J., Lazarus, A.L., Nam, D.H., et al., 2010. Mercury-associated DNA hypomethylation in polar bear brains via the LUminometric Methylation Assay: a sensitive method to study epigenetics in wildlife. Molecular Ecology. https://doi.org/10.1111/j.1365-294X.2009.04452.x.

Skinner, M.K., 2007. Endocrine disruptors and epigenetic transgenerational disease etiology. Pediatric Research. https://doi.org/10.1203/pdr.0b013e3180457671.

Skinner, M.K., 2011. Environmental epigenetic transgenerational inheritance and somatic epigenetic mitotic stability. Epigenetics 6 (7), 838-842. https://doi.org/10.4161/epi.6.7.16537.

Skinner, M.K., Guerrero-Bosagna, C., 2014. Role of CpG deserts in the epigenetic transgenerational inheritance of differential DNA methylation regions. BMC Genomics 15 (1), 692. https://doi.org/10.1186/1471-2164-15-692.

Skinner, M.K., Manikkam, M., Guerrero-Bosagna, C., 2010. Epigenetic transgenerational actions of environmental factors in disease etiology. Trends in Endocrinology and Metabolism. https://doi.org/10.1016/j.tem.2009.12.007.

Skinner, M.K., Manikkam, M., Haque, M.M., Zhang, B., Savenkova, M.I., 2012. Epigenetic transgenerational inheritance of somatic transcriptomes and epigenetic control regions. Genome Biology 13 (10), R91. https://doi.org/10.1186/gb-2012-13-10-r91.

Skinner, M.K., Manikkam, M., Tracey, R., Guerrero-bosagna, C., Haque, M., Nilsson, E.E., 2013. Ancestral dichlorodiphenyltrichloroethane (DDT) exposure promotes epigenetic transgenerational inheritance of obesity. BMC

Medicine 11, 228. https://doi.org/10.1186/1741-7015-11-228.

Smith, L.E., Carvan III, M.J., Dellinger, J.A., et al., 2010. Developmental selenomethionine and methylmercury exposures affect zebrafish learning. Neurotoxicology and Teratology 32 (2), 246-255. https://doi.org/10.1016/j.ntt.2009.09.004.

Sookoian, S., Rohr, C., Salatino, A., et al., 2017. Genetic variation in long noncoding RNAs and the risk of nonalcoholic fatty liver disease. Oncotarget 5 (0). https://doi.org/10.18632/oncotarget.15286.

Thompson, S a, White, C.C., Krejsa, C.M., Eaton, D.L., Kavanagh, T.J., 2000. Modulation of glutathione and glutamate-L-cysteine ligase by methylmercury during mouse development. Toxicological Sciences 57 (1), 141-146. http://www.ncbi.nlm.nih.gov/pubmed/10966520.

Tilson, H.A., MacPhail, R.C., Crofton, K.M., 1995. Defining neurotoxicity in a decision-making context. Neurotoxicology 16 (2), 363-375.

Tsuchiya, M., Ji, C., Kosyk, O., et al., 2012. Interstrain differences in liver injury and one-carbon metabolism in alcohol-fed mice. Hepatology 56 (1), 130-139. https://doi.org/10.1002/hep.25641.

Unoki, T., Akiyama, M., Kumagai, Y., et al., 2018. Molecular pathways associated with methylmercury-induced Nrf2 modulation. Frontiers in Genetics 9. https://doi.org/10.3389/fgene.2018.00373.

Vandegehuchte, M.B., Janssen, C.R., 2014. Epigenetics in an ecotoxicological context. Mutation Research: Genetic Toxicology and Environmental Mutagenesis 764-765, 36-45. https://doi.org/10.1016/j.mrgentox.2013.08.008.

Varga, M., Ralbovszki, D., Balogh, E., Hamar, R., Keszthelyi, M., Tory, K., 2018. Zebrafish models of rare hereditary pediatric diseases. Diseases. https://doi.org/10.3390/diseases6020043.

Waterland, R.A., Travisano, M., Tahiliani, K.G., Rached, M.T., Mirza, S., 2008. Methyl donor supplementation prevents transgenerational amplification of obesity. International Journal of Obesity. https://doi.org/10.1038/ijo.2008.100.

Weber, D.N., 2006. Dose-dependent effects of developmental mercury exposure on C-start escape responses of larval zebrafish *Danio rerio*. Journal of Fish Biology. https://doi.org/10.1111/j.1095-8649.2006.01068.x.

Weber, D.N., Connaughton, V.P., Dellinger, J a, et al., 2008. Selenomethionine reduces visual deficits due to developmental methylmercury exposures. Physiology and Behavior 93 (1), 250-260. https://doi.org/10.1016/j.physbeh.2007.08.023.

Weber, D.N., Klingler, R.H., Carvan III, M.J., 2012. Zebrafish as a model for methylmercury neurotoxicity. In: Methylmercury and Neurotoxicity. Springer, pp. 335-355.

Wilson, C a, High, S.K., McCluskey, B.M., et al., 2014. Wild Sex in Zebrafish: Loss of the Natural Sex Determinant in Domesticated Strains, vol. 198. https://doi.org/10.1534/genetics.114.169284.

Xin, F., Susiarjo, M., Bartolomei, M.S., 2015. Multigenerational and transgenerational effects of endocrine disrupting chemicals: a role for altered epigenetic regulation? Seminars in Cell and Developmental Biology. https://doi.org/10.1016/j.semcdb.2015.05.008.

Yee, S., Choi, B.H., 1996. Oxidative stress in neurotoxic effects of methylmercury poisoning. Neurotoxicology 17(1), 17-26.

Yorifuji, T., Kato, T., Kado, Y., et al., 2015. Intrauterine exposure to methylmercury and neurocognitive functions: minamata disease. Archives of Environmental and Occupational Health 70 (5), 297-302. https://doi.org/10.1080/19338244.2014.904268.

第 六 部 分

大数据和生物信息学

第二十九章

斑马鱼行为筛选的设计

Robert T. Gerlai[1]

29.1 为什么要筛选？

大多数学术研究者、基金评审人及基金会都不看重“非假设”驱动的研究。他们认为，基于坚实的基础和大量的初步数据，经过深思熟虑并且逻辑性强的研究才是推动科学进步的途径。这个论点很有道理。如果没有基于先前知识或者数据精心设计的假设，人们可能无法提出适当的实验设计，也可能无法收集有用的经验证据或者数据来决定假设是否正确。根据这一论点，非假设驱动的实验只不过是在黑暗中射击，击中目标的可能性极小！

然而，上述论点中有一个经常被人们遗忘的固有假设：这类研究的基础。假设驱动的研究只有在有足够坚实的基础，有充足的初步数据作为假设的依据时，才能有意义地进行。到目前为止，科学家们可能会产生疑问，我们已经对大脑如何工作有了很好的理解，所以假设驱动的研究确实是一个不错的方法。但是，有许多复杂的问题仍然没有答案。为了回答这些问题，我们不得不放弃假设驱动的方法，而采用另一种方法，即在黑暗中盲目射击，也称为无偏大规模筛选（unbiased large-scale screening）。后一种方法的支持者（如 Stein, 2003）强调，假设驱动的研究可能会受到狭隘视野的影响。换句话说，实验者可能会过早地决定关注什么，检测什么假设，以及忽略什么其他假设。这种判断是不合理的，因为他们可能还没有足够的数据。如何才能收集到足够的数据？可以通过全面的大规模筛选。本书有多个章节介绍了学习和记忆等复杂大脑功能的实例。从分子角度出发，在大约 400 个分子靶标（基因及其蛋白质产物）的基础上，科学家们已经针对该主题进行了数以千计的假设驱动研究（如 Sweatt, 2009）。然而，在约 25000 个已知的蛋白质

[1] Department of Psychology, University of Toronto Mississauga, Mississauga, ON, Canada.

编码基因中，至少有 50% 的基因可以在哺乳动物大脑中表达（Naumova et al., 2013），其中许多基因可能会参与神经元可塑性的基础过程，即促进学习和记忆的功能。在斑马鱼大脑中也发现了类似的基因表达数量（Pan et al., 2011）。简而言之，我们甚至没有在假设驱动的研究中触及表面。高通量和全面的筛选可以让我们识别和发现大多数基因的功能。关于大脑功能和功能障碍的其他几个复杂问题，也可以提出类似的论点。例如，尽管我们已经有很多证据证明了乙醇成瘾或其他乙醇相关疾病［包括胎儿乙醇谱系障碍（fetal alcohol spectrum disorders, FASDs）］的机制，但我们对乙醇影响大脑及其发育的潜在机制的理解还不足以研发有效的治疗方案，更不用说治愈方法了（Abrahao et al., 2017; Roberto and Varodayan, 2017）。从药理学的角度来看，乙醇长期以来被认为是一种极其复杂的药物（Pohorecky and Brick, 1988），因为它可以与大量分子靶标直接相互作用，并且这种相互作用依赖于浓度和给药方案（Ron and Barak, 2016; Abrahao et al., 2017; Roberto and Varodayan, 2017）。此外，考虑到乙醇对大量生化机制以及一系列发育事件的间接影响，我们可以理解为什么研发治疗乙醇中毒和 FASD 的方法并不简单。解决这种复杂问题的一种潜在方法是对可能改变大脑对乙醇反应的突变进行系统筛选，或者对具有类似乙醇效应的药物进行系统筛选。

29.2 我们应该筛选什么？

这个问题不涉及表型（在我们的例子中是行为）。我们稍后再讨论表型问题。这里的问题指的是一种操作，我们期望通过这种操作来诱导大脑的一些功能变化。正如我们在上面已经提到的，有两种根本不同的大脑功能操作是可以进行筛选的，或者对其筛选可能特别有用：突变和药物诱导效应。首先，我们将讨论前者。

在斑马鱼中可以有效地筛选突变，因为该物种繁殖力强、体形小、饲养容易，并且可以轻松高效地进行诱导突变。事实上，斑马鱼诱变筛选已成为生物学中最受关注的领域之一（Driever et al., 1996; Mullins et al., 1994）。从根本上说，有两种方法可以利用诱导突变来研究生物机制：正向遗传学（forward genetics）和反向遗传学（reverse genetics）。前者曾经是斑马鱼筛选的主要方法（例如 Malicki, 2000）。正向遗传学是指采用随机诱导突变的方法，即不知道哪些基因发生了突变，然后进行表型筛选以确定突变体并最终识别突变的载体，即基因。因此，正向遗传学被用于发现与目标表型相关且先前未知的基因。另一方面，当人们知道基因的核苷酸序列并专门针对（即操纵）该基因以发现其功能时，就会采用反向遗传学。因此，反向遗传学是一种假设驱动的方法，至少可以根据基因的序列或其他已知特征（如其表达模式）预测功能。而正向遗传学不仅是一种典型的非假设驱动方法，还是一种全面的且无偏倚的基因发现方法。

Patton 和 Zon（2001）设计了将化学诱变剂乙基亚硝基脲（ethyl nitroso-urea, ENU）用于随机诱变育种的方法，很好地说明了诱变研究需要高通量和大规模筛选的原因。在对斑马鱼（Knapik, 2000）和其他模式生物的正向遗传学研究中，ENU 是一种常用的化学诱变剂，因为在正确使用（以正确的浓度和正确的暴露时间）的基础上，它会均匀地诱导每个基因组的单点突变。育种方法主要包括以下内容。将雄性斑马鱼暴露在ENU溶液中，会导致其精子发生突变。为了清楚起见，我们称这些雄性为F0代。因为ENU不会对某个特定的靶序列产生突变，突变会随机发生在整个基因组。因此，已诱变雄性的每一个精子都可能携带不同的点突变。当这种突变的雄性与野生型（非突变型）雌性交配时，每个后代（F1代鱼）可能会以杂合子的形式携带不同的点突变，这种杂合子是因为只有来源于精子的姐妹染色体具有突变等位基因。然而，问题是，大多数点突变是隐性的，也就是说，可能不会以杂合形式表现出来。为了观察这种点突变的表型效应，需要将它们培育成纯合子。有人会认为，如果一条携带潜在突变的F1代鱼与另一条携带相同突变的鱼交配，就可以实现这一点。然而，我们还没有确定突变的位点或突变基因的序列，因此我们无法判断哪些鱼可能携带相同的突变。此外，在F1代中，发现两条携带相同突变的鱼是极不可能的，因为组成整个基因组的核苷酸数量巨大，而且ENU诱变是靶向基因组的随机位点。为了避免这个问题，人们需要用野生型（非突变）鱼繁殖出F1代，并产生F2代。在每个F2代中，对于给定的突变，50%的鱼是杂合子，50%是纯合的野生型。因此，在这一代鱼中，人们仍然无法观察到隐性突变带来的影响。为了使这种突变在表型上可观察到，需要将F2代与F1代个体（如这个F2代的父亲）进行回交，从而产生F3代。因为F2代中50%的鱼是野生型，所以50%的F2代鱼回交时不会导致F3的纯合突变。但是剩下50%的F2代鱼以杂合形式携带突变并与同样以杂合形式携带相同突变的父亲回交时，将产生纯合突变体（该杂交的F3代的25%）。回顾这种产生纯合ENU突变体的育种方法（Patton and Zon, 2001）说明了一个问题：必须将鱼大量且多代繁殖才能获得纯合突变体，从而在表型上可以观察到突变的影响。但是，我们只讨论了因单点突变生成具有纯合基因位点的鱼。斑马鱼基因组有数以万计的基因，这其中的每一个基因都可能以多种不同的方式发生突变。可以想象一下，对所有这些点突变进行上述杂交，经多代繁殖筛选才能获得目标纯合突变体。最重要的是，大多数表型特征都显示出一些变异性，因此需要一个合理的样本量（对于行为性状，通常是20到30条鱼）来可靠地识别突变效应。简而言之，全面的突变筛选可能需要测试数万条甚至更多的鱼。用生物医学研究中最受欢迎的实验室生物，即大鼠或小鼠来测试如此多的对象实际上是不可能的。事实上，很少有人能在啮齿类动物体内成功地使用这种策略。尽管对斑马鱼来说也不是一件小事，但上述突变筛选策略是可行的（例如，Baraban et al., 2007; Kegel et al., 2019），因为该物种的繁殖力很强（一条雌性斑马鱼每周可以产数百个卵），并且这种物种的小体形和群居性使其能够高效地饲养在小水箱中。总之，基于随机突变的正向遗传方法结合大规模筛选对斑马鱼而言是可行的。然而，筛选成功的关键不是诱变或育种策略，而是所采用的表型量化系统 / 方法，稍后我们将讨论这个问题。

还有一种也可用于斑马鱼大规模筛选的遗传方法，即CRISPR/Cas系统（Li et al., 2016）。从技术上讲，这个系统是一个反向遗传的方法。这种方法可以有效地靶向（敲除或敲入）所选择的目的基因。该方法在本书第十八章中有详细的描述，本章不作讨论。简而言之，这种方法可以应用于包括斑马鱼在内的许多物种，而且已经发展得足够简单和有效，该方法已经模糊了正向遗传学和反向遗传学之间的界限。也就是说，在CRISPR/Cas系统中，由于基因操作很容易实现，人们以预定的核苷酸序列为目标，虽然序列的数量可能非常大，但是还是可以采用这种方法进行筛查的，就像在正向遗传学中一样，但该方法不需要开展ENU突变研究中克隆定位所需要的劳动密集型后续连锁分析。

大规模筛选的第二个主要目的是药物（药用化合物）筛选。药用化合物，或者工业术语中的小分子是相当多的。不幸的是，基于预测化学结构（小分子）如何与复杂有机分子（如蛋白质）相互作用的智能药物设计仍然是一厢情愿的想法。因此，化学家创造的分子必须经过测试或筛选，才确定其在生物系统中的功效。斑马鱼是一种很好的体内系统，可以将斑马鱼暴露在含有某个化合物的溶液中（Gerlai, 2010; Kalueff et al., 2014）。一些小分子是水溶性的，但即使是不溶于水的化合物也可以通过使用二甲基亚砜（dimethyl sulfoxide, DMSO）来进行暴露。DMSO 是一种有机溶剂，通常用于将疏水性化合物溶于水中。并且已发现斑马鱼对 DMSO 相当耐受（Goldsmith, 2004）。如果化合物不能以这种方式溶解，可以采用腹腔注射法。化合物库包含成千上万种化合物，在这些化合物中，至少有一些化合物能够与它们的预期靶标相互作用，并且对于这些小分子子集来说，相互作用将以预期的方式进行。换句话说，因为我们无法预测小分子将如何影响生物系统的功能，我们必须大规模测试（筛选）大量小分子的功效。利用斑马鱼在神经行为遗传学中进行大规模药物筛选代表了新的研究方法，目前，首个证明这种方法可行性的例子已成功发表（Rihel and Schier, 2012）。

29.3 为什么要用行为表型作为筛查标准？

对行为分析的批评往往是行为的测量不可靠，因为影响因素太多。也许正因为如此，与生物学的其他领域不同，心理学有时被认为是一门“软”科学（不太精确）。然而，这正是对突变或药物改变大脑功能的效应感兴趣的神经科学家应该测试行为的原因之一：行为对许多因素高度敏感，包括遗传和药理操作。此外，行为分析不需要预先知道突变（或药物）在大脑中发挥作用的确切位置，以及它可能影响哪些电生理、生物化学或其他神经生物学过程。换句话说，这是一种非常好的非假设驱动的大规模筛选方法。其优势在于相对容易实现，而且成本低廉。

行为测试不需要昂贵的专用设备、高倍显微镜以及复杂的分子生物学工具包和试剂。行为分析尤其适用于斑马鱼，因为与啮齿动物不同，斑马鱼很容易诱导行为反应。斑马鱼是昼行性物种，即白天活动，晚上睡觉。它们感知的主要方式是视觉。因此，这些鱼对视觉刺激反应良好，对视觉刺激的传递和控制比其他形式的刺激容易得多。此外，因为我们人类的视觉系统也很发达，所以常规的消费类设备（电视或电脑监视器和摄像机）可以很容易地提供视觉刺激或记录基于视觉刺激的反应，而且价格低廉。总之，斑马鱼的行为分析可能是发现药物和突变对大脑功能改变影响的一种简单而廉价的方法。虽然它可能无法提供足够的机制细节，但行为分析是筛选此类影响的合适的第一步。关于为什么行为分析很重要，还有最后一点，它除了是一个相当好的研究工具，还被认为是治疗终点的测试。例如，在人类临床中，阿尔茨海默病患者不会关心其生物化学或神经元可塑性过程如何随着治疗而改变，他们所关心的是其记忆是否得到改善，以及记忆是否能在日常生活中更好地发挥作用。就大脑而言，行为分析确实是对疾病治疗终点最重要的测试。在过去的一些案例中，行为分析结果显示，即使研究人员不完全了解这种效果是如何实现的，也能显示出治疗的有益效果，例如最常用的抗抑郁药之一百忧解（Prozac）的治疗效果（Harmer et al., 2009）。

29.4 应认真推敲的行为测试概念

在开始解决这个实际问题之前，我们希望讨论一个概念上的重要问题。当查阅行为方面的文献时，人们可以经常查阅到研究者是如何“测量焦虑”或“量化学习和记忆”的。由于这种说法相当普遍，大多数读者可能不会再次思考它们的误导性，但我们必须强调的是这种说法代表了一种重要的且经常被忽视的问题。行为测试不能测量焦虑，也不能量化学习和记忆，它们所测量的只是作为时间函数的受试者的运动（运动反应）或位置。我们在这里建立的论点不仅仅是深奥的学术观点，认识到存在大量无法直接测量的复杂行为现象是非常重要的，它们只能基于多条信息得出结论，例如学习和记忆。在学习和记忆任务中的表现受损可能是由四个根本不同的原因引起的。第一，损伤可能确实是由异常的学习和/或记忆过程引起的（注意力、获得、巩固和/或记忆唤醒的改变）。第二，也可能是由运动功能紊乱引起的。在水迷宫试验中，突变小鼠可能无法到达目标平台的位置，不是因为它不记得，而是因为它无法游向目标平台。在恐惧条件反射任务中，正在接受药物处理的大鼠对训练过的音调信号仍然表现出活跃反应，这不是因为它不记得了，而是因为它无法保持静止（由于药物的作用而活动亢进）。在学习和记忆任务中改变表现的第三个原因也可能是一个微不足道的因素，即感知。如果突变小鼠失明，它将看不到水迷宫周围的视觉空间线索。如果接受药物处理的大鼠感觉不到疼痛，它不会对与足部电击相关的音调做出僵直反应。我们认为影响学习任务表现的最后一个因素是动机，如果动机发生改变，可能会导致学习和记忆的改变。上述例子通常适用于所有复杂的行为，不仅仅是学习和记忆，或者焦虑，当然也不是啮齿动物研究所独有的。我们如何分析突变或药物对复杂行为的影响？如何才能正确解释行为测试的结果？为了做到这一点，人们需要进行多种行为测试或测量多种行为参数，以分别识别上述因素，这就是行为测试组合（behavioral test batteries）发挥作用的地方。

29.5 行为测试组合，多重行为测量，重叠行为测试

行为测试组合可以更好地解释结果。为什么？因为除了一个单一的读数，比如关于学习成绩的读数，还可以独立测量可能影响学习成绩的某些其他因素。对于生物医学研究中最常用的实验室生物——小鼠，测试组合已经得到了很好的发展（Crawley and Paylor, 1997; McIlwain et al., 2001; Blanchard et al., 2003），利用概念上类似的方法，针对斑马鱼也开发了可以构成此类测试组合一部分的单独测试（Bruni et al., 2016）。例如，运动功能，包括耐力，可以在跑步机上测量小鼠，也可以在流动管中测量斑马鱼。可以使用视觉悬崖（visual cliff）实验测试小鼠的视觉感知，用视觉运动反应来测试斑马鱼。动机则更难直接测量。对于小鼠来说，实验者通常使用的测量方法是：为了一定数量的奖励而愿意按压杠杆的最大次数。例如，对于斑马鱼来说，在十字迷宫中，它们访问食物奖励臂的次数，可以作为寻找

和享用食物的动机的衡量标准。简单地说，许多用于小鼠的行为测试现在在斑马鱼中也有类似的方法（Kalueff et al., 2014）。

另一个关于如何改进对行为测试结果的解释的例子是，对相同的大脑功能进行多次测试，但这些测试有不同的表型呈现需求。继续以学习和记忆为例，如果实验者对检测小鼠海马体的功能改变感兴趣，可以进行 Morris 水迷宫空间学习任务测试，然后在情境和线索依赖的恐惧调节任务中测试携带相同突变或用相同药物治疗的小鼠。前一项任务与后一项任务具有不同的表型呈现需求，但两者都可以评估海马依赖性关联学习和记忆。例如，Morris 水迷宫需要视觉感知、主动运动反应（游到平台上），正确的反应会得到奖励（从迷宫中取出并放在温暖干燥的笼子里）。而恐惧条件反射任务（the fear conditioning task）使用听觉线索和各种其他方式，包括触觉和嗅觉线索，使得小鼠识别情境（电击发生的位置）。典型的行为反应表现为僵直，即被动反应而不是主动反应，疼痛是其行为表现的驱动力，例如，如果突变影响痛觉，这种损害将不会在 Morris 水迷宫试验中表现为关联记忆缺陷。同样，如果运动功能因突变而受损，则不会在恐惧条件反射任务中表现为记忆力减退。然而，如果这两项测试具有特定于任务的表型特征，突变动物在两个任务中均显示出相当的表型缺陷，则人们可以更有把握地得出结论，这种缺陷不是因为表型特征的改变，而是因为关联学习的改变。这两个任务，Morris 水迷宫、情境和线索依赖的恐惧条件反射组合试验的另一个方面是，它们有两个额外的特征可以帮助研究者解释结果：一是它们有非关系（非空间、非情境线索）的对照测试样式，二是它们都可以记录多种行为测量值，从而进一步帮助解释结果。Morris 水迷宫空间（隐藏平台）样式的对照任务是迷宫的可视化（或非空间）样式。恐惧条件反射任务的对照测试是音调线索探测。这些方面的任务已经在小鼠有关的文献中进行了详细讨论，因此读者可以参考这些出版物。然而，对于斑马鱼来说，我们还没有开发出足够多的行为范式，能够像在上述小鼠实验中那样，对表型进行同样强有力的对比。但这一概念仍然存在：具有重叠敏感性的多个任务利用特定的大脑功能，但有不同的表型呈现需求（动机、知觉和运动要求），能控制此类表型的任务，以及能够记录多种行为参数的任务，将有助于研究者设计相关行为分析试验以用于药物筛选或突变分析。

但是，应该如何组织此类测试或测试组合？该问题在啮齿动物相关文献中已有描述，从概念上讲，同样的解决方法也应该适用于斑马鱼。

29.6 测试组合的结构：自上而下与自下而上

有两种不同的方法可以组建测试组合，即自上而下或自下而上（Gerlai, 2002）。在自上而下的策略中，实验者从最复杂的目标表型开始。例如，如果一个人对关联学习感兴趣，他可以从空间或情境学习任务开始。如上所述，阳性目标、突变体或药物处理的动物在此类任务中表现不佳，可以进一步检查，以探索其表型受损的原因。后续研究可能包括对上述行为特征（动机、运动功能和感知）的测试。如果这些行为特征没有改变，我们可以进行后续的跟踪分析，进一步缩小可能导致关联学习任务表型受损的假设范围。这些分析应该包括对注意力、获得、巩固和记忆唤醒的测试。后三项测试很容易用药物进行，因为它们所需要的只是控制药物暴露的时间：在训练前给药可以检测获取，在训练后给药可以检测巩固阶段，在记忆探索试验前进行给药可以检测回忆（例如，Sison and Gerla, 2010）。注意力是获得记忆的重要前提，但测试起来有些复杂。对于啮齿类动物，已经开发了特定的测试方法来分析注意力，即 5 项选择反应时间测试（Muir et al., 1996）。该测试也适用于其他物种，包括斑马鱼（Parker et al., 2013; Fizet et al., 2016）。在上述情况下，与药物相比，突变更难测试，因为突变效应的时间不太容易控制。然而，已经精心设计了可诱导的基因打靶方法用于小鼠实验（Sauer, 1998），该方法也已经开始应用于斑马鱼实验（Pinzon-Olejua et al., 2017; Karttunen and Lyons, 2019）。

总而言之，在自上而下的方法中，实验者从最复杂和最重要的最终目标表型开始，如果出现阳性结果，则继续向下排除其他假设。这种方法的优点是，通常复杂的行为现象以及检测这些现象的行为测试对实验操作（药物或突变效应）高度敏感，因此实验者第一次就可能有很高的“命中率”。然而，这些命中的大部分可能是与目标表型（在我们的例子中为关联学习和记忆）无关的生物功能的改变，因此随着筛选的进行，将消除一些突变体或化合物。

自下而上的方法具有相反的理念。它从最基本的行为分析开始，然后发展到越来越复杂和更具体 / 狭义定义的表型，直到达到分析目标表型的实验目的。例如，这样的一组测试将从对运动功能的详细分析、对不同刺激方式的感知、对各种厌恶或喜爱的刺激做出反应的动机开始。随后，根据这些测试的结果，测试组合将转向对感兴趣的行为进行越来越复杂的分析。例如，如果目标表型是记忆，我们可以首先研究最简单的记忆形式，即习惯化和 / 或敏感化，然后转向简单的联想或操作性条件反射，来测试关联学习和记忆，最后将分析记忆（遗忘）的时间方面，包括保留间隔、消退和逆转学习。自下而上方法的优点是，随着筛选的进行，实验者将发现许多有趣的突变体或药物效应，尽管这些突变体或药物效应可能与目标表型没有直接关系，但仍可识别受各种药物和突变影响的大脑中的许多功能。换句话说，如果不想太具体，自下而上的方法是有用的，因为它允许通过分支来研究不同的目标表型。自上而下的方法虽然也允许分支，但筛选时更集中于预定的目标表型。

虽然可以通过设计筛选来减少重复测试，但在自上而下和自下而上的测试中，研究者必须对同一实验对象（突变或药物治疗动物）进行多次测试。对同一实验对象的重复测试存在复杂的测试间干扰问题。在这两种筛选方法中，重要的是组织测试的顺序，以尽量减少这种干扰。通常建议首先采用侵入性最小、厌恶性最小的测试，然后逐渐增加侵入性。McIlwain 等（2001）针对小鼠提供了如何在测试组合中进行行为测试的具体实例。尽管目前还没有针对斑马鱼的具体建议，但如何进行行为测试的一般原则也可能适用于斑马鱼。例如，在一组小鼠中，建议实验者从需要最少人工操作和 / 或与最轻压力或疼痛相关的测试开始。这也应该是斑马鱼行为测试组合的一个重要方面，因为斑马鱼的驯化程度远低于大鼠或家鼠，并且受人为操作的影响较大。事实上，一些研究小鼠行为的人认为，实验应该避免人为操作，并在它们各自的笼中测试动物的行为（de Visser et al., 2006）。饲养笼或者斑马鱼的水箱，可以配备大量刺激传递装置和行为反应监控摄像机，进行一系列的行为测试。对斑马鱼来说，采用这种方法的唯一复杂因素是该物种具有高度社会性。然而，鉴于视觉对斑马鱼来说是一种重要的感知方式，装有透明壁的小水箱可以使得单条斑马鱼在不受社会隔离有害影响的情况下进行单独测试。值得注意的是，这种家用水箱试验将允许同时分析大量斑马鱼的行为，因此可以实现大规模筛选所需的高通量。目前，尽管还没有任何关于这种筛选的公开报道，但是我们应该努力开发这种技术方法。

29.7　内表型

下面我们将讨论内表型的问题，即与重要的大脑功能和行为密切相关的特征，但可能比实际目标表型的测量更容易 / 更简单。使用内表型进行筛查的主要原因是简单。这种筛选往往不像我们上面描述的自上而下和自下而上那样全面和详细，但是它们的聚焦性很有吸引力，因为它们需要的工作量要少得多。如果内表型选择得当，这样的筛选可能会非常有效，能识别出参与或修饰复杂脑功能的关键基因或化合物，这些复杂脑功能是相关表型或中枢神经系统疾病的基础。这种内表型和筛选试验的一个例子是前脉冲抑制（prepulse inhibition, PPI）。在 PPI 中，较弱的刺激（前脉冲，通常是音调）会抑制生物体（通常是啮齿动物）对随后的强刺激（脉冲，通常是啮齿动物的吹气）的反应。PPI 是一种重要的感觉运动门控指标，精神分裂症患者的感觉运动门控功能受损（Braff and Geyer, 1990）。当然，PPI 受损并不等同于精神分裂症。PPI 确实无法捕捉与精神分裂症相关的一系列复杂变化，包括该疾病的阳性和阴性症状（Walker and Lewine, 1988）。尽管如此，在精神分裂症患者中发现的 PPI 损伤与其他症状之间的强相关性使大量研究人员相信 PPI 是一种极好的内表型。它可以用于精神分裂症的建模，以及使用动物来测试这种疾病的机制，并寻找治疗这种疾病的潜在改善效果。斑马鱼也适用于此类研究（Burgess and Granato, 2007）。

29.8　如何衡量行为?

我们要考虑的最后一点是如何量化行为。动物行为学家认为，测量行为的最佳方法是观察动物并测量行为谱的要素（Huntingford, 1984）。行为谱是指特定物种所特有的运动和姿势模式的完整集合。目前已经搜集了一些斑马鱼的行为谱（Kalueff et al., 2013）。行为要素的量化通常需要观察动物，并使用秒表或记录事件的计算机软件应用程序，测量动物执行行为的时间、行为发生的频率（次数）、行为首次发生的潜伏期以及行为的持续时间（每次行为开始时的持续时间）。有些人可能认为这种基于观察的行为量化方法具有主观性，相当不精确，但数十年的经验证据表明并非如此。即使有了复杂的计算机视觉应用，人脑似乎仍然是最精确和最复杂的模式识别设备之一，通过一些实践和标准化，基于观察的方法可以提供一组高度可靠和复杂的行为结果（Blaser and Gerlai, 2006）。然而，对于高通量和大规模筛选，坐在计算机显示屏前通过人工观察和测量动物的行为是不可行的。

另一种方法是使用计算机软件应用程序，允许在实验场地检测动物并以自动化方式记录其运动。本书的第三十章详细介绍了这种方法，这里我们只简单地讨论一下。如今，大多数能够以自动化方式测量动物行为的系统都采用视频跟踪，但过去的系统也采用了力传感器或光电管。市面上有许多复杂的视频跟踪系统（例如：www.noldus. com; www.viewpoint.fr/en/p/software/videotrack; https://www.harvardapparatus. com/smart-video-tracking-system.html; https://www.rsipvision.com/video-tracking-of-animal-behavior/）。这些系统特别适合斑马鱼，斑马鱼是一种白天活动的动物，这种物种的视觉感知很复杂，可以对视觉线索做出很好的反应，包括计算机动画视觉图像（Gerlai, 2017）。一些跟踪系统将行为反应的检测和量化与刺激的传递（视觉或其他类型，例如电击或食物传递）结合起来。这些系统可以测量动物运动（游泳）路径的许多参数，尤其擅长获取运动的定量测量。例如，或许实验人员无法判断实验中的斑马鱼是以 3.2cm/s 或 3.5cm/s 的速度游泳，还是以 142°或 124°的角度转弯，但量化此类行为对于视频跟踪系统来说根本不是问题。简而言之，当视频跟踪系统记录动物精确的 *x-y* 坐标，并从该数据矩阵输出大量衍生测量值时，可以获得大量关于受试者行为的信息，这里只提几个常用的跟踪参数，包括运动速度、运动位置、转角、加速 / 减速度、运动方向、到预设位置（点、线或区域）的距离。此外，除了这些测量处，大多数跟踪系统还可以输出超过特定用户设定阈值的时间或频率的数据。视频跟踪系统还可以输出个体内的大多数行为测量的时间变异性，这代表了动物移动 / 行为的一致性 / 不一致性。但是，这些系统不太擅长的是"识别"（检测和量化）复杂的运动或姿势模式，即测量行为谱的组成部分，尽管这对于现代软件应用来说已经不是什么问题了。例如，研究者可利用一些商业视频跟踪系统将斑马鱼跟踪参数的一些特征 / 定义组合起来。以斑马鱼拍打玻璃为例。行为谱中的这一行为现象通常是由于鱼类试图用力游过水箱的玻璃壁而产生的。当受试鱼试图接近或靠近奖赏刺激时，比如它自己的同类在测试水箱外游泳，会发生这种情况。当受试鱼试图逃离厌恶性刺激时，也会发生这种情况，例如，同域捕食者的动画图像投射到测试水池另一侧的计算机屏幕上。在这两种情况下，行为表现为快速移动和一系列靠近试验箱玻璃壁的快速转弯。在过去，只有当实验者观察动物的行为，并通过按键盘上指定的键将其手动输入计算机时，才能测量该行为（一种典型的事件记录软件应用程序方法）。然而，如今人们可以对视频跟踪软件应用程序进行编程，将往复移动作为更简单的跟踪参数的组合进行测量。例如，当游泳速度超过设定的阈值（比如 4cm/s）并以至少 180°的角度旋转，且频率超过每 5s 一次，现代跟踪系统就可以检测到斑马鱼的往返运动，并且这些响应发生在距玻璃壁 5cm 以内。将更简单的跟踪措施结合起来构建更复杂的行为的概念，也适用于静止、不稳定的运动、跳跃等行为，这些行为都是斑马鱼行为谱中过去需要人类观察记录的行为组成部分。

29.9 结论

鉴于众多神经生物学现象和中枢神经系统疾病的复杂性，可能需要无偏倚、全面、大规模的筛选来为后续假设驱动的研究奠定基础。这样的筛选有可能识别出大量的基因、蛋白质和生化途径，这些基因、蛋白质和生化途径是靶标现象或疾病表型的基础。然而，突变筛选和药物筛选是一个劳动密集型、昂贵且耗时的过程。斑马鱼因其实用的特点，包括其体形小和繁殖力强的特性，特别适合这种大规模筛选。虽然斑马鱼在行为神经科学和行为遗传学研究方面是一个相对新的物种，并且很少有人尝试将其用于行为筛查以检测药物和突变效应，但它在这方面的发展前景是可预期的。小鼠神经行为遗传学文献为如何构建和使用行为测试组合提供了很好的案例。这些方法的概念也适用于斑马鱼，我们的任务是真正地开发和试验这些方法。

（郭煊君翻译 石晶晶审校）

参考文献

Abrahao, K.P., Salinas, A.G., Lovinger, D.M., 2017. Alcohol and the brain: neuronal molecular targets, synapses, and circuits. Neuron 96, 1223-1238.

Baraban, S.C., Dinday, M.T., Castro, P.A., Chege, S., Guyenet, S., Taylor, M.R., 2007. A large-scale mutagenesis screen to identify seizure-resistant zebrafish. Epilepsia 48, 1151-1157.

Blanchard, D.C., Griebel, G., Blanchard, R.J., 2003. The Mouse defense test battery: pharmacological and behavioral assays for anxiety and panic. European Journal of Pharmacology 463, 97-116.

Blaser, R., Gerlai, R., 2006. Behavioral phenotyping in zebrafish: comparison of three behavioral quantification methods. Behavior Research Methods 38, 456-469.

Braff, D.L., Geyer, M.A., 1990. Sensorimotor gating and schizophrenia. Human and animal model studies. Archives of General Psychiatry 47, 181-188.

Bruni, G., Rennekamp, A.J., Velenich, A., McCarroll, M., Gendelev, L., Fertsch, E., Taylor, J., Lakhani, P., Lensen, D., Evron, T., Lorello, P.J., Huang, X.P., Kolczewski, S., Carey, G., Caldarone, B.J., Prinssen, E., Roth, B.L., Keiser, M.J., Peterson, R.T., Kokel, D., 2016. Zebrafish behavioral profiling identifies multitarget antipsychotic-like compounds. Nature Chemical Biology 12, 559-566.

Burgess, H.A., Granato, M., 2007. Sensorimotor gating in larval zebrafish. Journal of Neuroscience 27, 4984-4994.

Crawley, J.N., Paylor, R., 1997. A proposed test battery and constellations of specific behavioral paradigms to investigate the behavioral phenotypes of transgenic and knockout mice. Hormones and Behavior 31, 197-211.

de Visser, L., van den Bos, R., Kuurman, W.W., Kas, M.J., Spruijt, B.M., 2006. Novel approach to the behavioural characterization of inbred mice: automated home cage observations. Genes, Brain and Behavior 5, 458-466.

Driever, W., Solnica-Krezel, L., Schier, A.F., Neuhauss, S.C., Malicki, J., Stemple, D.L., Stainier, D.Y., Zwartkruis, F., Abdelilah, S., Rangini, Z., Belak, J., Boggs, C., 1996. A genetic screen for mutations affecting embryogenesis in zebrafish. Development 123, 37-46.

Fizet, J., Cassel, J.C., Kelche, C., Meunier, H., 2016. A review of the 5-choice serial reaction time (5-CSRT) task in different vertebrate models. Neuroscience & Biobehavioral Reviews 71, 135-153.

Gerlai, R., 2002. Phenomics: fiction or the future? Trends in Neurosciences 25, 506-509.

Gerlai, R., 2010. High-throughput behavioral screens: the first step towards finding genes involved in vertebrate brain function using zebrafish. Molecules 15, 2609-2622.

Gerlai, R., 2017. Animated images in the analysis of zebrafish behaviour. Current Zoology 63, 35e44.

Goldsmith, P., 2004. Zebrafish as a pharmacological tool: the how, why and when. Current Opinion in Pharmacology 4, 504-512.

Harmer, C.J., Goodwin, G.M., Cowen, P.J., 2009. Why do antidepressants take so long to work? A cognitive neuropsychological model of antidepressant drug action. British Journal of Psychiatry 195, 102e108.

Huntingford, F., 1984. The Study of Animal Behaviour. Springer, The Netherlands, p. 412.

Kalueff, A. V., Gebhardt, M., Stewart, A. M., Cachat, J. M., Brimmer, M., Chawla, J. S., Craddock, C., Kyzar, E. J., Roth, A., Landsman, S., Gaikwad, S., Robinson, K., Baatrup, E., Tierney, K., Shamchuk, A., Norton, W., Miller, N., Nicolson, T., Braubach, O., Gilman, C.P., Pittman, J., Rosemberg, D.B., Gerlai, R., Echevarria, D., Lamb, E., Neuhauss, S.C., Weng, W., Bally-Cuif, L., Schneider, H., the Zebrafish Neuroscience Research Consortium, 2013. Towards a comprehensive catalog of zebrafish behavior 1.0, and beyond. Zebrafish 10, 70-86.

Kalueff, A.V., Stewart, A.M., Gerlai, R., 2014. Zebrafish as an emerging model for studying complex brain disorders. Trends in Pharmacological Sciences 35, 63-75.

Karttunen, M.J., Lyons, D.A., 2019. A drug-inducible transgenic zebrafish model for myelinating glial cell ablation. Methods in

Molecular Biology 1936, 227-238.
Kegel, L., Rubio, M., Almeida, R.G., Benito, S., Klingseisen, A., Lyons, D.A., 2019. Forward genetic screen using zebrafish to identify new genes involved in myelination. Methods in Molecular Biology 1936, 185-209.
Knapik, E.W., 2000. ENU mutagenesis in zebrafishefrom genes to complex diseases. Mammalian Genome 11, 511-519.
Li, M., Zhao, L., Page-McCaw, P.S., Chen, W., 2016. Zebrafish genome engineering using the CRISPR-Cas9 system. Trends in Genetics 32, 815-827.
Malicki, J., 2000. Harnessing the power of forward genetics-analysis of neuronal diversity and patterning in the zebrafish retina. Trends in Neurosciences 23, 531-541.
McIlwain, K.L., Merriweather, M.Y., Yuva-Paylor, L.A., Paylor, R., 2001. The use of behavioral test batteries: effects of training history. Physiology & Behavior 73, 705-717.
Muir, J.L., Bussey, T.J., Everitt, B.J., Robbins, T.W., 1996. Dissociable effects of AMPA-induced lesions of the vertical limb diagonal band of Broca on performance of the 5-choice serial reaction time task and on acquisition of a conditional visual discrimination. Behavioural Brain Research 82, 31-44.
Mullins, M.C., Hammerschmidt, M., Haffter, P., Nüsslein-Volhard, C., 1994. Large-scale mutagenesis in the zebrafish: in search of genes controlling development in a vertebrate. Current Biology 4, 189-202.
Naumova, O.Y., Lee, M., Rychkov, S.Y., Vlasova, N.V., Grigorenko, E.L., 2013. Gene expression in the human brain: the current state of the study of specificity and spatiotemporal dynamics. Child Development 84, 76-88.
Pan, Y., Mo, K., Razak, Z., Westwood, J.T., Gerlai, R., 2011. Chronic alcohol exposure induced gene expression changes in the zebrafish brain. Behavioural Brain Research 216, 66-76.
Parker, M.O., Ife, D., Ma, J., Pancholi, M., Smeraldi, F., Straw, C., Brennan, C.H., 2013. Development and automation of a test of impulse control in zebrafish. Frontiers in Systems Neuroscience 7, 65. https://doi.org/10.3389/ fnsys.2013.00065.
Patton, E.E., Zon, L.I., 2001. The art and design of genetic screens: zebrafish. Nature Reviews Genetics 2, 956-966.
Pinzon-Olejua, A., Welte, C., Chekuru, A., Bosak, V., Brand, M., Hans, S., Stuermer, C.A., 2017. Cre-inducible site-specific recombination in zebrafish oligodendrocytes. Developmental Dynamics 246, 41-49.
Pohorecky, L.A., Brick, J., 1988. Pharmacology of ethanol. Pharmacology & Therapeutics 36, 335-427.
Rihel, J., Schier, A.F., 2012. Behavioral screening for neuroactive drugs in zebrafish. Developmental Neurobiology 72, 373-385.
Roberto, M., Varodayan, F.P., 2017. Synaptic targets: chronic alcohol actions. Neuropharmacology 122, 85-99.
Ron, D., Barak, S., 2016. Molecular mechanisms underlying alcohol-drinking behaviours. Nature Reviews Neuroscience 17, 576-591.
Sauer, B., 1998. Inducible gene targeting in mice using the Cre/lox system. Methods 14, 381-392.
Sison, M., Gerlai, R., 2010. Associative learning in zebrafish (*Danio rerio*) in the plus maze. Behavioural Brain Research 207, 99-104.
Stein, R.L., 2003. High-throughput screening in academia: the harvard experience. Journal of Biomolecular Screening 8, 615-619.
Sweatt, J.D., 2009. Mechanisms of Memory, second ed. Academic Press, p. 362.
Walker, E., Lewine, R.J., 1988. The positive/negative symptom distinction in schizophrenia. Validity and etiological relevance. Schizophrenia Research 1, 315-328.

第三十章

斑马鱼全生命周期行为表型分析的软件工具

Fabrizio Grieco[1], Ruud A. J. Tegelenbosch[1], Lucas P. J. J. Noldus[1,2]

30.1 引言

随着斑马鱼作为行为神经科学、药理学、毒理学和遗传学中的生物模型的兴起，研究人员对全面、灵活和易于使用的自动化系统的需求日益增加，它能够可靠地检测行为模式并从中提取定量数据。在本章中，我们主要关注图像分析软件，因为这是我们30年行业经验的主要领域，也是个人科学兴趣的主题。

跟踪软件利用目标分割和背景差分等方法将单个对象（在我们的例子中是鱼）从周围环境中分离出来。然后将位置（即身体质量中心的x、y坐标）在图像序列中集成以形成运动轨迹。从数字对象中提取的其他特征包括特定身体点的位置（鼻和尾尖）以及被检测对象的方向、形状和表面积。除非本章另有说明，我们使用EthoVision XT软件（www.noldus.com/ethovision）跟踪身体点并分析单个或多个受试体的轨迹［图30-1（B）～（F）］。重点将放在实际案例中，在这些案例中，研究者的目的是检测特定的行为模式，例如，转向或惊吓反应。然而，当受试者的位置和方向不相关时，例如仍在绒毛膜中的斑马鱼胚胎，该软件会在像素水平分析图像亮度的时间变化来检测运动［图30-1（A）］。视频跟踪软件的一般描述，见Dell等（2014）；针对斑马鱼的具体描述，参见Orger和de Polavija（2017）。

[1] Noldus Information Technology BV, Wageningen, the Netherlands.

[2] Department of Biophysics, Donders Institute for Brain, Cognition and Behavior, Radboud University, Nijmegen, the Netherlands.

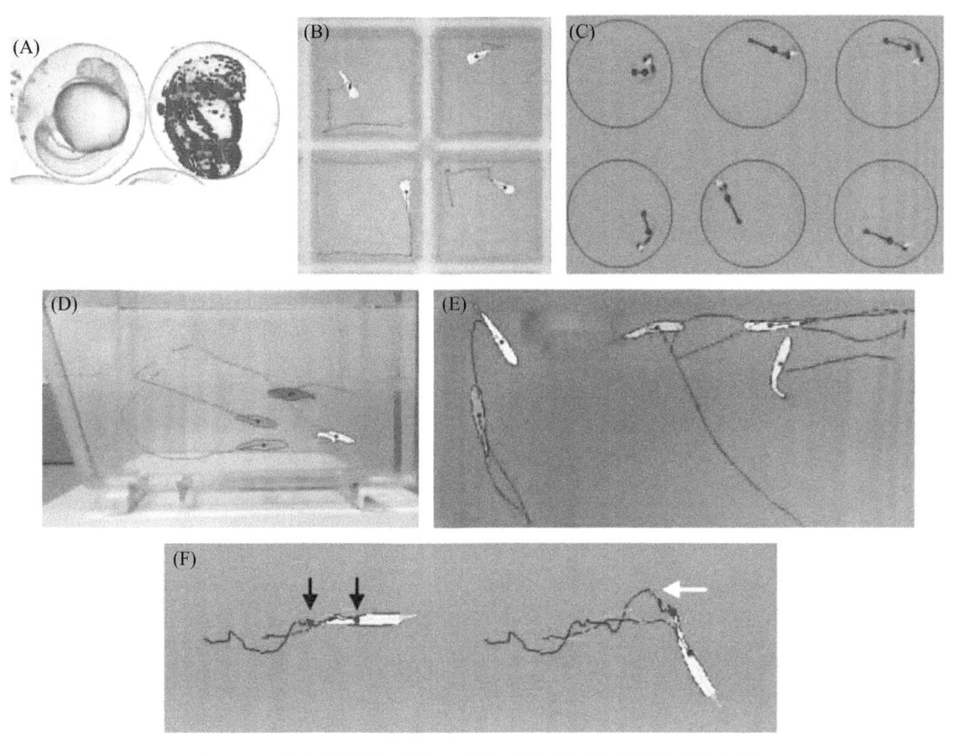

图30-1 **在不同实验装置中基于视频的斑马鱼行为观察示例**

（A）通过测量每帧亮度值（紫色突出显示）变化的像素数来检测胚胎的突发活动。（B）DanioVision系统中单条仔鱼的视频跟踪。每条仔鱼的中心点都被跟踪。（C）孔中单条仔鱼的视频跟踪；跟踪身体的三个点，即鼻点、身体中心和尾巴尖。（D）四条成年鱼在集群测试中的视频跟踪，正面视图。（E）另一次集群测试，这次从顶部拍摄鱼。（F）鱼在一个转弯过程中的中心点和尾端的轨迹，用黑色箭头表示；尾部各点形成的较长的路径用白色箭头表示；头部的方向用一个线段表示

30.2 追踪斑马鱼早期发育阶段的活动、心率、血流和肠（道）流

斑马鱼的胚胎发育非常迅速，第一次自发运动大约在17hpf时开始，在此期间胚胎重复地执行缓慢、交替的卷尾（Colwill and Creton, 2011a）。基于行为记录和遗传分析的研究已经确定了影响行为反应的基因（Muto et al., 2005; Smith et al., 2017）。未来行为分析面临的挑战之一是开发能够精确检测如卷曲等运动的系统（例如，完全卷曲与部分卷曲）。在本节中，我们简要描述使用DanioScope软件（www.noldus.com/danioscope）测量斑马鱼胚胎的突发活动。

与视频跟踪一样，斑马鱼胚胎的运动模式分析主要基于视频图像。然而，该软件不是跟踪特定的身体点，而是检测图像的时间变化，以确定动物运动的程度。如果动物是静止的，那么除了一些由背景产生的随机变化，视频中几乎所有的像素都保持相当稳定的亮度值。当动物开始运动时，一些像素的亮度值变化很快；有些像素变暗，有些变亮。移动越明显，像素变化越多。运动越快，描述这种变化的曲线越陡（见图30-2）。测量可以同

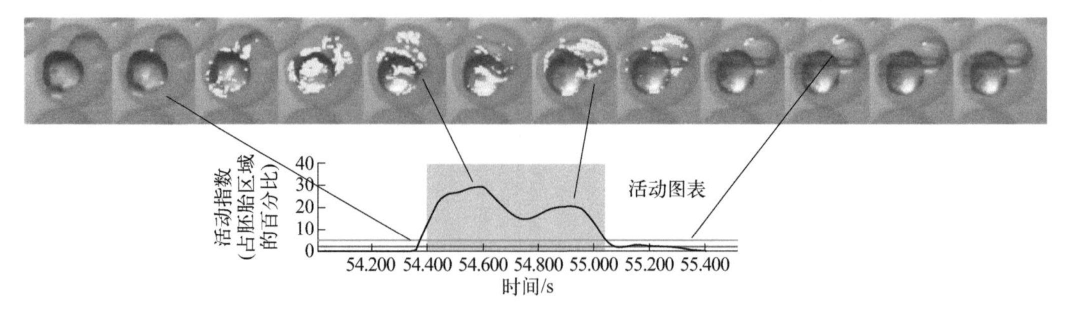

图30-2 **以斑马鱼胚胎的自发运动为指标测量活动值（activity）（另见正文）**

活动是图像中亮度（灰度值）从一个视频帧变化到下一个视频帧的像素数。上图：一个胚胎在绒毛膜中的视频。视频以25帧/s的速度录制（连续帧之间的时间为0.04s）。为了清晰起见，此处显示的是非连续帧，帧间间隔时间为0.12s。与前一帧相比，亮度发生显著变化的像素被高亮显示。第一帧（胚胎初期）和最后一帧（突发活动结束）的比较。下图：活动时间图，以占整个绒毛膜面积的百分比表示。注意活动的两个高峰，第二个代表尾部运动。颜色区域标志着运动的持续时间；通过比较活动值（%）和用户定义的阈值进行评估

时在多个胚胎中进行，而不需要移除绒毛膜。然后可以很容易地提取活动爆发次数、活动爆发持续时间和不活动时间的比例。例如，Swaminathan等人（2018）的报道显示，在20hpf时，*depdc5*基因敲除模型的自发卷曲活性降低。类似地，Riché等人（2018）敲除了与人类甘氨酸脑病相关的*gldc*基因，并显示*gldc*突变体在21hpf时降低了卷曲活性。在发展的后期，视频跟踪普遍用于测量运动距离、转向、活动期与不活动期等参数［如Ahmad et al.（2012）］。

视频图像中的像素变化原理也可应用于观察试验对象的特定局部，例如测量心率以及血管和肠道中的液体流量。不同之处在于，像素变化仅在焦点区域测量。对于心脏，像素变化的时间模式是周期性的；从离散傅里叶变换得到的频谱中提取出主频（即通过每分钟心跳次数计算的心率）（图30-3）。作为应用实例，我们报道了Grone等人（2016）的研究，他们创建了斑马鱼的合成蛋白结合蛋白1（STXBP1）突变系，该突变系与人类某些类型的癫痫相关。在许多差异中，在3dpf时，*stxbp1a*突变体仔鱼的心率低于同胞对照。其他商业软件，如De Luca等（2014）开发的ZebraBeat，也提供了通过视频分析量化心率的类似方法。ZeCardio软件（Cornet et al., 2017）分析了转基因斑马鱼仔鱼荧光心脏的图像（100hpf），并计算了校正的QT间期、射血分数（即心室每次收缩时泵出的血液比例）等

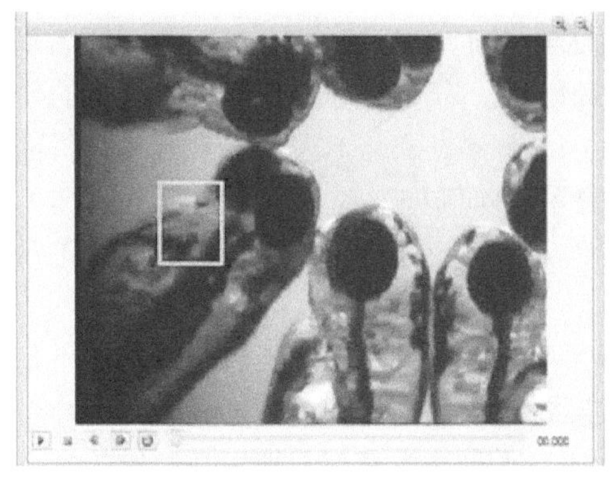

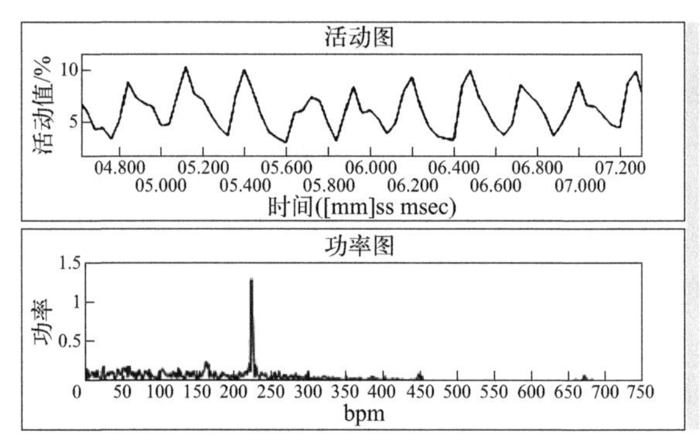

图30-3 **DanioScope心率分析软件截图**

左：多个斑马鱼仔鱼的视频，其中一条鱼的中心区域突出显示。该软件可以同时分析多条仔鱼。右上图：活动的时间变化（像素变化），说明周期模式。右下：显示峰值频率（心率）的功率谱，对应于223bpm

高级参数，并检测心脏骤停等事件。此外，视频中的像素变化也被成功地用于量化血管和肠道中的液体流动。

30.3 在孔板中用斑马鱼仔鱼进行高通量筛选

基于在受控测试环境中进行高通量遗传筛选的需求，科研人员开发了带有视频跟踪软件的观察室，如 DanioVision（www.noldus.com/daniovision）和 ZebraBox（ViewPoint）。在这里，我们主要介绍 DanioVision 观察室，它可以记录对视觉（闪光灯）、听觉、触觉（敲击孔板支架）、光遗传和化学刺激的行为反应，其中大部分是自动化的，在微孔板中可同时测试多达 96 条试验鱼（图 30-4）。

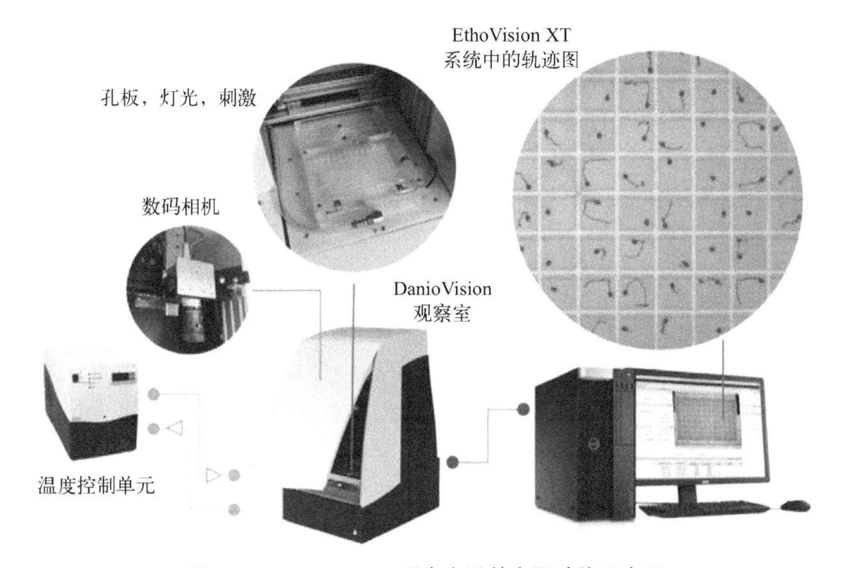

图 30-4 DanioVision 观察室及其主要功能示意图

左图：温度控制单元，用于在长期试验期间保持孔周围的水温恒定；箭头表示水流的方向。中图：DanioVision 观察室。右图：计算机和 EthoVision XT，以及 96 孔板中斑马鱼仔鱼运动特写。DanioVision 摄像机可将图像进行放大，以更详细地跟踪更少的孔

活力和节律（activity and rhythmicity）。运动活力和节律是通过受试者在试验期间或特定时间间隔内移动的距离来测量的。作为应用实例，我们报道了对生物钟基因（Ben-Moshe et al., 2014）、斑马鱼帕金森病同源基因敲除（Lopes da Fonseca et al., 2013）和帕金森病模型功能恢复（Cronin and Grealy, 2017）的研究。

惊吓反应（startle response）。例如，我们报告了暗闪光测试，在这个测试中，重复短时间关闭光源，产生周期性的间歇照明（Kim et al., 2017）。为了诱导僵直 / 惊吓反应，作者将孔板照亮 30s，然后关闭 30s（闪光黑暗条件）。这个过程重复了五次。以每 10s 时间内每条仔鱼移动的总距离来测量惊吓反应。在光照条件下，仔鱼表现出僵直反应，然后在黑暗条件下表现出惊吓反应。

当同时拍摄大量孔时，测试对象的表观体积小，限制了软件检测体形变化的能力，如 C 形弯曲。例如，用标准摄像机（1920×1080 像素）记录 96 孔培养皿中 4mm 的仔鱼可

能只有25～30像素长。然而，可以通过降低检测数量将仔鱼聚焦在24孔板中，从而获得更大的图像细节。然后，将合适帧率的拍摄与多个身体点的跟踪相结合，可以得到对行为反应的准确描述（见图30-6中的示例）。孔的大小不仅影响实验量，而且影响实验对象的行为。斑马鱼仔鱼在小孔里的移动比在大孔里的要少（Farrell et al., 2011; Padilla et al., 2011），当它们在小孔中饲养数天时，与在大孔中饲养的相比，它们更有可能出现骨骼畸形（Selderslaghs et al., 2009）。因此，在设计实验和解释数据时，应考虑测试孔的尺寸。

30.4 斑马鱼仔鱼回避行为的自动分析

越来越多的研究人员选择斑马鱼用于分析脊椎动物恐惧反应以及人类精神病理状况，如焦虑（Gerlai, 2010a, 2013）。目前，已经开发了一些行为事件用于筛选恐惧反应中的突变或药物引起的变化。斑马鱼的成鱼和仔鱼在许多刺激下会产生逃避和回避行为（Colwill and Creton, 2011b）。研究中的行为包括实验对象对捕食者的运动图像（Bass and Gerlai, 2008, Gerlai et al., 2009; Ahmed et al., 2011）、抽象形状（Ahmed et al., 2012）或机器人复制品（Ladu et al., 2015）的反应。此外，斑马鱼的仔鱼会从投射到孔板上的移动对象前游开，这可能是一种捕食者回避机制（Pelkowski et al., 2011; Barker and Baier, 2015）。

当实验者了解测试对象的反应行为时，如何自动测量恐惧反应以增加行为筛选的通量则是实验者需考虑的另一个问题。标准的视频跟踪参数已被广泛用于量化离开刺激点的时间，或到刺激点的平均距离，或到安全区域的距离。然而，逃跑和反掠食运动模式原则上可以通过个体短时速度增大（由于更频繁的快速游泳）、转弯角度变化（快速转弯）和游泳路径上更频繁、更急剧的转弯来检测。

在本节中，我们将重点放在一类测试上。在此类测试中，实验鱼（通常是仔鱼）将暴露在一个计算机化的二维（2D）对象（在大多数情况下是一个点，或“弹跳球”）面前（Barker and Baier, 2015; Colwill and Creton, 2010; Richendrfer et al., 2012; Bridi et al., 2017）。假设该物体可以模仿捕食者或捕食者的影子。在这种测试的另外一种形式中，设定该物体快速向鱼移动，并记录下鱼的反应。斑马鱼仔鱼会游离物体（逃逸或主动回避），并保持安全距离（被动回避）。我们的目的是展示视频跟踪系统在检测接近 - 回避事件和更详细地测量鱼的反应能力方面所具有的潜力。

在标准的双目标交互测试中，对运动物体和鱼进行跟踪。一般情况下，只跟踪中心点；但是，如果需要了解鱼的体型信息，例如，检测C形弯曲。就需要追踪它的鼻子和尾巴。使用Ethvision XT，我们定义了代表接近 - 逃逸事件顺序的变量（图30-5）。例如，目标靠近 > 鱼加速 > 鱼远离目标。为了定义这样的元素，我们使用标准的分析变量，比如加速，或者合并两个或更多个的变量。例如，目标靠近是一个具有两个可能值的状态变量，“on”是指当对象以高于特定阈值的速度移动时，以及与鱼的距离正在减少时；在其他所有情况下都是“off”。当鱼的加速度超过一定的值时，鱼加速就是主动的。鱼可能对靠近的目标有很好的反应，也可能没有反应；软件就可以很容易地计算出，例如，在这种方法之后有多少次出现了回应。

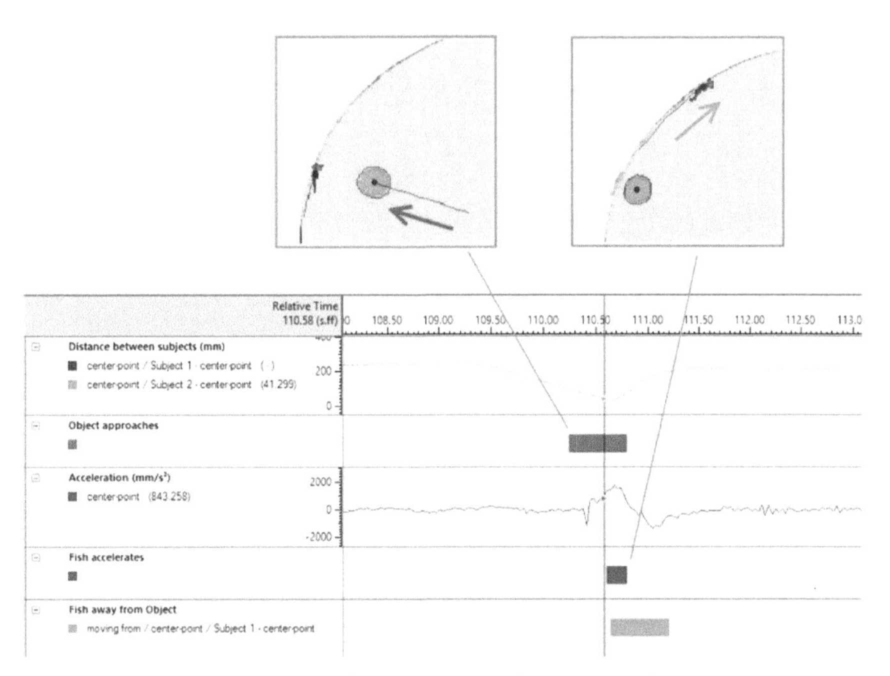

图30-5 **在捕食者-回避测试中检测的一系列事件**

上图：两个主要阶段，斑点靠近（左）和斑马鱼仔鱼的反应（右）。下图：变量发生的时间顺序图，从上到下：鱼与移动点之间的距离；基于圆点运动的状态变量；仔鱼加速；基于高加速度值的状态变量；基于仔鱼相对于圆点的运动的状态变量

深入分析的目的是提取响应过程中轨迹形状的信息。快速U形转弯可以通过观察主轴方向的时间变化来检测。此外，当软件正确检测到鱼的轮廓，并且成功跟踪了三个身体点（中心点、鼻子和尾巴），软件可以利用三个跟踪点连接的线段形成的角度找到C形弯曲（见图30-6）。

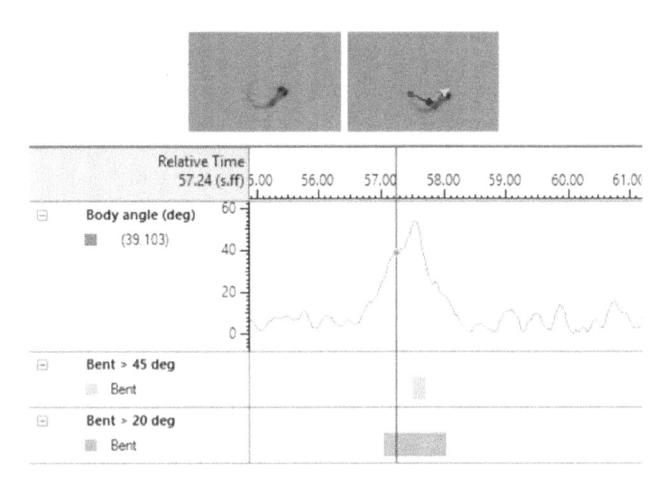

图30-6 **检测斑马鱼仔鱼C形弯曲**

左上图：弯曲时的仔鱼图像。右上图：与检测到的三个身体点相同的图像。下图：身体角度和两个状态变量的时间图，当身体角度超过一个阈值时得分。在计算状态变量之前，可以先对数据进行简化处理，去除身体角度的随机变化。通过不同的阈值（在这个例子中分别是45°和20°），我们可以检测到运动模式的不同阶段。例如，计算到最大弯曲的持续时间。在这个例子中，以高速（1000fps）拍摄实验对象；在不丢失帧数的情况下，通过减慢速度将视频转换为30帧/s。然后可以容易地将统计结果中的响应持续时间转换为实际值。原则上，如果可以的话，可以使用原始的高速记录来进行分析。视频由挪威分子医学中心的Nancy Saana Banono博士和Camila Vicencio Esguerra博士提供

30.5　在成年斑马鱼的常见测试中检测特定行为

在本节中，我们重点讨论两种常见的测试，新水箱潜水实验和T形迷宫。新水箱潜水实验是一种用于测试鱼暴露在新环境中的焦虑相关反应的范式（Cachat et al., 2010）。斑马鱼表现出明显的习惯化反应，类似于啮齿类动物的行为反应（Wong et al., 2010）。新水箱潜水实验中重要的终点是实验鱼在水箱下半部分的游动时间，从水箱底部到顶部的转换频率，以及僵直（freezing）行为的持续时间和频率（Egan et al., 2009; Cachat et al., 2010）。在新水箱潜水实验中，鱼通常是从侧面拍摄的，只跟踪它的几何中心。图30-7显示了已有的输出示例。从轨迹数据中，软件可以很容易地提取信息，如实验鱼从底部到顶部的

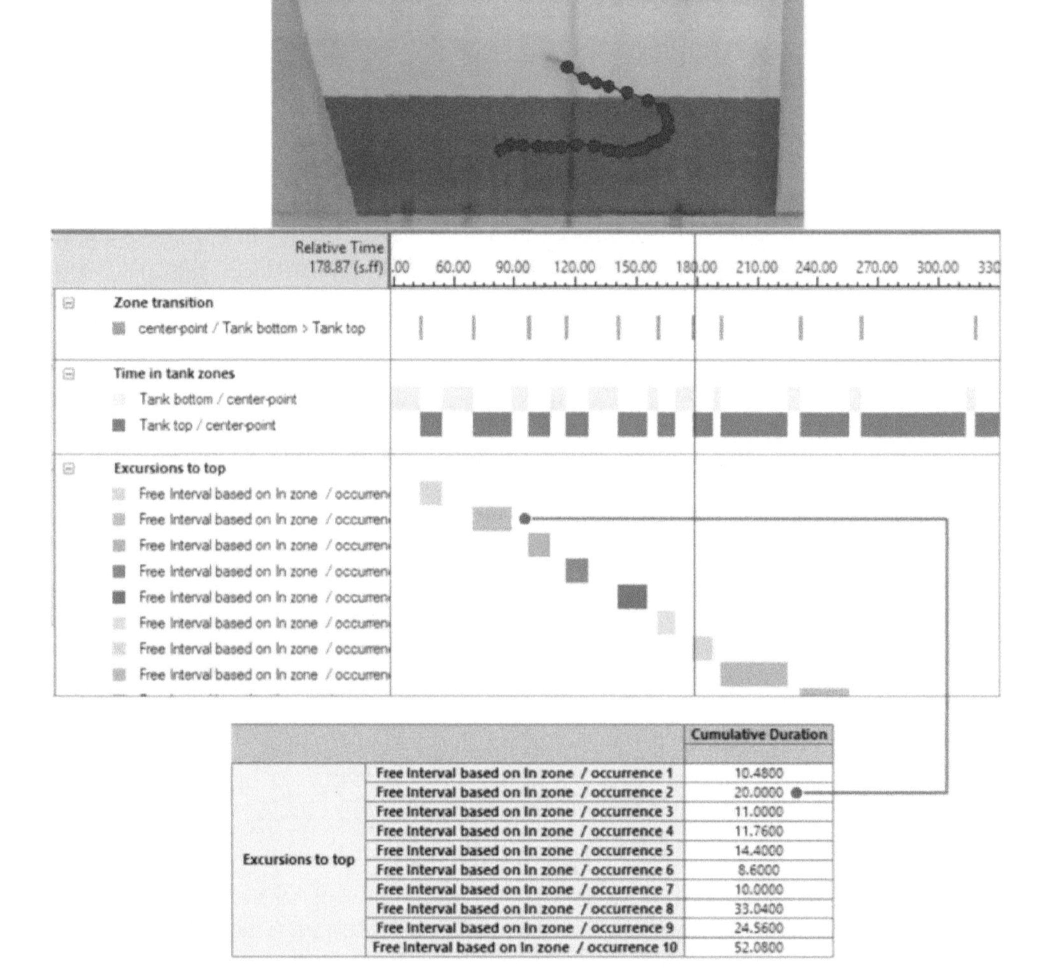

图30-7　EthoVision XT中新水箱潜水实验的截图

可以检测到从底部区域到顶部区域的过渡；每个间隔数据可以使用自由区间参数（Free Interval parameter）进行分离，并分段导出以供进一步分析

转换次数（以不同颜色显示），以及在底部区域和顶部区域所花费的总时间。在测试中，可以对底部 / 顶部驻留的独立实例进行详细分析（在第三个图中用不同颜色的条表示），例如，底部驻留的时间是否随着测试的进行而显著减少，以及在哪个时间段内会减少。与压力有关的静止期（僵直；定义可参照文献 Blaser and Gerlai, 2006）可以使用两种类型的视频跟踪输出变量进行有效评分，①在相邻的视频图像之间没有空间位移或存在非常短的空间位移（例如，Cachat et al., 2011a）或②在相邻的视频图像之间检测到的鱼的体型没有变化或变化非常小［Ethovision XT 中的 Immobility（静止）］（Cachat et al., 2010）。然而，这两种方法都不够灵敏，无法检测到像在僵直时观察到的频繁的眼部运动，因此不能排除类似睡眠 / 休息的行为（Kalueff et al., 2013）。此外，轨迹数据还可以以更直接的方式来表示行为的时间变化。例如，通过将轨迹数据转换为密度函数，可以创建鱼类位置的热图（图 30-8）。在 6min 的测试中，将为随后的三个 2min 间隔创建热图。随着测试的进行，鱼在水箱顶部停留的时间也会越来越长。

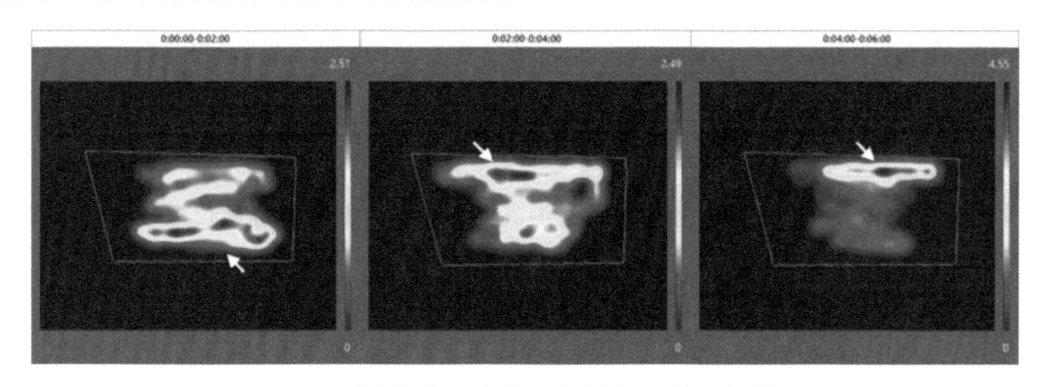

图 30-8 **在新水箱潜水实验中创建斑马鱼点位置的热图**

实验时间设定为 6min；数据被分成 2min 的间隔。随着时间的推移，需注意“热点”（用箭头表示）的位置从底部到顶部下发生了变化

T 形迷宫通常用于学习和记忆范式。鱼可以在迷宫的两条视觉上截然不同的臂中进行选择。例如，选择一种刺激，即迷宫壁的颜色或质地，与食物奖励相关联，而选择另一种刺激则没有奖励。在典型的实验中，斑马鱼对与食物奖励相关的刺激有明显的偏好，并且还能学习到辨别逆转（Colwill et al., 2005）。

在测试开始时，一扇断头台式的门被升起，将实验对象置于迷宫中。在门的两侧，存在两个区域，“释放室”和“启动区”（见图 30-9 中的①）。当鱼从“释放室”移动到“启动区”时，软件将开始跟踪。当鱼进入迷宫时，门立即降下。当鱼到达目标（诱饵）区域时，跟踪停止（图 30-9 中②）。然后，将鱼从迷宫中取出，软件准备好对“释放室”和“启动区”之间的下一条鱼的区域过渡进行检测（the next transition）。下一个跟踪测试场次可以自动启动，无需操作员的干预。当鱼在当前测试场次结束后返回到起始区域附近时，为了防止不必要的启动，必须进行区域过渡检测（transition）。此外，通过电动 T 形迷宫，软件可以控制门的升降。

在每一次测试中，软件系统都能标记出目标臂是否正确。由于这一点在以后的实验中可能会发生变化，因此在实验开始，实验方案会注明哪个臂是正确的目标，哪个不是，分析过滤功能根据方案来显示结果，例如特定目标手臂的数量和错误率。另一个常见的测量指

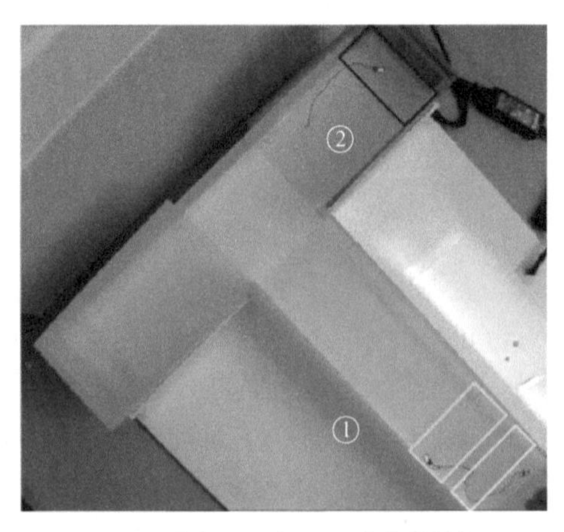

图 30-9 **成年斑马鱼 T-形迷宫的俯视图**

① 定义了“释放室（release chamber）”和“启动区（start）”两个区域，当鱼从前一个区域移动到后一个区域时，实验自动开始。② 在每个实验臂的末端定义一个区域以停止实验。该方案可以为每个实验指定一个臂作为停止区。参见正文

标是测量离开“释放室”区域或到达“目标区”的潜伏期（Colwill et al., 2005）。

Gould 等人（2011）描述了水箱内带有奖励盒的 T 形迷宫的改进版本。为了研究成年鱼的联想学习，还使用了四臂“+”字迷宫（Sison and Gerlai, 2010）。由于改进的迷宫具有与 T 形迷宫类似的结构，并且测试是通过高架摄像机进行的，因此可以参考应用以上概述的区域定义和行为终点。另一个版本的 T 形迷宫，设计用于药物敏感性测试，以评估实验鱼逃到有利的栖息地，包括目标箱（Darland and Dowling, 2001）。

30.6 成年斑马鱼的攻击行为检测

在斑马鱼中，攻击性的表现被定义为鱼竖起背鳍、尾鳍、胸鳍和臀鳍的一种姿势，通常伴随着波动的身体运动或尾鳍的小拍打（Gerlai et al., 2000）。攻击性表现经常与攻击行为交替出现。这是一种典型的短距离快速游向对手的行为，有时还伴有张嘴和咬的动作。

在本节中，我们将展示如何在视频跟踪中结合测量来记录运动模式，而不需要使用劳动密集型和缓慢的基于人为观察的手动事件记录方法。目前有不同的方法可用来检测攻击性（Way et al., 2015）。在这里，我们关注的是在镜子测试中可以观察到的身体和尾巴的运动。这个测试有不同的版本，取决于镜子是平行放置还是与鱼缸背面呈一定的角度放置（Blaser and Gerlai, 2006; Marks et al., 2005; Ariyomo and Watt, 2012; Moretz et al., 2007）。典型的观察终点通常以鱼实施攻击行为所花费的时间占比，或用于顶撞或啃咬镜子的时间占比表示，这些指标可采用人工计算获得（Gerlai et al., 2000; Marks et al., 2005; Pham et al., 2012; Teles et al., 2016）。我们将展示视频跟踪软件在捕捉和量化与攻击相关的行为方面的潜力。这一原则适用于类似的攻击性测试，例如，可将实验鱼暴露于多个同类个体、单个同类个体或虚拟鱼的电脑图像场景中。

为了检测尾拍，可以使用 EthoVision XT 中的三点跟踪方法，该方法利用基于参数的鱼形模型。该模型可强有力地形成对鱼的遮挡和身体接触，并且在相机置于头顶时是最佳的。在每个采样时间点提取三个身体点位：体中心点（质心），吻端点，尾尖点。在本节中，我们主要关注体中心点和尾尖点。从图 30-10 可以看出，当鱼有规律地游动时，体中心点和尾尖点的路径非常相似。然而，当鱼转了个弯，甚至只拍了一下尾巴，路径就会有所不同；尾尖点移动的距离比体中心点移动的距离长。我们在 EthoVision XT 中使用了一个可定制的 JavaScript 算法，读取体中心点和尾尖点从一个样本时间移动到下一个样本时间的距离。当尾尖点移动距离与体中心点移动距离的比值超过用户定义的阈值时，会检测到尾移动。通过多种方式可进行数据筛选，例如，使用移动的平均距离。也可以应用其他的筛选标准以确定鱼是否仍然停留在镜子前面。在这种情况下，体中心点移动的距离几乎为零，使得尾尖点移动距离与体中心点移动距离的比值高得离谱，在这一情况下，我们需要删除体中心点移动距离低于某一限制的数据点。为了有效地检测尾拍，视频的高分辨率和最低限度的反射是至关重要的，后者可以通过红外背光来实现。JavaScript 代码可以根据需求定制。

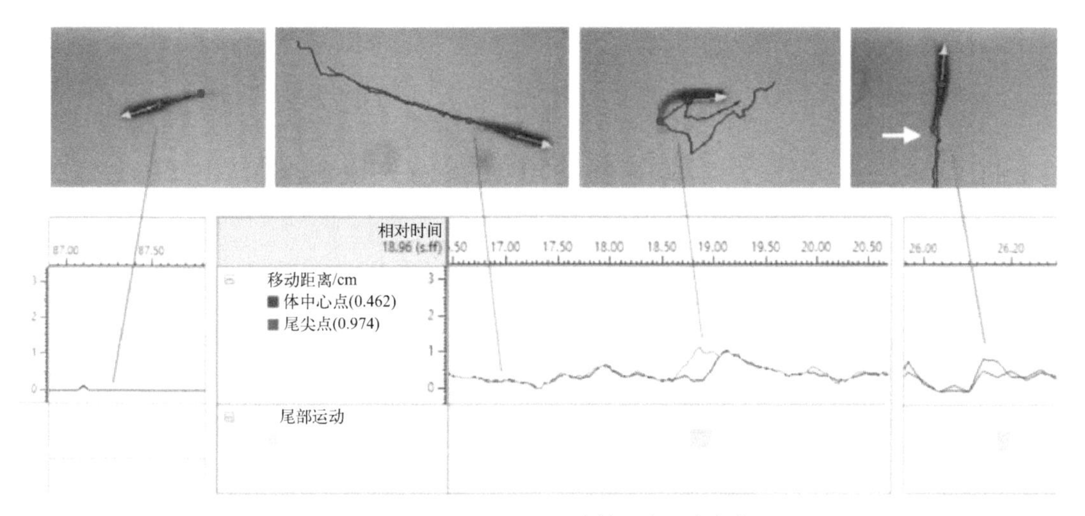

图 30-10　EthoVision XT 中的三点跟踪方法

上部：EthoVision XT 跟踪一条成年鱼的三个主体点：吻端点（三角形）、体中心点（圆形）、尾尖点（方形）的截图。从左到右：鱼不动；鱼慢慢地游着，大致呈直线；鱼在一系列不稳定的运动中转向；单尾跳动（箭头表示尾尖点相对于体中心点移动的额外距离）。中间：按体中心点（红色）和尾点尖（紫色）移动的样本（试验鱼）距离绘制的试验鱼游动时间图。下部：基于两个身体点移动距离的比例的自定义参数的时间图。当比值超过用户定义的阈值时，该参数的值为 on（用蓝色段表示）

一旦检测到了鱼尾运动，我们就选择那些与鱼的特定位置和方向相关的实例。为了达到这个目的，我们使用不同的头部指向区域来测量鱼相对于镜子的方向，并且只有当鱼距离镜子很近的时候才进行测量。头部方向也能用来区分鱼是否处于头朝前的姿势，也就是说，当鱼与镜子形成的角度约为 90°时表明头朝前；而当该角度小于 45°时，表明鱼体处于横向姿势。当鱼向镜子展示身体的左侧或右侧时，头部展示方向也不同（图 30-11）。

Gutiérrez 等人（2018）提出了一种检测斑马鱼攻击行为的类似方法，并将其应用于斑马鱼仔鱼。这个方法和我们的方法的主要区别是前者使用了游泳路径的弯曲度，而我们的方法是基于鱼的尾巴相对于体中心点的轨迹。

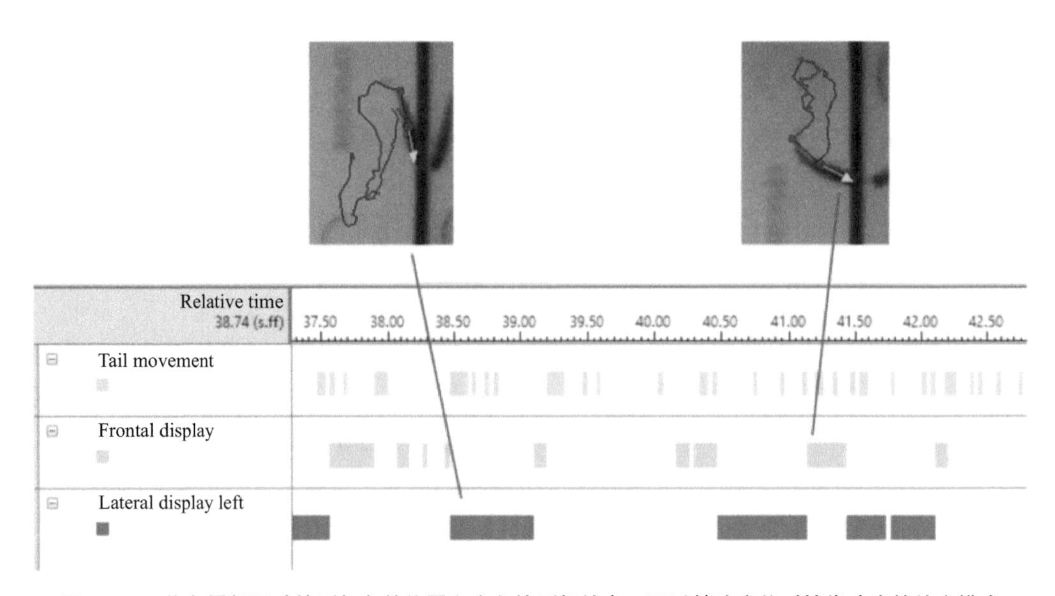

图30-11　**将鱼尾部运动检测与鱼的位置和方向检测相结合，可以精确定位对镜像响应的特定模式**

上图：截取自 EthoVision XT 的屏幕截图。下图：三个自定义状态变量的时间图，即尾部运动（如上图定义）、头朝前展示和横向展示。后两个变量基于鱼在刺激（镜子）附近的位置，并结合吻部的方向（EthoVision XT 中的头部朝向；参见正文）

30.7　成年斑马鱼群聚行为的量化

尽管有关斑马鱼行为的研究结果正在迅速增长，但关于群聚的研究仍存在很多未知，即在自然条件下群聚的基本特征尚未详细描述（Miller and Gerlai, 2011a）。在这里，我们专注于行为终点的一个子集，一旦将原始数据导出到电子表格中，就可以使用视频跟踪应用程序或借助对原始数据的简单操作来提取这些终点。有关群聚测试中使用的量化方法和典型行为终点的概述，以及与群聚特征相关的问题，参见以下文献：Saverino and Gerlai（2008），Green et al.（2012），Pham et al.（2012），Miller and Gerlai（2011b），Miller et al.（2012），Buske and Gerlai（2014）。

从基本的视频跟踪数据（即原始的 x, y 坐标）中可以提取鱼群凝聚力的相关行为指标。最近邻距离（nearest neighbor distance, NND）是任何一条鱼与距其最近的鱼之间的距离。在图 30-12 中，鱼 1 的中心（在上图的屏幕截图中用红点标记）和另一条鱼的中心之间的距离随时间绘图。要获得 NND，首先可以找到每个时间点的四个值中的最小值。在图 30-12 的示例中，鱼 1 在时间 70.0s 的 NND 约为 21.7mm。然后计算鱼 1 与其他鱼最小值的平均值，并对其余受试鱼也进行同样的计算。该报道通常需要计算出鱼的整个试验组的平均值。类似地，个体间距离（inter-individual distanc）（Miller et al., 2012）也可以通过对鱼群所有成员之间的距离进行平均计算（Green et al., 2012）。请注意，因为使用了整体平均值，所以即便在软件不跟踪单独标记的鱼时，也可以从中提取有意义的信息。在 Mahabir et al.（2013）、Séguret et al.（2016）、Kim et al.（2017）的研究中，报道了遗传背景或突变对鱼群凝聚力产生影响的案例。

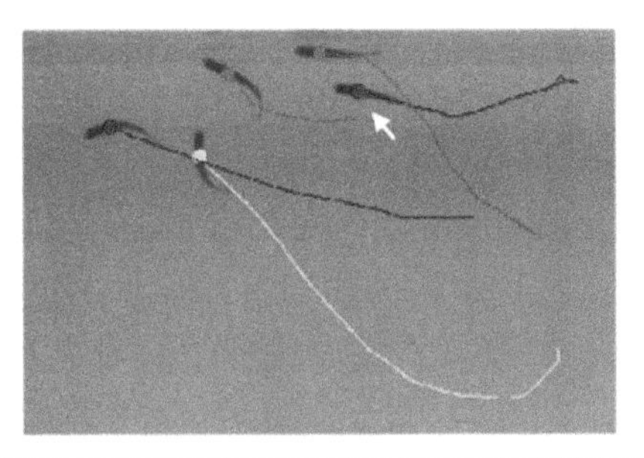

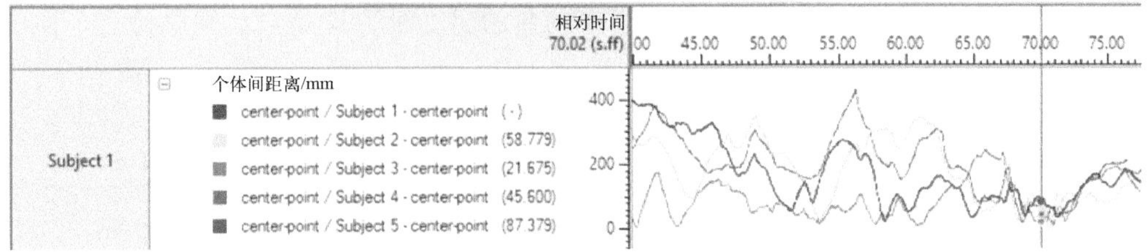

图 30-12　**受试鱼 1 与其他鱼之间距离的时间图（屏幕截图中的箭头表示受试鱼 1）**

计算受试鱼 1 和其他鱼之间的距离。为每条受试鱼都绘制了类似的图（即受试鱼 2 和所有其他鱼之间的距离等）

鱼群极化被定义为个体的前进方向偏离群体平均前进方向的程度，并以实验水箱中所有个体计算的平均运动矢量来测量（Miller et al., 2012）。极性指数是指基于一条鱼的前进方向角度和速度，这些数据很容易从视频跟踪软件中导出。如果所有或大部分鱼都朝同一方向游动（图 30-13）并以大致相同的速度游动，则该鱼群是高度极化的。Viscido 等人（2004）提出了一个无量纲的极性指数，该指数在 0（完全非极化）和 1（鱼完全对齐游动）之间变化。

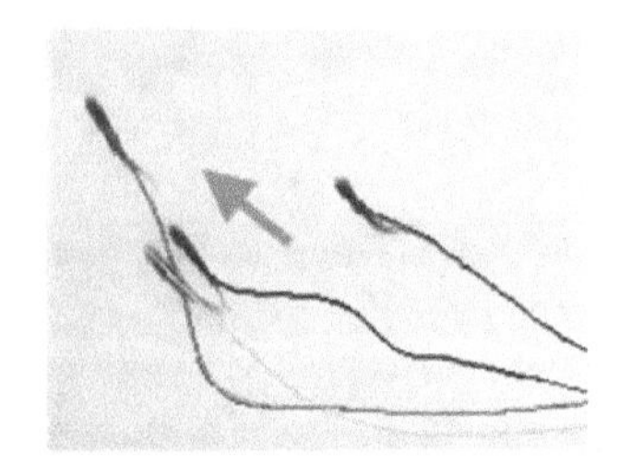

相对时间
486.81 (s.ff)
486.60　486.80　487.00　487.20　487.40

鱼头朝向(角度)
鼻尖点 (129.041) 0

鱼头朝向(角度)
鼻尖点 (133.847) 0

鱼头朝向(角度)
鼻尖点 (116.896) 0

鱼头朝向(角度)
鼻尖点 (130.954) 0

图 30-13　**使用 EthoVision XT 参数中的前进方向来测量鱼群极化**

前进方向测量的是一条鱼的运动矢量与参考轴（在本例中为水平轴）形成的角度。运动矢量在 x 轴上方时为正，在其他情况下为负。随着极化形成，鱼个体之间前进方向的差异很小。这种方法可以与参数“个体间的距离”相结合，来细化极化的定义当软件有效地检测到多个受试鱼游泳路径的转弯时，可以评估该转弯是否随机发生，或者一条受试鱼是否会跟随另一条受试鱼的转弯而转变，以及这种关联的强度（图 30-14）。相互关系可以揭示在哪个时间点转弯的鱼体间时滞相关性是最大的。当软件准确地跟踪鱼个体时，鱼很少或者几乎不变换位置的情况下，原则上软件是可以确定哪条鱼是领先的，哪条鱼是跟随的。鱼群内斑马鱼的个体识别已用各种方法解决了（Delcourt et al., 2018）。原则上，斑马鱼的颜色标记（Cheung et al., 2014）加上颜色跟踪可以进行自动个体识别，前提是该软件可以检测到背景上的颜色标记。关于鱼群个体识别追踪的综述，请参阅文献 Delcourt et al.（2013）。目前已成功应用于斑马鱼的另一种方法是自动视频检测个体皮肤的独特模式（指纹识别：Pérez Escudero et al., 2014）。

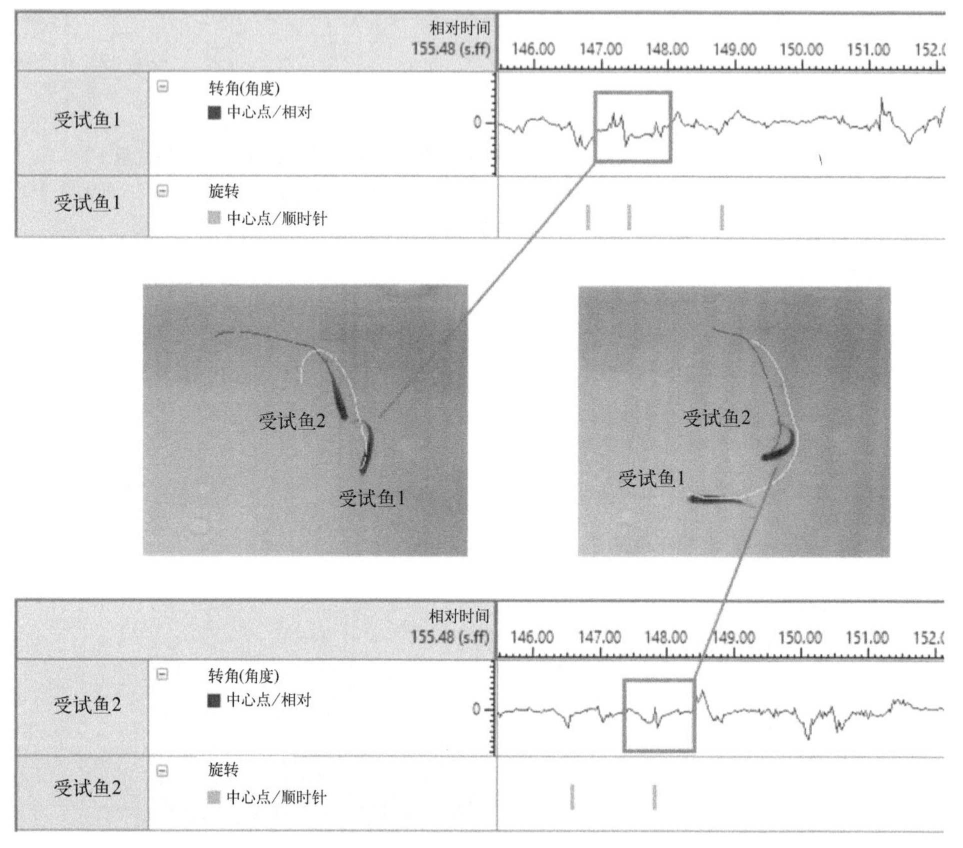

图30-14 **当正确检测到转弯时，可以测试不同受试者的转弯是否同时发生，即一个受试者在短时间内跟随另一个受试者**

上图和下图：转角（turn angle）和旋转（rotation）的时间图。后者是基于转角是否超过某个阈值的点类型事件

30.8 成年斑马鱼行为轨迹的三维跟踪

尽管将斑马鱼置于三维（3D）空间中游泳，但对其运动和行为的大部分研究都是使用摄像机视图进行的，结果是二维（2D）轨迹数据。最近的一项研究表明，与 3D 观察相比，

2D 数据隐藏了运动活动的差异，或者相反，高估了这种差异（Macrì et al., 2017）。下文我们将使用 EthoVision XT 附加模块 Track 3D 来说明 3D 跟踪技术的优势和潜力。

3D 跟踪系统的简化版本是使用两个相互正交的摄像头，以提供实验箱的俯视图和前视图。俯视图提供鱼的 x、y 坐标，而前视图提供 z 坐标（Cachat et al., 2011b）。这种情况描述的是鱼的近似位置，因为它没有考虑各种失真，包括由相机镜头引起的失真以及由水的折射率与空气的折射率不同引起的失真。为了对失真情况进行校正，研究者提出了各种 3D 校准方法；Track3D 使用了带有已知 3D 坐标标记的校准对象。在校准过程中，将物体放置在装满水的水箱内。然后，在给定由视频跟踪软件提取的 2D 坐标集的情况下，使用 2D 摄像机视图上的标记投影来重建鱼的 3D 位置（Stewart et al., 2015a）。如果摄像机的位置和方向相对于其跟踪的水量是不变的，则校准程序只需执行一次。为了将折射误差和三维重建的不确定性降至最低，摄像机应垂直于实验箱的两个相邻面放置，最好是相互正交。

Macrì 等人（2017）记录了成年斑马鱼在各种实验处理中的轨迹，并比较了 2D（水箱正面和顶视图）和 3D（结合摄像机视图中的 2D 数据后）测量的行为终点。他们发现，与 3D 测量相比，对 2D 测量中的许多终点（包括平均速度和平均角速度）的估计不足（作者假设 3D 代表了真实情况，因为它更准确地描述了鱼的真实轨迹）。造成这种差异的原因之一可参考图 30-15，该图显示了两个非连续视频帧中的鱼。使用俯视图时，通过这种方式获得的轨迹只能部分地描述潜水事件。两个视频帧中计算的鱼之间距离小于移动鱼间的实际距离。2D 跟踪由于运动模式的失真使测量值偏小（可能还存在组间差异）。这将导致对线性速度的低估（Macrì et al., 2017）。同理，从摄像机拍摄的 U 形转弯被“看”成是游泳方向约 180°的急剧变化，从而导致对角度变化和速度的高估（图 30-15）。

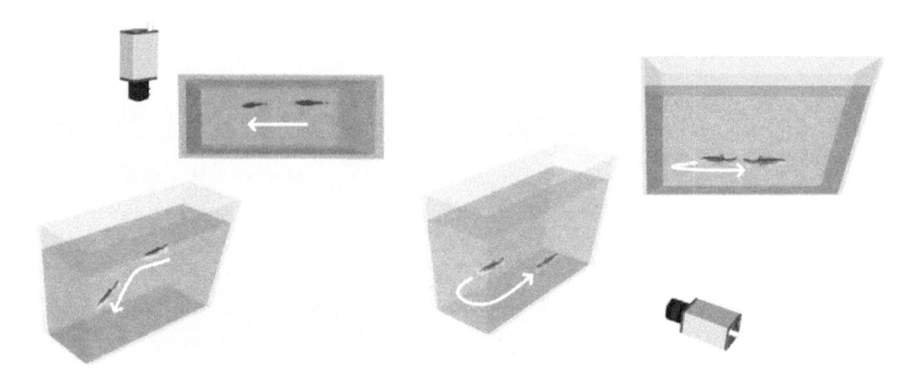

图 30-15 **在 2D 图像上进行的测量可能会低估距离或高估角度变化**

左图：潜水活动，用顶部摄像机拍摄，移动的距离和线速度被低估（上部插图）。右图：U 形转弯，用前置摄像头拍摄，跟踪数据显示方向发生了急剧变化，因此角速度被高估（上部插图）。另请参见文中的解释

此外，Macrì 等人（2017）也认为，实验处理的效果取决于是否使用 2D 或 3D 测量游泳轨迹。例如，当观察不同实验处理组对最大速度的影响时，与 3D 视图相比，2D 视图给出了特定处理的假阳性和假阴性结果。尤其是前视图存在问题时，频繁快速的水平运动可能会被掩盖。作者认为高比例的假阳性和假阴性结果也意味着 2D 与 3D 测量结果相比，3D 测量结果的统计能力更高。这种差异可以说明，通过 3D 跟踪，研究人员可以使用更少的受试鱼，这也是动物福利研究的基本要求（Slob, 2014）。然而，在我们看来，2D 跟踪在许多情况下仍然是有效的。例如，在 T 形迷宫实验中，由于水相对较浅，鱼在垂直

轴上移动不多，或者在新水箱潜水实验中，人们感兴趣的是从正面对受试鱼在底栖和顶栖之间的过渡进行评分。一般来说，只要不需随时了解鱼的确切位置及其轨迹，2D 跟踪就是一个很好的解决方案。

游泳路径的三维重建扩展了可测量的行为指标的范围。例如，苯环己哌啶（PCP）是一种致幻化合物，可触发斑马鱼的盘旋行为和“8 字形”游动模式。图 30-16 显示了实验处理对其中一项测量（3D 角度变化）的显著影响，这是由三个连续的斑马鱼位置形成的三维角度。与在一个平面内游泳的对照鱼相比，PCP 引起的不规则游泳模式的数值较大（图 30-16）。此外，与 PCP 处理组的鱼相比，麦角酸二乙酰胺（LSD）处理组的鱼的 3D 角度变化更小（Stewart et al., 2015a）。因此，通过分析此类 3D 测量，可以根据斑马鱼的游泳模式及其几何形状预测不同组神经活性化合物（或环境应激源，或基因型）的药理特征（Cachat et al., 2011b; Stewart et al., 2015b）。

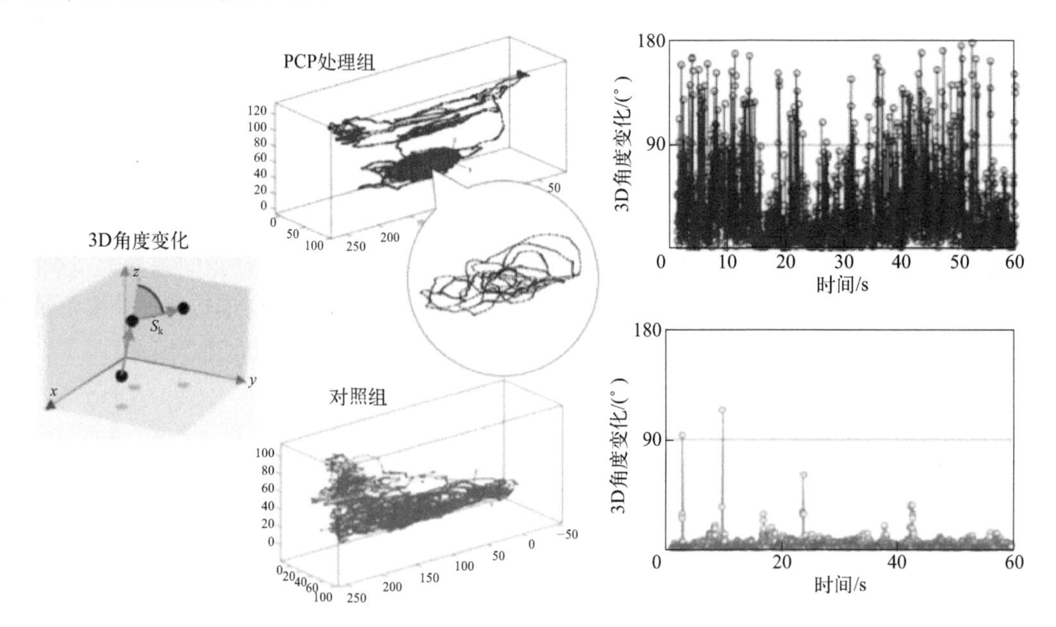

图 30-16　**对照鱼和经苯环己哌啶（PCP）处理组的鱼的 3D 测量结果有显著差异**

三维角度变化是 Track3D 计算出的运动参数之一，用于测量三维空间中的转弯程度。图中，在用 PCP 处理过的鱼中可观察到“8 字形”游动路径的放大视图。数据由 A.V. Kalueff 教授提供，图像由作者绘制。参见 Stewart, A.M., Gerlai, R., Kalueff, A.V., 2015. Developing higher-throughput zebrafish screens for in-vivo CNS drug discovery. Frontiers in Behavioral Neuroscience 9, 14

通常，3D 跟踪是通过两台摄像机从不同的位置拍摄鱼。其他方法仅使用一个摄像机，通过在水箱一侧（Pereira and Oliveira, 1994; Zhu and Weng, 2007）或与水面（Maaswinkel et al., 2013）成一定角度（通常为 45°）放置一面镜子来提供鱼的第二视图。单摄像头解决方案的优势在于，鱼的两个视图根据定义是同步的。一些研究者使用镜像系统开发了跟踪多条未标记斑马鱼的方法（Maaswinkel et al., 2013; Audira et al., 2018）。然而，当两条或更多条鱼的游泳路径交叉时可能会出现漏报个体互换率的现象。解决集群鱼类的阻塞问题（occlusions）至关重要，最近越来越引起研究人员关注（例如，Dolado et al., 2014）。此外，受试鱼可能会对其镜像做出反应（Maaswinkel et al., 2013）。研究人员应仔细考量镜子的位置和方向以及周围的照明设备，并验证该设置不会影响鱼的行为。

研究人员还提出了基于不同技术3D跟踪成鱼的方法。有些是基于红外光在不同深度水中的差异反射，可以估计鱼的垂直位置（Pautsina et al., 2015; Lin et al., 2018）。还有人提出了一种基于Microsoft Kinect I设备的方法（Saberioon and Cisar, 2016），但据我们所知，这些方法都没有在斑马鱼上进行过测试和验证。

30.9 结论

斑马鱼因其与人类具有高度的遗传保守性，且具有迅速发育、体形小、饲养成本低的优点，已经成为有效的基因和化学筛选脊椎动物模式生物。下文列出了我们认为未来面临的主要挑战。

高通量。在正向遗传学中，目标是识别与已知/特征表型（行为）相关的基因。因为通常不知道有多少基因会影响所研究的表型，所以需筛选的动物数量是正向遗传学要考量的一个关键因素。因此，一个主要挑战是增加行为分析的通量。在大多数情况下，任务通量仅取决于一个视频图像中可以容纳多少个测试箱（或用于测试仔鱼的孔）。视频跟踪系统获取和处理的视频必须是高分辨率的，才能拍摄大量的受试鱼并足够详细地记录每条鱼的行为。目前，96孔板中斑马鱼仔鱼的拍摄分辨率在120万到280万像素之间。4K的视频分辨率（约830万像素）有可能将通量系数提高至少6倍，从而实现一次实验中可以包含96×6=576条仔鱼。考虑到受试鱼的数量如此之多，只有在使用自动化系统来处理受试鱼和给予化合物时才能实现。在某些任务中，可以同时将多个个体暴露在测试环境中。例如，Gerlai（2010b）描述了Gómez-Laplaza和Gerlai（2010）开展的一项实验，其中多条鱼在探索实验期间暴露在复杂的迷宫中。在这些实验中，不需要跟踪和监测每条鱼，因此每天可以训练大量的个体。为了提高通量，任务的自动化也很重要。在研究社会偏好和捕食者回避时，刺激的呈现和受试鱼反应的记录越来越自动化，并可以由视频跟踪软件控制（例如，Brock et al., 2017；也参见下文）。

行为检测和分析。尽管斑马鱼的行为已经被研究了几十年，但对其行为研究的数量远低于对啮齿动物研究的数量。现在已经可以进行精确的基因操作，未来的一个挑战是将遗传方法与复杂行为的研究结合起来。在经典的行为跟踪中，生物学相关的运动和姿势被转换为可在实验组或时间点之间进行对比的测量指标。虽然可以用移动距离、速度和前进方向来进行这种对比，但我们也需要将行为分为离散的单元，如斑马鱼仔鱼的游泳和快速逃跑。在实验过程中，当观察一群鱼时，需要考虑在试验的特定时间，我们何时可以将一条鱼视为一群鱼的一部分。将受试鱼分配到一组的常见标准是基于距离（例如，身体长度的倍数）；然而，由于我们不知道个体间交流的极限距离，因此使用任何阈值都没有经验依据。此外，当研究仅限于特定的时间和空间尺度时，通常情况下，传统的模式识别算法无法处理更大的数据量，机器学习可能是一条实现途径（Soleymani et al., 2014）。事实证明，基于机器视觉的方法可以在没有监督的情况下（Wiltschko et al., 2015）识别啮齿动物的单一行为（van Dam et al., 2013）。我们相信对斑马鱼进行类似分析目前是可行的。

动物系统相互作用。要将神经回路功能与行为直接联系起来，操控神经功能并观察反应是至关重要的。通过光遗传学方法，人们可以上下调节动物的活动。然而，将自由活动动物的脑细胞暴露在光线下而不产生副作用仍然是一个挑战。例如，我们不能排除孔板上方的强烈光照本身可能会影响仔鱼的行为，该行为与光遗传刺激无关。Arrenberg 等人（2009）通过位于仔鱼头部上方的细光纤施加光刺激，一半的受试鱼行为受到了抑制。Chen 等人（2016）提出了一种替代方法，利用小分子或特定温度来激活斑马鱼仔鱼的神经元。

当研究社会行为时，未来的挑战之一将是使实验真正具有交互性，以在二元或多个体的背景下分析因果关系。视觉刺激缺乏物理输入并且是二维的。相比之下，模拟机器鱼的优势在于，可以更逼真地模仿斑马鱼的运动，并且可以通过编程来响应受试鱼的行为（Kopman et al., 2013; Ruberto et al., 2016; Kim et al., 2018）。研究者根据受试鱼与机器鱼的距离修改机器鱼的甩尾频率。他们发现受试鱼对机器鱼的吸引力程度取决于机器鱼的甩尾频率是否随着受试鱼的行为反应而变化。引入机器鱼等进行真正的交互式实验，根据测试对象的行为调节实验刺激，可能有助于解决有关感知、恐惧（Porfiri, 2018）、焦虑、社会偏好和集体行为等问题（Bonnet et al., 2018）。

30.10 补充数据

与本文相关的补充资料可在 https://doi.org/10.1016/ B978-0-12-817528-6.00030-9 在线查阅。

免责声明 / 利益冲突声明

下面列出姓名的作者提供了本手稿中讨论的主题或材料的详细信息，他们隶属于与这些主题或材料具有经济或非经济利益关系的组织或实体。F. Grieco 是行为研究顾问，R.A.J. Tegelenbosch 是产品经理，L.P.J.J. Noldus 是 Noldus Information Technology BV 的首席执行官（CEO），该公司是本章中描述的一些商业系统的开发商。

（郭煊君翻译　石晶晶审校）

参考文献

Ahmad, F., Noldus, L.P.J.J., Tegelenbosch, R.A.J., Richardson, M.K., 2012. Zebrafish embryos and larvae in behavioural assays. Behaviour 149, 1241-1281.

Ahmed, O., Seguin, D., Gerlai, R., 2011. An automated predator avoidance task in zebrafish. Behavioral Brain Research 216 (1), 166-171.

Ahmed, T.S., Fernandes, Y., Gerlai, R., 2012. Effects of animated images of sympatric predators and abstract shapes on fear responses in zebrafish. Behaviour 149 (10e12), 1129-1153.

Ariyomo, T.O., Watt, P.J., 2012. The effect of variation in boldness and aggressiveness on the reproductive success of zebrafish, *Danio rerio*. Animal Behaviour 83 (1), 41-46.

Arrenberg, A.B., Del Bene, F., Baier, H., 2009. Optical control of zebrafish behavior with halorhodopsin. Proceedings of the National Academy of Sciences 106 (42), 17968-17973.

Audira, G., Sampurna, B.P., Juniardi, S., Liang, S.T., Lai, Y.H., Hsiao, C.D., 2018. A simple setup to perform 3D locomotion tracking in zebrafish by using a single camera. Inventions 3, 11.

Barker, A.J., Baier, H., 2015. Sensorimotor decision making in the zebrafish tectum. Current Biology 25, 2804-2814.

Bass, S.L.S., Gerlai, R., 2008. Zebrafish (*Danio rerio*) responds differentially to stimulus fish: the effects of sympatric and allopatric predators and harmless fish. Behavioural Brain Research 186, 107-117.

Ben-Moshe, Z., Alon, S., Mracek, P., et al., 2014. The light-induced transcriptome of the zebrafish pineal gland reveals complex regulation of the circadian. Nucleic Acids Research 42, 3750-3767.

Blaser, R., Gerlai, R., 2006. Behavioral phenotyping in zebrafish: comparison of three behavioral quantification methods. Behavior Research Methods 38 (3), 456-469.

Bonnet, F., Gribovskiy, A., Halloy, J., Mondada, F., 2018. Closed-loop interactions between a shoal of zebrafish and a group of robotic fish in a circular corridor. Swarm Intelligence 12 (3), 227-244.

Bridi, D., Altenhofen, S., Gonzalez, J.B., Reolon, G.K., Bonan, C.D., 2017. Glyphosate and Roundup? Alter morphology and behavior in zebrafish. Toxicology 392, 32-39.

Brock, A.J., Sudwarts, A., Daggett, J., Parker, M.O., Brennan, C., 2017. A fully automated computer-based 'Skinner Box' for testing learning and memory in zebrafish. bioRxiv 110478.

Buske, C., Gerlai, R., 2014. Diving deeper into zebrafish development of social behavior: analyzing high resolution data. Journal of Neurosciences Methods 234, 66-72.

Cachat, J., Stewart, A., Grossman, L., et al., 2010. Measuring behavioral and endocrine responses to novelty stress in adult zebrafish. Nature Protocols 5, 1786-1799.

Cachat, J.M., Canavello, P.R., Elkhayat, S.I., et al., 2011. Video-aided analysis of zebrafish locomotion and anxiety-related behavioral responses. In: Kalueff, A.V., Cachat, J.M. (Eds.), Zebrafish Neurobehavioral Protocols. Neuromethods, vol. 51. Humana Press, New York.

Cachat, J., Stewart, A., Utterback, E., et al., 2011. Three-dimensional neurophenotyping of adult zebrafish behavior. PLoS One 6, e17597.

Chen, S., Chiu, C.N., McArthur, K.L., Fetcho, J.R., Prober, D.A., 2016. TRP channel mediated neuronal activation and ablation in freely behaving zebrafish. Nature Methods 13 (2), 147-150.

Cheung, E., Chatterjee, D., Gerlai, R., 2014. Subcutaneous dye injection for marking and identification of individual adult zebrafish (*Danio rerio*) in behavioral studies. Behavior Research Methods 46 (3), 619-624.

Colwill, R.M., Creton, R., 2010. Automated imaging of avoidance behavior in larval zebrafish. In: Zebrafish Neuro-behavioral Protocols Neuromethods, vol. 51. Humana Press, New York.

Colwill, R.M., Creton, R., 2011. Locomotor behaviors in zebrafish (*Danio rerio*) larvae. Behavioural Processes 86 (2), 222-229.

Colwill, R.M., Creton, R., 2011. Imaging escape and avoidance behavior in zebrafish larvae. Reviews in the Neurosciences 22 (1), 63-73.

Colwill, R.M., Raymond, M.P., Ferreira, L., Escudero, H., 2005. Visual discrimination learning in zebrafish (*Danio rerio*). Behavioural Processes 70 (1), 19-31.

Cornet, C., Calzolari, S., Miñana-Prieto, R., et al., 2017. ZeGlobalTox: an innovative approach to address organ drug toxicity using zebrafish. International Journal of Molecular Sciences 18 (4), 864-882.

Cronin, A., Grealy, M., 2017. Neuroprotective and neuro-restorative effects of minocycline and rasagiline in a zebrafish 6-hydroxydopamine model of Parkinson's disease. Neuroscience 367, 34-46.

Darland, T., Dowling, T.E., 2001. Behavioral screening for cocaine sensitivity in mutagenized zebrafish. Proceedings of the National Academy of Sciences of the United States of America 98 (20), 11691-11696.

De Luca, E., Zaccaria, G.M., Hadhoud, M., et al., 2014. ZebraBeat: a flexible platform for the analysis of the cardiac rate in zebrafish embryos. Scientific Reports 4, 4898.

Delcourt, J., Denoël, M., Ylieff, M., Poncin, P., 2013. Video multitracking of fish behaviour: a synthesis and future perspectives. Fish and Fisheries 14, 186-204.

Delcourt, J., Ovidio, M., Denoël, M., et al., 2018. Individual identification and marking techniques for zebrafish. Reviews in Fish Biology and Fisheries 28 (4), 839-864.

Dell, A.I., Bender, J.A., Branson, K., et al., 2014. Automated image-based tracking and its application in ecology. Trends in Ecology & Evolution 29 (7), 417-428.

Dolado, R., Gimeno, E., Beltran, F.S., Quera, V., Pertusa, J.F., 2014. A method for resolving occlusions when multi tracking individuals in a shoal. Behavior Research Methods 47 (4), 1032-1043.

Egan, R.J., Bergner, C.L., Hart, P.C., et al., 2009. Understanding behavioral and physiological phenotypes of stress and anxiety in zebrafish. Behavioural Brain Research 205 (1), 38-44.

Farrell, T.C., Cario, C.L., Milanese, C., Vogt, A., Jeong, J.H., Burton, E.A., 2011. Evaluation of spontaneous propulsive movement as a screening tool to detect rescue of Parkinsonism phenotypes in zebrafish models. Neurobiology of Disease 44 (1), 9-18.

Gerlai, R., 2010. Zebrafish antipredatory responses: a future for translational research? Behavioural Brain Research 207 (2), 223-231.

Gerlai, R., 2010. High-throughput behavioral screens: the first step towards finding genes involved in vertebrate brain function using zebrafish. Molecules 15, 2609-2622.

Gerlai, R., 2013. Antipredatory behavior of zebrafish: adaptive function and a tool for translational research. Evolutionary Psychology 11 (3), 591-605.

Gerlai, R., Lahav, M., Guo, S., Rosenthal, A., 2000. Drinks like a fish: zebrafish (*Danio rerio*) as a behavior genetic model to study alcohol effects. Pharmacology, Biochemistry and Behavior 67, 773-782.

Gerlai, R., Fernandes, Y., Pereira, T., 2009. Zebrafish (*Danio rerio*) responds to the animated image of a predator: towards the development of an automated aversive task. Behavioural Brain Research 201 (2), 318-324.

Gómez-Laplaza, L.M., Gerlai, R., 2010. Latent learning in zebrafish (*Danio rerio*). Behavioral Brain Research 208 (2), 509-515.

Gould, G.G., 2011. Modified associative learning T-maze test for zebrafish (*Danio rerio*) and other small teleost fish. In: Kalueff, A.V., Cachat, J. (Eds.), Zebrafish Neurobehavioral Protocols. Neuromethods, vol. 51. Humana Press, New York.

Green, J., Collins, C., Kyzar, E.J., et al., 2012. Automated high-throughput neurophenotyping of zebrafish social behavior. Journal of Neuroscience Methods 210, 266-271.

Grone, B.P., Marchese, M., Hamling, K.R., et al., 2016. Epilepsy, behavioral abnormalities, and physiological comorbidities in syntaxin-binding protein 1 (STXBP1) mutant zebrafish. PLoS One 11 (3), e0151148.

Gutiérrez, H.C., Vacca, I., Inguanzo Pons, A., Norton, W.H.J., 2018. Automatic quantification of juvenile zebrafish aggression. Journal of Neuroscience Methods 296, 23-31.

Kalueff, A.V., Gebhardt, M., Stewart, A.M., et al., 2013. Towards a comprehensive catalog of zebrafish behavior 1.0 and beyond. Zebrafish 10 (1), 70-86.

Kim, O.H., Cho, H.J., Han, E., et al., 2017. Zebrafish knockout of Down syndrome gene, DYRK1A, shows social impairments relevant to autism. Molecular Autism 8, 50.

Kim, C., Ruberto, T., Phamduy, P., Porfiri, M., 2018. Closed-loop control of zebrafish behaviour in three dimensions using a robotic stimulus. Scientific Reports 8, 657.

Kopman, V., Laut, J., Polverino, G., Porfiri, M., 2013. Closed-loop control of zebrafish response using a bioinspired robotic-fish in a preference test. Journal of the Royal Society Interface 10, 20120540.

Ladu, F., Bartolini, T., Panitz, S.G., et al., 2015. Live predators, robots, and computer-animated images elicit differential avoidance responses in zebrafish. Zebrafish 12 (3), 205-214.

Lin, K., Zhou, C., Xu, D., Guo, Q., Yang, X., Sun, C., 2018. Three-dimensional location of target fish by monocular infrared imaging sensor based on a Lez correlation model. Infrared Physics & Technology 88, 106-113.

Lopes da Fonseca, T., Correia, A., Hasselaar, W., van der Linde, H.C., Willemsen, R., Outeiro, T.F., 2013. The zebrafish homologue of Parkinson's disease ATP13A2 is essential for embryonic survival. Brain Research Bulletin 90, 118-126.

Maaswinkel, H., Le, X., He, L., Zhu, L., Weng, W., 2013. Dissociating the effects of habituation, black walls, buspirone and ethanol on anxiety-like behavioral responses in shoaling zebrafish. Pharmacology, Biochemistry and Behavior 108, 16-27.

Macrì, S., Neri, D., Ruberto, T., Mwaffo, V., Butail, S., Porfiri, M., 2017. Three-dimensional scoring of zebrafish behavior unveils biological phenomena hidden by two-dimensional analyses. Scientific Reports 7, 1962.

Mahabir, S., Chatterjee, D., Buske, C., Gerlai, R., 2013. Maturation of shoaling in two zebrafish strains: a behavioral and neurochemical analysis. Behavioural Brain Research 247, 1-8.

Marks, C., West, T.N., Bagatto, B., Moore, F.B.G., 2005. Developmental environment alters conditional aggression in zebrafish. Copeia 4, 901-908.

Miller, N.Y., Gerlai, R., 2011. Shoaling in zebrafish: what we don't know. Reviews in the Neurosciences 22 (1), 17-25.

Miller, N., Gerlai, R., 2011. Refining membership in animal groups. Behavioral Research 43, 964-970.

Miller, N.Y., Gerlai, R., 2012. Automated tracking of zebrafish shoals and the analysis of shoaling behavior. In: Kalueff, A.V., Stewart, A.M. (Eds.), Zebrafish Protocols for Neurobehavioral Research. Neuromethods, vol. 66. Humana Press, Totowa.

Moretz, J.A., Martins, E.P., Robinson, B.D., 2007. Behavioral syndromes and the evolution of correlated behavior in zebrafish. Behavioral Ecology 18 (3), 556-562.

Muto, A., Orger, M.B., Wehman, A.N., et al., 2005. Forward genetic analysis of visual behavior in zebrafish. PLoS Genetics 1, e066.

Orger, M.B., de Polavieja, G.G., 2017. Zebrafish behavior: opportunities and challenges. Annual Review of Neuroscience 40, 125-147.

Padilla, S., Hunter, D.L., Padnos, B., Frady, S., MacPhail, R.C., 2011. Assessing locomotor activity in larval zebrafish: influence of extrinsic and intrinsic variables. Neurotoxicology and Teratology 33 (6), 624-630.

Pautsina, A., Císař, P., Štys, D., Terjesen, B.F., Espmark, 2015. Infrared reflection system for indoor 3D tracking of fish. Aquacultural Engineering 69, 7-17.

Pelkowski, S.D., Kapoor, M., Richendrfer, H.A., Wang, X., Colwill, R.M., Creton, R., 2011. A novel high-throughput imaging system for automated analyses of avoidance behavior in zebrafish larvae. Behavioural Brain Research 223 (1), 135-144.

Pereira, P., Oliveira, R.F., 1994. A simple method using a single video camera to determine the three-dimensional position of a fish. Behavior Research Methods, Instruments, and Computers 26 (4), 443-446.

Pérez-Escudero, A., Vicente-Page, J., Hinz, R.C., Arganda, S., de Polavieja, G.G., 2014. IdTracker: tracking individuals in a group by automatic identification of unmarked animals. Nature Methods 11, 743-748.

Pham, M., Raymond, J., Hester, J., et al., 2012. Assessing social behavior phenotypes in adult zebrafish: shoaling, social preference, and mirror biting tests. In: Kalueff, A.V., Stewart, A.M. (Eds.), Zebrafish Protocols for Neurobehavioral Research. Neuromethods, vol. 66. Humana Press, Totowa, NJ.

Porfiri, M., 2018. Inferring causal relationships in zebrafish-robot interactions through transfer entropy: a small lure to catch a big fish. Animal Behavior and Cognition 5 (4), 341-367.

Riché, R., Liao, M., Pena, I.A., et al., 2018. Glycine decarboxylase deficiencyeinduced motor dysfunction in zebrafish is rescued by counterbalancing glycine synaptic level. JCI Insight 3 (21), e124642.

Richendrfer, H., Pelkowski, S.D., Colwill, R.M., Creton, R., 2012. On the edge: pharmacological evidence for anxiety-related behavior in zebrafish larvae. Behavioural Brain Research 228 (1), 99-106.

Ruberto, T., Mwaffo, V., Singh, S., Neri, D., Porfiri, M., 2016. Zebrafish response to a robotic replica in three dimensions. Royal Society Open Science 3, 160505.

Saberioon, M.M., Cisar, P., 2016. Automated multiple fish tracking in three-dimension using a structured light sensor. Computers and Electronics in Agriculture 121, 215-221.

Saverino, C., Gerlai, R., 2008. The social zebrafish: behavioral responses to conspecific, heterospecific, and computer animated fish. Behavioral Brain Research 191 (1), 77-87.

Séguret, A., Collignon, B., Halloy, J., 2016. Strain differences in the collective behaviour of zebrafish (*Danio rerio*) in heterogeneous environment. Royal Society Open Science 3, 160451.

Selderslaghs, I., Van Rompay, A., de Coen, W., Witters, H., 2009. Development of a screening assay to identify teratogenic and embryotoxic chemicals using the zebrafish embryo. Reproductive Toxicology 28, 308-320.

Sison, M., Gerlai, R., 2010. Associative learning in zebrafish (*Danio rerio*) in the plus maze. Behavioral Brain Research 207 (1), 99-104.

Slob, W., 2014. Benchmark dose and the three Rs. Part Ⅱ. Consequences for study design and animal use. Critical Reviews in Toxicology 44 (7), 568-580.

Smith, S.J., Wang, J.C., Gupta, V.A., Dowling, J.J., 2017. A novel early onset phenotype in a zebrafish model of merosin deficient congenital muscular dystrophy. PLoS One 12 (2), e0172648.

Soleymani, A., Cachat, J., Robinson, K., Dodge, S., Kalueff, A.V., Weibel, R., 2014. Integrating cross-scale analysis in the spatial and temporal domains for classification of behavioral movement. Journal of Spatial Information Science 8, 1-25.

Stewart, A.M., Grieco, F., Tegelenbosch, R.A.J., et al., 2015. A novel 3D method of locomotor analysis in adult zebrafish: implications for automated detection of. Journal of Neuroscience Methods 255, 66-74.

Stewart, A.M., Gerlai, R., Kalueff, A.V., 2015. Developing higher-throughput zebrafish screens for in-vivo CNS drug discovery. Frontiers in Behavioral Neuroscience 9, 14.

Swaminathan, A., Hassan-Abdi, R., Renault, S., Soussi-Yanicostas, N., Drapeau, P., Samarut, É., 2018. Non-canonical mTOR-independent role of DEPDC5 in regulating GABAergic network development. Current Biology 28, 1-14.

Teles, M.C., Oliveira, R.F., 2016. Quantifying aggressive behavior in zebrafish. In: Kawakami, K., Patton, E.E., Orger, M. (Eds.), Zebrafish—Methods and Protocols (Part Ⅲ). Springer, New York.

van Dam, E.A., van der Harst, J.E., ter Braak, C.J.F., Tegelenbosch, R.A.J., Spruijt, B.M., Noldus, L.P., 2013. An automated system for the recognition of various specific rat behaviours. Journal of Neuroscience Methods 218, 214-224.

Viscido, S.V., Parrish, J.K., Grünbaum, D., 2004. Individual behavior and emergent properties of fish schools: a comparison of observation and theory. Marine Ecology Progress Series 273, 239-249.

Way, G.P., Ruhl, N., Snekser, J.L., Kiesel, A.L., McRobert, S.P., 2015. A comparison of methodologies to test aggression in zebrafish. Zebrafish 12 (2), 144-151.

Wiltschko, A.B., Johnson, M.J., Iurilli, G., et al., 2015. Mapping sub-second structure in mouse behavior. Neuron 88, 1121-1135.

Wong, K., Elegante, M., Bartels, B., et al., 2010. Analyzing habituation responses to novelty in zebrafish (*Danio rerio*). Behavioural Brain Research 208, 450-457.

Zhu, L., Weng, W., 2007. Catadioptric stereo-vision system for the real-time monitoring of 3D behavior in aquatic animals. Physiology & Behavior 91, 106-119.

第三十一章

斑马鱼基因组测序计划：生物信息学资源

Kerstin Howe[1]

31.1 斑马鱼参考基因组测序计划简史

31.1.1 第一代测序

斑马鱼基因组测序计划始于2001年2月在英国剑桥郡辛克斯顿（Hinxton）桑格研究所（Wellcome Sanger Institute）召开的一场以斑马鱼为主题的会议。鉴于Tübingen品系斑马鱼在突变筛选中广泛使用，研究人员决定对其进行测序，并遵循已经成功应用于小鼠和人类的测序策略，例如，创建细菌人工染色体（bacterial artificial chromosomes, BAC）克隆载体和Fosmid基因组文库。利用限制性指纹图谱、放射性杂交和遗传图谱对这些克隆进行排序和定向，选择最小路径，进行测序和拼接。同时，对另一组斑马鱼仔鱼用Sanger技术（第一代DNA测序技术）进行全基因组鸟枪法（WGS）测序，并对测序结果进行组装。BAC克隆序列被用作基因组序列组装的可靠骨架，而WGS组装的重叠群被用来填补剩余的空白缺口。

多年来，越来越多的克隆使用Sanger和Illumina技术进行测序，减少了使用低质量WGS进行填补的必要性。得益于高密度基因图谱，研究者可以获得更好的基因图谱数据。单个双单倍体个体BAC和Fosmid基因组文库极大地解决了主要的组装问题，即为保留两种单倍型序列（单体复制）而对基因组区域进行的人工复制需要大量人工监管来纠正。斑马鱼基因组组装经过了Zv3到Zv9（GCA_000002035.2）的版本，直到Zv9才达到了满足发表要求的质量（Howe et al., 2013）。

[1] Wellcome Sanger Institute, Cambridge, United Kingdom.

31.1.2 基因组参考联盟

基因组参考联盟（the Genome Reference Consortium, GRC）是2008年由美国国家生物技术信息中心（the National Center for Biotechnology Information, NCBI）、欧洲生物信息学研究所（the European Bioinformatics Institute, EBI）、麦克唐纳基因组研究所（the McDonnell Genome Institute, MGI）和桑格研究所合作成立的，以确保人类和小鼠参考基因组组装（genome reference assemblies）的维护和持续改进。桑格研究所已经从各种斑马鱼基因组组装的管理中获得了丰富的经验，并提供和扩展其验证和管理平台gEVAL［geval.sanger.ac.uk（Chow et al., 2016）］用于人类和小鼠基因组组装。

Zv9公开发表后，斑马鱼的基因组装被添加到GRC参考组装集（reference assembly portfolio）中。参考集的改进持续进行，最终推出GRCz10（GCA_000002035.3）和GRCz11（GCA_000002035.4）版本。Zv9之后的大部分改进都是基于进一步的克隆测序以及光学和基因组图谱（BioNano, ionanogenomics.com）提供的远程绘图数据（long-range mapping data），这为连接重复区域、解决单倍型复制和排序/定向问题提供了理想的途径（Howe and Wood, 2015）（图31-1）。在Zv9和GRCz11之间，GRC提出并解决了近3000个基因组组装问题。

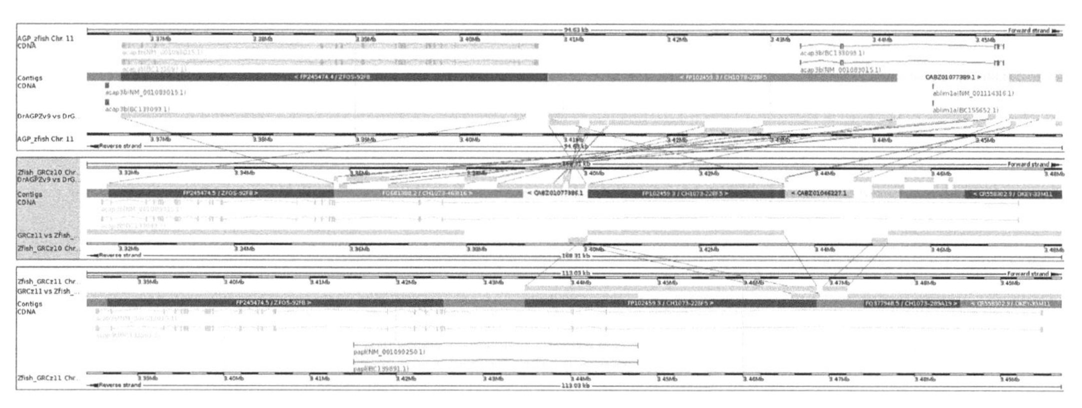

图31-1 *acap*3b区域从Zv9到GRCz11的改进

上图（Zv9）、中图（GRCz10）、下图（GRCz11）为*acap*3b区域排列的gEVAL屏幕截图。在Zv9中，*acap*3b基因存在于三个序列错误的片段中（红色和橙色基因结构，描绘的片段状态）和正链上。在GRCz10中，部分WGS片段重叠群（contigs）（黄色）被克隆序列（蓝色）所取代，并且这些组分重新排列（用深绿色勾勒的粉红色对齐区），得到一个反向的完整*acap*3b基因（绿色）。对于GRCz11，将该区域所有的WGS序列替换为克隆序列

GRCz11组装是第一个由主要组装（primary assembly）和额外的备用基因位点（alternate loci）组成的版本（图31-2A）。该主要组装包含25条染色体和尚未定位的序列。与相关主区域不同的已测序的单倍型代表Tübingen克隆被作为替代位点（alternate loci, ALT_REF_LOCI），这些克隆体在主要组装体的序列和位置可以与主要组装数据一起下载，克隆可以在染色体的基因组浏览器中显示，如Ensembl（www.Ensembl.org）。

31.1.3 持续的改进

从GRCz11开始，斑马鱼参考基因组已经完成了主动评估和改进阶段，GRC内部

维护已经从桑格研究所移交给了ZFIN（斑马鱼信息网，zfin.org）。ZFIN使用GRC工具收集用户报道的关于组装问题（assembly problems）的报告，并在必要时发布组装补丁（www.ncbi.nlm.nih.gov/grc/report-an-issue）。在编写本章时，没有计划对组装进行重大的结构更改，相反，任何更新将遵循建立的GRC模型，使用补丁发布次要的和非结构性改变的更新。补丁有两种形式，即修复补丁和新颖补丁。修复补丁可用来纠正序列，并通过替换错误区域而修订到下一个主要版本中。新颖补丁可将单倍型变异添加到主要组装中，并在下一个主要版本中成为ALT_REF_LOCI替代位点［图31-2（B），表31-1］。

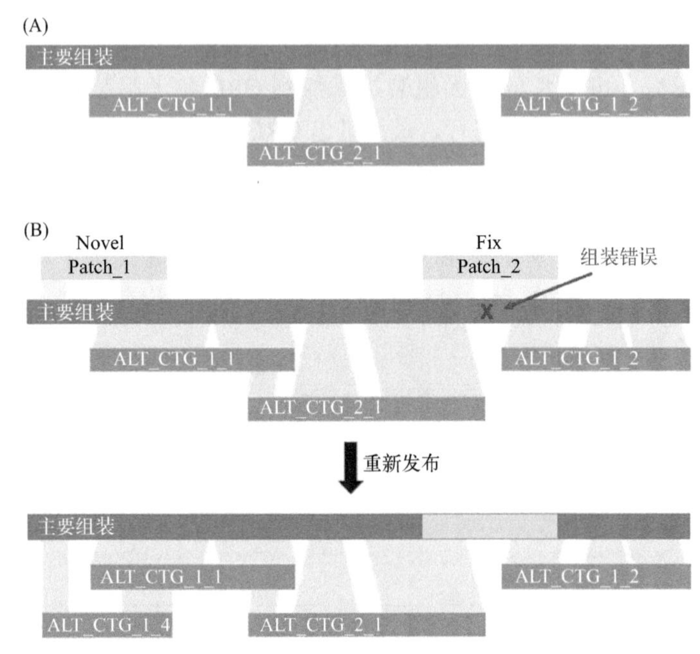

图31-2 **ALT_REF_LOCI和补丁及其与主要组装排列的示意图**

（A）三个假定的ALT_REF_LOCI基因座的排列，每个都代表主要组装区域中的交替单倍型。ALT_REF_LOCI被分组为非重叠区（例如ALT_CTG_1和ALT_CTG_2），以避免在重叠的单倍型中混合（例如ALT_CTG_1_1和ALT_CTG_2_1）。分配到同一个bin并不表示相同的单倍型起源，而只是避免重叠。GRCz11组装体总共包含930个ALT_REF_LOCI重叠序列。（B）在重新发布组装体时，修复补丁和新颖补丁经过不同处理分别被添加到主要组装体中

十多年来，欧洲和国际斑马鱼会议上都举办了斑马鱼基因组资源培训。表31-1包含了2018年研讨会的讲座和手册链接，以供进一步研究。

表31-1 **斑马鱼基因组参考汇编：URL**

主题	URL（统一资源定位符）
GRC斑马鱼主页	www.ncbi.nlm.nih.gov/grc/zebrafish
GRCz11基因库	www.ncbi.nlm.nih.gov/assembly/GCA_000002035.4
主要组装下载	ftp://ftp.ncbi.nlm.nih.gov/genomes/all/GCA/000/002/035/
对序列组装结果评估	www.ncbi.nlm.nih.gov/grc/zebrafish/data
桑格斑马鱼主页	www.sanger.ac.uk/science/data/zebrafish-genome-project
斑马鱼基因组网络资源教程2018	ftp://ftp.sanger.ac.uk/pub/grit/Madison_workshop_2018/

31.2 评估斑马鱼的基因组组装

组装一个基因组很不容易，特别是像斑马鱼这样具有大量重复和高杂合度的样本。因此，目前没有一个基因组组装能够包括所有变异或完全正确。但是，这不并会让用户在确定感兴趣的区域中是否存在问题时受制于基因组组装提供者。

GRC 在 www.ncbi.nlm.nih.gov/grc/zebrafish/issues 上提供了自 Zv9 以来在斑马鱼基因组组装中发现的所有问题的报告。这包括位置、受影响版本、对问题及其状态的描述，以及可能存在的解决方案。可以通过 issue ID 链接在本地序列浏览器上查看这些问题，并且也为 UCSC、Ensembl 和 NCBI 的基因组浏览器提供链接。

此外，GRC 还提供了基因组浏览器 gEVAL，其更侧重于组装质量而不是基因和特征注释。该浏览器将所有可用的数据集与给定的组装进行比对并标记出差异位点［geeval.sanger.ac.uk（Chow et al., 2016）］。用户可以通过检查基因缺失 / 排列错误的外显子、错位的 BioNano 图谱、错误的克隆重叠、克隆测序问题、克隆末端之间不寻常的距离或方向来识别当前和过去组装版本的潜在问题，这里仅列举几个可用的数据类型（图 31-3）。由于 gEVAL 提供了斑马鱼 BAC 和 fosmid 库中已测序克隆末端的比对，它还可以帮助识别连接某一特定特征或区域的克隆。数据可以通过其坐标来访问，例如从不同的浏览器导航，

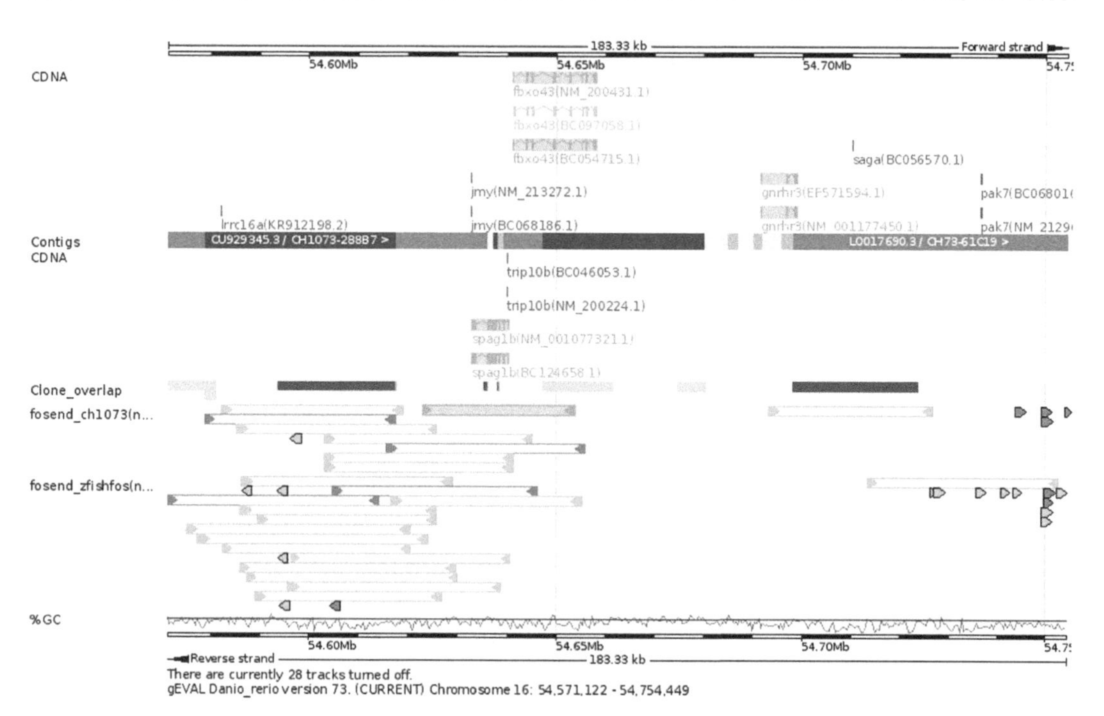

图 31-3 GRCz11 16 号染色体上的一个区域的 gEVAL 屏幕截图：问题区域实例

在克隆路径（蓝色）的剩余间隙中存在 WGS 重叠区（黄色）会导致三个基因的外显子缺失（以橙色表示为片段状态）。鼠标悬停可以提供相应 cDNA 比对的 ID 准确百分比和覆盖率，以及与外显子比对的链接。排列的 fosmid 端部基本在正确的距离和方向（绿色），但 *fbxo43* 和 *gnrhr3* 基因之间的区域未被覆盖。在 *gnrhr3* 区域的克隆重叠显示出超过 0.5% 的差异（红色），进一步证明该区域存在问题

或者通过文本搜索（例如，搜索基因名）。

任何与基因组组装有关的问题都可以（也应该）报告给 GRC，网址是 www.ncbi.nlm.nih.gov/grc/report-an-issue，以便进行调查和采取相关措施（表 31-2）。

表 31-2 组装评估：有用的 URL

主题	URL
问题报告	www.ncbi.nlm.nih.gov/grc/report-an-issue
GRC 问题跟踪	www.ncbi.nlm.nih.gov/grc/zebrafish/issues
GRCz11 gEVAL	geval.sanger.ac.uk/Zfish_GRCz11

31.3 斑马鱼基因组组装、基因注释和基因组浏览

GRCz11 上注释的斑马鱼基因组有两个来源：RefSeq（O'leary et al., 2016）和 Ensembl（Zerbino et al., 2018）。两者都是基于主要物种特异性 cDNA、RNAseq 和蛋白质数据与基因组装的比对，并使用基因模型预测软件来注释这些比对中正确的基因结构。Ensembl 受益于以往对基因模型进行的大量人工注释，这些注释可以解析许多不适合自动化方法的区域，例如包含扩展基因家族（如 NLR）重复排列的区域（Howe et al., 2016）。将这些人工注释的基因模型与自动注释的基因模型进行合并，形成了 Ensembl 斑马鱼基因组数据库。表 31-3 概述了这两种资源中目前可用的基因组。

表 31-3 GRCz11 主要组装的 Ensembl 和 RefSeq 基因注释的比较
（Ensembl release 95 和 NCBI Danio rerio release 106）

基因型	Ensembl 95	RefSeq 106
总基因数	32506	39241
编码蛋白数	25592	26197
假基因数	315	323
其他	6599	12721
总转录数	59876	48472

GRCz11 组装和斑马鱼基因模型可以在 UCSC、NCBI 和 Ensembl 三个主要的基因组浏览器中访问获取（表 31-4）。尽管浏览器在显示特征和对某些数据类型的偏好方面有所不同，但浏览器协议规定了它们都显示相同的序列和序列标识符。除了可用数据的转储，浏览器还提供搜索服务下载特定的数据集（例如，BioMart 和 UCSC table 浏览器，表 31-4）。当需要将数据集限制为特定的参数时，这些功能尤其有用。例如，用户希望获得具有 OMIM ID 的人类同源物的所有基因的 500bp 上游序列。

浏览器还允许用户附加自己选择的其他数据，其范围包括不同的文件格式到不同的跟踪枢纽（track hubs）和组装枢纽（assembly hubs）（genome.ucsc.edu/goldenPath/help/hgTrackHubHelp.html）。跟踪枢纽可以包含大量的全基因组数据包，并提供一种方便的方式来共享由用户控制的大型数据包，而不会给浏览器制作团队带来压力。GRC 会对斑马

表 31-4 斑马鱼基因注释和浏览：有用的网址

主题	网址
NCBI 浏览器	www.ncbi.nlm.nih.gov/genome/?term¼zebrafish
RefSeq 数据库	www.ncbi.nlm.nih.gov/genome/annotation_euk/Danio_rerio/106/
Ensembl	www.ensembl.org/Danio_rerio
Ensembl 下载	www.ensembl.org/info/data/ftp/index.html
BioMart	www.ensembl.org/biomart/martview
UCSC 浏览器	genome.ucsc.edu/cgi-bin/hgGateway
UCSC table 浏览器	genome.ucsc.edu/cgi-bin/hgTables
GRC 网址	ngs.sanger.ac.uk/production/grit/track_hub/hub.txt

鱼参考基因组装产生一系列的注释，包括正在审查的基因组问题、主要位点和备用位点或补丁之间的比对、克隆重叠连接认定、克隆序列异常和 BioNano 制图数据。这些数据每周更新一次，可以通过所有主要的基因组浏览器查看（图 31-4）。

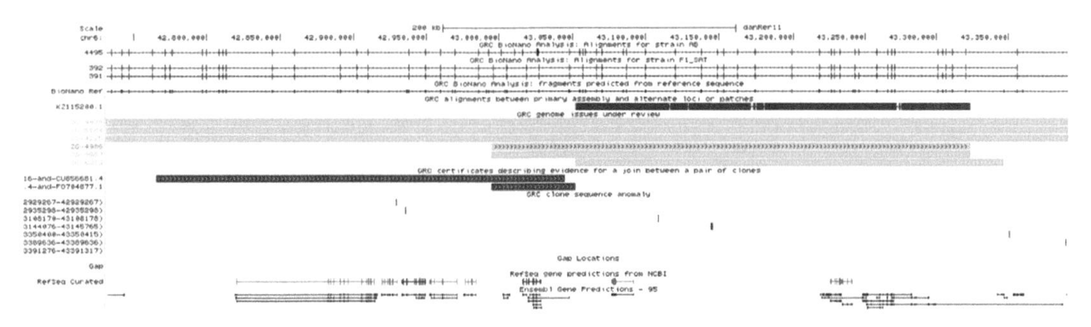

图 31-4 6号染色体上的斑马鱼装配区 UCSC 基因组浏览器截图

GRC trackhub 已经加载并显示（从上到下）BioNano 数据、主要组装和备用位点之间的序列、正在审查的基因组问题、克隆重叠图和克隆序列异常。此外，可以看到 gap 位置、curated Refseq 基因和 Ensembl 基因

31.4 序列比较分析

斑马鱼基因组与人类基因组具有高度的同源性（Howe et al., 2013），使得斑马鱼成为研究人类发育和疾病的模型。人类基因和基因组区域的功能可以通过研究斑马鱼来揭示，这是比较基因组学的一项工作，许多网络资源提供了基因组区域比对以及同源基因与同源染色体鉴定，极大地促进了这项工作的开展。

Ensembl 数据库比较分析提供了许多方法来研究基因组序列之间的关系，从基因家族和同源性列表到保守值和限定元素的各种比对与鉴定（www.ensembl.org/info/genome/compara）。基因组区域可以与一个或多个从 Ensembl 数据库中选择的物种对齐，并根据需要调整注释。感兴趣的基因组序列可以导出并用于创建比对和比对注释的其他资源，如 zPicture、PipMaker 和 Vista 等。zPicture 基于 Blastz 算法提供两个选择序列的在线比对，并提供各种可视化选项 [zPicture.dcode.org/ (Ovcharenko et al., 2004)]。“百分比标识图”

或 PipMaker 可以在线对 2 到 20 个序列进行比对，并生成图形概述，以强调用户自定义特性的保守性［pipmaker.bx.psu.edu/pipmaker/ (Schwartz et al., 2000)］。Vista 工具包提供了一套全面的应用程序，从在浏览器中预先计算的比对和 2 ～ 100 个用户自定义序列的在线对比到独立软件包［genome.lbl.gov/vista/index.shtml (Poliakov et al., 2014)］等。

在 Ensembl compara 或 Genomicus 中可以查看长程共线性（long-range synteny），该浏览器使用 Ensembl 数据来整理和描绘多个物种的共线性区域，并将它们与遗传数据集成在一起（图 31-5，表 31-5）（Nguyen et al., 2018）。

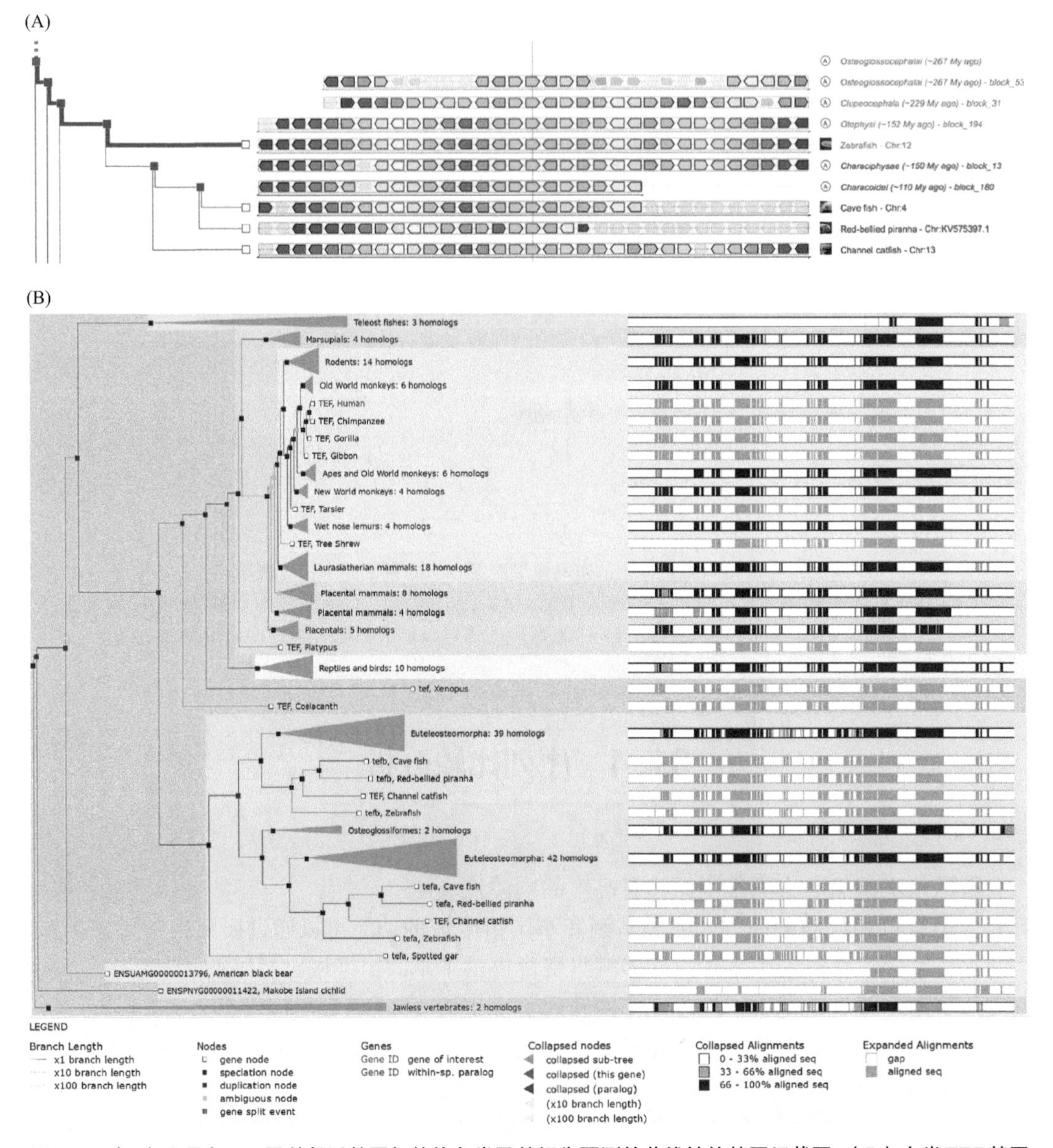

图 31-5 （A）斑马鱼 *tefa* 及其邻近基因与其他鱼类及其祖先预测的共线性的基因组截图；（B）人类 TEF 基因的 Ensembl 基因树（红色部分）

（A）图中，相应的基因名称出现在鼠标上方。（B）图中，重点研究其在斑马鱼中配对的同源基因 *tefa* 和 *tefb*，可以根据需要扩展和折叠分支，并将树结构和蛋白质序列导出到 Wasabi 查看器中，以便进行进一步编辑

表 31-5 比较基因组学：有用的 URL

主题	URL
Ensembl 数据库比较	www.ensembl.org/info/genome/compara
zPicture 数据库	zpicture.dcode.org/
PipMaker 数据库	pipmaker.bx.psu.edu/pipmaker
Vista 数据库	genome.lbl.gov/vista/index.shtml
Genomicus 数据库	www.genomicus.biologie.ens.fr/genomicus
ZFIN 数据库下载	downloads.zfin.org/downloads
Homologene 数据库	www.ncbi.nlm.nih.gov/homologene

通过 Ensembl 预测直系和旁系同源基因，有利于研究特定的同源关系。对于任何选择的基因，Ensembl 可以列出数据库中所有物种的直系和旁系候选基因，通过特定的比对和基因树，将目的基因和其他物种的同源基因置于进化环境中。至关重要的是，在像斑马鱼这样具有额外基因组复制的物种中，这支持了同源基因［由基因组复制引起的共享同一同源序列的同源基因，图 31-5（B）］的鉴定。有关同源性分析信息的更多资源可以在 ZFIN 的可下载列表（zfin.org/downloads）中找到，也可以在 NCBI 同源数据库中搜索（www.ncbi.nlm.nih.gov/ Homologene）。

31.5 变异

遗传变异是所有可遗传表型差异的基础。因此，对遗传变异的分析毫无疑问地与斑马鱼模型的使用相关联。基于单核苷酸多态性数据库 dbSNP（短遗传变异数据库，www.ncbi.nlm.nih.gov/snp）和染色体结构变异数据库 DGVa（基因组变异数据库，www.ebi.ac.uk/dgva/data-download）的集成，Ensembl 网络浏览器为斑马鱼变异分析提供了很多有用的资源。基因组学可以发现遗传性变异，并对其预测结果（包括表型）进行注释。

Ensembl 还提供了一些选项来研究由用户使用其他工具提供的序列。这其中包括变量记录器（转换不同的变量标识符），以及连锁不平衡计算器（a Linkage Disequilibrium Calculator）。最常用的工具是预测单个变异对基因和调节区域影响的变异效应预测器（Variant Effect Predictor, VEP），以及确定多个变量影响的 Haplosaurus。

Ensembl 变异分析（以及所有基因组浏览器和网络资源分析）的功能依赖于公开可用的高质量数据集。虽然公有领域的斑马鱼变异数据相当稀少，但可供人类使用的海量变异数据可通过上文提到的直系同源关系来解释在斑马鱼中的发现（表 31-6）。

表 31-6 变异：有用的 URL

主题	URL
Ensembl 数据库变异查询	www.ensembl.org/info/genome/variation
变异效应预测器网站	www.ensembl.org/info/docs/tools/vep

续表

主题	URL
Haplosaurus 网站	www.ensembl.org/info/docs/tools/vep/haplo
变异记录网站	www.ensembl.org/info/docs/tools/vep/recoder
连锁不平衡的计算器网站	www.ensembl.org/Homo_sapiens/Tools/LD
单核苷酸多态性数据网站	www.ncbi.nlm.nih.gov/snp
基因组变异数据库	www.ebi.ac.uk/dgva/data-download

31.6 鲋亚科鱼类基因组测序和组装

随着斑马鱼作为生物医学模型的普及和斑马鱼基因组组装的完成，人们对该物种及其近亲的生态和进化越来越感兴趣（Parichy, 2015）。鲋亚科（Danioninae）分支在形态特征上呈现出巨大的多样性，如色素图案、体形大小和骨骼结构。如果能掌握物种间的系统发育关系，这些性状的进化将更容易理解；然而这还不是决定性的（McCluskey and Postlethwait, 2015）。为了深入研究这些问题，并将鲋亚科分支用作进化发育生物学模型，人们开展了多个可生成几种斑马鱼品系及其亲缘基因组组装的测序项目。

其中一个项目由桑格研究所（www.sanger.ac.uk）与脊椎动物基因组项目（VGP, vertebrategenomes.org）合作，提供不同水平的测序，从简单的 Illumina 测序到分析变异，再到完全单倍型解析的组装，由多种数据类型所创建（表 31-7）。生成的数据提交到 INSDC 档案，完成的组装也可以在 VGP 的 GenomeArk 存储库（VGP.github.io/Genomeark/）访问。

表 31-7 桑格研究所目前的 Danioninae 测序项目
（所有组装都通过 gEVAL 评估完成，以消除明显的组装错误）

种属	组装水平
Danio rerio SAT	完全单倍型解析的 TrioCanu 组装 (PacBio, chromium, BioNano, HiC，基因图谱)
Danio rerio AB	Supernova 组装 + BioNano，基因图谱
Danio rerio Cooch Behar	Supernova 组装
Danio rerio Nadia	Supernova 组装
Danio aesculapii	Full VGP pipeline assembly (PacBio, chromium, BioNano, HiC)
Danio kyathit	Full VGP pipeline assembly (PacBio, chromium, BioNano, HiC)
Danio tinwini	Supernova 组装
Danio albolineatus	Supernova 组装
Danio choprai	Supernova 组装
Danio jaintianensis	Supernova 组装
Danionella dracula	PacBio 组装
Danionella translucida	Supernova 组装

致谢

本章大部分内容基于斑马鱼网络资源年度教程，由 Kerstin Howe 和 Jane Loveland 在欧洲和国际斑马鱼会议期间策划和授课。关于比较基因组学和变异的内容受到 Jane Loveland 整理和讲授的专题的启发。

（吴方晖翻译　赵丽审校）

参考文献

Chow, W., Brugger, K., Caccamo, M., Sealy, I., Torrance, J., Howe, K., 2016. gEVAL e a web-based browser for evaluating genome assemblies. Bioinformatics 32 (16), 2508-2510. https://doi.org/10.1093/bioinformatics/btw159.

Howe, K., Clark, M.D., Torroja, C.F., Torrance, J., Berthelot, C., Muffato, M., et al., 2013. The zebrafish reference genome sequence and its relationship to the human genome. Nature 496 (7446), 498-503. https://doi.org/10.1038/nature12111.

Howe, K., Schiffer, P.H., Zielinski, J., Wiehe, T., Laird, G.K., Marioni, J.C., et al., 2016. Structure and evolutionary history of a large family of NLR proteins in the zebrafish. Open Biology 6 (4), 160009. https://doi.org/10.1098/rsob.160009.

Howe, K., Wood, J.M., 2015. Using optical mapping data for the improvement of vertebrate genome assemblies. GigaScience 4, 10. https://doi.org/10.1186/s13742-015-0052-y.

McCluskey, B.M., Postlethwait, J.H., 2015. Phylogeny of zebrafish, a "model species," within *Danio*, a "model genus". Molecular Biology and Evolution 32 (3), 635e-652. https://doi.org/10.1093/molbev/msu325.

Nguyen, N.T.T., Vincens, P., Roest Crollius, H., Louis, A., 2018. Genomicus 2018: karyotype evolutionary trees and on-the-fly synteny computing. Nucleic Acids Research 46 (D1), D816-D822. https://doi.org/10.1093/nar/gkx1003.

O'Leary, N.A., Wright, M.W., Brister, J.R., Ciufo, S., Haddad, D., McVeigh, R., et al., 2016. Reference sequence (RefSeq) database at NCBI: current status, taxonomic expansion, and functional annotation. Nucleic Acids Research 44 (D1), D733-D745. https://doi.org/10.1093/nar/gkv1189.

Ovcharenko, I., Loots, G.G., Hardison, R.C., Miller, W., Stubbs, L., 2004. zPicture: dynamic alignment and visualization tool for analyzing conservation profiles. Genome Research 14 (3), 472-477. https://doi.org/10.1101/gr.2129504.

Parichy, D.M., 2015. Advancing biology through a deeper understanding of zebrafish ecology and evolution. eLife 4. https://doi.org/10.7554/eLife.05635.

Poliakov, A., Foong, J., Brudno, M., Dubchak, I., 2014. GenomeVISTAean integrated software package for wholegenome alignment and visualization. Bioinformatics 30 (18), 2654-2655. https://doi.org/10.1093/bioinformatics/btu355.

Schwartz, S., Zhang, Z., Frazer, K.A., Smit, A., Riemer, C., Bouck, J., et al., 2000. PipMakerea web server for aligning two genomic DNA sequences. Genome Research 10 (4), 577-586.

Zerbino, D.R., Achuthan, P., Akanni, W., Amode, M.R., Barrell, D., Bhai, J., et al., 2018. Ensembl 2018. Nucleic Acids Research 46 (D1),D754-D761.https://doi.org/10.1093/nar/gkx1098.

第三十二章

注册、标准化和交互性：斑马鱼神经解剖学在线资源综述

Arnim Jenett[1]

32.1 引言

虽然斑马鱼（*Danio rerio*）最初是作为遗传学和发育生物学研究领域的模式生物而建立的，但在过去的几十年里，从肿瘤学到生理学，再到学习、记忆和心理学的许多其他研究领域都开始使用斑马鱼进行研究。

最近的综述（Wyatt et al., 2015; Bradford et al., 2017; Ablain and Zon, 2013; Clarkson, 2016）和特刊（Burton and Palladino, 2010）都详细描述了这一进展。

因为这些新领域中的大多数（如果不是全部的话）都以这样或那样的方式将其结果与模型的神经解剖学联系起来，描述性和功能性神经解剖学领域即将成为研究物种之间相关性的关键领域，就像遗传学领域一样。

重建解剖学作为各种研究之间的桥梁纽带，最容易实现的目标之一就是开发标准化的神经解剖学图谱。这些在线数据库通过提供易于访问的已有知识源，帮助研究者设计和分析他们的实验。在项目的设计阶段，这些知识库可以帮助研究者在选择转基因动物或抗体来检测和研究表型时做出明智的决定，新结果可以与公共数据相关联，甚至共享。理想的状况是：所有这些信息来源，包括各种数据库以及新的数据都应该能融合到一个数据库中。然而现实中的情况是：数据模式和数据库结构之间的差异使得这种目标目前还无法实现。然而，目前正处于起步阶段的共享数据存储库和不同图谱之间的桥接注册的方法可能会给数据共享带来一些希望，即可以实现不同数据库中可互换数据的共享，这将促进不同平台和研究之间的数据关联。

[1] TEFOR Paris Saclay, CNRS UMS2010, INRA UMS1451, Universite Paris Sud, Université Paris-Saclay, Gif-sur-Yvette, France.

32.2 斑马鱼解剖数据的收集

随着斑马鱼研究的进展，有少数团队开始致力于系统地记录和描述斑马鱼的神经解剖学。

一般来说，这些图谱是以印刷品形式出版的（Bryson-Richardson et al., 2012; Holden et al., 2012; Wullimann, Rupp, and Reichert, 1996; Mueller and Wullimann, 2016, 2005），但是随着更快捷的互联网的应用，许多资源转移到网上。互联网当然有灵活与方便之处：可以随着时间的推移添加新数据。然而，与印刷书籍相比，这也给这些项目带来了维护负担：出版媒体的不断发展需要持续对在线资源进行维护。这可能就是许多最初很有前途的在线资源多年来停滞不前或完全关闭的原因。在本章，我们只关注目前（2019 年）已建立或即将发布的斑马鱼解剖学特别是斑马鱼神经解剖学在线资源。

32.2.1 ZFIN 网站上的斑马鱼解剖学

最早的斑马鱼解剖学在线项目之一是斑马鱼信息网（the Zebrafish Information Network, ZFIN），该网地址为 https://zfin.org/zf_info/anatomy/dev_atlas.html。这些含注释的斑马鱼早期胚胎亚甲蓝染色的横切面和矢状面切片是在 21 世纪之交建立的，可以认为是网上描述的所有其他资源的源头。该项目早期满足了对斑马鱼在线解剖资源的需求，并代表了未来资源精确性和一致性的理想水平。

第一种数字化解剖切片方法基于高质量的细胞分辨率图像。虽然可从两个方向上提供四个不同的胚胎和早期仔鱼阶段的图像，但也仅能表现出斑马鱼身体的一部分，注释有限。

在 ZFIN 解剖学版块的首页，该项目定义如下：“这些斑马鱼在四个发育不同阶段（24hpf，48hpf，72hpf，120hpf）的切片有助于了解斑马鱼胚胎的内部结构。环氧树脂切片采用亚甲蓝染色。拍摄的图像进行了数字化处理。读者可以找到胚胎的每个发育阶段的概视图，并通过链接（点击图像上的数字）可查到各部分图像。读者也可以下载高分辨率图像（JPEG，约 1MB），在大多数情况下，这些图像可以放大到单细胞核水平。”

32.2.2 斑马鱼血管解剖交互式图谱

在 ZFIN 创建解剖学版块的同一时期，美国国家卫生研究院（the National Institute of Health, NIH）尤妮斯•肯尼迪•施赖弗（the Eunice Kennedy Shriver）国家儿童健康和人类发展研究所（National Institute of Child Health and Human Development, NICHD）温斯坦实验室（the Weinstein lab）创建了第一个更专业的在线资源库：斑马鱼血管解剖交互式图谱（The Interactive Atlas of Zebraf ish Vascular Anatomy）（https://zfish.nichd.nih.gov/Intro%20Page/intro1.html）（Isogai et al., 2001）。这个在线资源库将激光共聚焦扫描显微镜拍摄的早期胚胎高分辨率 3D 荧光血管造影图像转换成带有轻微斜角镜头的视频（lightweight tilt-movies），由于其具有视差效应，非常有助于理解发育中的斑马鱼仔鱼血管系统的复杂解剖特征。该资源库提供了一种传统的数据处理方法，可将一个 3D 堆叠中每个叠片转换为一帧视频画面。但其未自带播放器，因此需要将视频下载后使用本地软件查看。由于它们所占的容量小，文件格式规范，软件和操作系统兼容性强，这是一种非常好的数据处理方法。

轻极简化风格的交互式斑马鱼血管解剖图谱入口页面将这些资源描述为："本网站是一个在线互动的'图谱'，三维细节展示了发育中斑马鱼的完整血管解剖结构。用户可以查看斑马鱼受精后 1 ～ 7 天的任何发育阶段的胚胎的任何区域的三维荧光血管造影图像。用户还可以访问针对这些血管造影制作的图表和描述性文本。这些图表详细说明了主要血管的连接方式和名称。"

32.2.3 斑马鱼大脑解剖

2010 年，伦敦大学学院（University College London）细胞和发育生物学系在 zebrafishbrain.org 网站（http://zebrafishbrain.org）启动了一个非常有前景的媒体收集项目。不幸的是，该资源目前（2019 年春季）正在重组中，因此处于离线状态。然而，由于媒体数据仍然可以通过 zebrafishucl.org 访问，该资源很有可能很快就会恢复。当 zebraf ishbrain.org 还在线的时候，网站首页显示如下描述："zebrafishbrain.org 将与传统神经解剖学图谱不同，因为它将使用多种媒体类型来呈现数据，包括激光共聚焦切片和重构、图解模型以及最终的 3D 和 4D 视频。目前，我们的方法是从处于多个发育阶段的不同转基因品系中收集高分辨率的共聚焦叠层图像，并使用 Improvision volocity 软件将这些图像重建为平面投影。"

32.2.4 FishFace/FaceBase 数据库

另一个早期采用互联网作为复杂解剖数据交流平台的是"细胞分辨率的交互式斑马鱼颅面发育图谱（The interactive atlas of zebrafish craniofacial development at cellular resolution）"——FishFace（https://www.facebase.org/fishface/home/）（Eames et al., 2013）。该数据库起初由俄勒冈大学（the University of Oregon）的坎摩尔实验室（Kimmel lab）基于荧光共聚焦激光扫描显微镜（confocal laser scanning microscopy, CLSM）和光学投影断层扫描（optical projection tomography, OPT）数据所构建，后来这些数据被整合到 facebase 中，facebase 是由美国国家口腔与颅面研究所（NIDCR）资助的合作联盟，旨在生成数据以支持颅面发育和畸形的前沿研究（来自 https://www.facebase.org/about/）。该资源提供了在公共信息基础设施中可访问的关于人类、小鼠、斑马鱼和黑猩猩的各种类型的数据，令人印象深刻。由宾夕法尼亚大学的 Shannon Fisher 和波士顿儿童医院的 Matthew Harris 组成的团队牵头启动了正常斑马鱼的颅骨发育资源的二期工作，他们的目标是利用发育后期的 microCT 和 CLSM 数据丰富现有的数据。将该标准图谱与从转基因细胞系筛选获得的数据进行比较，或许能够加深人们对临床上颅骨形成异常的遗传学基础的理解。

在网站首页，该资源的作者对 FishFace/FaceBase 数据资源作了如下定义："颅面骨骼的组成部分是如何产生、生长和重塑的？这个问题要借助分子遗传学和细胞生物学方法来解答，首先要对构成骨骼的发生事件和过程进行精确描述。斑马鱼的遗传学和基因组学已被充分认知，有利于发育的表型分析，且斑马鱼具有与所有脊椎动物都一致的发育模式，使得其成为研究颅面发育的强有力的动物模型。特别是利用现有的转基因方法，人们可以在活着的、完整的胚胎和仔鱼的长时间发育过程中，以细胞分辨率观察斑马鱼颅面骨骼组织结构，这种研究方法在准确性和敏感性方面都是无与伦比的。我们构建了这个颅面骨骼发育图谱。FishFace 可作为此类研究的指南。”

32.2.5 FishNet/ZFAP 网站

莫纳什（Monash）大学的布莱森 - 理查森研究组（Bryson-Richardson Research Group）于 2007 年建立了 FishNet（Bryson-Richardson et al., 2007）网站，于 2012 年创建了斑马鱼解剖学门户网站（http://zebrafish.anatomyportal.org/）（ZFAP）（Salgado et al., 2012a,b,c）。作为更广泛的解剖学门户网站的一部分，ZFAP 还包含鹌鹑发育数据。ZFAP 重点关注利用甲醛固定样本的自发光 OPT 成像实现斑马鱼从早期幼体（24h）到成年（17mm 体长）的发育过程可视化。该网站可在线浏览 18 个不同发育阶段的三维数据，界面非常直观，显示了横向、冠状、矢状三个切面的二维可视图。与当前样本配对的概览图像通过一组彩色编码的十字线可在样本中确定 2D 可视化的位置。通过点击概览图像，这些图像还可以用作选择切片集的输入工具，该切片集显示在 2D 查看器中。结合相邻的导航工具（允许沿着主轴逐步导航数据），其界面非常简单。切片的注释很丰富，有助于对 2D 数据的研究和理解。数据的 3D tif 下载选项也允许用户选择本地工具对其进行本地研究和 3D 可视化。总之，这个资源库提供了多种生物体的不同发育阶段广泛注释的 2D 视图。但是由于成像方法的限制，图像质量尚未达到细胞水平的分辨率，网站的数据通量和更新率较低。

解剖学门户网站的首页对这个项目有如下描述：“斑马鱼解剖门户网站（ZFAP）是一个解剖学资源库，提供胚胎发育的三维模型。它旨在提供有价值的模式生物参考资料，并通过不断添加新物种，为比较解剖学研究提供资源。”

ZFAP 对其项目描述如下：“斑马鱼解剖学门户网站（ZFAP）是了解斑马鱼解剖学特征的重要资源，包含了早期幼体（从 24h）到成年斑马鱼的三维模型。”

32.2.6 neurodata.io 网站

与本章描述的其他资源库不同，以 Joshua Vogelstein 为核心的团队开发的神经数据资源网（https://neurodata.io/）（Vogelstein et al., 2018）并不是一个现成的斑马鱼解剖图谱库，而是一个主题分类平台，用于共享大量 3D 数据包和可视化工具集。neurodata.io 目前基于 EM 数据（Hildebrand et al., 2017），以及刺激和行为依赖性神经活动的 6dpf 斑马鱼的全脑图谱（Randlett et al., 2015），对出生后 5.5 ～ 7 天的斑马鱼大脑中神经元进行稀疏重构。该项目的模块化设计可以集成各种数据模式和软件工具，用于注册、可视化和分析。该资源库也包含其他物种（果蝇、小鼠、秀丽隐杆线虫、人类）的数据，这些数据比斑马鱼物种的数据要多得多。该网站资源应用的大量技术对来访的科学家既是惊喜也是困扰：不同模式的数据可以直接获取，对大多数人来说，甚至连这些数据的注释工具都可以得到。然而，对于特定物种的数据包尚未登记到一个共同的坐标系中。目前，这限制了给定数据集在其特定工具集中的使用。然而，托管软件工具的严格开源理念，以及 Vogelstein 团队的持续努力，也推动了协同注册和通用可视化工具的进一步开发。

作为一个令人印象深刻且持续成长的新项目，建议读者经常浏览 neurodata.io，并在这个网站上跟踪研究进展。

https://neurodata.io/ 的自我设定目标如下："通过扩展与融合统计机器学习和大数据科学来解决我们这个时代最重要的脑科学和心理健康问题。"他们的使命宣言可能是实现这一目标的最佳途径："神经科学数据采集的革命正在对分析提出更高要求。我们利用云计算技术实现大规模神经数据的存储、探索、分析和建模，使全球科学家能够提出和测试大脑功能和功能障碍的理论。"

32.2.7 Bio-Atlas 数据库

目前宾夕法尼亚州立大学Cheng实验室创建的斑马鱼生物图谱数据库Bio-Atlas（http://bio-atlas.psu.edu/zf/progress.php）由 14 个不同发育阶段斑马鱼的 555 个（部分注释）高分辨率连续切片所组成，由苏木精和伊红（H&E）染色（Sabaliauskas et al., 2006)。这项研究数据分辨率非常高，能够很好地帮助人们了解斑马鱼作为一个有机整体的大体解剖结构。然而，遗憾的是，头部区域的强烈变形降低了这些原本完美的数据集在中枢神经系统（CNS）研究中的使用价值。

该团队近期优化了斑马鱼切片的固定和包埋方案（Copper et al., 2018; Lin et al., 2018），启动了使用同步加速 microCT 扫描进行斑马鱼整体的细胞分辨率成像的新项目（Ding et al., 2018, 2019; Xin et al., 2017)，可能很快就能够克服这一缺陷。第一个传统 H&R 染色与 microCT 方法比较的样本已经可以通过该项目的"样本比较"页面访问。

Bio-Atlas 项目组是这样描述该项目的：

斑马鱼虚拟幻灯片：第一个目标是提供基础结构，作为正常和异常斑马鱼组织数字化切片的中央存储库，允许以与现有解剖和表型本体相协调的方式进行渐进放大和标记。该项目的下一个目标是扫描收集的切片，并将数据库结构定义为可扩展的格式。

人类和其他虚拟切片：第二个目标，也可能成为主要目标，是获得人体解剖学和病理学的切片，以便与其他的模式生物数据进行比较。一个子目标是后期建立其他模型系统的切片。

MicroCT：我们正在利用斑马鱼来实现对整个毫米级生物体（以及后来的标本）进行表型组学（高通量表型研究）的 3D成像。因此，我们正在寻求实现微米级的细胞分辨率 X 射线断层扫描。使用同步加速 microCT，给斑马鱼创建了一个三维组织模型，称之为全细胞组织断层扫描（pan-cellular tissue tomography, PANCETTO)。我们的目标是重构图谱，可以看到任意方向的完整 2D 动物切片。这项工作是与芝加哥大学的 La Riviere 实验室和阿贡国家实验室（Argonne National Laboratory）先进光子源的物理学家合作进行的。

32.2.8 zTrap 和 ZeBrain 线上资源

日本国家遗传学研究所的Kawakami 实验室根据自己收集的转基因斑马鱼资料（Asakawa

and Kawakami, 2008; Kawakami et al., 2016）提供了两个在线资源库：zTrap（http://kawakami.lab.nig. ac. jp /ztrap/faces/image/ImageBrowse.jsp）和 ZeBrain（https://zebrain.nig.ac.jp/zebrain/page.do?p= welcome）。

通过优化日本青鳉（*Oryzias latipes*）的转座子 Tol2 在斑马鱼中转录，且将其导入种系，从而获得了该数据库的转基因斑马鱼（Koichi Kawakami et al., 2004; Kawakami, Shima, and Kawakami, 2000）。这些转基因鱼编码转录因子 Gal4FF，能够可靠地驱动各种用于 Gal4 阳性细胞可视化或功能操作的 UAS 效应器，这为各种解剖学和功能性实验开辟了道路。

zTrap 利用宽场（原位杂交）和荧光显微镜（Gal4FF; UAS:GFP），展示了该资源库的 1168 个转基因品系在仔鱼早期（受精后 0 ～ 5 天）的表达模式。使用 33 个解剖学本体（anatomic ontology）对表达模式进行了注释，并通过本地数据库以及 ensembl.org 中的文本和图像展示，对每一品系的遗传学进行了详细描述。

另一方面，ZeBrain 使用宽视场荧光显微镜和共聚焦激光扫描显微镜来可视化解剖的成年斑马鱼大脑中 GFP 的诱导表达模式。使用包含 187 个解剖学术语的本体对表达模式进行解剖学注释，包括 25 个神经束、神经连合或神经以及 162 个层次有序的脑区。该本体论可用于从数据库中检索 Gal4 品系。由于这个在线资源库当前处于工作进行中的状态，因此其提供的表达模式较少。

ZeBrain 的入口页面上有这样的描述："ZeBrain 是一个包含成年斑马鱼大脑（来自日本三岛国家遗传学研究所 Kawakami 实验室培育的转基因斑马鱼）中 GFP 表达模式的数据库。在 Kawakami 实验室，通过 Tol2 转座子介导的基因捕获和增强子捕获方法已经产生了大量的转基因斑马鱼，它们以时空限制的方式表达 Gal4FF 转录激活子。从这一转基因资源中，研究者收集了在成年大脑中有特异性 Gal4FF 表达的转基因鱼，并进行了切片分析。ZeBrain 数据库的建立是为了方便搜索这些有用的转基因斑马鱼品系，以寻找感兴趣的表达模式。"

32.2.9 AGETAZ 数据库

在卡尔斯鲁厄理工学院（Karlsruhe Institute of Technology）欧洲斑马鱼资源中心（the European Zebrafish Resource Center, EZRC）建立的"成年斑马鱼大脑基因表达图谱"（The atlas of gene expression in the adult zebrafish brain, AGETAZ）（https://itgmv3.itg.kit.edu/agetaz/index.php）提供了 1205 种不同转录因子原位杂交后的成年斑马鱼端脑横切面数据集（Diotel et al., 2015）。

准备聚类分析时，根据斑马鱼大脑图谱（Wullimann, Rupp, and Reichert, 1996）中定义的区域图对记录的表达模式进行注释，便于根据特定感兴趣区域内的表达模式检索一组转录因子。这种解剖学搜索在该资源库的"基因表达的聚类分析"部分能够非常清晰地呈现。

此外，这个数据库也可以通过基因名称、其人类同源基因或在 Zfin、UniGene、GenBank、RefSeq 或 Ensembl 数据库中定义的标识符进行查询。

对于每个包含的转录因子，该数据库提供了一个信息丰富的摘要页面，包括从名称、同义词、人类同源物和适用的基因本体术语等描述性数据，到外部数据库、微阵列和RNA 表达谱的链接列表（如果可用）。

这种与现有数据库环境的整合，为解剖驱动的转录因子研究到基因调控和一般生物信息学领域研究提供了一个很好的过渡，反之亦然。AGETAZ 网站上的介绍如下：“成年斑马鱼端脑基因表达图谱（AGETAZ）是一个成年斑马鱼（*Danio rerio*）端脑中转录因子（TF）和转录相关因子的表达图谱。该数据库是收集了斑马鱼胚胎中 TF 表达模式的鱼类因子数据库（fish factor database, FFdb）的后续，AGETAZ 旨在绘制这种模式生物成年端脑中表达的所有 TF，并且提供了一种快速的方法来发现你感兴趣的任何端脑区域（E 区）表达的 TF。因此，AGETAZ 可用于识别斑马鱼端脑中存在的不同细胞核和细胞类型的新标志物。”

32.3 标准化斑马鱼神经解剖学资源

虽然上述收集的数据为了解斑马鱼整体或特定解剖结构打开了方便之门，但单独的数据集只有通过其共享的染色、获取和可视化技术，以及适用的发育阶段进行协调。这阻碍了直接在屏幕上对不同数据集进行比较：这些资源中的一个数据集与另一个数据集的精确叠加是不可能的，必须在科学家头脑中进行。这是有问题的，因为它需要用户具有相当高水平的知识储备，而且同样重要的是，它为用户主观上的想法提供了充足的空间，这可能导致对所提供数据的曲解。

在下文中将介绍标准化斑马鱼神经解剖学的在线图谱，这些图谱在配准到一个通用坐标系后将其数据进行托管。由于经过配准（也称为“对准”），来自不同标本和 / 或成像模式的数据库可以以叠加方式交互显示，或抽象地通过其表达谱在通用坐标系空间内的分布进行查询。由于配准的处理和质量在很大程度上取决于所采用的技术和提供通用坐标系的模板，下面内容旨在对图谱构建中的这几个方面进行简要概述。

32.3.1 模板

标准神经解剖学图谱的核心是模板。作为标本神经解剖学的原型，它定义了坐标系，所有其他数据集都被配准到该坐标系统中。注意，在显微镜中，3D 图像由单个亮度测量（三维像素，体积像素）的立方矩阵（三阶张量，rank-3 tensors）组成，而在不同的标本中同样解剖位置的测量很少发生在这个张量内相同位置或方向上，因此模板作为通用坐标系的基础，其关键作用变得显而易见。

传统上，模板为数据库中选定的“最正常的数据集”。虽然这个选择过程可以通过算法优化，但最近的研究表明，统计学模板比基于单个样本的模板可以产生更好的种群配准结果（Joshi et al., 2004; Aubert-Broche et al., 2006; Dickie et al., 2017）。鉴于这项技术在标准化神经解剖学的其他物种领域得到了发展并被广泛接受［果蝇（Bogovic et al., 2018; Arganda-Carreras et al., 2018），小鼠（Dorr et al., 2008），人类（Xiao et al., 2017; Fonov et al.,

2011; Sanchez, Richards and Almli, 2012a, 2012b; Fillmore et al., 2015; Evans et al., 2012）和灵长类动物（Quallo et al., 2010, Seidlitz et al., 2017; orsten Rohlfing et al., 2012; Nadkarni et al., 2018; Van Essen and Dierker, 2007）]，因此，目前在斑马鱼群中也采用这种方法。

32.3.2 参考染色

神经解剖学的标准化一般基于参考模式的概念，参考模式在模板和给定图谱的所有数据集之间是通用的。理想情况下，这种省力、便宜、稳定的染色与表达模式的可视化或感兴趣的染色并行进行。其信号采集与感兴趣的信号同时采集，使得两个通道具有解剖相关性（内部相关性，IC）。因此，图谱数据集至少有两个通道：参考通道和感兴趣的信号通道。参考通道提供了与模板直接相关的解剖框架（外部相关性，EC）。可以假设所有标本的参考染色具有一致的形态，因此配准算法可以通过参考染色将不同数据集的解剖结构对准到模板的通用坐标系中。因此，配准的处理不是在感兴趣的数据上进行的（这些数据可能高度分散），而是在参考染色通道的基础上进行的。由此产生的转换随后被直接应用于实际信号通道，将其转换为图谱的通用坐标系。

对于斑马鱼，目前的参考染色方法有三种：转基因表达模式、免疫标记和染料染色。使用转基因品系作为参考模式的优点是可以直接与活体成像兼容，但需要在给定的转基因品系遗传背景下进行活体实验。据报道，转基因元件（转录因子、蛋白质）的表达有可能促进果蝇（Rezával et al., 2007; Pfeiffer et al., 2010; ramer and Staveley, 2003; Mawhinney and Staveley, 2011）和鱼类（Halpern et al., 2008）的神经退行性变化，如果转基因品系内的修饰对实验动物产生非预期的生理影响，或者由于转基因参考种群内发生基因修复或基因沉默，转基因表达模式的形态将随着时间的推移而发生变化，那么转基因背景可能会产生问题。与转基因表达模式相比，染料染色和免疫标记只适用于离体标本，这也限制了它们作为模板模式的通用性。因此，随着不同应用范围的扩大，对不同参考染色的需求也将持续存在（表 32-1）。

表 32-1 参考染色和模板类型

图谱	参考对象	分类	模板类型	操作方案
VibeZ	TOTO-3 或者 Sytox 染料	染色	单个大脑	Rath et al. (2012)
Brain Browser	tERK 抗体（Anti-tERK），vglut2a: dsRed 转基因鱼	免疫组化，转基因鱼	单个大脑	Marquart et al. (2017)
Z-Brain	tERK 抗体（Anti-tERK）	免疫组化	167 个标本平均值	Randlett et al. (2015)
TEFOR	DiI 膜染料，PCNA 抗体（anti-PCNA）	染色，免疫组化	15 个标本中位数	Arganda-arreras et al.(2018)
MPI	突触蛋白抗体（anti-synapsin）	免疫组化	12 个标本平均值	Kunst et al. (2018)

32.3.3 配准技术

在构建标准化神经解剖学图谱的过程中，最重要的步骤之一是将单个数据库配准到模板的通用坐标系中。

目前，标准化斑马鱼神经解剖学领域已建立了四种不同的方法：

• 计算形态测量工具包（computational morphometry toolkit, CMTK）（Rohlfing and Maurer, 2003）

• 高级标准化工具（advanced normalization tools, ANTs）（Avants et al., 2009）

• 开源通信软件包（elastix）（Klein et al., 2010）

• “扩展 yoUr 视图”工具包（“eXtend yoUr View” Toolkit, XuvTools）（Emmenlauer et al., 2009）

虽然 XuvTools 的开发似乎已于 2011 年停止，但其他三个工具仍在积极开发中（表 32-2）。

表 32-2 斑马鱼神经解剖学标准化在线图谱的配准方法

图谱	配准方法	实施方案
VibeZ	XuvTools	Ronneberger et al. (2012)
BrainBrowser	CMTK, ANTs	Marquart et al. (2017)
Z-Brain	CMTK	Randlett et al. (2015)
TEFOR	ANTs	Arganda-Carreras et al. (2018)
MPI	ANTs	Kunst et al. (2018)

32.3.4 VibeZ 网站

第一个标准化斑马鱼神经解剖学资源库是 2012 年建立的 VibeZ（http://vibez.informatik.uni-freiburg.de/）（Ronneberger et al., 2012），它是弗赖堡大学（University Of Freiburg）和弗里德里希 • 米歇尔生物医学研究所（Friedrich Miescher Institute for Biomedical Research）的合作成果——斑马鱼虚拟大脑探索器（Virtual Brain Explorer for Zebrafish, ViBE-Z）。ViBE-Z 是一个用于斑马鱼仔鱼大脑虚拟共定位研究的成像和图像分析框架，目前可用于 48hpf、72hpf 和 96hpf 的仔鱼（http://vibez.informatik.uni-freiburg.de/）。

基于使用稀疏自动标记检测和对齐进行图像配准的开源工具集“XuvTools”（Emmenlauer et al., 2009），VibeZ 提出将新用户数据在线配准到其解剖坐标系统中。然而，为了实现这一点，新数据的生成必须严格遵循既定的协议（Rath et al., 2012），包括对每个样本的背侧和腹侧进行成像，这将使采集时间加倍。在此之后，生成的数据必须上传到弗赖堡的 VibeZ 服务器，以便配准到图谱的通用坐标系中。这严重影响了数据资源的可用性，从而导致这个本来很有前途的资源库开发处于停滞状态。

32.3.5 BrainBrowser 网站

国家儿童健康和人类发展研究所伯吉斯实验室（Burgess lab at the National Institute of Child Health and Human Development）开发的在线网站 BrainBrowser（http://metagrid2.sv.vt.edu/wchris526/zbb/）（Marquart et al., 2015, 2017）被认为是目前最先进、最通用的幼年斑马鱼标准化神经解剖学 3D 图谱网站。网站界面非常简洁，提供了大量配准数据，包括 250 多个转基因品系的表达模式（其中 191 个由该实验室制备）（Marquart et al., 2015, 2017;

Tabor et al., 2019）和 25 种抗体的染色模式。所有的解剖数据都与相应的文献和 ZFIN 直接相关。解剖部位的图谱可以帮助用户在数据中找到自己的模式，空间搜索可以快速访问用户确定的感兴趣区域中的表达或染色模式。

作为网站界面的替代选择，人们也可以批量下载数据并使用 Zebrafish BrainBrowser 2.0 桌面版进行搜索。该解剖数据库的本地执行有助于将个人数据集成到其框架中，同时保持新数据的私密性。

这个数据库的唯一缺点是与 macOS 操作系统上的 Safari 浏览器不兼容。

在 BrainBrowser 网页上显示如下介绍："BrainBrowser 是一种三维可视化工具，用于显示 6dpf 幼体斑马鱼转基因和基因表达模式。目前，该数据库包括 133 个转基因、增强子捕获或免疫染色仔鱼的表达模式，每一个都以高分辨率扫描，并配准到参考脑图谱。该浏览器可以在接近单细胞的分辨率下对表达模式进行比较，并在几乎任何选定的神经元组中识别表达 Gal4 的转基因品系。"

32.3.6 Z-Brain 网站

最新启用的"Z Brain Atlas"（Z-brain）（https://engertlab.fas.harvard.edu/Z-Brain/#/home/）网站为用户提供了一个易于使用和非常活跃的 2D 查看器，可以查看 6 日龄斑马鱼仔鱼头部 309 种不同抗体染色或转基因表达模式（其中 47 种来自其自身成果）。与早期项目"Fish DB 计划"和早期 Z 脑图谱（Z Brain Atlas）相比，添加了许多额外的功能，展示了哈佛大学 Engert 实验室在这个资源项目中的进展。虽然显示的数据似乎与早期版本相同，但它们显然是经过同源对比增强的，因此可读性明显提高。当在通用坐标系中以伪彩色显示多达四种不同的表达模式时，这一点特别值得关注，有助于理解相关种群的潜在相关性或相互作用。对于这种可视化的微调，每个伪彩色通道的不透明度可以单独调整，允许叠加和比较强度差异很大的表达模式。稍有不足的是，不同通道伪彩色混合会导致至少一个通道为零的区域出现剪切伪影。在使用此资源时需要考虑到这一点，否则可能会导致对现有数据的曲解。

为了对表达模式进行解剖学注释并提供额外的定位，Engert 实验室将 6 日龄的斑马鱼模板划分为 294 个区域。在转基因表达模式的显示上，最多可以用预设的伪彩色覆盖四个不同的区域。这些区域还用于表达模式的注释，从而支持基于区域的品系查询。在当前的显示中，这种查询的选定结果（灰色）作为参考模式加载到查看器中，可以直接与最多三个其他表达模式叠加。由于表达模式的注释已经包含了"普通"术语，如"多巴胺"或"5- 羟色胺"，因此该工具在促进宿主表达和染色模式信息获取与研究方面可能会变得更加强大。

在这种情况下，有必要对 Engert 实验室的增强子诱捕品系资源进行说明。在构建时，该资源对 Engert 实验室 47 个增强子诱捕品系平均表达模式（群体模式）的最大密度投影进行了 38 个解剖区域的注释。在搜索特定区域的表达模式时，该资源可以快速且方便地访问 Engert 增强子诱捕品系库。每个品系都有其单独的页面，在该页面上可以根据翻译视频对表达模式进行更详细的研究：所有数据库的切片都被转换为视频的帧。从这些单独的页面可以直接在 Z Brain 图谱中打开表达模式以进行进一步的分析，这很好地将该资源集成到整个图谱框架中。

值得注意的是，这个资源非常易于使用共享功能。一个方便定位的“Share”按钮会生成一个 URL，它会在任何给定的浏览器中重新生成当前的画面。当沟通关于指定模式和 / 或区域组合的特定细节时，该功能是非常有用的。

此外，值得一提的是，Z-Brain 只能被认为是 Engert 实验室开发的更复杂项目——多尺度虚拟斑马鱼的中间步骤，该项目旨在将荧光显微镜和电子显微镜的不同分辨率的解剖数据与功能和连接组学数据的形态相结合，以完成斑马鱼仔鱼大脑的计算机建模。

在网站的“关于”页面，对 Z-Brain 的描述如下：“我们建立 Z-Brain 的目的是为斑马鱼仔鱼大脑的解剖和功能研究提供一个标准的参考大脑图谱（Randlett et al., 2015）。Z-Brain 是一个 3D 参考图谱，包含标准坐标空间内的神经解剖学标签和区域定义。Z-Brain 是基于 6 dpf Nacre/mitfa 突变仔鱼的抗 ERK 染色而建立的斑马鱼神经解剖学参考图谱。通过将不同解剖标记（markers）（抗体染色、转基因、染料染色）染色的鱼配准到这个共同的参考大脑中，我们创建了一个平台，可以在同一参考空间中探索很多的标签（labels）。目前我们已经在图谱上配准了 29 个标签。对于每个标签，我们配准了多条斑马鱼，然后将该标签表示为这些鱼的平均值，以便可视化这些神经元或特征的平均位置和染色。基于这些标记，我们还将大脑划分为 294 个区域。”

32.3.7 Tefor 斑马鱼图谱

Tefor Paris-Saclay 的斑马鱼图谱（https://zebrafish.tefor.net）最初是为成年果蝇的标准化神经解剖学［fruitfly.tefor.net（Arganda-Carreras et al., 2018）］而开发的，目前处于开发的早期阶段。该数据库拥有 219 个 7 日龄斑马鱼仔鱼的表达模式，目前显示了 4 个转基因品系表达模式的相关性和个体间的差异性。该技术展示平台是 Tefor Infrastructure 和 VRVis 生物医学图像信息学团队的合作项目，将在未来几年扩展成为一个神经发育图谱数据平台，其中包含从斑马鱼早期仔鱼阶段（5dpf）到成年阶段大脑的转基因表达模式、免疫组化染色模式和其他成像 / 可视化技术（例如荧光原位杂交）模式。此外，随着时间的推移，Tefor Infrastructure 作为服务提供商将不断地将其运行的高度专业化的研究数据注入数据库，从而持续丰富这一资源。

新开发的参考染色方案使得这一过程变得容易实现（Affaticati et al., 2017; Dambroise et al., 2017; Xavier et al., 2017）。来自 DiI 家族的孵育用荧光亲脂性阳离子吲哚碳菁染料为富含膜的结构（主要是中枢神经系统的白质）提供了一种廉价、强效和稳定的染色技术，并已经用于将单个数据集配准到图谱的通用坐标系中。这种染色技术也成功应用于其他物种（Affaticati et al., 2018）中，也为这些物种建立相关图谱创造了可能性。

由于 7dpf 以上的斑马鱼标本变得不再透明，因此，在过去几年中建立了一套组织清除方法，可应用于年龄较大的斑马鱼样本处理。

目前，Tefor 斑马鱼图谱提供了基于文本的查询，即可以根据注释查询到种属，同时也提供了空间查询，允许用户在图谱中选择感兴趣的自由区域，而不需要提供已有的知识或注释，并在该区域中可查询到其染色或表达模式。

32.3.8 Max-Planck 斑马鱼大脑图谱

马克斯 - 普朗克神经生物学研究所（Max-Planck Institute of Neurobiology）拜尔实验室（Baier lab）的 Max-Planck 斑马鱼大脑图谱（https://fishatlas.neuro.mpg.de/）描述了斑马鱼在受精后 6 天的标准化神经解剖学的三个不同方面：转基因表达模式、单细胞形态重建和空间填充区域化（space-filling regionalization）。虽然该在线资源库所有图像数据都源于共聚焦扫描显微镜图像，但它们是在两个抽象层次上呈现的。

在该资源的“转基因系”部分，用户可以在 2D 和 3D 水平上叠加和探索 376 个转基因系的平均表达模式，其中 85 个是开发者自己创建的产品（Förster et al., 2017; Scott and Baier, 2009）。在这些数据库的生成过程中，将同一转基因品系的多个共聚焦图片叠加（multiple confocal stacks）的灰度值配准到同一坐标系后组合成平均表达模式。因此，这些模式应该被视为“种群模式”，并且已经包括了转基因品系的生物系统固有的个体间变异。

与上述内容相比，“单细胞”（single cells）模块部分采用了更高层次的数据提取。多个刊物发表的数千个单个神经元染色的数据很复杂，将原始灰度值数据替换为球 - 棒形式表达的 3D 重建数据，有助于这些复杂数据的可视化（Miyasaka et al., 2014; Helmbrecht et al., 2018; Kunst et al., 2018）。该资源提供了 85 个分层结构的大脑区域的表面重建，这些区域可以单独或成组显示，并有助于根据单个神经元数据了解这些大脑区域之间的连接（表 32-3）。

表 32-3 在线解剖学资源表

类型	解剖学资源名称	发布时间
数据集	ZFIN anatomy	2000 年 4 月
数据集	The Interactive Atlas of Zebrafish Vascular Anatomy	2001 年 2 月
数据集	Zebrafishbrain	2010 年 5 月
数据集	Fishface	2011 年 8 月
数据集	FishNet	2007 年 8 月
数据集	ZFAP	2012 年 9 月
数据集	Neurodata	2015 年 9 月
数据集	Bio-Atlas	2015 年 12 月
数据集	Zebrain	2017 年 11 月
数据集	AGETAZ	2019 年 1 月
已注册	VibeZ	2012 年 7 月
已注册	BrainBrowser	2015 年 11 月
已注册	Z-Brain	2017 年 4 月
已注册	Tefor Zebrafish atlas	2018 年 11 月
已注册	Max-Planck zebrafish brain atlas	2018 年 12 月

32.4 结论与展望

本章所描述项目的理想前景之一是不同资源的交叉整合。幸运的是，这并不是遥不可及的，因为不同的标准化神经解剖学平台的开发者 / 用户们已经在进行友好地交流。

虽然这些资源最初是为了满足相应实验室的特定需求而开发的，但它们正变得越来越具有协作性。这已经体现在开发者正在使用桥接配准（bridging registrations, BR）将其他实验室的数据集成化（例如，BrainBrowser 网页中的 Z-Brain 数据和 MPI 图谱中的 Burgess 数据）。

在不同图谱的模板之间实现桥接配准，可以用相对较低的计算成本将数据从一种资源库传输到另一种资源库：由于给定数据库（dbA）的所有数据已经在同一个坐标系中，这个坐标系是由它们的模板（templA）定义的，因此可以很容易地通过对 dbA 的数据集应用 templA 到 templB 的桥接转换（bridging transformation），将它们转移到另一个数据库（dbB）中。但这种方法的缺点是，以这种方式导入的数据分辨率较低，并且与原始数据库相比，在目标数据库（coordsB）的坐标系中配准的效果较差，原因是需要进行两次转换（将数据集配准到 dbA，将数据集从 dbA 配准到 dbB），其中存在相关的错误传递和将数据重新采样到新模板的张量（tB）中。然而，这些问题可以通过将转换数据与原始的、未配准的数据一起提供给公共存储库来解决。这样，任何给定数据库的转换都可以根据其各自的配准参数（individual registration parameters）和桥接变换参数（bridging transformation parameters）来计算，而不需要双重配准和重采样。这种方法有可能使所描述的任何标准化神经解剖学资源中的每个数据库对任何其他人可用。

由 Volgelstein 实验室建立了数据（图像数据以及元数据，例如转换、协议、素材）和算法的集中数据存储库基础架构，这种方法是非常有应用前景的：根据所需的功能、科学背景或共同持有的数据，一个数据库生成的数据可以在另一个数据库中集成和使用。

这样的数据库越大，它就越有用。

这种方法潜在用途非常强大，因为它可以使实验室能够在没有标准化神经解剖学数据库的情况下，将自己的数据提供给这个数据库，并将自己的数据与不同的数据库关联起来。这不仅将大量的数据获取工作量分配给多个来源，而且还减少了使用的实验动物数量，因为在一项研究中获得的数据可以直接作为“已有知识”资源用于其他研究。通过这种方式，这些资源有助于斑马鱼研究的可持续性（3R），同时也有助于加速其发展，因为如果能够轻松访问已有知识数据库来重新确认自己的初步实验结果，那么就可以大大缩短初步研究的时间。

（吴方晖翻译 赵丽审校）

参考文献

Ablain, J., Zon, L.I., 2013. Of fish and men: using zebrafish to fight human diseases. Trends in Cell Biology 23 (12),584-586.

Affaticati, P., Le Mével, S., Jenett, A., Rivière, L., Machado, E., Mughal, B.B., Fini, J.-B., 2018. X-FaCT: xenopus-fast clearing technique. Methods in Molecular Biology 1865, 233-241.

Affaticati, P., Simion, M., De Job, E., Rivière, L., Hermel, J.-M., Machado, E., Joly, J.-S., Jenett, A., 2017. zPACT:

tissue clearing and immunohistochemistry on juvenile zebrafish brain. Bio-Protocol 7 (23). https://doi.org/10.21769/BioProtoc.2636.

Arganda-Carreras, I., Manoliu, T., Mazuras, N., Schulze, F., Iglesias, J.E., Bühler, K., Jenett, A., Rouyer, F., Andrey, P.,2018. A statistically representative atlas for mapping neuronal circuits in the *Drosophila* adult brain. Frontiers in Neuroinformatics 12, 13.

Asakawa, K., Kawakami, K., 2008. "Targeted gene expression by the gal4-UAS system in zebrafish." development. Growth & Differentiation 50 (6), 391-399.

Aubert-Broche, B., Evans, A.C., Collins, L., 2006. A new improved version of the realistic digital brain phantom. NeuroImage 32 (1), 138-145.

Avants, B.B., Tustison, N.J., Song, G., Gee, J.C., 2009. ANTS: open-source tools for normalization and neuroanatomy. Heanet. Ie X (Xx), 1e11.

Bogovic, J.A., Otsuna, H., Heinrich, L., Ito, M., Jeter, J., Meissner, G.W., Nern, A., et al., 2018. An unbiased template of the Drosophila brain and ventral nerve cord. bioRxiv. https://doi.org/10.1101/376384.

Bradford, Y.M., Toro, S., Ramachandran, S., Ruzicka, L., Howe, D.G., Eagle, A., Kalita, P., et al., 2017. Zebrafish models of human disease: gaining insight into human disease at ZFIN. Institute of Laboratory Animal Resources Journal 58 (1), 4-16.

Bryson-Richardson, R., Berger, S., Currie, P., 2012. Atlas of Zebrafish Development. Academic Press.

Bryson-Richardson, R.J., Berger, S., Schilling, T.F., Hall, T.E., Cole, N.J., Gibson, A.J., Sharpe, J., Currie, P.D., 2007. FishNet: an online database of zebrafish anatomy. BMC Biology 5, 34.

Burton, E.A., Palladino, M.J., 2010. Of fish, flies, worms and men: powerful approaches to neuropsychiatric disease using genetic models. Neurobiology of Disease 40 (1), 1-3.

Clarkson, M.D., 2016. Representation of anatomy in online atlases and databases: a survey and collection of patterns for interface design. BMC Developmental Biology 16 (1), 18.

Copper, J.E., Budgeon, L.R., Foutz, C.A., van Rossum, D.B., Vanselow, D.J., Hubley, M.J., Clark, D.P., Mandrell, D.T., Cheng, K.C., 2018. Comparative analysis of fixation and embedding techniques for optimized histological preparation of zebrafish. Comparative Biochemistry and Physiology. Toxicology and Pharmacology: CBP 208, 38-46.

Dambroise, E., Simion, M., Bourquard, T., Bouffard, S., Rizzi, B., Jaszczyszyn, Y., Bourge, M., et al., 2017. Postembryonic fish brain proliferation zones exhibit neuroepithelial-type gene expression profile. Stem Cells 35 (6), 1505-1518.

Dickie, D.A., Shenkin, S.D., Anblagan, D., Lee, J., Blesa Cabez, M., Rodriguez, D., Boardman, J.P., Adam, W., Job, D.E., Wardlaw, J.M., 2017. Whole brain magnetic resonance image atlases: a systematic review of existing atlases and caveats for use in population imaging. Frontiers in Neuroinformatics 11, 1.

Ding, Y., Vanselow, D.J., Yakovlev, M.A., Katz, S.R., Lin, A.Y., Clark, D.P., Vargas, P., et al., 2018. Three-dimensional histology of whole zebrafish by sub-micron synchrotron X-ray micro-tomography. bioRxiv. https://doi.org/10.1101/392381.

Ding, Y., Vanselow, D.J., Yakovlev, M.A., Katz, S.R., Lin, A.Y., Clark, D.P., Vargas, P., et al., 2019. Computational 3D histological phenotyping of whole zebrafish by X-ray histotomography. eLife 8. https://doi.org/10.7554/eLife.44898.

Diotel, N., Rodriguez Viales, R., Armant, O., Martin, M., Ferg, M., Rastegar, S., Uwe Strähle, 2015. Comprehensive expression map of transcription regulators in the adult zebrafish telencephalon reveals distinct neurogenic niches. The Journal of Comparative Neurology 523 (8), 1202-1221.

Dorr, A.E., Lerch, J.P., Spring, S., Kabani, N., Henkelman, R.M., 2008. High resolution three-dimensional brain atlas using an average magnetic resonance image of 40 adult C57Bl/6J mice. Neuroimage 42 (1), 60-69.

Eames, B.F., DeLaurier, A., Ullmann, B., Huycke, T.R., Nichols, J.T., Dowd, J., McFadden, M., Sasaki, M.M., Kimmel, C.B., 2013. FishFace: interactive atlas of zebrafish craniofacial development at cellular resolution. BMC Developmental Biology 13, 23.

Emmenlauer, M., Ronneberger, O., Ponti, A., Schwarb, P., Griffa, A., Filippi, A., Nitschke, R., Driever, W., Burkhardt, H., 2009. XuvTools: free, fast and reliable stitching of large 3D datasets. Journal of Microscopy 233 (1), 42-60.

Evans, A.C., Janke, A.L., Collins, D.L., Baillet, S., 2012. Brain templates and atlases. Neuroimage 62 (2), 911e922. Fillmore, P.T., Phillips-Meek, M.C., Richards, J.E., 2015. Age-specific MRI brain and head templates for healthy adults from 20 through 89 years of age. Frontiers in Aging Neuroscience 7, 44.

Fonov, V., Evans, A.C., Kelly, B., Almli, C.R., McKinstry, R.C., Louis Collins, D., Brain Development Cooperative Group, 2011. Unbiased average age-appropriate atlases for pediatric studies. Neuroimage 54 (1), 313-327.

Förster, D., Arnold-Ammer, I., Laurell, E., Barker, A.J., Fernandes, A.M., Finger-Baier, K., Filosa, A., et al., 2017. Genetic targeting and anatomical registration of neuronal populations in the zebrafish brain with a new set of BAC transgenic tools. Scientific Reports 7 (1), 5230.

Halpern, M.E., Jerry, R., Goll, M.G., Akitake, C.M., Parsons, M., Leach, S.D., 2008. Gal4/UAS transgenic tools and their application to zebrafish. Zebrafish 5 (2), 97-110.

Helmbrecht, T.O., Dal Maschio, M., Donovan, J.C., Koutsouli, S., 2018. Topography of a visuomotor transformation. Neuron 100, 1429-1445.e4.

Hildebrand, D.G.C., Cicconet, M., Torres, R.M., Choi, W., Quan, T.M., Moon, J., Wetzel, A.W., et al., 2017. Wholebrain serial-section electron microscopy in larval zebrafish. Nature 545 (7654), 345-349.

Holden, J.A., Layfield, L.J., Layfield, L.L., Matthews, J.L., 2012. The Zebrafish: Atlas of Macroscopic and Microscopic Anatomy. Cambridge University Press.

Isogai, S., Horiguchi, M., Weinstein, B.M., 2001. The vascular anatomy of the developing zebrafish: an atlas of embryonic and early larval development. Developmental Biology 230 (2), 278-301.

Joshi, S., Davis, B., Jomier, M., Guido, G., 2004. Unbiased diffeomorphic atlas construction for computational anatomy. NeuroImage 23 (Suppl. 1), S151-S160.

Kawakami, K., Asakawa, K., Hibi, M., Itoh, M., Muto, A., Wada, H., 2016. Chapter three e Gal4 driver transgenic zebrafish: powerful tools to study developmental biology, organogenesis, and neuroscience. Advances in Genetics 95, 65-87.

Kawakami, K., Takeda, H., Kawakami, N., Kobayashi, M., Matsuda, N., Mishina, M., 2004. A transposon-mediated gene trap approach identifies developmentally regulated genes in zebrafish. Developmental Cell 7 (1), 133-144.

Kawakami, K., Shima, A., Kawakami, N., 2000. Identification of a functional transposase of the Tol2 element, an Aclike element from the Japanese medaka fish, and its transposition in the zebrafish germ lineage. Proceedings of the National Academy of Sciences of the United States of America 97 (21), 11403-11408.

Klein, S., Staring, M., Murphy, K., Viergever, M.A., Pluim, J., 2010. Elastix: a toolbox for intensity-based medical image registration. IEEE Transactions on Medical Imaging. https://doi.org/10.1109/tmi.2009.2035616.

Kramer, J.M., Staveley, B.E., 2003. GAL4 causes developmental defects and apoptosis when expressed in the developing eye of *Drosophila* melanogaster. Genetics and Molecular Research: GMR 2 (1), 43-47.

Kunst, M., Laurell, E., Mokayes, N., Kramer, A., Kubo, F., Fernandes, A.M., Förster, D., Dal Maschio, M., Baier, H.,2018. A cellular-resolution atlas of the larval zebrafish brain. SSRN Electronic Journal. https://doi.org/10.2139/ssrn.3257346.

Lin, A.Y., Ding, Y., Vanselow, D.J., Katz, S.R., Yakovlev, M.A., Clark, D.P., Mandrell, D., Copper, J.E., van Rossum, D.B., Cheng, K.C., 2018. Rigid embedding of fixed and stained, whole, millimeter-scale specimens for section-free 3D histology by micro-computed tomography. Journal of Visualized Experiments 140. https://doi.org/10.3791/58293.

Marquart, G.D., Tabor, K.M., Brown, M., Strykowski, J.L., Varshney, G.K., LaFave, M.C., Mueller, T., Burgess, S.M.,Higashijima, S.-I., Burgess, H.A., 2015. A 3D searchable database of transgenic zebrafish Gal4 and Cre lines for functional neuroanatomy studies. Frontiers in Neural Circuits 9, 78.

Marquart, G.D., Tabor, K.M., Horstick, E.J., Brown, M., Geoca, A.K., Polys, N.F., Nogare, D.D., Burgess, H.A., 2017.High-precision registration between zebrafish brain atlases using symmetric diffeomorphic normalization. GigaScience 6 (8), 1-15.

Mawhinney, R.M., Staveley, B.E., 2011. Expression of GFP can influence aging and climbing ability in *Drosophila*. Genetics and Molecular Research: GMR 10 (1), 494-505.

Miyasaka, N., Arganda-Carreras, I., Wakisaka, N., Masuda, M., Sümbül, U., Sebastian Seung, H., Yoshihara, Y., 2014. Olfactory projectome in the zebrafish forebrain revealed by genetic single-neuron labelling. Nature Communications 5 (1), 3639.

Mueller, T., Wullimann, M., 2016. Atlas of Early Zebrafish Brain Development: A Tool for Molecular Neurogenetics, second ed. Elsevier.

Mueller, T., Wullimann, M.F., 2005. Atlas of Early Zebrafish Brain Development: A Tool for Molecular Neurogenetics. Elsevier.

Nadkarni, N.A., Bougacha, S., Garin, C., Dhenain, M., Jean-Luc Picq, 2018. Digital templates and brain atlas dataset for the mouse lemur primate. Data in Brief 21, 1178-1185.

Pfeiffer, B.D., Ngo, T.-T., Hibbard, K.L., Murphy, C., Jenett, A., Truman, J.W., Rubin, G.M., 2010. Refinement of tools for targeted gene expression in Drosophila. Genetics 186 (2), 735-755.

Quallo, M.M., Price, C.J., Ueno, K., Asamizuya, T., Cheng, K., Lemon, R.N., Iriki, A., 2010. Creating a populationaveraged standard brain template for Japanese macaques (*M. Fuscata*). NeuroImage 52 (4), 1328-1333.

Randlett, O., Wee, C.L., Naumann, E.A., Nnaemeka, O., Schoppik, D., Fitzgerald, J.E., Portugues, R., et al., 2015. Whole-brain activity mapping onto a zebrafish brain atlas. Nature Methods 12 (11), 1039-1046.

Rath, M., Nitschke, R., Filippi, A., Ronneberger, O., Driever, W., 2012. Generation of high quality multi-view confocal 3D datasets of zebrafish larval brains suitable for analysis using virtual brain explorer (ViBE-Z) software. Protocol Exchange.

https://doi.org/10.1038/protex.2012.031.

Rezával, C., Santiago, W., Ceriani, M.F., 2007. Neuronal death in *Drosophila* triggered by GAL4 accumulation. The European Journal of Neuroscience 25 (3), 683-694.

Rohlfing, T., Maurer, C.R., 2003. Nonrigid image registration in shared-memory multiprocessor environments with application to brains, breasts, and bees. IEEE Transactions on Information Technology in Biomedicine: A Publication of the IEEE Engineering in Medicine and Biology Society 7 (1), 16-25.

Rohlfing, T., Kroenke, C.D., Sullivan, E.V., Dubach, M.F., Bowden, D.M., Grant, K.A., Pfefferbaum, A., 2012. The INIA19 template and NeuroMaps atlas for primate brain image parcellation and spatial normalization. Frontiers in Neuroinformatics 6, 27.

Ronneberger, O., Liu, K., Rath, M., Rueb, D., Mueller, T., Skibbe, H., Drayer, B., et al., 2012. ViBE-Z: a framework for 3D virtual colocalization analysis in zebrafish larval brains. Nature Methods 9 (7), 735-742.

Sabaliauskas, N.A., Foutz, C.A., Mest, J.R., Budgeon, L.R., Sidor, A.T., Gershenson, J.A., Joshi, S.B., Cheng, K.C., 2006. High-throughput zebrafish histology. Methods 39 (3), 246-254.

Salgado, D., Marcelle, C., Currie, P.D., Bryson-Richardson, R.J., 2012c. The zebrafish anatomy portal: a novel integrated resource to facilitate zebrafish research. Developmental Biology 372 (1), 1e4.

Salgado, D., Marcelle, C., Currie, P.D., Bryson-Richardson, R.J., Carmen, E., Richards, J.E., Robert Almli, C., 2012a. Age-specific MRI templates for pediatric neuroimaging. Developmental Neuropsychology 37 (5), 379-399.

Salgado, D., Marcelle, C., Currie, P.D., Bryson-Richardson, R.J., Carmen, E., Richards, J.E., Robert Almli, C., 2012b. Neurodevelopmental MRI brain templates for children from 2 weeks to 4 years of age. Developmental Psychobiology 54 (1), 77-91.

Sanchez, C.E., Richards, J.E., Almli, C.R., 2012. Age-specific MRI templates for pediatric neuroimaging. Developmental Neuropsychology 37 (5), 379e399. https://doi.org/10.1080/87565641.2012.688900.

Sanchez, C.E., Richards, J.E., Almli, C.R., 2012. Neurodevelopmental MRI brain templates for children from 2 weeks to 4 years of age. Developmental Psychobiology 54 (1), 77-91. https://doi.org/10.1002/dev.20579.

Scott, E.K., Baier, H., 2009. The cellular architecture of the larval zebrafish tectum, as revealed by Gal4 enhancer trap lines. Frontiers in Neural Circuits 3, 13.

Seidlitz, J., Sponheim, C., Glen, D., Ye, F.Q., Saleem, K.S., Leopold, D.A., Ungerleider, L., Adam, M., 2017. A population MRI brain template and analysis tools for the macaque. NeuroImage. https://doi.org/10.1016/j.neuroimage.2017.04.063.

Tabor, K.M., Marquart, G.D., Hurt, C., Smith, T.S., Geoca, A.K., Bhandiwad, A.A., Abhignya Subedi, et al., 2019. Brain-wide cellular resolution imaging of Cre transgenic zebrafish lines for functional circuit-mapping. eLife 8.https://doi.org/10.7554/eLife.42687.

Van Essen, D.C., Dierker, D.L., 2007. Primer Surface-Based and Probabilistic Atlases of Primate Cerebral Cortex. https://doi.org/10.1016/j.neuron.2007.10.015 no. 34 cm.

Vogelstein, J.T., Perlman, E., Falk, B., Baden, A., Roncal, W.G., Chandrashekhar, V., Collman, F., et al., 2018. A community-developed open-source computational ecosystem for big neuro data. Nature Methods 15 (11), 846-847.

Wullimann, M.F., Rupp, B., Reichert, H., 1996a. Neuroanatomy of the Zebrafish Brain: A Topological Atlas. Neuroanatomy of the Zebrafish Brain: A Topological Atlas. Birkh[Ux9451]user Verlag.

Wullimann, M.F., Rupp, B., Reichert, H., 1996b. Neuroanatomy of the Zebrafish Brain. Birkhäuser, Basel. Wyatt, C., Bartoszek, E.M., Yaksi, E., 2015. Methods for studying the zebrafish brain: past, present and future. The European Journal of Neuroscience 42 (2), 1746-1763.

Xavier, A.L., Fontaine, R., Bloch, S., Affaticati, P., Jenett, A., Demarque, M., Vernier, P., Yamamoto, K., March 2017. Comparative Analysis of Monoaminergic Cerebrospinal Fluid-Contacting Cells in Osteichthyes (Bony Vertebrates). https://doi.org/10.1002/cne.24204.

Xiao, Y., Fonov, V., Chakravarty, M.M., Beriault, S., Al Subaie, F., Sadikot, A., Bruce Pike, G., Bertrand, G., Louis Collins, D., 2017. A dataset of multi-contrast population-averaged brain MRI atlases of a Parkinson's disease cohort. Data in Brief 12, 370-379.

Xin, X., Darin, C., Ang, K.C., van Rossum, D.B., Copper, J., Xiao, X., La Riviere, P.J., Cheng, K.C., 2017. Synchrotron microCT imaging of soft tissue in juvenile zebrafish reveals retinotectal projections. In: Alfano, R.R., Demos, S.G. (Eds.), Optica l Biopsy XV: Toward Real-Time Spectroscopic Imaging and Diagnosis, 10060:100601I. SPIE Proceedings. SPIE.

索　引

Y

Z

其他